Lecture Notes in Networks and Systems

Volume 161

The series "Lecture Notes in Networks and Systems" publishes the latest developments in Networks and Systems—quickly, informally and with high quality. Original research reported in proceedings and post-proceedings represents the core of LNNS.

Volumes published in LNNS embrace all aspects and subfields of, as well as new challenges in, Networks and Systems.

The series contains proceedings and edited volumes in systems and networks, spanning the areas of Cyber-Physical Systems, Autonomous Systems, Sensor Networks, Control Systems, Energy Systems, Automotive Systems, Biological Systems, Vehicular Networking and Connected Vehicles, Aerospace Systems, Automation, Manufacturing, Smart Grids, Nonlinear Systems, Power Systems, Robotics, Social Systems, Economic Systems and other. Of particular value to both the contributors and the readership are the short publication timeframe and the world-wide distribution and exposure which enable both a wide and rapid dissemination of research output.

The series covers the theory, applications, and perspectives on the state of the art and future developments relevant to systems and networks, decision making, control, complex processes and related areas, as embedded in the fields of interdisciplinary and applied sciences, engineering, computer science, physics, economics, social, and life sciences, as well as the paradigms and methodologies behind them.

**** Indexing: The books of this series are submitted to ISI Proceedings, SCOPUS, Google Scholar and Springerlink ****

More information about this series at http://www.springer.com/series/15179

Svetlana Igorevna Ashmarina ·
Valentina Vyacheslavovna Mantulenko
Editors

Digital Economy and the New Labor Market: Jobs, Competences and Innovative HR Technologies

Springer

Editors
Svetlana Igorevna Ashmarina
Applied Management Department
Samara State University of Economics
Samara, Russia

Valentina Vyacheslavovna Mantulenko
Department of Applied Management
Samara State University of Economics
Samara, Russia

ISSN 2367-3370 ISSN 2367-3389 (electronic)
Lecture Notes in Networks and Systems
ISBN 978-3-030-60925-2 ISBN 978-3-030-60926-9 (eBook)
https://doi.org/10.1007/978-3-030-60926-9

This Springer imprint is published by the registered company Springer Nature Switzerland AG
The registered company address is: Gewerbestrasse 11, 6330 Cham, Switzerland

Contents

Employment and the Labor Market of the Future: Current Trends

Role of Additional Education in the Context of Labor Market Volatility

G. N. Alexandrova(✉) and G. V. Glukhov

Samara State University of Economics, Samara, Russia
Alexandrova.gn@rambler.ru, Glukhov.g@mail.ru

Abstract. The article explores new trends that appear on the labor market and affect the structure of employment. The authors analyze the possible competencies that are necessary for graduates of universities of economic and legal specialties, taking into account new professions that will appear in the future. In order to be in demand on the labor market and not to remain without work, students need to engage in self-education and improve not only professional competencies, but also constantly improve their over-professional skills. The system of additional education at universities should be a useful assistant for students and teachers. The authors consider the system of continuing education as a flexible and dynamic structure that allows students to choose the course program themselves.

Keywords: Additional education · Competences · Profession · Labor market

1 Introduction

The rapid development of society in connection with the acceleration and improvement of technological processes and introduction of artificial intelligence elements into our daily and professional life have resulted in the emergence of new professions and, accordingly, the need for new specialists. Nowadays, many analytical articles have appeared and many predictions have been made about what the labor market should be in the near future, what specialists will be in demand and what workers and employees will soon be unnecessary for the country's economy. According to experts, the largest number of jobs will appear in the IT sector. Many human activities will in one way or another be associated with high-tech areas. In this regard, the hard and soft skills that employees must possess are changing. This fact poses serious challenges for schools and higher education institutions. The world is becoming unstable, everything in it is changing rapidly; people will no longer work all their lives at the same enterprise, as it was before. A university graduate, who studied for five years as an accountant, is surprised to understand that no one needs him as a specialist, since his duties are successfully carried out by a computer program, that replaced a whole department. So, sooner or later, it can happen to any profession. Therefore, modern reality requires employees and employers to be prepared for the fact that it will be necessary think constantly about acquiring new knowledge, competencies and skills. It is assumed, that all citizens of working age will have to study, improve their skills and acquire new knowledge all their lives [4].

S. I. Ashmarina and V. V. Mantulenko (Eds.): IPM 2020, LNNS 161, pp. 3–8, 2021.
https://doi.org/10.1007/978-3-030-60926-9_1

The aim of this work is to develop recommendations for designing and implementing additional education programs in non-linguistic universities. The goal setting involves solving the following tasks: analysis of the list of future professions compiled by the Agency for Strategic Initiatives and Skolkovo [1, 7] in order to establish new areas for training specialists in the future in accordance with the profile of the university; determine the set of necessary competencies for each area of training; develop a list of soft skills that are not included in the range of mandatory professional competencies; make a list of possible courses for students of economic and legal specialties in the framework educational of goals of the university.

2 Methodology

The study carried out by the Agency for Strategic Initiatives and Skolkovo showed which areas of professional training will be in demand in the future, and revealed the importance of developing the following skills: strategic thinking, creative thinking, systemic thinking and critical thinking. Based on the need for these skills, students should be taught the following competencies: the ability to solve complex problems; the ability to make decisions in situations of uncertainty, novelty and lack of information; the ability to create a new and original product; the ability to adapt and make complex decisions in a networked environment; the ability to analyze and highlight important issues among multiple facts and so on [1, 7]. At the same time, according to the results of a Stanford University survey, high school students cannot distinguish true information from fake [8], and an all-Russian study conducted by the Regional Public Center for Internet Technologies (ROCIT) in 2018 showed that digital literacy index of respondents went down by 14,7% compared to the previous year [6]. It follows that more than half of young people lack necessary digital skills. Experts of the Moscow School of Management Skolkovo analyzed the professional skills mentioned by Russian employers as the most important ones for future specialists. The most important according to employers are: multilingualism and multiculturalism, fluent English and knowledge of another language, understanding the national and cultural context of partner countries, inter-business communication skills, customer focus, systems thinking, sociability, programming of IT solutions, the ability to work with complex automated machines, with artificial intelligence [7]. The authors of this investigation used the following research methods: testing, experiment, and the problem task method. We used the following research tools: tests, questionnaires, case studies, surveys and specially designed experimental tasks that requires the solution of complex problems both individually and in a group. Modern challenges force higher education institutions to develop new strategies and adaptive practices to meet today's requirements and expectations of business and the state.

3 Results

In 2018, the Russian government announced the national project "Education", one of the tasks of which is the development of a socially responsible individual based on moral values, historical, cultural, and national traditions [5]. The announced project helped us to formulate the goal and objectives of our research that wourld work clearly and accurately. We carried out our study in line with the national education development strategy. The authors focused on studying the prospects for the development of additional education at universities that prepare students of legal and economic specialties. In this regard, we selected those skills and abilities that, in our opinion, require closer attention, since, firstly, they should be developed and maintained throughout life, and secondly, not sufficient time is allocated for these skills training. Analysis of labor market research, conducted by Russian and foreign experts, [1] allowed us to conclude that in the future many companies will become more customer-oriented. Many specialists will work with a specific person, taking into account their individual desires, needs and opportunities, there will be no universal schemes and proposals. For example, a personal brand manager will be engaged in the formation of the image of a specific person, his responsibility will be to represent this person in social networks as a good professional, organize various meetings, briefings and seminars for this purpose. Accordingly, in addition to the managerial professional skills, students need additional knowledge in psychology, since it is necessary to convince potential stakeholders of the professional superiority of an individual; linguistics, as it is necessary to write bright and persuaded texts, possibly in several languages; artistic creativity, as the manager will need to creatively present information, build up an attractive and vivid image of the client, take photos, think over the style of clothing and makeup.

In order to determine the themes of the courses, competencies, skills and abilities that need to be included in the course program, we conducted a study among students and potential employers. We pursued the study with the involvement of university students from the first to the fourth years. Overall, 485 students took part in the study. The study also involved middle managers - participants of training vocational courses. We conducted a survey among them to find out their opinion on the qualifications of university graduates, their level of creativity and critical thinking skills. To assess the creativity level of students, the authors of the article have chosen the following methods: testing, experiment, and the problem task method. We have chosen the following research tools: tests, questionnaires, and specially compiled experimental tasks that requires the solution of complex problems by the students, both individually and in a group. In the process of working with students of additional education courses, the main task was to find out what skills and competencies university graduates miss, when they apply for a job, and what skills the graduates will need in the near future. Because of the surveys and cases, we have found out that future applicants lack elementary literacy: they do not know how to write business letters, explanatory notes, advertising texts and reports. In addition, young professionals are often afraid of responsibility and do not know how to make strategic decisions. 60–65% of young specialists cannot speak fluent English, 95% do not know second foreign language. 75%–80% are not ready to work overtime even for an additional fee, 80–85% do not make any efforts to advance on the career ladder. We understand that these data are based more on the subjective feelings of the respondents, but we take them into account, as they, in any case, may be useful in compiling the content of future courses.

That part of the study, which we conducted with the involvement of undergraduate students, was aimed at studying the level of general erudition, creativity and the level of critical thinking of undergraduate students. The following research tools were selected:

questionnaires, tests and specially composed tasks. The questionnaires were designed in such a way as to find out the students' career ambitions, their professional expectations, their desire to find a job in their chosen specialty, and the amount of wages they would like to receive.

The purpose of the tests and special tasks was to determine whether the students have the over-professional competencies necessary for effective work in the future. Over-professional competencies are based on general erudition, education and upbringing of a person. Each employee should deal with perfection and acquisition of over-professional competencies independently. If a young employee has a good classical education, is educated and disciplined, he is able to map the educational path on his own and move in this direction successfully. However, most people cannot or do not have enough time to search necessary information on the Internet, choose the right literature, educational courses or educational institution to develop and gain new knowledge. Why do they need over-professional competencies? To answer this question, let us consider a specific example. On February 28, 2020, in the “Ninth Val” sauna complex in Moscow, the birthday celebration ended tragically - three people died and three more guests of the party were hospitalized in serious condition. The cause of the tragedy was dry ice, which organizers of the party threw into the pool to enhance the effect of evaporating water. The party organizers did not know that dry ice in water turns into carbon dioxide and does not rise up, but rather spreads above the surface of the water. The lack of elementary knowledge in physics and chemistry led to tragedy. Another example: in the early nineties, in Russia there was an advertisement for yeast-free bread. The commercials and articles described the advantages of yeast-free bread compared to the leaven. Nevertheless, any yeast dies in the process of baking bread, that is, all bread, regardless of whether the yeast was added there or not, is yeast-free. Thus, the advertising manager demonstrated his ignorance and misled potential buyers.

Such advertising, in our opinion, undermines the confidence of the client in the producer of goods. The advertising manager, as well as the entertainment services manager, lacked the knowledge essential for effectively coping with their direct professional duties. There are a huge number of such examples. Given the fact that demand in the labor market is constantly changing – old professions disappear, new ones appear instead – it is impossible to foresee all the necessary competencies in the university curriculum. The results of the study of creativity and critical level of thinking of SSEU students showed that: 76%, 71%, 68% and 59% of students of 1, 2, 3 and 4 courses, respectively, find it difficult to determine reliable sources of information. 81%, 73%, 64% and 58% of 1, 2, 3 and 4 courses students, respectively, cannot critically assess the importance of information they need for writing essays, reports and term papers. 30%, 45%, 64% and 90% of students of the 1, 2, 3 and 4 courses, respectively, showed a high coefficient of creativity when performing specially compiled creative tasks in their main areas of professional training. We evaluated the creativity according to the number, originality and novelty of the ideas offered by the students during fulfilment of assignments. Students of all four courses showed not very deep knowledge of Russian and foreign literature, the history of foreign countries, contemporary art, and the world culture.

4 Discussion

In our opinion, the system of additional education should fill all possible gaps in the knowledge of university graduates. Therefore, in 2018, the Language Training Center, together with the Department of Linguistics and Foreign Language Business Communication of SSEU, began serious research work in order to improve the quality of university education in general and to predict the prospects for its further development. Humanitarian education is closely connected with such categories as ethics, morals and moral principles of society, universal and national values, spirituality, the nation mentality, a subjective world picture [3]. Man cannot live only in the world of material objects, salaries and incomes. In our time, the production and providing of services are increasingly dependent on investors, that is, on those who have money. They form the final product. The quality of our lives depends on managers, lawyers and politicians. Society expects them to behave responsibly and take fair decisions.

According to the forecasts of futurologists, many companies in the near future will switch to new forms of work with their employees. Already, many employees work remotely, using computers and softwares. This will allow people to choose not only companies, but also countries. According to the forecasts of specialists, work in the usual sense will not exist anymore, permanent activity will be replaced by project activities on temporary basis [7]. Accordingly, the nature of the work can change, professional responsibilities can change, and with it the set of necessary competencies will alter. Therefore, it is necessary to prepare students for changing learning environment, to acquiring additional over-professional competences and get ready to self-study. During their study at the university they should master several specialties or majors. In this regard, there will be a need for additional knowledge on intercultural communication, for learning languages, customs, culture, history and mentality of other peoples. The set of competencies required in the labor market will constantly change in connection with the emergence of new professions [2].

Universities are expected to transmit knowledge of all kinds to students. A system of additional education, which already demonstrates great flexibility and customer focus, can help navigate the labor market and find a suited job. While developing course designs, we consider the wishes of students on what themes the courses cover, the number of hours, the form and methods of teaching and types of tasks. Recently, blended learning has been in greatest demand - the use of a distant and traditional ways of conducting classes [9]. We select teaching staff in accordance with the requests of students and the course topics. It is important to invite not only university professors, but also practitioners for organizing effective learning process. The isolation period and the measures taken by the governments of different countries in order to stop spreading Coronovirus-19 showed that part of the working population of the country and the world do not lose interest in self-development and the acquisition of new knowledge, both professional and humanitarian.

5 Conclusion

In connection with the latest trends in the development of society and the ongoing changes on the labor market, it is necessary to introduce new methods and forms into the higher education system. The task of classical university education is to provide students with basic fundamental knowledge in the main specialty of study. Basic curricula and university education programs cannot change quickly and adapt to new market conditions. This task and obligation should be delegated to universities. The system of continuing education at universities should not only transmit knowledge to students, but inspire them to self-study and self-development all their life. Modular educational courses using distance learning along with traditional classroom teaching with a professor are a good solution in the modern educational environment. In our opinion, a paradigm shift has occurred in the educational process. Modular courses can be created quickly as needed and as tasks become available, they can be easily combined and changed by replacing some modules and adding others. Further additional education is more flexible and dynamic, it responds better and faster to the challenges of the time and helps students acquire additional competencies. In addition, the course program is created taking into account the individual wishes of the student, that is, it puts the student in the center of training and is student-centered. The distance education system and blended-learning education can be useful for conducting vocational training. Its tailored programs meet the needs both employers and employees.

References

1. Agency for Strategic Initiatives: Informal education. https://www.asi.ru/leaders/initiatives/education_leaders/future_skills. Accessed: 24 June 2020
2. Beech, D.: Art and postcapitalism: Aesthetic labour, automation and value production. London: Pluto Press. https://doi.org/10.2307/j.ctvr0qv2p. (2019)
3. Downes, S.: Personal and personalized learning. https://us8.campaign-archive.com/?u=17ce08681f559814caf1359d3&id=fa1770e58d&e=6fb1272e29. Accessed: 14 June 2020
4. Helm, C.: Leadership dispositions: what are they and are they essential to good leadership. Academic Leadership **8**(1), 21 (2010)
5. Ministry of Education of the Russian Federation: National project "Education". https://strategy24.ru/rf/education/projects/natsionalnyy-proekt-obrazovanie. Accessed: 14 July 2020
6. ROCIT: The digital literacy index of Russians decreased by 14.7% compared to last year. https://rocit.ru/news/index-digital-literacy-2018. Accessed: 27 July 2020
7. Skolkovo: Atlas of new professions. http://www.skolkovo.ru/public/media/documents/research/sedec/SKOLKOVO_SEDeC_Atlas_2.0_Eng.pdf. Accessed: 27 June 2020
8. Stanford University: High school students are unprepared to judge the credibility of information on the Internet, according to Stanford researchers. https://ed.stanford.edu/news/high-school-students-are-unprepared-judge-credibility-information-internet-according-stanford. Accessed: 27 June 2020
9. Yamada, S.: Traditional apprenticeship as an educational and life experience: Life stories of young auto repair apprentices in Kumasi, Ghana. In: Ofosu-Kusi, Y., Matsuda, M., (Eds.), The Challenge of African Potentials: Conviviality, Informality and Futurity. pp. 181–208. Bamenda: Langaa RPCIG. https://doi.org/10.2307/j.ctv10h9fs0.12. (2020)

Socio-Psychological Model of PR Specialist Image in the Labor Market

K. A. Arzhanova[1(✉)], G. V. Dovzhik[2], and V. N. Dovzhik[2]

[1] Peoples' Friendship University of Russia, Moscow, Russia
kristina.arzhanova@gmail.com
[2] State University of Management, Moscow, Russia
{gv_dovzhik, vn_dovzhik}@guu.ru

Abstract. The research purpose is to develop a socio-psychological model of PR specialist's image and practical recommendations for the employer on using this model to form the image of a particular profession in the Russian labor market. The main tasks of this work include: identifying key components of the image of a public relations specialist and perception features of the specialty "public relations" by the method of a private semantic differential; reducing key variables and identifying factors that influence the formation of the image of a public relations specialist; building a socio-psychological model of this image. The methodological basis includes analysis, synthesis, schematization, abstraction, categorization, as well as empirical research methods (document analysis, survey method of particular semantic differential, associative experiment), data processing methods (factor analysis). The calculations were performed in the computer program SPSS Statistics. Factors that influence the formation of the image of a public relations specialist were identified: "wisdom", "effectiveness","sociability","Intelligence", "tact","strength". Based on them, a multidimensional socio-psychological model of the image of a PR specialist was built, where each dimension is one of the perception factors of the specialty in public relations.

Keywords: Image · Perception · PR specialist · Socio-psychological model

1 Introduction

The relevance of the topic of this research is the importance of studying the basis for building a socio-psychological model of forming the image of a public relations specialist by analyzing the perception of this profession by its future representatives – students of the field of training "Advertising and public relations". By studying the perception features and factors that influence the formation of the PR specialist's image among Russian students, it will be possible to adjust requirements of employers in order to attract a highly professional workforce in the Russian labor market and significantly improve the image of this specialty.

The sphere of public relations has recently undergone certain changes. This is due to the active development of new technologies and their gradual introduction into marketing communications. In this regard, there is a need to train specialists at a

S. I. Ashmarina and V. V. Mantulenko (Eds.): IPM 2020, LNNS 161, pp. 9–17, 2021.
https://doi.org/10.1007/978-3-030-60926-9_2

completely different level, who are ready to apply new technological solutions in practice and actively implement modern tools in their field of activity. In order to understand how future specialists perceive the image of their profession, this study was conducted. Understanding the peculiarities of image perception allowed us to develop a socio-psychological model of its formation. This model enables to evaluate the current image of a particular profession or adjust it from a socio-psychological point of view, using the considered factors. Korneychuk and Lukin list the following personal qualities of a public relations specialist: honesty, decency, sincerity, ability to set clear goals and achieve high results, love of work, responsibility, ability to communicate, establish social links [7]. The competencies for the position of PR specialist include the following: analytical and structured thinking, flexibility and creativity of thinking, dynamic thinking, memory, organization, planning and control, communication skills, ability to work in a team, stress resistance, self-confidence, observation and attention, high learnability and breadth of interests, discipline and normativity, special knowledge and skills. The degree of importance of a particular competence is not specified in the competence model.

The communication component of a public relations specialist is one of the main components of his work, especially when it comes to crisis situations. Choosing the right representative to communicate with the public during and after a crisis is a strategic decision that can have far-reaching consequences [8]. The following definition of image was taken as the basis for forming the image of a public relations specialist: image is a socio-psychological complex formation of individual, group or mass consciousness that creates an idea of the imaginary characteristics of an object based on subjective perception, which arose as a result of the action of socio-psychological mechanisms [1]. The definition of Dovzhik also states that this socio-psychological education includes semiotic, emotional-evaluative, cognitive components that have a relative constancy, dynamism, associativity, schematicity, openness, incompleteness and affect the behavior of people [2].

A general analysis of the degree of scientific development of the research topic (the image of a public relations specialist) makes it possible to conclude that, despite the presence of research in the field of socio-psychological mechanisms, the formation of the image of a public relations specialist has not been studied enough in both domestic and foreign socio-psychological science. There is no clear understanding of nature of the socio-psychological model of this mechanism.

2 Methodology

The novelty of the work lies in the fact that similar studies have not been conducted before to identify a model of the socio-psychological mechanism of a public relations specialist's image. The research hypothesis is: the socio-psychological model of the PR specialist's image is a multidimensional object, where each dimension is one of the factors of its perception. To develop a socio-psychological model for forming the image of a public relations specialist, a study was conducted in March - May 2020 among undergraduate students of 1-3 years studying in the field of "Advertising and public relations", using a number of methods.

The used theoretical methods include analysis, synthesis, schematization, abstraction, categorization. Empirical research methods applied in this research are document analysis, survey using the method of particular semantic differential, associative experiment. The research tools were systematically tested in a number of dissertations to identify the socio-psychological mechanism of image formation, including Dovzhik (in the study of socio-psychological mechanism of formation of the external image of the organization) [2] and Arzhanova who applied the given tools in the study of socio-psychological mechanism of formation of the political leader's image during the election campaign [1]. The data processing method is factor analysis. This method was used to construct a socio-psychological model for forming the image of a PR specialist. The calculations were performed in computer programs Microsoft Excel, SPSS Statistics. The specificity of the research subject is the justification for the choice of methods for collecting and analyzing the obtained data.

3 Results

An associative experiment was conducted to develop scales of a particular semantic differential. The sample includes 1-3 year students of the training direction "Advertising and public relations" (52 students). The age of respondents was from 17 to 21 years. As part of an association experiment, respondents were consistently asked to name associations, thoughts, and images that came to mind when reading questions: "What do you think is the positive image of a PR specialist?"; "What do you think is the negative image of a PR specialist? » ; "What is, in your opinion, a positive image in advertising?"; "What is, in your opinion, a negative image in advertising?"; "What is, in your opinion, the image of an ideal PR specialist?"; "What is, in your opinion, the perfect public image in advertising?". Then the adjectives were counted. The total number of adjectives was 306. Then the frequency of mentioning all adjectives was calculated and those that were repeated more than three times in the questionnaires were written out. The result was 72 adjectives that were ranked by frequency of mention. To construct scales of semantic differential, it was decided to select the most frequently mentioned characteristics that occurred more than three times. Thus, there are 21 adjectives (Table 1), repeated at least three times, which made up the scales-antonyms of the particular semantic differential used by us to develop a socio-psychological model for forming the image of a PR specialist. The order of scales was determined randomly so that similar concepts were not located in one part of the questionnaire. The possibility of a neutral response was eliminated by application of a four-point rating scale.

Table 1. Scales of the particular semantic differential

1	Cretaive	Not creative	12	Active	Passive
2	Sociable	Not sociable	13	Smart	Stupid
3	Educated	Not educated	14	Confident	Hesitant
4	Ethical	Cynical	15	Sly	Artless
5	Tactful	Tactless	16	Strong	Weak
6	Modern	Conservative	17	Calm	Scandalous
7	Fun	Serious	18	Purposeful	Passive
8	Responsible	Not responsible	19	Bright	Grey
9	Kind	Malicious	20	Competent	Not competent
10	Hard-working	Lazy	21	Interesting	Bored
11	Attentive	Inattentive			

(Source: authors).

The independent variable in the experiment was the concept of "Good PR specialist". Dependent variables were represented by semantic differential scales. The main stages of the experiment were organized as follows. The respondent in turn received a form for filling. Experiment participants noted on a four-point scale values of parameters between the poles (for example, creative – non-creative) that most accurately corresponded to their idea of the PR specialist's image. The data was processed using Microsoft Office Excel. A key image component for the PR specialist is the feature "sociable" (2,96).

With the aim of reducing the number of variables, which is 21 in the result of a particular semantic differential, and defining the structure of the relation between these variables, i.e. their classification, the authors conducted a factor analysis of data. The data was processed using the SPSS Statistics program. The principal component method using the varimax rotation method reduced the number from 21 explanatory variables to 6 factors with eigenvalues higher than 1.0. For interpretation, each factor included variables with loads on this factor from 0.50 and higher.

The Kaiser-Meyer-Olkin sample adequacy criterion allows us to draw a conclusion about the general suitability of the available data for factor analysis, i.e. how well the constructed factor model describes the structure of respondents' responses to the analyzed questions – a value that characterizes the degree of applicability of factor analysis to this sample. The results of this test range from 0 (the factor model is completely inapplicable) to 1 (the factor model perfectly describes the data structure). In our case, this indicator is equal to 0.548, which is a good result.

Bartlett's test of sphericity is a criterion of multivariate normality for distribution of variables. In addition to normality, the criterion checks whether correlations differ from 0. Statistic that determines the suitability of factor analysis for the Bartlett test is significance. At an acceptable level of significance (less than 0.05: an alternative hypothesis is accepted that the variables are correlated with each other), factor analysis is considered suitable for analyzing the sample under study. In the existing case, the significance of the test is equal to 0.000, therefore, the use of factor analysis is correct for the analysis of the sample under study. The full explained variance is shown in Table 2. Selection method is principal component analysis. The Cattell's rock scree criterion was applied too (1966).

Table 2. Full explained variance

Explained total variance									
Component	Initial eigenvalues			Extracting the sum of squares of loads			Rotation of the sum of squares of loads		
	Total	% variance	Aggregate %	Total	% variance	Aggregate %	Total	% variance	Aggregate %
1	4,845	23,071	23,071	4,845	23,071	23,071	3,378	16,084	16,084
2	2,655	12,645	35,716	2,655	12,645	35,716	2,351	11,194	27,278
3	2,055	9,785	45,501	2,055	9,785	45,501	2,308	10,993	38,271
4	1,739	8,280	53,781	1,739	8,280	53,781	2,037	9,698	47,969
5	1,418	6,752	60,533	1,418	6,752	60,533	2,013	9,587	57,556
6	1,227	5,841	66,374	1,227	5,841	66,374	1,852	8,818	66,374
7	,999	4,755	71,129						
8	,947	4,510	75,640						
9	,855	4,074	79,713						
10	,762	3,628	83,341						
11	,577	2,747	86,088						
12	,553	2,635	88,723						
13	,485	2,309	91,032						
14	,470	2,239	93,271						
15	,381	1,813	95,084						
16	,297	1,413	96,497						
17	,218	1,039	97,535						
18	,206	,982	98,518						
19	,139	,663	99,181						
20	,104	,496	99,677						
21	,068	,323	100,000						

(Source: authors).

The "stony scree" graph visually showed the number of factors: the 6 component/ factor shows a sharp decrease in the curve (Fig. 1).

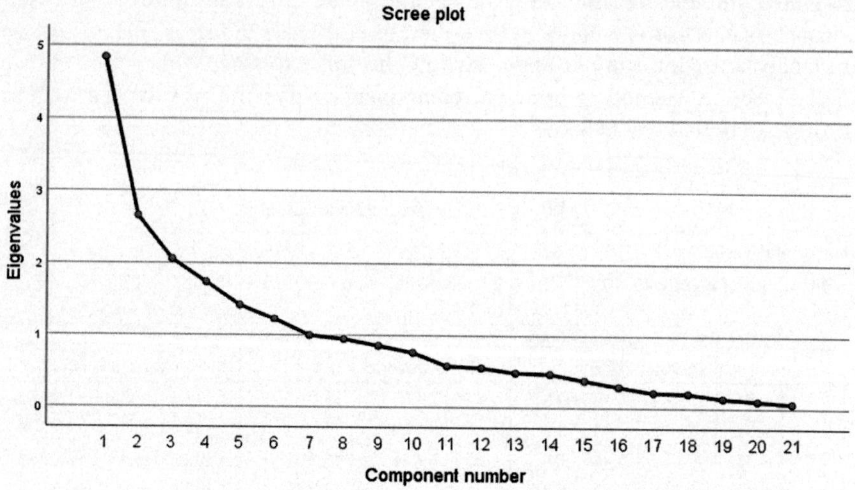

Fig. 1. Graph of "stony scree" (Source: authors).

In the resulting rotated matrix of factor coefficients, the results of classifying variables by factors were obtained. Using the automatic method of determining the number of factors, a factor model was constructed, in which 20 of the 21 variables were uniquely classified by 6 factors (Table 3). As a result, the variables were divided into the following groups (factors): 1 group (responsible, competent, hardworking, creative, attentive, modern); 2 group (interesting, bright, kind); 3 group (active, fun, sociable); 4 group (educated, smart); 5 group (sly, ethical, tactful, confident); 6 group (calm, purposeful, strong).

Table 3. Rotated component matrix

Rotated matrix of components (Rotation converged in 19 iterations)						
	Component					
	1	2	3	4	5	6
Responsible	,888					
Competent	,750					
Hard-working	,718					
Creative	,617					
Attentive	,613					
Modern	,540					
Interesting		,681				
Bright		,679				
Calm						,452
Kind		,579				

(continued)

Table 3. (*continued*)

Rotated matrix of components (Rotation converged in 19 iterations)						
	Component					
	1	2	3	4	5	6
Sly					,264	
Active			,765			
Fun			,652			
Sociable			,534			
Educated				,826		
Smart				,804		
Ethical					,795	
Tactful					,725	
Confident					,520	
Purposeful						,772
Strong						,674

(Source: authors).

The authors used the method of factor extraction: principal component method, rotation method: Varimax with Kaiser normalization. There were given conditional names for each of these factors: 1 factor - "wisdom"; 2 factor - "effectiveness"; 3 factor - "sociability"; 4 factor - "intelligence"; 5 factor – "tact"; 6 factor - "strength". Based on these factors, a model of the socio-psychological structure of the image of a public relations specialist was built. As a result of identifying factors that affect the image perception of a PR specialist, it became possible to develop a model for it. Below is a universal model of image perception for a public relations specialist (Fig. 2).

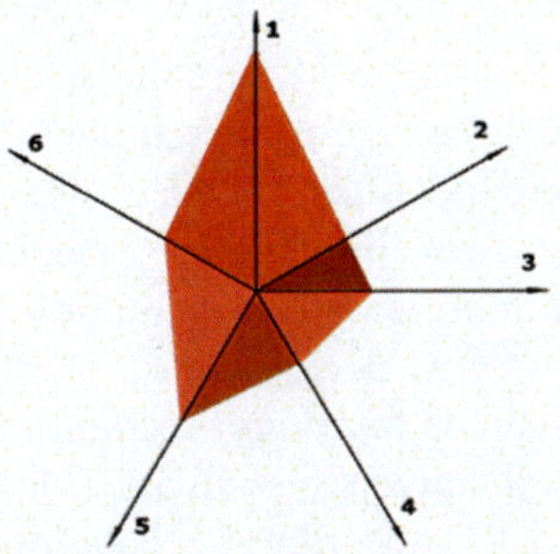

Fig. 2. Socio-psychological model of the PR specialist's image (Source: authors).

The model of the socio-psychological structure of the PR specialist's image can be represented as a multidimensional object, where each dimension is one of the factors of its perception. This figure schematically shows the structure of constructs corresponding to the independent variable "Good PR specialist", built on the basis of the factor analysis of results of a particular semantic differential. The polyline connects the points on the axes corresponding to the sum of the average values of the parameters of the PR specialist's image existing in the minds of respondents, distributed by the identified factors.

4 Discussion

Understanding the key components of the PR specialist's image was based on the Federal State Educational Standard of higher education in the direction of 42.03.01 "Advertising and public relations" (bachelor level), approved by order of the Ministry of education and science of the Russian Federation No. 997 of 11.08.2016 [3], according to which the field of professional activity of bachelors in the considered training field includes a number of components: 1) communication processes in the interpersonal, social, political, economic, cultural, educational and scientific spheres; 2) mass, business and personal communication techniques and technologies; 3) technologies of promotion of competitive properties of goods, services, commercial companies, non-profit and public organizations, state institutions and bodies, their positioning in the market environment; 4) public opinion [3]. As you can see, communication processes play a key role in the activities of these specialists. Further, the standard specifies types of professional activities that graduates are prepared for in this area, including communication activities as one of the key competences, in addition to organizational and managerial, project, advertising and information skills. Communication activity refers to the work on creating an effective communication infrastructure of the organization, providing internal and external communication, and participating in the formation and maintenance of corporate culture.

5 Conclusion

The process of working with the image of a public relations specialist will allow the image maker to create a particular image of a communicator. As a result of the study, the factors of its perception were identified, on the basis of which a model of the socio-psychological image structure was built. This model of the socio-psychological structure of the image can be represented as a multidimensional object, where each dimension is one of the factors of its perception.

The results of the research can be taken into account in the SRM strategy, since it is believed that a student-oriented university will be able to apply the correct implementation of its plans and goals. Accordingly, it helps to strengthen the academic culture so that students are provided with the necessary skills and knowledge for their career growth. SRM is a multi-pronged academic strategy that aims to use a chain of student satisfaction, retention, and loyalty for developing sustainable knowledge-based relationships with students [6]. According to some researchers, students are considered as main consumers [4] and partners due to the interactive and dynamic nature of higher education institutions [5].

According to the research conducted by Mingaleva, Deputatova, Akatov, Starkov, & Mitrofanova, business and organization managers, research workers, innovators should be aware of the importance of the knowledge creation process and its close connection with the innovation process. Managers of enterprises and organizations should contribute to the creation of an environment for the organization of a dynamic and effective knowledge formation process [9]. Accordingly, the data obtained during the research on the image of a PR specialist can be used for these purposes. Received research results show that the key component of the PR specialist's image is the quality "sociable" (2.96). The developed model could be useful for developing an HR strategy. For employer companies, opinions of future labor market participants (students) can serve as a starting point for adapting HR strategies in the digital environment [2, 10].

References

1. Arzhanova, K.A.: Socio-psychological mechanisms of forming the image of a political leader in the election campaign. Ph.D. thesis. Moscow: State University of Management (2016)
2. Dovzhik, G.V.: Socio-psychological mechanism of forming the external image of the organization. Ph.D. thesis. Moscow: State University of Management (2006)
3. Federal State Educational Standard of higher education in the direction of 42.03.01 "Advertising and public relations" (bachelor level), approved by order of the Ministry of education and science of the Russian Federation No. 997 of 11.08.2016. http://fgosvo.ru/uploadfiles/fgosvob/420301.pdf. Accessed: 09 March 2020
4. Gholami, H., Saman, M.Z.M., Mardani, A., Streimikiene, D., Sharif, A., Zakuan, N.: Proposed analytic framework for student relationship management based on a systematic review of CRM systems literature. Sustainability **10**(4), 1237 (2018). https://doi.org/10.3390/su10041237
5. Gholami, H., Saman, M.Z.M., Sharif, S., Zakuan, N., Abu, F.: Critical success factors of student relationship management. Sustainability **10**(12), 4527 (2018). https://doi.org/10.3390/su10124527
6. Gholami, H., Zakuan, N., Saman, M.Z.M., Sharif, S., Abdul Kohar, U.H.: Conceptualizing and operationalizing the student relationship management strategy: towards a more sustainable-based platform. J. Clean. Prod. **244**, 118707 (2020). https://doi.org/10.1016/j.jclepro.2019.118707
7. Korneychuk, Y.V., Lukin, N.A.: History of formation and specificity of the profession of PR-specialist. Young. Sci. **1**, 528–531 (2015)
8. Marsen, S.: Navigating crisis: the role of communication in organizational crisis. Int. J. Bus. Commun. **57**(2), 163–175 (2020). https://doi.org/10.1177/2329488419882981
9. Mingaleva, Z., Deputatova, L., Akatov, N., Starkov, Y., Mitrofanova, E.: Application of hadi-cycle for providing sustainability of processes of knowledge and innovation. Entrepreneurship Sustain. Issues **7**(2), 1628–1640 (2019). https://doi.org/10.9770/jesi.2019.7.2(58)
10. Vasileva, O.A., Dovzhik, G.V., Musatova, S.A.: Work motivational factors of Generation Z in the digital economy. In: Proceedings of the 2nd International Scientific and Practical Conference "Modern Management Trends and the Digital Economy: From Regional Development to Global Economic Growth" (MTDE 2020). Advances in Economics, Business and Management Research, vol. **138,** pp. 1018–1024. Atlantis Press, Paris (2020) https://doi.org/10.2991/aebmr.k.200502.168

Modern Trends of Russian Labour Market: Employers and Employees Expectations

S. A. Grishaeva[1(✉)], T. A. Beregovskaya[1], and E. V. van der Voort[2]

[1] State University of Management, Moscow, Russia
grishaeva@bk.ru, tbereg@gmail.com
[2] AB Sciex Netherlands B.V., Capelle Aan Den Ijssel, The Netherlands
katik@katik.nl

Abstract. This article discusses trends in the Russian labour market which reflect changes brought by digitization, changing gender roles, generational characteristics of the labour force, etc. Special attention is paid to the processes of flexibilization, which, in particular, result in such a phenomenon as "precarious employment". The authors consider the horizontal career as one of the trends, which allows maintaining motivation and involvement when reaching the "ceiling" of a vertical career, as well as the specifics of working in a remote format and the attitude of employers to this format of work. The conclusions in this paper are based on the data from a sociological research conducted by the headhunting company Kontakt InterSearch Russia in 2019-2020.

Keywords: Digitalization · Generation Z · Horizontal career · Labour market · Remote work

1 Introduction

The modern world is increasingly dynamic in its development and imposes new demands on social processes and areas of community life. Noticeable changes are also taking place in the sphere of labour and employment, affecting the employee-employer relationship, influencing their motivation to work, changing the requirements for professional competences, amount of labour, workers' ability to change, etc. The modern reality has shifted the most of activities into digital environment affecting all of the economy's fields. Intensive digital processes require fundamentally different skills. For example, Polyakova points out in her publication that digitalization has accelerated decision-making processes, which brings some benefits, but puts the individuality at risk [12]. A new generation of employees, the so-called generation Z, appears on the labour market and begins to play a significant role in it, the specifics of which have to be considered by employers. These and many other issues require a comprehensive analysis and conceptualization.

S. I. Ashmarina and V. V. Mantulenko (Eds.): IPM 2020, LNNS 161, pp. 18–24, 2021.
https://doi.org/10.1007/978-3-030-60926-9_3

2 Methodology

In this paper we used a number of sociological researches conducted by the head-hunting company Kontakt InterSearch Russia (https://kontakt.ru) in 2019-2020 and employed the methods of desk research:

1. "Expectations of employers and young professionals from each other". Where there were two audiences surveyed: students, and graduates of Moscow universities aged 18-23 (1,157 respondents), and owners, CEOs, and HR Directors of Russian companies (584 respondents) [16].
2. "Generation Z: myths and reality". Where there were the following surveyed audiences: CEOs, HR Directors, senior and middle-level managers of Russian companies (778 respondents) [17].
3. "Restarting Careers: What do CEOs do when their maximum is reached?" Where there were the TOP-managers surveyed of Russian companies: owners and CEOs, mainly medium and small businesses (373 respondents) [15].
4. "Additional education of managers. Required competencies". Where there were the TOP-managers of Russian companies from different surveyed industries (556 respondents) [14].
5. "Women and men in leadership positions: diversity, career development, family roles". Where there were the TOP-managers surveyed of Russian companies: Members of the Boards of Directors, General Directors, Directors with various functions (876 respondents) [9].
6. "Involvement, motivation, control of teleworking teams: what do employers do?" Where there were the HR-Directors and TOP-managers surveyed of Russian companies (452 respondents) [8].

3 Results

Our analysis of the above-mentioned studies revealed some trends distinct for modern Russian labour market. First of all, we would like to point out the considerable influence of such motivation factors for job search or job change as prospects for professional and career growth. Moreover, this trend is typical for various categories of employees: for young professionals (who work in junior positions and those who have just started integrating into the labour market), and high-level managers. In the research "Expectations of employers and young professionals from each other" the results show that young employees prioritize career prospects as the most important factors in their work (65% of respondents), followed by career development (56% of respondents), and then income (52% of respondents) [16]. Employers are in solidarity with young applicants, although to a somewhat lesser extent: career prospects are the most important factors for young professionals: 56 per cent of employers, career prospects are 51 per cent and income level is 37 per cent. [16]. A similar trend has also been observed in the study of career orientations of managers (CEOs) [15]. According to them, top-level employees sacrifice many things when changing their careers, including their income (43% of respondents) and their position (32% of respondents), but they still do not regret the decision they made. According to experts, the opportunity for development and professional growth is very important for managers, because the higher the degree of manager's involvement is, the easier and more effective it is for them to motivate their employees. However, it is difficult to do so [15].

Secondly, the horizontal type of career development is becoming increasingly relevant in contrast with the vertical one. As mentioned by the researchers, Grishaeva, Makarova: For many professionals in a mature age, the ceiling of a vertical career has been reached, and the volume of knowledge, skills and abilities has outgrown their current position [6]. In such a case one of the ways to motivate the employee to work would be the possibility of performing a switch to horizontal career. The data obtained from surveys conducted by Kontakt InterSearch Russia confirms this trend for both mature and young professionals. Thus, the company's experts note, that dynamic development and project activities are important for young professionals, which gives an opportunity to constantly learn something new and solve various tasks outside of their usual scope [16, p. 7]. Alikperova, Yarasheva, Vinogradova noted "… readiness of young people to change the type of professional activity" [1, p. 226]. Shafranov-Kutsev, Cherkashev, who have studied young people's focus on competition and entrepreneurship note that « … professional development, continuous and purposeful self-education, acquirement of professional, entrepreneurial, and business competences transform educational, innovative potential into a special competitive advantage ensuring competitive power [19, p. 118]». According to the results of a survey of 556 managers of Russian companies from different industries [9] 64% of respondents decided to get additional education in order to expand their functionality, i.e. to increase the possibilities of their professional realization also in related fields. 29% of managers are starting their own business after deciding on career changes [15]. It should be noted that the issue of horizontal careers is gender-specific [9]: 39% of female executives versus 36% of male executives believe that they have achieved the highest position in the company they currently work for, and 16% of female executives versus 5% of male executives stated that they do not want to grow in current company and plan to develop in another place/direction. In addition, women noted other factors as obstacles on their path to further development: uncertainty and fear of change to be the main obstacles to further development (22%), discrimination (17%), ignorance of prospects (12%) and consider lack of experience and knowledge (11%). At the same time, 48% of women said that nothing prevents them from growing further, while men gave such answer in 81% of cases [9].

Thirdly, flexibilization of processes is becoming widespread, resulting in, among other things, precarious employment, i.e. "non-standard labour relations that result in unstable social status for employees, an uncertain and unguaranteed degree of involvement in labour activities, unstable forms of distribution of surplus product and arbitrary remuneration", - Popov [13, p. 156]. This instability of the socio-economic situation requires a change in the behaviour of employees towards the transformation of their professional competencies and significantly changes the modern Russian labour market [2]. According to Shevchenko and Shevchenko, it is the education that is appropriate to the needs of the market that can increase the market value of an employee and their competitiveness [20]. This trend is understood by employers when hiring young professionals: two-thirds of respondents, employers' representatives, note that their companies have a development programme for young professionals, naming the following types: internship programmes (36%), trainings conducted by direct supervisors (23%), and management-trainee programmes (4%) [16]. This seems to be logical, because according to Hohlova and Hohlov "precarious employment, the so-called precarization, which is a main characteristic of the modern era, mostly affecting young people who have not yet had time to acquire a permanent profession" [7, p.90]. It is interesting to note that the assessment of specialized higher education is not equally evaluated by employers and young employees. As a mandatory requirement for young specialists, the presence of "only specialized education" is suggested by 40% of responded employers against 5% of responded young applicants [16]. 46% of employer's states that "orientation on the result" is the most demanded competence for young specialists [16]. Other competences, however, with named by a smaller number of employers, we also called in demand. Those are: "Client orientation" − 19% of respondents, "ability to work in a team" − 13%, "Analytical mindset" − 11%, "Leadership" − 8%, "Sociability" − 5% and "Stable performance at work" − 4%. As we can see, Soft Skills are the first to become in demand, but in combination with profiled education. Flexibilization processes also affect the leaders who have to improve their educational level constantly: "investing in education is one of the key areas for improving the competitiveness of employees", Golenkova, Kosharnaya, Kosharnyy [4]. The survey of managers showed that 95% of the surveyed managers received some form of additional education, which, according to the majority (82% of respondents), helped them in their career development [14]. The rating of the most demanded managerial competencies (according to the respondents) is as follows:

- ability to react very quickly to changes/stay ahead of changes – 86%,
- ability to look at the problem comprehensively – 75%,
- ability to work in the environment of an increase in budgets cuts – 50%,
- knowledge related to the digitization of business, understanding the principles of digital channels – 48%,
- ability to build work of a division/Department/company remotely – 27% (the study was conducted in 2019, before the coronavirus pandemic),
- ability to work according to the principles of holacracy – 11%.

As you can see, the most demanded competence is directly related to adaptation to the processes of flexibilisation. We see that all changes and trends in the market influence the requirements for the competences. Flexibility, the ability to quickly adjust strategy, multitasking, and the ability to learn quickly come to the fore. These qualities allow us to solve problems in the current crisis conditions: everything changes every day, new technologies are forced to appear, and we have to quickly adapt to these changing circumstances – says Tulubenskaya [11], partner at Kontakt InterSearch Russia.

Fourthly, a new generation of workers is entering the labour market, the so-called generation Z, which has grown up in the digital age and, due to these circumstances, has some specific characteristics. The appearance of this new generation coincides with technological changes in the economy, and "if these two processes (technological structure change and generation change in the economy) occur asynchronously, it threatens social stability", says Sadovaya [18]. There are many recommendations in the literature to employers on how to behave with members of the generation Z. However, there are studies that question many of these recommendations [3]. According to the results of a survey of 778 respondents (CEOs, HR Directors, senior and middle-level executives of Russian companies) [15], it turned out that most of the characteristics of young employees are not related to the generational theory, but are specific to all young professionals due to their lack of experience. The majority of respondents sufficiently agreed that generation Z has a more pronounced focus on work-life balance compared to the employees over the age of 30 (an average score of 7.5 on a 10-point scale), they strive for instant success (7.4), and flexible schedules (7.3), and that it is more difficult for them to deal with routine tasks (7.1) in contrast with older employees. However, when working with new generations of employees, it is certainly necessary to take into account the areas and directions in which young professionals feel most organically – for example, they have advantages over others in areas such as digital communication, social media marketing, programming and information systems administration.

Fifthly, digitalization, and now the coronavirus pandemic, have outlined another trend in the labour market – “working remotely”. According to the results of a study conducted by Kontakt InterSearch Russia between April 6 and April 15, 2020, [8], 20% of the companies are ready to leave their employees to work remotely even after the restrictions are lifted. At the same time, a significant share of respondents (30%) note that the efficiency of employees after switching to remote mode has decreased, 35% of respondents believe that their efficiency in this regime has not changed, and 18% of respondents note that it has increased on average. It is safe to assume that they are the backbone of those companies that are willing to leave their employees working outside the office for good. Of course, employers understand that telework requires other tools of motivation and control: according to Kontakt InterSearch Russia partner “… careful work with each employee, constant monitoring of burnout and loyalty levels comes to the fore” [8, p. 4].

According to the respondents, the most frequently used control tools “on remote”, are the following:

- focus on micromanagement: creating tasks for the day and monitoring their completion during the day – 42% of respondents,
- regular calls – 39% of respondents,
- chat chec – 31% of respondents.

It is interesting to note that the next most popular response about the used control instruments used (it was chosen by 21% of the respondents) – “not controlled at all”. In our opinion, this can be explained by the fact that evaluation (and, consequently, control) is carried out based on the results of work in any case. As noted by Staseva partner of Kontakt InterSearch Russia: “Remote communication leaves a crunch from the most important working questions, on which it is easier to assess the real involvement of an employee in the performance of their functions” [8, p. 4].

4 Discussion

The list of presented trends is not exhaustive. There is a noticeable imbalance between the education received and the immediate needs of the labor market, leading to youth unemployment [1]. The data obtained by Kontakt InterSearch Russia in the study "Expectations of employers and young professionals from each other" confirm the existence of this imbalance [16]: 52% of the surveyed managers name the lack of professional knowledge among graduates as the main difficulty in selecting young professionals. Assessing the level of knowledge that University students receive, employers do not show optimism: 73% of respondents believe that "Universities do not provide applied skills"; 56% of respondents believe that "Universities do not keep up with changes in technologies and approaches", and 21% of respondents do note that "Universities only provide a document necessary for applications". Only 8% of respondents are optimistic ("Universities provide an excellent base").

Another labour market trend discussed in the academic literature is a more active policy on employment of people with disabilities [11]. The trend we have already mentioned, caused by flexibilisation combined with the trend towards distant work, leads to an increase in the number of freelancers, whose specific work activity is also strongly condemned in the scientific community.

Strebkov, Shevchuk point out that freelancing is often more positively assessed in the ordinary mind than it should be, as an opportunity to be freed from organizational hierarchies and to strike a balance between work and life [21]. But not everything is so optimistic: By "choosing independence and autonomy, freelancers are trapped in flexible employment. According to the apt expression of Gorz, self-exploitation and self-sale of "man-enterprises", innumerable "I-JSC" take over the place of exploitation" [5, p. 13].

5 Conclusion

The trends identified in the Russian labour market will allow to creating forecasts that meso-, and micro-level of labour relations management. In our opinion, digitalization is the dominant factor of changes in these levels. It changes the expectations of employees and employers in relation to each other, the nature of the labour relationship, and labour legislation. At the same time, the processes of digitalization interact with many other social changes, which are not always correlated with each other. In our opinion, the speed of social changes in the modern world leads to a constant modification of all types of social relations in general and labour relations in particular. Therefore, constant monitoring of trends in the sphere of labour and employment is the most important area of interest for scientists in various areas and managers.

Acknowledgements. We express our gratitude to the company Kontakt InterSearch Russia for fruitful cooperation and research materials.

References

1. Alikperova, N.V., Yarasheva, A.V., Vinogradova, K.V.: Motivating young people's labor behavior as an opportunity for implementing financial strategies. Econ. Changes: Facts, Trends, Forecast **12**(1), 226–240 (2019)
2. Bobkov, V.N., Kvachev, V.G., Novikova, I.V.: Precarious employment in the regions of russian federation: sociological survey results. Econ. Reg. **14**(2), 366–379 (2018)
3. Bogacheva, N.V., Sivak, E.V.: Myths of the « Generation Z». NRU HSE, Moscow (2019)
4. Golenkova, Z.T., Kosharnaya, G.B., Kosharnyy, V.P.: Influence of education on improved competitiveness of employees in the labour market. Integr. Educ. **22**(2), 262–273 (2018)
5. Gorz, A.: The immaterial knowledge, value and capital. HSE, Moscow (2010)
6. Grishaeva, S.A., Makarova, E.V.: Determinants of a professional career crisis in the period of maturity and the possibilities of overcoming it. Bull. Univ. **1**, 162–167 (2018)
7. Hohlova, M.G., Hohlov, I.I.: Youth in the russian labour market. World Econ. Int. Rel. **62** (9), 88–96 (2018)
8. Kontakt InterSearch Russia: Involvement, motivation, control of teleworking teams: What do employers do?: https://www.cossa.ru/articles/add/Kontakt_.pdf. Accessed: 30.06.2020. (2020)
9. Maryina, A.: Almost half of female managers have experienced gender discrimination at work – survey. https://rb.ru/news/discrimination-gender/.: 30 June 2020
10. Mitkevich, A.: Only multitaskers will survive. What skills are needed to get a job in a crisis. https://www.forbes.ru/forbes-woman/403023-vyzhivut-tolko-mnogostanochniki-kakie-navyki-nuzhny-chtoby-ustroitsya-na-rabotu. Accessed: 30 June 2020
11. Olkhina, E.A., Medvedeva, E.Y., Dmitrieva, E.E., Kashtanova, S.N., Kudryavtsev, V.A.: Employers on employment of persons with health impact assessment and disabilities: opinions and facts. Perspect. Sci. Educ. **40**(4), 377–392 (2019)
12. Polyakova, N.L.: New horizons of the theory of society at the beginning of the XXI Century: from postindustrialism to neocapitalism. Moscow State University Bulletin. Series 18. Sociol. Politic. Sci. **2**, 83–115 (2016)
13. Popov, A.V.: From precarious employment to precariat. Soc. Stud. **6**, 155–160 (2020)
14. RBC: Managers of the Russian Federation receive additional education mainly at their own expense. https://pro.rbc.ru/demo/5c6579039a794742acb7ebe3?from=newsfeed. Accessed: 30 June 2020
15. RBC: Most CEOs in Russia are thinking about changing their careers. https://pro.rbc.ru/demo/5cdeaf209a794788cc2f7590. Accessed: 30 June 2020
16. RBC: Parade of expectations: what employers and young professionals expect from each other. https://pro.rbc.ru/demo/5d14b5b79a79473e8ed2e706. Accessed: 30 June 2020
17. RBC: What generation Z myths do employers believe. https://pro.rbc.ru/demo/5d761d129a79471db05c3e33. Accessed: 30 June 2020
18. Sadovaya, E.S.: The digital economy and the new labour market paradigm. World Econ. Int. Rel. **62**(12), 35–45 (2018)
19. Safranov-Kuzev, G.F., Cherkashov, E.M.: Youth focus on competition and entrepreneurship. Sociol. Stud. **4**, 117–123 (2020)
20. Shevchenko, I.O., Shevchenko, P.V.: Gender characteristics of precarity. Sociol. Stud. **9**, 84–94 (2019)
21. Strebkov, D.O., Shevchuk, A.V.: The trap of flexible work schedule: How unusual working patterns influence the work-life balance of freelancers. Monitor Public Opinion: Econ. Soc. Changes **3**, 86–102 (2019)

Prospects for the Development of State Personnel Policy in Russia

S. Kalyugina, T. Gridina, and D. Chuprova(✉)

North-Caucasus Federal University, Stavropol, Russian Federation
s.kalyugina@gmail.com, grita25@yandex.ru,
dina.chyprova@mail.ru

Abstract. This research was aimed at studying the state of state personnel policy in Russia and extrapolating the future state of the country's labor potential. The objectives of the study were to study the state of the economy and analyze changes in the labor market from the 1990 s, when the socio-economic development of society was slowed down as a result of irrational use of human resources of the country, to the present time, when the labor market is adapting to the market economy. Research methods include comparative, system analysis, abstract-logical, economic-statistical, sociological. The main research results are: analysis of the implementation of the state personnel policy should start from the study of measures developed at the federal level to improve the efficient, ensure of labor resources, the activation of the employment policy, creating a system of state social support for unemployed citizens. The study focused on the Russian Concept of State Personnel Policy, which was developed in 2001. However, the Concept was never adopted. Currently, it is necessary to create a unified legal and conceptual framework for the effective implementation of the state personnel policy.

Keywords: Competition · Labor resources · State personnel policy

1 Introduction

HR policy is a pre-planned course of action that sets out the fundamental principles and strategies for working with human resources. The HR policy helps to form rules for how an organization should act in relation to its staff and helps to implement a consistent approach to management [3]. Changes that took place in our country in the 90 s, in the intellectual, physical, social, technical and technological spheres, led to significant losses in the country's human resources potential [6]. There were practically no effective mechanisms for regulating labor and personnel relations at the state level As a result of irrational use of the country's human resources, the socio-economic development of society was hindered. The country's potential was significantly weakened, due to the deterioration of social indicators and a decrease in the prestige of professional highly intelligent personnel. The largest changes affected the areas of industry, construction, transport and agriculture.

For Russia, the solution of these issues is extremely important, since our country is the penultimate place in the ranking countries in terms of labor productivity and is

S. I. Ashmarina and V. V. Mantulenko (Eds.): IPM 2020, LNNS 161, pp. 25–31, 2021.
https://doi.org/10.1007/978-3-030-60926-9_4

about 2 times behind the average level [7]. In the regions of the Russia, in addition to measures to improve labor productivity, it is planned to create a new employment system that will meet the emerging needs of job seekers and employers [10].

2 Methodology

The relevance of the existence of a personnel policy and the analysis of the problems contained in its development were analyzed by representatives of the classical school of management. The most significant principles of scientific management were outlined by Taylor in 1911. Based on his experience, he identified four postulates of management: replacing methods of empirical work with methods based on the scientific study of tasks; scientific selection and development of each employee; and dividing the work equally between managers and workers [12]. Another founder of personnel policy is Fayol. According to the concept developed by him, key management functions are based on observing the principles of division of labor, power and responsibility, and moving towards the same goal [1]. The transition to a market economy has significantly changed the fundamental principles and content of Russian personnel policy.

In our research, we analyzed the labor market and made a forecast of the development of Russia's labor potential, taking into account the implemented programs to ensure employment of the population. In our research, we analyzed the labor market and made a forecast of the development of Russia's labor potential, taking into account the implemented programs to ensure employment of the population. The changes currently taking place in the Russian economy make it necessary to determine long-term aspects of personnel policy based on science-based planning. The emerging federal standard for a new type of employment center involves the development of unified requirements for centers, optimization of processes, and improving the efficiency and quality of provided services. The effectiveness of project activities can be determined based on the program indicators. However, statistics do not allow you to use these indicators in a short time and are often a year late. To avoid this, we suggest using monitoring and operational control to track national projects.

3 Results

The state's policy on the labor market in the 2000 s was passive and concerned primarily issues in the field of employment promotion and was aimed at supporting non-competitive individuals and persons with disabilities. The main measures were quotas for jobs; preferential taxation and subsidies for employers providing jobs to disabled people; organization of seasonal employment, and others [11]. Over time, the country's economy began to stabilize. The labor market has changed. The composition of the employed and unemployed began to change by age, educational, and professional characteristics.

The state is carrying out large-scale work to create a labor potential that would best contribute to combining the goals and priorities of the state, the enterprise and the employee [5]. In 2001, the draft concept of state personnel policy included principles aimed at the development of democracy. The principles of openness and professionalism, legality and social accessibility were declared.

1. Later, the Order of the Government of the Russian Federation of 06.05.2003 N 568-R « About the concept of actions in the labor market for 2003-2005 » approved the Concept of actions in the labor market for 2003-2005, where the key task was defined as the implementation of measures in the social and labor sphere, allowing in the future in a qualitatively new economic environment to increase the efficiency of employment of the population [9].

The Concept of actions in the labor market for 2003-2005 laid down the main directions and priorities of the activities of state authorities to implement state policy in the field of employment, aimed at increasing the flexibility of the labor market, developing effective employment, improving the system of employment management, creating conditions for reducing unemployment and providing social support to unemployed citizens.

The Concept of actions in the labor market for 2003-2005 was implemented as part of anti-crisis measures in the economy and provided a reduction of tension in the labor market: reducing the unemployment rate, maintaining about 2.0 million jobs, creating about 4.0 million additional jobs, including more than 306.8 thousand permanent jobs (Fig. 1) [9].

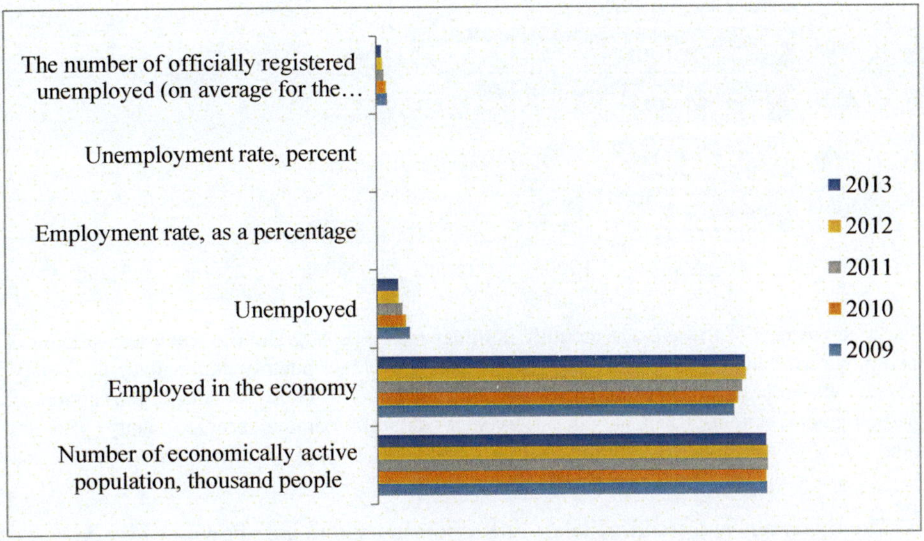

Fig. 1. Number of economically active population, employed and unemployed in 2009-2013 (Source: authors based on [2]).

As can be seen from Fig. 1, Russian researchers use two indicators of the unemployment rate to characterize the labor market: the general unemployment rate and the officially registered unemployment rate in employment agencies. Statistical agencies calculate the level of total unemployment according to the methodology of the International labour organization. Using this method of calculation leads to the fact that the level of officially registered unemployment is several times higher than the real one. The reason for the excess of the overall unemployment rate is that the majority of unemployed citizens do not resort to the help of state agencies for employment due to the low level of offered financial support and difficulties in registering the status of an unemployed person in employment centers.

At the same time, important achievements of this period were the growth of the average monthly nominal wage by 3.5 times, as well as maintaining the standard of living of unemployed citizens while increasing the maximum amount of unemployment benefits by more than 1.5 times (Fig. 2).

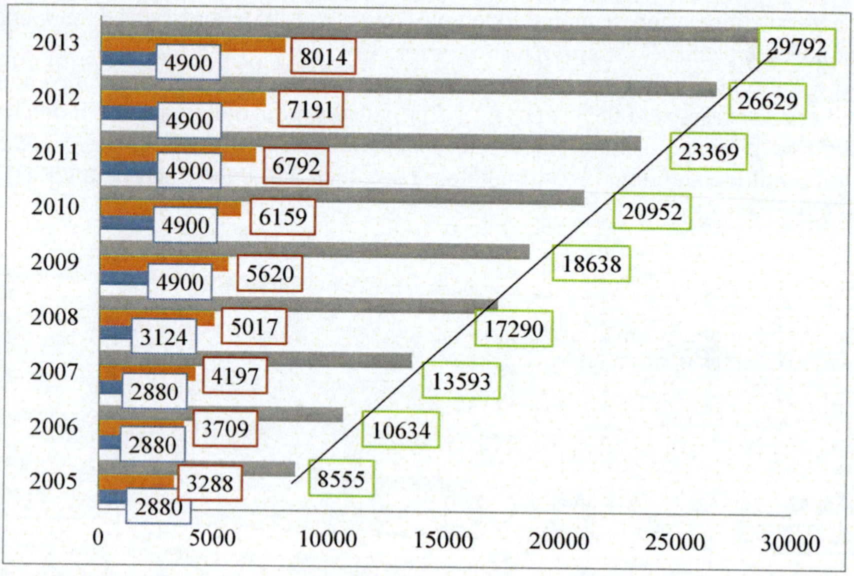

Fig. 2. Dynamics of the average monthly nominal accrued salary and the maximum amount of unemployment benefits for 2005-2013. (RUB) Grey - average monthly nominal accrued salary, Orange - the minimum subsistence level for able-bodied citizens, Blue - maximum amount of unemployment benefit, Black – Linear (average monthly nominal accrued salary) (Source: authors).

The dynamics and trend of labor utilization indicators are shown in Fig. 3.

In 2018, with the support of the state, measures to increase labor productivity and support employment were implemented at 200 enterprises in 16 regions of the Russia. At the same time, 70% of enterprises achieved a 10% increase in labor productivity, revenue-by 13%, and the tax base on profits-by 18%. In the country's economy increased labor productivity is by 3.4% compared to 2015 (Fig. 4).

This has become a serious prerequisite for the development of a National project "Productivity and employment support" in 2019. In 2019, 1,025 enterprises from 36

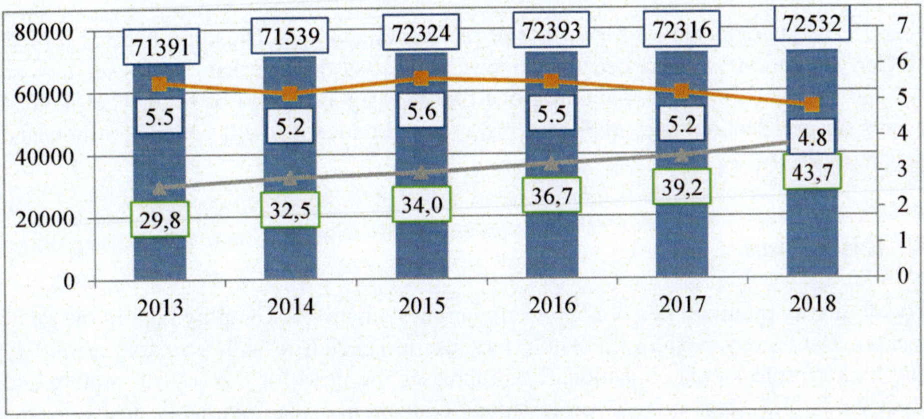

Fig. 3. Indicators of the use of labor resources: Blue - Number of people employed in the economy, thousands, Grey - average monthly nominal accrued salary, thousand rubles, Orange - Unemployment rate, % (Source: authors based on [10]).

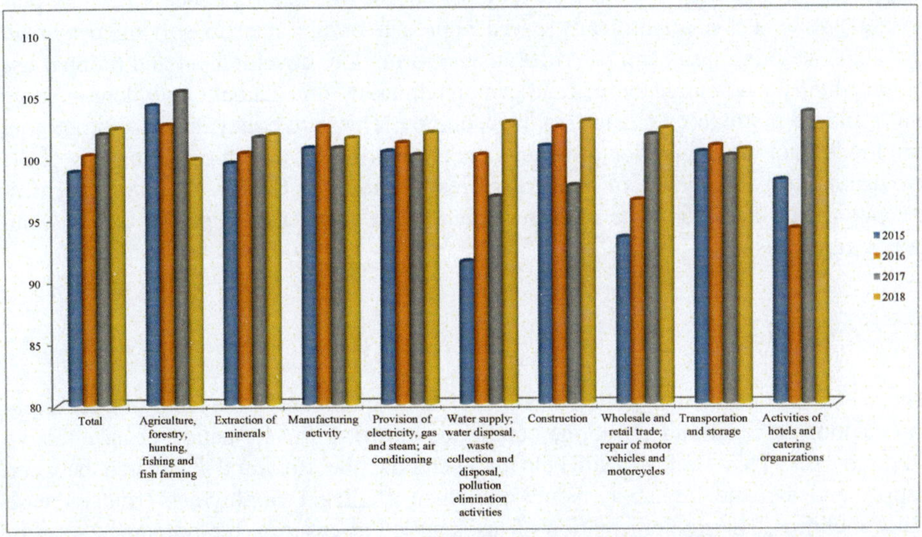

Fig. 4. Growth (decline) rates of labor productivity by individual types of economic activity (Source: authors based on [10]).

regions of the Russian Federation in 341 areas of activity participated in the implementation of the national project [8]. By 2024, the project will involve all regions of the Russian Federation, covering at least 10,000 large and medium-sized enterprises [4]. Big questions provide the economy with human resources and increasing employment is also expected to be solved in the framework of another national project "Demography" through Federal projects "Senior citizens" and "Promoting women's employment - the creation of conditions of preschool education for children aged up to three years." The state does a lot for people of pre-retirement age. Thus, the Federal project "Older generation" was created for citizens who are at the age that entitles them to an old-age insurance pension, including one that is assigned ahead of time. The planned events on professional and additional professional education will cover 450 thousand people in 2024.

4 Discussion

Today, highly qualified personnel have the highest chances of employment in the labor market. Data analysis shows a steady increase in specialists with primary, secondary and higher professional education. This increases the need for theoretical training and changes in the standards of professional education. This invariably leads to the development of social partnership between the state, enterprises and educational institutions in the field of training workers, taking into account both the requirements of employers and civil society to train employees. It became obvious that the labor market is adapting to a market economy, and government assistance in the form of individual measures will not help to adjust all its proportions. We need a concept of state personnel policy, i.e. a scientifically based approach of the state to solving personnel problems, defining tasks and priorities for the formation, development and rational use of the country's human resources and personnel, taking into account both demographic shifts and deep structural changes in the economy. The state policy aimed at promoting employment of the population proclaims the basic principles of unity and quality of the provision and performance of public services provided by the law on employment of the population in accordance with the federal state standards of public services and public functions.

5 Conclusion

The declared goals of state policy that affect the development of the labor market imply the creation of legal and economic conditions aimed at the formation of an effective labor market. This strategy will help to overcome the structural mismatch between supply and demand for labor, reduce the share of illegal employment, and increase labor mobility. The implementation of tasks aimed at ensuring the implementation of the right of citizens to protection from unemployment, improving the efficiency of regulating the use of labor resources and ensuring the protection of labor rights of citizens, and introducing a culture of safe work should contribute to the achievement of these goals. For each of these tasks, the state authorities have outlined areas of activity: stimulating the development of employment, creating conditions for extending the terms of employment, ensuring compliance with established norms and rules in the field of labor market regulation and labor relations, improving the effectiveness of monitoring and supervision of their implementation. Thus, the analysis of employment promotion programs adopted in the Russian regions shows that the development of regional policy in the labor markets of the regions was influenced by the following factors: either the regions reacted to the possibility and need to develop regional programs determined at the federal level. The convergence of recent total and officially registered unemployment indicates that the measures taken in the framework of national projects and state programs are the result of the change to the actual regulation of the labour market and personnel policy.

Acknowledgements. The study was carried out at the expense of a grant from the Russian Foundation for Basic Research (project No. 18-010-00732).

References

1. Fayol, H.: General and industrial management. Martino Fine Books, Eastford, CT (2013)
2. Federal State Statistics Service: Human resources. http://rosstat.gov.ru/labour_force. Accessed: 14 July 2020
3. Fursov, V.A., Krivokora, E.I., Savchenko, I.P., Gorlova, E.B.: Methods of assessing the state of the labor potential of the region. Res. J. Pharm. Biol. Chem. Sci. **9**(6), 1955–1960 (2018)
4. Fursov, V.A., Krivokora, E.I., Strielkowski, W.: Regional aspects of labor potential assessment in modern Russia. Terra Economicus **16**(4), 95–115 (2018). https://doi.org/10.23683/2073-6606-2018-16-4-95-115
5. Giltman, M.A., Obukhovich, N.V., Tokareva, O.E.: The impact of the labour market policy on registered unemployment. Public Adm. Issues **4**, 51–76 (2017)
6. Kalyugina, S., Pyanov, A., Strielkowski, W.: Threats and risks to Russia's intellectual security in the context of global globalization. J. Institutional Stud. **12**(1), 117–127 (2020). https://doi.org/10.17835/2076-6297.2020.12.1.117-127
7. Mukhoryanova, O., Hercegova, K., Kalyugina, S., Tretyakova, O.: Socio-economic labor relations and personnel security in multicultural environment. Terra Economicus **16**(4), 130–143 (2018). https://doi.org/10.23683/2073-6606-2018-16-4-130-143
8. National project "Productivity and employment support" dated 24.12.2018. http://government.ru/info/35567/. Accessed: 25 June 2020
9. Order of the Government of the Russian Federation of 06.05.2003 N 568-R « About the concept of actions in the labor market for 2003–2005 ». https://rulaws.ru/goverment/Rasporyazhenie-Pravitelstva-RF-ot-06.05.2003-N-568-r/. Accessed: 22 June 2020
10. Strielkowski, W., Astachova, E., Larionova, N.: Assessment of personnel risks and threats in the context of state security: A neo-institutional perspective. J. Institutional Stud. **11**(3), 170–181 (2019)
11. Strielkowski, W., Kalyugina, S., Mukhoryanova, O.: Managing human potential and youth social entrepreneurship in peripheral regions. Manag. Res. Pract. **11**(3), 41–51 (2019)
12. Taylor, F.W.: The principles of scientific management. Dover Publications, New York (1997)

PR-Support of Sports Club as Way to Attract Funding and Labor

O. A. Kulikova[1], A. A. Komarova[1](✉), and Z. S. Khussainova[2]

[1] State University of Management, Moscow, Russia
kilka2002@yandex.ru, komarova_a@inbox.ru
[2] Buketov Karaganda State University, Karaganda, Republic of Kazakhstan
zhibekhll@mail.ru

Abstract. PR in sports is aimed at creating a positive image of an organization, the reputation of a sports club, or even an individual athlete, as well as to attract public attention to various sporting events and events in the field of sports. Domestic commercial sports is trying to find new sources of funding to facilitate the process of effective development of sports organizations. Among the sources of income for the clubs is implied ticketing, profit from the sale of television rights, merchandise and sponsorships. A set of specialized PR communications and tools is used to attract the attention of a disinterested audience and increase the popularity of a sports organization among fans. These are sports fans who occupy one of the leading positions in financing the activities of a particular sports organization in any country of the world a decade ago and now. Accordingly – the more fans, the better funding for club members, as well as more people who want to work in a well-known and attractive club.

Keywords: Hockey · Loyalty cards · PR support · Social networks · Staff loyalty

1 Introduction

The audience develops a positive attitude to sports events, the sports organization acquires a positive image with the help of thoughtful and planned public relations activities, and thanks to the efforts of internal press services and marketing departments, sports has become one of the most profitable areas of sponsorship and profit. The history of the Dynamo Moscow hockey club brand is inextricably linked with the formation of Russian hockey, as well as the transformation of sports events into a mass and popular product. Hockey has long been a special part of show business, like all professional sports. Professional hockey clubs in the country today are focused not only on sporting achievements, but also on profit. Gradually, the audience began to come to the arena not only for the sake of one hockey game, but also to participate in entertainment events, competitions and activities before and during the match itself, to buy souvenirs and attributes of the club in the lobby, take a photo and get to the big media club in the arena. Fans are one of the most reliable and stable tools for club financing. The commercialization of high-performance sports (elite sports) has made sport a significant player in regional and national economies and political economy.

S. I. Ashmarina and V. V. Mantulenko (Eds.): IPM 2020, LNNS 161, pp. 32–38, 2021.
https://doi.org/10.1007/978-3-030-60926-9_5

Sports are the most important public organizations in the social community and social events that unite groups from youth to professional competitions. Sports events compete with other entertainment events, as well as with each other for the attention of fans all over the world. Hockey, however, is the most expensive of the major international team sports. Indoor ice arenas are very expensive to build and operate, and equipment from skates, uniforms, and fall protection costs of the player more than in any other team sport. Ice hockey has a dominant position historically in countries such as Canada, Russia, Finland, Sweden, and the Czech Republic, which is due to the annual performances of the Russian team at the world championships, where the "Red machine" – the nickname of the Russian national ice hockey team, received by the team in Soviet times – is traditionally among the favorites.

According to a study by the global measurement company Nielsen, the hockey audience in Russia is 57 m.p. (or 57%) [8]. This is the number of respondents who are "interested" or "very interested" in this sport, with the highest level of interest expressed by 26 million Russians (or 26% of respondents): they can be considered active hockey fans. The potential audience of hockey in Russia is 83.3 million people (83%) in general - those who are at least a little interested in hockey and follow hockey news [8]. The Kontinental hockey league was created in Russia in 2008 to develop and popularize ice hockey. A project is independent from national federations and the Russian ice hockey Federation in particular. This was preceded by many years of work by Fetisov and Medvedev, and the result was an agreement between the Russian ice hockey Federation and the Kontinental hockey league that the Russian ice hockey championship would be organized by the KHL structures.

The Kontinental hockey league is a domestic and international league created for the development of ice hockey in Russia, as well as in the countries of Eurasia. Currently, the league includes 19 Russian clubs and 6 foreign ones, represented by teams from the Republic of Belarus, Latvia, Kazakhstan, Finland, Slovakia and China. The Kontinental hockey league is one of the leading sports club leagues in Russia. Recognition of the KHL in the country is at the level of 83%, but only 26% of respondents said that they follow the performances of their favorite club or matches on a regular basis. It is also worth noting the features of the so – called "salary cap" - a certain amount for which clubs form a team. There is a reduction in the "ceiling", aimed at increasing transparency, improving the quality of commercial and sports components and equalizing the financial capabilities of clubs in the domestic league, according to the five-year development strategy announced by the league. This will make the sports competition more spectacular, and the result of the match unpredictable, as the dominance of several single-digit leading teams has been established in recent years, the exit to the playoffs and the victory in the championship of which is predicted before the start of the season. The current situation is explained by the presence of a much larger budget, which allows you to form a team of the strongest players, offer hockey players more favorable financial conditions for moving to your club.

The "salary cap" was set at 850 million rubles for one club per season in the 2018/2019 season, and it was also possible to exceed it by paying the league a 20% additional fee. "The ceiling" will be reduced to 800 million in season 2019/20, exceeding it will cost 30%. Starting from 2020, a strict ceiling of 900 million rubles will be finally introduced, while the possibility of an additional payment if it is exceeded will be excluded. Conditions of economic instability require rapid transformation and adaptation to new conditions [9].

2 Methodology

The paper uses a secondary analysis of a number of sociological studies conducted by Nielsen. A full-scale survey conducted by Nielsen contains responses from more than 4,700 respondents. The methodology takes into account the territorial zoning of Russia. The general totality consists of citizens of the Russian Federation aged 16 to 65 years. We also analyzed internal documents - seminars on marketing and communications of the Kontinental Hockey League held in 2017, 2018 and 2019. A sociological survey was conducted in 2019 among the club's subscribers in social networks using an online questionnaire. The sample consisted of 372 respondents. Active and potential fans were studied in the aspects of ticket costs and attributes, their typical lifestyle, income group, social activity, attitude to sponsorship and the club. Content analysis of Dynamo club's media behavior in social networks (Facebook Instagram, Twitter, YouTube, Telegram) was carried out. Statistics of the club's subscribers, in particular their demographic data, have been studied.

3 Results

In order to correctly determine the direction of the club's development, the possibility of attracting staff and funding, and to place emphasis on working with fans we need to study our audience The research will be useful not only for the sports organization and the league, but also for potential sponsors: this is how they can evaluate the effectiveness of possible cooperation and investment. After all, the digital age requires us to transmit high quality information through social networks and other digital platforms [1]. Social network analysis was chosen and a traditional study was conducted as a way to study the audience of the Dynamo Moscow hockey club – an online survey was launched, which was attended by subscribers of the Dynamo Vkontakte group. In Russia, the Internet is still growing, including at the expense of a more adult segment of the audience, which suggests that we need to increase interaction in the virtual space [2]. The software allows you to quantify the social and demographic characteristics of users and their behavior in social networks [3]. The "geography" line shows the cities where subscribers are concentrated, in descending order of popularity in statistical data. This is Moscow and Saint Petersburg in most cases, followed by Latvia's Riga, Novosibirsk and Minsk.

A large proportion of subscribers to club accounts are men, but a fairly large percentage of women are interested in club news. We can conclude that hockey remains a team sport that is interesting for both men and women. The average age of a Dynamo fan varies from 25 to 44 years. There is a hypothesis that the average fan of the club is a family man who occasionally attends matches with his family. The average age of the audience that is interested in hockey is 30–40 years according to the survey.

Most of this audience is represented by married men who most often go to hockey with their family, and sometimes with friends. Thus, attending hockey matches is the same form of leisure as going to the cinema, theatre.

The portraits of the target audience were created using the obtained results. Loyal fans attend all or most of the regular season matches and all the club's playoff games, attend pre-season matches at the sports base or games held within the framework of the Moscow Mayor's Cup, purchase a season ticket for several seasons in a row, and monitor the club's Internet resources. Hockey is interesting as a sports competition, following not only Dynamo, but also other clubs of the Continental and National Hockey League. The whole family supports the club. Attends away games of the club in major cities: Saint Petersburg, Kazan, Minsk. They often buy club attributes on the days of home matches. The average receipt is 1200 rubles. They use the attributes and products of the club in everyday life, on match days they are located in the central stands. Income per family member is 50 thousand rubles. Have higher education. Married with two children. Are rational, would like to save money and participate in the incentive program.

Enthusiasts and fans of the weekend attend some matches, usually with top clubs, regard hockey as a show, go with family or friends. Occasionally they buy the club's products, usually when new products appear or the club produces limited - edition products-attributes with a special design, dedicated to the anniversary of the club or an important event, the team's entry into the playoffs, and so on. Subscribed to the several pages of Dynamo in social networks. The team's guest matches are watched on TV. They are receiving or have already received a higher education. Most fans note that they have an average income (78%). So, fans identified three factors that will attract them to attend a hockey match: the team's sports results (42%), the ticket price (38%), and the opportunity to have fun not only watching the competition. Dynamo fans noted among other things that the privileges available to users of the loyalty program will be able to attract an even larger audience to hockey competitions. They are interested in discounts or increased cashback for purchasing tickets and season tickets for team matches, access to club events: meetings of the team and management with fans, attending team training sessions, post-match press conferences, etc. Bonuses when paying for souvenirs in branded stores will allow loyal fans to use the services of the club on favorable terms, which will be a consequence of their motivation to cooperate.

The fans also want the club to pay great attention to its partners. At the moment, Dynamo's partners are companies that do not have a useful property for fans and are not relevant for the target audience. Cooperation with such organizations has a formal meaning and is provided by the banner space system on the club's website in exchange for monetary compensation. The signing of a cooperation agreement with a chain of restaurants or cafes, car brands, grocery stores and electronics, as well as fitness clubs provided the brand's appeal to the target audience and guaranteed the promotion of the club for people unfamiliar with the club's activities, as noted above.

4 Discussion

The main task of marketing and PR in sports is to establish a link between achievements in sports tournaments and performance outside the competition process. The target groups of the hockey club include the media, general sponsors, partners, fans and internal audience – the players themselves and employees of the club and the league. In order for the club to be able to pay for "expensive" players, it is necessary to work seriously with fans and sponsors, so you can see the specifics of the tasks of managing the club. The product produced by a sports organization is the final result of the competition. Marketing and PR work is based on what is happening on the ice. At the same time, the primary basis is the sale of unpredictability of the result and inextricably linked emotions and sensations.

While sport is considered and most often describes itself as a mass type of leisure organization, it is also the most personalized. Thousands of fans come to the stadium and, in fact, choose one product, but each needs a personal approach. Dynamo is proposed to restart and start implementing a new loyalty program to solve the problem of the club's poor image, to combat the established opinion of fans about unsatisfactory work with fans, as well as to increase revenue from commercial activities. This is necessary to increase engagement. It is necessary to inform the staff that scientific research accompanies all activities from planning to creating and promoting the product [10].

5 Conclusion

Digital technologies, and first of all, Internet technologies not only significantly transform the business process itself, but also modify its infrastructure and market conditions, when introduced into business: the market, demand features, consumer and competitor behavior, and so on [7]. The future development of corporate information systems based on advanced business analytics depends to a large extent on the level of maturity of business processes in the organization and the readiness of personnel for innovations [4]. In cooperation with VTB Bank, the club will issue multi-service bank cards "Our movement" for the fans of the Dynamo Moscow hockey club. The goal of the card is to make communication between fans and the club not only interesting, easy and memorable, but also as profitable as possible.

Fan card "Our movement" is a debit card. Owners will be able to pay for purchases with its help, including using innovative contactless payment technology, or make it into a mobile device and use it as a digital card, as well as withdraw cash from an ATM, make payments via internet banking. The cards have a special design with the club logo.

"Our movement" is the first card for a sports club in the history of VTB and the first Dynamo card issued on a bank card using modern technologies. Fans will be able to save on ticket purchases with its help, the club, in turn, will increase the turnover from sales of season tickets and products, and partners "Dynamo" and VTB Bank will get new customers. Thus, we work with three main target groups of the club: fans, partners and the general sponsor. This bank card can be used as a fan ID, and you can use it to record a season ticket to attend matches at the VTB arena sports complex. It also presents a bonus card for participation in the loyalty program of the hockey club.

It will provide a number of privileges for loyal fans. These include exclusive events of the club, having accumulated a certain number of points, fans will be able to participate in meetings with fans and the team, visit the team training, participate in post-match press conference and see the work of journalists from the inside, to visit the club and see the training process and communicate with their favorite Dynamo players, participate in photo and autograph sessions, and view the match from the VIP or press box. This also includes special events and promotions dedicated to the birthday of the "Our movement" cardholder. The program participant will be able to spend the accumulated bonuses on buying tickets, season tickets and hockey attributes with autographs of idols.

If a member of the loyalty program is a season ticket holder, then they not only secure their own place on the podium, but also get the opportunity to be one of the first to purchase a certain number of tickets for matches with top clubs, as well as buy a place in case the team reaches the playoff stage. In this way, the club will be attract a new audience that may have never attended a hockey match, and support the demand for Dynamo's services and products. It is recommended to implement a CRM system, Fanats Relationship Management, to properly work with the data that the club will be able to get later after working with the loyalty program.

CRM has become an indispensable tool for marketers in modern marketing. CRM marketing is always based on unique collected and analyzed information about consumers of a product or service [5]. The system records the history of relationships with customers, information about their preferences, and the pool of collected information forms the features and actions that need to be performed to maintain effective interaction. Recently, specialized CRMS for PR specialists have started to appear on the world market, which have attracted the attention of sports industry workers. This is not just software and computer software at the moment, it is a full-fledged communication tool that simplifies and systematizes communication with the audience.

The fan base is not limited to the usual list of names, phone numbers, and e-mail addresses in the Fanats Relationship Management system. Each contact has a personal card, which displays the history of relations with the fan, whether they have a subscription or information about participation in the loyalty program, information about purchases made by them, their interests, and activity regarding the club's mailing lists. This data organization facilitates teamwork with the database: all employees of the PR Department are aware of the specifics of interaction with a particular fan. It is a loyal fan who brings money to the club and at the same time wants to get the services he needs from it, in the form of a sporting event, as well as goods, in the form of souvenirs or branded attributes.

In its turn, the system can integrate into several subsystems and contain information not only about the audience of fans, but also information about journalists and media, as well as partners of the organization. Thus, the system provides templates for press releases and messages sent to both fans and journalists, depending on the tasks set. Fanats Relationship Management allows you to make thousands of contacts. One of the important advantages that the CRM system will give is feedback on the mailing list, since it will allow you to evaluate the effectiveness of the mailing list and provide information about the number of opened and read emails, the viability of mailboxes, and so on.

Sending out the latest news from the club's life will make life easier for many fans, which will increase their loyalty. Dedicated fans spend a large amount of time on Internet resources, searching for the latest news. If the mailing list is targeted, then the fan will get the impression that the club's representatives tell the news personally to him and if they share such information with him, then he feels that he belongs to the common cause. The mailing tool can also be used to invite fans to traditional meetings of fans and teams, and based on responses to make predictions about the attendance of such events. The automated Fanats Relationship Management system will solve a number of important strategic tasks: increase revenue from interaction with fans, attract new audiences and increase the loyalty of existing ones, as well as provide information and analytics. All these measures will attract strong players and increase the number of jobs in the club, as well as increase the loyalty of employees of the organization. When employees feel an internal similarity with the company, its values, goals, and feel satisfaction with their involvement, this is one of the indicators of successful integration of the employee into the corporate culture of the club [6].

References

1. Gault, F.: User innovation in the digital economy. Foresight STI Govern. **13**(3), 6–12 (2019)
2. Grishaeva, S.A., Kolosova, O.A., Kulikova, O.A.: Concepts of PR promotion in social networks. In: Proceedings of the 1st International Scientific and Practical Conference "Step into the Future: Artificial Intelligence and Digital Economy" pp. 215–221. State University of Management, Moscow (2017)
3. Karpikova, I.S., Artamonova, V.V.: Attracting audience to digital media using gamification elements. Quest. Journal. Theory Pract. **7**(4), 599–614 (2018)
4. Komarova, A., Tsvetkova, L., Kozlovskaya, S., Pronkin, N.: Organisational educational systems and intelligence business systems in entrepreneurship education. J. Entrepr. Educ. **22**(5), 445 (2019)
5. Komarova, A.A., Stepankov, G.A.: Using CRM marketing tools in B2B and complex markets. Modern Sci. Actual Prob. Theory Pract. Ser. Nat. Tech. Sci. **2**, 83–87 (2020)
6. Kulikova, O.A.: Culture shock: Socio-psychological adaptation to a new cultural environment. Bull. Univ. **5**, 190–195 (2017)
7. Moreeva, E.V., Komarova, A.A., Dianina, E.V.: Digital marketing in the tourism business. Labor Soc. Relat. **3**, 114–128 (2020)
8. Nielsen: Hockey fans: Features of the 57-million audience in Russia. (2018). https://www.nielsen.com/ru/ru/insights/news/2018/bolelshchiki-hokkeya-v-rossii-kto-oni.html. Accessed 22 June 2020
9. Popov, E.V., Omonov, G.K., Naumov, I.V., Veretnikova, A.U.: Trends in the development of social innovations. Terra Economicus **16**(3), 35–59 (2018)
10. Zaigrajkina, I.N., Ostapenko, R.I.: Innovative activity and innovative staff in the modern economy. Perspect. Sci. Educ. **2**(26), 68–71 (2017)

Assessment of the Labor Potential of the Stavropol Region

A. Yu. Lukhyanova(✉), S. A. Fedorova, and A. A. Prostyakov

North-Caucasus Federal University, Stavropol, Russia
anyrik@mail.ru, kull4@yandex.ru,
prostyakov.alexandr@yandex.ru

Abstract. The research purpose is assessing the labor potential of the Stavropol region based on a special methodology. The following components were identified as structural elements of the labor potential: demographic; economic; educational and qualification; psychophysiological; innovative. When calculating the labor potential, the following indicators were used: the total birth rate; the total mortality rate; the coefficient of migration growth (retirement) of the population; the ratio of marriages and divorces; age dependency ratio; employment rate (%), unemployment rate (by ILO methodology, %), labour productivity index, population with money incomes below the subsistence minimum in percentage of the total population, the average monthly nominal wage per employee in the full range of organizations (ruble); life expectancy at birth (years), availability of hospital beds (per 10 thousand people of the population), the provision of population with medical workers (per 10 thousand people of the population), expenditures of the consolidated budgets of the regions of the Russian Federation on healthcare (million rubles); the number of organizations performing R&D, the number of personnel engaged in R&D, internal R&D costs, the innovative activity of organizations and the volume of innovative goods, works, services, etc. The main sources of analysis were official statistic data.

Keywords: Evaluation of labor potential · Labor potential · Method of estimating the employment potential

1 Introduction

The unique feature of the Stavropol territory is its geopolitical position. It is part of the North Caucasus Federal district, which includes 6 republics. The region belongs to the "poles of growth" regions. These are small regions with low investment risk in terms of population and economic power, which will reach a ceiling in their development in the next 10–15 years. The neighboring Republic of North Ossetia-Alania belongs to the "problem regions", which, if their investment attractiveness increases, may become also one of the growth points and one of the steadily developing regions of the Russian Federation. Dagestan, Kabardino-Balkaria, Ingushetia, Chechnya, Karachay-Cherkess Republic belong to "regions of special attention". Thus, the Stavropol region, on the one hand, can become an attractive region for energetic and talented labor migrants from other regions of the North Caucasus Federal district, and, on the other hand,

S. I. Ashmarina and V. V. Mantulenko (Eds.): IPM 2020, LNNS 161, pp. 39–47, 2021.
https://doi.org/10.1007/978-3-030-60926-9_6

migration processes taking place in the region can contribute to the spread of ideas of extremism and nationalism among young people, carry the threat of conflict situations and significantly affect the labor market.

2 Methodology

A single methodology for assessing the labor potential has not been developed or approved. We have considered "Methodology of system analysis of the composition and structure of the labor potential of the Institute of Strategic Studies of the Republic of Bashkortostan" [15]; "Methodology of the Institute of Socio-Economic and Energy Problems of the North Komi science centre of Ural branch of Russian Academy of Sciences" [10]; "Methodology of the Institute of Socio-Economic Problems of Population Russian Academy of Sciences" [7]; "Methodology of the Institute of Socio-Economic Territories Development of Russian Academy of Sciences" [6]; "Methodology for assessing the actual state of the labor potential of Russian regions of the Southern Federal University" [5]. Each of the proposed methods has a number of positive and negative aspects. We decided to use the methodology developed by Kryshtaleva to assess the actual state of the labor potential of Russian regions developed by the Southern Federal University [5]. The issues of labor potential assessment have also been covered in the works of many other researchers [1–4, 13].

3 Results

The following components were identified as structural elements of the labor potential: demographic; economic; educational and qualification; psychophysiological; innovative. In this case, to bring the indicators of each component into a comparable form, we used the linear scaling method, which consists in displaying the values of each indicator in the range from 0 to 1, preserving all the proportions between individual values. The calculation takes into account the direct or inverse relation between the change in the indicator and the labor potential. The following formulas were used to calculate private indexes of the development of labor potential components:

– if there is a direct relation between the variables,

$$I_{xi} = \frac{x_i - x_{min}}{x_{max} - x_{min}} \quad (1)$$

– if there is an inverse relation between the variables:

$$I_{xi} = \frac{x_{max} - x_i}{x_{max} - x_{min}} \quad (2)$$

It should be noted that when calculating the integral index of labor potential at the regional level, a large array of statistical data needs to be processed. However, it should be taken into account that a comprehensive assessment is influenced by one part of the indicators that have a direct impact on the standard of living and the economic sphere (for example, consolidated budget revenues, the value of the gross regional product, etc.) and characterize the current level of economic and social development; the other part directly characterizes the quality of labor resources and does not allow us to directly judge the overall level of socio-economic development (for example, the number of graduates of universities and colleges, providing the population with beds, doctors, etc.). The methodology widely uses dynamic series of considered parameters with a fairly representative research horizon (12 years), which allowed us to obtain relevant values of labor potential indexes.

The research horizon is chronologically limited for the following reasons. The lower and upper borders of the series (2006 and 2017) are determined by the lack of open statistical data on the "index of changes in total labor costs" in the context of regions for further calculation of the "index of labor productivity in the regional economy" [14].

When calculating the demographic component, we used such indicators as total birth rate; overall mortality rate; rate of migration growth (disposal) of the population; the ratio of marriages and divorces (number of divorces per 1,000 marriages); the age dependency ratio. During the analyzed period, both positive and negative trends were revealed. So for the studied 12 years, the number of births per 1000 people per year increased by 0.1%, the mortality rate decreased by 0.2%, but the outflow of population due to migration increased (the migration growth rate (for 10 thousand people) decreased by 8 p.p.). The number of divorces per 1000 marriages increased by 3 units, as well as the demographic load factor (for 1000 people of working age, there are persons of non-working age) by 110 units. The changes have had a negative impact on the dynamics of the composite index for the demographic component of the labor potential of the Stavropol region (Fig. 1). There was a decrease of −0.0952 points.

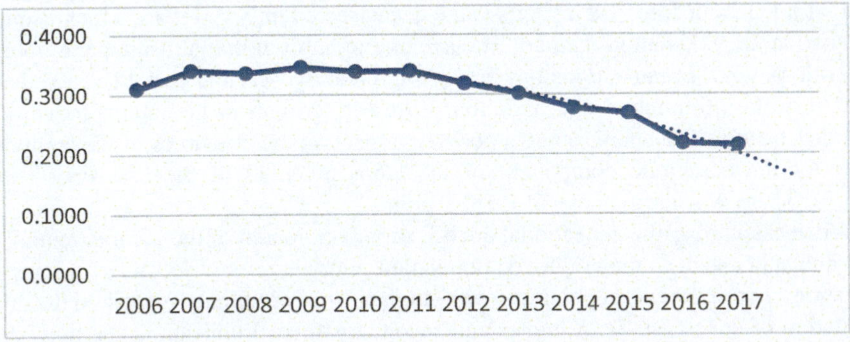

Fig. 1. Dynamics of the composite index for the demographic component of the labor potential (Source: authors).

Next, we considered elements of the economic component. This includes such indicators as the employment rate (%), the unemployment rate, the labor productivity index, the number of people with monetary incomes below the subsistence minimum, as a percentage of the total population, the average monthly nominal accrued salary per employee for a full range of organizations (in rubles). There is a positive trend of reducing the unemployment rate by 3.6%. However, this indicator will not decrease sharply for a number of reasons. One of the reasons is that the relevance of professions and the labor market are changing their structure. Many specialties lose their relevance, and in the market system, the process of retraining specialists occurs unsystematically and over a long period of time. Another reason is that the education system is disconnected from production and business in general, it complicates and increases the information gap where there is no feedback. In this regard, initially the training of specialists is carried out at a low level, which, in turn, reduces the competitiveness of domestic and regional labor resources. This indicator is calculated as the ratio of the number of unemployed people in a certain age group to the number of workers in the corresponding age group, calculated as a percentage. The data were obtained from sample surveys of the labor force presented in open sources [12]. Since 2017, the survey covers persons of the age of 15 and older (until 2017 – persons at the age of 15–72). The number of people with monetary incomes below the subsistence minimum also decreased by 8.3%, while it should be noted that during the considered period, the subsistence minimum increased by 2.8 times [14]. This figure is based on data on the distribution of population by average per capita money income and is a result of their commensuration with the minimum subsistence level; percentage of the total population. Data for the subjects of the Russian Federation up to 2018 and for Russia as a whole up to 2013 are calculated using the value of the macroeconomic indicator of the average per capita monetary income of the population, determined in accordance with the Methodology for calculating the balance of monetary income and expenditures of the population [9, 11]. Data for Russia for 2013–2018 are calculated in accordance with the Methodological provisions for calculating indicators of monetary income and expenditures of the population [12]. The employment rate is also calculated using data from sample labour force surveys. Its value decreased slightly (−1.1%), which is due to a change in the calculation method. The average monthly nominal wage per employee for a full range of organizations has grown significantly over the past 12 years, due to changes in the minimum wage, inflation processes, growth in the labor productivity, and other factors. The changes had a positive impact on the dynamics of the composite index for the economic component of the labor potential of the Stavropol region (Fig. 2). There was an increase of 0.1851 points.

When calculating the educational qualification component of the labor potential, we used indicators such as the employed population with higher and secondary vocational education, unemployed population with higher and secondary special education, graduation of specialists with higher and secondary special education, the number of educational institutions of higher and secondary special education, expenditures of consolidated budgets of regions of the Russian Federation on education.

The most significant indicator of the dynamics of the labor quality is the education level of the population in the region. Professional education in the modern economic world plays a role of a catalyst for many socio-economic processes. It is a significant social institution that shapes the social-economic environment and determines the ways and directions of society development. The analysis of data for the considered period shows that the education level of the employed population has significantly increased:

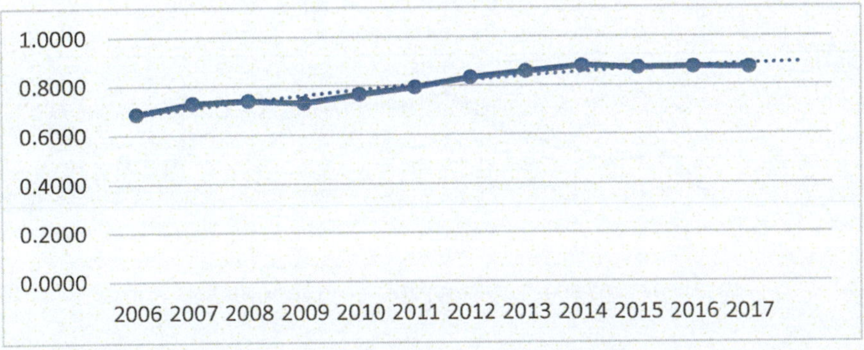

Fig. 2. Dynamics of the composite index for the economic component of the labor potential (Source: authors).

higher education by 10.4%, secondary vocational education by 13.9%. But the fact that the level of unemployed people with higher education has increased significantly in 2013 is alarming. We associate this phenomenon with the deterioration of the economic situation in the country and, as a result, the processes of reducing administrative and managerial personnel in the commercial sector and optimizing state institutions in the region. In addition, the output of specialists with higher and secondary special education decreased by 13.8 thousand people, which is due to a decrease in the number of educational institutions and their branches of higher and secondary special education by 39 units in the region.

Moreover, the number of educational institutions of higher education and scientific organizations that carry out educational activities under bachelor, specialist, master programs and their branches has decreased by half. And the expenses of the consolidated budgets for education from 2006 to 2017 increased by 3.5 times and amounted to 316.36 million rubles The current changes had a negative impact on the dynamics of the composite index for the educational and qualification component of the labor potential of the Stavropol region (Fig. 3). It declined by 0.0457 points. The authors identified some problems of lack of human resources for the regional labor market, namely, low qualification of personnel with technical and engineering skills and specialties based on technical and professional education; lack of knowledge on progressive advanced technologies; lack of scientific personnel in technical, engineering specialties and innovative management; insufficient harmonization of professional standards with educational standards.

When calculating the psychophysiological component of the labor potential, the authors used the following indicators: morbidity rate (people) (per 100 thousand people, registered diseases in patients with a diagnosis established for the first time in their life), life expectancy at birth (years), availability of hospital beds (per 10 thousand people), the provision of medical workers (per 10 thousand people), expenditures of consolidated budgets of regions of the Russian Federation on health care (million rubles).

The analysis of data for the years 2006–2017 showed morbidity with the diagnosis established for the first time increased by 15.6% for population living in the service

Fig. 3. Dynamics of the composite index for the educational and qualification component of the labor potential (Source: authors).

area of the medical institutions (for 100 thousand persons of the population), the expenses of consolidated regional budgets of the Russian Federation on health care increased by only 8%. In our opinion, this is a negative trend, since the issues of prevention and early detection of diseases are currently one of the priority tasks that the President of the Russian Federation sets for the Government. The provision of hospital beds decreased by 6.9 units for 10 thousand people, and there is a slight increase - by 5.2 people - in the provision with doctors per 10 thousand people. While life expectancy at birth has increased (number of years, which one person from generation born in a given year has to live, if throughout all life of this generation, the mortality rate at each age remains the same as in the year for which the indicator is calculated. The changes did not negatively affect the dynamics of the composite index for the psychophysiological component of the labor potential for the Stavropol region (Fig. 4). The indicator increased by 0.0592 points and tends to increase in the future.

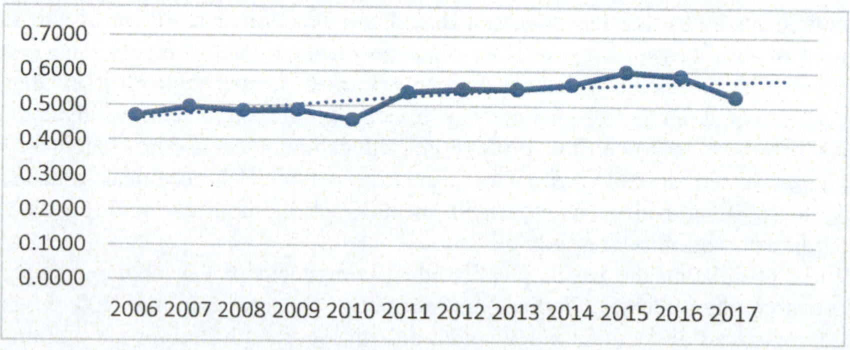

Fig. 4. Dynamics of the composite index for the psychophysiological component of the labor potential (Source: authors).

The final component of the Stavropol region's labor potential is the innovation component. The elements of the innovation component not only characterize the labor potential of the region, but also ensure stable economic growth of the region and its stability. In its calculation, such indicators as the number of organizations performing research and development, the number of personnel engaged in R&D, internal costs for research and development, innovative activity of organizations and the volume of innovative goods, works, and services were used. The calculation of criteria for determining the organization's innovation activity for 2018 has changed due to the release of a new edition of the international guide on the statistical measurement of innovation, implemented by the OECD in conjunction with Eurostat [8]. During the analyzed period, the number of organizations that performed research and development in the Stavropol region has increased by 3.2 times, and the number of personnel engaged in R&D has increased by 40%. An increase in internal spending on research and development by 5 times contributed to an 88-fold increase in the production of innovative goods, works, and services. The innovative activity of organizations, i.e. the change in the share of organizations that carried out technological, organizational, and marketing innovations in the reporting year, in the total number of surveyed organizations, was changing abruptly and as a result, in 2018 has decreased by 0.2% compared to 2016. The revealed changes had a positive impact on the composite index for the innovative component of the labor potential of the Stavropol region (Fig. 5), the size of which increased by 0.4994 p.

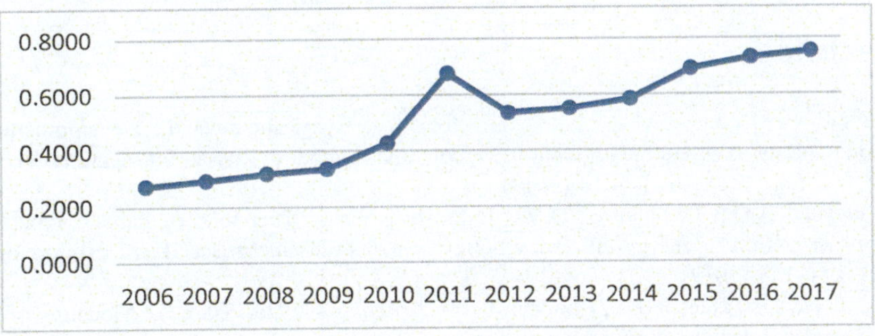

Fig. 5. Dynamics of indicators included in the innovation component of the labor potential (Source: authors).

4 Discussion

The labor potential index obtained with the help of the considered methodology needs to be further specified on the basis of additional collection of statistical information and comparison of existing methods. In addition, the use of this index in the context of reflecting regional aspects has a number of disadvantages, in order to eliminate which, it is quite reasonable to develop a special regional methodology for calculating the labor potential index, which can take into account all the features of life processes in the regions. However, the attempt to calculate and analyze the components that make up the labor potential index indicates the need to improve not only the methodological tools, but also the strategic policy of socio-economic development of the region in the perspective of transition to a new category of regions with a very high labor potential index.

5 Conclusion

The labor potential index, which is based on the aggregation of values of nine indicators that allow a comprehensive assessment of its state in the region, showed that it largely depends on the level of socio-economic development. Thus, in the Stavropol region there are problems in the remuneration level, employment, working conditions, as well as the capacity of the labor market. The region's labor market index can vary from 1 to 100. So, at the end of 2018, the labor market index in the Stavropol region was 54.4, which is 1.8 points higher than in 2017. Despite the positive dynamics, the characteristic feature of the region is still a low labor market index in comparison with other regions of the Russian Federation. Stavropol region occupies the 47th position in the rating of regions in the Russian Federation. The region is becoming unattractive for the labor potential employed in such areas as science, education, and healthcare. There is an outflow of the population with a high level of qualifications, which significantly leads to a negative dynamics of the labor potential index of the Stavropol region.

Acknowledgements. The study was conducted within the framework of the RFBR project no. 18-010-00887A "Comprehensive assessment of the impact of youth migration on the labor market of the border regions of the North Caucasus".

References

1. Fursov, V., Rudnev, A., Malinina, O., Alekhina, E., Kushnareva, I.: Development of a methodology to assess the regional labor potential in modern Russia. Int. J. Innov. Technol. Explor. Eng. **9**(1), 1613–1618 (2019)
2. Gaboyan, A.G., Podgajnov, D.V., Belousov, V.M.: The role of labor potential in implementation of energy efficiency policy at industrial enterprises. Terra Economicus **17** (1), 174–188 (2019)
3. Harmider, L., Taranenko, I., Honchar, L., Ovcharenko, O., Dotsenko, G.: Modeling of labor potential as a factor of influence on the region competitiveness. Montenegrin J. Econ. **15**(2), 111–125 (2019)
4. Kruhlov, V.V., Tereshchenko, D.A.: Public-private partnership as tool for developing regional labor potential. Sci. Innov. **15**(6), 5–13 (2019)
5. Kryshtaleva, T.Yu.: Assessment method of the state of labour potential in regions of the Russian Federation. World Econ. Manag. **17**(3), 35–46 (2017)
6. Leonidova, G.V., Panov, A.M.: Labor potential: territorial aspects of quality status. Prob. Territor. Dev. **3**, 60–70 (2013)
7. Migranova, L.A., Toksanbayeva, M.S.: Quality of labor potential of Russian regions. Population **2**, 102–120 (2014)
8. OECD: Oslo manual 2018 (2018). https://www.oecd.org/science/oslo-manual-2018-9789264304604-en.htm. Accessed 21 June 2020
9. Order of Federal State Statistics Service from July 2, 2014 N 465 "On approval of Methodological provisions on calculation of indicators of monetary income and expenses of the population" (2018). http://www.consultant.ru/document/cons_doc_LAW_167107/. Accessed 21 June 2020
10. Popova, L.A., Terentieva, M.A.: Employment potential of the Russian North. Arctic North **4**, 51–69 (2014)

11. Resolution of the State Statistics Committee of the Russian Federation of 16.07.1996 N 61 "On approval of methods for calculating the balance of monetary incomes and expenditures of the population and the main socio-economic indicators of the population's standard of living" (1996). https://legalacts.ru/doc/postanovlenie-goskomstata-rf-ot-16071996-n-61/. Accessed 21 June 2020
12. Rosstat: Application of the labor cost indicator in the calculation of labor productivity indexes (2011). https://www.gks.ru/bgd/regl/B11_04/IssWWW.exe/Stg/d07/2-zatrat.htm. Accessed 21 June 2020
13. Simonova, M., Kolesnikov, S., Spravchikova, N.: Interregional aspects of employment as a factor in the formation of the labor potential. Smart Innov. Syst. Technol. **139**, 425–434 (2019)
14. State Statistics: Number of employees (2020). https://www.fedstat.ru/indicator/34061. Accessed 21 June 2020
15. Yakshibayeva, G.V.: Labor potential of the Republic: quantitative and qualitative characteristics. Econ. Manag. **3**, 57–65 (2007)

Social Barriers to Innovation in Higher Education: Key Stakeholders' Perception

I. A. Martynova(✉)

Samara State University of Economics, Samara, Russia
miasseu9@gmail.com

Abstract. In this paper the results of an ascertaining experiment are presented. The experiment took place in Samara State University of Economics, Russia. The purpose of this study was to determine the individual barriers to technological innovations within higher educational institutions and to find possible remedies that could overcome the resistance to innovation among key internal stakeholders of universities. The methodological approach taken in this study is a mixed methodology based on content analysis of students' interviews and mathematical statistics. The cohort of students, who previously took part in the distance learning projects, was examined for their negative perceptions of technological innovations. The students' responses were then analysed and, the factors that could block the innovation were selected. The data analysis revealed the prevalence of non-affective barriers over the affective ones. This study suggests that innovation managers, who are responsible for implementing distance learning at universities, should closely monitor social factors that could block innovative projects, and support sustainable communication process with the internal stakeholders.

Keywords: Barriers to innovations · Noncognitive skills · Perception · Pedagogical innovation · Stakeholders

1 Introduction

Traditionally, higher education is particularly difficult to make more productive because of its complexity. That is why innovations are of special importance for the sphere of education. It is stated that innovations in education are intended to raise productivity and efficiency of learning and improve learning quality [13]. It should be noted, that this statement concerns proactive innovations, which are targeted to the strategic development of organization. Among proactive innovations technological ones play a paramount role. In education distance learning is directly associated with technological innovations. Distance learning is carried out remotely through the use of telecommunications, namely web resources, e-mail, video communication, etc. The existing body of research on distance learning suggests that it has a number of advantages over face-to-face training:

- operational - no time and space limits, instant feedback,
- informational - availability of remote educational arrays,
- communication - efficiency of interaction of subjects of training,

S. I. Ashmarina and V. V. Mantulenko (Eds.): IPM 2020, LNNS 161, pp. 48–54, 2021.
https://doi.org/10.1007/978-3-030-60926-9_7

- pedagogical – interactivity and greater motivation for learning,
- psychological – comfortable conditions, removal of psychological barriers of face-to-face communication,
- economic-saving on transport costs, maintenance of premises, reduction of paperwork,
- ergonomic – individual schedule and pace of learning and the right equipment.

However, using a new technology in organizational activities also implies social changes, since there the formation of new standards for behavior and activities takes place. Educational innovation concerns all stakeholders and requires their active involvement and support. Students are considered to be the most significant social stakeholder group that has direct influence upon the university's operations and has great relevance in the implementation of innovative policies [2, 15]. However, depending on their perceptions of the innovative project, students can become project assets or project problems. It is stated in Hueske and Guenter that attitudes, that could drive the mode of behavior, belong to individual or social barriers, which can 'hamper, delay or block innovation' [4]. According to Lašáková, Bajzíková, and Dedze, who determined three clusters of barriers to innovation within the higher educational institutions, the barriers that are examined in this paper belong to 'people-related issues' [8]. Although extensive research has been carried out on the barriers to innovation in higher education [1, 6, 8], no study exists that concentrates on the resistance to innovations within the higher educational institutions from the part of its internal stakeholders, students in this case.

The scene starter for this study was the sub-result of the previous educational experiment that took place at Samara State University of Economics in 2016–2019 [9]. During the experiment the students were instructed in the mode of blended learning course. The major objective of the experiment was to balance the equilibrium between the applicants' weak language background preparation and final educational goals for baccalaureate graduates. The blended English course was designed to support poor second language readers and equip them with the skills necessary for mastering Content and Language Integrated Learning course at the next stage of their university studies. The efficiency of this approach was proved and later verified at the second stage of the research in 2019. But one provocative sub-finding emerged at the verification stage of the experiment. The question on the attitude towards the distance mode of learning was included in the interview on the efficiency of the results of the blended course. Although the majority of the respondents acknowledged the efficiency of the course, only less than a half of respondents demonstrated positive perception towards distance learning mode, while the attitude of 49% was negative. The same observations were found in the previous research that has demonstrated that students continue to prefer traditional instructional methods, and their preferences may represent a barrier to innovation in education [1]. The authors also note, that, «*Given the assumed insufficiency of traditional methods, it is especially important to consider how to facilitate increased interest in less traditional, more effective innovative methods*» [1, p. 152]. Those observations gave rise to concern about the social aspects that could affect the innovations within the higher education institutions or even block their implementation. As long as students are the key stakeholders in the higher educational institutions, the analysis of the factors, that influence the decrease in their interest towards the innovation, could determine the track of interaction between the higher educational institution and its major stakeholders. It should be aimed at creating a sustainable future. Thus, this paper investigates the factors that are the most impactful and relevant to the negative students' perception of distance learning. Additionally, it speculates on the possible improvements that might positively impact the ways students treat innovative developments in higher educational institutions.

2 Methodology

The methodological approach taken in this study is a mixed methodology based on ascertaining experiment, content analysis of students' interviews, and mathematical statistics. Participants, who formed the cohort, belonged to the second and third year of baccalaureate programs in Samara State University of Economics. The cohort consisted of students (N = 87), who previously took part in a series of educational experiments with distance learning that was conducted at Samara State University of Economics. During the previous study it was noticed that students explain their negative attitude to distance learning mode by the necessity to study autonomously and independently out of the teacher's supervision. Based on those findings and combining them with the conclusions of Heckman and Rubinstein, who had demonstrated the quantitative importance of non-cognitive skills in educational attainment, the interview, which contained open questions, was conducted and then analyzed with the content analysis procedure that allows selecting the utterances meaningful to the research problem [3]. The study design uses the methodological framework based on the taxonomy of individual differences among the second language students adapted from Oxford [10], who suggested several attitude levels that can directly affect behaviors. Based on comprehensive literature review [6, 12] some other factors that could negatively affect an educational innovative project were added to the interview questions. The results are analyzed and interpreted in the next section of this paper.

3 Results

The first set of questions was aimed to examining the motives of students to study a foreign language. The motivation barrier had to be excluded from the sample, as motivation 'technically means the condition of being moved to action' [11, p. 275], and the lack of motivation automatically makes the study irrelevant. As it can be seen from the Table 1, all respondents demonstrated high level of motivation toward learning a foreign language. Interestingly, the majority of respondents reported having the instrumental motivation, which is less preferable than integrative motivation in case of language learning.

Simple statistical analysis was used during the second stage of ascertaining experiment to examine what individual factors could negatively influence the perception of distance learning. Table 2 illustrates the breakdown of individual barriers to educational innovation into categories according the frequency of their mentions in the

Table 1. Motivation towards the second language learning in the sample

Type of motivation	Number of respondents	Percentage
Integrative motivation - the desire to learn a language to integrate oneself with the target culture	12	13.79
Instrumental motivation - the desire to learn the language in order to get a better job or meet a language requirement	75	86.2

Source: author.

respondents' interviews. The themes identified in the interview responses are presented in the left column of the table.

Table 2. The proportion of individual barriers to innovation

Individual Barrier to Innovation	Frequency of mentions (%)
Lack of self-discipline/laziness	90.8
Anxiety on the lack of timely 'alive' feedback	62.6
Time stress	41.37
Computer anxiety	21.83
Ambiguity anxiety	9.19
Competition anxiety	5.74

Source: author based on [14].

From the data in Table 2, it is apparent that the majority of social barriers are of the affective nature. It is argued that these factors play significant role in influencing academic procrastination among students [12]. Moreover, stress and anxiety can 'cause motivation to plummet and attitudes to turn negative' [10]. Only self-discipline does not belong to the emotional domain in this list, but it has the biggest share in terms of percentage. It is acknowledged, that while distance learning offers more freedom for learners, it also requires planning of their own self-development and high self-discipline [6]. Self-discipline is a non-cognitive skill that is typically acquired at the initial stages of education. As it is stated in Gorbunovs, Kapenieks, Cakula self-discipline is a more complicated and difficult way to achieve success than in a case of motivational approach, but this is the most reliable way [6].

Based on these results the individual barriers were then divided into two bigger groups of affective or emotional barriers and non-affective ones. Comparing the type of motivation and the individual barriers, that has been reported, an interesting correlation was determined, which is shown in Table 3.

From this table, it can be seen that by far the respondents with instrumental motivation reported both emotional and self-discipline reasons to resist innovative projects, while the students with integrative motivation demonstrated lack of self-discipline in the majority of cases. This is a rather surprising outcome, as it is traditionally expected that integrative drivers motivate learners to persistence, and high activity level, that in turn should provoke self-discipline. It could be assumed that

Table 3. Generalized typology of individual barriers to innovation

Type of individual barrier to innovation	Number of respondents with integrative motivation	Number of respondents with instrumental motivation
Affective barriers	2	75
Non-affective barriers	10	69

Source: author.

pleasure and satisfaction from learning driven by intrinsic integrative motives allow that group of respondents to avoid emotional stress experienced by the second group of students. Based on the obtained data, it was determined that the most obvious and the most influential barrier to innovation among the students of higher education is their lack of self-discipline. In an attempt to outline the ways for finding the remedy to overcome the barriers, the students were again interviewed to determine the deeper reasons for resistance to innovation. At this stage of the experiment the students were tested for their locus of control. The results are summarized in Table 4.

Table 4. Locus of control in the cohort of respondents

Locus of control	Ratio of students (people/%)
External	79/88.2
Internal	8/11.8

Source: author.

The data in the table reveals that the majority in the cohort has the external locus of control, that means that the 'place one attributes control of one's life-is totally external (e.g., fate, luck, chance, the government, God, the Devil) as opposed to at least partially internal' [10]. Self-discipline in its turn implies that some degree of control exists within.

Taken together, these results suggest that the lack of self-discipline among the students could be the most compelling barrier for technological innovation in higher educational institutions. The next section of the paper, therefore, moves on to discuss the possible solutions of how to deal with this barrier in the process of innovation.

4 Discussion

The present study was designed to determine the barriers to innovative technological projects within a higher educational institution that can arise from its key internal stakeholders, namely students. With respect to the first research question, it was found that irrespective of sufficient motivational level there is a barrier that is represented by insufficiently developed non-cognitive skill of self-discipline among university students. These findings raise questions regarding the ability of higher educational institutions to overcome this barrier. It could be suggested that lack of self-discipline among students can become a compelling circumstance for a higher educational institution in an attempt to implement innovative policies. Prior studies that have noted the importance of non-cognitive skills generally tend to treat discipline as an unconscious process. According to Watkins, "Within education, emphasis is given to discipline as a form of subjection as opposed to a force that can also equip students with the capacity to learn" [14, p. 545]. Moreover, in reviewing the literature, it was found that this non-cognitive skill is formed at the initial stage of education, i.e. in primary or elementary school, and in modern school the stricter disciplinary codes characteristic of instructional pedagogies are largely viewed as anachronistic [14]. Several reports have also shown that university students in Russia lack the autonomous learning skills that have to be formed and developed in the secondary school [9]. And, finally this situation becomes even more complicated if we consider that innovative behavior does not imply adaptation, it implies the formation of own personality, self-development, which is hardly can be expected from the students with the lack of self-discipline. However, contrary to the apparent complexity of the task, there are several novel studies that suggest a key to solve the problem. According to Krskova, Breyer, Baumann, Wood university students from multiple faculties and at different stages of academic progression understand discipline in higher education as 'internally driven as opposed to being enforced externally' [7]. Thus, there is a conscious perception of this concept that could be appealed to when innovation managers design the strategy for implementing innovative projects. Another important issue to consider in implementing innovations within a higher educational institution is the careful and reasonable choice of communicator with the valid set of skills and abilities for effective and sustainable interaction with the stakeholders. For instance, in the framework of higher education the facilitator should be able to apply mental contrasting methodology [5] to promote self-discipline or should know what external stimuli could impact motivation to self-development.

5 Conclusion

This study was undertaken to determine the individual barriers to technological innovation from the part of key internal stakeholders within the higher educational institutions. The second aim of this study was to review the literature to examine the prospects for eliminating those barriers. Content analysis of cohort interviews combined with the simple statistical calculations revealed two major groups of barriers, from which the most controversial one was identified. Based on the analysis of the relevant research the possible strategical suggestions that present potential answers for innovation managers in dealing with the social aspect of innovation in higher educational institutions are made. The current data highlight the importance of social aspects of innovation in higher education. The findings of this research provide insights on problems with technological innovations in higher education. Although the generalizability of the results is subject to certain limitations, they support the idea that ensuring appropriate strategies, facilitating services and support for students should be a priority for innovation managers of higher educational institutions.

References

1. Bower, E.A., Girard, D.E., Wessel, K., Becker, T.M., Choi, D.: Barriers to innovation in continuing medical education. J. Cont. Educ. Health Prof. **28**(3), 148–156 (2008)
2. Chapleo, C., Simms, C.: Stakeholder analysis in higher education: a case study of the University of Portsmouth. Perspectives **14**(1), 12–20 (2010)
3. Heckman, J.J., Rubinstein, Y.: The importance of non-cognitive skills: Lessons from the GED testing program. Am. Econ. Rev. **91**(2), 145–149 (2001)
4. Hueske, A.K., Guenther, E.: What hampers innovation? External stakeholders, the organization, groups and individuals: a systematic review of empirical barrier research. Manag. Rev. Q. **65**(2), 113–148 (2015)
5. Ilyuk, V.V.: The methodological approach to managing innovative project stakeholders. Prod. Org. **4**(71), 38–55 (2016)
6. Gorbunovs, A., Kapenieks, A., Cakula, S.: Self-discipline as a key indicator to improve learning outcomes in e-learning environment. Procedia Soc. Behav. Sci. **231**, 256–262 (2016)
7. Krskova, H., Breyer, Y., Baumann, C., Wood, L.: An exploration of university student perceptions of discipline: Introducing F.I.R.S.T. discipline principles. High. Educ. Skills Work Based Learn. **10**(1), 61–82 (2019)
8. Lašáková, A., Bajzíková, Ľ., Dedze, I.: Barriers and drivers of innovation in higher education: Case study-based evidence across Ten European Universities. Int. J. Educ. Dev. **55**, 69–79 (2017)
9. Martynova, I.A.: Development of oral reading fluency in foreign language through a blended course in a non-linguistic university. Kazan Pedag. **1**(138), 128–135 (2020)
10. Oxford, R.: Who are our students? A synthesis of foreign and second language research on individual differences with implications for instructional practice. TESL Canada J. **9**(2), 30–49 (1992)
11. Oxford, R.L.: Language learning styles and strategies: concepts and relationships. Iral **41**(4), 271–278 (2003)
12. Rahardjo, W., Juneman, S.Y.: Computer anxiety, academic stress, and academic procrastination on college students. J. Educ. Learn. **7**(3), 147–152 (2013)
13. Serdyukov, P.: Innovation in education: what works, what doesn't, and what to do about it? J. Res. Innov. Teach. Learn. **10**(1), 4–33 (2017)
14. Watkins, M.: Discipline, consciousness and the formation of a scholarly habitus. Continuum **19**(4), 545–557 (2005)
15. Yaro, I., Arshad, R., Salleh, D.: Relevance of stakeholders in policy implementation. J. Public Manag. Res. **3**(1), 1–16 (2017)

Problems of Financial Support of SME Development in the Russian Federation

A. A. Prosvetova[1(✉)] and K. B. Gerasimov[2(✉)]

[1] Samara State University of Economics, Samara, Russia
mir.189@yandex.ru

[2] Samara National Research University, Samara, Russia
gerasimov.kb@ssau.ru

Abstract. The article examines issues of financial support mechanisms for the small and medium-sized business sector in Russia at the level of the entire country and at the level of federal subjects. Special attention is paid to the system of state bodies dealing with the financial development of small businesses in the Russian Federation, including from the organizational and legal point of view. Based on the survey, information about the problems of financial support for SMEs was structured and described. This is important not only for the development of SMEs in general, but also for the development of the country's labor market. The authors focus on the need to support and bring to a new development level of the country's economic sector, as well as the development of anti-crisis measures, promoting the growth of the market sector, which determines special attention to issues of SME development support, taking into account the foreign experience.

Keywords: Development factors · Entrepreneurship · Financial support · Labor market · Problems · Small and medium-sized enterprises

1 Introduction

The small and medium-sized enterprises (SME) sector is an integral and very important part of the entire market system. SMEs have a direct impact on a country's gross national product, its quality and structure. There is also a directly proportional relation between the level of SME development and the level of economic growth. SMEs have a huge number of working enterprises, which in turn provide jobs for a large part of the population. The result of these enterprises is production, on average, half of the country's GDP. The relevance of this research confirms the need to overcome consequences of the financial and economic crisis in the country. It is particularly important that, in the event of new or existing sanctions, SMEs should be stable in their work, and not lose the efficiency, while establishing new economic contacts, including with foreign partners.

Today, in our country, the problems of small business development are in the center of attention and discussions at various government levels. There is no single and standardized approach to defining methods and forms of SME support. The lack of state financial support can lead to weakening of the small business sector and an

S. I. Ashmarina and V. V. Mantulenko (Eds.): IPM 2020, LNNS 161, pp. 55–64, 2021.
https://doi.org/10.1007/978-3-030-60926-9_8

aggravation of social and economic life spheres [1]. Developed countries to a large extent solve the unemployment problem due to small business. The percentage of jobs in such countries provided by small businesses varies from 50 to 90%. Thus, by developing small businesses, the state ensures the growth and development of the market economy. SME, as a whole system, is one of the main satellites of the state's economy. Reducing or increasing this system will significantly reduce the economic development.

2 Methodology

The base of scientific research methodologies used in the work includes the system method of scientific knowledge, methods of induction and deduction, the method of dialectics, the method of analysis using logic, the system-functional, formal-legal method, analysis of statistics reports, normative legal acts and other documents. Financial assistance measures can be implemented in accordance with the legislation of the Russian Federation at the expense of budgets of the subjects of the Russian Federation, local budgets by providing subsidies, budget investments, state and municipal guarantees for the obligations of small and medium-sized businesses and organizations that form the infrastructure for SME support [2].

It is fair to say that the main financial source for supporting the business activities is money supply of a borrowed nature. One of the bodies providing assistance to SMEs is the National Guarantee System, which was established by order of the Ministry of Economic Development of the Russian Federation. This system is represented by a list of guarantee organizations that provide the necessary assistance to SMEs by providing affordable and preferential lending resources.

Let's consider the main instruments of the state financial support for the development of small businesses [9]:

- preferential microloans for business. By creating a quick access for businessmen to borrowed funds, there are small-scale lending enterprises (micro-credit organizations) in the regions of the Russian Federation that operate at already established rates. These collateral loan rates range from 3 to 6.25% per annum, and for non-collateral loans – from 6.25 to 15% per annum. This rate is linked to the key interest rate of the Russian Central Bank,
- program of granting incentives for lending to SMEs. Under this program, the loan can be issued in the amount of 500 thousand rubles to 1 billion rubles, although it is worth noting that in the practical application of this program, this figure is determined by the issuance of funds in the amount of 10 million rubles. In the process of making a decision to approve or refuse a loan, banks are guided solely by their own motives, based on their own legal acts. The government in turn does not set any requirements for obtaining financing from banks,
- a program for providing equipment leasing on benefits. The essence of this program is that the equipment that enterprises belonging to the SME sector leased was provided to them at a low rate of 6% and 5% for Russian equipment and equipment of foreign origin, respectively. The transaction amount is determined by the

framework from 5 million to 200 million rubles with an advance payment of at least 10% of the transaction amount [6],

- loans for industrial companies. The Industrial Development Fund, through consideration of incoming applications, provides borrowed funds to SMEs for their modernization and creation of new production facilities with the condition of project financing. To apply for participation in the program, you should write a project on business organization, which would include the introduction of new profitable methods of production,
- guarantees and credit guarantees for the SME sector. The provision of collateral security is a prerequisite for banks engaged in lending to SMEs. As a rule, the issue of collateral is more difficult for SMEs than for a manufacturer with a large scale and a large number of employees. To solve this problem, regional organizations have been created to help SMEs guarantee loans for small and medium-sized businesses. They help you get a loan by giving guarantees to the bank up to 25 million rubles, covering collateral up to 70%.

Bank financing is crucial for the economy functioning, as it is an important source of finance to support the development of SMEs [14]. Of course, all these measures of the state support for SMEs have their own limitations, but despite them, they are designed and can provide significant assistance to entrepreneurs [3]. In general, assistance provided with monetary funds, including programs of subsidies, concessional loans, etc., is in practice the most popular and effective form of state support. It should also be taken into account that the organization of the financial state support process is a complex structure of interaction between the state apparatus and its infrastructure organizations on the basis of a regulatory framework. One of the priority goals of financial support for the SME sector is regional development, so the geographical location of an economic entity in many cases plays a significant role in obtaining financial assistance from the state, as well as the level of income development of the population or other regional features of doing business. Regions with a low level of socio-economic development, or regions that are located on geographically difficult terrain, have some privileges. The presence of adverse weather and metrological conditions can also influence the state's decision to give, for example, a grant or subsidy. Economists argue that although large firms have a greater concentration on research and development, it is small firms that are responsible for most important inventions and innovations in fact [4]. Attracting external financing will be easier for an entrepreneur with a complex research project that offers many ways to commercialize its results [12].

3 Results

Sustainable development of small and medium-sized enterprises requires well-functioning financing of small and medium-sized enterprises with easy access to capital for investment [11]. 2020 has been declared the year of entrepreneurship in Russia. Today, the problem of SMEs in Russia has become a "byword". The reasons for the deplorable situation were and still are high tax rates for SMEs, bureaucratic delays, the high cost of loans (refusal to lend), and the lack of sufficient, and most importantly, the right level of the state support.

SMEs are at a standstill, as demand for commodity production and services is declining every year because of lower incomes for Russians. At the same time, production costs and the fiscal burden are growing. All this leads to a drop in the profitability of SMEs and subsequent bankruptcy. The federal media, unfortunately, do not openly talk about problems in the small business sector, and some data, in general, are significantly embellished.

This year, we conducted a study of the state of SMEs in Russia for 2019, which helped us to analyze the current economic situation. 3,698 SMEs with revenues of up to 350 million rubles, i.e. representatives of small businesses, were surveyed. The questions that were asked to entrepreneurs related to the following topics:

- assessment of the economic situation in the country and in the industry,
- assessment of the purchasing power of the population in the country or in the industry,
- financial position of the company (future plans, profit, revenue, development plan, success criteria),
- sales and the number of customers, as well as the frequency of purchases, the size of the average receipt,
- the situation in the field of import and export.

Questions were also asked about the business profile, location of the company, age of the entrepreneur, sphere of activity, number of employees.

The main criterion used to assess the situation in SMEs is the so-called "small business optimism index". It is calculated according to its methodology and its values can vary from minus 100 units (negative mood prevails) to plus 100 units (positive mood prevails).

Unfortunately, in 2019, the index was 32 points, remaining at the level of 2018. Small businesses were more negative only in 2015 during the crisis (the sentiment index was minus 38 points).

If we compare it with the end of 2017 and the beginning of 2018, the pessimism of businessmen has increased and companies are set for further deterioration of the economic situation in the country. They predict a further decline in the purchasing power of the population. The Ministry of Finance confirms these concerns and draws attention to the decline in investment activity.

The results of the survey of entrepreneurs are:

- 86% say that the current economic situation in the country hinders business development; 82% note a decrease in purchasing power. Companies also note a decrease in profits, a decrease in the amount of the average receipt and the number of customers,
- the tax burden has increased significantly, including because of the increase in VAT to 20%. The budget was affected by the cost of implementing cash register equipment. Competition with large companies has increased. The worst situation is in the areas of retail trade and construction,

- mandatory labeling of certain categories of goods introduced in 2019 also contributed to the deterioration of the situation. It is one of the top 5 problems that concern small businesses, as it entails additional costs,
- the number of respondents who perceive state support as a driver of small business has decreased. The current programs have not given a noticeable result, heads of small businesses do not believe the words of the state about reducing the burden on business, because they have experienced the opposite effect.

According to representatives of SMEs, development can only be helped by reducing the tax burden, increasing consumer demand, and easing the requirements of banks to issue loans. Let's look at the main problems of small business in more detail.

The Decline of Purchasing Power. People are buying less. These situation was caused by low wages, a decrease in the purchasing power of the ruble and a high level of credit. It is also worth noting that today economic growth is achieved mainly due to the high level of consumer lending.

The Central Bank believes that if banks had not issued 1.5 trillion rubles of loans to the population over the past year, economic growth would have remained zero [10]. Most Russians are forced to make purchases on credit, as they do not have sufficient funds of their own. As a result, the debt burden increases, the level of real disposable income decreases, and the amount of funds that a person can spend decreases.

High Tax Burden. The tax burden has continued to grow since 2017. In 2019, 46% of the surveyed businessmen were concerned about this problem. In 2020, the tax burden will increase by another 4.9%, which, according to entrepreneurs, will lead to disastrous consequences for small businesses.

The VAT increase led to significant price increases, customer churn and a decrease in the amount of the average check. Then the government prepared the indexing of deflators coefficients, this will result in increase of imputed income tax to the patent system of taxation and trading fees by 4.9%. In addition, the amount of mandatory contributions of SMEs to the Pension Fund and the Mandatory Health Insurance Fund (MHIF) will rise too. Contributions to the Pension Fund will increase by 10.5%, and to the MHIF – by as much as 22.4%. However, not all SMEs can take these amounts into account when paying their income taxes. As a result, SMEs continue to operate in unfavorable external conditions, and internal growth reserves are almost exhausted [15].

Lack of Working Capital and Lack of Profitable Loans. Another factor that hinders production is the low level of capitalization of entrepreneurs themselves and the high cost of providing loans by banks. According to economic scientists, only 25% of entrepreneurs in the Russian Federation are fully satisfied with their business conditions, as well as their current situation.

Entrepreneurs who are just trying to organize their business have the hardest time, because the procedure for obtaining a loan is too difficult for them, and sometimes even impossible. Such a small percentage of loans to start-up businesses has the following reasons:

- negative impact of foreign policy on the country's economic situation,

- significant reduction of banks in the Russian Federation as a result of the revocation of licenses by the Central Bank (the result was a reduction in the supply and choice of investments in the market),
- lack of necessary collateral (banks are reluctant to issue loans to entrepreneurs who cannot provide it with collateral or surety).

Administrative barriers also make a negative contribution to business activity. Red tape, a huge amount of necessary documentation, a large number of state inspections, increased supervision and control by the state – all this is a hindrance to doing business, which does not contribute to its expansion, but rather the opposite.

According to the owners of enterprises and organizations, in order to facilitate their activities, it is necessary to make a convenient and simpler procedure for registering a new enterprise, make changes to the terms of their registration, as well as the procedure for licensing and obtaining certificates, etc. After all, the calmness of an entrepreneur has a direct impact on his further prosperity as a successful businessman and the success of his production.

Currently, there are 197 control and supervision bodies in Russia that perform 236 functions and monitor compliance with business requirements. If you compare it with European countries, for example, in the UK, where the restaurant is regulated by 18 standards, we have more than 6 000 requirements for this sector. In order to create the most comfortable environment for small business development, it is necessary to reduce the number of regulations and requirements, among other aspects.

Cash Control Discipline. "Online cash registers will finally ruin business" - economic experts said in 2017 [8], when the Federal law required the vast majority of Russian entrepreneurs to switch to cash registers [5]. Despite the positive attitude of the tax authorities, the situation with the installation of cash register equipment was not so effective.

First of all, this entailed additional expenses, sometimes "incompatible with the life" of some SMEs. And it's not just about spending on the cash registers. Entrepreneurs also had to pay for the fiscal storage device, Internet connection, payment for services of the fiscal data operator, additional equipment, cash software, and services of technical centers.

In addition, some SMEs faced time and financial costs for training staff to work with online sales registers. Another disadvantage in the use of cash registers, according to businessmen, is technical failure in the operation of equipment, as a result of which incorrect data gets to the fiscal services (which threatens with fines).

Marking. Entrepreneurs did not approve of the idea of product labeling, which will negatively affect their business development:

- every sale point and checkout area needs to be equipped with 2D scanners, and this is quite expensive, especially for small retail stores with a profitability of 10–20 thousand per shift,
- you need to connect to the electronic document management system, and the cost of the operator's services on average is 3,000 rubles per year,

- every entrepreneur dealing with a marked product should purchase a qualified electronic signature to work with consignment notes and other documents in the electronic document management system,
- a commodity accounting system has to be installed on the cash register equipment to account for marked goods. Many developers charge a fee for the right to use such software. This also involves additional costs.

As for manufacturers who are required to label their products, they will have to equip each production line with marking equipment. The average cost of devices is about 250,000 rubles. In addition, manufacturers need to buy stamps at a price of 50 kopecks per piece. Naturally, the manufacturer will include all these expenses in the cost of the product.

Constant changes in the tax legislation are another reason for the negative mood of entrepreneurs. Because of constant innovations, they do not have confidence in the future. At any time, the government can change tax rates and introduce requirements that will be an unbearable burden for businesspeople to comply with. If we return to the situation with the abolition of the single tax on imputed income, this will undoubtedly lead to a reduction in the number of legal entities that will not be able to find an alternative for this tax regime. To avoid this, the government will need to provide measures to replace the regime.

The prosperity and expansion of the small business sector in the Russian Federation over the past ten years has been hindered by obstacles that can be eliminated with building a successful economic policy based on the experience of previous years. As the real business practice shows, not all state support measures for SMEs are correct and necessary.

In the current economic environment, it is necessary and possible to develop a state program to support SMEs that would take into account specific features of small economic entities, and would also apply to all production sectors. First of all, it means reduction of the tax burden and provision of relief in terms of the use of cash registers. Individual lending programs should be developed with low interest rates and less stringent requirements.

4 Discussion

In order to create and approve a set of measures to help SMEs, it is not enough to take into account only the experience of own country in this matter. Considering information about the state of this sector in other countries, and how they act in the framework of solving tasks we are interested in, can be very useful and effective when making a decision.

Today, when there is a process of economic globalization, the interaction of bodies that provide assistance to SMEs is not a complex and difficult process, but rather the opposite. The interaction of these organizations with each other provided an opportunity to look at existing support measures for expanding the scope and improving the quality of SME development from another perspective, to understand advantages and disadvantages of these measures, and to re-evaluate the contribution that SMEs can make to the economic and social life of the country.

In developed countries, small and medium-sized enterprises serve as the basis for stable production development [7]. As the analysis of the experience of developed countries shows, the state plays an important role in creating favorable conditions for stimulating innovation activity in all sectors of the economy [13].

In countries where the economic structure is market-based, assistance in the development of SMEs occurs in many areas of activity. Due to the different historical development of the economy of different countries as a whole, in advanced economically developed countries, the awareness of the SMEs' importance occurred much earlier than in our country, so the process of stimulating and increasing the share of this sector by authorities of these countries began earlier. It should be mentioned that SMEs work closely with national, regional and local authorities.

The experience of a country as advanced in production and technology as Japan, where a significant share of the country's production belongs to small entrepreneurs, is very informative. The Small and Medium Enterprise Agency (SMEA) is an Agency owned by the Japanese Government. SMEA is responsible for controlling and managing the entire business support area. SMEA is also authorized to implement the Japanese Government's global plan for small and medium-sized businesses. One of the tools for fulfilling such powers is joint work with scientific and research centers, major organizations, both private and public.

Of course, we cannot ignore the experience of our major trade and economic partner - China. The Chinese Government considers active promotion of SME development and growth to be a priority in the sphere of economic regulation. The Chinese system of small business management differs from the Russian one that is focused more on the "collective management", which implies the absence of a centralized administrative apparatus, but there are controls bodies across industries and specialized management (coordination centre for external cooperation of SMEs), the mandate of which includes the regulation of export turnover produced by small and medium producers and the regulation of output of producers outside the country.

The experience of the United States of America as a country with one of the most developed economies in the world shows that in 1953, the Small Business Administration (SBA) under the President of the USA was established to help SMEs. The fundamental goals of the SBA are:

- financing of the SME sector,
- assistance to entrepreneurs in obtaining state orders by organizations,
- analysis and verification of innovative business ideas,
- provision of technical support and consulting services.

All the above-mentioned countries rely on each other's experience to some extent. It is not possible to fully apply a particular policy of supporting SMEs in one country to another, since they differ both in their geographical location, development characteristics, and national and ethnic affiliation. But based on this experience, we can identify the strengths and weaknesses of Western mechanisms and try to adjust them to our country.

5 Conclusion

The need to support and bring to a new development level of the country's economic sector, as well as the development of anti-crisis measures, promote the growth of the market sector determines special attention to the issues of supporting SMEs. Today, the state clearly understands the significant relations between the development levels of the national economy as a whole and the SME sector. The country cannot grow its economy without a well-functioning main sphere.

The prosperity level of entrepreneurs engaged in the SME sector, like a mirror, reflects the real position of the state not only on the world stage, but also its internal regional development, social situation, and most importantly, it shows all its economic potential, which should be maintained at a high level in a country like Russia. The SME sector in the country is impressive in terms of the number of different types of production, the number of employees engaged in production, profitability indicators, the development level, and equipment in general.

The Government of the Russian Federation is developing and implementing a wide range of measures to maintain the sustainability, a high level of functioning of SMEs. The most necessary and common type of assistance among entrepreneurs is monetary financing. It is obvious that all programs have their shortcomings because of the design complexity or the presence of a large number of criteria, but they are all very effective, allowing many entrepreneurs to stay afloat. Of course, there are still unresolved problems, which are noted by entrepreneurs themselves who ask first of all to reduce the tax burden and give relief in terms of the use of cash registers. Individual lending programs should be also developed with low interest rates and less stringent requirements. In order to solve all the tasks that arise in the sphere of small business, the authors studied the experience of other countries. The research shows that each country has its own special SME factor, which they focus on in implementing support activities. The analysis of foreign experience will allow to create a new, more advanced system of state support for small businesses in Russia, and possibly make changes to the existing organizational and legal structure, taking into account all the collected experience of the own economy and economic systems of other countries.

References

1. Anoshkin, A.N.: Problems of small and medium-sized businesses in Russia associated with the economic policy of the state. Azimuth Sci. Res. Econ. Manag. **5**(4), 15–18 (2016)
2. Bessonova, A.A., Rybakova, Yu.V.: Small business in Russia: current state and development prospects. Azimuth Sci. Res. Econ. Manag. **7**(4), 270–272 (2018)
3. Bobrova, O.S., Tsybukov, S.I., Bobrov, I.A.: Entrepreneur Handbook. Yurayt, Moscow (2020)
4. Das, B., Hui, X., Sha, S.J.: Investment policies that support SME self-development. Hum. Syst. Manag. **37**(1), 15–25 (2018)
5. Federal law of 22.05.2003 54-FZ. On the use of cash registers for cash payments and (or) settlements using electronic means of payment (2003). http://www.consultant.ru/document/cons_doc_LAW_42359/. Accessed 28 June 2020

6. Gracheva, K.M., Peter, E.V.: State support for the development of small business in the Russian Federation. Sci. Almanac **7–1**, 12–16 (2018)
7. Gumar, N.A., Telagussova, E.U., Serikbaev, S.K., Seisenbaeva, A.M.: Financial support for the development of the SME sector in Kazakhstan. J. Soc. Sci. Res. **3**(Special Issue), 121–125 (2018)
8. Klerk: Will online sales registers ruin small and medium-sized retail businesses in Russia (2017). https://www.klerk.ru/boss/articles/461484/. Accessed 28 June 2020
9. Koroleva, E.A., Smulov, A.M.: Banking credit and consulting support for small and medium-sized enterprises in Russia. Pero Publishing, Moscow (2015)
10. Markelov, R.: The Central Bank spoke about the impact of a surge in lending on economic growth (2019). https://rg.ru/2019/06/28/cb-rasskazal-o-vliianii-vspleska-kreditovaniia-na-rost-ekonomiki.html. Accessed 28 June 2020
11. Qamruzzaman, M., Jianguo, W.: SME financing innovation and SME development in Bangladesh: an application of ARDL. J. Small Bus. Entrepr. **31**(6), 521–545 (2019)
12. Scott, T.J., Scott, J.T., Link, A.N.: Commercial complexity and entrepreneurial finance. Econ. Innov. New Technol. **26**(5), 489–500 (2017)
13. Shokhan, R., Karipova, A., Khoich, A., Kabdullina, G., Kudaibergenov, N., Niyetalina, G.: Economic levers of regulation of entrepreneurship. Entrepr. Sustain. Issues **7**(2), 1567–1581 (2019)
14. Wonglimpiyarat, J.: The new challenge of financing innovative economic growth through SME development in the People's Republic of China. Technol. Soc. **46**, 49–57 (2016)
15. Zhurba, A.S., Sotnikov, I.V.: Small business development in modern Russia. Sci. Soc. Econ. Law **1**, 154–159 (2020)

Current Issues of Legal Personnel Training in Telemedicine

A. Sidorova(✉) and E. Kalashnikova

Samara State University of Economics, Samara, Russia
an.sido@bk.ru, kalashnikova-helen@yandex.ru

Abstract. Nowadays, telemedicine has become an integral part of healthcare. Such services are provided both in Russia and foreign countries. Funding in this area increases every year. The economic and social value of these digital technologies in medicine is widely recognized. The worldwide coronavirus pandemic has only accelerated the implementation of telemedicine technologies and changed the attitude of citizens to online consultations. However, even now there are some problems. The study is devoted to analyzing one of them. The purpose of the study was to highlight the problematic aspects of training and retraining of legal personnel for healthcare institutions that provide telemedicine services. To do this, we have analyzed and summarized the factors that affect the requirements to lawyers who provide legal support of telemedicine services. The article highlights the areas of knowledge, necessary for lawyers that provide services in the field of telemedicine technologies. The result of the research will be the development of a proposal to reform the educational process of training lawyers in the field of telemedicine.

Keywords: Competence · Education · Healthcare institution · Informatization · Lawyer · Telemedicine

1 Introduction

The market economy, globalization and Informatization in all spheres of public life [9], the development of telemedicine technologies impose fundamentally new requirements to the legal personnel of health institutions, the organization and content of their training and retraining, and professional development. In the field of telemedicine, special requirements are imposed to the skills and qualifications, competencies and knowledge of legal personnel of healthcare institutions. High-quality functioning and development of any medical organization, especially in the field of telemedicine, is impossible without proper legal support and improvement of legal education. On the one hand, the task is to train a sufficient number of qualified legal advisers (lawyers). They know the specifics of healthcare in general and the specific rules for providing various types of medical care. On the other hand, telemedicine technologies require special knowledge about the specifics of personal data processing, ensuring the safety of personal data and the safety of personal data storage media; the creation and operation of information systems, data transmission, information protection from unauthorized access, etc.

S. I. Ashmarina and V. V. Mantulenko (Eds.): IPM 2020, LNNS 161, pp. 65–70, 2021.
https://doi.org/10.1007/978-3-030-60926-9_9

2 Methodology

Scientific research is the form of existence and development of science. Any scientific research begins with the definition of its purposes that determine the tasks. The purposes and objectives of the research determine its methodology. The correct choice of a set of knowledge methods in a particular subject area allows you to formulate reliable conclusions. The set of methods allows you to show the real situation of the studied area. The method of scientific knowledge for us is a method of theoretical or experimental study of a phenomenon or a process that contributes to the discovery of objective laws of reality. This research should be qualified as theoretical and applied. This has determined the set of its methodological tools. The main methods of this research are theoretical methods, in particular, it is analysis and idealization. Their use made it possible to identify problematic aspects of training lawyers in the field of telemedicine and to construct a progressive model of the educational process in this aspect. In addition, the authors also used empirical methods: observation, methodical analysis.

3 Results

The development of telemedicine technologies has a global scale. In Russia, its rapid development has began after 2018. However, even today there are disadvantages: a limited range of medical services provided by the nomenclature of medical services is used; insufficient development of the regulatory framework; lack of interaction between federal and regional ministries on the implementation, regulation and development of telemedicine; lack of a payment model for telemedicine services [7]. There are also problems that lie not only in telemedicine and law, but also in pedagogy.

High-quality functioning and development of any medical organization, especially in the field of telemedicine, is impossible without proper legal support and improvement of legal education [10]. On the one hand, the task is to train a sufficient number of qualified legal advisers (lawyers). They know the specifics of healthcare in general and the specific rules for providing various types of medical care. On the other hand, telemedicine technologies require special knowledge about the specifics of personal data processing, ensuring the safety of personal data and the safety of personal data storage media; the creation and operation of information systems, data transmission, information protection from unauthorized access, etc.

The Russian system of professional legal personnel training has a long history and established traditions. Decree of the President of the Russian Federation of 26.05.2009 N 599"On measures to improve higher legal education in the Russian Federation" has defined some measures to improve the training of legal personnel [2]. This concerns the development and implementation of a mechanism for public accreditation of federal, state and non-state educational institutions of higher professional education that train legal personnel, and provides for the development and public discussion of federal state standards of higher professional education in the field of training (specialty) "jurisprudence", which provide for increasing the volume of the practical part of the main educational program, improving the licensing procedures for educational activities.

Despite the fact that the legal literature suggests that it is necessary to stop the recruitment of students to law faculties of non-specialized universities [5], we believe that healthcare requires lawyers with special knowledge that can be obtained when studying at the law faculty of a medical university or specialization in the field of "Medical law" in specialized law schools. We agree with the statement "medical law is formed as a complex branch of legislation, including the norms of many branches of Russian law that regulate relations in various fields of medical activity - from health care management to specific relations between medical institutions and citizens [8]. In foreign countries, interest in medical law also does not weaken [3, 6].

Forms of professional training, retraining and advanced training of employees, the list of necessary professions and specialties are determined by the employer, taking into account the opinion of the representative body of employees in accordance with article 372 of the Labor code of the Russian Federation for the adoption of local normative acts [4].

When determining the content of training and further training of lawyers of the health institutions in terms of development of telemedicine it is necessary to provide understanding of the development trends of telemedicine; development of principles and rules for the development of telemedicine; application of standards and other normative legal acts in the sphere of public health and health protection including telemedicine; readiness to development of the local regulations of the medical organization based on the principles and requirements established by applicable standards, and clinical guidelines and other normative legal acts; skills formation of professional judgment in the qualification, the classification and assessment of the significance of various aspects of the telemedicine activities of the health institutions in the legal field; understanding of socio-technical processes of the telemedicine activities.

The basis for the development of educational programs for training, retraining and advanced training of legal personnel in the field of healthcare and telemedicine is the national legislative framework that regulates the use of telemedicine technologies and combines national legislation that is directly or indirectly related to the process of telemedicine interaction; national protocols and standards of health care; national and international harmonized engineering and telecommunications standards.

Changing approaches to training and retraining of legal personnel of healthcare institutions in telemedicine is determined by many factors: federal law on telemedicine; limits of out the telemedical consultations, the current regulations; the absence of approved procedures for registration of protocols; patients rights in remote communications (information booth and Internet website of the institution, forms of consents/refusals; respect the patient's right to receive information about his health, familiarization with medical documentation and providing it on request); medical confidentiality of the personal data security (reasons for disclosing information without a patient's consent, differentiation of personal data of the patient, the proper execution of the patient consent and/or his representatives to the processing of patient' personal data depending on the category of patient' personal data, the reasons for their treatment without a patient's consent, all medical records of the patient at the request of supervisory authorities, law enforcement agencies, court investigation, etc.; the possibility of video recording of the patient and by the patient, including hidden; transfer of research results and analysis by email & SMS); the proliferation of remote communications, including monitoring the health status of the patient; the adoption of a uniform format and recognition of the status of electronic medical records of the patient; the possibility of realization patient's right to receive anonymous medical services; the specifics of registration of contractual relations with patients in the provision of services within the mandatory and optional medical insurance (LPA and AMV) and reimbursement for provided medical services; specifics of payment of medical services provided under contract with an individual and honors the patient/consumer of services from the customer/payer; the possibilities of failure in the treatment/of treatment (including hospitalization) under the rule of law and algorithm design appropriate medical documentation; the possibility of failure of the patient and/or legal representative from intensive care taking into account patient's right to refuse medical intervention, including resuscitation, provided their disengagement with euthanasia.

Modern training, retraining and advanced training programs for legal personnel of healthcare institutions in the context of telemedicine development should focus on the formation of relevant key competencies considering the requirements for practice-oriented educational programs. Fundamental problems of legal responsibility of doctor and medical organization, defining new requirements for the competences, skills and qualifications of the legal professionals of healthcare institutions in the development of telemedicine are the following: legal responsibility of doctor and medical organization, which defines the responsibilities of doctor and medical organization in front of the patient, its forms, limitations and maintenance; legal security of doctor and medical organization; minimize legal risks; the possibility of criminal responsibility of medical personnel; the differences of professional and official offences "iatrogenic" crimes, medical malpractice; liability of physicians engaged by the court as third parties; the possibility of recourse actions within the institution.

4 Discussion

In the formation educational programs work, programs of retraining and advanced training of legal workers, healthcare institutions it is necessary to analyze very specific questions for ensuring the patient's right on secure medical services; status of orders and standards of care, clinical guidelines; safety criteria of medical care defects, its difference from the shortcomings in medical services; the procedure of internal control of the quality and safety of medical activities in recommendations issued by Roszdravnadzor. In the framework of practical training it is necessary to consider the currently claims of patients to medical organizations for various reasons (poor quality of medical aid, defects of healthcare, causing harm to life and health, the fact of a medical error, negligence of medical personnel, etc.); possible pre-trial settlement of the dispute and the options for compromise agreements; litigation practice in civil and criminal cases, particularly article 238 of the Criminal Code of the Russian Federation [1] (providing services without meeting safety requirements), the analysis of specific criminal and civil cases in the field of medical care, modern litigation practice, new trends in litigation, "strong" and "weak" positions of medical organizations in the role of responder in the judicial process; the possibilities of modern evidence-based medicine, independent examination, forensic medical examination; the adoption of reasonable risk and extreme necessity as conditions of exemption.

The main directions of improving the system of training and advanced training of legal personnel are: orientation of educational programs of higher educational institutions, as well as retraining programs for professional lawyers to an in–depth study of the order of medical care and legal regulation of telemedicine; orientation of educational programs of higher educational institutions, as well as programs for retraining of professional lawyers in healthcare and in telemedicine; monitoring the quality of educational programs of higher educational institutions, as well as programs for retraining of professional lawyers in healthcare institutions; updating and development of educational and methodical support of the educational process for lawyers in the field of medicine and telemedicine technologies.

5 Conclusion

Interaction of legal personnel of healthcare institutions with medical personnel of a medical organization should be due to the specifics of the functions and requirements division to the quality and safety of medical services. Medical personnel of the medical organization fully responsible for the quality and safety of medical care (including quality criteria and practices), compliance with standards and procedures of providing standards and clinical guidelines; results of examination of medical care quality in providing requirements of the state quality control and safety of medical activities.

References

1. Criminal Code of the Russian Federation from 13.06.1996 N 63-FZ (2020). http://www.consultant.ru/document/cons_doc_LAW_10699/. Accessed 05 July 2020
2. Decree of the President of the Russian Federation of 26.05.2009 N 599 "On measures to improve higher legal education in the Russian Federation" (2009). http://www.consultant.ru/document/cons_doc_LAW_88097/. Accessed 05 July 2020
3. George, R.P., Tollefsen, C.O.: The natural law foundations of medical law. In: Phillips, A. M., de Campos, T.C., Herring, J. (eds.) Philosophical Foundations of Medical Law, pp. 46–64. Oxford University Press, Oxford (2019)
4. Labor Code of the Russian Federation from 30.12.2001 N 197-FZ (ed. from 24.04.2020) (2020). http://www.consultant.ru/document/cons_doc_LAW_34683/. Accessed 05 July 2020
5. Matyusheva, T.N.: Improving the quality of legal education in solving the problem of the formation of the rule of law and civil society. Mod. Law **11**, 68–72 (2015)
6. Phillips, A.M., de Campos, T.C., Herring, J.: Philosophical Foundations of Medical Law. Oxford University Press, Oxford

7. Sidorova, A., Bezverkhov, A., Yudin, A.: Prospects for the development of the Russian telemedicine market: legal aspect. In: Ashmarina, S., Mantulenko, V. (eds.) Current Achievements, Challenges and Digital Chances of Knowledge Based Economy. Lecture Notes in Networks and Systems, vol. 133, pp. 789–796. Springer, Cham (2021)
8. Tsomartova, F.V.: New medical technologies - "Riddles" for law. J. Russ. Law **3**(267), 169–174 (2019)
9. Zakharova, M.V., Voronin, M.V.: Legal science in the challenges of the external environment: from the national past to the clash with the paradigms of globalization. Bull. Perm Univ. Legal Sci. **43**, 19–45 (2019)
10. Zhuravlev, M.S.: Interoperability as a factor in the development of law in the field of e-health. Law. J. High. Schl. Econ. **3**, 98–116 (2019)

Current Trends in the Labor Market Transformation Under the Influence of Environmental Factors

E. S. Smolina[1](✉), M. V. Greshnova[1], and A. S. Ryzhova[2]

[1] Samara State University of Economics, Samara, Russia
ekaterinsmolin@yandex.ru,
mariyagreshnova2019@yandex.ru

[2] Kazan Federal (Volga Region) University, Kazan, Russia
a.s.ryzhova@yandex.ru

Abstract. The first half of 2020 has become interesting and unique in terms of the changes that have taken place. The declared pandemic has affected all areas of life in both developed and developing countries, where cases of coronavirus infection have been detected. The labor market, which was previously influenced by factors and products of technological progress, as a result of the manifestation of crisis phenomena was among the vulnerable areas. In this case, crisis phenomena are understood as the adoption of quarantine regimes and general self-isolation, restriction or complete suspension of the functioning of economic entities (small, medium and large businesses). As a result of the impact of many factors, the labor market is transformed, the priorities of popular professions are changed, some of them are "dying", and some are just beginning to appear, becoming widespread. In this paper, based on the study of expert opinions (labor market) and analysis of information, the trends of the labor market are identified and potential changes in the requests of employers are described, which are interesting for potential applicants to study.

Keywords: Changes · Labor market · Pandemic · Robotization · Transformation

1 Introduction

Scientific and technological progress, the fourth industrial revolution, digitalization and robotics, the Internet of things – all these terms have entered our lives and have begun to change the conditions of the world's functioning, increasingly penetrating into various spheres of human life and society, including the influence exerted on the labor market, under the influence of factors, its transformation is taking place. In 2020, in addition to the development of technologies, the declared coronavirus pandemic had an impact on the economy and on certain areas, in particular, on the labor market. As a result of various events, new trends appear and spread, as well as changes in the interests and behavior of employers and job seekers.

S. I. Ashmarina and V. V. Mantulenko (Eds.): IPM 2020, LNNS 161, pp. 71–76, 2021.
https://doi.org/10.1007/978-3-030-60926-9_10

One of the trends of modern life is the spread of digitalization. It leads to an increase in the value of candidates who were able to adapt and learn to work in a digital format. Now remote work and flexible hours are becoming more and more popular, the coronavirus pandemic has also contributed to such conditions, companies have decided to transfer employees to such format, and employees have tried new conditions. New realities have provided an urgent mass transfer of employees to remote work. Companies that have the necessary resources have been able to change their methods of work.

The current trends in the labor market are:

- use of big data for various purposes (in particular, forming a value proposition for customers), cloud technologies. This leads to increased demand for mathematicians, programmers, and system architects [3],
- increasing the role of online sales leads to an increase in demand for hiring sales managers,
- remote work (reducing the need to rent office space and additional fixed costs,
- automation and simplification of routine processes leads to the replacement of people with “robots”.

As a result of these changes, the demand for professions changes, some are no longer in demand, and others, on the contrary, come to the fore.

In order to be in demand on the labor market both now and in 5–10 years, you need to make the effort and take the necessary steps to find significant areas, learn new methods of work, get additional knowledge that can be used in the future, choosing a set of educational programs and getting the necessary knowledge and competence.

2 Methodology

In 2018, “PwC” company prepared an analytical material on the labor market future based on global trends (megatrends) [7].

In the labor market these trends stand out:

- development of technologies, automation and robotics, making life easier and increasing productivity, a threat to social stability,
- demographic shifts – aging of the population, continued work of elderly employees,
- increasing the role of cities in the choice of employment, the impact of cities on the formation of jobs due to urbanization,
- economic problems of developing countries,
- climate change and depletion of natural resources [7].

Some scientists claim that there are not many areas where people will not be replaced by robots in the near future [4]. In health, education, household, and some other areas in which it is still difficult to completely exclude human participation and performance of certain actions, that is, it is difficult to refuse using of human intelligence. Despite this statement, it should be said that information mechanisms are and will be used in these areas with varying degrees of application intensity, and it is possible to use new robotic technologies in the processes.

Scientific materials publish the results of analytical work of labor market researchers with an assessment of current changes and a forecast of the future. Kolontay, professor of the Higher School of Finance and Management of the Russian Presidential Academy of National Economy and Public Administration (RANEPA), analyzing the labor market trends, wrote on an important trend, which is a reduction in the need for labor power with the growth of production (the jobless growth term) [5]. Robotization leads to a decrease in the need for the number of employees, as people begin to replace labor power with the "machine".

Considering the future of the labor market, we study how the demand for professions will change. We must understand what will be irrelevant in the future in order to use the time resource efficiently at the present time. The Future of Jobs study, published by the world economic forum in 2016, reported that between 2015 and 2020, 7.1 million jobs will disappear, most of them are "concentrated in routine white collar office", while 2 million new jobs "in Computer and Mathematical and Architecture and Engineering related fields" will be created [9]. Such changes are influenced by the socio-demographic situation and the development of new technologies. We will be able to see and evaluate the results of these changes in 2021.

In 2020, new trends caused by the spread of coronavirus began to appear, including remote work. Professions that may become popular in the near future as a result of the development of the labor market are highlighted. Based on current conditions and future forecasts, we can say that automation and robotics are among the main factors affecting the labor market.

3 Results

The world is constantly changing, both in favorable and unfavorable way, the division is conditional. Changes under the influence of external and internal factors contribute to development. The higher the level of development, the more serious changes begin to occur, the speed of implementation becomes higher. The labor market is also dynamic, it is influenced by technological factors. If earlier in Russia a lot of students have tried to get a degree in economics or law, now it becomes clear that the supply exceeds the demand, you need to look for more popular professions that guarantee future employment. In 2018, it was reported that a significant part of unemployed graduates (about half) have degrees in economics and law [8].

In order to be useful to employers and competitive in the labor market, you should learn new things. In the Sobesednik.ru review professions that may cause high demand in the near future were named: sanctionist, urbanist, Big Data analyst, specialist in religious conflicts and gerontologist [1].

Over the next 5 years, there may be more and more staff reductions due to robotization and automation. Payment terminals, Internet banks replace operators, self-service ticket offices, ticket terminals eliminate the need to maintain a full-time cashier, robotic car washes also do not require human participation, you can continue to list the areas of work in which a person can be displaced by a computer mechanism, as it is already happening in many areas. An adverse consequence is an increase in the number of unemployed. Major Russian companies announced large-scale cuts, Sberbank reported a reduction in accountants from 33,000 to 1,600, and then even to 600 people, Russian Railways announced the dismissal of almost 30 thousand employees by 2025, explaining this by the introduction of new technologies.

The coronavirus pandemic also have its impact on the labour market. Short-term demand is predicted for some professions, but they will be relatively new for Russia: an organizer of VR/AR broadcasts, an online doctor, an anti-crisis manager, a time manager, an IT consultant, and other professions that help to set up remote processes, optimize business, and take care of health. Additionally, well-known professions can gain popularity:

- financial consultant. In the second quarter of 2020 in Russia, there is an increased interest in the topics of personal and corporate finance management, ensuring solvency, and maintaining current and future financial stability,
- online store manager, as well as content manager will be in demand as a result of the spread of online trading to attract and interact with customers,
- psychological crisis phenomena affect not only the economy, but also relations. This fact requires the help of professionals to resolve conflicts and vulnerable situations,
- programmer. Digitalization will inevitably affect many companies, in order to use technologies and optimize business processes, it will be necessary to attract specialists for the initial configuration of data and in the future to provide support.

Quarantine restrictions make business representatives who are not yet represented in the online space to think on the need to take steps towards digital transformation. Currently, there is a trend for remote employment. The need to rent large offices is decreasing. After an office employee has "tried" remote work, there is no guarantee that they will have a desire to return to the office – because this generates additional costs of both time and money. From the point of view of the employer, it should be said that the lack of need for offices can also reduce the monthly fixed costs of maintaining the office, and the weight advantage will be that there are no borders for admission, if previously managers chose geographically convenient employees, now it is possible to find employees from other cities without their moving, which means that there are advantages for each side in remote employment. It is important to ensure information security and preserve the company's trade secrets, as well as find ways to motivate employees to work effectively.

After professions become unnecessary, some people begin to retrain and even reach career "heights" in a new direction, while the other part, due to low flexibility and lack of adaptability, remains for a long time without a job and a source of income, not realizing that there no longer will be former demand and you need to move on. Modern specialists are becoming more mobile and adaptive, striving to gain new knowledge and experience. Changes and uncertainty when changing the sphere of activity are not so much frightening. You need to constantly focus on learning and becoming professionals in the right areas. Robots can displace many professions in the labor market from simple to more complex, but the need for highly qualified specialists with a high level of knowledge and developed intelligence will remain.

4 Discussion

Digital transformation is a modern phenomenon of our life that affects business structures. The reason for changes in modern life is the development of technologies in all areas. The topic of labor market transformation has been in the focus of researchers for many years, presenting results in the form of analytical reviews and forecast materials on future changes. There are various scenarios for the development of the labor market. In the presented review materials and development models, some experts note the adverse side of the changes, concluding that automation and robotization will cause mass reductions and increase in unemployment, which, in turn, will harm the countries economies and exacerbate social tensions. To a greater extent, such shifts in the labor market will adversely affect developing countries, since the spread and use of IT technologies has so far been much lower than in developed countries.

In Russia, there is now a gap between supply and demand, that is, employers need one thing, and applicants are interested in another, interests differ, and a solution is needed. We need to track trends in order to be successful both from the point of view of the company and each individual, we need to focus on the needs of society caused by a number of factors that are constantly changing.

One of the current trends in the labor market is remote employment, which has its pros and cons. According to NAFI, the number of remote workers in the European Union is about 17%, in Japan, Sweden, Finland and the United States in some industries it reaches almost 40%. It is noted that for the employer, in addition to the advantages (saving on some items of fixed costs), remote work carries a number of disadvantages, for example, it is more difficult to control an employee when he is not near, which may lead to a decrease in productivity and deterioration in the quality of performed work [6], as a result, you need to look for motivation. By studying the topic of labor market transformation, you could understand what you need to be in demand and what steps you need to take to be in demand in the labor market for professional competencies in the future.

5 Conclusion

Professions of the future, if you have the inclination and interest, you need to learn now, this will give you an advantage when the need arises. The main "engines" of changes and transformation of the labor market are automation, robotics and digitalization, they contribute to the development of technologies in companies, which is why professions associated with performing routine operations are becoming less popular. Digital transformation should affect the processes of production, personnel management, and interaction with customers. Most of all, digitalization will be relevant for large and medium-sized businesses in both B2C, B2B, and B2G segments [2].

In addition to the development of information technologies, the labor market is also affected by socio-demographic changes, economic changes (economic transformation), urbanization, and others. One of the factors influencing the labor market in 2020 was the coronavirus pandemic, which contributes to the acceleration of the development of optimization and informatization processes due to the remote work processes development. Now we could see that in the future the world will be more technological, and some of the process will be performed automatically. Robots will not completely replace people, but they will replace some professions. We need to be in demand today, tomorrow, in 5 and 10 years. In order to do that, it is important to be professionals in the areas that the modern world needs.

References

1. Akhmirova, R.: Five professions that will be in demand in the near future (2017). https://sobesednik.ru/tehnologii/20171208-professii-budushego. Accessed 30 June 2020
2. Fomichev, K.: Go digital or die: Digitalization of business as an inevitability (2017). https://nand.ru/professional-information/and_library/20565/. Accessed 30 June 2020
3. IOT: Digitalization sets trends in the labor market (2018). https://iot.ru/promyshlennost/tsifrovizatsiya-zadaet-tendentsii-na-rynke-truda. Accessed 30 June 2020
4. IQ Consultancy: What professions will be in demand in 5–7 years (2017). https://www.iqconsultancy.ru/articles/kakie-professii-budut-vostrebovany-cherez-5-7-let/. Accessed 30 June 2020
5. Kolontay, M.M.: How to change the labor market in the next 20 years: Trends and opportunities (2017). http://mkolontay.ru/publications/275. Accessed 30 June 2020
6. NAFI: Russian companies are in no hurry to transfer employees to a remote work format (2019). https://news.myseldon.com/ru/news/index/212167194. Accessed 30 June 2020
7. PWC: The future of the labor market (2019). https://www.pwc.ru/ru/publications/workforce-of-the-future-rus.pdf. Accessed 30 June 2020
8. Vaschenko, V.: We do not need a lawyer: Universities got criticized for bachelor's training (2018). https://www.gazeta.ru/social/2018/01/29/11629591.shtml. Accessed 30 June 2020
9. World Economic Forum: The future of jobs. Employment, skills and workforce strategy for the fourth industrial revolution (2018). http://www3.weforum.org/docs/WEF_FOJ_Executive_Summary_Jobs.pdf. Accessed 30 June 2020

Evolution of the Labor Market: Challenges of the Millennial Generation

T. V. Suvalova[1(✉)], A. V. Troitskiy[1], and G. Sh. Zhaxybayeva[2]

[1] State University of Management, Moscow, Russia
{tv_suvalova, av_troitskiy}@guu.ru
[2] Buketov Karaganda State University, Karaganda, Republic of Kazakhstan
zhaksy@mail.ru

Abstract. The research purpose is to analyze changes in the labor market based on the needs and values of a new generation of young professionals. The authors consider the theory of generations, changes in the interaction between employers and candidates related to the development of digital technologies, open communications, the application of flexible schedules and remote work. The results of sociological surveys of professional aspirations, career preferences and expectations of millennials are summarized. The list of bonuses provided by Russian companies in addition to wages is formed. The most popular compensations are marked. Priority factors have been established for Russian young professionals when choosing an employer company. It is revealed that opportunities for professional development, participation in innovative projects, flexible working hours that allow combining work and private life (work-life-balance) are more relevant for young people than monetary values and the capital accumulation. An overview of the indicators of a successful career. It is noted that most young people are ready to invest in the own additional professional education, participate in international projects and internships. The issues of attracting young and talented candidates are considered in this paper too.

Keywords: Career · Labor market · Social mobility · Success indicators · Youth

1 Introduction

The Soviet economy underwent serious changes in the mid-eighties of the XX century. The Iron Curtain and the "era of stagnation" were replaced by "Perestroika" and "Glasnost". After the collapse of the Soviet Union and the reorganization of the public sector, stability in the forms of state orders and state budgeting was replaced by competitive market relations, entrepreneurial activity, and the development of the own businesses. The labor market has also undergone drastic transformations. The authoritarian leadership style has been replaced by democracy, leadership development, and soft skills. The image of the "ideal" manager and the employee's behavioral model have changed. Instead of orders and obedient execution, partnership relations and interaction between managers and subordinates, the development of delegation of authority and responsibility become priority. Rapid development of IT technologies has

S. I. Ashmarina and V. V. Mantulenko (Eds.): IPM 2020, LNNS 161, pp. 77–84, 2021.
https://doi.org/10.1007/978-3-030-60926-9_11

contributed to changes in business processes, communications, working conditions, and accordingly to requirements to education, competencies and skills of personnel. Under the current conditions, a new generation has grown up with attitudes to social freedom, open communications, independence, personal development, and willingness to open own business.

The theory of generations, which is popular today, appeared in the early 90's of the previous century. This theory is based on the dominant values in the time context. It should be noted that the time frame is quite conditional. Representatives of generation X belong to the "era of stagnation" and to the time period from 1965 to 1985. Generation Y or perestroika generation, or millennials belong to the period from 1985 to 2003. The word "millennials" derivates from the word "millennium", it means a turn between millennials. Unlike their predecessors, those born during this period strive for individual unlimited multi-dimensional development. Characteristic features of millennials are freedom of choice, thoughts and actions, the absence of prohibitions and dependencies, restrictions [4]. Generation Z, born from 2004 to the present, are active Internet users and adherents of online communication. The most adults of them in 2020 are about 20 years old [14]. In recent years, there has been debates about the birth of a new generation (Generation Alpha), a generation of "artificial intelligence" that was born since 2015 [1].

The study of the labor market on the example of generation Y is relevant due to their age (25-35 years old). Prospects for the development of the economy, politics, labor and market relations are formed by the younger generation. The formation of the Y' personality took place in the era of digital technology development and integration of digital transformations in all business spheres. Y' life priorities are intellectual performance and passion for sports (promotion of healthy lifestyle), careful attitude to the planet and its natural resources.

Nowadays, there are opportunities for self-employment, flexible working hours and remote work, the activity of social networks has increased as a means of communication and broadcasting of corporate information, we also have new opportunities for self-development and self-realization. Work functions are being transformed, and new professions are being developed.

2 Methodology

The younger generation has a high degree of personal and social freedom. The age of political and business leaders is rapidly rejuvenating. Key positions are held by thirty – to thirty-five-years-old managers. In the labor market, there is a growing demand for young talents with multi-potential, broad horizons and bold ambitions [3]. Betting on young people is gaining more and more relevance.

According to the experts of the HAYS study "Generation Y and the labor market", the young Russian generation is characterized by social mobility, the desire for development and, consequently, to increase personal income [11]. The most relevant criteria for choosing an employer for young people is the degree of job satisfaction (interesting work), and career prospects (opportunities for promotion).

The analysis of the labor market evolution is based on the results of labor market research and a review of wages in Russia, the current needs of the younger generation, and the prospects for remote work, conducted in the period from 2017 and 2019 [9, 10, 12]. The study of professional aspirations, moods and expectations from the career of generation Y (millennials) was conducted by HAYS in 2019 using a social survey of more than 1000 young Russian professionals aged from eighteen to thirty years, working in various business areas: consulting, manufacturing, finance. The proportion of respondents was 51% for men and 49% for women. Regional representation is expressed in the following percentages: the Central federal district – 25.8%, the North-Western federal district – 9.5%, the Southern federal district – 16.9%, the Volga federal district – 20.5%, the Ural federal district – 8.6%, the Siberian federal district – 14.0% and the Far Eastern federal district – 4.7%. The majority of respondents (44%) have more than three years of work experience. More than a third (38%) work full-time for at least 35 h a week, and only 11% work part-time for up to 35 h a week [11].

The purpose of the survey was to identify preferences, aspirations, and expectations from current or future work, as well as to form a portrait of an ideal manager, promising and talented employee. The research issues cover the impact of factors such as communication in the workplace, the degree of independence, opportunities for flexible schedules and remote employment on the labor market. Special attention in the survey was paid to the amount of material remuneration that makes up the compensation package, various bonuses and options.

3 Results

The statistical analysis of sociological studies of the world labor market for 2018-2019 showed that most organizations have expanded their staff. Especially rapid growth is observed in the information technology industry. It is worth noting that preference is given to young candidates. In the information sphere, the "fight for talents" is relevant. If we look at foreign experience, then a striking example is the leading technology companies in Silicon Valley. Their personnel policy is aimed at attracting talented university graduates. Amazon, Apple, Google, Cisco Systems, and Microsoft offer high starting salaries for newcomers, flexible schedules, remote work opportunities, free 24-h meals, recreation areas, and other bonuses.

In the bonus package, more than 50% of Russian employers include: medical insurance, payment for private medical services, additional payment for sick leave, a company car or compensation for using personal transport for official purposes, and phone payments. More than 30% of companies include in the list of bonuses provided parking compensation, or directly parking, food payment, provision of a sports complex on the territory, full or partial compensation for fitness, corporate training, additional vacation days, discounts on the purchase of products or services, life insurance, incentive payments. About 20% of organizations provide financial support for additional professional education and reimburse expenses for using public transport. Less than 10% of employers offer payments for anniversaries, birth of a child, corporate kindergarten and remuneration based on promotions, assistance in employment when leaving. The list and percentage of bonuses provided by Russian companies is shown in Table 1.

Table 1. List of bonuses provided by Russian companies

Bonuses	Percentage, %
Medical insurance, surcharges for private medical services	58%
Provision of official transport	54%
Payment for mobile communications	52%
Compensation payment for parking	43%
Food compensation	41%
Fitness, yoga, swimming pool compensation	35%
Life insurance	33%
Additional vacation days	34%
Incentive payments quarterly, annual (based on the results of work)	33%
Discounts on products or provided services	32%
Corporate training	31%
Compensation for additional education	19%
Compensation of expenses for using public transport	17%
Assistance in employment	9%
One-time payments for an anniversary or the birth of a child	9%
Equity participation	8%
Corporate kindergarten	2%

Source: authors.

In order to attract talents and retain key professionals, in addition to competitive salaries, bonuses and social packages, employers offer corporate training with subsequent careers, expanded opportunities to make decisions independently [15]. Priority factors when choosing an employer for Russian young professionals are: interesting work with opportunities for professional development, innovative projects, flexible schedules that allow you to combine work and personal life (work-life-balance), a workplace equipped with modern technology, a pleasant and beautiful office (workspace). A third of respondents noted the importance of independent work and the lack of strict control by the direct manager, the ability to express their own opinion, a sense of support from colleagues and management. Young people pay attention to remuneration. In addition to wages, a third of respondents expressed interest in a variety of flexible benefits and bonuses. The priority factors of working conditions are presented in Table 2. Due to the possibility of selecting multiple answers, the final percentage is more than 100%.

Table 2. Priority factors of working conditions for young professionals

Factors of working conditions	Percentage, %
Interesting work with opportunities for professional development, projects	73%
Flexible working hours and remote work	43%
Equipped with modern technology workplace	38%
Opportunity to work independently	36%
Regular corporate training	35%
Competitive salary	35%
Bonus package and flexible benefits	33%
Opportunity to work at home	31%
Well-maintained office space	29%
The popularity of employer's brand	25%
Convenient location of the company	24%

Source: authors.

Flexible schedules are very popular among young professionals. Almost half of respondents noted the possibility to choose the time of arrival to work and leaving. It is also a desirable option to work at home or remotely at least once a week. This trend is related to the growing need for a combination of work-life balance. The concept of success is inextricably linked with the career growth [2]. The millennial generation highlights the following indicators of a successful career: high wages that provide opportunities for capital accumulation and financial investments, interesting work that brings pleasure, the opportunity to have a work-life balance, professional mobility and acquisition of various competencies, managerial position, high status in the company, professional recognition, image of a highly qualified specialist.

These indicators determine the desire of young specialists for professional and personal development. Most of them are willing to invest their personal funds in additional education [19]. A significant number of young people are interested in internships abroad and participation in international projects.

4 Discussion

Theoretically, the millennial generation can be classified as "young professionals" by age criteria. According to Federal Law No. 98-FZ of 28.06.1995 "On state support for youth and children's public associations", this category includes citizens under the age of 30 years [6]. Decree of the President of the Russian Federation of 14.10.2012 No. 1380 "On improving the effectiveness of state support measures for employees of organizations of the military-industrial complex of the Russian Federation" has an impact on young engineering and technical workers, specialists and highly qualified workers of enterprises of the military-industrial complex of the Russian Federation under the age of 35 years [5]. It should be noted that the company is entitled to determine the status of a young specialist in its regulatory documents. Modern competition encourages promising companies to search for and carefully select talented young people.

Issues of attracting young and talented employees are studied by such large companies as Forbes Insights, HeadHunter, Hays [7]. For example, in the USA, Japan, Canada and Germany employment agencies monitor perspective candidates. In France, companies are willing to pay for internships and professional training programs for young professionals. In the United States and Western Europe, the "virtual organization method" is used to train new employees during the adaptation period. In her research, Ivanovskaya considers event-recruitment [13]. Guzhavin notes the role of programs for the selection and development of young employees [8]. Lobacheva believes that the potential and effective work of young professionals brings maximum benefits to organizations. We can conclude that in modern conditions, the most popular in the labor market are specialists at the age of 26–30.

The authors note the need to reveal the entrepreneurial potential of the millennials generation for the development of Russian business. Therefore, in order to attract and retain professionals of the millennial generation, it is important for the company management to take into account motivating factors and success indicators that are attractive to young candidates indicated in this study.

5 Conclusion

The study illustrated some rapid changes in the labor market related to young candidates. With the development of digital technologies, the horizons of business communications have expanded, work processes have been optimized, and the functionality of employees has changed [18]. In these conditions, we have a growing need for the ability to react flexibly and quickly, make "working" decisions, which is inherent in young professionals. The results of sociological research indicate the expansion of the staff primarily by attracting young people [16]. The company's management offers internships with subsequent employment, corporate training, competitive wages, and an extended bonus programs. Most employers provide health insurance, compensate for food, mobile communications, transportation expenses, and visits to sports complexes.

Directly for young professionals themselves, the priority indicators when choosing an employer organization are opportunities for professional development and participation in promising projects [20]. The interest in flexible hours and remote work is being updated. However, it is worth noting that not all employers agree with the feasibility of this option. An employee who works from the home office or comes to work several times a month is not considered a full-fledged member of the team, has difficulties interacting with colleagues, and is not involved fully in the company's life. The undeniable advantage of a flexible schedule is work-life-balance. This is an optimal combination of workloads and free time for personal life. And for the younger generation, compliance with this balance is an actual criterion for choosing an employer company. The development of technologies and means of communication contributes to the demand for "flexible working hours".

The platform for a successful career is getting new knowledge, regular training and self-development. The younger generation is well aware of the value of knowledge and is ready to invest in additional self-education. This is why candidates prefer companies with a developed institute of corporate training. The traditional indicator of choosing a future employer is the salary level. Young people prefer to combine the financial factor with job satisfaction. Money is a necessary but not a sufficient condition. The essence of the work is of paramount importance. Interest and variety of functions are put above monetary values.

References

1. Afisha Daily: What is the difference between boomers and Millennials in Russia (2019). https://daily.afisha.ru/brain/13620-chem-otlichayutsya-bumery-i-millenialy-v-rossii/. Accessed 09 June 2020
2. Anderson, H.J., Baur, J.E., Griffith, J.A., Buckley, M.R.: What works for you may not work for (Gen) me: Limitations of present leadership theories for the new generation. Leadersh. Q. **28**(1), 245–260 (2017)
3. Baker Rosa, N.M., Hastings, S.O.: Managing millennials: looking beyond generational stereotypes. J. Organ. Change Manage. **31**(4), 920–930 (2018)
4. Bencsik, A., Horváth-Csikós, G., Juhász, T.: Y and Z generations at workplaces. J. Competitiv. **8**(3), 90–106 (2016)
5. Decree of the President of the Russian Federation of 14.10.2012 N 1380. On improving the effectiveness of state support measures for employees of organizations of the military-industrial complex of the Russian Federation (2012). http://base.garant.ru/70242012/#ixzz6PsPBi8KQ. Accessed 18 June 2020
6. Federal Law No. 98-FZ of 28.06.1995. On state support for youth and children's public associations (1995). http://base.garant.ru/103544/#ixzz6PsPoosUu. Accessed 18 June 2020
7. Forbes Insights: Talent acquisition and onboarding. 4 steps to a winning candidate experience (2019). https://i.forbesimg.com/forbesinsights/oracle_HR_talent_acquisition/Oracle_TalentBrief1.pdf. Accessed 12 June 2020
8. Guzhavin, Y.: Role of corporate programs for a set of the young experts. Econ. Entrepreursh. **4–1**(45), 498–501 (2014)
9. Hays: Remote work (2018). https://hays.ru/res/remote-work/. Accessed 07 June 2020
10. Hays: Hays Salary guide 2018 (2018). https://hays.ru/res/salary-guide-2/. Accessed 07 June 2020
11. Hays: Generation Y and the labor market. Research of professional aspirations, moods and expectations from the career of a generation Y in Russia (2019). https://hays.ru/wp-content/uploads/2019/02/HAYS-Y-REPORT_RUS_XS.pdf. Accessed 08 June 2020
12. HeadHunter: Salary trends: Results of 2019 and employers' plans for 2020 (2019). https://hh.ru/article/25966. Accessed 08 June 2020
13. Ivanovskaya, L.V.: Modern approaches to attract young professionals to work in companies. Univ. Bull. **10**, 194–199 (2015)
14. Jenkins, R.: Generation Z versus millennials: the 8 differences you need to know (2017). https://www.inc.com/ryan-jenkins/generation-z-vs-millennials-the-8-differences-you-.html. Accessed 10 June 2020
15. Kashtanova, E.V., Ashurbekov, R.A., Maslennikov, I.A.: Russian labour market outlook. IOP Conf. Ser. Mater. Sci. Eng. **753**, 082026 (2020)

16. Konovalova, V.G., Laas, N.I., Svistunov, V.M., Starun, N.V.: The role of social work in improving of the quality of employees' life. In: Mantulenko, V. (ed.) International Scientific Conference "Global Challenges and Prospects of the Modern Economic Development". The European Proceedings of Social & Behavioural Sciences, vol. 57, pp. 708–719. Future Academy, London (2019)
17. Lobacheva, A.S.: Young specialists' labor activity stimulation at the enterprise. Hum Resour. Intell. Resour. Manage. Russia **2**(6), 58–62 (2013)
18. Mitrofanova, E.A., Konovalov, V.G., Mitrofanova, A.E.: Analysis of perspectives and risks of the corporate well-being programes. In: Ashamrina, S.I., Mantulenko, V.V. (eds.) Proceedings of the International Scientific Conference "Global Challenges and Prospects of the Modern Economic Development". The European Proceedings of Social & Behavioural Sciences, vol. 79, pp. 1392–1399. Future Academy, London (2019)
19. Parakhina, V., Ustaev, R.: The impact of the innovative potential of human capital on regional economic development: analytical aspects. Newslett. Expert Council **1–2**(12–13), 107–115 (2018)
20. Sviridova, S.V., Kablashova, I.V., Shkarupeta, E.V., Khmeleva, G.A.: Imperative of human capital management transformation in the development of economic systems. In: Mantulenko, V. (ed.) International Scientific Conference "Global Challenges and Prospects of the Modern Economic Development". The European Proceedings of Social & Behavioural Sciences, vol. 57, pp. 1355–1362. Future Academy, London (2019)

Changes in the Content of Labor and Their Social Consequences: Discussion Issues

V. V. Chekmarev[1] and O. A. Bulavko[2(✉)]

[1] Kostroma State University, Kostroma, Russia
Tcheckmar@ksu.edu.ru
[2] Samara State University of Economics, Samara, Russia
vikigorl63@mail.ru

Abstract. The article deals with debatable issues related to social consequences of changes in the content of labor caused by the development of socio-economic relations. It is shown that changing the content of labor determines the need to review characteristics of employees and workplaces, as well as the employment process itself. It is determined that changes in the content of labor is a direct challenge to the education system both in terms of content and organization (and especially in terms of resource provision). The author state the emergence of many new questions to the education economics as a scientific discipline and as an element of ensuring the educational production. Additional arguments are presented to support a position, according to which fundamental changes in the modern economy occur under the influence of fundamental shifts in the content of labor. Contradictions arising in the social structure in connection with changes in the content of labor are revealed.

Keywords: Changes · Creative class · Labor · Personnel

1 Introduction

The modern era, which is often called the fourth industrial revolution, is characterized by deep contradictions that permeate the entire "social fabric", but the initial changes, in our opinion, lie still in the content of labor and the qualitative transformations that occur at this level. All technologies are ultimately derived from human qualities and the transformation of mainly reproductive labor, characteristic of industrial society, labor that is primarily creative in its nature. These are key changes that cause corresponding shifts in the structure of the economy and in the social structure of our society, and even in politics and ideology [6].

The very question of the impact of changes in the labor nature on the social organization of society was raised by a group of sociologists at the beginning of the XX century, but without taking into account changes in the content of labor that have now appeared. Speaking about these fundamental changes, in our opinion, it is necessary to emphasize the legitimacy of the discussion that has been going on in the journal "Sociological Research" over the past years. It is mainly related to the analysis of new social phenomena, such as the formation of a mass creative class, on the one hand, and

S. I. Ashmarina and V. V. Mantulenko (Eds.): IPM 2020, LNNS 161, pp. 85–91, 2021.
https://doi.org/10.1007/978-3-030-60926-9_12

the "proto-class" of the "precariat", on the other hand, which is transformed into a new class. That is analyzed in a series of articles by Toshchenko and his monograph [17].

In turn, a series of works by Buzgalin and the discussion caused by these publications are devoted to the topic of genesis of the "creatosphere" and employees engaged mainly in creative activities [1, 2]. In our view, both processes are objectively conditioned, and are caused by a change of what is called in Marxism the "productive forces", meaning by this the active processing units of the means of production and man. And in this case, a man is acting not only as a labor force, but also as a creator, capable of creative change in the material body, social relations, and culture. These changes are at the heart of the transformation that is taking place in the economic and social space. But is that all? To answer this question, we should pay close attention to the discussion that is currently being conducted on the pages of both economic and philosophical literature, as well as in sociological research. We are talking about studies of the relations between market, planned, private and public forms of management, and more broadly-public life. This discussion was rather rigidly positioned in the text devoted to issues of the intellectual private property [13, 16, 18].

2 Literature Review

First of all, several circumstances should be noted. First, changes in the content of labor should be considered as a source of conflict between the development of productive forces and production relations and, at the same time, as a factor in the formation of a post-capitalist society. Secondly, changes in the content of labor are a catalyst for finding new solutions for social mechanisms regulating employment and unemployment, as well as "building new business conferences". Two publications on this topic in the journal "Sociological Research" by Buzgalin [1] caused a lively and objectively determined discussion in the publications of Ryazanov [15], Yakovleva [19] and other authors [4, 7, 11, 12, 14]. On the one hand, there was criticism, and on the other hand, support for the general thesis that characterizes a vector of strengthening social principles in the modern economy.

At the same time, it should be taken into account that there is another position that is associated with the justification of the need and possibility of developing mainly private, in some cases individual entrepreneurship and economic activity, determined by the progress of productive forces and the development of primarily creative content of labor [5]. The arguments of this side deserve close attention, because we really have a wide set of examples in countries from India to the United States (including Russia) of how a person's creative potential allows him to turn into an independent economic actor, get a status of a freelancer, while creating significant public wealth and providing a fairly high level of well-being for himself. However, the reverse side of this coin, rightly shown in the works of Toshchenko, indicates that the socio-economic factors in this area are characterized by volatility, instability, lack of social guarantees [17] (starting from employment guarantees, ending with guarantees of social security and income), the lack of social class community, political and ideological representation, and a number of negative consequences due to this process. This research work, on the whole, is devoted to the discussion of Buzgalin' concept, presented in the journal

"Sociological Research" [1] and other publications by this author. These publications are related to the limits of neoliberal economics and not by accident dedicated to the bicentennial birth anniversary of Marx [2].

Buzgalin shows that the modern system of industrial relations leads ultimately to a dead end and does not have the potential for further progress of productive forces [1, 2]. This conclusion about the system of industrial relations, which the researcher calls the global hegemony of capital, in our opinion, sounds excessively harsh and unambiguous, although certain arguments that are hidden behind this position, as well as arguments formulated above by Toshchenko, have considerable grounds [17]. First of all, this concerns such an aspect as the contradiction in the system of public relations, the sphere that is key for the economy and society based on creative activity. We are talking about the educational sphere, where the creative potential of a person is formed. The authors of this article link the problem of changes in the content of labor, the emergence of a creative class on the one hand, and the precariat on the other hand, with processes taking place in this area [3]. The choice of education as a key space that requires research is due not only to one of the authors' previous scientific specializations, but also to the role that this sphere plays in the economy of the XXI century and in the social development. We will point out only a few key points.

Qualitative changes in the content of the employee's work, which began in the middle of the XX century, led to the fact that the qualification requirements have changed no less radically. Today, more than half of employees of the most economically developed countries are people who have the potential for creative activity, which, unfortunately, is not always realized in practice, which, however, is rightly pointed out by the researchers mentioned above. This is a state when in today's economy, not just 12-16 years of education is in demand for almost all workers of a new generation and beyond, but also regular retraining of employees every 5 years. In the case we also consider the inclusion of non-working pensioners in the educational and cultural process, an example of which can be found in Belarus and even in Moscow. This phenomenon practically empirically proves the transformation of education into a sphere on which the current progress of productive forces is based, which requires the closest attention, although it is characterized as a component of the service sector, where educational services are produced, etc.

This point of view, which is widely presented in the economic literature, in the literature on economics and education, in our opinion deserves the most thorough criticism, since education is a more complex and deep process than the production of commercial services in a relatively narrow market [3]. By its specific weight, if it is calculated exclusively in monetary terms, which is not so significant when compared with the market for oil, gas, or simulacrum goods. If we look at the content of this process, then, as it has already been noted, this sphere forms the main productive force in the modern economy and the economy of the era of creative revolution in the first place. And this view of education raises the question of the extent of socialization in this area. In Buzgalin's articles and polemics on this issue in the texts of Klyuev, Ryazanov, Yakovleva and other authors, they consider the need to define measures in the development of the private and public sectors of culture in a broad sense, including not only art, as it is customary today, but also all areas that create cultural phenomena: and this includes science and education [10, 15, 19]. And here the question arises: what can be a criterion for distinguishing public and private beginnings in the development of the cultural sphere in general and education in particular? Buzgalin's message is quite unambiguous, but at the same time it is one-sided [2]. For him, the larger and wider the public sector develops, the better it is for the progress of human and the productive forces. This statement is not a new one, it is characteristic of the Marxist spectrum of social thought and has been reproduced for many years [8, 9].

3 Methodology

In our opinion, this formulation of the considered question requires a significant correction, and we would like to polemize with Professor Buzgalin, emphasizing that real economic processes are closely related to the interaction of social structures, classes, proto-classes and more local social groups, as well as political processes. In our opinion, such a view, a view through the prism of the social structure of society on the relation between the public and private subspaces of culture and, in particular, education, deserves the closest attention. The formation of the creative class causes by no means unambiguous trends in the growth of the public sector. As we have already noted, the creative class in real conditions of the modern economy and the modern political space is primarily focused on winning its relatively privileged position in the society, which is largely objectively determined. In the course of this research, methods of distribution of financial consequences of social risks were applied. The authors used methods of comparison and variational calculation, functional dependencies to solve considered socio-economic problems.

4 Results

The qualification level, the amount of social wealth created by the creative class, undoubtedly exceeds the potential of industrial workers, and workers engaged in the habitual reproductive labor. These objectively determined claims result in subjective social aspirations of this class, and in the corresponding political position, which, as a rule, supports not so much egalitarian, but elite trends in the development of social structures, education, access to culture, etc. And these intentions of the creative class, which have significant objective grounds, create powerful prerequisites for the formation of an elite model of education.

This model is directly or indirectly focused on the wealthiest segments of society and requires the use of a wide range of commercial principles, on the one hand. Examples of this in education are well known: elite universities, which involve directly paid education (this applies primarily to the elite universities of the USA), relying on the ability of parents to provide a high level of spending on special individual training for students through tutoring or other methods, which are then widely used (tutoring, additional counseling) and teaching at elite universities. On the other hand, this model does not allow to use fully the potential of a wide range of "ordinary" citizens; and the economy of the creative revolution era requires not only a narrow circle of highly qualified professionals that make up 10-15% of the society elite, but also the mass ingenuity of almost every man capable of labor in the economy of the mid-twenty-first century.

These workers are shaped by today's educational system. Here we are talking not only about the sphere of our own material production, engineers, highly qualified employees, technologists, economists, etc., but also about those areas that will create spaces, which Buzgalin called the "creatosphere" [1]. We speak about those employed in the field of education for all and through life, which will require in the near future up to 20% of those employed in the modern economy; about the sphere of forming a healthy person, which will require about the same amount of employment, if we consider that this sphere is not only the sphere of medicine, but also the sphere of forming a healthy lifestyle, including mass public sports for all ages and other components. This is a sphere which involves the formation of an adequate natural environment and solving problems of combating environmental pollution. Cleaning up the natural environment is another huge layer of social and creative work, which is necessary for today's economy no less than the production of steel, robots and new generations of smartphones. Finally, it is culture itself in a narrow sense as an art that can and should become the property of almost every person. Its prologue is, in our opinion, the experience of the Soviet Union, even in the first stage of its new economic policy, when in a relatively poor country, nevertheless, dozens of workers' clubs were created, centers for people's inclusion in poetic, theatrical, artistic, sports, scientific creativity, and this network covered millions of workers, and even the peasantry.

5 Conlusion

Summing up our reflections on the potential of the creative class and the contradictions associated with it, the authors would like to emphasize the reverse side of the creative revolution medal, with which this text begins. Technological challenges of the XXI century determine the main vectors of socio-economic development of countries, which are associated with the transition to a post-industrial economy, an economy based on knowledge, and an intellectual economy. Its technological appearance is a new sixth technological order. The intellectual economy, the main features of which are born during the transition to the sixth technological order, generally defines a new paradigm for managing economic development. The focus is not so much the development of competitive industries, but rather the creation of conditions for the harmonious, sustainable development of the economy and the system of social values and new moral principles, which are based on the idea of social justice and ensuring the well-being of the Earth's inhabitants for future generations.

At the same time, we should say about the formation of a mass layer of people who find themselves on the sidelines of the social development process. While maintaining a commercially-oriented and elitist education system, on the one hand, and introducing market, private principles in public life and the economy, on the other hand, a mass layer of people is now formed who are not in demand for either mass industrial labor or creative activity. And, as a rule, these are people with a fairly high level of humanitarian education and cultural potential. Some of them turn into freelancers with partial non-permanent, but employment, and some turn into a steadily reproducible (the further, the more large-scale) layer of lumpens, threatening to become the trouble that Marx once wrote about in relation to the XIX century, describing mass stagnant unemployment and the formation of related "lumpen estates". These are contradictions that should be discussed when analyzing the creative revolution and the qualitative changes in the content of labor, as well as the transformation of the social structure that results from them. Using the concepts of "social", "asocial" and "anti-social", and a new methodological base, without finding answers about how to resolve these contradictions, the ongoing changes in the nature of labor will most likely be idealized by proponents of the society development (including economy).

References

1. Buzgalin, A.V.: Creative economy: private intellectual property or ownership by everybody of everything? Sociol. Res. **7**, 43–53 (2017)
2. Buzgalin, A.V.: The decline of neo-liberalism (to the 200th birth anniversary of Karl Marx). Voprosy Ekonomiki **2**, 122–141 (2018)
3. Chekmarev, V.V.: New political economy in the context of expectations of the realization of the predictive function of economic science. Issues of Political Econ. **3**, 20–30 (2015)
4. Cherepantseva, Y.S.: The prefectariat and its influence on the development of labor relations. Proceedings of the Orenburg Institute (Branch) of the Moscow State Law Academy **1**(31), 140–145 (2017)
5. Dolzhenko, R.A.: Preconditions and consequences of changes in the scope and nature of labor in the context of innovation economy formation. Bull. Tomsk State Univ. Econ. **3**(35), 77–88 (2016)
6. Grigoriev, G.: The sky is under repair. Saint-Petersburg: "Palmira" (2017)
7. Gunko, N.N., Dinukova, O.A.: Impact of education on the structure of youth employment. In: Mantulenko, V.V. (Ed.) Proceedings of the 18th International Scientific Conference "Problems of Enterprise Development: Theory and Practice". European Proceedings of Social and Behavioural Sciences, vol. 82, pp. 874–880. European Proceedings, London (2020)
8. Inozemtsev, V.L.: 1985: Reminders of the present. Free Thought **9**, 5–16 (2010)
9. Kara-Murza, S.G.: A lost mind: Intelligentsia on the ashes of Russia. Eksmo, Moscow (2012)
10. Klyuev, A.: Music: the way to the absolute. the quests of the absolute. Russian J. Philosophical Sci. **5**, 124–138 (2016)
11. Kogut, O.Y., Janshanlo, R.E., Czerewacz-Filipowicz, K.: Human capital accounting issues in the digital economy. In: Ashmarina, S., Mesquita, A., Vochozka, M. (eds.) Digital Transformation of the Economy: Challenges, Trends and New Opportunities. AISC, vol. 908, pp. 296–305. Springer, Cham (2020). https://doi.org/10.1007/978-3-030-11367-4_29
12. Korneeva, T.A., Potasheva, O., Tatarovskaya, T.E., Shatunova, G.A.: Human capital evaluation in the digital economy. In: Ashmarina, S., Mesquita, A., Vochozka, M. (eds.) Digital Transformation of the Economy: Challenges, Trends and New Opportunities. AISC, vol. 908, pp. 66–78. Springer, Cham (2020). https://doi.org/10.1007/978-3-030-11367-4_6
13. Makhmudova, I.N., Bogatyreva, I.V., Ilyukhina, L.A.: Evaluation of administrative work efficiency in the context of digitalization. In: Mantulenko, V.V. (Ed.) Proceedings of the 18th International Scientific Conference "Problems of Enterprise Development: Theory and Practice". European Proceedings of Social and Behavioural Sciences, vol. 82, pp. 625–63. European Proceeding, London (2020)

14. Marinosyan, K.: The new strategy of nation-states in the context of globalization. Part I. Russian J. Philosophical Sci. **3**, 7–15 (2016)
15. Ryazanov, V.T.: Creative labor and ownership of everyone for everything: opportunity and problems. Sociol. Res. **3**, 138–143 (2018)
16. Schekoldin, V.A., Bogatyreva, I.V., Ilyukhina, L.A.: Digitalization of labor regulation management: new forms and content. In: Ashmarina, S.I., Vochozka, M., Mantulenko, V.V. (eds.) ISCDTE 2019. LNNS, vol. 84, pp. 137–143. Springer, Cham (2020). https://doi.org/10.1007/978-3-030-27015-5_17
17. Toshchenko, Z.T.: The birth of a new social class. Bull. Russian Acad. Sci. **86**(3), 231–236 (2016)
18. Troshina, E.P., Mantulenko, V.V.: Influence of digitalization on motivation techniques in organizations. In: Ashmarina, S.I., Vochozka, M., Mantulenko, V.V. (eds.) ISCDTE 2019. LNNS, vol. 84, pp. 317–323. Springer, Cham (2020). https://doi.org/10.1007/978-3-030-27015-5_38
19. Yakovleva, N.G.: Education in russia: public good or commercial service? Sociol. Res. **3**, 149–153 (2018)

Economic Consequences of Current Precarious Employment, Alignment Paths and Development Trends

K. S. Chernousova(✉)

Samara State University of Economics, Samara, Russia
samara.oxy.travel@gmail.com

Abstract. The economic crisis associated with the epidemiological situation in the other countries and in Russia in particular has a powerful impact on the level of employment now, and further, according to analysts, the situation will worsen. The unemployment rate is rising, along with the level of employment falling. The role of small business is not only to maintain balance in the labor market, create new jobs and absorb excess labor, but also to develop the economy and ensure the revenue of the state budget. However, to optimize their costs, employers go into the shadows. This problem has always existed; tax control measures have made it possible to solve it for the formal sector of the economy, though it still remains among small businesses. There is a need to develop, improve the efficiency of employment as well as research and optimization of precarious employment. The paper analyzes the category of "precarious employment" conducting a study of both employment and precarious employment in various businesses, identifies promising areas against precarization of employment. Proposals have been made to determine the category of precarious employment in small businesses, to increase the level of employment in certain sectors over the long term.

Keywords: Employment · Precarious employment · Small and medium businesses

1 Introduction

The issue of legalizing jobs and white wages in the segment of medium-sized and small businesses has been a long-standing issue. According to various estimates, in the "shadow sector", there are from 14 million to 16 million people, that is, approximately every tenth citizen, the scale of the shadow sector in the Russian Federation in different definitions and according to different methods has a significant spread, from 23 to 40% of GDP, it is difficult to take into account the scale of shadow employment and the problem of "double counting" when tax evasion. The fight against shadow employment and the legalization of jobs are the main goals of the national project for the development of small and medium-sized enterprises. The economic crisis caused by the coronavirus epidemic did not so much create new difficulties, but rather exacerbated the need to solve the existing systemic problems of small and medium-sized businesses - taxation, employment, precarious employment, security, and a number of other issues [2].

S. I. Ashmarina and V. V. Mantulenko (Eds.): IPM 2020, LNNS 161, pp. 92–100, 2021.
https://doi.org/10.1007/978-3-030-60926-9_13

This work is devoted to the problem of employment in the economic crisis caused by Covid-19, and in the future, to develop proposals for increasing the level of employment and reducing the level of precarious employment in the segment of small and medium enterprises, as well as among the self-employed category.

The economic crisis associated with the epidemiological situation in other countries and in Russia in particular has a powerful impact on the level of employment now, and further, according to analysts, the situation will worsen. The unemployment rate is rising, along with the level of employment falling. At the same time, the level of competition for the workplace has also grown. According to employment services, on the average there are seven resumes per vacancy, which is much higher than usual averages [7]. Tourism, marketing, education, restaurant business are areas of activity that are primarily affected by the effects of coronavirus in all developed countries. Small business came under a very serious blow.

In modern crisis conditions, the role of revenue in the state budget is increasing. The leading position among the revenues of any state is tax revenues. Revenues from small and medium-sized businesses are the basis of the US budget and a number of developed European countries. That is, the role of small business is not only to maintain equilibrium in the labor market, create working places and absorption of excess labor, but also in the development of the economy ensuring the revenue of the state budget. Therefore, the development of small business is also one of the directions of state policy of the Russian Federation. In 2020, according to Forbes, the tax burden on small businesses will increase by 4.9% [9]. Taxation of labor, or rather, income from wages and other forms of income from labor activity consists of taxes that must be paid by both the employer and the employee himself, this is a tax on income of individuals whose tax payers are recipients of income from labor activity and insurance contributions, tax burden on which lies with the employer. Taxation of wages and other income from employment significantly affects the financial condition of the organization, the total amount of tax payments may be more than 40% of the wage fund itself. To optimize their costs, employers go into the shadows. This problem has always existed, tax control measures have largely solved it for the formal sector of the economy, however it remains among the representatives of small and medium-sized businesses. At the same time, the coronacrisis aggravated the long-overdue problems of the self-employed part of the population, small and medium-sized businesses of Russia. The fall in employment for the second quarter of 2020 in Russia, according to various estimates, ranged from 1 to 25%, depending on the region. Unemployment has grown.

There is a need to develop, improve the efficiency of employment, research and optimization of precarious employment. The paper analyzes the views of scientists on the category of "precarious employment", conducts a study of employment and precarious employment in small and medium-sized businesses, as well as in the formal sector of the economy in the current crisis, identifies promising areas of struggle against precarization of employment. Proposals have been made to determine the category of precarious employment in small businesses, to increase the level of employment in certain sectors over the long term.

2 Methodology

In the study of the selected problem, methods of observation and collection of facts on employment and precarious employment of the population in the segment of small and medium-sized businesses, as well as self-employed categories were used.

The study was based on the principles of dialectical logic: the unity of historical and logical, the unity and difference of form and content. The methodological basis of this work was a system-structural analysis. The work used the approaches and provisions of the classical and institutional economic theories. Theoretical generalizations are based on the work of domestic and foreign scientific economists in the areas of entrepreneurship and employment; on the theory of the labor market and precarious employment; on the development of small and medium-sized businesses.

The study was based on such classical scientific studies and theories as the concept of labor value by Smith and Ricardo [15], the nonequilibrium concept of employment by Keynes [8]. In addition, the works of Russian scientific economists on this issue were studied. In domestic studies of the concept of employment, a special place is occupied by the work of Popov and Leonidova [17]. It should be noted that a very significant contribution to the formation of the concept of employment management was made by such scientists as, Bulgakov, Manannikova, Sayapin, [4], Odintsova, [14], Simonova, Zhukova, [19], Schekoldin, Bogatyreva, Ilyukhina, [20], Simonova, Sankova, Mirzabalaeva, Shchipanova, Dorozhkin [18], Tsurkan, Sotskova, Aksinina, Sukhanova, Shirnaeva, Lyubarskaya, Tkacheva, Mokronosov, [21]. Works of economists Aleshkova, Greshnova, Smolina, Popok, [1], Tuguskina, Rozhkova, Taktarova, Salnikova [22], Molotkova, Kulikov, Kudryavtseva, Pecherskaya, [12], Kot, Spanagel, Belozerova [10] devoted to the study of innovation and digitalization of the labor market. The issues of taxation of the labor market and entrepreneurship are presented in the works of Nazarov, Mikhaleva, Chernousova, [13], Piskunov, Manyayeva, Tatarovskaya, Bychkova [16] are devoted to internal control and management in small businesses.

Among foreign studies of this problem, the works of such scientists as Doren, Benz [5] Dunford and Devine [6], Maxson [11]. The empirical base of work was Rosstat statistical reporting data, periodical press materials, official Internet sites, the methodological and regulatory framework of the Russian Federation, as well as economic studies on market economies, small and medium enterprises, self-employed citizens, employment and precarious employment.

3 Results

As a target function, the total economic effect of the reconstruction of objects can be used, calculated according to the described method. The crisis of 2020 seriously affected all sectors of the economy in the Russian Federation, however, small and medium-sized businesses require special attention. As for employment in this economic segment, it can be said that employment here, as a rule, is in the "shadow". Employees in this segment have precarious work. And although this economic phenomenon is widespread, it does not find legislative reflection. The question arises of the definition of "precarious employment". Let's consider how leading experts on this issue define this category. Bobkov defines it this way: precarious employment is the loss by an employee of a typical, familiar civil law society employment relationship, the basis of which was an employment contract (standard employment) [3].

Popov and Leonidova consider the concept of precarious employment as a separate scientific category, does not see negative consequences for labor relations in it, since he believes that any standard employment may have elements of precarious employment [17]. And even standard employment mechanisms may not always give full protection to employees. Therefore, precarious employment can also be considered a form of labor relations between the employer and employees.

As a rule, workers and the employer are forced to choose non-standard labor relations due to certain reasons. The manifestation of unstable forms of employment does not always lead to a decrease in the quality of labor relations, and in some cases even makes them better [17]. Thus, it can be said that precarious employment in small business is a form of relations in the labor market that does not have certain legislative consequences for either the employer or hired personnel, and which is usually unsystematic.

Consider the reasons for the widespread phenomenon of "precarious employment" in small and medium-sized businesses. In our opinion, the key reason is that the employer must transfer more than 30% of the salary fund to the state budget, in addition, a tax is added on the income of an individual, the tax payer of which is the recipient of labor income, however, the obligation to pay, according to Russian legislation also falls on the employer. Thus, the total tax burden on wages is more than 40%. For small businesses this is a very high cost. The tax burden is simply unbearable, and the remaining problems arise from this, including employment issues, "said the small business ombudsman in St. Petersburg during a roundtable in May 2020 [2]. To work in such conditions, part of the organizations moves into the informal sector. Often, not only hired employees are forced to accept precarious work, but the employer himself can be issued only half or even a quarter of the rate. For this reason, the exact number of victims of the crisis is difficult to name here, since many organizations use wage labor that is not supported by a proper legal basis.

4 Discussion

It is possible to draw a conclusion about a real decrease in the level of employment with information from labor exchanges. According to the Ignatova, in the first week of quarantine alone, the number of unemployed increased by 264, and in April the unemployment rate amounted to 3.4% of the total labor force [7]. Compared to April 2019, the number of employed people (million people) in the age range of 15 years+ in our country in April of this year 2020 decreased by 1,154,000 people, or 1.6%. The number of unemployed at the age of 15+ increased by 732,000 people or by 20.6% (in April 2019, the number of unemployed was 3,554,000).

The unemployment rate in April 2020 amounted to 5.8% of the total able-bodied population. Based on the data presented in Fig. 1, it can be said that the largest drop in employment occurred in the Republic of Ingushetia, the Kabardino-Balkarian

Republic, the Altai Republic, the Republic of Buryatia, the Republic of Tyva, and the smallest - in the Central Federal District and the Yamalo-Nenets Autonomous District. The change in the unemployment rate by region is shown the Fig. 1.

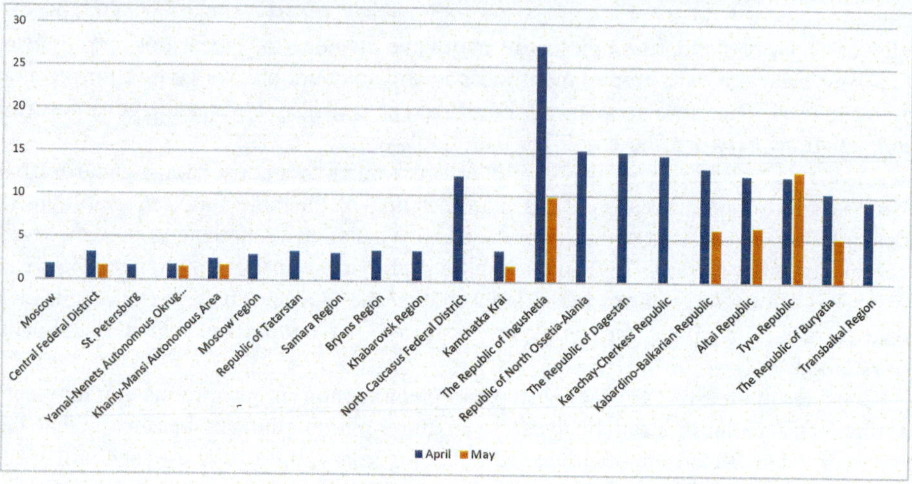

Fig. 1. Dynamics of the number of unemployed in May and April to the regions of the Russian Federation (in % of the total labor force) (Source: author).

In the future, according to experts, the numbers will go up. A similar situation is developing in other countries. According to the International Labor Organization, as a result of strict quarantine measures, about 300 million people will lose their jobs in the near future, and more than 1.5 billion people on Earth will be left without a livelihood. And this is an optimistic forecast. For example, in India, 122 million people were already unemployed, the unemployment rate reached unprecedented levels there and amounted to 27.1%. In the countries of the Eurozone, the International Labor Organization makes a forecast for a reduction of 12 million people (this is 7.8% of the total number of employees), in the Arab states they risk losing their jobs up to 5 million (8.1 %of the workforce). As for our country, here experts predict from 15 million to 25 million unemployed. The number of unemployed in the world, according to the end of the first half of 2020, by country, is presented in the Fig. 2.

If we assume that every unemployed person could receive a salary of 12130 rubles, then we can say that the budget loss for April alone amounted to more than a billion rubles, and in addition to losses, the state pays funds for the payment of benefits. This convoy raises an acute question: how to reduce unemployment and raise the level of employment. And we can say that for some areas, the coronary crisis can serve as a serious impetus for growth. For example, for organizations of the tourism business, the question arises of a reorientation from external tourism to domestic tourism, which will enable the economic development of the hotel business and transportation. A competent approach will make it possible to increase the occupancy rate of the hotel, sanatorium and resort base of Russian resorts, and will open up the possibility for the

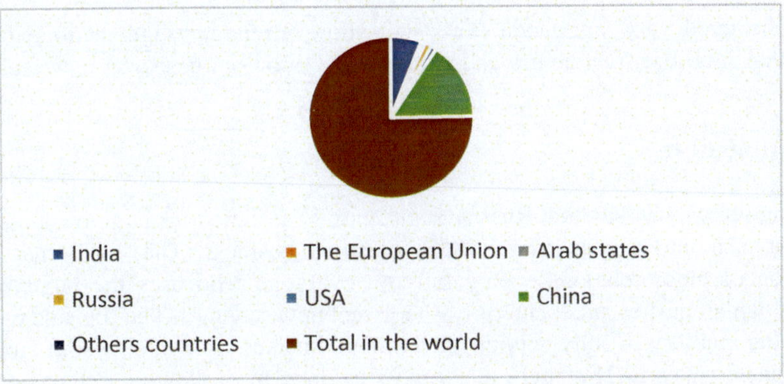

Fig. 2. The number of unemployed in the world by country in 2020 (million people) (Source: author).

development of new tourist sites that have not been studied much before. The mass nature of domestic tourism will make it possible to optimize prices for domestic vacations. All this will not only save jobs in this segment, but also create conditions for creating new jobs, that is, increase the level of employment. Thanks to the crisis, the additional and non-state segment also has the opportunity to open up new prospects for growth. In the context of remote work, many people have additional free time previously spent, for example, on the road. The population has the opportunity to learn, improve their qualifications, improve their knowledge and skills, and learn new ones. It will be possible to invite people from different regions and from different countries. A competent transition to online platforms will allow not only to survive, but to develop this sector. For organizers of conferences and exhibitions, it becomes possible to expand the geography of attracted participants. Reducing travel time for many people will be additional positive points for participating in such events.

Speaking about the consequences of the coronacrisis, it should be noted that there are areas of activity in which only in March 2020 the number of employees increased significantly, among them, first of all, medicine and pharmaceuticals can be distinguished, and construction is also singled out among the sectors that cope with the crisis. If the reason for the increase in employment in the sectors related to medicine and pharmaceuticals in the current conditions is understandable, then, in my opinion, the increase in employment in construction is caused by the inability to use foreign labor, the labor of citizens of neighboring countries due to closed state borders.

Enormous help for all sectors can be government support measures: material payments, interest-free loans to affected sectors, tax exemptions and others. The excitement, which will become an incentive for changing the consciousness and structure of the labor market, attracts employers to the legalization of jobs. In particular, our organization was recognized as affected by the consequences of the Covid-19 pandemic. In an employment relationship, there can only be an employment relationship formalized by law. The help did not extend to any "gray", "shadow schemes". In those cases when some areas of activity simply stopped their work, they should have been considered very significant. This will allow the larger segment to redistribute difficulties, mobilize, restructure and continue the duration, based on new realities.

5 Conclusion

Summing up, we can say that Russian legislation is designed for large business, and it does not take into account the realities of small business. The tax burden on the regulation of labor relations evenly falls on everyone who uses the labor of hired workers, which pushes small enterprises to invent new schemes. The flip side of this, in turn, is the inability to fully support government business that affects both employees and business owners. The study allowed us to draw the following conclusions. The 2020 pandemic caused by the Covid-19 virus exacerbated labor market problems throughout the world, and in particular in the Russian Federation. As a result of measures to eliminate the consequences of the pandemic, about 300 million people will lose their jobs in the near future, and more than 1.5 billion people on Earth will be left without a livelihood. In Russia, the number of unemployed is estimated at 25 million people.

First of all, the negative consequences affected representatives of small and medium-sized businesses, especially in such areas as the hotel business, tourism, fitness and sports activities, catering, the activities of road freight transport and transportation services, additional and non-governmental education, and household services, production of printed materials. It was these areas in the first place that always tried to bring their labor resources into the shadows, and the current crisis exacerbated the situation. The concept of precarious employment finds its manifestations not only in medium-sized and small businesses, but also in the formal economy, that is, it is quite common. This leads to a decrease in the economic efficiency of labor taxation, and, consequently, to tax losses of the revenue part of the budget of the Russian Federation.

The concept of "precarious employment" is disclosed in the work as an unsystematic form of relations in the labor market that does not have certain legislative consequences for either the employer or hired personnel. The most important reasons for the growth of precarious employment in the Russian Federation are the heavy tax burden for employers and for recipients of income from labor activity, as well as the lack of a tax culture. The rejection of precarious employment is also necessary for the labor resources of hired workers, but also for the owners of medium-sized and small businesses. He gives to employees wage guarantees of social and labor protection, pension and medical care, and others, and to the employer - state support. It is the organizations that showed real employees at the time the pandemic started that received significant government support: paying off bank loans to pay salaries to employees, subsidies for salaries, and others. This proved the necessity and feasibility of legalizing staff employment. The study showed that in spite of the fact that the consequences of the Covid-19 crisis somehow affected all areas of activity. Affected all continents, however, there are industries that have already begun to overcome its negative effects. The paper suggests some ways to preserve employment in the affected sectors of the economy: changing conditions, reorienting the market, expanding the geography of activity, accessing online platforms and others. For each segment, for each enterprise he has his own. If structural changes in the segment of small and medium-sized enterprises do not happen, then an increase in informal employment, an increase in unemployment, a reduction in the total number of jobs across the country is a small fraction of what small businesses can expect in the near future. Undoubtedly, this will lead to the mass transfer of some of the employees to the category of unemployed, self-employed or precariously employed, to the growth of the shadow sector, a decrease in the country's economic growth rate, a decrease in budget revenues, and, consequently, a lack of tax revenues in the budget. Structural changes in the economy, in the quality of supply and demand in the labor market should contribute to a smooth exit from the current crisis. Enhanced legal opportunities for the use of precarious labor relations and flexible employment will also help stabilize the labor market. By legalizing the labor market, the number of entrepreneurs can be increased by 2 million people, which can be considered a serious positive step.

References

1. Aleshkova, D.V., Greshnova, M.V., Smolina, E.S., Popok, L.E.: Research of efficiency of tax stimulation of innovative entrepreneurship. In: Ashmarina, S.I., Vochozka, M., Mantulenko, V.V. (eds.) ISCDTE 2019. LNNS, vol. 84, pp. 80–84. Springer, Cham (2020). https://doi.org/10.1007/978-3-030-27015-5_10
2. Babaeva, V.: Covid-19 and small business: The problem is not in the crisis, but in the system.https://www.dp.ru/a/2020/04/28/COVID-19_i_malij_biznes. Accessed: 25 June 2020
3. Bobkov, V.N.: Unstable employment in the russian federation: state and directions of decline. Population **2**, 91–104 (2019)
4. Bulgakov, G.V., Manannikova, O.N., Sayapin, A.V.: To the question of the impact of small business on the growth of employment in modern conditions. Socio-Econ. Phenomena Process **14**(105), 114–118 (2019)
5. Doren, B., Benz, M.R.: Employment inequality revisited: predictors of better employment outcomes for young women with disabilities in transition. J. Spec. Educ. **31**(4), 425–442 (1998)
6. Dunford, B.B., Devine, D.J.: Employment at will and employment discharge: a justice perspective on legal actions following termination. Personnel Psychol. **51**(4), 903–934 (1998)
7. Ignatova, O.: Named the regions with the largest and lowest number of unemployed. Rossiyskaya Gazeta, **75**(8129). https://rg.ru/2020/06/27/reg-szfo/nazvany-regiony-s-naibolshim-i-naimenshim-chislom-bezrabotnyh.html. Accessed: 25 June 2020
8. Keynes, J.: General theory of employment, interest and money. Eksmo, Moscow (2007)
9. Kononova, V.: Small business and taxes: Who whom. Forbes. https://www.forbes.ru/biznes/385605-malyy-biznes-i-nalogi-kto-kogo. Accessed: 25 June 2020
10. Kot, M.K., Spanagel, F.F., Belozerova, O.A.: Problems of Digital Technologies Using in Employment and Employment Relations. In: Ashmarina, S., Mesquita, A., Vochozka, M. (eds.) Digital Transformation of the Economy: Challenges, Trends and New Opportunities. AISC, vol. 908, pp. 227–234. Springer, Cham (2020). https://doi.org/10.1007/978-3-030-11367-4_21
11. Maxson, J.: Employment express: an innovative approach to improve employment outcomes. J. Visual Impairment Blindness **91**(5), 1–3 (1997)

12. Molotkova, N.V., Kulikov, N.I., Kudryavtseva, Yu.V., Pecherskaya, E.P.: (Innovative technologies as a new social challenge in the labor market. In: Mantulenko, V.V., (Ed.), Proceedings of the International Scientific Conference "Global Challenges and Prospects of the Modern Economic Development. European Proceedings of Social and Behavioural Sciences, vol. **57,** pp. 1259–1268, London: Future Academy (2019)
13. Nazarov, M.A., Mikhaleva, O.L., Chernousova, K.S.: Digital transformation of tax administration. In: Ashmarina, S.I., Vochozka, M., Mantulenko, V.V. (eds.) ISCDTE 2019. LNNS, vol. 84, pp. 144–149. Springer, Cham (2020). https://doi.org/10.1007/978-3-030-27015-5_18
14. Odintsova, E.V.: Unsustainable employment in Russia: Quantitative identification in the sphere of employment. In: Raskov, D.E., Kadochnikov, D.V., (Eds.), Proceedings of the VII International Conference Labor and Leisure, p. 83, Saint Petersburg: Astericon (2019)
15. Petty, V., Smith, A., Ricardo, D.: Anthology of economic classics. Moscow: Moscow. (1993)
16. Piskunov, V.A., Manyayeva, V.A., Tatarovskaya, T.E., Bychkova, E.Y.: Risk-oriented internal control: the essence, management methods at small enterprises. IEJME-Math. Educ. **11**(7), 2710–2731 (2016)
17. Popov, A.V., Leonidova, G.V.: The spread of precarious employment as a limiter of Russia's economic growth. Sci. Works of the Free Econ. Soc. Russia **212**(4), 270–293 (2018)
18. Simonova, M.V., Sankova, L.V., Mirzabalaeva, F.I., Shchipanova, D.Y., Dorozhkin, V.E.: Assesment problems and ensuring of decent work in the Russian regions. Int. J. Envir. Sci. Educ. **11**(15), 7608–7626 (2016)
19. Simonova, M.V., Zhukova, A.A.: Informal employment in the labor market: Analysis, trends. Bull. Voronezh State Univ. Series: Econ. Manage. **2**, 82–89 (2018)
20. Schekoldin, V.A., Bogatyreva, I.V., Ilyukhina, L.A.: Digitalization of Labor Regulation Management: New Forms and Content. In: Ashmarina, S.I., Vochozka, M., Mantulenko, V.V. (eds.) ISCDTE 2019. LNNS, vol. 84, pp. 137–143. Springer, Cham (2020). https://doi.org/10.1007/978-3-030-27015-5_17
21. Tsurkan, M.V., Sotskova, S.I., Aksinina, O.S., Sukhanova, E.I., Shirnaeva, S.Y., Lyubarskaya, M.A., Tkacheva, O.N., Mokronosov, A.G.: Influence of the participatory budgeting on the infrastructural development of the territories in the Russian Federation. Int. J. Envir. Sci. Educ. **11**, 7684–7702 (2016)
22. Tuguskina, G.N., Rozhkova, L.V., Taktarova, S.V., Salnikova, O.V.: The role of human capital in the digital economy. In: Mantulenko, V.V., (Ed.), Proceedings of the International Scientific Conference "Global Challenges and Prospects of the Modern Economic Development. European Proceedings of Social and Behavioural Sciences,vol. **57,** pp. 960–968, London: Future Academy (2019)

Analysis of the Population Replacement Component in Demographic Development of Russia

O. F. Chistik(✉)

Samara State University of Economics, Samara, Russia
yurijchistik@yandex.ru

Abstract. The relevance of statistical analysis of demographic development in Russia is due to negative trend of natural rate of population growth, low birth rate, and high mortality as observed in last two decades. A solid information and analytical approach to the study of negative phenomena in demographic development is necessary for development of effective measures and management decisions on different levels to ensure the positive demographic change in Russian Federation. In this work, we study processes associated with the population replacement component in demographic development of Russia. Analyzing the demographic development, we used general scientific and statistical research methods such as: dialectics, comparison, time series analysis, the method of generalizing indicators, the method of multidimensional grouping. Stages of demographic development were identified based on its component analysis and quantitative assessment of the natural population decline replacement. A system of indicators to allow the development of an integral indicator and to characterize the population replacement component of demographic development has been formed. As the result of the study, three groups of regions were formed using the "Pattern" method: having low, medium and above-average demographic development based on the integral estimate ranking of regions.

Keywords: Demographic development · Demographic dynamics · Pattern method

1 Introduction

The prospects for economic competitiveness, geopolitical stability and socio – economic development depend largely on the demographic problem solution, which is one of the most acute problems nowdays. The continuing population decline due to low fertility and very high mortality level in developed countries, aging and reduction of the working-age population, accompanied by low internal mobility and ineffective migration component has a negative impact on Russia's future development.

The issues of spatial demographic development and analysis of its main factors at the regional level are highlighted in scientific works [2, 6]. The author reveals the concept of "demographic dynamics" [12]. The research defines the dynamics of the working-age population as the most important factor of economic potential [7].

S. I. Ashmarina and V. V. Mantulenko (Eds.): IPM 2020, LNNS 161, pp. 101–107, 2021.
https://doi.org/10.1007/978-3-030-60926-9_14

The author links the perspectives of human capital development with the growth of the economy [16]. This analysis provides further consideration by the researcher [5] of the issues of discrimination between men and women in the labor market. Researchers target their analysis on the issues of mortality and life expectancy [4, 8]. Problems and contradictions of the current demographic policies dealing with population reduction are also considered in other studies [1, 3, 9, 11, 13–15]. The researcher analyzes the types of financial support or regions while implementing priority national projects [10].

Population change is caused by the processes associated with population replacement and migration components. Demographic problems urge the study of demographic development in order to optimize the replacement component of the population change process. The subject of our research is to outline the quantitative patterns of demographic development in Russia. The research object is demographic development in the Russian Federation and its regions. The purpose of the research is to study the processes of population replacement component of demographic development in Russia.

2 Methodology

In our analysis of the demographic situation in the regions, there general were used scientific and statistical research methods such as: methods dialectics, comparison, time series analysis, multidimensional grouping method (the "Pattern" method of expert evaluation), tabular method of forming and studying the information array. Our study is focused on the important assessment of the effective public policy measures implementation in terms of national demographic development [11, 12]. The unresolved demographic problems of individual regions urge to focus on strong information and analytical support for management decisions in the demographic sphere. Processing the information array was carried out on the basis of application packages "STATISTICA" and "Microsoft Excel".

3 Results

Demographic development and undoubtedly sustainable natural population growth are of strategic importance. In our work we conducted a thorugh analysis of the main population replacement component of demographic dynamics. To do this, we set a task: to establish the stages of demographic development and give a quantitative assessment of the replacement of the natural population decline by the migration growth surplus based on the use of the replacement index (Table 1).

Period I (2001-2012). It should be noted that Russia went through the first decade of the XXI century and the following years 2010-2012 with a natural decline of population. At the same time, migration growth did not provide for substantial population replacement until 2009. In the period 2001-2008 the natural decline of population was almost 6 million people, thus, on average, the reduction of more than 740.8 thousand people per year occured with an average annual replacement of only 25.8%.

Table 1. Indicators of demographic dynamics in the Russian Federation for 2001-2018

Time Period, years	Natural growth, decline (−)	Migration growth	Total growth, decline (−)	Replacement Index, %
	Average per year, thousand people			
I.2001–2012	−545,7	299,2	−246,5	54,8
2001–2008	−740,8	190,1	−550,7	25,8
2009–2012	−124,4	516,2	391,8	4,2 times
II.2013–2015	28,8	1037,2	1066,0	–
III.2016–2018	−120,9	199,2	78,3	1,6 times
Including2018	−224,5	125,5	−99,0	55,9

Source: author.

During this period, there were processes of varying growth in the number of births and decrease in the number of deaths. From 2009 to 2012, there has been a steady increase in the number of births and an unstable decline in the number of deaths. The predominance of migration growth over natural decline allowed for population growth in this period. This clearly marked the process of depopulation, characterized by a steady excess of the number of deaths over births. Natural population decline is stable and in a long-term trend, but its rate has gradually slowed down.

Period II (2013-2015). In 2013, there was a decrease in the number of births and a significant decrease in the number of deaths, which allowed for a natural increase in the population by 24.0 thousand people. Since 2014 there has been a decline in the number of births and an unstable dynamic in the number of deaths. At the same time, the excess of births over the number of deaths in 2014 and 2015 provided a natural increase.

Period III (2016-2018). In 2016 and 2017 a significant decrease in the number of births with a corresponding decrease in the number of deaths has led to a natural decline, overlapped by migration substitution. In 2018, there was a decrease in the population size by 99.0 thousand people. It was caused by a significant decrease in the number of births, outpacing the decrease in the number of deaths compared to the previous period. This led to a natural decline of population, while its replacement by migration growth contributed only 55.9%. Analyzing demographic development, it is important to take into account its territorial features in Russia. The study of demographic development should focus on population replacement component. In this regard, to assess the degree of differentiation among regions by the level of demographic development, a nonparametric "Pattern" method is being used. This method allows giving an integral characteristic of phenomena that has a multidimensional character.

Due to the multidimensional nature of the process under consideration, a system of indicators has been formed as initial data that characterize the replacement component of demographic development: the indicator of the population structure by gender, relative indicators of birth rate, total and child mortality, working-age mortality, life expectancy, marriage prevalence and divorce rates. The original matrix was converted according to the established algorithm into a matrix of normalized values, and then, based on the partial coefficients derived, the integral indicator numerical values were determined. Note that the "Pattern" method is a type of expert methods. Based on a content analysis of the values of integral (general) indicators, we have identified 3 groups of regions: with a low level of demographic development, with an average level and above the average (Table 2).

Table 2. Typology of regions of the Russian Federation based on the integral indicator of the level of demographic development in 2018.

Level of demographic development, index	Number of regions		Composition of groups
	In total	%	
Low 0,46–0,54	69	80,2	Regions: Belgorod, Bryansk, Vladimir, Voronezh, Ivanovo, Kaluga, Kostroma, Kursk, Lipetsk, Oryol, Ryazan, Smolensk, Tambov, Tver, Tula, Yaroslavl, Arkhangelsk, Vologda, Kaliningrad, Leningrad, Murmansk, Novgorod, Pskov, Astrakhan, Volgograd, Rostov, Kirov, Nizhny Novgorod, Orenburg, Penza, Samara, Saratov, Ulyanovsk, Kurgan, Sverdlovsk, Chelyabinsk, Irkutsk, Kemerovo, Novosibirsk, Omsk, Tomsk, Pskov, Astrakhan, Amur, Magadan. Jewish Autonomous region, Chukotka Autonomous Okrug
Average 0,54–0,61	14	16,3	Republics: Kabardino-Balkar, North Ossetia, Tatarstan, Tuva, Sakha (Yakutia) Region: Krasnodar. Regions: Tyumen, Sakhalin. Cities of national significance: Moscow, Saint Petersburg, Sevastopol. Autonomous districts: Nenets, Ugra, Yamalo-Nenets.
Above the average 0,61 and more	3	3,5	Republics: Dagestan, Ingushetia, Chechen Republic

Source: author.

The largest group is the one with a low level of demographic development, it includes 69 regions (80.2%) with 62 regions having a natural population decline. At the same time, 42 regions have high mortality rates above 13%. And in 30 regions, there is a low level of birth rates - below 10%. In 35 regions of this group, infant mortality rates are higher than 5.1%. Currently in the European Union the infant mortality rate is extremely low and has a tendency to further decrease, approaching the biologically justified minimum level of 2-3%. 59 regions have high mortality rates for people of working age, and 12 regions have an average life expectancy below 70 years.

The second group, with an average level of demographic development, includes 14 regions (16.3%). 3 regions have high infant mortality rates (the Republic of Tuva,

Yamalo-Nenets Autonomous district, and the city of Moscow). In two regions of this group there is a natural decline in population: the city of Sevastopol and the Sakhalin region. In the Republic of Tuva, with a high birth rate (20.2%), there is a high infant mortality rate (9.4%), and a low average life expectancy (66 years).

The group with above-average demographic development level includes 3 regions of the North Caucasus Federal district. These republics are characterized by high birth rates, medium life expectancy, and low overall mortality rates, as well as mortality rates for people of working age. However, the infant mortality rate remains high for all regions (6.1% - 7.8%).

The analysis of territorial differentiation of demographic development allows to determine that a significant part (69 entities) of the Russian Federation have a low level of demographic development. The situation is relatively favorable in 3 regions of the North Caucasus Federal district (republics: Chechen, Dagestan, Ingushetia) due to their ethno-cultural features.

4 Discussion

Demographic development is a complex, multi-component phenomenon and calls for quantitative and qualitative assessment of demographic processes that occur in a particular territory. For our demographic situation analysis we used data from Rosstat. For research we employed active up-to-date methodological tools: the method of multi-dimensional grouping, the tabular method, integral indicators and time series methods. Our study of the population replacement component includes the stages of demographic development characteristics based on a quantitative assessment of the natural population decline. It's clear, the main problem in this matter a modern society may face is to achieve a stable, ongoing excess of birth rate over death rate. In our case, the question is whether Russia will be able to achieve demographic growth in the long run. The starting point is to conduct a comprehensive quantitative and qualitative study of demographic processes in the regions of the Russian Federation. We have proposed a system of indicators that describe the population replacement component of demographic development. This system serves as the basis for obtaining integral estimates of the level of demographic development in the regions. According to the "Pattern" method, 3 groups of regions with different levels of demographic development were formed based on the ranking of the obtained integral estimates. Each group identified similar problems in demographic development of its member regions.

5 Conclusion

Statistical analysis of the population birth rate matched by mortality rates revealed the long period of natural population decline (2001-2012) changed into a short period (2013-2015) of natural population growth. The recent period (2016-2018) of natural decline is associated with a decrease in the birth rate greater then the decrease in mortality. To maintain the Russian population and increase its number, that is to ensure national demographic security, there should be taken measures to slow down the natural decline and provide stimulus for further natural population growth.

Demographic development in the regions of the Russian Federation is of vital strategic importance. Applying the method of multidimensional classification based on the proposed system of indicators allowed us to obtain integral estimates and form three groups of regions sorted by the level of demographic development (low, average, above average). The current situation in the first group is of prime concern. The first group with a low level of demographic development has 62 regions with a natural population decline. This group is characterized by high rates of mortality in the entire population: infant mortality, mortality of working-age people, and a low birth rate. In 12 regions of this group, the average life expectancy does not exceed 70 years.

The problem of sustainable population replacement can be solved by stimulating the growth of the birth rate and a permanent reduction of the death rate of the population by means of informational, methodological and analytical support. The proposed informational and methodological approach can serve as a basis for developing effective policy and making effective management decisions at the federal and regional levels to ensure progressive demographic development.

References

1. Borjas, G.J.: Does welfare reduce poverty? Res. Econ. **70**(1), 143–157 (2016)
2. Fomin, M.V.: Demographic forecasts and correctness of spatial development statistics for Siberia and the Russian Far East. Population **3**, 15–29 (2019)
3. Hardy, B., Smeeding, T., Ziliak, J.P.: The changing safety net for low-income parents and their children: Structural or cyclical changes in income support policy. Demography **55**(1), 189–221 (2018)
4. Herr, M., Arvieu, J.J., Ankri, J., Robine, J.M.: What is the duration of life expectancy in the state of frailty? Estimates in the SIPAF study. Euro. J. Ageing **15**(2), 165–173 (2017). https://doi.org/10.1007/s10433-017-0438-z
5. Kalabikhina, I.E.: Parental responsibilities and employment discrimination. Popul. Econ. **1** (1), 89–116 (2017)
6. Karmirgodieva, A.A., Cherkesova, E.Y.: Influence of demographic factors on socio-economic development. Concept **17**, 249–253 (2016)
7. Kayukov, V.V., Melchakova, Y.L.: The relationship of demographic processes with the state of the economy. In: Kaynova,: G.A.(Ed.) Economics, Management, Finance: Proceedings of the III International Conference, pp. 177–180. Mercury, Perm (2014)
8. Lariscy, J.T., Hummer, R.A., Rogers, R.G.: Cigarette smoking and all-cause and cause-specific adult mortality in the United States. Demography **55**(5), 1855–1885 (2018)
9. Lebihan, L., Mao Takongmo, C.O.: The impact of universal child benefits on family health and behaviours. Res. Econ. **72**(4), 415–427 (2018)
10. Milchakov, M.V.: Features of financial support for regions in the implementation of national projects Scientific-research financial institute. Financial J. **3**, 22–37 (2019)
11. Rudnitskaya, A.P., Novikov, E.A.: The main directions of formation, problems and tasks of demographic policy in modern Russia. PolitBook **1**, 43–56 (2015)
12. Rudnitskaya, A.P., Quartnikov, P.V.: The role of civic activity in the formation of value orientations of modern youth. CITIZE **1**, 24 (2018)

13. Rybakovsky, O.L.: Demographic dynamics of Russia: Basic concepts, indicators, results for 1946-2017. Population **21**(4), 14–22 (2018)
14. Slack, K.S., Kim, B., Yang, M.Y., Berger, L.M.: The economic safety net for low-income families with children. Child Youth Serv. Rev. **46**, 213–219 (2014)
15. Sleptsova, E.V.: Evaluating the effectiveness of demographic policy in Russia. Econ. Sustain. Dev. **3**(19), 194–199 (2014)
16. Tkachenko, A.A.: Russian socio-demographic perspectives: problems and solutions. Econ. Taxes Right **12**(20), 6–18 (2019)

Role of Universities in the Infrastructure to Support Small and Medium-Sized Business

S. I. Ashmarina(✉) and G. M. Murzagalina

Samara State University of Economics, Samara, Russia
asisamara@mail.ru, gulnazmur@yandex.ru

Abstract. The purpose of this work is to analyze activities of existing infrastructure facilities that support small and medium-sized businesses in the Russian Federation within the framework of the theoretical paradigm of national projects and state programs, to identify contradictions between the declared goals (tasks) and the actual achieved results, and to justify the role of universities in this process. Universities are considered as scientific and training centers for development of entrepreneurial culture and creation of entrepreneurial spirit. In this regard, the author justifies the need to review existing approaches to the entrepreneurship support, in particular in the youth environment, taking into account functions that higher education institutions implement and their potential in this area.

Keywords: Entrepreneurship · Entrepreneurial culture · Small and medium-sized business · Universities

1 Introduction

The small and medium-sized entrepreneurship (SME) support infrastructure in Russia today is a very complex system of elements built into a single structure, the main task of which is to create a favorable environment for creating a new or developing an existing business. SME support infrastructure is represented both at the federal and regional levels (Fig. 1). Despite the common infrastructure goals, there are some differences between federal and regional level elements.

It should be noted that elements of the federal support level are usually aimed at creating favorable conditions for the development of existing businesses, or at forming new entrepreneurs by creating, adopting and implementing special state documents, financing business initiatives, grant-making and supporting the SME interests. The support directions and SME development directions are also indicated in national projects, in particular the project "Small and medium-sized businesses and support for individual business initiatives" [8]. The national project passport was developed by the Ministry of Economic Development of the Russian Federation in compliance with Presidential Decree No. 204 of May 7, 2018 "On national goals and strategic objectives for the development of the Russian Federation for the period up to 2024" [12]. Targets of the national project are: increase in the number of employees in small and medium enterprises, including individual entrepreneurs up to 25 million people by the end of 2024; increasing the share of small and medium entrepreneurship in the GDP to 32.5%

S. I. Ashmarina and V. V. Mantulenko (Eds.): IPM 2020, LNNS 161, pp. 108–116, 2021.
https://doi.org/10.1007/978-3-030-60926-9_15

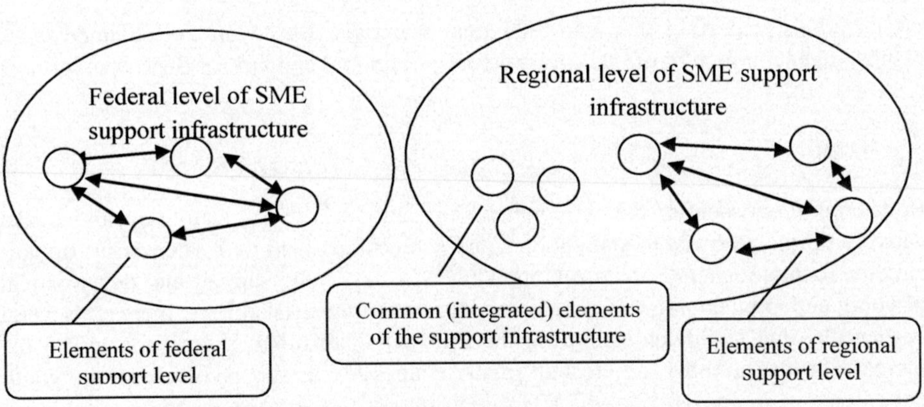

Fig. 1. Levels of SME support infrastructure (Source: authors).

by the end of 2024; increasing the share of exports of small and medium-sized enterprises, including individual entrepreneurs in the total volume of non-oil exports to 10% by the end of 2024. National projects should become a powerful incentive for the development and support of entrepreneurship at all levels.

There is also a local level of SME support operating in the regions of Russia, which is engaged in direct interaction with business representatives, young entrepreneurs and people who want to organize their own business. It should be noted that the organizations represented in the regional context that are working for the SME development can be grouped into separate categories of support: property, financial, information, consulting and training assistance groups.

2 Methodology

The research purpose of this article is to show the role of higher educational organizations in the infrastructural and other forms of SME support based on the analysis of real practice of state and non-state support for small and medium-sized businesses in Russia and determination of contradictions that exist in this area. The main research tasks include:

- study public and non-public sectors of the infrastructure support for small and medium-sized businesses in the Russian Federation,
- consider real available infrastructure elements of SME support for young people in regions,
- investigate the interaction mechanism of infrastructure elements for SME support,
- analyze the role of higher educational institutions in this infrastructure in the Russian Federation.

The research methods include analysis of foreign practice of SME and youth entrepreneurship support [2, 5–7, 9, 10]. We also studied the SME support infrastructure in all regions of the country, relying on official electronic resources of organizations, reporting data and analytical materials. Based on the obtained work results, a general picture of the infrastructure of regional support for SME was formed.

3 Results

In general, the system of the entrepreneurship support is dynamically developing and actively promoted by both state bodies and subjects, as well as by non-profit organizations, foundations and organizations. However, as for the support and development of youth entrepreneurship and development of entrepreneurial culture, there are several bottlenecks that significantly impede the full realization of measures directed to the increase in the number of created small businesses, active involvement in youth entrepreneurship and growth of SME's contribution to the gross income.

The maximum concentration of young people, namely people of the age of 15 to 30 years, occurs in the system of higher and secondary education. In universities, educational institutions of secondary vocational education and schools should create foundations of entrepreneurial knowledge, skills and competencies [1]. These institutions should be a place where entrepreneurial culture is formed, and the opportunity to try yourself in business, create your first product or implement entrepreneurial initiatives is maximized. Obviously, the more fully the entrepreneurial culture and practice are formed, the more potential entrepreneurs will start their business after graduation. It is obvious that this youth environment need information, organizational, infrastructure and financial support for the formation and development of an entrepreneurial culture and foundations for entrepreneurial activity. It is in relation to this category that it is the most appropriate to talk about the popularization of entrepreneurship, since the decision to start a business, as the real business practice shows, occurs at this age, the proof of this is the world statistics on the issue. Thus, the role of universities as subjects of support and development of youth entrepreneurship is great. Accordingly, it would be logical to assume that the role of universities, as a place of concentration of young people who are the most ready to start entrepreneurial activity, in the SME support infrastructure, should be decisive [3, 4].

However, according to the analysis, the main efforts to support and develop entrepreneurship in the existing paradigm are aimed at existing entrepreneurs who are starting their own business or are already developing it. It should be noted that, of course, students and schoolchildren can also take part in them, but, according to statistics, it looks so spotty that it is difficult to talk about significant results of this participation. Universities, in fact, are removed from the general structure of the subjects-organizers of this activity and subjects–recipients of funding for the implementation of these activities, in contrast to all other infrastructure elements. In addition, the main grant holders in the business support system are not always ready to actively cooperate with universities in the implementation of support programs. Most of the activities aimed at the development of entrepreneurial competencies at universities and the development of entrepreneurial culture are carried out on an initiative basis and at the expense of universities' own funds.

At present, in relation to the subjects of the higher education system, namely, speaking about universities, strict requirements are imposed by the state and the founder. They are indicated in monitoring indicators for evaluating the performance of universities and national projects, such as "Education" and "Science", as well as in indicators for evaluating the performance of universities necessary for assessing their compliance with accreditation procedures. The need for the formation of entrepreneurial competencies and the development of youth entrepreneurial culture are not identified as essential in these requirements.

At the same time, they are designated as basic ones for other entities – regional ministries, organizations that provide business development, foundations, etc. Thus, there is a certain dissonance between the existing need for organizational, informational and financial support for the development of youth entrepreneurial culture and entrepreneurial competencies, setting of appropriate tasks and the allocation of resources for the implementation of tasks for the development of youth entrepreneurship at the state level [11].

The crucial role of universities in the SME support infrastructure and direct support of youth entrepreneurship is determined by several factors. First, the university is represented by a contingent of young people, most of them fit into the age range of 16 to 25 years. Entrepreneurial initiatives that arise at this age are directly related to the category of "youth entrepreneurship" and are initiatives that most of all need special support and development. There are several important reasons: the age of potential entrepreneurs is the most susceptible to lay foundations of entrepreneurship and develop entrepreneurial culture; the number and concentration of the relevant audience are maximum; the level of perception and capabilities to create an entrepreneurial culture is rather high.

Secondly, universities have created an educational environment to support and teach the basics of entrepreneurship through the implementation of educational programs on entrepreneurship (in different variations), formation of team activities through practice and in other formats. Third, universities themselves create various types of business support infrastructure elements on their ground. For example, many universities have their own business incubators, business accelerators, and coworking centers. Thus, the university has the greatest significance in the overall SME support infrastructure, and should become its important element, since it is the central link between the most active segment of the population (youth) and the SME support infrastructure itself.

4 Discussion

The existing infrastructure is mainly focused on interaction with entrepreneurs who have some experience, while at the same time it is not focused on the cooperation with young entrepreneurs. Universities, if they are aimed at creating entrepreneurial culture among students, can increase the share of people interested in starting a business, that is, increase the flow of customers for the existing business support infrastructure.

Conditionally, the mechanism for the SME support in its general form can be divided into two areas: the public sector and the non-governmental sector. The public infrastructure sector has two levels: federal and regional. The non-governmental sector is represented linearly by one level. The components of each level were described earlier. However, there is a relationship between many of the components of the infrastructure, which is expressed in certain principles of interaction. The principles of interaction between elements of the federal and regional levels of SME support are based on their main functions. The federal level of SME support has the following functions of its activities:

1. Conducting research aimed primarily at obtaining objective data on the SME development and youth entrepreneurship in Russia, in order to identify positive and negative processes. It should be noted that research is usually carried out with the active participation of higher educational institutions by allocating funds for state tasks, grants and competitions of thematic research papers.
2. On the basis of the obtained data, including results of scientific research, the formation of draft legal acts, federal laws, strategies and state programs for the development of entrepreneurship. In the formation of such a level of normative legal acts, the forces of industry ministries and higher educational institutions are involved, among others.
3. Forming a policy and stimulating innovative youth entrepreneurship, creating a positive image of entrepreneurship. This function is assigned to the Federal Agency for Youth Affairs "Rosmolodezh" (specifically to its subordinate institution - Federal state budgetary institution "Russian center for youth entrepreneurship assistance"). This center carries out measures to promote and encourage entrepreneurship among young people, allocating grants for these purposes.
4. Financial incentives and support for entrepreneurship, by allocating funds to various funds and banks whose activities are aimed at supporting SME.
5. Credit and financial assistance for SME representatives by providing targeted funds to support certain areas of business, including among young people. Financial resources are allocated, including in the form of subsidies.

Elements of the regional level of SME support provide targeted support to entrepreneurship in accordance with category of support they belong to: property, financial, information, etc. Some elements of the regional SME support infrastructure operate in accordance with the federal laws that have been developed, adopted and define activities of specific types of SME support facilities. These objects include: chambers of commerce and industry, employers' associations, non-profit partnerships, large cooperatives and microfinance organizations, consumer cooperatives, and special economic zones. In this aspect, the essence of interaction between elements of the federal and regional sectors of the SME support infrastructure is reduced to coordinating activities of elements of the regional sector on the part of the federal sector through legal regulation [14].

Higher education institutions deserve special attention in the SME support infrastructure. Let's take a closer look at the role of universities in supporting youth entrepreneurship. Based on the rationale given above, the university is an integral part of the SME support infrastructure. Speaking about the university as a part of the whole system, first of all, we should pay attention to functions performed by it [13].

First, universities are actively involved in research on the state of entrepreneurship, both on a national scale and in the context of regions. The scientific potential of universities plays an important role in forming a common understanding of trends and processes occurring in the field of entrepreneurship.

Secondly, the university, as a place of training for potential and real entrepreneurs, can independently form an internal support infrastructure. This applies to the organization of separate structural units that contribute to the development of students' entrepreneurial skills, the implementation of innovative projects and support for entrepreneurial initiatives.

It is worth noting an important aspect: the university can support both the state infrastructure for small and medium-sized enterprises and create its own internal infrastructure for supporting youth entrepreneurship.

Analysis of the universities' structure in the Russian Federation has shown that, as a rule, to meet the goals of supporting youth entrepreneurship in the country's universities, there are such structural divisions as: student business incubators, clubs of directors, business accelerators, student business clubs, technology parks, coworking centers, centers for the development of youth entrepreneurship, small innovative enterprises [4, 13]. Each of these types of structural divisions of the university, whose activities are aimed at supporting youth science, can be organized within the university, and supported by its management. However, not all Russian universities have such a wide list of possible elements of intra-university infrastructure to support entrepreneurship. As a rule, universities have one or two types of support infrastructure objects. The most popular form of youth entrepreneurship support at universities is students' business incubators.

Speaking about the university's place in the state infrastructure for supporting entrepreneurship, it should be noted that first of all it depends on the type of interaction of the university with other objects. Almost all elements of the state infrastructure for the entrepreneurship support do not have direct interaction with universities. There is a buffer zone filled with funds, committees, organizations-operators of entrepreneurship support programs, etc., whose activities, in relation to the development of youth entrepreneurship, are not associated with universities or with the main pool of potential young entrepreneurs. Unfortunately, quite often the interaction of the university with other elements of the SME support infrastructure does not bring significant benefits for beginning entrepreneurs or potential entrepreneurs (the youth student audience).

5 Conclusion

The study of the support infrastructure for the development of entrepreneurship in general and youth entrepreneurship in particular allowed us to draw the following conclusions:

1. The SME support mechanism is a complex environment formed by elements whose main function is to support business activities. However, considering the infrastructure as a single system, we can say that the distribution of its functionality has significant limitations, and is not able to fully promote the development of youth entrepreneurship in the regions. The infrastructure is focused on serving mainly

experienced entrepreneurs. In particular, youth entrepreneurship does not have full access to the full range of services offered by the infrastructure organizations.
2. Regions do not have an identical set of SME support infrastructure facilities. Each region of the Russian Federation itself forms an element base, which is why we observe a heterogeneous picture, expressed in the imbalance of opportunities offered by the SME support infrastructure for entrepreneurs.
3. The established mechanism for the entrepreneurship support in the regions has a set of different functions aimed at the development of certain areas. At the same time, there are no organizations (that are part of the overall infrastructure for supporting SMEs) whose activities are aimed at promoting entrepreneurship, in particular among young people. The infrastructure element that is the closest to the implementation of promotional functions is represented by regional universities.
4. Taking into account the fact that universities play a leading role in the formation of a positive image of entrepreneurship in the regions, the study revealed a weak interaction of elements of the regional SME support level with universities in these issues. In fact, universities are not built-in elements of the entrepreneurship support, performing functions of supporting entrepreneurial youth initiatives based on their own projects and at their own discretion.
5. The presented elements of SME support, which are part of a single infrastructure, are aimed at development of business initiatives, which they partially cope with. However, it is worth noting that this infrastructure is aimed at supporting entrepreneurship at the earliest stages, but does not have a positive impact on the formation of entrepreneurial initiatives themselves, entrepreneurial culture, and does not form a favorable environment for the cultivation of entrepreneurial spirit.
6. The SME support infrastructure does not have a single governing body, but is a network of disparate elements united by a single principle. At the same time, infrastructure is affected by external factors, among which we highlight such as legislative (federal and regional legislation), economic (the influence of economic factors and trends), technological (the deep impact of digitalization aspects on the development of entrepreneurship).

The role of universities in forming entrepreneurial culture is undeniable and significant. Moreover, it is the university that is able to most widely reach young people to form foundations of entrepreneurial thinking and increase entrepreneurial activities. This thesis is based on a number of arguments.

First, universities are engaged in this activity or should lay foundations of entrepreneurial competencies. Given that not all students are initially intended to do business after graduation, it is the university and its entrepreneurial culture that should be aimed at developing entrepreneurial competencies through educational technologies and the practice of engaging in business processes to increase the overall economic, financial culture of young people.

Secondly, the university is able to influence the formation of a positive image of entrepreneurship among students and schoolchildren through the implementation of special programs, conferences, events and the educational process. The main task in this aspect is to cultivate the interest of schoolchildren and students in entrepreneurship, instilling skills for generating entrepreneurial ideas and methods for their promotion and acceleration.

Third, the university can act as a platform for supporting and developing entrepreneurs by organizing its own facilities to support entrepreneurial initiatives (business clubs, business centers, case clubs, etc.).

Fourth, the university is able to build practice-oriented approaches to teaching students, which contributes to the formation of the entrepreneur's personality, and the acquisition of the necessary experience, the lack of which is often a serious reason for receiving support from some elements of business support (in particular financial).

In connection with all of the above, we can talk about the key role of universities in the process of forming an entrepreneurial culture and increasing the level of entrepreneurial activity, and the need to change the infrastructure architecture building models for supporting entrepreneurship and youth entrepreneurship. When developing the SME support infrastructure, it is advisable to distinguish between measures to support entrepreneurship in general and support and development of youth entrepreneurship, including the system of management, control and organization of these measures and activities. This is based on the following aspects:

- entrepreneurs who carry out their activities and require support are concentrated in different areas and are geographically separated; potential entrepreneurs from the youth environment are mainly concentrated in educational institutions of different levels,
- measures to support existing entrepreneurs and potential entrepreneurs (youth environment) differ both organizationally and financially; it follows from this that the implementation of existing support measures, which in many cases are implemented for a wide audience, usually requires greater personalization and consideration of the specifics and requests of the audience,
- universities should become an important independent part of the infrastructure for supporting and developing youth entrepreneurship,
- financial, organizational, informational, and other forms of support and development of youth entrepreneurship should be carried out in a targeted way in universities and other educational organizations, bypassing buffer organizations that distribute resources,
- it is important to assign responsibility to universities (at the state level) for implementing measures to develop entrepreneurial competencies, promote entrepreneurship, and create an entrepreneurial culture.

References

1. Akhmetzyanova, M.M., Galimova, A.S.: Participation of young people in small business. Young Sci. **1**(48), 80 (2013)
2. Alaref, J., Brodmann, S., Premand, P.: The medium-term impact of entrepreneurship education on labor market outcomes: experimental evidence from university graduates in Tunisia. Labour Econ. **62** (2020). https://doi.org/10.1016/j.labeco.2019.101787

3. Ashmarina, S.I., Kandrashina, E.A., Izmailov, A.M., Mirzayev, N.G.: Gaps in the system of higher education in russia in terms of digitalization. In: Ashmarina, S., Mesquita, A., Vochozka, M. (eds.) Digital Transformation of the Economy: Challenges, Trends and New Opportunities. AISC, vol. 908, pp. 437–443. Springer, Cham (2020). https://doi.org/10.1007/978-3-030-11367-4_43
4. Ashmarina, S.I., Mantulenko, V.V.: Regional development of small and medium-sized businesses: value and behavioral aspects. In: Proceedings of the 13th International Scientific Conference European Forum of Entrepreneurship 2020 "Protectionism – protection or destruction of national economies?", pp. 46–52. NEWTON College, Prague (2020)
5. Brixiová, Z., Ncube, M., Bicaba, Z.: Skills and youth entrepreneurship in Africa: analysis with evidence from Swaziland. World Dev. **67**, 11–26 (2015). https://doi.org/10.1016/j.worlddev.2014.09.027
6. Kantis, H.D., Federico, J.S., García, S.I. Entrepreneurship policy and systemic conditions: evidence-based implications and recommendations for emerging countries. Soc.-Econ. Plann. Sci. (2020).https://doi.org/10.1016/j.seps.2020.100872
7. Morris, M.H., Santos, S.C., Neumeyer, X.: Entrepreneurship as a solution to poverty in developed economies. Bus. Horizons **63**(3), 377–390 (2020). https://doi.org/10.1016/j.bushor.2020.01.010
8. National project "Small and medium-sized businesses and support for individual business initiatives" (2018). https://base.garant.ru/72185938/. Accessed 03 Aug 2020
9. Nguyen, A.T., Do, T.H.H., Vu, T.B.T., Dang, K.A., Nguyen, H.L.: Factors affecting entrepreneurial intentions among youths in Vietnam. Child Youth Serv. Rev. **99**, 186–193 (2019). https://doi.org/10.1016/j.childyouth.2019.01.039
10. Olugbola, S.A.: Exploring entrepreneurial readiness of youth and startup success components: entrepreneurship training as a moderator. J. Innov. Knowl. **2**(3), 155–171 (2017). https://doi.org/10.1016/j.jik.2016.12.004
11. Petrov, N.A., Mikhaylov, A.M.: Features of state regulation of small business financing in the russian federation. In: Ashmarina, S.I., Mantulenko, V.V., Vochozka, M. (eds.) Engineering Economics Week 2020. LNNS, vol. 139, pp. 183–190. Springer, Cham (2021). https://doi.org/10.1007/978-3-030-53277-2_21
12. Presidential Decree No. 204 of May 7, 2018 "On national goals and strategic objectives for the development of the Russian Federation for the period up to 2024" (2018). http://kremlin.ru/acts/bank/43027. Accessed 03 Aug 2020
13. Shirokova, G.V., Bogatyreva, K.A., Beliaeva, T.V., Laskovaia, A.K., Karpinskaia, E.O.: Global university entrepreneurial spirit students' survey national report Russia 2018 (2018). http://www.guesssurvey.org/resources/nat_2018/GUESSS_Report_2018_Russia.pdf. Accessed 29 Jan 2020
14. Sukhanova, E.I.: Small business in the regions of Russia: statistical analysis and modeling. In: Ashmarina, S.I., Mantulenko, V.V., Vochozka, M. (eds.) Engineering Economics Week 2020. LNNS, vol. 139, pp. 443–453. Springer, Cham (2021). https://doi.org/10.1007/978-3-030-53277-2_53

Necessary Conditions of Inclusive Development of Territories for Labor Market

R. R. Gilfanov[1(✉)], G. A. Chedzhemov[2], and N. A. Igoshina[2]

[1] Center of Advanced Economic Studies of the Academy of Sciences of the Republic of Tatarstan, Kazan, Russia
rus.kamtent@mail.ru

[2] Samara State University of Economics, Samara, Russia
irongerman@mail.ru, bume63@mail.ru

Abstract. In the article, the author analyzes a number of controversial issues related to the definition of the essence of inclusive development of territories, its nature, as well as determining the necessary level of centralization of management of the process of inclusive development of the territory. Approaches to determining the territory from the position of inclusiveness of its development are analyzed, the question of the positive and negative impact of growth on the development of the territory in the strategic perspective is considered.

Keywords: Decentralization of territory development management · Economic marginalization of the territory · Growth for the poor · Inclusive development · Territory

1 Introduction

Experts are increasingly discussing the possibility of inclusive development of territories and regions. In this regard, there are several lines of discussion, namely:

- what is considered a territory that the concept of inclusive development applies to,
- what type of development can be considered inclusive,
- what level of centralization of the process of inclusive development management is necessary to achieve maximum results.

The term "territory" comes from the Latin territorium. Historically, this concept is associated with the concepts of power and domination. The current definition of "territory" depends on social, political, cultural, and linguistic contexts. There are two approaches: Anglo-Saxon and French. The Anglo-Saxon approach is based on the idea of expanding capitalism, forming territorial forms of regulating expanding trade. The French approach is closer to the semantic field of public policy and is based on the institutional environment.

Territory development policies are often seen as "universal" policy that focus on infrastructure development (roads, railways, access to water and sanitation, electrification) and financial support mechanisms and instruments aimed at attracting large firms to the least developed territories. This development policy responds to the needs expressed by citizens and often leads to attracting investment in infrastructure as a

S. I. Ashmarina and V. V. Mantulenko (Eds.): IPM 2020, LNNS 161, pp. 117–121, 2021.
https://doi.org/10.1007/978-3-030-60926-9_16

means of improving market access in remote areas or helping areas that lack physical, financial and other capital. However, policies that focus on infrastructure development and investment mechanisms are often implemented without much consideration for territories specifics. The "one size fits to everyone" policy does not allow achieving sustainable development [1, 2].

Indeed, many studies indicate that despite their positive impact, infrastructure development policies often lead to polarization and economic agglomeration. Despite the economies generated by these agglomerations, they are also the root of growing economic marginalization. Many peripheral regions of Europe [4, 8, 10] and China, where significant investments have been made in infrastructure, illustrate this marginalization. Similarly, development policies based on financial incentives that encourage firms to settle in less industrialized regions have shown their limitations [9, 11].

Indeed, firms found themselves in regions that were unable to absorb such investments due to a lack of competent human resources and institutional deficiencies. And even mastered investments that lead to growth cause mixed results. Growth is usually associated with progress and development, but growth can also be a source of inequality, especially if inequality is geographically localized, or if it excludes certain groups of the population, for example, on the basis of ethnic origin. In the short term, inequality can fuel growth, for example by allowing industry to benefit from low wages for unskilled staff.

2 Literature Review

In the long term, poverty and inequality can become a serious obstacle for growth, these processes are described in detail in the works of Lanchovichina, and Lundstrom [6], Berg and Ostry [3]. Redistributing results as a way to reduce poverty and inequality can contribute to sustainable growth over time [7]. The term "growth for the poor" quickly came to the fore in literature. There are two approaches to defining the term "growth for the poor". The first approach takes into account the impact on poverty in absolute terms, that is, "growth for the poor" is such when it contributes to reducing the level of poverty. The second approach defines the impact of growth on poverty in relative terms, meaning that "growth for the poor" is when the incomes of the poorest grow faster than those of the rest of the population, which leads to a decrease in inequality.

Based on the first definition, a policy of targeted redistribution in favour of the poorest can achieve inclusive growth if one thinks within the boundaries of the first definition. But this approach does not work within the second definition. Within the second definition, reducing inequality requires increasing the income of the poorest social groups, through improving their skills and access to higher-paying jobs, thus ensuring that the territory's development is inclusive.

There are several approaches to the definition of "inclusive development". Thus, White notes that the concept "inclusive" has at least six meanings in the literature [12]:

- more equal distribution of income,
- the reduction of absolute poverty,

- internalization of external growth factors,
- greater equality in access to opportunities created by growth (access to education, financing, etc.),
- reducing inequality between territories,
- increasing the role of developing countries in the management of international financial institutions.

For research purposes, only the first four are of interest. The first two relate to the definition of growth for the poor. The next two are broader concepts that define the complexity of measuring "inclusivity".

3 Methodology

The following research methods were used: scientific abstraction, comparative analysis, system-structural approach, analysis and synthesis, induction and deduction, analysis of theoretical literature, statistical grouping method, stochastic and simulation modeling method. Together, they made it possible to achieve the stated goals of the work. There was also an analysis of scientific publications on the impact of ongoing socio-economic shifts on the inclusive development of territories.

The theoretical relevance of the study lies in the expanded understanding of the need for inclusive development of regions and factors that can influence its course. This collection of information can serve for further study of the problem of regions inclusive development.

4 Results

It should be noted that the representation of a region as a system allows us to analyze it through an important feature of the system - the presence of behavior - action, changes, functioning, etc. This underlies in the ability of a region (as an object) to develop. By behavior we mean the process of purposeful change in the state of the system in time, provided that the behavior is realized exclusively by the system itself (the region), based on its own goals. The behavior of each system is determined, firstly, by the structure of the low-order systems that make up this system, and secondly, by the presence of equilibrium (homeostasis) signs that determine the state (states), which is preferable for the system (region). Therefore, most often, the behavior of systems is described in terms of the restoration of these states when they are disturbed as a result of changes in the environment. Third, the behavior of a system (region) is determined by the growth (development), which is considered as the most important component of behavior.

The Commission on growth and development's report on strategies for sustainable growth and inclusive development notes that the concept of inclusiveness includes equity, equity and protection in access to high-tech jobs. In general, skilled employment is a central point of reference in determining inclusive growth [13].

Inclusivity is often defined based on one or more outcomes (for example, income growth), but it can also include the process leading to these outcomes. Inclusive growth is defined by both the speed and structure of growth. Sustained growth is necessary to reduce poverty, but not sufficient to ensure that growth is inclusive. If, for example, growth is localized geographically or poorly diversified, its results will certainly affect only a small part of the population. On the other hand, growth can create many jobs that may be unproductive or destroyed over time.

Thus, inclusiveness includes a productive dimension that should be both static and dynamic. Inclusivity also includes a structural transformation of the economy that allows a broader population:

- find employment opportunities where the skills of the population can be fully used,
- adapt to technological progress and, consequently, to the process of creating and disappearing jobs that it entails,
- create new employment opportunities,
- make the best use of available resources.

5 Conclusion

Inclusiveness also refers to the process that generates growth, and can be seen in terms of participation in the decision-making process leading to public policy choices. Thus, inclusive growth can also be seen as a process of participation, which is important because it relies on the efficient development of existing resources and the development of resources over time. Therefore, it is a question of identifying potential sources of growth and the constraints that affect it, which implies not only a combination of public policy at the macro - and micro-economic levels, but also the existence of good management.

The question on how the decentralization of management affects the results of inclusive growth of the territory remains debatable. So, Jutting, et al. conducted a case study regarding 19, which showed that the impact of decentralization on the level of development inclusiveness, in some cases, was positive and negative in others [5].

The nature of the impact of management decentralization on the effectiveness of inclusive development of the territory is largely determined by the quality of local governance, the level of accountability of local government, as well as the active civil participation in public decision-making, and the development of partnerships between public and private communities. Thus, the quality of management and country characteristics are very important for assessing the impact of decentralization on the degree of economic growth inclusion.

References

1. Ascani, A., Crescenzi, R., Iammarino, S.: Regional economic development: a review (2012). http://projects.mcrit.com/foresightlibrary/attachments/article/1236/WP-1.3.pdf. Accessed 27 June 2020
2. Barca, F., McCann, P., Rodriguez-Pose, P.A.: The case for regional development intervention: place-based versus place-neutral approach. J. Regional Sci. **52**(1), 134–152 (2012)
3. Berg, A., Ostry, J.: Equality and efficiency: is there a trade-off between the two or do they go hand in hand? Financ. Dev. **48**(3), 12–15 (2011)
4. Dall'Erba, S., Le Gallo, J.: Regional convergence and the impact of European structural funds over 1989–1999: a spatial econometric analysis. Papers Regional Sci. **87**(2), 219–244 (2008)
5. Jutting, J., Kauffmann, C., Mc Donnell, I., Osterrieder, H., Pinaud, N., Wegner, L.: Decentralization and poverty in developing countries: exploring the impact (2004). https://ideas.repec.org/p/oec/devaaa/236-en.html. Accessed 27 June 2020
6. Lanchovichina, E., Lundstrom, S.: Inclusive growth analytics: framework and application, policy research working paper, No. 4851. World Bank, Washington (2009)
7. Lopez, J.H.: Pro-poor growth: a review of what we know (and of what we don't know). World Bank, Washington (2004)
8. Puga, D.: European regional policies considering recent location theories. J. Econ. Geogr. **2**, 373–406 (2002)
9. Trigilia, C.: Sviluppo senza autonomia: effetti perversi delle politiche nel mezzogiorno. Il Mulino, Bologna (1992)
10. Vanhoudt, P., Mathä, T., Smid, B.: How productive are capital investments in Europe? Eur. Invest. Bank Papers **5**(2), 81–106 (2000)
11. Viesti, G.: Come nascono i distretti industriali. Laterza, Rome (2000)
12. White, W.R.: Policy Debate: How do You Make Growth Moreinclusive? OECD Publishing, Paris (2012)
13. World Economic Forum: The inclusive growth and development report 2017 (2017). http://www3.weforum.org/docs/WEF_Forum_IncGrwth_2017.pdf. Accessed 27 June 2020

Protection of Economic Human Rights in the European Court of Human Rights

S. N. Revina(✉) and D. S. Zemlyanikin

Samara State University of Economics, Samara, Russia
29.revina@mail.ru, Dima.zemlyanikin@mail.ru

Abstract. Today, European society actively uses rights and freedoms established by law, including the right to international protection. Due to the growing number of complaints to the European Court of Human Rights (hereinafter - the ECHR), as well as increasing confrontation between national and international legal systems, the research topic is relevant. The research object is relations that arise as a result of violations of human rights and freedoms. The research subject is the international protection of economic human rights. The purpose of this research is to highlight the role of the ECHR in protecting economic human rights in the context of the modern world space development. The authors conclude that the ECHR plays a leading role in the protection of economic human rights at the international level. This is confirmed by the growing demand for the protection of violated economic rights. The activities of the ECHR show certain shortcomings of national legal systems, the need for constant monitoring of this issue, or the adoption of measures to reform a particular institution.

Keywords: Economic human rights · ECHR · International protection of economic rights · Property rights

1 Introduction

The ECHR is an international court that considers complaints related to violations of human rights and freedoms, which in turn are provided by the European Convention on Human Rights (hereinafter - Convention) [4]. To better understand the role of the ECHR in protecting economic rights, it is necessary to be aware of what economic rights are enshrined in the Convention. For example, economic rights are more widely disclosed in the Constitution of the Russian Federation [3] than in the Convention. Accordingly, if the complaint sent to the ECHR does not contain a violation of the Convention, it is rejected.

In accordance with Article 1 of the Additional Protocol to the Convention [4], protected economic rights include property rights. The norm of this article provides that individuals and legal entities have the right to respect for their property. It is also provided that the state can only deprive a person of the right to property in the interests of society. The state exercises control over the use of property and ensures the payment of taxes. But when protecting property rights, rights and claims arising from them are

S. I. Ashmarina and V. V. Mantulenko (Eds.): IPM 2020, LNNS 161, pp. 122–126, 2021.
https://doi.org/10.1007/978-3-030-60926-9_17

also protected. To understand this judgment, it is necessary to refer to international judicial practice and analyze decisions made by the ECHR.

2 Methodology

In this work, the following methods of scientific research were used. The dialectical method allowed us to consider activities of the ECHR for the international protection of economic human rights in the context of constant development of public relations. Such methods as structural-functional and system-based ones have made it possible to study elements of constitutional and legal guarantees of human rights and freedoms and their protection, as well as to consider them as a single mechanism that is an integral part of the legal state. The formal legal method was used in the analysis of the ECHR decisions related to the protection of economic human rights which are provided by the Convention. The statistical research method allowed to process digital information and interpret statistical data.

3 Results

The Convention protects property rights in the classical sense, that is, such property rights as disposal, use and possession. However, the ECHR considers complaints that arise from these legal relations in a broader sense, for example, complaints concerning long-term non-enforcement of a court decision on property disputes, complaints related to unreasonably high fines and taxes, complaints concerning damage to property as a result of military exercises, etc. Let's analyze the content of the ECHR's rulings regarding the protection of property rights provided by the Convention.

In accordance with the decision of the ECHR in the case "Vladimirova v. the Russian Federation" of 10.04.2018, the applicant sent a complaint to the ECHR, which stated that, as a director of the company "Aquilon", she was engaged in the sale of food products [5]. In 2001, a transaction was to take place, according to the results of which the shipped goods should remain by the buyers, and the seller company had to receive its profit. But a criminal case was opened against the buyers on suspicion of fraudulent actions, and the shipped goods were not returned to the seller, as they were material evidence. In 2007, the criminal case was closed, and the buyers' guilt was not proved. In 2005, the applicant was awarded compensation for the caused damage, but it was not paid until 2007. Given these facts, the ECHR ruled that Article 6 of the Convention and article 1 of the Additional Protocol to the Convention were violated, due to the fact that the court's decision was not enforced for 2 years. 1,500 Euros were awarded as compensation for non-pecuniary damage.

In the judgment of the ECHR in the case "Gyrlyan v. Russia" of 09.09.2018, the applicant sent a complaint to the ECHR, which contained the following information [6]. In 2014, the applicant travelled by plane from Ukraine to Russia carrying 100,000 U.S. dollars in a hand baggage obtained from the sale of real estate. Upon arrival at the airport in Russia and passing the hand baggage check, the applicant reported the presence of the specified amount of money, thinking that its declaration should be made after checking the hand baggage. However, the money was withdrawn from the applicant, and an attempt to appeal the actions of the customs officers was not successful. In this case, the ECHR ruled that there is no doubt that the applicant is the owner of the confiscated money amount, which means that this fact confirms the interference of the state in the use of the applicant's property. But this intervention is legal according to the Additional Protocol to the Convention (Paragraph 2, Article 1) [4], since this is regulated by the national legislation, namely Article 16.4 of the Administrative Code of the Russian Federation [1].

However, since these funds were obtained as a result of a sales contract, there is no reason to believe that the applicant obtained them illegally. A violation of the applicant was not timely declaration of funds. Thus, the ECHR ruled that Article 1 of the Additional Protocol to the Convention was violated because of the excessive punishment. The court awarded compensation for material damage in the amount of 73,000 Euros and compensation for non-material damage in the amount of 1,500 Euros.

In accordance with the decision of the ECHR in the case "Tkachenko v. Russia" of 20.03.2018, the court found that Article 1 of the Additional Protocol to the Convention was violated [7]. The applicant's complaint stated that in 1999 he had privatized part of the house, which was located in the city center, but the land under the house was municipal. In 2003, according to one decree of the city head, the land and the entire house were transferred to the company Z for building an apartment building. In 2004 the decision of the national court terminated the ownership of part of the applicant's house, and he was offered a house with a plot of land in the suburbs. In 2005, the applicant was evicted by the private company M, which later developed the land plot. The ECHR ruled that the applicant had not been fully granted the right to property protection, and awarded 5,000 Euros for material damage and 5,000 Euros for non-material damage.

Having analyzed a number of decisions of the ECHR in cases involving violations of Article 1 of the Additional Protocol to the Convention, we can identify some property types that fall under these legal relations:

1. Property that exists in its natural form and to which the right of ownership is registered.
2. Property seized by the government.
3. Unrealized income that could have been received by the property owner.
4. License that provides certain commercial activities, since the revocation of the license may violate the right to respect for property;
5. Intellectual property products;
6. Shares with economic value;
7. Social benefits, such as pensions, etc.

In order to most accurately understand the significance of the ECHR in cases of property rights protection, let's consider a number of statistics data that were taken from the Annual report 2019 of the European Court of Human Rights, which is published on the official website of the ECHR [2]. According to data for 2019, the ECHR received a total of 59,800 complaints that is 166 complaints per day on the average. According to the report, the largest number of complaints was sent against Russia (25% of the total number of complaints) (Fig. 1).

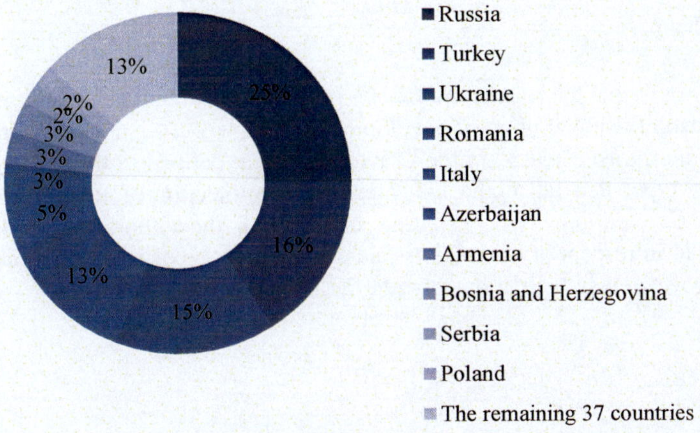

Fig. 1. Percentage of complaints received to the ECHR by country in 2019

However, it should be noted that in 2019, 198 ECHR rulings were issued in respect of Russia from the large list of complaints. It means that most of the complaints are not accepted for consideration for one reason or another. According to the report, 10% of all complaints are related to violation of property rights, which also confirms the significance and relevance of this issue and the need to address this problem.

4 Discussion

According to the research results, it can be concluded that the role of the ECHR in the international protection of economic rights is enormous. This is due to the fact that countries that have ratified the Convention allow their citizens, if all national mechanisms for protecting violated rights and freedoms have been exhausted, to apply for international protection in the ECHR.

Having analyzed a number of ECHR rulings, the authors note that most of the complaints sent to the international court are not only of a restorative nature, which implies the restoration of violated rights and material compensation. They also make the national government to take measures for structural changes, if the complaints show systemic problems.

The statistical indicators indicated in the study also emphasize the significance of this scientific work. In 2019, on average, the ECHR received 166 complaints per day, these are quite serious figures, and every fourth complaint is against Russia. From the 59,800 complaints, 10% are related to violations of intellectual property rights, which is also a fairly serious indicator.

5 Conclusion

This study highlights the role of the ECHR in the international protection of economic human rights in the contemporary development of the world legal and economic space. The economic rights specified in the Convention were defined in this study, as well as the property types that fall under the international protection of economic rights provided in the Convention were considered. In this work, the authors analyzed the ECHR rulings issued in respect of complaints related to violations of economic human rights, as well as reviewed statistical data on the ECHR activities in 2019.

References

1. Administrative Code of the Russian Federation (2001). http://www.consultant.ru/document/cons_doc_LAW_34661/. Accessed 20 Aug 2020
2. Annual report 2019 of the European court of human rights (2019). https://www.echr.coe.int/Documents/Annual_report_2019_ENG.pdf. Accessed 20 June 2020
3. Constitution of the Russian Federation (1993). http://www.consultant.ru/document/cons_doc_LAW_28399/. Accessed 20 June 2020
4. Council of Europe: European Convention on Human Rights (2010). https://www.echr.coe.int/Documents/Convention_ENG.pdf. Accessed 20 Aug 2020
5. ECHR Ruling in the case "Vladimirova v. Russia" of 10 April 2018 (2018). http://base.garant.ru/72143658/. Accessed 20 Aug 2020
6. ECHR Ruling in the case "Gyrlyan v. Russia" of 09 September 2018 (2018). http://prof-sommer.ru/postanovlenie-evropeyskogo-suda-po-delu-gyrlyan-protiv-rossii-zhaloba-no-3594315-ot-9-oktyabrya-2018. Accessed 20 Aug 2020
7. ECHR Ruling in the case "Tkachenko v. Russia" of 20 March 2018 (2018). http://prof-sommer.ru/postanovlenie-evropeyskogo-suda-po-delu-tkachenko-protiv-rossii-zhaloba-no-2804605-ot-20-marta-2018. Accessed 20 Aug 2020

HR Management in Digital Conditions: IT for Creating a Healthy Work Environment and Managing Well-Being

Management of Personnel Health and Well-Being in the Context of Distance Employment

N. I. Arkhipova(✉), S. V. Nazaikinsky, and O. L. Sedova

Russian State University for the Humanities, Moscow, Russia
Prorektor207@gmail.com, svyatoslavrsuh@gmail.com,
olga-sedova@yandex.ru

Abstract. The purpose of the study is to identify the main problems of managing the health and well-being of personnel in the conditions of distance employment and to identify the ways to solve them. To achieve this goal, we defined the following main tasks: to identify the main factors of influence of distance employment on the health and well-being of personnel; to study the practices of management of health and well-being of personnel in organizations; to identify the areas of improvement for the programs of management of health and well-being of personnel. The main problems and the ways to solve them were described as a result of the analysis and generalization of research data from international and Russian consulting companies, expert opinions on the research topic, the experience of Russian companies in managing the health and well-being of personnel, and surveys.

Keywords: Distance employment · Health · Management · Personnel · Well-being

1 Introduction

In recent years, experts have considered the health and well-being of employees to be the main condition for increasing the productivity and labor efficiency of the personnel. There are several definitions of this concept in the literature. Within the framework of this article, we will use the definition of the World Health Organization, where health is understood not only as the absence of diseases but also as the state of physical, mental (emotional and intellectual), social and spiritual well-being [3]. It should be noted that the concept and essence of programs for managing the health and well-being of workers in organizations have undergone significant changes in recent years. Thus, at the beginning of the 21st century, the well-being programs were aimed mainly at ensuring the physical and mental well-being of employees. Now they include ensuring emotional, social, financial, and professional well-being.

According to Towers Watson consulting company, comprehensive programs for managing the health and well-being of workers should include three main blocks [12]. The first block should be aimed at preserving the health of workers (health screening, healthy lifestyle programs, sports activities, etc.); the second - at their financial well-being (financial literacy programs, financial advisory, savings management, etc.); the

S. I. Ashmarina and V. V. Mantulenko (Eds.): IPM 2020, LNNS 161, pp. 129–134, 2021.
https://doi.org/10.1007/978-3-030-60926-9_18

third - at meeting the individual needs of the employee. Thus, the programs for managing the health and well-being of employees are aimed not only at creating optimal working conditions but also at the well-being "in the work itself", regardless of the form of employment.

According to the «Global Human Capital Trends - 2020» study by Deloitte consulting company, about 80% of respondents marked "well-being as a part of work" as one of the most important modern trends in personnel management [4]. The issues of managing the health and well-being of workers are particularly relevant in the context of distance employment.

2 Methodology

The methodological basis of the study is the methods of logical and comparative analysis, expert assessments, opinion poll, survey. In recent years, the problem of the well-being of workers has become the subject of both theoretical and empirical research. The paper analyzes the research results of international and Russian consulting companies, expert opinions, as well as the experience of Russian companies in managing the health and well-being of personnel. We have analyzed the impact of modern digital technologies on the health and well-being of employees, including the remote workers. The research work is based on the concepts and hypotheses presented in the works of Russian and foreign researchers, in particular, the concept of work and personal life balance, the concept of zero injuries, the concept of well-being, the concept of a digital workplace, the theory of generations, etc.

3 Results

The Russian practice of managing the health and well-being of personnel is traditionally aimed at creating safe working conditions, reduction of industrial injuries and occupational diseases. This is because, on the one hand, non-compliance with safety rules in production processes leads to a reduction in the number of employees due to injuries and occupational diseases, and, on the other hand, to a decrease in labor productivity and financial costs.

In recent years, the Russian Federation has developed more than one hundred regulatory documents on labor protection for various types of economic activity and types of work. A special assessment of working conditions has been introduced, and the Russian Ministry of Labor and Social Protection is constantly monitoring the application of labor protection legislation by organizations. All this helps to preserve the health of employees.

Many Russian companies implement health management technologies, which include not only regular medical examinations, vaccination of employees, and recreational activities, but also the study of medical risks for staff and the development of a plan of preventive measures. Large Russian organizations, such as Lukoil, Gazprom, Severstal, Sberbank, and others, have developed and implemented the employee well-being programs.

The benchmarking of the personnel well-being programs implementation has shown that the majority of Russian organizations are focused on maintaining the physical health of workers and reducing injuries and occupational diseases at work, while in foreign companies, the well-being programs are comprehensive and aimed at improving physical, emotional, and mental condition of the personnel [1].

In November 2019, the department of organizational development of the Russian State University for the Humanities conducted a study of the effectiveness of implementing the personnel well-being program in a large domestic commercial and industrial company for the production and sale of office and hotel furniture, which employs 956 people. The study was conducted by interviewing employees using a specially developed questionnaire on the level of satisfaction with the well-being program which has been implemented in the company since 2017.

Following the well-being program, each employee of the company within the planned budget (60 410 rubles), along with the standard social package (voluntary medical insurance, health and life insurance, compensation for the cost of sports activities, compensation for the cost of food, discounts on housing purchases) has the opportunity to participate in events and promotions aimed at increasing knowledge about maintaining health, anti-smoking and anti-obesity campaigns, etc.

For two years of implementing the well-being program in the company, the number of its participants increased from 75% to 88.9%, which indirectly confirms its effectiveness. At the same time, 77% of the participants of the well-being program noted that they were mostly satisfied with its content, however, 30% of the respondents indicated the need to expand the number and focus of events and actions. Also, more than 30% of respondents are not satisfied with the standardized character of the well-being program which does not take into account the state of health, marital status, and other individual characteristics of employees. Therefore, we can conclude that enhancing the effectiveness of well-being programs can be associated, on the one hand, with an integrated approach to its development, and on the other hand, with the individualization of the approach to each employee.

The problem of managing the health and well-being of workers is particularly relevant in the context of alternative forms of employment, including distance employment of workers. For example, according to the results of the PWC consulting company study involving 1,200 company executives and HR managers from 79 countries as respondents, the spread of distance employment is the most important condition for a company's competitiveness in the labor market [9]. The forms of distance employment of workers have become widespread in the context of the pandemic associated with the spread of COVID-19.

In the context of the introduction of additional preventive measures and sanitary standards by the authorities, a large number of Russian organizations employees were transferred to a remote working regime. According to experts, 2020 has become "the world's largest experiment in distance working".

Employers and HR specialists face new challenges in organizing effective work of personnel and in managing the health and well-being of the workers. It should be noted that Russian organizations have accumulated a certain experience in managing workers in remote employment conditions. For example, the remote work format was quite widespread in marketing, PR, IT, accounting, etc. At the same time, experience has proven that in the context of the mass transition of workers to the remote mode of work, many organizations were not ready for effective personnel management, as well as for ensuring the well-being of their employees.

First, in the mode of stationary work in the office, a special assessment of the working conditions of employees is carried out and their workplaces are supplied with professional technical equipment and licensed software, while in the conditions of distance employment, the employees often use the equipment available in their houses. In March 2020 the Mail.ru Group company surveyed 12464 employees transferred to the remote mode of work. The results of a survey showed that most of them used personal equipment (one or more technical facilities at the same time): smartphone (62%), computer (51%), tablet (10%). The most common services used for work were email (91%), messengers (62%), and video call services (23%) [7]. The diversity of the equipment, its insufficient power, and low-speed Internet are often the reason for increasing the complexity of the work performed.

Second, the organization of the workspace and the workplace itself in most cases does not meet sanitary and hygienic standards and requirements, which negatively affects the health of workers.

Third, in the context of distance employment, the balance between work and personal life is often upset. On the one hand, employees can independently plan their work and rest hours, and on the other hand, they have become "available" to their managers almost 24 h a day. These conditions increase the intensity of labor, the amount of work, and the requirements on such personal qualities of employees as reliability, organization, concentration, etc. [8]. The information load and constant contact with the employer lead to "burnout" of employees, lower involvement and productivity [10].

Fourth, remuneration systems depending on the achievement of goals or the implementation of KPI have become widely used in management practice. To ensure financial well-being, employees increase their working hours, perform more job duties, etc. This leads to increased competition between employees, as well as the emergence of workaholics. Workaholism causes a high level of emotional stress, professional burnout, the appearance of chronic fatigue syndrome, and the imbalance of work and personal life.

Fifth, in the conditions of distance employment, the communication of employers with their colleagues is limited, which can lead to a deterioration in their psychological well-being [5].

All this leads to a deterioration in physical health, increased levels of stress, depression, etc. It should be noted that the well-being of employees in the context of distance employment depends not only on the efforts of the employer but also on the formation of "self-management" competencies which include the ability of personnel to manage their energy, resources, time, emotional intelligence.

4 Discussion

The specifics of organizing work in distance employment conditions make it necessary to transform the health and well-being management of personnel in companies. According to the results of the study conducted by the authors, the main factors of workers' health and well-being deterioration in distance employment conditions are the imperfect organization of the workspace and workplace, non-compliance of working conditions with sanitary and hygienic standards, upset work-personal life balance, limited communication.

Managing the health and well-being of personnel in distance employment requires joint development and implementation of programs by both the employer and the employee. The employer should provide the employee with a virtual workplace. It is advisable to organize the workplace of an employee working in a remote format following the "Digital workplace" concept according to which the employee is a consumer of the necessary "content and services here and now" [11]. In this case, it is advisable to determine the standard composition of the virtual digital workplace for each position. It is worthwhile to include a training and development module, as well as a knowledge base in the digital workplace to ensure the professional well-being of the employees, i.e. the possibilities of their development, career advancement, and professional self-realization.

In the process of practical implementation of personnel well-being programs, specialized services and tools should be used. For example, the use of fitness trackers allows to track the level of activity of the employee and give recommendations on its adjustment. The Welltory service allows measuring the level of stress and energy of the employee based on the analysis of data from 200 fitness applications and gadgets. The Waterbalance app reminds to drink water. The Lifeaddwiser mobile app evaluates the strengths and weaknesses of the employee and develops personalized recommendations for improving productivity, etc.

The employee needs to master the skills of using tools that allow enhancing the work effectiveness and formation of digital, professional and social competencies [2]. Thus, in the conditions of distance employment, there is a trend of "shifting focus from paternalistic care for employees to partnerships" when employees share responsibility for their development and well-being with the employer [6].

5 Conclusion

One of the main trends in personnel management in modern conditions is ensuring the health and well-being of personnel. This led to the development and implementation of corresponding management programs in modern companies. The effectiveness of the personnel health and well-being management programs can be ensured only if there is an integrated approach to their development and implementation. Personnel health and well-being management programs should be aimed at ensuring physical, financial, social, emotional and professional well-being. Along with ensuring physical and financial well-being, the management of emotional (ability to adapt to constantly changing conditions) and professional (self-realization in the profession, career advancement, development) well-being is of particular importance in the conditions of distance employment.

Another way to improve the effectiveness of personnel health and well-being management programs is an individual approach to their development and implementation, i.e., taking into account the needs and individual characteristics of each employee. Therefore, the first step in developing the personnel health and well-being management programs should be to assess the level of well-being of employees and their needs [6]. The specific features of distance work organization make it necessary to improve programs for managing the personnel health and well-being using modern digital tools. The development and implementation of the personnel health and well-being management programs in organizations can ensure the balance of work and personal life of employees, increase their engagement and satisfaction, and contribute to the accumulation of experience through the use of new methods of work.

References

1. Arkhipova, N.I., Nazaikinskii, S.V.: Human resources potential of the Russian economy: risks and opportunities. RSUH/RGGU Bull. Ser.: Econ. Manag. Law **4**(2), 154–165 (2019). (In Russian)
2. Arkhipova, N.I., Sedova, O.L.: Application of digital-instruments in the staff selection and screening in the organization. RSUH/RGGU Bull. Ser.: Econ. Manag. Law (2), 9–22 (2018). (In Russian)
3. Constitution of World Health Organization, adopted on 22 July 1946 (1946). https://www.who.int/ru/about/who-we-are/constitution. Accessed 20 June 2020
4. Deloitte: Global Human Capital Trends – 2020 (2020). https://www2.deloitte.com/kz/ru/pages/human-capital/articles/human-capital-trends_msm_moved.htm. Accessed: 12 Apr 2020
5. Konobevtsev, F., Laas, N., Gurova, E., Romanova, I.: Remote work: technologies and experience of the organization. Vestnik Univ. **7**(1), 9–17 (2019). (In Russian)
6. Lifeaddwiser: The value of corporate well-being (2020). https://lifeaddwiser.com/blog/wellbeing-conference. Accessed 12 June 2020
7. Mail.ru Group: Hi-Tech Mail.ru found out how many Russians work from home (2020). https://www.searchengines.ru/hi-tech-mail-ru-udalenka.html. Accessed 17 Apr 2020
8. PWC: Millenials at work. Reshaping the workplace (2011). https://www.pwc.com/co/es/publicaciones/assets/millennials-at-work.pdf. Accessed 15 May 2020
9. PWC: Preparing for tomorrow's workforce, today Insights from a global survey of business and HR leaders (2018). https://www.pwc.com/gx/en/people-organisation/pdf/pwc-preparing-for-tomorrows-workforce-today.pdf. Accessed 13 July 2020
10. Schwarz, J., van Berkel, A.: The overwhelmed employee. Simplify the work environment (2014). https://www2.deloitte.com/us/en/insights/focus/human-capital-trends/2014/hc-trends-2014-overwhelmed-employee.html. Accessed 15 June 2020
11. Training Center "Competences": difficulties in implementing the digital workplace (2019). https://hr-media.ru/kejs-slozhnosti-vnedreniya-tsifrovogo-rabochego-mesta-sotrudnika-digital-workplace/. Accessed 14 June 2020
12. Towers Watson: Wellbeing Diagnostic. https://www.willistowerswatson.com/en-US/Solutions/products/wellbeing-diagnostic. Accessed 25 June 2020

Contractual Arrangements Between Providers and Consumers of Digital Technologies in Space Industry

E. K. Belyaeva[1], D. Yu. Ivanov[1], and S. V. Domnina[2,3](✉)

[1] Samara University, Samara, Russia
Belyaeva1301@gmail.com, ssau_ivanov@mail.ru
[2] Samara State University of Economics, Samara, Russia
swdomnina@mail.ru
[3] Samara State Institute of Culture, Samara, Russia

Abstract. Currently, almost all spheres of human activity are being transformed under the influence of information and telecommunication technologies. The development of the world space market contributes to the emergence of new market segment of the digital technologies and services provision. In order to ensure and develop a clear digital economy infrastructure, it is necessary also to optimize the interaction of market participants in the sphere of space activities. The purpose of the research is to develop an economic model of formation of contract optimal parameters for interaction between the providers and consumers of digital technologies in space industry and to give an overview of the satellite services market state. Accounting the contract nature of space market participants interaction, we apply the basic principles of the contract theory. We present a model of interaction between providers and digital services consumers, which allows to align the interests of each side of contract. We consider the model in terms of symmetric and asymmetric information. We find the optimal menu of contracts that maximizes provider's profit.

Keywords: Contract theory · Decision-making model · Digitalization · Digital economy · Digital technologies · Optimal contract

1 Introduction

The beginning of the XXI century got marked by the breakthrough advancements in digital technologies, revolution in space of information and acceleration of economic globalization processes [6]. The development of the digital economy is impossible without the use of modern space technologies: the transmission of data over long distances requires clear satellite support. Space market concentrates a variety of segments, including the provision of digital services such as satellite television, remote sensing, mobile and Internet services, etc. [13, 20]. Its specificity is the explosive development and consumption of new advanced technologies. Currently, projects on unmanned aerial and remotely piloted vehicles are being actively developed, their implementation requires the use of spacecraft. Thus, the practical implementation of projects in both digital and space economy requires the organization of participants'

S. I. Ashmarina and V. V. Mantulenko (Eds.): IPM 2020, LNNS 161, pp. 135–142, 2021.
https://doi.org/10.1007/978-3-030-60926-9_19

interaction [11]. This problem can be solved by the construction of contractual arrangements between participants of the digital services and technologies provision segment of the space market.

2 Methodology

The model of contractual arrangements between providers and consumers of digital technologies in space industry is based on adverse selection model of two types of agents in the context of the contract theory [4, 8, 9, 15]. The case of asymmetric information assumes the imposition of incentive restrictions for the presented equation system. The digital technologies consumers' utility functions are selected in such a way that condition of strict single crossing is met [3].

3 Results

Obviously, the development of the digital economy is closely connected with the possibilities of using the results of space activities, as the transmission of information over long distances is available because of the development of the satellite Internet. Digital services are one of the components of the space market – satellite services segment, which includes four main areas: mobile satellite services (mobile data and voice), earth observation, fixed satellite services (transponder agreements and managed services), consumer satellite services (satellite TV, radio, broadband, Internet TV). Figure 1 presents the revenue dynamics of the satellite services segment.

According to the Satellite Industry Association Report, the revenue from the satellite services segment in 2018 amounted to 126.5 billion USD, while satellite TV services accounted for 74% of all satellite services revenues, nearly 40% of global revenues attributed to U.S. [20]. The volume of the fixed satellite services segment in 2018 amounted to 17.9 billion USD, and the consumer satellite services segment – 102,4 billion USD. The income of the satellite services segment in 2018 reached 46% of the total income of the world space market – that makes this segment the largest among the other world space market segments. The compound annual growth rate of the satellite services segment revenues over the five-year period is estimated as 7% [13]. The comparison of the «downstream» (activities associated with the provision of space services and products to the end users) and «upstream» (activities associated with space systems and space vehicles construction, development and launch) segments of the space market shows that the downstream drives the space economy with nearly 5050 companies engaged in contrast to the upstream with 40 companies engaged [13].

Let us consider an interaction of the provider of digital technologies in space industry and its consumers. Digital technologies providers within the space market will be understood as companies engaged in providing such digital services as mobile, fixed, consumer satellite services and Earth observation. Let us simulate two cases: symmetric and asymmetric information. Digital technologies consumers differ in certain information characterizing them, which, following the terminology of the contract theory, is called their type, and in our case will be defined as consumer's value of the

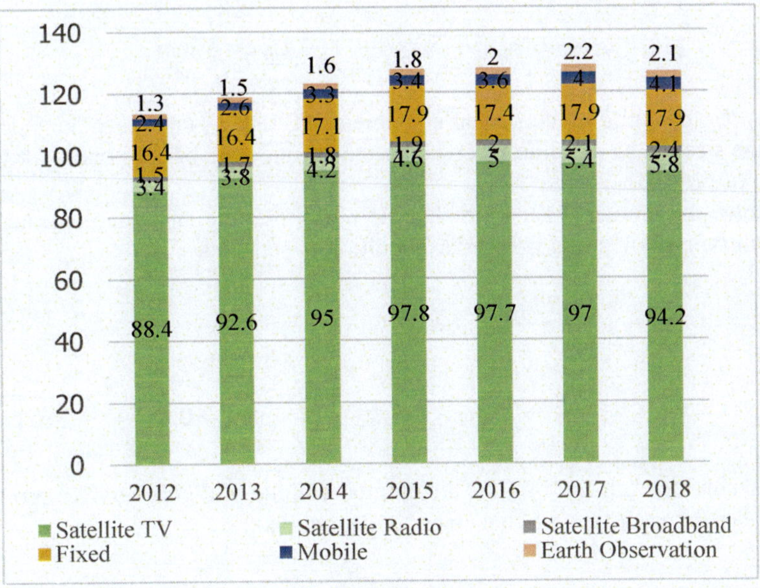

Fig. 1. The revenue dynamics of the satellite services segment, billion USD (Source: authors).

digital services. In case of symmetric information the type of the digital technologies consumers is known to the provider. Having this information, he takes a decision on what price policy he should determine. In case of asymmetric information, the provider is not able to differentiate the types of the consumers. In order to increase the contract policy effectiveness in such conditions he has to design the mechanism that is based on the digital technologies consumers type identification principle.

Let us assume that consumers of digital technologies are divided into two types: θ_1 and θ_2, moreover, the share of consumers of the first type is π, and the second - $(1 - \pi)$. The target function of the digital technologies provider is:

$$\Pi = (p - c)D(p) \to \max_{p}, \tag{1}$$

where $D(p)$ is the demand for a certain volume of digital service;

c stands for costs of the provider for the unit of service offered to consumers.

The provider can offer a contract (x, p) for both types of consumers in case of symmetric information and individual contracts (x_i, p_i), $i = 1, 2$ in case of asymmetric information, where x_i stands for the volume of the provided digital services for each type of consumer, p_i is the price of the digital service conventional unit. The consumers of digital technologies maximize their utility.

The target functions of the first and second type consumers will be as follows:

$$u_1 = \theta_1(a_1x_1 + b_1x_1) - px_1 \to \max_{x_1}, \tag{2}$$

$$u_2 = \theta_2(a_2x_2^2 + b_2x_2) - (1-\mu)px_2 \to \max_{x_2}, \tag{3}$$

where a_1, b_1, a_2 and b_2 – numerical coefficients, a_1 and a_2 take negative values;

μ - the size of the discount received by the second type of consumers for the large volumes of orders.

To solve the problem of maximizing the utility of a digital technologies consumer of the *i*-th type (2)–(3), the first-order condition is satisfied:

$$\frac{\partial u_1}{\partial x_1} = 2\theta_1 a_1 x_1 + \theta_1 b_1 - p = 0, \tag{4}$$

$$\frac{\partial u_2}{\partial x_2} = 2\theta_2 a_2 x_2 + \theta_2 b_2 - (1-\mu)p = 0. \tag{5}$$

When solving each Eq. (4)–(5), the optimal volume of digital services provided by the provider for each type of consumers is determined:

$$x_1 = \frac{p - \theta_1 b_1}{2\theta_1 a_1}, \tag{6}$$

$$x_2 = \frac{(1-\mu)p - \theta_2 b_2}{2\theta_2 a_2}. \tag{7}$$

The total demand for digital services in case of two types of the consumers will be described by a system of expressions:

$$D(p) = \begin{cases} \dfrac{\theta_2 a_2 \pi(p - \theta_1 b_1) + \theta_1 a_1(1-\pi)((1-\mu)p - \theta_2 b_2)}{2\theta_1\theta_2 a_1 a_2}, 0 \le p \le \theta_1 \\ (1-\pi)\cdot\dfrac{(1-\mu)p - \theta_2 b_2}{2\theta_2 a_2}, \theta_1 < p \le \theta_2 \\ 0, p > \theta_2 \end{cases}. \tag{8}$$

The finding of the price of the digital services conventional unit involves the substitution of the corresponding demand expression (8) in the provider's objective function (1) and its further differentiation at a price. Thus, if both markets are served the price will look like:

$$p_{2m} = \frac{\theta_2 a_2 \pi(\theta_1 b_1 + c) + \theta_1 a_1(1-\pi)(\theta_2 b_2 + c(1-\mu))}{2(\theta_2 a_2 \pi + \theta_1 a_1(1-\pi)(1-\mu))}. \tag{9}$$

After substitution of the price of the digital services conventional unit (9) in the target function of the digital technologies provider (1), we get the expression of the provider's profit from both types of consumers:

$$\Pi_{2m} = \frac{(\theta_1 a_1((1-\pi)(\theta_2 b_2 - c(1-\mu)) + \theta_2 a_2 \pi(\theta_1 b_1 - c))^2}{-8\theta_1 \theta_2 a_1 a_2(\theta_1 a_1(1-\mu)(1-\pi) + \theta_2 a_2 \pi)}. \tag{10}$$

The optimal price that should be set in case of serving only the second type of consumers will look like:

$$p_{1m} = \frac{\theta_2 b_2 + c(1-\mu)}{2(1-\mu)}. \tag{11}$$

The provider's profit from the second type consumers market will be as follows:

$$\Pi_{1m} = \frac{(1-\pi)(\theta_2 b_2 - c(1-\mu))^2}{-8\theta_2 a_2(1-\mu)}. \tag{12}$$

So, as a result, the optimal parameters of contracts between providers and consumers of digital technologies were found under symmetric information.

Let us consider the second version of the information structure - asymmetry. In the modified target function of the provider, we will include the following limitations, by which, in terms of the theory of contracts, the principle of identifying agent types (in our case, two types of consumers) is realized:

$$\Pi = \pi(p_1 - c)x_1 + (1-\pi)(p_2 - c)x_2 \rightarrow \max_{p_1, p_2, x_1, x_2} \tag{13}$$

$$\begin{cases} \theta_1(a_1 x_1^2 + b_1 x_1) - p_1 x_1 \geq 0 & (14) \\ \theta_2(a_2 x_2^2 + b_2 x_2) - p_2 x_2 \geq 0 & (15) \\ \theta_1(a_1 x_1^2 + b_1 x_1) - p_1 x_1 \geq \theta_1(a_2 x_2^2 + b_2 x_2) - p_2 x_2 & (16) \\ \theta_2(a_2 x_2^2 + b_2 x_2) - p_2 x_2 \geq \theta_2(a_1 x_1^2 + b_1 x_1) - p_1 x_1 & (17) \end{cases}.$$

Inequalities (14)–(15) are conditions for restricting participation, due to their introduction into the mechanism for determining optimal contract parameters, for both consumers it becomes unprofitable abandoning the transaction. The conditions for the restriction of compatibility on incentives (16)–(17) guarantee that both consumers will prefer contracts with individual parameters designed specifically for their type by obtaining greater utility in choosing their "own" contract rather than "someone else's". At the optimal point, the restriction (14) becomes equal. Restrictions (15)–(16) at the point of the optimal contract are ineffective and therefore they cannot be included in the optimization task. Then the expressions that describe the price per unit of digital services, which is individual for each type of consumer, will look like:

$$\Pi = \pi(p_1 - c)x_1 + (1-\pi)(p_2 - c)x_2 \rightarrow \max_{p_1, p_2, x_1, x_2} \tag{18}$$

$$\begin{cases} p_1 = \dfrac{\theta_1(a_1x_1^2 + b_1x_1)}{x_1} \\ p_2 = \dfrac{\theta_2(a_2x_2^2 + b_2x_2) - \theta_2(a_1x_1^2 + b_1x_1) + \theta_1(a_1x_1^2 + b_1x_1)}{x_2} \end{cases}. \tag{19}$$

Further, the prices found are substituted one by one in the target function of the provider (13), after which a first-order condition is made to determine the volume of digital services. We get the optimal offer of the volume of digital services for consumers of both types [10]:

$$x_1 = \frac{\pi(c - b_1\theta_1) + (1 - \pi)(\theta_2 b_1 - \theta_1 b_1)}{2(\pi\theta_1 a_1 - (1 - \pi)(\theta_2 a_1 - \theta_1 a_1))}, \tag{20}$$

$$x_2 = \frac{c - \theta_2 b_2}{2\theta_2 a_2}. \tag{21}$$

Thus, optimal values of contract parameters for digital technologies consumers under symmetric and asymmetric information were found.

4 Discussion

The formation and development of the global digital economy is considered in the works of Khitskov et al. [17], Silkina [21], etc. The impact of digital economy on various economic spheres is reflected in the works of such authors as Abdikeev [2], Bajgoric and Moon [7], Dementiev et al. [16], Veselovsky et al. [22]. Formation of digital economy in Russia is studied, for example, in works by Babkin et al. [6] and Romanova [19]. Contract as a tool of coordination of economic agents' actions and interests is described in the works of Aase [1], Araujo and Moreira [3], Ardalan et al. [4], Bossaerts et al. [15], Zhou et al. [23], etc. The problem of adverse selection is viewed for labor markets, insurance, credit market, cars. The authors consider the implementation of the agency theory principles for the space market participants' interaction [11, 12]. The study of contract theory is considered in works of Asimit et al. [5], Boonen [14], Quiggin and Chambers [18], etc.

5 Conclusion

We developed a model of contractual interaction between providers and consumers of digital technologies in space industry in terms of symmetric and asymmetric information. We found the optimal contract that maximizes provider's profit. The application of the "adverse selection" model allows taking into account the economic relations' uncertainty associated with information asymmetry. The model of the providers and consumers of digital technologies contract interaction allows to determine such target contract parameters as the volume of digital services and the price per unit of digital services. Such an approach to the contract policy formation helps to align interests of the space market participants and increase the efficiency and profit of the company that offers the contract.

References

1. Aase, K.K.: Optimal insurance policies in the presence of costs. Risks **5**(3), 46 (2017). https://doi.org/10.3390/risks5030046
2. Abdikeev, N.M.: Cognitive technologies of business processes management in digital economy. In: Proceedings of 2017 Tenth International Conference Management of Large-Scale System Development, MLSD 2017, pp. 1–3. Institute of Electrical and Electronics Engineers (IEEE), Moscow (2017). https://doi.org/10.1109/mlsd.2017.8109583
3. Araujo, A., Moreira, H.: Adverse selection problems without the Spence-Mirrlees condition. J. Econ. Theory **145**(3), 1113–1141 (2010). https://doi.org/10.1016/j.jet.2010.02.010
4. Ardalan, F., Almasi, N.A., Atasheneh, M.: Effects of contractor and employer's obligations in buy back contracts: case study of oil exporting country. Entrep. Sustain. Issues **5**(2), 345–356 (2017). https://doi.org/10.9770/jesi.2017.5.2(13)
5. Asimit, A.V., Bignozzi, V., Cheung, K.C., Hu, J., Kim, E.S.: Robust and Pareto optimality of insurance contracts. Eur. J. Oper. Res. **262**(2), 720–732 (2017). https://doi.org/10.1016/j.ejor.2017.04.029
6. Babkin, A.V., Burkaltseva, D.D., Vorobev, D.G., Kosten, Yu.N.: Formation of digital economy in Russia: essence, features, technical normalization, development problems. St. Petersburg State Polytech. Univ. J. Econ. **10**(3), 9–25 (2017). https://doi.org/10.18721/je.10301
7. Bajgoric, N., Moon, Y.B.: Business model for digital economy era: a framework based on the Churchman's theory of design integrity. Int. J. Bus. Syst. Res. **11**(3), 284–308 (2017). https://doi.org/10.1504/IJBSR.2017.085466
8. Balbás, A., Balbás, B., Balbás, R.: Good deals in markets with friction. Quant. Financ. **13**(6), 827–836 (2013). https://doi.org/10.1080/14697688.2013.780132
9. Balbás, A., Balbás, B., Balbás, R., Heras, A.: Optimal reinsurance under risk and uncertainty. Insur.: Math. Econ. **60**, 61–74 (2015). https://doi.org/10.1016/j.insmatheco.2014.11.001
10. Belyaeva, E.K.: The economic-mathematical models of the optimal contracts construction on the space services market. Manag. Econ. Syst.: Sci. Electron. J. **6**(100) (2017). http://uecs.ru/index.php?option=com_flexicontentandview=itemsandid=4461. Accessed 21 Oct 2018
11. Ivanov, D., Belyaeva, E., Kurilova, A., Popkova, E.: Models for the interaction between space services providers and manufacturers of space vehicles. Entrep. Sustain. Issues **5**(4), 846–857 (2018). https://doi.org/10.9770/jesi.2018.5.4(10)
12. Belyaeva, E., Bogatyrev, V., Klevtsov, D., Kurilova, A.: Formalization of the structure of the space market and models of interaction of its participants. Qual.-Access Success **19**(2), 76–87 (2018)
13. Bochinger, S.: Space as business (2017). https://www.roscosmos.ru/media/files/docs/2017/SpAsBus/1_bocindzer.euroconsult.-.roscosmos.1.ru.pdf. Accessed 21 Oct 2018
14. Boonen, T.J.: Nash equilibria of over-the-counter bargaining for insurance risk redistributions: The role of a regulator. Eur. J. Oper. Res. **250**(3), 955–965 (2016). https://doi.org/10.1016/j.ejor.2015.09.062
15. Bossaerts, P., Ghirardato, P., Guarnaschelli, S., Zame, W.R.: Ambiguity in asset markets: theory and experiment. Rev. Financ. Stud. **23**(4), 1325–1359 (2010). https://doi.org/10.1093/rfs/hhp106

16. Dementiev, V., Evsukov, S., Ustyuzhanin, V., Ustyuzhanina, E.: Impact of digital economy on the transformation of models of business activity organization within corporations. Espacios **38**(48), 10 (2017)
17. Khitskov, E.A., Veretekhina, S.V., Medvedeva, A.V., Mnatsakanyan, O.L., Shmakova, E.G., Kotenev, A.: Digital transformation of society: problems entering in the digital economy. Eur. J. Anal. Chem. **12**(5b), 855–873 (2017). https://doi.org/10.12973/ejac.2017.00216a
18. Quiggin, J., Chambers, R.G.: Bargaining power and efficiency in insurance contracts. Geneva Risk Insur. Rev. **34**, 47–73 (2009). https://doi.org/10.1057/grir.2008.15
19. Romanova, O.A.: Industrial policy priorities of Russia in the context of challenges of the fourth industrial revolution. Econ. Region **14**(2), 420–432 (2018). https://doi.org/10.17059/2018-2-7
20. SIA: State of the satellite industry report (2017). https://www.sia.org/wp-content/uploads/2017/07/SIA-SSIR-2017.pdf. Accessed 21 Oct 2018
21. Silkina, G.Yu.: Information and communication technologies in ensuring of innovative development. In: Soliman, K.S. (ed.) Proceedings of the 29th International Business Information Management Association Conference - Education Excellence and Innovation Management through Vision 2020: From Regional Development Sustainability to Global Economic Growth, pp. 1165–1176. International Business Information Management Association, Vienna (2017)
22. Veselovsky, M.Y., Pogodina, T.V., Ilyukhina, R.V., Sigunova, T.A., Kuzovleva, N.F.: Financial and economic mechanisms of promoting innovative activity in the context of the digital economy formation. Entrep. Sustain. Issues **5**(3), 672–681 (2018). https://doi.org/10.9770/jesi.2018.5.3(19)
23. Zhou, R., Li, J.S.H., Tan, K.S.: Modeling longevity risk transfers as nash bargaining problems: methodology and insights. Econ. Modell. **51**(C), 460–472 (2015). https://doi.org/10.1016/j.econmod.2015.08.019

The Impact of Digitalization on Employee Engagement

E. O. Gasparovich[1], E. V. Uskova[1](✉), and E. V. Dongauzer[2]

[1] Ural Federal University, Ekaterinburg, Russia
geol605@yandex.ru, uskova80@mail.ru
[2] Ural State Pedagogical University, Ekaterinburg, Russia
dong-elena@yandex.ru

Abstract. The article discusses the relationship between the concepts of "digitalization" and "employee engagement", describes the factors that affect the involvement of personnel. Based on the identified relationships, the authors suggest that the degree of digitalization of the enterprise can be identified as a separate factor that determines the degree of employee engagement. The current state of digitalization of the Russian economy at various levels: state, industry, and individual enterprise level is described. The authors conclude that at the moment digital transformation is the most effective way to increase the efficiency of individual enterprises. Further, the mechanism of influence of the digital transformation of the enterprise on the level of employee engagement is described from various sides. The authors also make suggestions on measures to increase employee engagement in the context of increasing digitalization of Russian enterprises.

Keywords: Digitalization · Digital transformation of the economy · Employee engagement · Factors of employee engagement · Impact of digitalization on employee engagement

1 Introduction

In the modern economy, there is a transition to universal digitalization. The age of digital information is coming to all spheres of society, including social and economic. The sphere of human resource management is no exception; it is also subject to the merging of digitization processes. The federal project "Personnel for the digital economy", adopted in the development of the program "Digital economy of the Russian Federation", provides for the implementation of measures to assist citizens in mastering the key competencies of the digital economy until the end of 2024 [10, 11]. In the era of digitalization authors discussed the need for training qualified personnel from various sides [7].

The implementation of the above-mentioned federal project involves the transformation of all levels of the education system, including the training and retraining of highly qualified personnel who meet the requirements for the key competencies of the digital economy. Accordingly, by 2024, graduates of educational institutions, as well as employees with developed digital economy competencies, who are focused on working

S. I. Ashmarina and V. V. Mantulenko (Eds.): IPM 2020, LNNS 161, pp. 143–150, 2021.
https://doi.org/10.1007/978-3-030-60926-9_20

in new economic and technological conditions, will "enter" the labor market. The research purpose is to study the impact of digitalization as a factor of employee engagement. The research hypothesis: digitalization as a factor will have a direct productive impact on the level of employee engagement and increase the efficiency of employee engagement.

2 Methodology

"It is important to attract very talented people, since the brightest employees create a much greater comparative value" - this statement, expressed by Michaels, Handfield-Jones, and Axelrod, is necessarily relevant at present [9]. At the same time, taking into account the guidelines for the development of digitalization, this statement may be changed and supplemented in the context of the topic under consideration: it is important to attract very talented people with developed digital economy competencies, since they will create much more.

However, the problem, in our opinion, is the question of whether Russian companies will be ready to fully "use" and continue to develop the "digital" competence of their employees and prospective candidates for employment by 2024. Companies with state participation, under the methodological control of the state, create and implement innovative development programs that ensure that their strategic development takes into account national digitalization projects. Other companies, in our opinion, should ensure a competitive advantage in the fight for talents with key components of the digital economy by switching to digitization standards.

Armstrong and Taylor express the opinion that increasing the employee engagement is possible by ensuring a favorable working environment (working conditions) [2]. In the context of the universal development of digital technologies, we can assume that the conditions for providing internal services to employees with the help of digitalization tools, that is, quickly and with the required quality, without excessive bureaucracy, also can be attributed to the basic factors of labor satisfaction. It is worth noting that motivation is the main component of professional interest and the reason for career growth and promotion of employees in the company [6]. Besides, the digital transformation of enterprises significantly affects the level of technological development of the enterprise, its reputation in society and in the labor market, as well as its corporate culture. That is why we believe that the degree of development of the company in the field of digitalization can be considered not just as one of the factors of labor satisfaction, but as a factor of increasing employee engagement. The research uses methods of statistical analysis, generalization, synthesis of scientific literature and personnel management practices on the research problem.

3 Results

Recently, such areas of development as computerization and informatization, were the most relevant for our society. The process of computerization involved the introduction of electronic computing technology in all sectors of production and economy and informatization is the use of information technologies to solve economic problems. What is the fundamental difference between the processes of digitalization and these areas? The first concept of "digitalization" was used in 1995 by the American computer scientist Nicholas Negroponte (University of Massachusetts).

For the purpose of considering the impact of digitalization on employees and their loyalty to the company, we will consider the concept of digitalization in a broader sense of the word, not just as the transition of information to digital form, but also as the process of transferring to the digital environment of business processes (activities and functions) previously carried out by personnel.

Digitalization is possible at various levels of the economy: at the state, industry, and individual company levels. At the state level, the program "Digital economy of the Russian Federation" was formed. At the moment, the state of digitalization of the domestic economy as a whole can be characterized as follows:

- there are some advantages in the digitalization field, but there is a number of problems and disadvantages in this area (a special feature of the Russian mentality-suspicious attitude to new technologies and methods of work, etc.; insufficient digital qualification of graduates of domestic educational institutions [15]; a system of remuneration established at domestic enterprises that equalizes the salary levels of employees of complex professions and employees with typical skills and abilities, etc.),
- opportunities for digital transformation exist in all industries,
- the digitization process faster and easier for the most vulnerable sectors related to information and telecommunication technologies (ICT),
- the lowest level of digitalization is observed in the production sector, due to the high inertia of industrial enterprises that use outdated technologies not only in the production itself, but also in the management sphere.

All subjects of the Russian Federation have planned to finance digital transformation in the budget. In the leading regions, large-scale digitalization programs are being implemented ("Smart city" in Moscow, "Informatization Program" in the Republic of Tatarstan, etc.). In the lagging regions of the Russian Federation, these are mainly regional acts on the creation of working groups to increase the level of digitalization, decrees of regional heads on the creation of research and technology parks, etc.

If we talk about industry digitalization, then, as noted earlier, the greatest development of digitalization has been in the areas related to ICT. These are services, telecommunications, software development, communications, trade, construction, medicine, the financial sector of the economy – insurance and banking segments, transport, media business, automotive industry, e-Commerce, energy, housing and public administration.

In general, certain parts of the required digital infrastructure are already functioning in Russia. For example, software in the field of cybersecurity has been developed (see Kaspersky Lab products), the rate of Internet penetration in households is increasing every year, there are very strong solutions in the field of industrial robotics and automation, etc.

In modern conditions, digital transformation is probably the main way to improve the efficiency of individual companies' economic activities. It should be borne in mind that digitalization processes for companies are global changes that cannot but affect the staff. The question of how to avoid "unrest" in organizations in connection with the ongoing digital transformation is topical now. Most importantly, on the contrary, make digitalization a factor that increases employee engagement.

Before studying the mechanism of influence of the process of digital transformation of enterprises on the degree of employee engagement, let's consider what the concept of "employee engagement" is. The concept of "engagement" was first used by Thompson. In his opinion, in order to fully convey business values to the consumer, the employee must not only understand these values, but also fully share them [14]. Prior to the concept of engagement, the terms "commitment" and "loyalty" to the enterprise were mainly used.

Later, the term "engagement" was used in social psychology as an assessment of the degree of adaptation of people to a certain role. Over time, the study of employee engagement has become used in the field of human resource management of enterprises to diagnose the attitude of personnel to their work and the organization as a whole. Thus, for the purposes of this article, we will refer to engagement as the emotional, intellectual, and physical state that motivates staff to perform their duties in the best possible way. Engagement seems to be the highest level of a person's commitment to the company, when an employee is heart-sick for their company and strives to work as best as possible.

Indicators of employee engagement are their desire to actively participate in all work processes (including decision-making), their enthusiasm and initiative, involvement in the social life of the team, etc. It's obvious "when employees are satisfied with the employee engagement initiatives practiced by their organizations, they are more engaged with their jobs and it subsequently, contributes to their job satisfaction as well" [4, p. 242].

Gallup researchers found a relationship between employee engagement and business indicators, such as company profitability, customer satisfaction, product (service) quality, employee turnover, productivity, and others [3]. Organizations with high levels of employee engagement were characterized by a low level of employee turnover, as well as low rates of labor discipline violations. In turn, the indicators of profitability and earning power ratio were significantly higher for enterprises with a high involvement of personnel.

The significant positive impact of employee engagement on the economic results of enterprises revealed in the course of research confirms the need for a thorough study of all factors that affect employee engagement. In order to increase the level of employee engagement, the employer must find sources of internal motivation for staff, that is, form such relationships in the organization and such working conditions that employees are internally motivated to demonstrate the best of what they are capable of. Material incentives are no longer the only means of motivation.

Thus, employee engagement depends on various factors:

- the degree of accessibility and openness of information (employees ' awareness) about the company, priority goals and tasks, decisions made, etc.,

– encouraging initiative and creativity on the part of the management,
– involvement of staff in decision-making (taking into account the opinions and suggestions of employees),
– compliance of the corporate culture of the enterprise with modern trends in the economy and labor market, motivational profile of employees,
– the content of the work of specific employees (the presence of a large volume of routine monotonous tasks significantly reduces the employee's involvement),
– the company's reputation in the eyes of various categories of the population (consumers, potential employees, shareholders, etc.),
– availability of opportunities and prospects for career and professional growth in the company, etc.,
– employer brand and internal PR [1].

4 Discussion

As already mentioned, we believe that the degree of digitalization of the enterprise should be considered as a separate factor in increasing employee engagement. Let's look at the degree influence mechanism of the organization's digitalization on employee engagement in more detail. First, digital transformation involves creating a more comfortable digital environment for employees in the services of employers, as well as increasing the reliability and openness of data about the company and its activities.

According to a study by VMware, the largest developer of virtualization software, the degree of comfort of the digital environment for enterprise staff is correlated with the rate of business growth [15]. For example, Sievert and Scholz note that "the use of internal social media fosters stronger employee engagement" [12, p. 894]. Organizations with a more favorable digital environment were characterized not only by higher business growth rates, but also by a more developed organizational culture, as well as a more stable workforce. Thus, an internal digital environment that is comfortable for employees is also an important condition for high employee engagement.

In addition to transforming employer services, digitalization implies a certain technological revolution. Undoubtedly, many employees will feel a sense of pride from working with the most modern technologies, from the awareness of the company's advanced position in technological and digital processes. At the same time, without proper attention, including to the processes of personnel training, a high degree of digitalization of the enterprise can turn into a factor of instability.

According to data for 2016, published at the Disposal of the Russian Federation Government, the value of the international digital economy and society index (I-DESI), Russia ranks 6th out of 16, yielding in the development of the digital economy to Australia, Canada, USA, Japan, The European Union, etc., while outstripping Turkey, China, Chile, Mexico and Brazil. This index takes into account a wide range of factors, but assigns significant importance to digital skills and human capital. In terms of human capital, Russia is significantly behind Australia, Iceland, New Zealand, South Korea, and others [8].

Currently, the share of employees whose responsibilities are directly related to the development and implementation of digital technologies is only 2% of the total employed population. This indicator corresponds to a low share of the digital economy in the GDP structure. At the same time, in the leading countries of digital transformation, this indicator is approximately 4%. Russia is also behind the leaders in terms of employment in science-intensive and high-tech industries. In Russia, this figure is estimated at 5.5%, and, for example, in Germany-it is close to 10% [5].

Thus, the actual number of employees who are “at the peak” of the digital transformation of the Russian economy at the moment is quite small, which partly explains the low level of staff involvement in Russian enterprises.

5 Conclusion

Based on the above, we can conclude that the level of digitalization will be one of the determining factors of loyalty and engagement of generation Z employees (these are people who were born since 1995 (according to some versions - since 2003) to the company, as they have digital skills from birth. There is a special investigation of variables of employer attractiveness important for generation Z. The Steckl, Simshäuser, and Niederberger consider that companies are united in their strategic and targeted communication of engagement for employee’s health [13].

Further, digitalization of the enterprise significantly improves the content of the work of many categories of employees, frees them from the need to perform a number of routine and monotonous functions. At the same time, we should not forget about the need for timely training and retraining of personnel, in particular when a number of professions, mainly requiring average qualifications, disappear due to digital transformation.

To begin with, it is obvious that organizations need to develop measures to attract and retain promising employees with a high level of qualification in the digital sphere. Within this direction, targeted training programs can be implemented in higher education institutions, the requirements for new employees entering the enterprise are adjusted, the criteria for including employees in the personnel reserve are revised, and opportunities for remote work are prepared.

This is not all. Let’s imagine a company that develops measures to attract and retain staff with key competencies of the digital economy, but at the same time does not create a convenient digital environment for staff. With this approach, none of the strategies for staff development will allow you to ensure the necessary level of staff involvement for the successful functioning of the company. At a minimum, the staff will not be satisfied with either the employer or the job, and will not be able to demonstrate the high productivity that they are capable of.

We are confident that the high level of digitalization of the company is, and in the future will be, the most significant argument for specialists in choosing a job. An employer with a highly developed digital economy structure will definitely have advantages in attracting talent, and, due to the high efficiency of employees with key “digital” competencies, higher staff involvement.

Another direction of measures for digitalization of the company to maintain staff involvement should be the development of the digital economy competencies of existing staff. When creating a digital environment in a company, in our opinion, one of the main priorities is professional retraining and advanced training of personnel in the field of the digital economy. First of all this direction is important in order to prevent the loss of key knowledge for the company, which is shared by staff with missing "digital" competencies. It is quite possible that in the near future we will talk about "reverse" or mutual mentoring, when young people who came to the company and received basic key competencies in the field of digital economy will train existing staff. At the same time, of course, they adopt practical knowledge and experience.

In order for employees of other generations to remain engaged, companies must provide them with timely training and professional development. In addition to the enterprises' own funds, there are also state education programs, for example, a program for citizens of pre-retirement age for the period up to 2024. This program provides free education (including programs "Network and system administration", "Internet of things", "Industrial automation") for citizens aged 51–55 years for women and 56–60 years for men.

Thus, in accordance with the goal, the influence of digitalization as a factor of staff involvement is studied. The research hypothesis was confirmed: digitalization as a factor has a direct productive impact on the level of employee engagement and improving the efficiency of personnel management. As a result, digital transformation is an important way to improve the efficiency of domestic enterprises in particular, and the Russian economy as a whole.

References

1. Akhmetshin, E.M., Ilyina, I.A., Kulibanova, V.V., Teor, T.R.: 'Employee engagement' management facilitates the recovery from crisis situations. In: Proceedings of the 2019 IEEE Communication Strategies in Digital Society Seminar, pp. 50–55. IEEE, New Kersey (2019)
2. Armstrong, M., Taylor, S.: Practice of Human Resource Management, 14th edn. Piter, Saint Petersburg (2018)
3. Beck, R., Harter, J.: Why good managers are so rare (2014). https://hbr.org/2014/03/why-good-managers-are-so-rare. Accessed 21 June 2020
4. Delina, G., Samuel, P.E.: A study on the interrelationships between employee engagement, employee engagement initiatives and job satisfaction. Int. J. Bus. Excell. **20**(2), 242–268 (2020)
5. McKinsey: Digital Russia: A new reality (2017). https://www.mckinsey.com/ru/~/media/McKinsey/Locations/Europe%20and%20Middle%20East/Russia/Our%20Insights/Digital%20Russia/Digital-Russia-report.pdf. Accessed 21 June 2020
6. Gasparovich, E., Tokareva, Y., Tokarev, A.: Gender trajectories of professional and personal development as a condition of effective management. In: Proceedings of the 3rd International Conference on Advances in Education and Social Sciences, pp. 992–999. International Organization Center of Academic Research, Istanbul (2017)
7. Ivanova, I.A., Odinaev, A.M., Pulyaeva, V.N., Gibadullin, A.A., Vlasov, A.V.: The transformation of human capital during the transition to a digital environment. J. Phys: Conf. Ser. **1515**(3), 032024 (2020)

8. Kolodyazhnyi, S., Korshunova, S., Perevozchikova, L.: Problems of adaptation of technical university graduates in the formation conditions of new economic reality. In: Solovev, D.B., Savaley, V.V., Bekker, A.T., Petukhov, V.I. (eds.) Proceeding of the International Science and Technology Conference "FarEastCon 2019". Smart Innovation, Systems and Technologies, vol. 172, pp. 61–68. Springer, Cham (2020)
9. Michaels, E., Handfield-Jones, H., Axelrod, E.: The War for Talent. Mann, Ivanov and Ferber, Moscow (2015)
10. Passport of the Federal project "Personnel for the digital economy": Minutes of the Presidium of the Government Commission on digital development, the use of information technologies to improve the quality of life and conditions for conducting business activities dated 28 May 2019, no. 9(2019). https://digital.gov.ru/uploaded/files/pasport-federalnogo-proekta-kadryi-dlya-tsifrovoj-ekonomiki.pdf. Accessed 21 June 2020
11. Program "Digital economy of the Russian Federation". Order of Russian Federation the Government of 28 July 2017 No. 1632-p. (2017). http://static.government.ru/media/files/9gFM4FHj4PsB79I5v7yLVuPgu4bvR7M0.pdf. Accessed 21 June 202
12. Sievert, H., Scholz, C.: Engaging employees in (at least partly) disengaged companies. Results of an interview survey within about 500 German corporations on the growing importance of digital engagement via internal social media. Public Relat. Rev. **43**, 894–903 (2017)
13. Steckl, M., Simshäuser, U., Niederberger, M.: Attraction of Generation Z: A quantitative study on the importance of health-related dimensions at work. Pravent. Gesundheitsforderung **14**(3), 212–217 (2020)
14. Thomson, K.: The Employee Revolution: Rise of Corporate Internal Marketing. Pitman Publishing, London (1990)
15. WMware: VMware research: the engine of digital transformation-employees (2018). https://www.vmware.com/ru/company/news/releases/2018/vmware-research-digital-transformation-engine-employees-041218.html. Accessed 21 June 2020

Professionally Oriented Training in the School-University System

M. D. Goryachev[1], A. A. Popov[2], and V. V. Mantulenko[3](✉)

[1] Samara State University of Social Sciences and Education, Samara, Russia
mdg16@mail.ru

[2] Primakova's Regional Gymnasium, Razdory, Odintsovo District, Moscow Region, Russia
popovsky_91@mail.ru

[3] Samara State University of Economics, Samara, Russia
mantoulenko@mail.ru

Abstract. The article investigates a problem of organizing professionally oriented training in the system "school-university". The authors identified approaches to the study of the continuity principle and defined their objectives; analyzed the essence of the continuity principle in professional-oriented education in the system "school-university"; specified scientific understanding of the concept of "readiness of high school graduates to continue their education in high school"; justified the structure of school graduates' readiness to continue their education in a higher education institution and determined the content of its components. The results of the experimental work allow making conclusions that the implemented organizational and pedagogical conditions contribute to increasing the level of school graduates' readiness to continue their education, successful adaptation in new conditions, and thus provide professionally-oriented training in the "school-university" system.

Keywords: Continuity principle · Professionally oriented education · School-university system · Readiness to continue education · Organizational and pedagogical conditions · Social adaptation

1 Introduction

Orientation of Russian education to the competence paradigm of continuous human education throughout life implementing successive basic educational programs and various additional educational programs, as well as the interpretation of the continuity of basic educational programs as the main factor in ensuring the unity of the educational space of the Russian Federation speak about the need for organizational and informative interrelation of professionally oriented education at each level of this system (preschool, primary, basic, secondary, vocational, higher) and stage (primary, basic, secondary (full), higher, postgraduate) [5, 9].

Professionally-oriented education in the system "school-university" is considered as a process of organizing professionally-oriented activity, in which its goals, objectives, forms, content and methods are aimed at the development of professionally-oriented

S. I. Ashmarina and V. V. Mantulenko (Eds.): IPM 2020, LNNS 161, pp. 151–161, 2021.
https://doi.org/10.1007/978-3-030-60926-9_21

knowledge, skills and social experience [1, 3]. The essence of professionally-oriented training is to integrate general education programs of special disciplines into the educational process in order to obtain additional professional knowledge for students and form professionally significant personal qualities.

2 Literature Review

In modern research, continuity is considered from the point of view of various aspects: as a principle, process and method of resolving the contradiction between the special tasks of higher education and the general nature of training in secondary schools [6]; pedagogical category [7]; the basis of relation between links of step-by-step teacher training, when the university, based on the school experience acquired by students, creates conditions for further accumulation of knowledge and skills, their improvement in professional activities [15]. Studying continuity as a transition from one educational stage to another, it is necessary to take into account differences in the development levels of subjects in the pedagogical process. Hodusov and Kononova analyzed the basics of continuity in teacher's training through the formation of its methodological culture [8]; Zhukovsky considers issues on continuity in the system "school-military university" [16]. The continuity of the educational process in the system "school-university" affects the content of training, its forms, methods, strategies of interaction of subjects of educational activity, changes in the personality of students, helps to organize and build phases of the educational process in a holistic hierarchy.

In our study, continuity is a principle that directs the teacher to ensure the continuity in the acquisition, accumulation and transformation of social experience by students. It (the principle) is implemented through scientific and practical justification of subjects' goals, content, activities in the educational process, the choice of forms, methods and means of teaching that allow students to provide an individual educational trajectory, adjust it at all educational levels (from general secondary (full) education to higher), taking into account personal and social preferences. Continuity in education is provided by professional self-determination of individuals – the most important element of the individual educational trajectory. In this regard, the question arises about the value "core" of continuity, the content of which, according to Esaulova, is formed by universal values: morality, spirituality, humanity. Continuity is an important methodological principle, which is based on the productive experience of previous generations, is designed to form a new system of values and knowledge with the leading role of education [4].

Considering the continuity principle in the context of the system of professionally-oriented entraining, note the need to analyze relations in the content of educational programs, to make sure that each subsequent program is built based on the previous one. In modern conditions, there are curricula with a large number of subjects that lead to overload of students. Therefore, it is necessary to exclude filling general education programs with material that makes the learning process formal. For students with different levels of training and personal maturity, it is necessary to provide for the variability of educational programs, which allows individualizing and differentiating the learning process. The continuity principle implies a link between stages of personality development, formed during the transition from one educational stage to another. This principle is possible at various stages between educational social roles and statuses, knowledge and skills, values, motives, norms of educational activity, "I-concepts" of the individual. A lot of scientists emphasize the multidimensional nature of continuity in the educational context. Smantser identifies four levels of continuity. The first (high) describes patterns of human development in the system of continuing education; the second (generally pedagogical) describes the basis of this system. The third level is a didactic principle that allows students to fully perceive disciplines they are learning. The fourth level (methodical) reveals features of continuity in the organization of training in a specific subject [15]. In our study, we consider continuity as a principle that reflects regularities of reforming the content structure of educational material and optimizing teaching methods which are aimed at overcoming contradictions of the linear-discrete nature of learning process and characterize changes in the ways of implementing these regularities, depending on educational goals, up-bringing and development of students. It is important to note that continuity in modern professional education should be interpreted as a pedagogical principle and phenomenon.

3 Methodology

The analysis of scientific literature allows us to attribute continuity to the number of terms included in the categorical apparatus of pedagogy, as well as to the main pedagogical principles of the educational process organization. In our study, it seems appropriate to formulate a definition of professionally-oriented education in the "school-university" system based on the continuity principle. The recipient – the person to whom general education training is directed, is the trainee (student): in this study, they are school children and university students. When determining continuity based on the description of the object of activity, the organization of training, educational activity, educational subject, and the personality of a student are characterized. Based on this, it seems logical to consider the content and organization of general and professional education as an object of continuity. The principle of continuity affects the content, forms, methods of teaching, strategies of subjects' interaction in the framework of educational activity, changes in students' personality, contributes to the systematization and construction of stages of the educational process in an integral hierarchy.

The educational system, built on the continuity principle, gives teachers an opportunity to choose the most convenient and adequate means, forms and methods of teaching that contribute to the disclosure of the student's personality. Education based on the continuity principle provides a chance to adjust the educational program at all its levels. The organization of professionally-oriented education in the "school-university" system implies the creation of conditions for the transition of an individual from one educational organization to another, one of which is psychological and pedagogical support for students in a new educational environment. It means inclusion of the individual in socially significant activities, organization of external influence on students stimulating their internal activity, ensuring adaptation in the team. The implementation of professionally-oriented training in the school-university system based on the continuity principle implies some interrelated stages, through which the teacher creates appropriate organizational and pedagogical conditions within the socio-pedagogical aspect.

In this work, we consider the first stage that includes two areas contributing to the implementation of professionally-oriented education in the school-university system. The first is working with the teaching staff. To manage the student's progress from secondary general education to higher education, it is necessary to review the work of teachers. It becomes mandatory to familiarize them with the logic and laws of professionally-oriented education in the "school-university" system based on the continuity principle: a set of forms, methods, means of its organization, provide them with specialized knowledge about interaction ways and organization tools within the study group or between individual students.

The second direction is the following work with students: recruitment of students of secondary education organizations for pre-professional training courses and summer programs; diagnostics of students' adaptation to the environment of a new educational organization, their abilities, professional opportunities and interests; development of students' cognitive interest and need to enrich their social experience; formation of a positive attitude to activities; inclusion of students in various types of activities at educational organizations; determination of the content of students' and teachers' activities in terms of identifying a set of methods, forms, and means of teaching that reflect the sequence of stages of the educational process. This complex ensures the integrity of training and education, provides a smooth transition from one educational stage to another; determination of resources of the educational organization which are necessary for the implementation of professionally-oriented training in the "school-university" system (availability of places and time for educational and extracurricular activities).

It should be noted that this direction is implemented through individual and group work. Special attention should be paid to the following forms: consulting school graduates in order to help them to choose a profession; conversations, discussions that form knowledge, skills, motives, professional interests, inclinations, and a positive attitude to future professional activities. These forms of work with students provide professionally-oriented education in the "school-university" system, and vocational guidance work carried out with students will form their idea of own qualities, norms and rules of education at the higher educational institution and will act as an organizational and pedagogical condition that is implemented by the teacher in the educational environment.

Let us consider the following actions of the second direction which in the socio-pedagogical aspect highlight a number of organizational and pedagogical conditions of professionally oriented training in the system "school-university": development of schoolchildren' s cognitive interest and the need to enrich their social experience, the formation of positive attitudes towards educational activities, inclusion of students in various activities at educational institutions, diagnostics of pupils' adaptation to the new environment of the educational organization, their abilities, professional opportunities and interests. The formation level of cognitive interest, as an integral characteristic of the personality, expressing its positive and emotional attitude to the cognitive process, to its emotional, intellectual and communicative-volitional capabilities, determines a lot of characteristics of students and affects the motivation to learn. Cognitive interest plays a special role in determining the overall orientation of the student's personality and is revealed through the overall level of its development, character, relation to other motives, its connection with further life plans and forms a positive attitude to the activity.

4 Results

In our study, we consider a special type of cognitive interest that affects the student's educational motivation – interest in an academic subject, which is defined as the orientation of an individual to the process of mastering knowledge, skills and abilities in a particular area. It should be noted that the interest in the subject is determined by specifics of educational activities, especially at the middle and senior school levels.

The set of formed cognitive interests of a student determines the educational motivation – the desire to get new knowledge, and its stimulation should be considered as one of the organizational and pedagogical conditions when considering the socio-pedagogical aspect of professionally-oriented training in the "school-university" system. The inclusion of students in various types of activities on the basis of educational organizations creates opportunities for creative development and self-realization of each individual, which involves subject-subject interaction (teacher-student), cooperation and active creative activity in the classroom.

Attention should be paid to the adaptation of students to conditions of higher education institutions, since the creation of appropriate conditions by the teacher implies adaptation to them. Difficulties of freshmen in adapting to university conditions are associated not with the poor quality of school training, but with the fact that the student has not formed the following qualities: readiness to learn, the ability to independently acquire knowledge, self-control, individual cognitive characteristics, time management. Orientation of students to master skills of independent study and rational organization of self-training can be considered as one of the organizational and pedagogical conditions within the socio-pedagogical aspect of professionally-oriented education in the "school-university" system.

The research of theoretical aspects of students' adaptation to conditions of higher education allowed us to conclude that the most urgent problem is social adaptation. It should be considered as a continuous communication process in which students adapt to each other, develop ways to interact with the social environment. At the same time, the ultimate goal is to integrate the individual into the system of social relations, including the consolidation and development of interpersonal communication skills, the development of behavior patterns that reflect the system of norms and values and determine behavior in this social group. Social adaptation is a process and result of an individual's active adaptation to conditions of social environment – acceptance of its physical and social requirements in the absence of internal discomfort. Communication is a significant factor of a student's social adaptation at the stage of inclusion in a new social environment. It ensures that the individual learns certain patterns of behavior adopted in a given social group, follows them, and is controlled by adults. In addition, communication is a part of joint educational activities, meeting the need for informal interaction with peers, recognition and understanding, that is so necessary for young people.

A significant factor for the success of social adaptation is the development of a unified university policy in the above areas, including psychological assistance and the activation of student self-government. This will not only speed up the solution of the main tasks, but also interest young people in the social life of a particular educational institution. This will have a beneficial effect on the development of communication skills and work with people, and form an active life position. Therefore, in our opinion, it is necessary to diagnose the adaptation of students to the environment of a new educational organization at school in order to determine the development of interests, requests, inclinations, and predict the further social development of the student's personality. The implementation of professionally-oriented education in the "school-university" system based on the continuity principle ensures the creation of organizational and pedagogical conditions allowing the student to move from school to the social environment of the higher education institution and successfully adapt to it.

An important aspect of studying professionally-oriented education in the school-university system based on the continuity principle and allocation of organizational and pedagogical conditions of professionally-oriented training in the school-university system is a set of pedagogical approaches used in the process of its (professionally-oriented education) implementation. These are system-based and person-centered approaches. As characteristic of the system-based approach in the framework of our research, we consider the organization of professionally-oriented training in the "school-university" system based on the continuity principle as a system with its own structure, and its entry into a higher-order system.

According to Serikov, the personality-oriented approach should be considered as a construction of a special pedagogical process with its own goals, tasks, content, technologies, which orients individuals to (self-)development of their personal qualities [14]. The person-centered approach to the study of students' motivation allows taking into account individual characteristics of each student, implies learning individualization (the teacher adapts the pedagogical impact to the student, modifies the educational process in accordance with particular specific characteristics). In our study, the student from the position of a person-centered approach is presented as an active person who has the opportunity to choose in the educational process, as well as the own system of views. Training in the framework of a person-centered approach provides an increase in the level of motivation for learning, self-development and development of the student's personality based on his individual aptitudes, abilities, interests, value orientations and subjective experience; allows the student to carry out cognitive educational activities.

An important factor determining the success of training includes motives and interests. Therefore, a necessary organizational and pedagogical condition for professionally-oriented education in the "school-university" system is to stimulate the educational motivation of students, which implies the desire to get new knowledge. The success of educational activities is impossible without students' activity, so managing the motivational sphere of the individual, awakening the interest in obtaining knowledge, increasing the responsibility and self-discipline through self-organization in the process of social adaptation are equivalent tasks of university training.

According to some experts, the relevance of forming students' motivation for educational activities at the stage of the future profession choice is explained by the constant change in the content of education, the problem of forming students' needs for independent acquisition of knowledge and the development of an active life position [11]. Teachers often face a lack of focus and regularity in the educational activities of students and the need to stimulate these qualities. A student who is properly attuned to educational activities should feel the need to get an education, and be aware of the importance and necessity of future professional growth.

It should be noted that high motivation in training is the basis for its success, and therefore contributes to the implementation of professionally-oriented education in the "school-university" system based on the continuity principle. As part of a person-centered approach, the teacher should apply various ways to increase students' motivation in the own pedagogical practice. A variety of used forms, a positive predisposition of a teacher, a favorable psychological atmosphere in the classroom can form a stable motivation for students to acquire knowledge.

The person-centered approach to the formation of students' skills of independent educational work and ways of rational organization of self-training is revealed through the continuity as a principle of organizing independent work of secondary school students and university students. Independent educational work in the framework of this approach is to activate the creative activity of students by the teacher, since the teacher affects the formation of professional qualities of the individual and builds this work taking into account the individual characteristics, capabilities and interests of their students. Various types of independent educational work (teaching, training, drilling, repeating, developing, creative and control) are used by teachers in the process of curricular and extracurricular work.

The analysis of scientific approaches to the organization of professionally-oriented education in the "school-university" system allows us to make a general conclusion that we are talking about building an educational process aimed at forming the readiness of school graduates to continue their education at the higher educational institution. It (readiness) includes professional self-determination, qualities and personality traits that will ensure successful training at the university for a school graduate. Readiness for a certain activity is an indicator of the educational level of the individual and the main condition for his/her productive activity, as it implies meaningful and instrumental mastering of the activity structure– acceptance of the goal, possession of skills, motivation, etc.

According to Saveleva, school graduates have a low level of readiness to master university educational programs. Many people are characterized by a low level of cognitive activity, poor general educational and intellectual skills, methods of independent cognitive activity, insufficient motivation to learn, real interest in the chosen profession [13]. In addition, not everyone has such formed qualities as purposefulness, independence, determination, sociability, etc. Readiness is a state of personality characterized by dynamism and staging. Under certain conditions, "readiness-state" becomes "readiness-quality". In the scientific literature, readiness is usually considered as a state that a person experiences when starting a familiar, non-demanding job.

We defined the readiness of school graduates to continue their education at higher education institutions as an integral personal formation, which is a disposition to acquire, improve and use the received subject and meta-subject knowledge, skills and abilities in further education. It (readiness) is an indicator of professionally-oriented education in the school-university system. We identified components of school graduates' readiness to continue their education in higher education (activity, motivational and cognitive). We also determined a structure of such readiness, including criteria and indicators tracked by results of student's activity, which allowed us to mathematically calculate the formation of the considered personal formation and present it as a system of levels: low, medium, high. The structure of readiness of school graduates to continue their education at higher education institutions (components and criteria) can be clearly presented in the form of Table 1.

Table 1. Structure of graduates' readiness to continue their education at the university

Components	Criteria	Indicators
Motivational	Students' motivation and emotional attitude to learning	- personal meaningfulness of learning, the ability to set goals; - cognitive and social motives, external and internal motivation, the desire to achieve success and to avoid failure
Cognitive	Subject knowledge, ways, methods, techniques necessary for studying at the university	The student knows: - basic concepts of the main sections of the school course and their properties; - system of functional concepts, functional language and symbolism; - visual AIDS for interpretation, argumentation, and illustration; - studied concepts, results, methods for solving practical problems and problems from related disciplines
Activity	Cognitive and regulatory universal educational actions (UEA) required for studying at the University	The student is able to: - to set a new learning task (goal setting); - to determine the sequence of intermediate objectives in the light of the final result; - to make a plan and sequence of actions (planning); - to compare methods and a result of the action with the standard in order to detect deviations and differences (control); - to make necessary adjustments and additions to the plan and methods of action in case of discrepancies with the standard; - to highlight and be aware of what has already been learned and what is still to be learned; - to be aware of the quality and level of the material assimilation (assessment); - to formulate educational goals; - to search and highlight information; - to identify essential and non-essential features in the analyzed concept; - to select grounds and criteria for comparing and classifying objects; - to establish a causal connection; - to build a logical chain of reasoning, evidence; - to put forward and justify hypotheses; - to solve problems of creative and exploratory nature

Source: authors.

Means of professionally-oriented education in the system “school-university”, which are implemented through organizational-pedagogical conditions, forms and readiness of school graduates to continue their education at higher education institutions, are: the variety of lesson types, business games, learning-research activity of students, group work, pair work, various types of diagnostic tasks (tasks “error search”, “deliberate error”, a task to find information in the proposed sources, “labyrinths”, organizing chain mapping-supports, working with cards, tables, drawing and recognizing diagrams), mutual control and mutual verification.

Primary diagnostics revealed a low formation level of cognitive (author’s diagnostic work in mathematics in the format of a unified state exam), motivational (modified questionnaire of Ch.D. Spielberger) [12] and activity (author’s diagnostic tasks) components in the structure of school graduates’ readiness to continue their education. Based on the data of primary diagnostics, a general conclusion is made: the majority of students showed a low or average level of readiness to continue their education at the university, which indicates a weak need of students to continue their education, the lack of clearly expressed values and motives for educational activities, and a limited assessment of their personal qualities.

The results of the forming experiment showed an increase in values for all indicators. The dynamics of indicators in each component is noted: in the cognitive component – from knowledge of the basic concepts of the main sections of the school course, the system of functional concepts, functional language and symbols to the ability to apply the studied concepts, results, methods for solving practical problems and problems from related disciplines, work with information (search, analysis, interpretation), define concepts, establish analogies, create generalizations, classify, establish cause-and-effect relations, build logical reasoning, inference (deductive, inductive, by analogy) and draw conclusions (high results in 85% of students); in the motivational component – from the personal meaningfulness of learning and the ability to set goals to the predominance of cognitive or social motives, external and internal motivation, the desire to achieve success or avoid failure, the implementation of educational motives in behavior (high results in 80% of students); in the activity component – from the weak formation of the tracked skill within the framework of formation of cognitive and regulatory UEAs (some mastered operations do not pass into a complete action) to the demonstration of confident and effective use of this skill by students without extra effort (high results in 70% of students).

5 Conclusion

A comparison of the data obtained during the initial diagnosis and the forming experiment showed a positive dynamics of growth in the values of all indicators. This fact is confirmed by statistical analysis conducted to compare results of the experimental and control groups in the framework of cognitive and motivational components based on the results of the forming experiment. The obtained empirical value $t = 3{,}3$ ($t = 3{,}7$), respectively, based on the results of repeated diagnostics aimed at identifying the level of mathematical training, is in the zone of significance (at $p \leq 0{,}05$, TCR = 2) (according to Student’s t-test [2]). Thus, the implementation of organizational and pedagogical conditions in the learning process provides an increase in indicators of the components formation of participants' readiness in the experimental group to continue their education at the university (the results are shown in Table 2).

The last stage of the experimental work (analytical) (method of socio-psychological adaptation of Rogers and Diamond [10]) showed that students with high and medium levels of readiness are more successfully adapted to higher education conditions, and therefore, a professionally-oriented education is provided in the "school-university" system. From 54 participants in the experimental group who showed a high (50 students) and average (4 students) level of readiness, almost all demonstrated the appropriate level of adaptation at the university (52 first-year students with a high level of adaptation, which is 96% of the total number of respondents).

Table 2. Dynamics of school graduates' readiness to continue their education at higher education institutions (experimental group) *

Levels	Readiness components (%)					
	A	M	C	A1	M1	C1
Low	33	0	25	0	0	0
Average	45	80	55	30	20	15
High	22	20	20	70	80	85

Source: authors.

*(Abbreviations indicate the readiness components at the beginning (letter) and at the end (letter + unit) of the experiment: (A (A1) – activity; M (M1) – motivational; C (C1) – cognitive).

The results of the experimental work allow us to conclude that the implemented organizational and pedagogical conditions contribute to increasing the readiness level of school graduates to continue their education at the university, successful adaptation at the higher education institution.

References

1. Aimicheva, G., Kopeyev, Z., Ordabayeva, Z., Tokzhigitova, N., Akimova, S.: A spiral model teaching mobile application development in terms of the continuity principle in school and university education. Educ. Inf. Technol. **25**(3), 1875–1889 (2019). https://doi.org/10.1007/s10639-019-10051-z
2. Azevich, A.I., Alekseeva, S.I.: Teaching experiment and means of descriptive statistics. Bull. Peoples' Friendship Univ. Russia Ser.: Inf. Educ. **2**, 57–65 (2016)
3. Cooper, B., Cowie, B., Campbell, C.: A New Zealand collaborative university–school partnership: applying the STEPS framework. In: Hobbs, L., Campbell, C., Jones, M. (eds.) School-based Partnerships in Teacher Education, pp. 209–218. Springer, Singapore (2018). https://doi.org/10.1007/978-981-13-1795-8_11

4. Esaulova, M.B.: Continuity as a condition for the effectiveness of professional training in higher education for college graduates. In: Bugashev, S.I., Minin, A.S. (eds.) Proceedings of the International Scientific Conference "Humanities in Modern Higher Education as the Basis of Intercultural Interaction", 10 December 2018, pp. 68–70. Publishing house of SPbSU of industrial technologies and design, Saint Petersburg (2018)
5. Federal law "On education in the Russian Federation" of 29 December 2012 N 273-FZ (2012). http://www.consultant.ru/document/cons_doc_LAW_140174/. Accessed 13 July 2020
6. Godnik, S.M.: Methodological paradoxes in studying the concept of object-subject transformation of a personality. Bull. Voronezh State Univ. Ser.: Probl. High. Educ. **2**, 77–83 (2015)
7. Guschina, K.G., Kustov, Y.A., Statsuk, S.V.: Continuity principle as a backbone factor of the continuous nourishing of the quality culture in the students in the system "school-university". Bull. Volga State Univ. Named After V.N. Tatishchev **2**(18), 79–88 (2015)
8. Hodusov, A.N., Kononova, S.A.: The development of vocational education in the institutions of professional education. Provinc. Sci. Notes **1**(9), 11–15 (2019)
9. Kandrashina, E.A., Izmailov, A.M., Mantulenko, V.V., Mirzayev, N.S.: Theoretical aspects of gaps formation in the system of Russian higher education. In: Mantulenko, V.V. (ed.) Proceedings of the I International Scientific Conference "Global Challenges and Prospects of the Modern Economic Development", The European Proceedings of Social and Behavioural Sciences (GCPMED 2018), vol. 57, pp. 1135–1142. Future Academy, London (2019). https://doi.org/10.15405/epsbs.2019.03.114
10. Okonechnikova, L.V., Wilhelm, A.M., Wilhelm, A.V.: Study of the relationship of socio-psychological adaptation and emotional intelligence of first-year students. Pedagog. Educ. Russia **7**, 154–160 (2019)
11. Pryadyokho, A.N., Pryadyokho, A.A.: Motives of learning as means of stimulating pupils' cognitive activity. Bull. Bryansk State Univ. **1**, 55–60 (2013)
12. Raigorodsky, D.Ya. (ed.): Practical Psychodiagnostics. Methods and Tests. Publishing House "BAHRAKH-M", Samara (2001)
13. Saveleva, N.N.: Design of the system of adaptation of first-year students to higher education (direction-pedagogy). Ph.D. thesis, Vladivostok (2007)
14. Serikov, V.V.: Experience of scientific and pedagogical school of personal development education. Bull. Voronezh State Univ. Ser.: Probl. High. Educ. **2**, 11–18 (2018)
15. Smantser, A.P.: Theory and practice of realisation of continuity of training pupils and students. BSU, Minsk (2011)
16. Zhukovsky, V.P.: Continuity of educational activities in the system "school-military university". Doctor thesis, Saratov (1999)

Features of Personnel Training/Retraining in the Conditions of Digital Transformation

T. V. Gromova(✉)

Samara State University of Economics, Samara, Russia
gromova73@yandex.ru

Abstract. Today, no one disputes about the prerequisites of digital transformation, because it has already been happening. It would be strange to assume that businesses will be able to work according to the old model, with traditional processes and the same efficiency. Everybody is aware education to be an essential part of the economy, and the degree of education in society is compared with parameters of economic development and welfare of people. The purpose of the research is to improve effectiveness of personnel training/retraining for work in the conditions of digital transformation. The tasks are: to analyze the impact of digital transformation on labor market; justify the need for training/ retraining of personnel for these activities in the context of digital learning; consider stages of training personnel using distance education technologies; analyze participants' interaction in the educational process. Empirical (observation, comparison), theoretical (analysis, synthesis, modeling) and diagnostic methods (of the condition and causes) research methods were used. The article analyzes the impact of digital transformation on labor market and the trends of the latter's development; the necessity of staff retraining for activities in digital transformation; the stages of personnel training, and the interaction of educational process participants in distance education.

Keywords: Digital transformation · Digitalization of education · Distance learning · Labor market · Staff retraining

1 Introduction

Digitalization, digital transformation (DT) — the tendency has at present been resolutely recognized by representatives of global business in various spheres and is the most essential focus for business improvement at the moment. The opportunity of this trend is merely gigantic. At present, in line with certain estimations, more than 60% of the global biggest businesses have now been operating on their DT policy [5]. Furthermore, there has been a great upsurge in amount of requests for such services from Russian companies over the past year.

The main obstacle to successful business transformation is the lack of readiness of companies and personnel - lack of motivation and often lack of experience. In the process of business transformation, the main components also change: the approach to conducting business changes – businesses move from a designed marketing model to a customer-oriented system, methodologies are reborning from project management to product

S. I. Ashmarina and V. V. Mantulenko (Eds.): IPM 2020, LNNS 161, pp. 162–169, 2021.
https://doi.org/10.1007/978-3-030-60926-9_22

management, the organizational structure is being transformed, and there is a need for continuous development, training and professional development of employees [9].

Along with new technologies, new positions are being created in companies, such as Digital Director, Data Director, Data processing Specialist, Transformation Director, and others. The structure of companies is also changing – there are completely new departments, for example, the Transformation Team, which includes specialists from various areas of business activity.

Digital transformation is no longer a service of consulting companies, but an inescapable technological process that global businesses are going through to adapt to the latest realities of the digital economy.

The Russian economy will also undergo major transformations in the coming years. Many businesses and firms will leave the market, others will lose their profits, and completely new businesses will take their place. The hardest hit due to the introduction of new technologies will be the labor market. There are now about 7 billion people living in the world, of which about 3.5 billion are able-bodied persons older than 15 years. Simultaneously, there are only 1.2 billion jobs available [3].

Humanity is on the verge of the highest unemployment rate in its history. To stay alive, everyone will have to adapt to the new conditions: the state, private enterprises and employees. At the start of 2019, the level of joblessness was approximately 5.2% in Russia that is roughly 4 million people. It might give the impression the rate of unemployment to be rather little. Nevertheless, it is used merely to officially recorded people without jobs. Indices of self-employed remain more significant. In keeping with numerous estimations, from 15 to 22 million Russian people catch salaried jobs in so called "grey", i.e. illegal segment [3]. And since the staff will have to work in the conditions of digital transformation, it would be logical to train/retrain them in the conditions of digitalization with the use of distance and communication technologies. In this article, we will focus on some aspects of staff (hereinafter referred to as listeners/trainees) training, in particular at the stages of training, the interaction of participants in the educational process in distance learning (DL).

2 Methodology

The investigation basis consists of: scientific papers; psychosomatic literature; theories and reports at scientific conferences, e-resources on the subject.

Empirical (observation, comparison), theoretical (analysis, synthesis, modeling) and diagnostic methods (diagnostic analysis of the condition and causes, in particular the impact of digital transformation on the labor market and its development trends) research methods were used in the study.

Key areas of digital transformation include the following:

- customer focus,
- partnership and collaboration,
- use of data,
- search and implementation of innovations,
- value management,
- strategy of personnel management and digital culture [8].

As it was noted earlier, it is advisable to conduct training/retraining of personnel for activities in the context of digital transformation in the context of digital learning using distance and info communication technologies (ICT) [6]. The use of ICT in higher education system allows, first, to ensure equal access to educational resources (electronic libraries, cases, educational platforms) regardless of the listener's place of residence, health status, etc. secondly, it forms the ability of listeners to build independent educational strategies, and also contributes to the development of skills for independent work with information, which increases the analytical potential of listeners [1].

3 Results

Digital transformation in the economy and business directly depends on digital transformation in higher education [7]. In this section of the article will focus on some of the characteristics of the personnel training process (hereinafter listeners/trainees/learners), in particular at the stages of learning using ICT, interaction of participants of educational process in distance learning (DL).

3.1 Stages of Training

Distance learning, carried out with the help of ICT, is a process of mediated communication between teachers and trainees, the goal of which is organized gaining of information, competences and capacities specified by academic standards [4].

The procedure of cooperative work of trainees and trainers in the DL is offered in the shape of a sample comprising five basic stages: 1) acquaintanceship, interaction and incentive; 2) information exchange; 3) understanding; 4) application; 5) knowledge building and improvement.

First, let's provide the general characteristics of the model. The necessary prerequisite for effective activity of listeners in DL system is their motivation, the desire to receive education (advance their abilities) via DL (the first stage). This phase relates as well to the acquaintanceship of trainees and trainers. Since that time, the trainer starts escorting trainees: defining their requirements, problems, and degree of their readiness. During the second phase, trainees learn the data delivered by the DL teacher or autonomously seek out the content concerning the training course. Didactic backing is delivered via advising at this phase. The third and fourth phases contain activity on the integration of the data acquired. At the fifth phase, the learned information and skills are summarized.

During every phase, trainees require definite technical abilities. Every phase needs definite abilities from DL trainer as well. The interactivity scale shows the degree of interactivity that can be expected at each stage. When organizing practical classes, one must keep in mind that during the first stage trainees interact, as a rule, only with a few listeners. Following the second phase, the amount of listeners that they cooperate with and periodicity of interactions gradually increases, although the fifth stage involves a large share of independent work of listeners.

Let's look at features of every phase in in details.

Stage 1: Acquaintanceship, interaction and incentive. Through the learning process, and particularly at the start of it, when the listener is faced with possibly unfamiliar systems and modes of activity, it is important for him/her to be aware of the support from the DL teacher. During the first phase, incentive is very significant. The result of teaching mostly relies on the trainees' view of training forecasts, aims, and probable challenges. The mission of the trainer is to support trainees get used to it and propose assistance. It is here that the trainer encourages team interrelation and reciprocal help groups to be structured on his initiative The trainer ought to provide the most comprehensive data on the training course on the whole and its details. It is essential to clarify the importance of homemade and last tasks in the course, the needs for them. That is the trainer should "eliminate" probable listeners' inquiries in advance. The first phase stops after trainees become acquainted with the group and begin e-mailing primary letters to other trainees and the teacher.

Stage 2: Information exchange. Listeners get used to a new learning environment; a society of people working on common tasks arises. When listeners master this stage, become familiar with the etiquette of communication and technology, they are ready to perceive and search for information. Speaking about communication at this stage, it is worth noting that the teacher should make every effort to ensure that the listeners have a desire to communicate. Group discussions often show how quickly and smoothly the group works, the level of mutual understanding of participants. At this stage the teacher is to encourage the development of mutual respect among listeners, to eliminate imminent conflict, to help listeners to engage in better communication.

One of the demands of support in DL states that the course states that the course and training manuals must be self-sufficient in terms of information availability. Often, the course content is provided to listeners in the form of well-made and ready printed materials, in text format on the internet. Listeners independently master theoretic resources, advising with the trainer if required. The trainer, in turn, ought to inspire trainees to seek out extra data, search for essential connections and resources. Abilities to seek for material, choose the essential one at this phase may not be quite established, so the trainer ought to manage this skill. At the start of this phase, the trainer should consider means and practices of group-discussions, partaking in computer sessions, etc.

Stage 3: Understanding. Trainees recreate the gained knowledge and spread it to answer common learning problems, and they start cooperate more dynamically. To realize this aim, tutorials are considered to be efficient, where debates, role-playing games, and case studies should be planned under the supervision of a trainer and numerous active learning means which may eliminate holes in the data of the theory should be used.

At this stage, it is important that the teacher correctly manages the listeners' learning activities. It may be justified to attract listeners as "experts" - different people have different experience, knowledge from different fields, and this experience, vision and ideas can contribute to better assimilation of theoretical knowledge, problem solving, etc. The teacher should direct listeners to analyze their own activities (Which sections caused the greatest difficulties and why? What skills do the listeners lack?). It is likely that this stage will be held in full-time mode, at the so-called installation session.

Stage 4: Application. Trainees solve problems connected with using their knowledge to solve non-standard problems. They verbalize their understanding of the theme being investigated. What they study is not so much a product as a mental original practice, comprising the give-and-take of views, their evaluation, and the skill to fix them through debates or pair work. Learning is not mere active, it is interactive.

Working in pairs and groups allows solving tasks, considering different points of view, model, coming to a common opinion, etc. The listener is involved in the discussion and interaction with other listeners and experts in the communication process. At this stage a lot depends on teacher's behavior – controlling or participative style, dependence on/her own experience, etc. A complete variety of active educating means should be applied at this phase. Reflection of activity is also very essential.

Stage 5: Knowledge building and improvement. At this stage, there is a kind of generalization of the knowledge and skills acquired by listeners. They are presented the chance to alter their situation, improve their attitude and use it in their future work as specialists. Knowledge building happens when trainees study a task, arrive at some view, argue it, and estimate their positions, i.e. participate in reflection.

There may be a situation when listeners will need less support and assistance from a teacher-consultant. During the phase, the trainer does not offer prepared answers; listeners become more authors, not just information carriers. At this stage, the ability to think critically and evaluate the results of independent activity is formed. Listeners and teachers can use the so-called constructivist approach, which encourages listeners to think independently and explore the process of knowledge accumulation [4].

During the 5th stage reflection also plays a significant role: analysis and results of training course and its efficiency, the influence of technology on the educational process. Tutors, in turn, ought to be prepared to provide trainees with activities and tasks that improve critical reflecting (for instance, commenting other listeners' written work; defending final projects, the topics and conditions of which were agreed at previous stages). During the phase, it is worthwhile to perform a business play as one of the laborious types of active training. During the 4th and 5th phases, there is no teaching in the form of lessons. DL offers listeners an opportunity to autonomously search and analyze data, build their knowledge. The teacher is supposed to promptly react, summarize, assess, that is supervise the learning work of listeners. If in the process of the whole training trainees are given ongoing adequate assistance from the tutor, then the process of transferring from one phase to another and reaching more advanced stages will be rapid and efficient.

3.2 Collaboration of Partakers in Educational Process in Distance Learning

Coordination of the teacher's activities with the participants of the educational process is of great importance in the organization of training/retraining of personnel to work in the conditions of digital transformation. In their professional activities, teachers have to interact with the following persons involved in the learning process: the course leader, authors of training courses, technical staff, administrators, other teachers, including the teacher- consultant, listeners.

Let's look at some of the relationships between participants in the educational process. Relationships with the course leader. The DL teacher should report periodically to the course leader to develop these relationships. Remember that both of you have specific expectations from each other. The ideas about these expectations are given in Table 1.

Table 1. Mutual expectations of the distance learning teacher and the course leader

Your expectations from the course leader	The course leader's expectations of you
You assume that your course leader will	Your course leader assumes that you will
Keep in touch with you	Contact him
Discuss with you the content of the course	Master this course and develop your own ideas on the activities of the DL teacher
Let you know about resource limits and what he expects from you	Do what is expected of you
Assist you in improving your business	Constantly improve yourself as a teacher
Give you feedback on the results of your work	Welcome feedback from him and respond effectively to it
Provide you with necessary information	Ask him to provide you with the necessary information and apply it
Assist you in solving emerging problems	Report and discuss your problems as they arise

Source: author.

Relationships with the course authors can be conducted in face-to-face (instructing) and distance mode, their goal is to get explanations (if necessary) and improve the course. You can get a variety of data, including information about students, from the administrator, who should also be friendly with the DL teacher. You both organize the training process with the support of the administrator. Relationships of a DL teacher with the teacher - consultant. For a DL teacher, especially a beginner, such relationships are invaluable: getting practical support, answers to many different questions at the initial stage, the teacher-supervisor will check the quality of your comments on the written tasks of students, and help you adjust your professional activities. Considerations about these expectations are given in Table 2.

The nature of the relationship of DL teacher to other members of DL system depends largely on the efficiency of the educational process, and yet, since the teacher is one of the main participants in the educational process, the educational process will be as successful as he himself knows the training material, owns a situation, is educated and competent. Intensive interaction with other participants will enable the teacher to contribute to fruitful work and mutual assistance among the teaching staff of the course and strengthen their position in the professional environment.

Table 2. Mutual expectations of the distance learning for teacher and the teacher- consultant

Expectations of the teacher - consultant from the DL teacher	DL teacher' expectations from the teacher - consultant
Carefully studies the feedback and applies it to their professional development	Analyzes his/her activities objectively and thoroughly
Informs the teacher about how effective the consultant's comments were for the teacher	Starts and maintains, along with tracking, other contacts (social networks, email, etc.)
Treats him/her like a colleague	Treats him like a colleague
Ready for joint professional development	Ready for joint professional development

Source: author.

4 Discussion

At the moment, few doubt that we are experiencing the era of the digital revolution. Digitalization, digital transformation - these are trends that are already firmly fixed in the realities of today. The main economic trend of the current era is the digital transformation of business. This concept was thought for a long period exclusively as transformation and storage of common information in digital format and transferring of media from analogue to digital one [2]. Now digital technologies are being actively introduced into all spheres of business and traditional industries that have not modified for years. It first of all influences the labor market: according to the World Economic Forum robots will ruin more than 75 million global jobs in the near future, though they will generate 133 million new jobs [11]. The DT is built on certain know-hows: industry and business procedures robotization; manmade intellect; many-channel gathering of information, its study and use both for forecasting and satisfying consumers' request and wishes etc. At the moment everybody ought to understand that such concepts as "profession", "stable employment", "career" in the classical sense of the word, no longer exist. And never will be. Moreover, this does not depend on the qualifications of the employee. This applies to the situation both in the whole world and in Russia. There are examples when a former private security guard became a system administrator, completely and radically changing his profile of activity. And there are examples when investment bankers have been unemployed for years, because there are practically no investment banks left in Russia, but they continue to hope for a miracle.

The general pattern is as follows. For "blue-collar" workers, the amount of jobs reduced by near 20%. There remain fewer and fewer positions for low-skilled employees which makes it more difficult for them to gain their money. There has been redeployment among "white-collar" workers. Like the "blue ones", the number of jobs also decreased by 20%, but new ones came to replace those that disappeared. The status quo has roughly remained, but the situation has required many people to retrain, get a second higher education, and generally not sit on an even keel. Therefore, the key skill now is constant self-education and obtaining new competencies, creating a flexible set of them that allows you to adapt to any changes [10]. That is why the study was aimed at solving these problems.

5 Conclusion

Summing up, we can conclude that the digital transformation has significantly affected the labor market, causing the need for enterprises to train/retrain personnel. The paper analyzes the impact of digital transformation on the labor market; substantiates the need for training/ retraining/advanced training of personnel for new activities; proves the feasibility of such retraining in the conditions of digital learning; considers the stages of training using distance educational technologies; analyzes the interaction of participants in the educational process in distance learning. The current situation requires a corresponding restructuring of business, and, on the other hand, changes/modernizes education system, both of which will undoubtedly cause a number of transformations, which we will witness in the very near future.

Acknowledgements. The author of the article expresses her appreciation to the Department for support of publication activity of Samara State University of Economics for the comprehensive assistance and support provided.

References

1. Akimova, O.B., Shcherbin, M.D.: Digital transformation of education: timeliness of educational and cognitive independence of students. Innov. Projects Programs Educ. **1**, 27–34 (2018)
2. Albukhitan, S.: Developing digital transformation strategy for manufacturing. Procedia Comput. Sci. **170**, 664–671 (2020)
3. Forbes: How digital transformation will change the labor market in Russia? https://yandex.ru/turbo?text=https%3A%2F%2Fwww.forbes.ru%2Fkarera-i-svoy-biznes%2F371537-kak-cifrovaya-transformaciya-izmenit-rynok-truda-v-rossii. Accessed: 10 June 2020
4. Gromova, T.V., Belousov, A.I.: Functional components of the model of teacher's pedagogical activity in the distance learning system. Bull. SSAU(National Research University) **14**(2), 248–260 (2015)
5. Guinan, P.J., Parise, S., Langowitz, N.: Creating an innovative digital project team: levers to enable digital transformation. Bus. Horiz. **62**(6), 717–727 (2019)
6. Jackson, N.C.: Managing for competency with innovation change in higher education: examining the pitfalls and pivots of digital transformation. Bus. Horiz. **62**(6), 761–772 (2019)
7. Limani, Y., Hajrizi, E., Stapleton, L., Retkoceri, M.: Digital transformation readiness in higher education institutions (HEI): The case of Kosovo. IFAC-Papers OnLine **52**, 52–57 (2019)
8. Merge, I., Edelmann, N., Haug, N.: Defining digital transformation: Results from expert interviews. Gov. Inf. Q. **36**, 101385 (2019)
9. Ryzhkov, V.: What is digital transformation? https://komanda-a.pro/blog/digital-transformation. Accessed: 10 June 2020
10. Santos, H., Batista, J., Marques, R.P.: Digital transformation in higher education: the use of communication technologies by students. Procedia Comput. Sci. **164**, 123–130 (2019)
11. WEF: Robots 'will create more jobs than they displace. https://www.bbc.com/news/business-45545228. Accessed: 15 May 2020

"Digital Trade Union" in the Personnel Motivation System

O. A. Dzhulai[1(✉)], O. G. Savchenko[2], and D. N. Savinskaya[3]

[1] Municipal Educational Institution Kindergarten "Muravishka", Tolyatti, Russia
sseu313@mail.ru

[2] Samara State University of Economics, Samara, Russia
savoed15@rambler.ru

[3] Kuban State Agrarian University (Named After I.T. Trubilin), Krasnodar, Russia
savi_dinki@mail.ru

Abstract. In the article, the authors analyze the issues of modernizing the system of personnel motivation by using the achievements of digitalization. An example for analysis is the "Digital trade union" project, which is used by trade unions to motivate their staff. The study used methods such as analysis, observation, and comparison. There are three main blocks of motivation, based on which methods of motivation are developed: legal motivation; socio-psychological motivation; material motivation. It is concluded that such a project as "Digital trade union" can significantly affect the system of personnel motivation in the modern world.

Keywords: Digital trade union · Personnel motivation · Trade union

1 Introduction

At present, trade union organizations are beginning to play an increasing role in the life of employees. Trade unions are called upon to protect the social and economic rights of employees, to act as a motivator for the team, and to ensure that all social guarantees are fulfilled for the staff. The role of trade unions was particularly evident in the current situation caused by the pandemic, when many businesses were forced to close for quarantine and send their employees to down time. At the same time, most of the closed enterprises that do not belong to state structures were not able to pay their employees for non-working days. At such times, it is the unions that act to protect the rights of employees. However, this is not the only significant function of these organizations.

It is considered that the function of a trade union as a motivator is one of the most important functions for ensuring a favorable psychological climate in the collective. The quality of work of trade union organizations can be considered in terms of the activity of the staff in their work, the quality of business communications in it, as well as information transparency in terms of informing the staff of current labor legislation and new opportunities that can be used by members of the trade union [8]. We can say that from the point of view of trade unions, motivation can be interpreted as a system of

S. I. Ashmarina and V. V. Mantulenko (Eds.): IPM 2020, LNNS 161, pp. 170–174, 2021.
https://doi.org/10.1007/978-3-030-60926-9_23

forms and methods of work that would contribute to the growth and strengthening of the positions of trade union structures and, as a result, improve the quality of the workforce.

The need for trade unions has been reflected in various periods of society's existence [10]. It is believed that trade unions are able to take a position in society in which they will stand on an equal footing with employers and public authorities. The advantages of trade union organizations on this matter are confirmed by the larger number of its members, which allows them to increase their own influence in society.

According to the Federal law "On trade unions, their rights and guarantees of activity" of 12.01.1996 N 10-FL, the key to the life of trade union organizations is currently to ensure the preservation and growth of the number of trade union members [5]. These moments can preserve the influence of trade union organizations in the long term. The size of a trade union organization and its financial base make it possible to assess the organization's capabilities in terms of representation and protection of the interests of represented employees at various levels: the organization, the industry, the subject of the country or the country as a whole. In this regard, a new important task for trade union organizations is to improve the forms of staff motivation to improve the working climate and attract new members to their ranks [11].

Article 377 of the Labor Code of the Russian Federation emphasizes the voluntary participation of employees in the activities of trade unions, so an employer or any representative of a trade union organization cannot forcibly oblige employees to join a trade union organization [7]. This further emphasizes the importance of attracting new and existing employees to the trade union ranks with the help of new motivational forms and measures.

Most of the motivation problems are solved within the framework of a specific labor collective, which is related to the specifics of the trade union organization in the organization under consideration at that time. The main motivation mechanism used for a long period is the collective agreement. This document regulates the activities of a trade union within a specific organization. It can be concluded that the main purpose of motivating trade union membership is to attract all employees who have not yet joined the union to their ranks, through disclosure of the forms of activity of trade union organizations and demonstration of the advantages of trade union membership.

2 Methodology

In the framework of the study, the main general logical methods were used, consisting in analyzing the received information and using a systematic approach. At the same time, empirical methods such as observation and comparison were also used. These methods were used when considering the theoretical and methodological basis of the study. We analyzed the research of domestic and foreign scientists on this topic, the main legal acts, as well as available data on current methods of motivation used in trade unions, as well as innovations among them. Such an analysis will clearly show the development of the motivation system, taking into account the development of digitalization of society and the universality of new technologies in relation to various branches of human activity.

3 Results

When implementing measures to motivate trade union membership, trade unions should take care of the growth of the number of trade union organizations and, consequently, their strength and financial stability. At the same time, it is necessary to take into account the external and internal conditions of motivation, as well as use newly emerging technologies to meet the realities. Among the external conditions of motivation, there is a link with the image of the trade union organization not only directly in the labor collective, but also among various social partners. After all, with a lack of image, many social partners will not want to cooperate with the organization and present preferences to employees of a particular labor collective. This moment is reached through numerous forms:

- organization of meetings of labour collective with participation of the chairman and members of the staff necessary to ensure the information transparency concerning the results of trade union works,
- organization of periodic reporting meetings and meetings on the work results,
- organization of dissemination of information about the activities of the trade union using stands, newspapers, leaflets, websites and other resources of the trade union association. Currently, in the age of digitalization, it is electronic methods of transmitting information that are widely used, which greatly simplifies and speeds up this process,
- organization of an open day and much more.

Currently, the most commonly used methods of motivating staff are the organization of cultural and leisure activities, as well as the organization of a thorough system of labor protection and safety [11]. However, these methods are losing their relevance among young representatives of the labor collective, which requires trade union organizations to use the most relevant methods of motivation and attracting employees now.

Basic methods of motivation provide for the creation of a special section in the collective agreement that regulates benefits and preferences that can only be used by members of trade unions. To modernize the motivation system, it makes sense to rely on thematic blocks that represent the most promising areas for the activities of trade unions. These blocks are legal motivation (regulation of legal, contractual and other relations within the workforce); socio-psychological motivation (career, training, incentive system, rest and recovery); material motivation (remuneration, compensation, bonuses and incentive payments).

In 2019, within the second block "Socio-psychological motivation", the project "Digital trade union" was developed, which began to be implemented in 2020 [1]. The main objectives of the project are:

1. Switching to a new dual-use electronic union card.

2. Creating conditions for additional social support for trade union members using discount and bonus programs combined with an electronic trade union ticket.

3. Introduction of a system of electronic registration of trade union members, automated collection and processing of statistical reports in the trade union, which should ultimately contribute to the organizational strengthening of the trade union, increase the efficiency of its activities.

The task of the chairmen of trade union committees of each institution was to fill in the register and enter into the database all the necessary information on employees of their institution, including members of the trade union. After a trade union member is entered in the database, they must register in the AIS (automated information system) and receive an electronic trade union ticket. A trade union member can download the mobile app and register for the bonus program. For an ordinary trade union member, the digital trade union project provides an opportunity to use the bonus program.

In addition, an employee under the program gets the opportunity to register in the "Profcards" system using the trade union ticket number. This incentive system allows you to get a refund from 1 to 30% of the money spent on goods and services. When you purchase through online stores, bonuses are awarded automatically. Purchases in offline stores can be accounted for by scanning barcodes on the receipt in the mobile app you are using. Users of the program will be able to withdraw any amount starting from 1 ruble to any bankcard.

Considering the experience of foreign countries, it should be noted that some of them already have a system to support the use of digital technologies in the activities of trade unions [12]. The European socio-economic committee has long noted the relationship between digital transformation and trade union development strategies [4]. The impact of digitalization on staff working conditions is actively studied, which largely attracts the attention of the main social partners. All this confirms the need to use digitalization products in improving the system of personnel motivation.

4 Discussion

Booth in his work notes the inefficiency of the information system, which is currently used to attract new employees to trade unions, and the modernization of this system will help to reduce staff turnover and increase the efficiency of their work [2]. Brand and Niedermoser say that trade unions are designed to change the balance between staff welfare and the growth of capitalism in the direction of improving the quality of employees' life [3].

The German experience shows that the existence and development of trade unions can be carried out without the use of membership fees. Despite this, trade unions continue to fully perform their duties, which is very attractive for employees [6]. Some authors note that trade union organizations have to take into account the specifics of the organization's sphere of activity when organizing a motivation system [9]. One of the ways of motivation can be identified and the impact of trade union organizations on tax legislation and wages of employees belonging to trade union organizations [13]. Such a comprehensive analysis of the activities of trade unions and their methods of motivation confirms the relevance of the study of these issues.

5 Conclusion

The importance of digitalization for various spheres of human life is increasingly confirmed in the course of society development. At the same time, it makes sense to talk not only about the impact of digitalization on each individual, but also about the impact on entire businesses and industries. As mentioned earlier, one of the main resources of the enterprise is the labor force, which must be constantly stimulated and motivated to get the desired result. Such motivators are trade unions, which are forced to periodically adjust existing ones and develop new ways of motivation.

Digitalization makes it possible to use its advantages when upgrading any processes or activities, and motivation is no exception. Many countries already use digital products in the system of motivation of trade union organizations. So in Russia, this is reflected in the development of the project "Digital trade union", which combines the achievements of digitalization and methods of motivating and stimulating staff. Such a product can significantly influence the system of personnel motivation in the modern world.

References

1. All-Russian Trade Union of Education: Digital trade union. https://www.eseur.ru/digital_union/. Accessed: 11 June 2020
2. Booth, A.: The economic behavior of trade unions. Discussion paper № 670. (2012)
3. Brand, U., Niedermoser, M.K.: The role of trade unions in social-ecological transformation: Overcoming the impasse of the current growth model and the imperial mode of living. J. Clean. Prod. **225**, 173–180 (2019)
4. Europäischer Wirtschafts-und Sozialausschuss: Digital transformation and the trade unions strategy. https://www.eesc.europa.eu/de/node/65062. Accessed: 10 July 2020
5. Federal Law "On trade unions, their rights and guarantees of activity" of 12.01.1996 N 10-FL. http://www.consultant.ru/document/cons_doc_LAW_8840/. Accessed: 11 June 2020
6. Goerke, L., Pannenberg, M.: Trade union membership and dismissals. Labour Econ. **18**(6), 810–821 (2011)
7. Labor Code of the Russian Federation of 30.12.2001 N 197-FL (as amended on 25.05.2020). http://www.consultant.ru/document/cons_doc_LAW_34683/. Accessed: 15 June 2020
8. Makarova, V.A., Goncharov, M.A.: Place and role of trade unions in the life of modern Russian society. Poly. Netw. Electron. Sci. J. Kuban State Agrarian Univ. **121**, 511–524 (2016)
9. Räthzel, N., Uzzell, D.: Trade unions and climate change: The jobs versus environment dilemma. Glob. Environ. Change **21**(4), 1215–1223 (2011)
10. Soshnikova, T.A.: Ways to increase the role of trade unions in modern Russia. Moscow State University for the Humanities, Moscow (2014)
11. Troshina, E.P., Mantulenko, V.V.: Influence of digitalization on motivation techniques in organizations. In: Ashmarina, S.I., Vochozka, M., Mantulenko, V.V. (eds.) ISCDTE 2019. LNNS, vol. 84, pp. 317–323. Springer, Cham (2020). https://doi.org/10.1007/978-3-030-27015-5_38
12. TUC Digital: Double-down on reps, not digital. https://digital.tuc.org.uk/. Accessed: 10 July 2020
13. Wang, Y., Quattara, K.S.: Employment double dividend hypothesis with the presence of a trade union. Econ. Lett. **193**, 109273 (2020)

Professional Education Digitalization (Example of Physical Culture)

S. I. Zizikova[1], Yu. Yu. Kareva[2(✉)], and I. L. Matasova[3]

[1] Volga State Academy of Physical Culture, Sports and Tourism, Kazan, Russia
zizikova@yandex.ru
[2] Samara State University of Economics, Samara, Russia
Kareva-19911987@mail.ru
[3] Samara Branch Moscow City University, Samara, Russia
inna_gu@mail.ru

Abstract. The article deals with the problem of Russian education digitalization. The digitalization of all social and economic spheres is a very fast process. Digital changes are like a rapid tsunami. Scientists and specialists in the field of physical culture and sports pay considerable attention to the application of new innovative technologies in physical culture and sports training, including the introduction of new technologies in various educational programs. Currently, most universities are equipped with computer classes, interactive whiteboards and other multimedia equipment; they can use digital technologies and tools for educational purposes. A large number of electronic educational resources, interactive environments, collections of video, audio resources, aggregator platforms, automated information systems, etc. have been developed and successfully used in the education system. The purpose of the article is to study the problem of digitalization of education and personnel support in the sphere of physical culture and sports. Using the example of physical culture, we show options for implementing digital technologies in the training and competition process, as well as consider the transformation of the personnel structure of physical culture and sports in the era of digitalization of society. The paper uses methods of theoretical analysis and generalization of research data.

Keywords: Digital technologies · Digitalization · Physical culture · Sport

1 Introduction

All areas of human life today are affected by major changes. A huge role in these transformations belongs to the rapid development of scientific and technological progress, the global digitalization of the leading countries of the world community, the penetration of information technologies into all spheres of our life, regardless of their specification. Education is one of the areas where digitalization plays a major role [4, 5]. Accordingly, the concepts of education, as well as learning strategies, were revised [13]. Approaches to education and universities that must produce people who meet the requirements of the digital work environment are changing [8]. Digitalization can be the key to increasing students' access to education for sustainable development [1].

S. I. Ashmarina and V. V. Mantulenko (Eds.): IPM 2020, LNNS 161, pp. 175–183, 2021.
https://doi.org/10.1007/978-3-030-60926-9_24

In these conditions, it is necessary to encourage young people to develop their digital skills, as well as digital and information literacy [6].

The term “digitalization” arose from the strengthening of information and communication technologies. Some scientists consider this concept as a translation of information into numbers and, at the same time, simultaneously infrastructural, managerial, behavioral, and cultural components of the content of education.

Digitalization can be considered one of the main approaches to the use of digital resources in the transformation of not only education, but also the economy. At the same time, the role of technologies and processes is supposed to be reallocated in order to improve the information and educational environment. Digitalization leads to the intensification of production processes [11].

Digital technologies are the future of humanity. They are based on speed and versatility, which makes them so popular in all areas of human life. A discrete system based on methods of encoding and transmitting information, which allows you to perform a variety of tasks in the shortest possible time, has found its application in sports. Monitoring and analysis of the athlete’s actions is based on the latest achievements of scientific thought - from nano-electronics to molecular biology. This helps to achieve accuracy in everything from the organization of training to the process of treatment and recovery of athletes after injuries, including doping tests. The use of digital sports technologies is typical not only for sports of the highest professional achievements, but also refers to physical culture. it is easier to arouse interest in sports and physical culture among young people if innovative methods and digital technologies are introduced into the training process. For this purpose, the Ministry of Sports of Russia issued order No. 971 dated 25.11.2019 “On approval of the concept of digitalization of the state system of training and management in the field of physical culture and sports of the Ministry of Sports of the Russian Federation for the period 2019-2024” [10]. The introduction of digital technologies in the sphere of physical culture and sports requires retraining of personnel, training them in the use of modern means, forms and methods of organizing physical culture and sports work. The overall goal of the concept is to transform the working processes of public administration in the field of physical culture and sports in order to significantly increase their efficiency by switching to a digital management model and an electronic format of interaction.

2 Methodology

The purpose of this research is to identify modern trends in the use of digital technologies in education. When writing the article, we used the method of theoretical analysis and generalization of data from scientific works of domestic and foreign specialists in recent years. In total, more than 70 literary sources have been studied, the main ones are listed in the list of references. The methodological basis of the research was the works of leading experts in the field of digitalization, physical education, sports training, health and adaptive physical culture. On this basis, the features of the use of digital technologies in various spheres of physical culture and sports activities were determined: educational, physical culture and recreation, training, and competition.

The article shows the relationship between the processes occurring in the field of education and the digitalization of society.

3 Results

The process of digitalization, which began with the economy, gradually affects other areas of human activity. Education is no exception in this regard. Until recently, information technologies were actively introduced into the practice of educational activities. This level of development is successfully completed and started the next level of digitalization. This process is associated with the development of the Internet and its rapid spread. The advent of the Internet has created a new world-a virtual world that has expanded the possibilities of communication through social networks, online games, forums, chats, etc. Over time, the emergence of a hybrid world was observed–the merging of the real and virtual.

The penetration of the Internet and the development of mobile technologies are significantly changing the modern economy, penetrating deeply into public life and work. If previously most work tasks were solved using a desktop computer, the young generation Y (born after 1980) and the young generation Z (born after 2000) cannot even imagine life without their smartphones. For today's youth, searching for information on the Internet and surfing in the virtual world is routine. 99% of modern students have smartphones, 92% have their own computer, 84% download music from the Internet, 76% use instant messaging systems, 44% participate in social networks, 34% receive all news only via the Internet and almost do not watch TV (less than once a week), 28% have their own blogs, and 20% create their own web pages, 18% are online. Today, the technology of digitalization is changing very quickly. And the speed of their change increases all the time, and they, in turn, change everything else with which they interact. It is very important to understand this process and adapt to it accordingly. Rapidly changing, it brings both new benefits and additional challenges to society as a whole and to education in particular. In this regard, new approaches to education, wider use of information technologies, and more active involvement of students are needed [7].

Smart education is a fundamentally new educational environment. It is an Association of teachers, students and knowledge from all over the world. Smart is a property of an object that characterizes the integration of two or more previously unrelated elements into this object, which is performed using the Internet. A smart goal is a successful strategy for turning even the most fantastic task into an implementation category. The technology of intellectual education is independent, motivated, flexible, technical, based on independent judgment, enriched with resources and technical means of learning.

Accordingly, a smart teacher (student) is a participant in the educational process, constantly using technological innovations and the Internet to achieve a new quality of the educational process that meets the requirements of a smart society. In this regard, there is a new concept – the network student, which is characterized by: flexibility in learning in an interactive environment; individualization of the program for each student; multidisciplinary (inter-disciplinarity) of training programs; the ability to learn at any time and in any place based on free access to content around the world; the opportunity to work, which is most relevant for most modern students.

Within the framework of the international interdisciplinary conference on engineering education, the world's leading scientists highlighted the areas of digital learning: "design and deployment of cloud-based learning and cloud-based research environment of an educational institution, joint open courses of the educational community, and slide Wiki platforms".

The relevance of using cloud technologies in education is that they not only serve as tools, but also provide qualitatively new opportunities for learning, developing skills for independent learning, and contribute to the creation of new forms of education and upbringing.

Currently there are three most common ways to work with cloud technologies in educational institutions:

- distance learning,
- joint project work of students,
- working together on documents.

In Internet technologies of distance learning, there are several main directions, they are shown in the figure.

Figure-distance learning directions [3].

Educational cloud services are used not only in distance learning, but also in traditional forms of learning. They allow you to manage the processes of the virtual space of the university. Virtual space is an environment aimed at supporting distance learning, as well as managing and optimizing the business processes of the University itself.

Cloud technologies should be considered as a means of consolidating internal subsystems and creating a virtual environment that provides interaction between teachers and students, making the following processes available:

- posting news, announcements and event announcements,
- conduct webinars and online conferences,
- creating virtual laboratories on the Internet,
- exchange of electronic messages between users centrally or by separate categories,
- remote interaction with students, including the provision of educational materials in electronic form, testing, online consultations, information about the schedule of exams and classes,
- electronic interaction with applicants, including consulting, informing, remote registration of applicants' applications [2].

Leventsov and Mukhanova, investigating possible directions of digitalization of education, note that this process should not be limited to simple translation of familiar textbooks into electronic form, digitization of document flow in educational institutions or open Internet access for students and teachers. According to the authors, a mandatory condition for the organization of digital educational space in higher education should be a change in the approach to both training and emerging competencies, which will be in demand by the market in 10-15 years [9].

Pritchina agrees with them, noting that digital transformations that have affected all spheres of human activity have led to a rethinking of the tasks of higher education. In his opinion, within the digital space, society needs graduates of a new "digital format", which means that professional competencies in the field of digital technologies should become the dominant vector of the educational process [12]. For example, the use of gadgets and specialized programs was tested by manufacturers in two leading higher educational institutions in Russia – the first Moscow State Medical University named after Sechenov and the Bauman Moscow State Technical University. A competition was held for students between two higher educational institutions. The goals of the competition project were as follows: increasing students' motor activity, involving young people in a healthy lifestyle, digitizing the self-control diary.

The result of the project was overall activity and increased power in the conditions of increased physical activity (67% of people), more than half of the participants 58% began to monitor their diet using an electronic food diary.

Competitions in motor activity using the sensor technology of the controller were held in 2018 by the federal center for organizational and methodological support of physical education. The Ministry of Education and Science of the Russian Federation, together with the Federal Institute for Educational Development, the Russian State University of Physical Culture, Sports, Youth and Tourism, and Ubisoft, provided active support for this project. Educational institutions from each federal district took part in the competition. Preliminary and regional finals were held. The experiment showed that the use of interactive digital technologies in physical education classes provides not only increased interest in the subject, but also reduces the neuropsychiatric stress in students, but also improves their functional state and accelerates the processes of adaptation of the body to educational activities.

4 Discussion

Currently, there is a widespread introduction of digital technologies in the field of physical culture and sports. On the one hand, the trend of using analytical technologies to support productivity improvements (for the sake of winning) continues and increases. For example, during the 2014 FIFA World Cup, the successful performance of the German national football team was supported by SAP, a German software provider. This example shows how SAP programmers, based on the use of the "March insight" software, allowed the team to identify existing shortcomings, make timely adjustments and improve the effectiveness of competitive activities.

On the other hand, the use of digital technologies in sports also extends to areas such as the organization and management of sports teams and their stakeholders, access to and interpretation of sports information, and the invention of new tools and strategies that would otherwise be impossible. This requires an urgent increase in the "digital" education of sports professionals such as directors of sports clubs, team leaders, managers, psychologists, and sponsors.

Finally, digitalization has led to the creation of a new sport –eSports, which has profound implications for the very nature of the sports field. In Sweden, this became clear after the Swedish eSports Association applied for membership in the Swedish

Sports Confederation. Despite the rejection of the request of the Swedish Sports Confederation came to the conclusion that they need to review what the sport is. The creation of eSports requires the training of new specialists: coaches, judges, competition organizers, and service personnel of eSports competitions.

Competition in sports is now largely linked to competition on digital assets. For example, in Formula 1, digitalization combined with a new formula for hybrid powertrains with internal combustion engines paired with energy recovery systems led to an era of software competitiveness. Accordingly, the team has changed from primarily mechanical and aerodynamic engineers and technicians to programmers and data analysts. In addition to the technical progress of physical equipment and the inclusion of its components, IP researchers are also concerned with the ever-increasing importance of data and information generated, integrated and analyzed using various digital tools, both software and hardware. Admittedly, this trend can be traced back to the time of the seminal work of Michael Lewis, which presents the procedure for completing the Oakland baseball team based on computer analysis of morphological and weight indicators of athletes-candidates for the team. Thus, thanks to the use of Internet of Things technologies, such as sensors, Wearables and analytical tools, what used to be "hidden" information about the technological component of the team (for example, the level of physical skills of players, data on the performance of the game), has become more accessible, tracked and visible not only to the general public, but also, more importantly, to representatives of competing teams. For example, in 2014, MLBAM (the owner of a baseball club) in the United States introduced a "camera-and radar-based system" for all MLB team stadiums, which allows you to track real-time data related not only to baseball (such as movement speed), but also the performance of each player [14]. Organizing and conducting events of this kind requires special training not only for programmers, but also for sports operators, doctors, and statisticians.

The world of sports has been constantly changing over the years, and the use of digital technology is just one of the areas that have had a significant impact on the development of many sports. For example, thanks to the "Polar Team 2" athlete monitoring system, coaches will get unique information about the players' capabilities, planning different types of training based on their heart rate to optimize them and prevent overtraining and eliminate injuries to athletes. "Polar Team 2" will provide the coaching staff with the ability to record and study data on the physical fitness of the entire team in real time, taking into account the age, weight, stress level and maximum heart rate of each individual player when developing specialized training modes. Based on data collected from players' heart rate monitors, coaches can individually assess how much energy a player has expended, how much damage their body has suffered, and how much recovery time they will need to maximize their performance on game day and throughout the season. The above points to the need to attract specialists in the field of medicine (cardiologists, traumatologists, nutritionists, etc.), as well as sports psychologists.

Hawk-Eye technology is the name of a computer system that tracks the trajectory of a ball. It is used in international cricket and tennis, and many other sports are also considering using this technology. This system is also being trialled in soccer as part of the assessment the goal line. The Premier League of football in the UK has agreed to introduce goal-line sensors after receiving approval from football lawmakers. The system being developed by the British company Hawk-Eye was supposed to give the final decision on whether the ball crossed the line. Hawk-Eye Technology uses a camera that takes 600 frames per second on the goal line, with the information analyzed by a computer and sent to the referee's headset or device on his wrist. In 2015, Hawkeye technology was also used by Rugby officials at the 2015 Rugby world Cup to improve decision-making by the official televised match, as well as help with player safety.

Finally, as mentioned above, the digitalization of sports is also driving the growth and popularity of eSports, which, from a technological point of view, is the commercialization of the physical skills and knowledge that sports manufacturers possess when creating digital products (often in the format of digital games). For example, the popularity of FIFA football can be attributed to its simulation of the real world, where players live and control players whose skills are duplicated from data collected on their live counter parts. This area will attract additional attention, given the great potential of VR (virtual reality) and AR (augmented reality) technologies in creating simulated game scenarios that are close to the real experience [14].

5 Conclusion

At the present stage of higher school development, there are processes of accelerating the introduction of digital technologies in the field of education, including physical education. However, the quality of implementation of these processes depends on the competent use of the personnel of university departments. In the field of physical culture, the transformation of digital technologies in the educational process proceeds in four main directions:

1. Educational process. With the help of digital technologies, students are transferred theoretical knowledge, methodological and practical skills, and control over their assimilation, which is carried out in the form of monitoring of physical fitness, motor activity and psychophysical state of students in the course of physical culture and sports. For the successful implementation of this direction, teachers of adepartment should not only have perfect computer skills, but also have the skills to develop electronic educational and methodological documentation, visual AIDS, test tasks.

2. The process of sports training. Achieving high results in modern sports is possible only through the use of advanced forms, tools and methods of training. This information is obtained on the basis of online viewing and analysis of training and competitive activities of leading domestic and foreign teams. The personnel of teams (coaches, medical personnel, masseurs, statisticians, etc.) must possess digital technologies that allow assessing the current functional state of athletes (caliperometry, bioimpedance, etc.), as well as develop long-term plans for their training, taking into account the data of medical and pedagogical control.

3. The process of competition. The competition referees should be able to apply various video analysis programs that allow online evaluation of the performance of an athlete (for example, the quantity and quality of technical and tactical actions of a particular player or team in the game), various start-finish control systems, electronic systems of judging and timing, etc.

4. Health-improving and adaptive physical culture. Digital technologies in this area are aimed at creating computer programs of health and rehabilitation orientation. The purpose of these programs is not only to improve the health of students, but also to identify abnormalities in the state of health at an early stage and make appropriate adjustments to previously developed programs. The process of health-improving and adaptive physical culture activities should be carried out by highly professional physical culture instructors together with medical professionals.

References

1. Ahel, O., Lingenau, K.: Opportunities and challenges of digitalization to improve access to education for sustainable development in higher education. In: Leal Filho, W., Salvia, A.L., Pretorius, Rudi W., Brandli, L.L., Manolas, E., Alves, F., Azeiteiro, U., Rogers, J., Shiel, C., Do Paco, A. (eds.) Universities as Living Labs for Sustainable Development. WSS, pp. 341–356. Springer, Cham (2020). https://doi.org/10.1007/978-3-030-15604-6_21
2. Ananchenko, I.V.: Cloud technology in higher education. Modern Sophisticated Technol. **5**, 48–52 (2015)
3. Bulgakova, D.M., Parfirova, A.A., Tuchkova, A.S.: Application of cloud technologies in the educational process. Modern Scientific Research and Innovation,vol. **3**. http://web.snauka.ru/issues/2017/03/79865. Accessed: 09 July 2020
4. Crittenden, W.F., Biel, I.K., Lovely, W.A.: Embracing digitalization: Student learning and new technologies. J. Mark. Educ. **41**(1), 5–14 (2019). https://doi.org/10.1177/0273475318820895
5. Kazaishvili, A.: Managerial skills can be provided by the universities in the digital era. Quality - Access to Success **20**, 89–94 (2019)
6. Kmecová, I.: Digitization, digital technology, and importance of digital technology in teaching. In: Ashmarina, S.I., Vochozka, M., Mantulenko, V.V. (eds.) ISCDTE 2019. LNNS, vol. 84, pp. 526–537. Springer, Cham (2020). https://doi.org/10.1007/978-3-030-27015-5_63
7. Kondakov, V.L., Kopeikina, E.N., Voloshina, I.G., Usatov, A.N.: Main directions digitalization in the higher education in the world and Russia (Based on the example of physical culture and sports). Univers. J. Educ. Res. **7**(10B), 47–52 (2019). https://doi.org/10.13189/ujer.2019.071810
8. Krichevsky, M., Martynova, J., Budagov, A.: Methods of machine learning in the master's educational program. In: Rudoy, D., Murgul, V., (Eds.), Proceedings of Innovative Technologies in Environmental Science and Education (ITESE-2019), E3S Web of Conferences, Vol. **135**, pp. 03069. Les Ulis: Web of Conferences. https://doi.org/10.1051/e3sconf/201913503069. (2019)
9. Leventsov, V.A., Mukhanova, N.V.: Quality of education in the epoch of the digital economy. In: Proceedings of the Saint Petersburg International Economic Forum, pp. 77–79. Saint Petersburg: Publishing House of Polytechnical University. https://doi.org/10.18720/spbpu/2/id18-80. (2018)
10. Order of the Ministry of Sports of the Russian Federation No. 971 of 25.11.2019 "On approval of the concept of digitalization of the state system of training and management in the field of physical culture and sports of the Ministry of Sports of the Russian Federation for the period 2019–2024.". http://www.consultant.ru/cons/cgi/online.cgi?req=doc;base=EXP;n=617284#06651133236754201. Accessed: 09 July 2020

11. Petrova, N.P., Bondarev, G.A.: Digitalization and digital technologies in education. World Sci. Cult. Educ. **5**(78), 353–355 (2019). https://doi.org/10.24411/1991-5497-2019-00138
12. Pritchina, L.S.: Digitalization and new economic education. Pedagogical Educ. Sci. **2**, 121–124 (2018)
13. Vogt, T., Rehlinghaus, K., Klein, D.: School sport facing digitalisation: a brief conceptual review on a strategy to teach and promote media competence transferred to physical education. J. Phys. Educ. Sport **19**, 1424–1428 (2019). https://doi.org/10.7752/jpes.2019.s4206
14. Xiao, X., Chian Tan, F.T., Lim, E.T.K., Henningsson, S., Vatrapu, R., Hedman, J., Van Hillegersberg, J.: Sports digitalization: An overview and A research agenda. Paper presented at the ICIS 2017: Transforming Society with Digital Innovation. (2018). https://pdfs.semanticscholar.org/c5b9/fabd8ac3a7f8a80931719f49df5ba7fd4878.pdf. Accessed: 09 July 2020

Application of IT Technologies in Personnel Management in the Era of Digitalization

N. V. Kozhukhova[1](✉), J. V. Veselova[2], and S. V. Chekuldova[2]

[1] Samara State University of Economics, Samara, Russia
kuka_55_@mail.ru
[2] Samara State Transport University, Samara, Russia
veselova-uv@yandex.ru, Chekuldova-sv@mail.ru

Abstract. The article explains the need to use IT technologies in personnel management in the era of digitalization. The basic mechanisms of embedding in personnel management system are analyzed and the basic results improving administrative processes, such as automation of working space of the head, creation of a uniform platform for the analysis of variants of typical administrative decisions in a mode of real time, use of an artificial intelligence in creation of reference norms of work and processes of training and development of talents are characterized. Emphasis is placed on best practices in the application of technology.

Keywords: BI-analytics · Digital technologies · Human resources management · IT technologies · Pilot project

1 Introduction

In the age of rapidly developing technologies, company owners and business managers who have reacted quickly to emerging trends are acting decisively, in real time, considering HR as a new source of value. Their carte blanche includes digital tools for working with data, methods of analytics and forecasting statistics, artificial intelligence and other novelties of IT technologies not only in HR management, but also in the company as a whole. Other managers are limited to minor changes (using data analysis, etc.) or persistently continue to implement a static approach of the «stone age», which has dangerous consequences in the era of digital change.

2 Methodology

The task of providing management and analytical control over the company's activities through a unified IT environment of interaction, document flow management, analysis and optimization of interaction processes, planning of human, material and financial resources is now a concern for companies and organizations in all areas of activity [2]. Choosing the best information-analytical tools of IT technologies for embedding in the business environment of the company, top management, first of all, sets the following purposes and tasks. The main purposes are: 1) contribute to improving the quality and

S. I. Ashmarina and V. V. Mantulenko (Eds.): IPM 2020, LNNS 161, pp. 184–192, 2021.
https://doi.org/10.1007/978-3-030-60926-9_25

validity of expert and analytical recommendations and management decisions in the company itself and in conjunction with the rapidly transforming economic community; 2) reduce costs in the work of the organization; 3) develop information technologies in Russia taking into account international experience [1, 4–8, 10].

Tasks of embedding:

- comparative multi-criteria analysis of functionality and technical characteristics of information-analytical tools presented on Russian and foreign markets within the framework of embedded IT environment,
- opportunities to demonstrate technological platforms and analytical tools aimed at improving the efficiency of management solutions online and off-site,
- selection of the best information and analytical tools in the IT environment, taking into account proven global trends,
- planned opportunities to develop proposals and recommendations on areas and spheres of application of the best information and analytical tools in the business environment of the company.

The main estimated indicators include:

- functionality (a complex of implemented functional capabilities in relation to the subject area and the scale (level) of tasks to be solved),
- complex of technical and technological characteristics (including general functionality of the tool without reference to the subject area),
- ergonomics,
- flexibility (possibility to use the result for solving tasks in companies of different branches of economy and social sphere, in the interests of different groups of consumers without software modifications),

level of technical support.
There are also additional ones:

- number (scale) of implementations on the Russian market taking into account the vector of Western technologies development,
- estimated cost of minimum standard configuration implementation,
- approximate average cost of ownership per year (for typical configuration per user).

For example, the use of Russian IT- platform Malahit: BI is based on the collection of data from different information systems, data aggregation and operational presentation to specialists and management staff to support management and personnel decisions.

- creation of various information panels,
- support for a wide range of visual elements - widgets (panels, bookmarks, carousels, tables, charts, traffic lights, speedometers)
- creating widget containers,
- using various widget controls,
- setting up and analysis of key indicators (quality, production, personnel management, project work, finance, logistics, sales, procurement),
- industrial discipline control,

- flexible configuration of indicators and alerts for key events,
- construction of statistical models to identify critical resources, predict equipment failure,
- ready-made configurations - possibility to create customized configurations for individual functional tasks (personnel management, quality control, project management, KPI management, resource control),
- designer of information panels - support for a large set of visual elements for the development of information panels, with the possibility of nested calls for data detail (drill-down),
- access rights delimitation - each user can have his own set of information panels and available information to work with, according to the role model,
- Malakhit: BI system is a Russian development and is included in the unified register of Russian programs for electronic computers and databases,
- low cost of ownership, minimal development costs for new information panels,
- easy training of staff, no need for expensive specialists,
- The system is not demanding on hardware resources, there are no restrictions on the server and terminal equipment used,
- easily integrates with the hardware architecture of the implementation site,
- high quality of implementation: specialists - system developers are connected to the implementation process, in case of necessity changes to the existing version of the product are promptly made,
- the platform is implemented on modern web-technologies.

3 Results

Turning to the experience of top management going «in step with the times» , it can be noted that an important aspect of HR management is information and its timely submission, both from the sender to recipients and back. This can be facilitated by the development and use of «Mobile Office for the Executive» , i.e. the automation of his workspace, taking into account digital trends in business processes and advanced BI-analytics. Today the managing director wins quickly if he/she makes the right decisions. Top managers and managers need to make decisions based on huge volumes of information from various sources. In turn, large volumes of information cannot be effectively used in «manual mode» : adaptation to changes in real time is impossible, large expenditure of resources (development of models, taking into account all influencing factors, etc.), low speed of decision-making, time spent on data processing.

In order for the demands of modern challenges of economy and market to meet the vision and needs of the company, it is necessary to develop and clearly define their «want» to owners and management of companies. Most often, the functionality of such a software package may include: Visual presentation of information in the form of charts, traffic lights and dashboards, updating information in real time, notification of managers on the trigger values of key indicators, the ability to refine mathematical models and data showcases, enhanced functionality, secure access to data from any device, even from a mobile phone or tablet. The most important is the ability to predict results, which will take only 2–3 s with the help of interactive filters that will help to set the forecast conditions and instantly get the required information. The ability to get accurate answers to questions, for example, how will the efficiency of measures load the staff in relation to last month, how will the capacity increase with a decrease in the given coefficients, which is associated with the change in value per one violation of intervals last month, why the dynamics of the number of employees serving the equipment decreased, what is the dynamics of losses on the horizontal inefficiency, what is the deviation from the real trajectory of the month, etc. The algorithm of work assumes formation of a uniform database including more than 100 sources of information, cost of resources, technical condition of the equipment, tariffs, personnel expenses, activity of a business position in the market, with use of developed mathematical models which will be reflected in a show-window of the data. The architecture of the system consists of ERP-system of resource management (SAP.1C), BI - system of business analytics (Foresight Platform. Oracle. PosgreSQL…), DBMS-system of database management (MS SQL Server), ESB - service bus of the enterprise (file resources, SAP PI), then everything will be reflected on the server where synchronization and authorization will be carried out, and further output to mobile devices (mobile application, mobile framework and SQLite database). Analytical BI - platform for visual data analytics, as well as mobile solutions for business process automation is convenient from the experience of advanced companies. It will allow you to connect all the necessary collected sources of personnel data, ensure the processing of the required volumes of information, support the necessary speed of work with the interface manager.

Advanced BI - analytics for effective management of large data aggregates and updates data in real time and displays interactive data on the tablet manager. Naturally, to embed it into the system, you will need to form a pilot project and perform all related calculations. The pilot project should include:

1. Formation of the decision concept and terms.
2. Development of the solution architecture.
3. Connection of information sources.
4. Design of mathematical models and data showcase.
5. Pilot operation and evaluation of results.
6. Scaling up to other units.

The focus of the project is on the principle of prioritization. For this purpose, the main figures about the company are collected, then the main focus of competencies, then the key difficulties in making decisions when working with large volumes of data are identified, and in the final analysis of the technological platform on which the solution "Mobile Office of the Head" is based. The main requirements to the final product, basic values and benefits of implementing an advanced analytics system are formulated, the main development trends and key development initiatives are studied. The basis for the structure-values and benefits from the introduction of BI-solution in the structure of personnel management and enterprise management as a whole.

Functional requirements for BI-application:

- preferably autonomous operation of the application,
- application screens open in 5–7s with the update included,
- inclusion of users by sending PUSH-notifications to mobile devices,
- need to draw the user's attention to the problem area there is a system of traffic lights and trends,
- diverse types of business analytics: comparison with competitors, per diem indicators, quarterly indicators and annual results, plan-fact analysis, etc.,
- application of analytical documents answering questions, why indicators are such, etc.

The standardization of the work is also rather time-consuming process because of the considerable volumes of information processing [1]. It also requires a complete transformation taking into account the minimization of time expenditure and losses [1].

Another example that can be cited from the Russian practice of IT implementation in HR is the use of artificial intelligence, which helps monopolist companies to create reference labor standards. This is facilitated by the fact that many leading Russian enterprises have already developed and started to gradually implement a digital transformation strategy. Its implementation will give companies an economic effect of up to 160 billion rubles, and the country's economy - 400 billion rubles.

For example, on the Russian railways, the adopted strategy provides for the creation of eight digital platforms in key areas of activity: transportation, logistics, infrastructure, commerce, human resources management, non-production processes, transport composition. The implementation of these platforms will require the development of fifty-five projects, including the inclusion of Big Data technologies and quantum computing models in them. This year several pilot projects have already been launched, namely, the creation of «digital twins» of infrastructure objects, with their help it will be possible to predict and prevent failures based on Big Data. The pilot project on implementation of software robots for process automation of the main computer center allowed to increase labor productivity by 20%. The possibilities of artificial intelligence began to be used in rolling stock management - machine vision allows the locomotive to recognize arrows, traffic lights and other infrastructure objects, to notice obstacles on the tracks. The same technology can be used to identify and optimize the cost of working time, to create reference labor standards.

One of the basic principles of increasing productivity is the elimination of unreasonable working time costs in the performance of the production process [2]. At present, it is problematic to objectively assess these losses due to the human factor: the quality and accuracy of data largely depend on the speed of reaction of an observer, his or her vision, ability to timely record all changes at different stages of the operation. In order to accumulate enough information for further analysis and optimization, it is necessary to see the performance of the same task by other employees or in another division. To compile objective norms of work, it is necessary that time parameters at different stages coincide, but they are not repeated with the necessary share of stability because of the arising «error» at the performer or at the observer. This complicates the process of optimizing working time costs.

As now possibility of introduction of a computer vision for automation of process of working out of standards of work and their revision in case of change of technologies is worked out. Neural networks have been gaining popularity recently, and the interest in their use among customers is growing. Embedded next-generation machine vision systems based on deep neural networks can be used to do business process on analysis of working time costs in structural subdivisions of company branches. With the help of machine vision, the system is expected to record the beginning and end of a production operation, recognize it and the tools used in the process, identify employees, calculate the time of work. All data of video observation will be automatically loaded in the Unified Corporate Automated Human Resources Management System (right now the information is entered manually by engineers of services on the organization and labour standardization). Further the system will analyze video files for deviations from time and technology norms, automatically calculate norms on average or better result and formulate reference technological maps of production processes. This technology can be a step towards creating a digital work organization in the workplace.

In the future it is supposed that technologically-normal cards can be sent directly to the executor of works on a mobile device. The chain of actions will look so: results of diagnostics of the car-defectoscope in the deciphered form come in the mobile device of the road master of distance of a way, after that it forms in a digital template an outfit on performance of works and sends it to the foreman and fitters of a way of brigade on urgent works. Then a technologic-standardization map of work performance, path characteristics at the place of work and digital instructions on occupational safety are uploaded to the mobile device of the executors, which are digitally signed by the employees. In the process of task fulfillment in the augmented reality mode, hints on technology and norms of work fulfillment time are displayed on safety glasses. At the same time, video recording of work performance is performed with automatic detection of errors or violations by the employee. The dispatcher or master receives information about the execution of the work along with it. Automatically formed calculation of piece-work earnings or part of the bonus for the performance of work with the display of dynamics as the work is performed, this calculation comes to the performer on a mobile device.

4 Discussion

Another digital initiative, most often considered by both foreign colleagues and Russian researchers and practitioners, is the creation of a single platform for the analysis of variants of standard management solutions in real time in the field of HR [7, 9]. The existing approaches to analytics and HR data management do not correspond to the modern principles of digital platforms [2, 3]. There are no system mechanisms for building predictive models, no visualization of digital data. The speed of data array processing and decision making is low [3]. This pushes companies to create a control center with an IT block to develop an analytical platform that will allow the information to be presented in a user-friendly way.

The experience of foreign companies reveals the need for the analytical platform to aggregate the information from all sources in a single information space, and the formation of reports occurs in different directions with varying degrees of detail. Correct visualization of information allows to increase the speed of perception and the efficiency of management decisions in the field of HR. The companies set a task of full integration of Unified Corporate Automated Human Resources Management System (now information is manually entered by the engineers of labor organization and rationing services) with industrial information systems as then there is a possibility of automatic (without participation of executors) counting of the standard number of employees, indicators of rewarding, Labour Payments Fund and other labor indicators. In a context of megatrends in questions of work with the personnel with use of advanced IT-technologies there are expedient questions: how to train the personnel to new technologies, how to react to resistance to changes?

Digital megatrends will change business, processes and people around the world. The expectations of Generation Z employees are focused on the consumption of digital technologies that an employer is expected to have. Business demands are focused on the need to provide quality HR services, while maximizing values and minimizing costs. Analyzing foreign experience, at the stages of automation, integration, involvement and productivity, there are changes not only in HR functions, but also in the leadership role (Table 1).

Many companies use SAP Talk - real real-time dialogues - in the training process. By announcing this program, companies strengthen a culture of continuous dialogue, and focus on the continuous development and growth of their employees. SAP Talk includes the following components that are easily assimilated by the staff: Tasks/Activities, Achievements, Feedback/Development. The trend towards minimalism in the presentation of their thoughts has been set for a long time. Long texts are becoming less and less popular, and less and less readable. The aim of Small SAP Talk is to optimally identify an important topic and present it in a concise manner. In the Learning experience, the use of IT is based on certain requirements for the offered products. Let's consider them. Customised Convenience: the offered services and decisions should be flexible, i.e. be able to be recreated and will change under expectations and requirements of employees, new technologies and innovative decisions, external calls.

Everyday Magic: each product should include many possibilities and ways of application, and it is important that it is constantly improved and modernized. End-to-End Experience: Companies expect a complete and ready-made solution that solves problems or provides a need. Instant Everything: access to information and internal products must be constant −24/7, in one click. Service in «360°» format. Quiet Simplicity: today, it's important not only the content of the product, but also how simple and quick the experience of its use will be.

Table 1. Stages of change

Digitalization of HR management system	Deepening and coordinating talent development	Study on the quality of management of participation, commitment, culture and results based management	Team and usability management
HR processes are automated	One platform for all HR processes	Correct placement of employees, performance motivation and development	Advanced IT environment tools integrated into self-organization and self-learning processes
-Technologized through the implementation of IT blocks standard, template, streamlined processes (payroll, primary documentation requiring routine processing, etc.); - HR system cycle - complete, closed and controlled by managers with the participation of employees	-The processes are organized and their coordination is optimized; - The possibility of independent use of the data platform has been taken into account; - algorithms for embedding new talent development processes have been prescribed, taking into account advanced IT tools	-HR-cycle is organized under independent management within the structure; - the IT platform is linked to the business goals and built into this HR system as a business partner (full analysis is performed and the vector of changes is supported); -Analysis and calculation of HR contribution to the company's value formation (for each employee)	- Companies set tasks that require creativity and willingness to cooperate with artificial intelligence systems, a global goal - universal development; - IT environment HR - compact, convenient and maximally optimized for the mission; -Vector of HR development, within the complexity and polarization of the surrounding space, maintaining a culture with a projection for the future
Automation	Integration	Involvement	Productivity

Source: authors.

5 Conclusion

Changes in the world are so rapid that in the routine of current affairs are not always managers and specialists can keep track of current information. The information flow is endless and employees' resources are limited. In such a paradigm it is important to think one step forward, have a clear vision of business development and be ready to face the challenge of the future. The challenges of global change for HR: set goals and objectives in terms of changes in the economy and new trends for business [11], digital transformation, new technologies and super-technologies, speed and depth of change, endless information flow, new generation of employees and their expectations, new business models, changes in customer experience, cultural changes. All the above mentioned should contribute to actualization of HR functions and development of HR as a new source of company value.

References

1. Bogatyreva, I.V., Kozhukhova, N.V., Hakobyan, D.A.: Management of labor productivity/wage ratio as a factor of Samara region economic development. In: Mantulenko, V., (Ed.), Eurasia: Sustainable Development, Security, Cooperation. SHS Web of Conferences, **71** (04006). Les Ulis: EDP Science. (2019)
2. Bogatyreva, I.V., Kozhukhova, N.V., Zheleznikova, E.P.: Analysis of the current state of labour productivity in Russia. Econ. Labour **5**(3), 683–698 (2018)
3. Bogatyreva, I.V., Ilyukhina, L., Simonova, M.V., Kozhukhova, N.V.: Estimation of the efficiency of working time use as a factor of sustainable increase of labor productivity. In: Mantulenko., V, (Ed.), Proceedings of the International Scientific Conference "Problems of Enterprise Development: Theory and Practice". SHS Web of Conferences, **62,** 06002. Les Ulis: EDP Science. (2019)
4. Haag, S., Bauerdick, C., Campitelli, A., Anderl, R., Abele, E., Schebek, L.: A framework for self-evaluation and increase of resource-efficient production through digitalization. Procedia CIRP **72**, 850–855 (2018)
5. Hicran, H.C., Arif, E: Job assignment model in intelligent transportation systems: Istanbul electric tram tunnel administration. In: Zehir, C., Erzengin, E., (Eds.), Proceedings of the 7th International Conference on Leadership, Technology, Innovation and Business Management, **75,** (pp. 10–26). London: European Proceedings (2019)
6. Kozhukhova, N.V., Adykova, D.B.: To the question of increasing the efficiency of personnel work in the organization. Sci. XXI Century: Actual Dir. Develop. **1**(1), 299–303 (2019)
7. Movsesian, V.E., Kozhukhova, N.V.: Theoretical aspects of labour management in conditions of innovative economy. Probl. Enterp. Develop. Theor. Pract. **1**, 121–124 (2017)
8. Pechová, J.: Personnel trends in a globalised world. In: Kliestik, T, (Ed.), Proceedings of the 19th International Scientific Conference Globalization and its Socio-Economic Consequences 2019 – Sustainability in the Global-Knowledge Economy. SHS Web of Conferences, **74** (01024). Les Ulis: EDP Science (2020)
9. Simonova, M.V., Kozhuhova, N.V.: Strategies for obtaining added value in developing technological innovations. In: Ashmarina, S.I., Vochozka, M., Mantulenko, V.V., (Eds.), Digital Age: Chances, Challenges and Future. Lecture Notes in Networks and Systems, **84**, (pp. 128–136). Cham: Springer. (2020)
10. Schallmo, D.R.A., Williams, C.A., Christopher, A.: Digital Transformation Now! Guiding The Successful Digitalization of Your Business Model. Springer, Cham (2018)
11. Stark, J.: Digital Transformation of Industry. Springer, Continuing change. Cham (2020)

Evaluating the Effectiveness of HR Management Departments Based on Cluster Analysis

M. A. Kolotilina(✉), A. A. Korobetskaya, and V. K. Semenychev

Samara State University of Economics, Samara, Russia
ms.kolotilina@bk.ru, kornast@yandex.ru,
valeriysem@icloud.com

Abstract. The article examines problems of evaluating the effectiveness of personnel management services (departments) of JSC "Russian Railways" branches based on the statistical method. Special attention is paid to the selection and justification of criteria that ensure selective use of data sets for cluster construction. The classification of criteria is analyzed depending on HR metrics. The authors focus on the clustering algorithm, the k-means method. The use of statistical clustering method allows analyzing the effectiveness of personnel management departments of JSC "Russian Railways". The analysis of HR policy and HR management strategy allowed us to create a set of HR metrics that take into account the main directions of the company's HR processes. Each department has a specific development, which is explained by the production conditions, organization and management features. However, it is possible to speak about regularities when these features are repeated in the aggregate of services (departments). It is the objectivity of features that helps to distinguish homogeneous groups of services (departments).

Keywords: Clustering · HR metrics · HR management · K-means method · KPI indicators · R language

1 Introduction

Modern tools with modeling capabilities allow to monitor the main performance indicators of personnel management services (departments), make optimal management decisions based on a comprehensive analysis of personnel information, as well as to develop, realize and monitor the implementation of corporate human resource management strategies. Predictive Analytics is of particular importance for JSC "Russian Railways", where processing and understanding of a huge amount of data is required and there are high risks when making decisions. Statistical training is an indispensable set of tools for extracting useful information from large and complex data sets [4]. The KPIs' values are not always used effectively, but they can be used to optimize operational processes and improve technical and economic indicators.

The methods of modeling and forecasting used by the authors include linear regression, classification, clustering, repeated sampling, regularization, decision trees, support vector machines, and others. Why are they limited to clustering? Examples of

S. I. Ashmarina and V. V. Mantulenko (Eds.): IPM 2020, LNNS 161, pp. 193–200, 2021.
https://doi.org/10.1007/978-3-030-60926-9_26

practical implementation of methods are realized using R, an extremely popular open-source statistical computing environment. When forming an assessment of the effectiveness of personnel management departments, there is usually a situation when several indicators of HR metrics are proposed, which form an assessment of the effectiveness of the company's HR policy [8].

According to Alifanov, the following ratios are reflected among the performers of the considered problem: 60% of HR specialists do analytics manually in Excel; 35% of HR professionals work without regular HR analytics and data management processes; 15% of HR specialists who have all HR data store this information in one integrated system; 3% of HR professionals use Predictive Analytics tools based on machine learning [1]. Thus, obvious is the difficulty to accurately and promptly assess the effectiveness of the personnel department performance: that is, to determine to what extent the company's system of indicators analysis allows to ensure greater labor productivity, increase profitability, and reduce the time to achieve strategic goals [2].

Therefore, it is reasonable to use clusterization, which means a wide range of methods designed to detect clusters of data. When clustering observations from a certain data set, separate groups are selected so that observations within each group are similar to each other, and observations from different groups differ markedly from each other. To specify this task, we have identified a number of features that characterize the activities of personnel management departments. When analyzing the subject area, we identified 24 attributes (Table 1). Based on the results of the analysis of JSC "Russian Railways", the authors obtained clusters of combined groups based on similarity of features. The competitive moment helps the personnel management departments of JSC "Russian Railways" branches to improve key performance indicators and the quality of work.

2 Methodology

The RStudio development environment based on k-means clustering was chosen as a tool for analyzing key indicators. This method is a simple approach for splitting some data set *k* into separate, disjoint clusters. To perform clusterization using this method, we have to specify the desired number of clusters. The k-means method is that clustering is good if the intra-cluster variation is minimal ("within-cluster variation"). In the next step, the average feature values in each cluster are constants that minimize the sum of square deviations, and in the second step, redistributing cluster labels can only improve. As the algorithm runs, the resulting clustering will gradually improve until the result no longer changes; the target function will never increase. Stabilization of the clustering result means that the local optimum has been reached. The name of the method is determined by the fact that at the centroids step, average values of observations are calculated as the assigned ones to each cluster. The results depend on initial clustering of observations at the first step of the algorithm [2].

Table 1. The main KPI's

Name of the indicator	Characteristics of indicators
The number of enterprises with less than 97% staffing of the main production groups (only 1 group)	The companies' share with the main production groups is less than 97%
	Staffing levels excluding employees who are on leave for child care
	Staff turnover including those dismissed by agreement of the parties
Personnel reserve	Turnover of heads of the nomenclature of f (N, NZ), FR)
The recruitment and appointment of candidates	Percentage of employees who have more than 50 days off
	Share of long-term vacancies in the nomenclature f, (N, NZ), FR
	Effectiveness of the personnel reserve of the nomenclature f (N, NZ), FR
	The share of released managers of the nomenclature of f (N, NZ), FR with less than 1 year of work experience
Qualitative composition	Replacement of positions requiring higher education by practitioners
	Replacement of positions requiring secondary professional education by practitioners
	The number of employees with an inappropriate level of education in engineering and technical positions
	Number of employees with an inappropriate education profile in engineering and technical positions
Personnel assessment	Percentage of employees who have a corporate competency rating and are in the basic reserve
	Percentage of employees who have a corporate competence rating and are in the corporate reserve
Staff training and development	Implementation of the plan for training and retraining of workers, (%)
	Training at the training center
	% implementation of the workers' professional development plan
	Implementation of the plan for improving the skills of managers and specialists, (%)
	The number of workers related to traffic safety with an overdue period of professional development
	The share of managers and professionals with the expired period of training in mandatory programs
Young professionals and youth	Percentage of target employees with higher education who have been employed for more than 1 year
	Percentage of young people under 35 years of age, %
Personnel administration	Percentage of those who leave their positions in the first year of work (dismissal on the employee's initiative, including retirement), by agreement of the parties, and for violation of discipline
	Percentage of employees reinstated by a court decision from the whole list of employees

Source: authors based on [9].

3 Results

Evaluating the performance of HR management departments of JSC "Russian Railways" branches requires an analysis of the main KPIs of HR management efficiency and management policies. Key indicators for cluster analysis are presented in (Table 1). Key performance indicators of personnel management are determined based on the general recommendations and practices of companies [9].

The presented indicators characterize the effectiveness of the HR policy and HR management strategy. The indicators are determined in accordance with the methodology for determining the effectiveness of the personnel management services (departments) of JSC "Russian Railways" branches.

Each of the KPI indicators is taken into account as a cumulative total. The indicator is recorded quarterly on the last day of the month, for each department of HR management of JSC "Russian Railways" branches. JSC "Russian Railways" pays great attention to the strategy of personnel development and improvement of strategic indicators. The task is to conduct clustering of 26 personnel management departments of JSC "Russian Railways" branches, each of which is characterized by certain attributes (Table 1). Quarterly data were collected for 26 personnel management departments of JSC "Russian Railways" branches for the 2018 quarter II and 2019 quarter II.

The advantage of this method is that it is also an easy way to evaluate not only whether you belong to a cluster, but also how close you are to other services (departments) in the same cluster. This allows you to identify one or more of the most "typical" services (departments) that are close to the center of the cluster and should be studied in more detail. To combine objects into clusters, you have a similarity measure. The most important characteristics of a cluster structure are the shape of the cluster and the size of the clusters (expressed in the number of objects).

In the k-means method, the clustering algorithm is constructed in such a way that it seeks to minimize the total square deviation of objects from cluster centers:

$$V = \sum_{i=1}^{k} \sum_{x \in S_i} (x - \mu_i)^2 \rightarrow min, \quad (1)$$

where k is a number of clusters, Si – obtained clusters, μ_i – cluster centers.

As a distance measure between objects, the Euclidean distance in the space of certain quantitative features is used, and the centroid method is applied to determine the distance between clusters.

The algorithm performs an iterative procedure for recalculating cluster centers until stable clusters are formed. However, the result depends on the selected starting centers. They are usually selected randomly, but can also be set by the researcher based on the nature of the clustered objects and existing hypotheses about their grouping.

In this study, the analysis of the performance of HR departments of JSC "Russian Railways" branches is carried out using the clustering method, the k-means method in the RStudio development environment, the window of which is shown in Figs. 1 and 2. For the k-means method, the k-means function from the standard statistics package was used.

```
K-means clustering with 3 clusters of sizes 13, 7, 6

Cluster means:
  ykomplektovannost.97 ykomplektovannost  tekychest dolivakancii  dolirykov zamejeneipraktiki zamejeneipraktikicredne_tex doli_rab_inener doli_rab_ne_coot_inener
1           0.05170069         0.6760955 0.24876030  -0.02850810 -0.2160111        -0.4755517                 -0.02980363      -0.7681105             -0.2180015
2           0.34542408        -1.3504999 -0.53762804  -0.06380385 -0.1673853         0.7283121                 -0.48010200       1.2398478             -0.3116018
3          -0.51501292         0.1107096 0.08825208   0.13620537  0.6633070         0.1806646                  0.62469353       0.2177504              0.8358720
  doli_rabot_komp doli_rabot_komp_baza vupoln_plana ychebnuecentru vupoln_plana_povush vupoln_plana_povush_kvalfikacii doli_rabotnikov_bezop_dvej doli_rykov_cpecial
1      -0.5327527           -0.0243051   -0.8530535     -0.4341714          -0.4877128                      -0.1231660                  0.5216921          0.3678396
2      -0.5153107           -1.1954490    0.4553545     -0.3914557          -0.4612125                      -0.1065807                  0.5384879          0.3843505
3       1.7554934            1.4473516    1.3170357      1.3974029           1.5947923                       0.3912039                 -1.7585688         -1.2453948
  долядоli_celivikov doli_molod  doli_yvol
1        0.008853139  0.5484245 -0.3894503
2        0.029014491 -1.3487727  0.2986177
3       -0.053032041  0.3853152  0.4954217

Clustering vector:
      Д       Т      ДИ    КДМВ     ДРП     ДПМ     НТЭ     ДТВ    РДЖВ    ОЦОр     ИВЦ    ДИТО    КДПО      НС      ТР      ДМ   ТЦФТО      НА     АХЦ     КЦМ     ДСС
      3       3       1       1       1       1       1       3       3       1       1       1       1       1       1       1       1       3       3       2       2
   ДАВС   КУЦПК БАЗА 47     ДЭЗ    КХТЛ
      2       2       2       2       2

Within cluster sum of squares by cluster:
[1]  53.77825  47.27058 174.15521
 (between_SS / total_SS =  45.0 %)

Available components:

[1] "cluster"      "centers"      "totss"        "withinss"     "tot.withinss" "betweenss"    "size"         "iter"         "ifault"
> |
```

Fig. 1. Result of cluster analysis based on data for 2018, the development environment window RStudio. Source: authors.

To ensure the stability of clustering results, it is necessary to select features that form a space that is close to orthogonal one, i.e. those that are loosely related to each other. In addition, clustering results are negatively affected by the large dimension of the feature space. The KPI indicators (Table 1) that characterize the effectiveness of the implementation of the HR strategy were considered. The next step is to determine the number of clusters and initial cluster centers that the clustering result depends on. The number of clusters from 2 to 4 was considered, and random points and/or random services (departments) were assigned as cluster centers. For each option of the starting conditions, the algorithm was run several times to determine the stability of the resulting clusters. As a result, 3 clusters were allocated. For each of the remaining 26 departments, the cluster membership was determined in different years and a measure of proximity to the cluster center was calculated ("typicality" of departments). The percentage of the cluster-explained spread of values in this split is 45% in 2018 and 45.8% in 2019. After analyzing the composition of each cluster and the characteristic values of features, we identified each of them as follows. Clusters based on a set of homogeneous features of departments are presented in Tables 2 and 3.

```
K-means clustering with 3 clusters of sizes 4, 20, 2

Cluster means:
  ykomplektovannost.97 ykomplektovannost   tekychest dolivakancii  dolirykov zamejeneipraktiki zamejeneipraktikicredne_tex doli_rab_inener doli_rab_ne_coot_inener
1           -1.0676976        0.81908404  0.40119214  -0.02850810 -0.2160111       -0.38950364                   0.3171958     -0.31337255            -0.09971772
2            0.1545039       -0.03321288 -0.02647562  -0.04086161 -0.1989921       -0.05419939                  -0.1874081     -0.06532511            -0.25076161
3            0.5903563       -1.30603924 -0.53762804   0.46563233  2.4219432        1.32100116                   1.2396890      1.27999621             2.70705154
  doli_rabot_komp doli_rabot_komp_baza vupoln_plana ychebnuecentru vupoln_plana_povush vupoln_plana_povush_kvalfikacii doli_rabotnikov_bezop_dvej doli_rykov_cpecial
1        1.755493            1.7143654    1.5007316      1.0172961           1.1201993                       1.7742335                 -1.7585688        -1.87374306
2       -0.526648           -0.4342055   -0.3951107     -0.4192209          -0.4784377                      -0.1173612                  0.5275706         0.37361844
3        1.755493            0.9133240    0.9496439      2.1576166           2.5439784                      -2.3748554                 -1.7585688         0.01130174
  долядоli_celivikov doli_molod   doli_yvol
1        -0.42651301  1.0918195 -0.09972118
2         0.01590961 -0.1155946 -0.14862651
3         0.69392990 -1.0276934  1.68570744

Clustering vector:
      Д       Т      ДИ    КДМВ     ДРП     ДПМ     НТЭ     ДТВ    РДЖВ    ОЦОр     ИВЦ    ДИТО    КДПО      НС      ТР      ДМ   ТЦФТО      НА     АХЦ     КЦМ     ДСС
      1       1       2       2       2       2       2       1       1       2       2       2       2       2       2       2       2       3       3       2       2
   ДАВС   КУЦПК БАЗА 47     ДЭЗ    КХТЛ
      2       2       2       2       2

Within cluster sum of squares by cluster:
[1]  44.86189 181.44034  44.60011
 (between_SS / total_SS =  45.8 %)
```

Fig. 2. Result of cluster analysis based on data for 2019, the development environment window RStudio. Source: authors.

Table 2. Directorates divided by cluster according to data for 2018

Group	The name of the services (departments)
Group 1	Infrastructure directorate, Kaliningrad directorate of motor car rolling stock, Kuibyshev directorate for track repair, Kuibyshev directorate for operation of track machines, Kuibyshev directorate for energy supply, Kuibyshev regional general service center, Samara information and computing center, Logistics directorate, Samara communications directorate, Kuibyshev directorate for repair of traction rolling stock, Kuibyshev directorate for terminal and warehouse complex management, Kuibyshev territorial center for corporate transport services
Group 2	Kuibyshev center of metrology, Directorate of social sphere, Directorate of emergency recovery facilities, Kuibyshev training center for professional qualifications, Base 47, Kuibyshev directorate for the operation of buildings and structures, Kuibyshev chemical and technical laboratory
Group 3	Transport directorate, Business management service, Administrative and economic center, Kuibyshev heat supply directorate, Kuibyshev regional railway directorate of railway stations

Source: authors.

Table 3. Directorates divided by cluster according to data for 2019.

Group	The name of the services (departments)
Group 1	Transport directorate, Kuibyshev heat supply directorate, Kuibyshev regional directorate of railway stations
Group 2	Infrastructure directorate, Kaliningrad directorate of motor car rolling stock, Kuibyshev directorate for track repair, Kuibyshev directorate for operation of track machines, Kuibyshev directorate for energy supply, Kuibyshev regional general service center, Samara information and computing center, Logistics directorate, Samara communications directorate, Kuibyshev directorate for repair of traction rolling stock, Kuibyshev directorate for terminal and warehouse complex management, Kuibyshev territorial center for corporate transport services, Directorate of social sphere, Directorate of emergency and recovery facilities, Kuibyshev training center for professional qualifications, Base 47, Kuibyshev directorate for the operation of buildings and structures, Kuibyshev chemical and technical laboratory
Group 3	Business management service, Kuibyshev center of metrology, Administrative and economic center

Source: authors.

4 Discussion

The well-known solutions to the problem of personnel management in large companies by foreign and domestic scientists are considered and taken into account [3, 5–7, 9]. These studies investigate managerial innovations in the field of personnel that determine the increase in the company's innovative potential, identify a set of personnel innovations that implement a process approach to management, and establish their relationship with the company's innovation performance indicators. Our research deals with the problem of analyzing the effectiveness of the personnel management services (departments) of JSC "Russian Railways" branches as a necessary condition for the development of a new type of economy in the conditions of digitalization of the Russian Federation. Modern analytics tools have been identified that allow HR specialists to conduct a multi-sided analysis of KPIs. The authors offered a new look at a well-known problem.

5 Conclusion

In the company of JSC "Russian Railways" in 2018 the analysis of the activities was carried out using the rating system. The rating allows to select services (departments) based on the calculation of the indicator rank. The rating is always a one-dimensional scale, and clusters are built in a 24-dimensional (in our case) space of all specified attributes. The k-means method is the most popular clustering method. Our research suggests using this method to evaluate the effectiveness of human resources management services (departments). This method allows you to solve a practically significant problem and formulate new recommendations for HR services. The k-means method in the R environment enabled to divide HR management services (departments) into 3 clusters based on homogeneous characteristics. This makes it easier to evaluate the effectiveness of human resources management services (departments). By accessing any of the 3 clusters, you can see a set of departments that are similar in terms of KPI performance.

Acknowledgments. The reported study was funded by RFBR, project number 20-010-00549.

References

1. Alifanov, K.: People and digital. HR-Partner **3**, 16–19 (2019)
2. Chadwick, C., Li, P.S.: HR systems, HR departments, and perceived establishment labor productivity. Hum. Res. Manag. **57**(6), 1415–1428 (2018)
3. El-Kassar, A.N., Singh, S.K.: Green innovation and organizational performance: The influence of big data and the moderating role of management commitment and HR practices. Technol. Forecast. Soc. Chang. **144**, 483–498 (2019)
4. James, G., Witten, D., Hastie, T., Tibshirani, R.: An introduction to statistical learning with applications in R. Springer, New York (2017)
5. Matusova, D., Gogolova, M.: Using of new management approaches in the field of personal marketing in the transport company. IOP Conf. Ser.-Mater. Sci. Eng. **245**, 042027 (2017)
6. Simek, D., Sperka, R.: How robot/human orchestration can help in an HR department: a case study from a pilot implementation. Organizacija **52**(3), 204–217 (2019)
7. Turriago-Hoyos, A., Thoene, U., Arjoon, S.: Knowledge workers and virtues in Peter Drucker's management theory. Sage Open, **6**(1) (2016). http://sgo.sagepub.com/content/6/1/2158244016639631. Accessed 24 06 2020
8. Vienni, Bachtiar, M.: Analysis of performance measurement at HR-GR department using the balance scorecard method. In: (2017). IOP Conference Series-Materials Science and Engineering, vol **277**, p. 012006 (2017)
9. Zamecnik, R.: Human resources controlling as a tool for measuring human resources key performance indicators. In: Pastuszkova, E., Crhova, Z., Vychytilova, J., Vytrhlikova, B., Knapkova, A., (Eds.), Proceeding of the 7th International Scientific Conference on Finance and Performance of Firms in Science, Education and Practice, (pp. 1681–1695). Zlin: Tomas Bata University. (2015)

Professional and Public Accreditation of HR Management Educational Programs: Prospects and Challenges

V. G. Konovalova[1](✉), M. A. Fedotova[2], and Inh Bình[3]

[1] State University of Management, Moscow, Russia
vg_konovalova@guu.ru

[2] Moscow Aviation Institute (National Research University), Moscow, Russia
fedotova-ma@yandex.ru

[3] Center for Chinese-Russian Humanitarian Cooperation and Development, Beijing, China
bide9368454@hotmail.com

Abstract. In order to strengthen control of the educational programs market by the professional community, the decision of the council on professional qualifications in the field of personnel management, basic regulation for conducting professional and public accreditation of educational programs in the field of personnel management has been adopted. The article reveals the essence and objectives of professional and public accreditation as a necessary condition for achieving the goals of education quality assessment; analyzes the structure of the main processes and groups of participants in professional and public accreditation. The method of conducting accreditation expertise is presented; the features of applying accreditation criteria and indicators are revealed, taking into account the experience of the conducted expertises. An issue related to the implementation of accreditation procedures that require further clarification are highlighted. Based on the work results, groups of tasks for further development of professional and public accreditation were identified, including: 1. normative and methodological development, 2. use of criteria and indicators for assessment of programs, and 3 increasing opportunities for applying its results.

Keywords: Accreditation expertise · Accreditation of educational programs · Accreditation professional and public · Independent assessment · Professional education · Quality of education

1 Introduction

Professionally-public accreditation received legislative regulation with the adoption of the Federal law of 29.12.2012 № 273-FL "On education in the Russian Federation" [6] in accordance with specified accreditation "Is the recognition of the quality and level of training of graduates who have mastered an educational program of each organization performing educational activities that meet the requirements of professional standards, labor market requirements specialists, workers and employees of the corresponding profile". The development of professional and public accreditation of educational

S. I. Ashmarina and V. V. Mantulenko (Eds.): IPM 2020, LNNS 161, pp. 201–209, 2021.
https://doi.org/10.1007/978-3-030-60926-9_27

programs, as world practice shows, should become an important mechanism for maintaining healthy competition in the market of educational services, a factor in ensuring the quality of education both in Russia and internationally [2, 3, 12, 17].

2 Methodology

The purpose of professional and public accreditation is to improve the quality of education based on and taking into account the opinions of expert practitioners representing the professional community and participating in the processes: designing and improving the content of implemented educational programs; identifying the best practices of educational organizations; informing the public about the best educational experience that meets modern standards of education quality. The national council for professional qualifications under the President of the Russian Federation has developed documents regulating the basic principles of professional and public accreditation, the procedure for its conducting, selecting, monitoring and controlling the activities of organizations that carry out accreditation, etc.

The Ministry of Science and Higher Education of the Russian Federation has created a system for monitoring professional and public accreditation, which contains registers of accreditation organizations and organizations that carry out accreditation expertise. Today, we have already accumulated some experience in professional and public accreditation of professional educational programs in the direction of "Personnel Management", which took into account the existing foreign experience and approaches to emerging problems solving in the implementation of such programs [1, 5, 9, 11, 14, 16].

In order to strengthen control in the educational programs market by the professional community, the council for professional qualifications in the field of personnel management approved the basic regulation for conducting professional and public accreditation of professional educational programs in the field of personnel management, standard documentation is prepared, the list of experts approved, and the accreditation organization is determined [4].

The structure of the main processes of professional and public accreditation includes the interaction of three participants: subjects of the educational services market (main and additional educational programs), the accreditation organization, and an expert group. Further promotion of professional and public accreditation of professional educational programs in the field of personnel management is possible only if there are clear and understandable conducting regulation, appropriate methodological tools, as well as if the establishment of close cooperation between educational organizations and employers and their partnership motivation, which determined the objectives of this study.

3 Results

Let's look at the main processes of professional and public accreditation (Fig. 1):

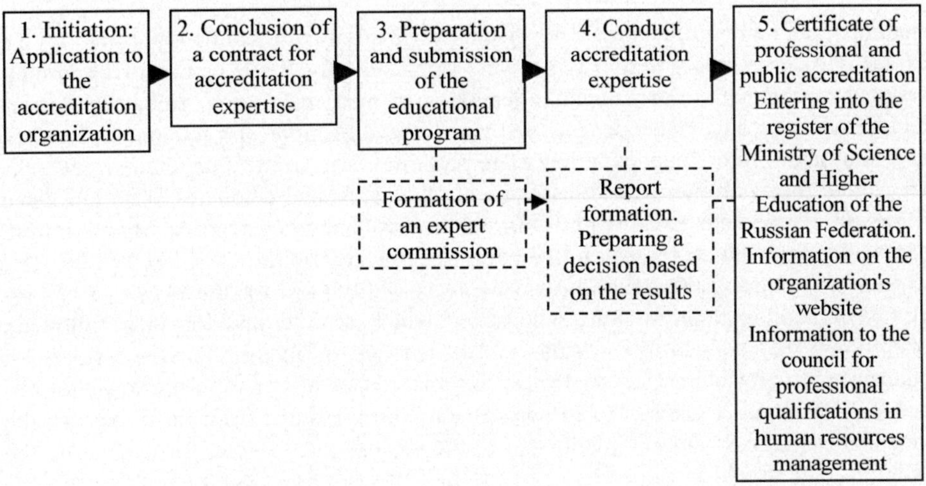

Fig. 1. General procedure for conducting professional and public accreditation (minimax scheme for the customer) (Source: authors).

- subjects of the educational services market make a request to the accreditation organization and agree on the documents necessary for the expertise of the professional educational program (the form and content of documents, for example, a certificate on the program as a whole, orders to open the program, teaching staff, training and evaluation tools, etc.),
- the accrediting organization enters into a contract for expertise, forms an expert group, makes additional requests for materials on the program (if necessary) and issues certificates of accreditation (based on the results of the expertise),
- the expert group forms a report on the results of the expertise, implements additional expertise procedures, including making requests to representatives of the professional community (if necessary).

The basic component of the educational programs expertise when conducting accreditation in the field of personnel management is the use of criteria and indicators for assessment of programs. Initially, the criteria are defined by the National Agency of the Qualifications Development [15], the council for professional qualifications in the field of personnel management has developed a procedure for assessment indicators [4]. In this article, we will share the experience of assessing and analyzing indicators, taking into account the experience of the conducted expertise. In general, the accreditation expertise is based on seven criteria that are assessed through groups of indicators (the assessment features are related to the level of the educational program: higher or additional education and the group of criteria used for assessment).

Criterion 1: The results of the graduates of the educational program passing a professional exam in the form of an independent assessment of qualifications (if there is an independent assessment of qualifications for the corresponding qualification).

This criterion is considered as one of the most important for assessment and making a decision: the higher the specified indicator (for example, as a percentage of all students), the more grounds for a positive assessment of the accredited program by the expert. However, this criterion is not very applicable for the expertise of basic educational programs; since students generally cannot meet the mandatory formal requirements of an independent assessment of qualifications (professional education and minimum professional experience are required). Because of this, the current tasks are to find forms of intermediate assessment of students and graduates, or to introduce forms of intermediate assessment of formed professional competencies, which, in turn, can be used for the accreditation procedure for this criterion.

On the contrary, this criterion can be successfully used for the expertise of programs of additional professional education, which is determined by the following conditions: the availability of data on the passage of independent assessment of qualifications procedures for graduates, the analysis of information on how the procedure is planned as a result of the professional program, the decision to pay for the independent assessment of qualifications (at the expense of the participant or the program organiser).

Criterion 2: Compliance of the training results stated in the educational program (in the form of professional competencies) with professional standards and other qualification requirements established by federal laws and other regulatory legal acts of the Russian Federation.

The analysis can be carried out based on the content of the educational program, starting with documents confirming the program legitimacy (opening order, curriculum, set of educational materials, etc.). In documents the relevance of the terms of their approval (dates, signatures), compliance with standards and qualification requirements (indication of compliance with the state educational standard, professional and other standards in the program text), including the level of qualification are checked. According to this criterion, the analysis of programs of higher and additional professional education practically does not differ.

Criterion 3: Compliance of curriculums, work programs of academic subjects, courses, disciplines (modules), evaluation materials and procedures with the planned results of the educational program development (competencies and learning outcomes).

This criterion reflects an important substantive assessment of the program on compliance with professional standards, and the training materials on compliance and sufficiency of professional training. Conducting the expertise, professional training and specialist experience in the reviewed area is very important (e.g., program selection, assessment or staff development and the qualified expert understanding in this area). The main point of the examination is similar for higher and additional professional education: it is necessary to determine in which topics, sections, and disciplines professional competencies are formed. Expertise can be started with assessment of the compliance of the specified parameters for the sections of the program and their presence in the relevant training materials (topics, questions, tasks, etc.). When conducting an assessment, the expert expresses an opinion on the extent to which the training topic (section, discipline, etc.) on the training material and evaluation tools allow to form professional competencies corresponding to generalized labor functions/labor actions. Practice shows that at this stage of the expertise, in some cases, additional external expertise may be required, but the procedure for such actions has not yet been developed. The solution may be to form a pool of experts in certain professional areas and the possibility of contacting them for expertise on special "narrow" topics and issues.

Criterion 4: Compliance of personnel, material and technical, information and communication, educational and methodological and other resources that directly affect the quality of graduates training, the content of professional activities and professional tasks for which the graduate is preparing. The indicators used to assessment (indicating the threshold value) are set for programs of additional professional education and affect the quality of the program teaching staff, including: the proportion of teachers involved in the training on educational programs and with a higher education (professional training) (30%); the proportion of teachers trained in relevant organizations or involved in consulting activities for an accreditation application of educational programs for the last three years (30%), etc.

The current assessment procedure includes checking the staffing of the educational program. If the normative standards for staffing are set by the corresponding state educational standard, the parameter cannot be lower than the normative standard, and in a situation of lack of practice on the indicator, the expert has the right to indicate the lack of information on the parameter and not take it into account in the general conclusion on the program.

Criterion 5: there is a demand for the educational program, the demand for graduates of professional educational programs by employers (the latter is only for higher education programs).

The availability of demand for the program is assessed by experts initially on the basis of information provided by the educational organization. For higher educational programs, this is usually the information obtained from a survey of graduates and confirmed by certificates on the program. As a basis for making a decision and assessment of the criteria, it should be recommended to conduct interviews with employers who are ready to confirm the success of employment. Information confirming that the program has students working on their specialty may be the criterion method of assessment on programs of additional professional education. In practice, experts faced situations when the indicator for assessment of the program of advanced training could not be set, which was noted in the expert opinion, respectively, the criterion was assessed partially (for example, by the indicator: the percentage of graduates who were employed within a year after the end of the program (70%), from them - employed in the specialty (50%).).

Criterion 6: Confirmed participation of employers (in the design of the educational program, including the planned results of its development, evaluation materials, curriculums, work programs; organization of students project work; development and implementation of practice programs, the formation of planned results of their passage; the development of topics of final qualifying works that are significant for the relevant areas of professional activity (only for higher education programs)).

The procedure for assessment of this criterion immediately included differentiation of indicators for programs of higher and additional professional education. The practice of conducting expertise shows that confirmation of this criterion begins with checking the availability of reviews from the employer on the educational program, indicating the positions that were agreed with this criterion. Since for higher education programs the emphasis is on project work and practices, it is advisable to look at the contracts and the corresponding reports of students' practices, to record what percentage of the topics of final qualifying works were completed in agreement with employers (for example, how much the quality of the final work satisfies the employer and affects the assessment of the possibility of offering a subsequent employment for a graduate). In addition, it is possible to provide interviews with employers, during which it is recommended to ask questions on what professional competencies are important for the employer and how much the educational program allows to master them.

Considering the expertise practice of additional education programs for evaluation by this criterion, it is also advisable to clarify the following question: whether the program is universal (developed as an offer to the open market of educational services) or developed for a specific customer and the corresponding labor functions. As practice has shown, it is recommended to clarify information on the extent to which the evaluation tools of the program (tests, cases, questions, tasks, and other evaluation tools) were agreed with the customer and developed for a specific professional standard and labor functions.

Criterion 7: Assessment of criteria is carried out taking into account the interaction of the organization engaged in educational activities with other organizations, including in the framework of the network form of educational programs implementation. At the moment, there are no specific indicators for this criterion assessment, but taking into account the experience of experts in the educational services market, we can recommend paying attention to the following:

- the practice of implementing educational programs in the field of personnel management with the use of a network form has not yet found an active distribution in the market. There is also no experience of such programs expertise yet,
- potentially for assessment, we can consider whether the educational organization and its partners have certificates, etc., confirming the assessment of programs by employers and partners, including international certification,
- in general, additional indicators for a positive assessment of accreditation expertise can be: the presence of certificates from employers'communities, professional associations, data on students' victories in professional skill competitions, etc.,
- when implementing programs using the network form, the presence of such achievements should be considered by all participants of the interaction.

To develop the possibility of using this criterion, we can recommend selecting a list of possible accreditations and certificates that will be analyzed and positively considered according to the criterion. All these criteria are evaluated comprehensively, based on the orientation of a specific educational program to a specific labor market and the needs of key employers.

4 Discussion

During the testing of the accreditation procedure and its methodology, a number of areas were identified that require further clarification:

1. The period of reporting for the expertise. Current solution: use of information on the implementation of a professional educational program for the last three years (or since the beginning) of the program. According to experts, the rules of accreditation should provide for an increase in the period for submitting information on the implemented program, if necessary.
2. Cost of independent assessment of qualifications for accredited programs. Current solution: experience is being developed. According to experts, in order to spread professional and public accreditation in the educational services market, the accreditation regulations should provide incentives for paying the cost of passing an independent qualification assessment for graduates of accredited programs.
3. In a number of situations, additional special expertise of modules/disciplines may be required (for example, when conducting a check on the criterion of compliance with the declared labor functions/actions, competencies and training results). Current solution: there is no practice. According to experts, the regulation of accreditation should provide for forms and procedures of attracting additional experts/obtaining additional expertise on controversial and/or specific issues.
4. List of documents for request on the accredited program. Current solution: information requested by the expert group. According to experts, the regulations for professional and public accreditation should provide a list of necessary documents and forms of references not only for the main professional educational programs, but also for programs of additional professional education, taking into account their specifics and implementation features.
5. Save a conclusion on the results of professional and public accreditation when changing the content of the program. Current solution: there is no practice. According to experts, the regulation of professional and public accreditation should provide for the boundaries of changes in the professional educational program, in which the expert opinion retains legitimacy.
6. Deviation of one or more indicators from the threshold value by 10–15%. Current solution: there is no practice. According to experts, it is advisable to analyze the question of whether the threshold value can be exceeded for different indicators.
7. Additional opportunities for using the results of professional and public accreditation. Current solution: there is no practice. According to experts, it is advisable to provide for the formation of a rating of professional educational programs based on the results of professional and public accreditation (including programs of additional professional education).
8. The composition of the expert group. Current solution: accounting of mandatory requirements for the procedure. Possible development: it is worth considering the task of allocating expert classes depending on the specifics of the programs [7].

5 Conclusion

Among the most obvious results of professional and public accreditation are the trust of the target audience to the quality of services provided by the educational organization; the ability of graduates who have completed programs that have passed professional and public accreditation to receive a certificate of their high professional qualifications recognition. The opinion of experts representing the professional community allows us to identify the shortcomings of the educational program and opportunities for its improvement. However, since professional and public accreditation is not a mandatory procedure, most of the mechanisms for its implementation are still being defined and are being formed.

Among the nearest regulatory and methodological tasks of professional and public accreditation of educational programs in the field of personnel management: correction of normative documentation taking into account the results of trial accreditations and development of methodological recommendations for accreditation. Considering the experience of conducting professional and public accreditation, we can talk about the tasks of developing and refining indicators according to the evaluation criteria, namely: 1) maintaining the accreditation specifics for higher and additional professional education and even strengthening it for a number of indicators (for example, additional consideration of the importance of developing a program of additional education in accordance with the goals of the employer); 2) dividing criteria and indicators into main and additional when conducting evaluation procedures (for example, further consideration of the feasibility of introducing the weight of evaluation indicators), especially taking into account the experience where in some cases not all indicators can be established.

It is important to emphasize that one of the obstacles to the activation of independent quality assessment processes in the field of professional education is the weak motivation for conducting professional and public accreditation both in the educational community and among employers involved as experts, and the entrenched view on its subjectivity. In this regard, we should pay attention to the following tasks for the development of professional and public accreditation and the possibility of using its results. Today, higher education institutions with the status of national research universities in Russia consider an important component of assessment of their professional status to obtain a certificate of professional and public accreditation according on relevant professional standards. And here it is worth noting two aspects: the employer's assessment of training programs relevance and compliance with the "standards" of the profession. The need to consider such an assessment when allocating "targeted training" places is also discussed. Increasing the significance by considering the results of professional and public accreditation as an additional "plus" when passing the state accreditation procedures. There are suggestions on this issue on the possibility of accounting the accreditation results in terms of assessment of employers representatives involved in the implementation of the main educational program as a partial or full-fledged replacement for the questionnaire.

Ranking according to the results of professional public accreditation results on the official websites of councils for professional qualifications and educational institutions (e.g. admissions officers of universities, retraining centers) can be seen as the evidence confirming the quality of programs by the employer and their compliance with professional standards.

World experience also shows that mechanisms for accreditation of educational programs and certification of specialists should be developed in a coordinated manner, for which it is necessary to develop criteria and procedures that ensure continuity and comparability of assessments of the educational content quality and the level of specialists' qualification [8, 10, 13]. In many countries, a mandatory condition for certification of a specialist's qualification is the completion of a university program that has passed professional and public accreditation. This experience is also useful for specialists in the field of personnel management.

References

1. Alzafari, K., Ursin, J.: Implementation of quality assurance standards in European higher education: does context matter? Qual. High. Educ. **25**(1), 58–75 (2019)
2. Bendixen, C., Jacobsen, J.C.: Accreditation of higher education in Denmark and European Union: from system to substance? Qual. High. Educ. **26**(1), 66–79 (2020)
3. Bendixen, C., Jacobsen, J.C.: Nullifying quality: the marketisation of higher education. Qual. High. Educ. **23**, 1–15 (2017)
4. Council for Professional Qualifications in the field of Personnel Management: professional and public accreditation (2018). http://sovethr.ru/poa-sovethr/. Accessed 14 April 2020
5. Dakovic, G., Gover, A.: Impact evaluation of external quality assurance by institutional evaluation programme. Qual. High. Educ. **25**(2), 208–214 (2019)
6. Federal Law of 29.12.2012 No. 273-FZ On education in the Russian Federation. http://www.consultant.ru/document/cons_doc_LAW_140174/. Accessed: 30 June 2020 (2012)
7. Fedotova, M.A., Binn, I.: Tasks of professional and public accreditation of educational programs in the field of personnel management. In: Durakova, I.B., Taltynov, S.M. (Eds.), Management in Management Training Programs, pp. 112–114. Voronezh State University, Voronezh (2019)
8. Ferran, J.Z.C.: Business school accreditation in the changing global marketplace: a comparative study of the agencies and their competitive strategies. J. Int. Educ. Bus. **9**(1), 52–69 (2016)
9. Lagrosan, S.: Quality through accreditation. Int. J. Qual. Serv. Sci. **9**(3/4), 469–483 (2017)
10. Manatos, M., Huisman, J.: The use of the European standards and guidelines by national accreditation agencies and local review panels. Qual. High. Educ. **26**, 1–18 (2020)
11. Margarov, G., Mitrofanova, E., Konovalova, V., Mitrofanova, A., Trubitsyn, K.: Effectiveness of implementing practice-oriented higher education programs based on stakeholders. In: Ardashkin, I.B., Martyushev, N.V., Klyagin, S.V., Barkova, E.V., Massalimova, A.R., Syrov, V.N. (Eds.), Proceedings of the International Conference on Research Paradigms Transformation in Social Sciences. The European Proceedings of Social and Behavioural Sciences, vol. 35, pp. 906–914. Future Academy, London
12. Marshall, S.: Quality as sense-making. Qual. High. Educ. **22**(3), 213–227 (2016)
13. Matei, L., Iwinska, J.: Quality Assurance in Higher Education. Central European University, A practical handbook. Budapest (2016)
14. Mitrofanova, E.A., Konovalova, V.G., Mitrofanova, A.E.: Opportunities, problems and limitations of digital transformation of HR management. In: Mantulenko, V.V. (Eds.), Proceedings of the International Scientific Conference Global Challenges and Prospects of the Modern Economic. The European Proceedings of Social and Behavioural Sciences, vol. **57,** pp. 1717–1727. London: Future Academy
15. National Agency of the Qualifications Development: Professional and public accreditation is a guarantee of the training quality (2018). https://nark.ru/news/za-professionalno-obshchestvennuyu-akkreditatsiyu-.php. Accessed 14 April 2020
16. Salto, D.: Quality assurance through accreditation: when resistance meets over compliance. High. Edu. Q. **72**, 78–89 (2018)
17. Ulker, N., Bakioglu, A.: An international research on the influence of accreditation on academic quality. Stud. High. Educ. **44**(9), 1507–1518 (2018)

Organization of Labor Activity at Construction Enterprise During the Spread of Coronavirus

M. V. Lovcheva(✉)

State University of Management, Moscow, Russia
lovchevamv@mail.ru

Abstract. The article reflects the results of the practical implementation of measures to organize the work of the personnel of construction organizations, which were developed and implemented during the spread of the new coronavirus infection (COVID-19). According to the author, the Russian Government has chosen the right strategy to combat COVID-19, and as a result, the damage caused by coronavirus infection was less in our country than in many other countries. To develop a system of measures that have been successfully implemented at a construction enterprise, the author studied the general for all employers and special industry (for construction) regulatory documents and methodological recommendations developed by the authorities of the Russian Federation, Federal Service for Surveillance on Consumer Rights Protection and Human Wellbeing, the Ministry of Health of Russia, the Ministry of Construction of Russia, the Government of Moscow, The Government of the Moscow Region, the Association "All-Russian Non-State Non-Commercial Organization - the All-Russian Industry Association of Employers "National Association of Self-Regulatory Organizations Based on Membership of Persons Carrying Out Construction", and on their basis a system of organizational and control measures was prepared and implemented into the work of the enterprise, presented in the article.

Keywords: Building · Coronavirus · Labor resources · Personnel · Personnel management

1 Introduction

5.7% is the contribution of the construction industry to the Russian economy. In the industry today there are two hundred seventy-eight thousand organizations, which employ almost five million people. Every year, the professional training system produces about two hundred thousand young specialists with higher, secondary and primary professional education. To provide more accurate compliance with the requirements of employers for the qualifications of industry specialists in the field of construction, eighty-eight professional standards have been developed and approved, the Unified Certification System for managers and specialists of the construction industry, 38 councils for professional qualifications in construction, and 56 centers for qualification assessment are operating in 36 cities of Russia. In 2017, the national registers of specialists in the field of construction, design, engineering surveys became

S. I. Ashmarina and V. V. Mantulenko (Eds.): IPM 2020, LNNS 161, pp. 210–220, 2021.
https://doi.org/10.1007/978-3-030-60926-9_28

fully operational. A major reform of the town-planning legislation, the construction financing system and the construction self-regulation system is ongoing.

At the same time, the problem of labor resources in the industry is very acute, according to the Deputy Prime Minister of the Russian Federation Khusnullin [17], the industry still lacks about two million builders.

Personnel problem in the industry has become even more acute during the spread of the new coronavirus infection COVID-19. So, according to a survey of 229 developers from 59 regions of the Russian Federation, which was carried out by the Institute for the LLC "Development of the Construction Industry" in the period from 14/04/2020 to 26/04/2020 at the request of the National Association of Builders (National Association of Self-Regulatory Organizations Based on Membership of Persons Carrying Out Construction) [12] as part of the implementation of paragraph 2.7 of the priority measures plan (actions) to ensure sustainable economic development in the context of a worsening situation due to the spread of a new coronavirus infection, approved by the Government of the Russian Federation on March 17, 2020 [14], developers named a number of issues that have arisen most acutely during the pandemic:

- the end of work of the majority of construction contractor organizations, manufacturers of construction materials, suppliers of construction materials and equipment,
- changes in the cost and delivery time of building materials,
- complication of interaction with banks, construction financing,
- a high degree of bureaucratization of procedures in construction (building permits, commissioning permit, obtaining authorization documentation, approval, etc.),
- reduced demand for real estate.

In the field of personnel, 56% of developers experienced some kind of problems, and 26.4% named these problems significant. Problems were notified both in terms of personnel, and in terms of organizing the work of personnel during a pandemic, and in terms of organizing and financing preventive measures, as well as in connection with an increase in cases of panic in collectives and the outflow of migrant workers, which in many construction organizations constitute the largest share in the number of workers):

- 36.2% of developers noted that there were difficulties with the purchase of protective equipment (masks, antiseptics, etc.),
- 24.5% indicated that there are no clear rules for working in the new conditions with high requirements of regulatory authorities,
- 21.4% noted difficulties with the delivery of personnel to workplaces,
- 15.3% experienced a shortage of foreign workers,
- in 13.5% of collectives there was fear and panic among workers. However, in this case it should be noted that a reasonable amount of fear is inevitable [16] and even necessary.

Contractor organizations (organizations carrying out construction) experienced similar problems - interruptions in financing from customers (developers), sale interruptions, shipment, delivery of domestic building materials and equipment, interruptions in the sale, shipment, delivery of imported building materials and equipment, suspension of competitive bidding procedures for concluding new contracts; and a number of problems related to personnel. According to the survey "Impact of the coronavirus pandemic on the activities of contractors in construction" conducted by the LLC "Construction Industry Development Institute" in the period from 07/04/2020 to 27/04/2020 at the request of the National Association of Builders among 3564 contractors from 65 regions of the Russian Federation [11], 43.4% of construction organizations experienced problems due to the prohibition of personnel access to the construction site; 21.5% had difficulties with the delivery of personnel to the construction site. In 9.0% of organizations, foreign workers (migrants) did not come to the construction site, and in 3.8% of organizations, foreign workers were forced to leave.

2 Methodology

It should be noted that the theory and practice of personnel management has not yet developed the methodology for organizing the work of personnel in the context of the rapid spread of a serious disease transmitted by airborne transmission, since the world has faced such a situation for the first time. Therefore, it is so important to generalize and publish all the practical developments of this difficult period in order to preserve and replenish the experience of organizing the work of personnel in conditions of mass morbidity of such scale [2].

The Government of the Russian Federation, Federal Service for Surveillance on Consumer Rights Protection and Human Wellbeing, Federal Service for Labour and Employment, the Ministry of Health of Russia, the Ministry of Construction of Russia, as well as the National Association of Builders and the National Association of Designers and Surveyors as the situation with the spread of a new coronavirus infection developed, provided construction organizations with documents and guidelines that allowed them to organize the work of enterprises in the new conditions [3–10, 15].

On the basis of these documents, local enterprises in an operational mode developed and adopted local regulations governing working hours, the mode of personnel employment, as well as the rights, duties and responsibilities of workers and employers in the new conditions. During the spread of the new coronavirus infection, construction and repair works were suspended, with the exception of medical facilities construction, as well as the construction and maintenance of the metro, railway, ground, public transport and airports. From May 12, 2020 in Moscow and from May 18, 2020 in the Moscow region, all construction organizations are allowed to resume work. As of the end of June, one constituent entity of the Russian Federation has extended restrictions on the work of construction organizations until July 12. In 16 regions, the work of construction organizations is allowed ensuring a number of restrictions (passes, lists), and in other regions, all restrictions have now been lifted, provided that preventive measures are taken.

3 Results

For the purposes of this publication, it is not necessary to elaborate on the measures that have been taken at all enterprises, including construction companies, that are common to all employers, namely, the ban on collective meetings and mass events, the transfer of all employees who can be transferred to remote working mode, measures against employees over 65 years of age and employees of other ages from risk groups, restrictions on business trips, the need to use protective equipment and disinfectants. But the organizational and control measures taken by employers in the construction industry to organize the work of employees of construction companies in order to prevent the spread of a new coronavirus infection are discussed below in more detail, since they have industry specifics and have been developed on the basis of industry orders and recommendations.

Organizational and control measures for organizing the work of the personnel of a construction organization in the conditions of mass spread of infection can be conditionally divided into six enlarged blocks [10, 18].

The First Block - Arrangements for Organizing Access to the Construction Site:

- introduce access control to the construction site, construction camp or industrial premises,
- keep records of the arrival and departure of employees in the form of a journal (report card), indicating the date and time of entry and exit, the presence or absence of symptoms of ARVI (acute respiratory viral infection), data on body temperature, a survey of the employee' health and a survey of the health of individuals living together with the employee,
- restrict access to the construction site for individuals not involved in the performance of construction works and construction maintenance,
- require visitors and employees entering or leaving the construction site to ensure social distance, including by dividing the time of entry and exit of teams, to observe a social distance of at least two meters in the case of a delay in the entry/exit process and the emergence of a waiting situation,
- recommend drivers from transport vehicles, when delivering building materials to a construction site, not to leave the car cab, if the technology of building materials unloading allows it. If it is impossible to stay in your transport vehicles, tell the driver to wash or disinfect his hands, monitor that the driver (freight forwarder) keeps the social distance with the workers involved in unloading at the construction site,
- do not allow individuals without a pass to enter the construction site.

In order to reduce the probability of mass accumulation and the intersection of personnel at the entrance and exit of the construction site, the work schedules of individual teams and construction sites were adjusted in such a way as to exclude the intersection of workers from different departments (shifts) when entering and exiting the construction site. Employees of checkpoints are required to control the order of entry and exit to construction sites in accordance with schedules that exclude the mass accumulation of personnel and the intersection of workers at the entrance and exit, and keeping the distance established by special markings between employees while waiting for entry and exit from the construction site. Additional control of the situation at the beginning and end of the working day (shift) was carried out by the security service, monitoring the situation through video cameras at the entrance to the construction site, as well as a specially appointed controller (steward). Every week, each construction site summed up the results and received additional points to the CTU, if the controlling authorities (both internal-steward, site manager, and external-Federal Service for Surveillance on Consumer Rights Protection and Human Wellbeing, Stroynadzor) did not detect violations of the regime and did not give an order.

Activities for Organizing the Workflow are the Second Block.
Common measures for the organization of the work process – for all employees:

- isolate work teams from each other when performing work,
- ensure minimal contact between employees during the working day (shift),
- cancel optional works that require physical contact of employees, if it is impossible to cancel - to perform work in gloves and personal respiratory protection,
- organize the disinfection of reusable work tools, the interior of the transport vehicles and construction equipment between the use of different workers.

All weekly workshops that are typical for the construction process (usually involving the customer, general contractor, contractors, labor protection, construction supervision, designers, author supervision, etc.) were held using the videoconferencing service (zoom.ru), and each interested party of construction carried out monitoring of the work on the site in turn, in order to avoid a one-time congestion of people and reduce the risk of infection. Protocols were kept exclusively in electronic form, and electronic images of documents were accepted for execution.

Additional measures for organizing the work process – for migrant workers:

- ensure that soap, hand sanitizers, and thermometers are available in places where migrant workers are accommodated,
- ensure daily disinfection of accommodation facilities for migrant workers, with the use of disinfectants to sanitize all surfaces,
- isolate migrant workers when performing work, except the cases where such isolation will lead to a violation of the production technology,
- if the presence of a new coronavirus infection (COVID-19) is confirmed, place the specified workers in quarantine for 14 days in a specially designated place.

In order to protect the staff, it was decided to test for covid those foreign citizens who have just come to work.

The Third Block - Measures to Monitor the Health of Employees:

- carry out morning and evening (at the entrance and exit from the construction site) visual inspection for signs of ARVI, temperature control, survey of employees' health condition with entering the results in the journal (report card) of arrival and departure of employees,

- conduct daily non-contact body temperature measurement of employees and visitors at the entrance to the construction site, as well as at the end of the working day (shift),
- carry out disinfection control of sanitary and amenity facilities and cabins of the construction town during the working day (shift),
- restrict access to work of employees with signs of ARVI, who got signs of ARVI in the course of work,
- keep a register of employees with the identified symptoms of ARVI, monitoring the following information: full name; date; characteristics of health (satisfactory, slight malaise, severe malaise); body temperature; cough; runny nose, loss of smell; headache, chills; other symptoms; individuals with whom the individual contacted (full name); signature of the employee indicates that the information was provided voluntarily and recorded correctly from his words,
- if an employee of the enterprise has an increased body temperature (37.0 °C or higher) and (or) signs of ARVI, to suspend the employee from work and arrange for the employee the transportation to the place of residence for self-isolation for a period of at least fourteen days. Notify a sick employee on the need to immediately seek medical care at home,
- if an employee is diagnosed with a "coronavirus infection", check the isolation of all employees who came into contact with this individual after the onset of symptoms, and their compliance with the period of self-isolation at the place of residence for at least fourteen days,
- if the worker of the construction site was diagnosed with the "coronavirus infection", the responsible person appointed by the head of the company immediately notifies the territorial authority of Federal Service for Surveillance on Consumer Rights Protection and Human Wellbeing on the employee diagnosis "coronavirus infection", including address and name of object of capital construction, the address of the place of the sick employee isolation, the date of occurrence of ARVI symptoms of people that were in contact with the person after the diagnosed ARVI symptoms, addresses of their isolation places,
- organize keeping records of all employees with identified symptoms of respiratory diseases,
- if an employee has an increased body temperature (37.0 °C or higher) and/or signs of respiratory diseases, suspend the employee from work and arrange for the employee to be taken home for self-isolation for a period of fourteen days,
- ensure that separate rooms are available to isolate workers if they have increased body temperature or symptoms of respiratory illness before the ambulance arrival.

The Fourth Block - Measures to Ensure Personal Hygiene of Employees and Disinfection of Premises:

- instruct employees on the prevention of the spread of a new coronavirus infection (COVID-19) in accordance with the instructions of Federal Service for Surveillance on Consumer Rights Protection and Human Wellbeing with mandatory recording in the appropriate journal,

- issue information materials (memos) to each construction site employee on symptoms of coronavirus infection (COVID-19) and measures to prevent infection at the entrance to the construction site and in public areas,

- control availability on the construction site:

1. Disposable paper towels, towel bins, or disinfectants.
2. Means of individual protection for individuals admitted to the construction site.
3. UV bactericidal irradiators (recirculators).
4. Five-day stock, detergents, disinfectants, means of individual protection, gloves, rags, cleaning equipment, based on the estimated need.
5. In places for washing and disinfecting hands, soap, disposable paper towels, trash cans for towels, or disinfectant in the case of a temporary interruption or restriction of water supply.

- organize on the construction site and monitor in the construction town and production space:
- regularly empty trash cans with used disposable hand towels and utilize the used disposable hand towels,
- centralized collection of used disposable masks. Before placing them in waste collection containers place them in tight packaging and in 2 plastic bags,
- provide the individuals admitted to the construction site, as well as to production, household, office premises, administrative and household premises with personal protective equipment (masks, gloves, etc.),
- ensure control over employees located on construction sites and in construction towns in order to prevent their movement outside of the construction site and construction towns, except for movement from their place of residence to their place of work and back,
- when performing work (if it is not related to the technological process), as well as when waiting for entry and exit from domestic premises, buildings, structures, observe a distance of 2 (two) meters between employees,
- organize places for washing and disinfecting hands at the entrance to the construction site, in places where food is taken, in toilet rooms, and in public areas.

In addition to the measures listed above, construction organizations are recommended to oblige every employee who starts working after the resumption of work of construction enterprises to:

- exercise daily self-monitoring of your health status,
- move from the place of residence alone, if possible using personal transport. When you are in transport, keep a social distance of at least two meters,
- inform the responsible individuals who keep the journal of monitoring the health status of construction site workers, reliable information on their own health status,
- wash or disinfect hands when entering and exiting the construction site, before eating, before and after using the toilet, after contact with objects that have been used by others,
- exclude handshakes, hugs and other contact forms of communication during the period of preventive measures for COVID-19,

- when self-detection of increased body temperature (37.0 °C and more) and (or) signs of ARVI:
- do not go to work,
- immediately inform the immediate supervisor of your condition, and also the head of the HR department or general director of the company,
- provide a regime of self-isolation for a period of not less than fourteen days,
- request medical assistance at home.
- Employees whose duties include cleaning in office, work and household premises are charged with the following duties:
- carry out daily disinfection of the construction site premises and common areas with the use of disinfectants 2 times a day in the morning and in the evening,
- conduct regular (every 2 h) cleaning of contact surfaces (door handles, buttons, switches, handrails, railings, telephone equipment, keyboards, office equipment, tables, locks, taps, sinks, toilets, etc.) using disinfectants,
- every two hours to ventilate the production areas of the headquarters building and common areas,
- make appropriate notes in the contact surface and room disinfection journal.

Technical and administrative staff of building organizations (PTO engineers, cost consultants, designers, personnel officers, accountants, lawyers, etc.) were quickly put in remote work with the necessary working conditions (connection to the software, the compensation of additional costs for mobile communication and Internet), with the exception of work producers, heads of construction sites, project managers. Their construction sites continued work employment or faster than others received permission to easing restrictions related to prevention of a new coronavirus infection spread. It was necessary to make adjustments to the schedules of work production and delivery of construction materials to construction sites, taking into account the requirements for the permissible number of workers for each type of construction work.

4 Discussion

Like many other enterprises, construction organizations had a very difficult time during the period of restrictive measures. However, according to a survey of developers [12], 72% of the surveyed construction organizations managed to retain almost all staff, about 15% dismissed some of their employees, and less than 2% dismissed the entire staff. If we talk about the size of construction organizations that did not survive the restrictive measures, these are microenterprises and small businesses. Among these organizations, the lowest percentage of those who retained all staff, and the highest percentage of layoffs. More than 80% of enterprises of large and medium-sized businesses in construction have retained their staff without significant losses in numbers (correctly organized remote work, introduced flexible working hours, complying with legal requirements, revised the holiday schedule, shift mode, duration of shifts, etc.). If we analyze data on the subjects of the Russian Federation, of which there were organizations that took part in the survey, then the leader in the number of organizations that had to lay off more employees is Moscow and the Moscow region, Saint

Petersburg, and the Khanty-Mansi Autonomous Okrug - Yugra. Also, many construction organizations noted difficulties and problems with organizing work by migrant workers — problems with entering the territory of the Russian Federation, problems with extending mandatory documents for work, problems with the fact that some of the staff was deported.

5 Conclusion

Thus, thanks to the efforts of management bodies, as well as the discipline and responsible behavior of employers and employees of construction organizations, there were no cases of mass diseases and outbreaks of infection in construction organizations in Russia during the period of the coronavirus infection spread. The situation and results of today confirm that the construction industry personnel had sufficient "resources necessary to participate in pandemic self-government (knowledge of how and when to shop, availability of people who can help, the nearest hospital having enough respirators, etc." [13]. The experience gained during the pandemic, as well as the results of the dedicated work of military builders, allowed both employees and employers and authorities to take a new look at the labor resources of the industry, to think about the strategic prospects of personnel support for construction and the quality of training of qualified labor for construction organizations in the Russian Federation. The global coronavirus pandemic (COVID-19) has already had a huge impact and will undoubtedly have profound consequences for many next years [1], and no one will come out of the crisis without losing something [19].

References

1. Balog-Way, D.H.P., McComas, K.A.: COVID-19: Reflections on trust, tradeoffs, and preparedness. Journal of Risk Research, In Press (2020). https://www.tandfonline.com/doi/full/10.1080/13669877.2020.1758192. Accessed 01 June 2020
2. Bryce, C., Ring, P., Ashby, S., Wardman, J.K.: Resilience in the face of uncertainty: Early Lessons from the COVID-19 Pandemic. Journal of Risk Research, In Press (2020). https://doi.org/10.1080/13669877.2020.1756379. Accessed 22 June 2020
3. Decree of the President of the Russian Federation On declaring non-working days in the Russian Federation dated 25.03.2020 No. 206 (2020). http://www.consultant.ru/document/cons_doc_LAW_348485/. Accessed 20 June 2020
4. Decree of the President of the Russian Federation of 11.05.2020 No. 316 On determining the procedure for extending measures to ensure the sanitary and epidemiological well-being of the population in the subjects of the Russian Federation in connection with the spread of a new coronavirus infection (COVID-19) (2020). http://www.consultant.ru/cons/cgi/online.cgi?rnd=410DFCB1B6E87DDBE69691C2747130BDandreq=docandbase=LAWandn=352133andstat=refcode%3D16876%3Bindex%3D0#oa0j3kbr6o. Accessed 20 June 2020

5. Decree of the President of the Russian Federation dated 02.04.2020 No. 239 On measures to ensure the sanitary and epidemiological well-being of the population on the territory of the Russian Federation in connection with the spread of a new coronavirus infection (COVID-19) (2020). http://www.consultant.ru/cons/cgi/online.cgi?req=docandbase=LAWandn=349217andfld=134anddst=1000000001,0andrnd=0.653442326359952#09453038710275816. Accessed 20 June 2020
6. Decree of the President of the Russian Federation of 28.04.2020 No. 294 On the extension of measures to ensure the sanitary and epidemiological well-being of the population on the territory of the Russian Federation in connection with the spread of a new coronavirus infection (COVID-19 (2020). http://www.consultant.ru/cons/cgi/online.cgi?req=docandbase=LAWandn=351539andfld=134anddst=1000000001,0andrnd=0.1171375320457595#004435075431923008. Accessed 20 June 2020
7. Letter of the Ministry of Construction Industry, Housing and Utilities Sector of the Russian Federation dated 03.04.2020 No. 13158-IF/03 On recommendations for the prevention of the spread of coronavirus infection for organizations in the construction industry (2020). http://nostroy.ru/news_files/2020/04/07/01/GetAttachmentCashedImageStream%20(14).pdf. Accessed 20 June 2020
8. Letter of Federal Service for Surveillance on Consumer Rights Protection and Human Wellbeing from 10.03.2020 no 02/3853-2020-27 On measures for the prevention of new coronavirus infection (COVID-19) (with Recommendations for the prevention of new coronavirus infection (COVID-19) among employees) (2020). http://www.consultant.ru/document/cons_doc_LAW_347459/. Accessed 20 June 2020
9. Letter of Federal Service for Surveillance on Consumer Rights Protection and Human Wellbeing from 18.04.2020 N 02/7329-2020-27 On carrying out preventive and disinfection measures for employees of the construction industry (with MR 3.1/2.2.0172/2-20. 3.1. Prevention of infectious diseases. 2.2. Labour health. Recommendations for the prevention of new coronavirus infection (COVID-19) among construction workers. Guidelines) (2020). http://www.consultant.ru/cons/cgi/online.cgi?req=docandbase=LAWandn=350801andfld=134anddst=1000000001,0andrnd=0.7655392409223332#048416180473402615. Accessed 20 June 2020
10. Methodic of measures aimed at preventing the spread of a new coronavirus infection (COVID-19) when contractors-members of self-regulating organizations perform works on construction, reconstruction, major repairs, demolition of capital construction projects, as well as in places where people are allowed to perform these works, approved by President of National Association of Builders 06.04.2020 (2020). http://nostroy.ru/news_files/2020/04/05/%D0%BC%D0%B5%D1%82%D0%BE%D0%B4%D0%B8%D0%BA%D0%B0%20%D0%BF%D1%80%D0%BE%D1%82%D0%B8%D0%B2%D0%BE%D0%B4%D0%B5%D0%B9%D1%81%D1%82%D0%B2%D0%B8%D1%8F%20%D0%BA%D0%BE%D1%80%D0%BE%D0%BD%D0%B0%D0%B2%D0%B8%D1%80%D1%83%D1%81%D0%BD%D0%BE%D0%B9%20%D0%B8%D0%BD%D1%84%D0%B5%D0%BA%D1%86%D0%B8%D0%B8%20%D0%BD%D0%B0%20%D1%81%D1%82%D1%80%D0%BE%D0%B9%D0%BF%D0%BB%D0%BE%D1%89%D0%B0%D0%B4%D0%BA%D0%B0%D1%85%20-%2006.04.2020.pdf. Accessed 22 June 2020
11. National Association of Builders: Survey Impact of the coronavirus pandemic on the activities of contractors in construction, conducted by LLC Institute for the development of the construction industry in the period from 07.04.2020 to 27.04.2020 (2020). https://nostroy.ru/nostroy/situation_center/analitics_data/Опрос%20(Подрядчики%2027.04.2020).pdf. Accessed 15 June 2020

12. National Association of Builders: A survey of developers in the Russian Federation conducted by the LLC "Construction Industry Development Institute" in the period from 14.04.2020 to 18.04.2020 (2020). https://nostroy.ru/nostroy/situation_center/analitics_data/Опрос%20(Застройщики%202этап%20-%2014.04-27.04.2020).pdf. Accessed 22 June 2020
13. Nygren, K.G., Olofsson. A.: Managing the Covid-19 pandemic through individual responsibility: The consequences of a world risk society and enhanced ethopolitics. Journal of Risk Research. In Press (2020). https://www.tandfonline.com/doi/full/10.1080/13669877.2020.1756382. Accessed 22 June 2020
14. Plan of priority measures (actions) to ensure the sustainable development of the economy in the context of the deterioration of the situation due to the spread of a new coronavirus infection (approved by the Government of the Russian Federation 17.03.2020) (2020). http://www.consultant.ru/cons/cgi/online.cgi?req=docandbase=LAWandn=348153andfld=134anddst=1000000001,0andrnd=0.6021272823433681#014087510286109417. Accessed 21 June 2020
15. Recommendations to employers towards the application (distribution) on workers mode of working days from 30 March to 3 April 2020 in connection with the decree of the President of the Russian Federation from March 25, 2020 No. 206 On the Declaration in the Russian Federation of non-working days, approved on the meeting of the operational headquarters for the prevention of introduction and spread of new coronavirus infection in the Russian Federation 26.03.2020 (2020). https://mintrud.gov.ru/labour/relationship/379. Accessed 20 June 2020
16. Ruiu, M.L.: Mismanagement of Covid-19: Lessons learned from Italy. Journal of Risk Research. In Press (2020). https://www.tandfonline.com/doi/full/10.1080/13669877.2020.1758755. Accessed 22 June 2020
17. Russian Future. National projects: Khusnullin: to meet the goals of the national project for housing commissioning, another 2 million builders are needed (2020). https://futurerussia.gov.ru/nacionalnye-proekty/husnullin-dla-vypolnenia-celej-nacproekta-po-vvodu-zila-nuzny-ese-2-mln-stroitelej. Accessed 22 June 2020
18. Sharfuddin, S.: The world after Covid-19. Round Table **109**(3), 247–257 (2020)
19. Standard for organizing work on construction sites in order to prevent the spread of a new coronavirus infection (COVID-19), approved by order of the Deputy Chairman of the Government of the Moscow region on 22 04 2020 No. 28-p (2020). http://www.consultant.ru/law/review/207655546.html/. Accessed 22 June 2020

Features of Organizational Culture of Russian Companies Transformation Under Conditions of Digitalization

V. M. Svistunov(✉), G. P. Kuzina, and V. V. Lobachev

State University of Management, Moscow, Russia
{svistunov, vvl}@guu.ru, gpkuzina2009@yandex.ru

Abstract. The purpose of the study is to identify the organizational cultures that have developed in Russian companies and develop recommendations for their targeted transformation in the context of digitalization of the economy. The article presents the study results on identifying the type of existing and preferred organizational cultures in 12 Russian companies of various activity fields and development levels. Based on the results of the analysis of the existing organizational cultures in companies, it was concluded that they are more likely to show signs that do not take into account the system characteristics of the company, changes in the external environment and the influence of competitors. The calculation of correlation coefficients for existing and preferred organizational cultures in the context of digital transformation has shown that a high level of correlation is typical only for a small number of large export-oriented companies. Small and medium-sized businesses are characterized by a low level of correlation between existing and preferred organizational cultures.

Keywords: Company employees · Digital transformation · Organizational culture · Organizational culture profile · Value system

1 Introduction

Each company has its own organizational culture, which distinguishes it from other organizations with a unique set of values, beliefs, rituals, symbols, artifacts, heroes, norms and rules of employee behavior. Modern trends, globalization and digitalization, dictate the need for organizational changes that lead to the transformation of the organizational culture of companies. At the same time, changes often encounter hidden or obvious resistance from employees and are even doomed to failure if they contradict the established organizational culture [11]. The purpose of the study is to identify existing and preferred organizational cultures for employees of Russian companies, as well as to develop recommendations for their transformation in the conditions of digitalization of the economy, taking into account the above factors.

The organizational culture of large Russian companies often has deep development traditions since the socialism time. Considering changes in the priorities of companies' activities in the post-soviet period, their organizational cultures have been transformed. The transition to market-based economic conditions determined profit-making and

S. I. Ashmarina and V. V. Mantulenko (Eds.): IPM 2020, LNNS 161, pp. 221–229, 2021.
https://doi.org/10.1007/978-3-030-60926-9_29

market competition as the main goals of the company's functioning, which led to the transformation of not only the old value system, but also behavioral models [2, 7]. The situation is somewhat different for medium-sized and small enterprises created mainly in the post-soviet period. They are more homogeneous in organizational culture, although more often the organizational culture was formed spontaneously and now exists in an implicit form, without being the subject of close attention from the leadership. Based on the results of the analysis, the following problems were identified: organizational culture is not formalized, and leadership, in most cases, does not identify it, despite such symptoms as high staff turnover, dissatisfaction with relationships in the team, conflicts, low level of competitiveness, etc. [10, 12].

2 Methodology

An earlier study of the reasons for the failure to implement modern innovative management methods in industrial enterprises allowed the authors to identify a low level of trust to leadership from the employees of the enterprise as a fundamental reason [13]. It was hypothesized that the level of trust is directly related to the organizational culture of the enterprise, which does not motivate employees to use new innovative methods of work. Therefore, it is very important to determine which type of organizational culture is most typical for modern Russian companies and the direction of its transformation in the conditions of the economy digitalization.

12 organizations of various fields of activity were selected as research objects: industrial enterprise, commercial bank, energy company, university, container transport operator, advertising agency, medical equipment sales company, engineering service company, IT organization, real estate agency, design bureau, marketing company. Employees of these companies (a total of 438 people) were involved in the research. Among these people, surveys were conducted to identify signs of existing and preferred corporate cultures in the conditions of digitalization.

The methodology developed by Cameron and Quinn was chosen as a tool to assess the current corporate culture of the company, its desired state and to develop recommendations for the transition from the current state to the preferred one. The method involves evaluating four types of cultures: clan, adhocratic, market, and bureaucratic, which manifest themselves in companies with varying degrees of intensity. The orientation of the organizational culture depends not only on the company's strategic goals, but also on the company's place and role in society.

The typology of organizational cultures developed by Cameron and Quinn was the basis for their assessment using the OCAI tool on six criteria: the most important characteristics of the organization, the overall leadership style in the organization, employee management, the connecting essence of the organization, strategic goals, and success criteria [3]. Based on the results of assessing the degree of influence of different types of cultures on each of the six criteria, profiles of existing and preferred organizational cultures are constructed. The Pearson correlation coefficient was used to determine the degree of closeness of the relationship between them. Based on a systematic approach, taking into account the place and role of the company in society and the nature of the dominant processes, recommendations for the transformation of organizational cultures were developed.

3 Results

Table 1 presents the average indicators for evaluating the types of organizational cultures of the studied commercial organizations, obtained by interviewing employees. For the existing and preferred organizational culture, there are two most characteristic types that determine their profile.

Table 1. Assessment of the profile of the average organizational culture of 12 commercial organizations

	Existing, %	Preferred, %	Growth rate
Clan	**27**	**28.1**	**1.04**
Adhocracy	20	**26.4**	**1.32**
Market	25	23	0.92
Bureaucracy	**28**	22.5	0.8
Profile	Bureaucratic-clan	Clan-adhocratic	–

Source: authors.

As you can see, the average characteristic of the existing organizational culture is bureaucratic-clan (although the difference between the bureaucracy and the clan is only 1%). This type of culture does not motivate, but rather slows down the process of digitalization. Bureaucracy in the context of digitalization demonstrates resistance to changes, suppresses initiative, and clan system creates conditions for joint responsibility, sabotage, and collusion against any organizational changes.

The preferred organizational culture is considered by employees to be clan-adhocratic. Since the organizational cultures of large companies differ significantly from those of medium and small businesses, the analysis of corporate cultures was performed separately for each of these two groups.

Large business organizations include 4 companies: an industrial enterprise, a commercial bank, a university, and a design department. Organizations of medium and small businesses include 8 research objects: an energy company, a container operator, an advertising agency, a company selling medical equipment, an engineering service company, an IT organization, a real estate agency, and a marketing company.

Table 2 presents the results of the assessment of large organizations organizational cultures.

The analysis shows that the average profile of their existing organizational culture coincides with the average profile of all 12 research objects and is bureaucratic-clan and adhocratic-clan for the preferred one. The current culture is aimed at regulating and standardizing the interests of the clan, while the preferred one is aimed at activating the creative potential to increase the level of satisfaction of the clan interests. In the preferred culture, the level of bureaucracy should be reduced by 30%, and the level of

Table 2. Results of organizational cultures evaluation of 4 large companies, %

	Existing organizational culture				Profile*
	Clan	Adhocracy	Market	Bureaucracy	
Industrial enterprise	20	12.5	**34.2**	**33.3**	M-B
University	**42.5**	21.2	6.2	**30.1**	C-B
Commercial bank	21.7	22.5	**24.2**	**31.6**	B-M
Design bureau	13.3	23.3	**26.7**	**36.7**	B-M
Average culture	**24.38**	19.88	22.82	**32.92**	B-C
	Preferred org. culture				Profile*
	Clan	Adhocracy	Market	Bureaucracy	
Industrial enterprise	**26.7**	38.3	24.2	10.8	A-C
University	**45**	10	15	**30**	C-B
Commercial bank	15	20	**25**	**40**	B-M
Design bureau	**21.7**	**43.3**	16.7	18.3	A-C
Average culture	**27.1**	**27.9**	20.22	24.78	A-C

* - A-adhocracy, B-bureaucracy, C-clan, M-market.
Source: authors.

adhocracy should increase by 40% and be approximately similar to the clan system. The level of marketability should decrease by 12%. Thus, the average profile for large companies corresponds to the general trend across the entire sample. As we can see from the analysis results, there is a market component in the organizational cultures of large companies. The exception is the university. The first dominant is the market component of an industrial enterprise, i.e. the culture of an industrial enterprise is aimed at satisfying the interests of the market, while the bureaucracy as the second dominant shows that the products and their production technology are standardized. This culture can be recognized as satisfying the requirements of digitalization [6]. At the same time, the majority of the survey participants supported reducing the level of market and bureaucracy and increasing the level of adhocracy and clan-system.

According to the expressed preferences, the level of adhocracy in an industrial enterprise should increase by more than three times. In a market environment, this is appropriate, since external focus and differentiation are enhanced, but a decrease in marketability with an increase in clan system is negative. Adhocratic-clan culture will direct the creative potential to meet the interests of the clan. For the university, the preferred culture coincides with the existing clan-bureaucratic one. The preferred organizational culture of a commercial bank coincides with the existing bureaucratic market culture. Employees do not perceive the market requirements as the main motivating factor in the organization's activities, internal control-oriented values prevail, and they are satisfied with the existing pseudo-effective culture. At the design bureau, the levels of adhocracy and clan system are growing at the highest rate, which indicates that employees understand the need to search new non-standard solutions, implement innovations, increase the level of team cohesion and loyalty to the organization. The market-adhocratic culture is more preferable. Tables 3 and 4 provide data on the organizational cultures of small and medium-sized businesses.

Table 3. Results of assessment of existing organizational cultures of medium and small businesses

Type of organization	Clan	Adhocracy	Market	Bureaucracy	Profile
Energy company	**22.5**	12.5	**45**	20	M-C
Estate agency	**33.3**	20.8	**35**	10.9	M-C
Engineering service company	15.8	19.2	**20.8**	**44.2**	B-M
IT-company	15.3	**35.8**	**36.6**	12.3	M-A
Container operator	**31.6**	10.9	15	**42.5**	B-C
The company selling medical equipment	**38.4**	13.3	17.5	**30.8**	C-B
Advertising agency	**33.4**	23.4	**26.6**	16.6	C-M
Marketing company	**44.6**	25.4	**17.8**	12.2	C-A
Average org. culture	**29.36**	20.16	**26.79**	23.69	C-M

Source: authors.

Table 4. Results of assessment of preferred organisational cultures of small and medium businesses companies

Type of organization	Clan	Adhocracy	Market	Bureaucracy	Profile
Energy company	**30**	16.6	25.8	**27.6**	C-B
Estate agency	28.3	**33.3**	**28.4**	10	A-M
Engineering service company	14	**31**	**37**	18	M-A
IT-company	**28.6**	32.8	22.8	15.8	A-C
Container operator	**42**	19	**21**	18	C-M
The company selling medical equipment	**29**	18	7	**46**	B-C
Advertising agency	**30**	20	5	**45**	B-C
Marketing company	**27.5**	18.33	**28.33**	**25.84**	M-C
Average org. culture	**28.67**	23.63	21.92	**25.78**	C-B

Source: authors.

The profile of the average existing organizational culture of medium and small businesses organizations is clan-market. These are competing characteristics. Clan system, as the first dominant, puts the interests of the clan above the interests of the market, i.e. the market serves as a mean of satisfying the interests of the clan, creating an illusory appearance of working for the market.

The most effective are the market-adhocratic organizational culture of the IT company and the market-clan culture of the energy company, since they are aimed at the market, and correspond to the type of companies that are under development. The profile of the average preferred organizational culture belongs to the clan-bureaucratic one, i.e. the interests of the clan are fixed in norms and regulations, which most likely reflects the real state of affairs and hinders the organizations development. According to the authors, the transformation of organizational culture should take into account the place and role of the organization in the digital economy, the nature of consumed

Table 5. Profiles of existing and preferred organizational cultures of companies and the close relationship between them

Type of organization	Profile of existing org. culture	Profile of preferred org. culture	Coefficient of Pearson	The closeness of the connection
Industrial enterprise	M-B	A-C	0.704	High
University	C-B	C-B	0,733	High
Commercial bank	M-B	M-B	0.972	High
Design bureau	M-B	C-A	0.056	Weak
Energy company	M-C	C-B	0.151	Weak
Estate agency	M-C	A-M	0.465	Moderate
Engineering service company	B-C	M-A	0.079	Weak
IT-company	M-A	A-C	0.230	Weak
Container operator	M-C	C-M	0.048	Weak
The company selling medical equipment	C-B	M-C	0.444	Moderate
Advertising agency	C-M	B-C	0.181	Weak
Marketing company	C-A	M-C	0.004	Weak
Average culture of all 12 organizations	B-C	C-A	0.083	Weak
Average culture of 4 large organizations	B-C	A-C	0.163	Weak
Average culture of 8 small and medium-sized organizations	C-M	C-B	0.017	Weak

Source: authors.

resources, the range of provided products or services, and the specifics of the dominant internal processes. Thus, market-bureaucratic organizational culture is appropriate for industrial enterprises producing standardized products with low freedom of choice, and market-adhocracy organizational culture is appropriate for release of non-standard, unique products and high freedom of choice.

For organizations with a dominant innovation process and a high freedom of choice (university, IT company, design bureau, engineering service company, advertising agency, marketing company), a market-oriented organizational culture is preferable. Organizations with a predominance of infrastructural processes (commercial banks, container transport operators, medical equipment companies) should preferably focus on a market-oriented and adhocratic organizational culture. In organizations that are dominated by operational processes with low freedom of choice (energy companies and real estate agencies), it is recommended that the preferred organizational culture is market-bureaucratic. Based on the conducted research, profiles, existing and preferred organizational cultures were compared for all companies, with the calculation of the Pearson coefficients values showing the closeness of the relationship (the level of correlation) between them. The results are presented in Table 5.

As the calculations show, all companies can be divided into three groups: with a high, moderate and weak correlation between the established and preferred organizational culture. The group with a high level of correlation includes the three largest organizations: an industrial enterprise, a university, and a commercial bank. This is due to the fact that these are mature organizations that present themselves as socially responsible, have a development strategy, shared values, and the organizational culture is perceived and supported by employees who have an understanding of the leadership's actions to implement modern IT products [4, 14].

There are two organizations of small and medium businesses in the group with a moderate level of correlation: the real estate agency and a company selling medical equipment. Their organizational culture is in the formation process. The main values of organizations are not formulated and notified to employees, there are no clearly established, understandable rules of behavior, there is no understanding by employees of the prospects for the development of the organization and their place in it.

The group with a weak correlation level includes seven organizations, including a large business - a design bureau; medium and small businesses: an energy company, an IT company, an advertising agency, an engineering service company, a container operator, and a marketing company. The organizational cultures of these organizations are either not fully formed, or are unproductive and need more detailed study and transformation [1].

4 Discussion

There are several methods for evaluating corporate culture, including those of Cameron and Quinn [3], Trompenaars and Kuberg [15], Hofstede and McCrae [5], Krasovsky [9]. Consulting companies have developed and offer their numerous methods and services for conducting audits and improving organizational culture. Not every company can afford such research because of the high costs and the risk that these costs will not pay off, but the organizational culture cannot simply be copied from another organization [8]. It should be understood that nowadays organizational culture must conform to the time requirements and be considered from system positions, involving some unification, the scope of which depends on the nature of the processes and freedom of choice in making management decisions.

Table 6. Existing, preferred and recommended organizational culture profiles of the studied companies

Type of organization	Profile of existing org. culture	Profile of preferred org. culture	Recommended profile in the digitalization conditions
Industrial enterprise	M-B	A-C	M-B/M-A
University	C-B	C-B	A-M/M-A
Commercial bank	M-B	M-B	M-B
Design bureau	M-B	C-A	M-B/M-A
Energy company	M-C	C-B	M-B
Estate agency	M-C	A-M	M-B
Company of engineering service	B-C	M-A	M-A
IT-company	M-A	A-C	M-A/A-M
Container operator	M-C	C-M	M-B
The company selling medical equipment	C-B	M-C	M-B
Advertising agency	C-M	B-C	M-A
Marketing company	C-A	M-C	M-A

Source: authors.

5 Conclusion

Based on the results of the analysis, it was concluded that signs of organizational cultures that are not aimed at innovation dominate most companies, do not take into account changes in the external environment and the digitalization of the economy. Table 6 presents the existing, preferred, and recommended organizational culture profiles for the studied companies. The complete overlap between the existing, preferred and recommended cultures can only be noted in the commercial bank; other companies have significant differences. This indicates the need to determine the vector of organizational cultures transformation.

There is no single recipe for transforming the company's organizational culture. Since each organization must have its own identity, it is necessary to constantly study and improve its organizational culture. At the same time, it is necessary to approach this issue from a systemic perspective, consider the place and role of each organization in the system of interaction with other organizations, the nature of consumed resources, produced products and services, and the nature of the dominant processes. According to the authors, in the conditions of globalization and digitalization of the economy, the market dominant is the most preferable for commercial organizations. The adhocracy, bureaucracy and clan should only strengthen it.

References

1. Astafeva, O.V., Pecherskaya, E.P., Tarasova, T.M., Korobejnikova, E.V.: Digital transformation in the management of contemporary organizations. In: Ashmarina, S.I., Vochozka, M., Mantulenko, V.V. (eds.) ISCDTE 2019. LNNS, vol. 84, pp. 382–389. Springer, Cham (2020)
2. Bogdanov, S.V., Nechaev, S.Y.: Comparison of the indices of strategic competitive initiatives of the Russian banking and metallurgical business under modern global economic recession conditions. Russ. Metall. (Metally) **6**, 651–656 (2019)
3. Cameron, K., Quinn, R.: Diagnostics and Change of Organizational Culture. Peter, Saint Petersburg (2001)
4. Godin, V.V., Terekhova, A.E.: Digital ecosystems as a form of modern business transformation. In: Becker, J., Matveev, M., Tartukhin, V. (eds.) 1st International Conference of Information Systems and Design. CEUR Workshop Proceedings, vol. 2570, no. 19. CEUR, Aachen (2020)
5. Hofstede, G., McCrae, R.R.: Culture and personality revisited: linking traits and dimensions of culture. Cross-Cult. Res. **38**(1), 52–88 (2004)
6. Ivanov, I., Lukyanova, T., Orlova, L.: Digitalization as a driver of innovation for industrial enterprises. In: IOP Conference Series: Materials Science and Engineering, vol. 753, no. 7, p. 082023 (2020)
7. Kot, M.K., Spanagel, F.F., Belozerova, O.A.: Problems of digital technologies using in employment and employment relations. In: Ashmarina, S.I., Mesquita, A., Vochozka, M. (eds.) Digital Transformation of the Economy: Challenges, Trends and New Opportunities. Advances in Intelligent Systems and Computing, vol. 908, pp. 227–234. Springer, Cham (2019)
8. Kraev, V.M., Tikhonov, A.I.: Risk management in human resource management. TEM J.: Technol. Educ. Manage. Inf. **8**(4), 1185–1190 (2019)
9. Krasovsky, Yu.D:. Resource constructors of simulation modeling in sociology: methodology, theory, practice. Management, **1**(11), 31–41 (2016)
10. Mitrofanova, E., Konovalova, V., Mitrofanova, A.: Methodical approach to conflict management in the system of social and labour relations in the organization. In: Solovev, D. B., Salavey, V.V., Bekker, A.T., Petukhov, V.I. (eds.) Proceeding of the International Science and Technology Conference “FarEastCon”. Smart Innovation, Systems and Technologies, vol. 172, pp. 397–406. Springer, Cham (2020)
11. Mory, L., Wirtz, B.W., Götte, W.: Factors of internal corporate social responsibility and the effect on organizational commitment. Int. J. Hum. Resour. Manage. **27**(13), 1392–1425 (2016)
12. Rothenberg, S., Hull, C.E., Tang, Z.: The impact of human resource management on corporate Social performance strengths and concerns. Bus. Soc. **56**(3), 391–418 (2017)
13. Svistunov, V.M., Kuzina, G.P., Lobachev, V.V.: The level of trust of the company staff as the factor of the increase of social and economic systems management efficiency. In: Nkitov, S.A., Bykov, D.E., Borovik, S.Yu., Pleshivtseva, Yu.E. (eds.) XXI International Conference “Complex Systems: Control and Modeling Problems (CSCMP)”, pp. 799–802. IEEE, Paris (2019)
14. Temnyshov, I.A., Belyaev, A.M.: Business modeling in business operations of manufacturing company. In: IOP Conference Series: Materials Science and Engineering, vol. 753, no. 8, p. 082027 (2020)
15. Trompenaars, F., Kuberg, P.H.: 100 Key Models and Management Concepts. Mann, Ivanov and Ferber, Moscow (2019)

Harmonization of the Employee's Career in Digital Economy

N. Sotnikov(✉) and S. Sotnikova

Novosibirsk State University of Economics and Management, Novosibirsk, Russia
n.z.sotnikov@edu.nsuem.ru, s.i.sotnikova@nsuem.ru

Abstract. The purpose of the article is to substantiate the strategic imperatives of the harmonization of the career dual nature on the basis of the concept of time management. To achieve this purpose, the following research tasks were solved: career time architectonics that allows us to understand its duality has been justified, a methodological approach to evaluating the harmony of a employee's career based on the concept of time management has been proposed and tested; strategic alternatives to the harmonization of careers based on individual employee choices of an occupation, education, life partner, friends, children, etc. have been systematized. The research methodology is based on the theoretical analysis of the results of scientific works and the empirical data characterizing the size and structure of the employees' career time at commercial banks in Novosibirsk during six years. The scientific novelty of the article lies in the formation of the author's conceptual approach to the harmonization of an employee's career: the author's interpretation of the career time concepts and career harmony, the essence of the architectonics of career time, a system of indicators for assessing career harmony, justified methodological approach to career harmonization has been proposed and tested.

Keywords: Business career · Career harmony · Duality of career · Employee's career · Harmonization of a career · Time management

1 Introduction

Under contemporary conditions of VUCA economy, one of the cost effective and long term solutions to the problem of overcoming the staff alienation from the process and results of work, stimulating the sense of responsibility and professional pride in their competitiveness in the labour market, is implementation of a professional approach to a career that allows employees to be more productive, free and economically independent personalities.

At present, in Russian and foreign works an employee's career is no longer defined as an informed subjective judgment of the employee about their future employment, but as an awareness of it as a common system development and human movement in various fields [4, 13], the process of formation of an individual in social life, the progress in discovering their potential, carried out time-sequentially, stage by stage [22], the way of life, providing stability in the flow of social life [3, 5, 27], etc. In other

S. I. Ashmarina and V. V. Mantulenko (Eds.): IPM 2020, LNNS 161, pp. 230–239, 2021.
https://doi.org/10.1007/978-3-030-60926-9_30

words, nowadays the career is regarded as a way of self-actualization of the employee, defining it in a complex hierarchy of human relationships and social roles in work, home, family, leisure.

In this context, an employee's career is defined as an integrated formation of a *business career* [14] (or a career in working life, in the market system of division of labor) and *personality career* [28] (or off-duty [21], derivative [16], household [14], supporting [15], "career in a personal life" [27]). The interaction of business career and personality in the "work-family-leisure" system is associated with the desire of the employee to achieve, on the one hand, material well-being and competitiveness in the labour market; on the other hand, social welfare, spiritual well-being in the personal life.

Both business and personality careers are important for the employee, which they want to enjoy, control and manage to achieve personal goals. In ideal world, equality is needed between these careers: a person strives for harmony and prosperity in the present and future in both personal and work life. However, in real life, such equality is difficult to achieve. It is possible to develop a situation in which the business career does not correspond, but contradicts the personality one, as a result, hidden conflicts and imbalances appear, unpredictable situations and problems arise [4, 7, 20, 23]. In this regard, the focus of the research is to identify the imperatives of achieving the harmony in business and personality career, allowing to improve the quality of working life of the employee, increase their competitive advantage in the labour market.

To achieve this objective, the following research tasks were *solved:* career time architectonics that allows us to understand its duality has been justified, a methodological approach to evaluating the harmony of an employee's career based on the concept of time management has been proposed and tested; strategic alternatives to the harmonization of careers based on individual employee choices of an occupation, education, life partner, friends, children, etc. have been systematized.

2 Methodology

The scientific novelty of the study is the development of theoretical and applied bases of achievement of harmony of business and personality careers based on the concept of time management in the conditions of informatization and digitalization of national economy. *The subject of the study* is the time for an individual career of an employee taking into account the prospects (strategy) of the organization's development. *The object* is commercial banks in the city of Novosibirsk in 2013–2018. The study was conducted on a random quota multistage sample, which was used as a micromodel of the survey object, formed on the basis of statistical data (quota parameters). The research methodology is based on the theoretical analysis of the results of scientific works, which can be called "personal time management as a firm management" [1, 8, 17, 18]. This approach raises the problem of choosing priorities and offers principles for building a rational organization of personal time: coherence (consistency), balance, focus, humanity, flexibility. A person is considered as a "complex organization", whose management system should have all the elements of classical management theory – strategy, mission, management accounting, marketing, etc.

This approach makes it possible to increase the structure of methods for harmonizing the dual nature of a business career in order to increase personal efficiency, as well as to borrow from the classical management of the enterprise regularities. The research was conducted using statistical methods. The object is commercial banks in the city of Novosibirsk in 2013-2018. The study was conducted on a random quota multistage sample, which was used as a micromodel of the survey object, formed on the basis of statistical data (quota parameters).

3 Literature Review

The reference to the concept of "harmony of the employee's career" is not limited to the terminological makeover of the existing concepts (work life balance [6, 11, 12, 19], work and leisure (rest) [10], "satisfaction with their functioning at work and at home" [2, 3, 24], and suggests a methodological reorientation to its socio-economic value. This reorientation is in the fact that the harmony of the employee's career is such an interaction of opposite, mutually exclusive types of it, which allows the employee to receive personally significant benefits in various life and work situations. It is an expression of will, conscious personal activity of the employee, their vision of socio-economic value of business and personality career and to achieve social sustainability in work and personal (private) life through investments (direct and indirect) into the career. In other words, the achievement of career harmony involves not only "adjustment" (adaptation) of the needs and resources of the employee to the market (current and strategic) goals, but also the formation of a full-fledged career, desirable, bringing satisfaction and joy to work and into personal (private) life.

Considering that business and personality careers are carried out in time, measured in terms of days, weeks, months, years, etc., the harmony of career can be understood in the temporal dimension of achieving social (personal and labour) stability. In this context, the understanding of career harmony is associated with an objective assessment of the employee's temporary contribution to the management of their career. To date, there are debatable provisions, important from the theoretical and methodological points of view, related to the content and structure of time for the employee's career. The existing studies do not consider the concept of "career time", its content and structure.

4 Results

4.1 Time of Business and Personality Careers: Concept and Structure

Career time should be understood as the time spent at the employee's discretion on the development and self-design of life activities, self-actualization of professional and personal potentials, as well as self-destruction of the career. From this point of view it is necessary, first of all, to allocate *time of business career* and *time of personality career* related to the development and use of physical and mental abilities of the worker respectively in working hours, out-of-work and free time (Table 1).

Table 1. The career time of an employee: types of time for the employee's business and personality career

Type of career	Kind of career	Sphere of career manifestation	Form of career manifestation	Kind of time for career	Economic content of career time	Examples of career time
Business career	Professional	Working life	Productive (efficient) work in the organizational structure	Working hours	Time of the professional abilities development	Time of study and professional development. Self-education time Time of volunteer (social) work
	Working				Time of spending of physical and mental forces	Fixed time. Overtime
				Out of work time		Performance of official functions outside the workplace in the household: consultations on working issues by the phone with colleagues, revision of reports, consideration of projects, etc.
Personality career	Mundane	Personal (private) life	Productive (efficient) work in the household			Time to make household items Time of work on a personal subsidiary farm
			Domestic service			Time for cooking, cleaning rooms and furniture, clothes and shoes, laundry, children, etc.
	Leisure		Leisure activities	Free time	Time to meet physical, intellectual and social needs	Time of study and professional development Self-education time Time of volunteer (social) work Time of entertainment (including visits to cultural institutions, communication with family, friends, etc.) Time for physical education, sports, hobby

Source: authors.

Time of business career is working (both fixed and overtime) and out-of-work time associated with the preservation or acquisition of some desired personal success in terms of clearly defined positions, posts, statuses, roles, perceived as a result of achieving the demanded quality of working life [25, 26]. An employee, organizing and developing a business career, acts not only as a performer of a socially prescribed professional role, but also as an autonomous, creative, responsible person capable of self-regulation, self-determination and self-development [28]. In this regard, it is necessary to distinguish the time of *work career and professional career.*

The time of personality career is an out-of-work time connected with implementation by the person of productive (efficient) and service work in a household, and with satisfaction of physical, intellectual and social needs during free time. An employee, managing a personal career, achieves stability in the flow of personal (private) life through the formation, firstly, of a certain system of principles of socialization and self-realization associated with their life experience and activities in the household, i.e. mundane career; *secondly*, personal, physical, spiritual development in their free time, i.e. *leisure career.*

Given the proliferation of artificial intelligence, virtualization of work processes, self-organization of labour, distribution of electronic documents, replacing the leadership types of controls by a calculated and reasoned solutions based on computer processing of information, BigData, etc., the boundaries of classification time for career groups does not have clearly defined contours, but rather the area of growth and enrichment of one group characteristics at the expense of the other groups. Ultimately, the range of types of business career and personality time is expanding.

For example, as the market for digital solutions and digital management services expands, the creative work of most modern employees extends beyond the normal working day and working week [9]. This is the performance of official functions by an employee during out of work time: completion of reports, urgent execution of tasks on weekends and holidays, discussion of work issues on the phone with colleagues, etc.

4.2 Career Harmony: Methodical Approach to Its Assessment (Example of Bank Employees)

The achievement of a full-fledged career, desirable, bringing satisfaction and joy in work and in personal (private) life, ultimately depends on the use of scarce resources – time. The amount of time for business and personality careers are not arbitrary. In spite of the fact that each employee defines them according to the abilities, motives, quality of life, but in general they depend on needs and opportunities of society and the market, approach to the person, work and property. In other words, business and personality careers have necessary market value.

In terms of time management concept, career is *presented as a fusion* of different types of career into a unified whole, allowing a person to achieve the desired quality of life under the changing conditions inside and outside organization realities so that the time for working career decreases, and the time for personality career increases, but does so in such a way that the time for mundane career is reduced more than increased time for professional and leisure career.

Since the harmony of the employee's career has a versatile nature, it is difficult to name a single indicator characterizing its measure. As its indicators, it is advisable to *use indicators of balance, independence, proactivity and career success* (Table 2).

Table 2. Indicators of harmony of business and personality careers depending on the strategy of the banking organization (based on the results of the working day photographs), ratio

Indicators for assessing employee's career development	Banks with strategies					
	Entrepreneurial		Dynamic growth		Profitability	
	Back office	Front office	Back office	Front office	Back office	Front office
The ratio of the employee's career balance $C_{individual} = \frac{T_{personality}}{T_{business}}$	0.50	0.68	0.55	0.86	0.53	0.95
Private business career independence ratio $I_{bu \sin ess} = \frac{T_{business}}{T_{business} + T_{personality}}$	0.67	0.59	0.64	0.54	0.65	0.51
Private personality career independence ratio $I_{personality} = \frac{T_{personality}}{T_{business} + Tpersonality}$	0.33	0.41	0.36	0.46	0.35	0.49
Integral individual's career independence ratio $(I = I_{business} \times I_{personality})$	0.223	0.241	0.229	0.249	0.226	0.250
Career proactiveness ratio $P = \frac{T_{leisure} + T_{prof}}{T_{working} + T_{mundane}}$	0.22	0.21	0.29	0.19	0.25	0.22
Career success ratio $S = \frac{T^{successful}_{personality} + T^{successful}_{business}}{T_{personality} + T_{business}}$	0.67	0.41	0.43	0.56	0.71	0.87

Source: authors.

The obvious indicator characterizing the quantitative agreement between the time for business $T_{business}$ and personality $T_{personality}$ careers is *the indicator of balance*. Ideally, this figure should tend towards 1: ($C_{individual} \rightarrow 1$). This means that the employee pays equal attention to achieving both business career goals and personality career ones. This career situation (Table 2) is more or less typical for front office workers in organizations with a strategy aimed at profitability ($C_{individual} = 0,95 \rightarrow 1$). Under other organizational development strategies, there is a situation when business career is in conflict with personality career ($C_{individual} < 1$), resulting in the hidden contradictions and disparities of career, problems and barriers to socialization and self-realization.

In case of "business career is above all" ($C_{individual} < 1$) an employee considers business career as a key factor of social well-being and financial independence, sets strict restrictions in private life, consciously chooses the infringement of personality career, primarily, leisure career, which is perceived by them as a lack of freedom and discomfort resulting from this dependence. As can be seen from Table 2, this is a typical career situation in banking organizations: $(0,5 \leq C_{individual} \leq 0,86)$.

To characterize the quantitative side of an employee's career harmony, *a career independence ratio can be used,* characterizing the involvement of an employee in a particular type of career relative to different periods of working life. The involvement of an employee in the business and personality career depends on the obvious existing options to achieve their personally significant benefits through either their professional development, positioning their influence, power, authority, status, competencies in the professional environment in a specific inside and outside organizational realities $I_{business}$, or meeting physical, intellectual and social needs during working hours and free time in organizational structures. $I_{personality}$. Under the law of equilibrium, to ensure career harmony, time resources must be equally invested into business and personality careers, i.e.$(I_{business} \geq 0,5)$, $(I_{personality} \geq 0,5)$ and $(I \rightarrow 0,25)$. In principle, the career situation in terms of independence was favourable, allowing employees to solve the dual task of material and social well-being: $(0,223 \leq I \leq 0,25)$.

The relationship between business and personality career is much more complex and contradictory than the quantitative proportionality of time for a business career and time for a personality one. Career allows a person to achieve recognition of his uniqueness, significance for other people, for society as a whole [13]. In this regard, it is advisable to use the *P proactivity ration* as an indicator of career harmony, reflecting the degree of employee responsibility for personal development, competitiveness and quality of life.

It is obvious that the reactive career is the most common type of career among bank employees, regardless of the organization strategy: $(0,19 < P < 0,29)$ (Table 2). It is associated with an employee's preferential perception of their business career as the only existing way of socialization and self-realization. Business career, being the most significant, overshadows not only leisure, but also often mundane career. Consequently, there is a narrowing of the circle of communication, restriction of personal relationships and friendly contacts, lack of hobbies and entertainment, as well as an increased likelihood of forming a dysfunctional climate in the family, the inability to arrange a personal life and other social problems.

The indicator of career success is an important subjective characteristic of career harmony. This *ratio* reflects the subjective assessment of satisfaction with the time spent on career achievements at work $T_{business}^{successful}$ and in life $T_{personality}^{successful}$. Front office employees at banks with a profitability strategy consider their career to be the most successful (Table 2) $(S = 0,87)$ and back office employees of banks with an entrepreneurial strategy $(S = 0,41)$ think of their career as the least successful.

4.3 Strategy of Career Harmonizing: Concept and Types

Career harmonization strategy is a long-term course of employee's impact on the amount and structure of time spent on career in order to achieve the desired quality of life in the changing conditions of modern reality. This is a set of premeditated principles of behaviour, covering 4Ps of career marketing complex: jobs (product), time investment (price), resources (place), brand (promotion). The following strategies can be applied: *strategy of personal strategy balance, double employment, double career of the family, resource saving and career (business and/or personality) downshifting* (Table 3).

Table 3. Strategies to achieve harmony between business and personality careers (% of respondents)

Time spent on a business career	Time spent on a personality career		
	Low	Average	High
High	Strategy of a personality career downshifting (23%)	Strategy of a double career of the family (10%)	Double employment strategy (9%)
Average	Career resource saving strategy (7%)	Double employment strategy (11%)	Strategy of a double career of the family (12%)
Low	Double employment strategy (6%)	Career resource saving strategy (5%)	Business career downshifting strategy (17%)

Source: authors.

The strategy of double employment (typical for 15% of the respondents) is based on the fact that business and personality careers are designed to mutually balance each other: the employee is forced to give their best both at work and in the household ("the employee is both the breadwinner, and the householder"). In this context, an employee chooses from a certain number of more or less acceptable jobs and personal positions the one that allows him to obtain personally significant benefits in life and at work [27].

The strategy of career resource saving (typical for 12% of the respondents) assumes that the employee, assessing the available resources (abilities, motives, time, money) and career opportunities (market system of division of labour, competition), seeks to achieve a certain prolonged personal benefit in their working or personal life depending on the stage of the life cycle.

The strategy of a double career of the family (typical for 22% of the respondents) involves strengthening the autonomy of spouses and children, the importance of self-realization for each member and building on this basis complementary relationships in the social cultural environment of the family, facilitating self-development and certain personally significant benefits in various aspects of life to each family member, without losing family integrity.

The strategy of career downshifting (typical for 40% of the respondents) assumes achievement of career harmony by means of voluntary abandonment of self-realization in different socially significant spheres of life.

5 Conclusion

Theoretical significance of the study lies in the further development of the theory of duality of an employee's career on the basis of time management, namely: conceptual approach to career time architectonics, reflecting its duality, is formed; socio-economic essence and concepts of "an employee time for career", "career harmony" are defined; system of indicators to measure career harmony is proposed and approved; methodical approach to formation of harmony strategy within the concept of time management is justified. The theoretical provisions and conclusions of the study of career duality allow us to understand the general nature and underlying causes, sources, driving forces of the processes of its organization, development and destruction, alternatives to synchronization of the interests of labour market participants. The practical significance of the study is determined by the possibility of using the results of the study for the formation of personnel policy in organizations with a focus on increasing personnel competitiveness, improving the quality of life, boosting the employer's image. Applied aspects of the research are universal, i.e. they can be taken as a basis by any organization regardless of field of activities and territorial affiliation, organizational and legal form, etc.

References

1. Arkhangelskiy, G.A.: Organization of Time: From Personal Efficiency to the Development of the Company. Piter, Moscow (2008)
2. Başlevent, C., Kirmanoğlu, H.: The impact of deviations from desired hours of work on the life satisfaction of employees. Soc. Indic. Res. **118**(1), 33–43 (2013). https://doi.org/10.1007/s11205-013-0421-9
3. Booth, T., Murray, A.L., Overduin, M., Matthews, M., Furnham, A.: Distinguishing CEOs from top level management: a profile analysis of individual differences, career paths and demographics. J. Bus. Psychol. **31**(2), 205–216 (2015). https://doi.org/10.1007/s10869-015-9416-7
4. Bullinaria, J.: Agent-based models of gender inequalities in career progression. JASSS **21**(3), 7 (2018)
5. Buravcova, N.V.: Modern approaches to career and careerism. SMALTA **1**, 39–42 (2014)
6. Clark, S.C.: Work/family border theory: a new theory of work/family balance. Hum. Relat. **53**(6), 747–770 (2000). https://doi.org/10.1177/0018726700536001
7. Cohen, J.R., Dalton, D.W., Holder-Webb, L.L., McMillan, J.J.: An analysis of glass ceiling perceptions in the accounting profession. J. Bus. Ethics **164**(1), 17–38 (2018). https://doi.org/10.1007/s10551-018-4054-4
8. Covey, S.R.: The 7 habits of highly effective people (2019). http://englishonlineclub.com/pdf/Stephen%20R.%20Covey%20-%20The%207%20Habits%20of%20Highly%20Effective%20People%20[EnglishOnlineClub.com].pdf. Accessed 22 June 2020
9. Demina, V.V.: The boundaries of the working and leisure time for the employees today. Rossiysk. Predprinimatelstvo **12**(10), 30–36 (2011)
10. Demina, V.V.: Time maneuvering as a feature of modern economic system. J. Moscow State Regional Univ. Ser.: Econ. **4**, 13–20 (2018)
11. Greenhaus, J.H., Collins, K.M., Shaw, J.D.: The relationship between work-family balance and quality of life. J. Vocat. Behav. **63**(3), 510–531 (2003)

12. Grzywacz, J.G., Carlson, D.S.: Conceptualizing work-family balance: implications for practice and research. Adv. Dev. Hum. Resour. **9**(4), 455–471 (2007). https://doi.org/10.1177/1523422307305487
13. Karpov, A.V., Subbotina, N.V.: The research of personality determinants of career preferences of managers. Bull. Yaroslavl state Univ. Named After P. G. Demidov Ser. Hum. Sci. **4**(22), 77–80 (2012)
14. Kibanov, A.Y.: Basis of Personnel Management. INFRA-M, Moscow (2015)
15. Kotomanova, O.V.: Professional career as a path of personal development of a woman. Bull. Buryat State Univ. **5**, 102–106 (2013)
16. Kozlova, O.P.: Organizational and economic bases of regulation of career of scientific and pedagogical employees of the higher school. Doctor's thesis. Novosibirsk State University of Economics and Management, Novosibirsk (2010)
17. Kulikova, V.N.: Make Time Work for You. Centrpoligraf, Moscow (2008)
18. Morgenstern, J.: Technologies of Effective Work. Dobraya Kniga, Moscow (2006)
19. Mustapa, N.S., Noor, K.M., Mutalib, M.A.: Why can't we have both? A discussion on work-life balance and women career advancement in Malaysia. J. Asian Financ. Econ. Bus. **5**(3), 103–112 (2018). https://doi.org/10.13106/jafeb.2018.vol5.no3.103
20. Ono, H.: Career mobility in the embedded market: a study of the Japanese financial sector. Asian Bus. Manag. **17**(5), 339–365 (2018). https://doi.org/10.1057/s41291-018-0042-x
21. Orel, A.A., Sidorova, D.G.: The major factors influencing on career development and on advance of a career ladder. Electronic Scientific Journal "APRIORI". Ser.: Hum. Sci. **2** (2014). https://cyberleninka.ru/article/n/osnovnye-faktory-vliyayuschie-na-razvitie-kariery-i-prodvizhenie-po-kariernoy-lestnitse. Accessed 22 June 2020
22. Romanov, V.L.: Social Self-organization and State. RAGS, Moscow (2003)
23. Salmela-Aro, K., Upadyaya, K.: Role of demands-resources in work engagement and burnout in different career stages. J. Vocat. Behav. **108**, 190–200 (2018). https://doi.org/10.1016/j.jvb.2018.08.002
24. Smotrova, T.N., Gritsenko, V.V.: Satisfaction with different life aspects as an indicator of social and psychological adaptation success of fellow countrymen in Russia. . Izvestiya Saratov Univ. Ser. Educ. Acmeol. Dev. Psychol. **6**(1), 53–59 (2017). https://doi.org/10.18500/2304-9790-2017-6-1-53-58
25. Sotnikov, N.Z., Sotnikova, S.I., Mikhailova, E.M.: Business career personnel: modern management models of Russian banking organizations. In: Mantulenko, V.V. (ed.) Proceedings of the International Scientific Conference "Global Challenges and Prospects of the Modern Economic Development". European Proceedings of Social and Behavioural Science, vol. 57, pp. 1302–1315. Future Academy, London (2019). https://doi.org/10.15405/epsbs.2019.03.132
26. Sotnikov, N.Z.: Business career as mechanism of formation of competitiveness of commercial bank employees. Doctor's thesis. Novosibirsk State University of Economics and Management, Novosibirsk, Omsk (2017)
27. Sotnikova, S.I.: Methodological approaches to investigation of career of worker: possibilities and limitations. Vestnik NGUEU **2**, 142–154 (2014)
28. Sotnikova, S.I.: On the career space in the Russian society. Proc. Irkutsk State Acad. Econ. **3**, 46–52 (2014)

Check for updates

Formation of Stress Competence of Customs Officials

K. V. Trubitsyn(✉), O. Y. Kalmykova, and Y. N. Gorbunova

Samara State Technical University, Samara, Russia
Trubitsyn.KV@samgtu.ru, oukalmiykova@mail.ru,
080505@mail.ru

Abstract. The purpose of the study is to develop methodological and practical recommendations for the formation of stress competence of customs officials, which involves improving the system of educational work in customs authorities and implementing conflict-related counseling of officials in the performance of personnel management functions in customs authorities. The tasks of the diagnostic stage of the study included the identification of stress factors of the organizational environment of customs authorities, diagnostics of the level of organizational stress of officials, identification and assessment of personnel risks, determining the comfort of the organizational environment and diagnostics of coping strategies of officials (2014–2019). Empirical methods, such as pedagogical experiment, observation, questioning, testing, analysis of products of educational and professional conflictological activity of university professors, methods of quantitative and qualitative analysis of the results obtained and methods of mathematical statistics were used in the study. The results of various sociological studies conducted of customs officials are presented.

Keywords: Conflict counseling · Professional stress · Stress factors · Stress competence

1 Introduction

Organizational changes in the customs system have a negative impact on the socio-psychological climate in the team: the stress level in the organizational environment is increasing; job satisfaction and the level of officials' attitude are decreasing, which, accordingly, leads to a decrease in efficiency and productivity. The reasons of stressful situations in the customs authority may be as follows: unsatisfactory working conditions, low professional or stress competence of both ordinary officials and managers, problems of labor motivation, violations of intra-group norms of behavior of individuals, etc.

To achieve the goals of functioning and strategic development of the customs authority through consistent harmonization of discrepancies in the implementation of management, economic, innovation and production situations, it is necessary to form a strategy for managing the professional stress. The effectiveness of implementing the strategy in management practice depends not only on the level of professional competence, but also on the level of stress competence of customs officials.

S. I. Ashmarina and V. V. Mantulenko (Eds.): IPM 2020, LNNS 161, pp. 240–248, 2021.
https://doi.org/10.1007/978-3-030-60926-9_31

The head, together with the specialists of the personnel service of the customs authority, should have the tools to diagnose the labor behavior of officials, carry out a set of preventive and educational measures. The existing facts of operational and service misconduct and violations of professional ethics in relation to their colleagues or participants in foreign trade activities, and a decrease in the level of general culture of civil servants cause an urgent need to develop preventive and educational measures aimed at increasing the level of stress competence of officials.

Socio-economic instability in society creates an urgent need to study the impact of stress factors of the organizational environment on the efficiency and productivity of personnel. In modern research in the field of personnel management, it is important to take into account the relationship between performance, satisfaction and incentives [7]. Nowadays, in addition to professional skills, customs officials in the Russian Federation also need trainings that prioritize the following: development of skills such as leadership, communication, interpersonal skills, logical thinking, problem solving, working as a team member and ethical judgments, stress tolerance, and involvement in the organizational culture of customs.

Therefore, on the one hand, there is a need to improve the standards of training of future customs specialists by creating special training programs for customs guidance that meet the international standards of PICARD accreditation and ensure the formation of a meta-competence profile of education in the field of customs administration [20]. But, on the other hand, the lack of proven tools for making decisions on adapting existing professional standards as a model for achieving strategic and systemic goals is a limiting factor in the implementation of the standard in the field of customs affairs [17, 19].

The head and officials of customs authorities should be prepared to implement professional tasks in a stressful organizational environment and promptly implement personnel measures to minimize the consequences of stressful situations. These personnel measures can be classified as “personnel needs” of customs authorities, which requires the development and implementation of specialized programs [13]. This will allow one to prepare a specialist who meets the modern requirements of the employers, as well as capable to adapt to changing conditions [4, 19].

2 Methodology

The methodological basis of the research included concepts, approaches, principles in the study of social conflicts, their nature and management. The following methods were used in the study: the method of assessing the stress level in the organizational environment; the method of diagnosing the comfort of the behavioral space of the organization [15, 16].

The formation of stress competence of customs officials involves solving multi-level tasks that allow the integration of *Customs* students and customs officials:

– at the university level: analysis of psychological and pedagogical conditions for the formation of students’ stress competence; formation of the concept of scientific and

methodological support for the discipline *Conflict and Stress Management in Customs Authorities*,
- at the level of customs authority: formation of stress competence of managers and officials; development of training programs for psychological self-regulation of emotional states; implementation of a system of conflict counseling (creation of the *Center for Conflict Counseling* on the basis of partner educational organizations) etc.,
- at the level of university and customs authority: testing of educational and methodological support used in the educational process of the university and in the process of conflict counseling of customs officials; development of recommendations to improve the level of stress competence of managers and officials; formation of training programs in customs authorities [9].

The process of forming the stress competence of officials should be aimed at mastering the following competencies:

- methods for diagnosing the sources and causes of increased social tension in the workplace,
- methods for the diagnostics of stress factors that cause an increase in the level of professional stress of managers and officials,
- strategies for the prevention of occupational stress of managers and officials,
- programs for prevention of addictive behavior of managers and officials,
- programs for prevention of corrupt behavior of officials; coping strategies of managers and officials,
- methods and programs for preventing the syndrome of emotional (professional) burnout of officials,
- regulatory and documentation support of the personnel strategy for managing conflicts and stresses in customs authorities.

The level of formation of stress competence components, namely, readiness for constructive conflict resolution and prevention of professional stress can be assessed in the framework of training sessions, during the current assessment of the professional activities of customs officials. When developing a system for monitoring the stress competence level of officials, it is necessary to select indicators that meet the goals and objectives of the study, on the basis of the general functions of the conflict and stress management strategy in customs authorities.

The study of this problem involves the solution of many tasks, one of which is the filling of educational content that methodically provides the process of forming the stress competence of customs officials. The authors have developed a *Bank of educational and methodological materials* that includes a set of conflict-related tasks, that was formed taking into account the specifics of the professional activities of customs officials.

Conflict-related tasks are used both in the educational process at the university and in the process of conducting various training sessions in the customs authorities. Like a number of other authors who study the current conditions of the educational process, the authors of the paper emphasize the need to improve the quality of professional training of University students in the context of modernization of the educational process [13, 19]. In order to improve the process of forming the stress competence of a specialist in the field of customs, educational and methodological materials on the methodology of personnel risk management in customs authorities are being developed [5, 14].

3 Results

Stress monitoring was carried out to determine the stress factors of the customs authority's organizational environment, the level of satisfaction with working conditions and the officials' emotional state [15]. In the period from 2014 to 2017, the amount of stress factors affecting officials was at an average level and did not reduce the labor productivity on average for the customs authority, but in 2018 it was already at a critical level and could initiate disorganization of labor processes. Assessment of the emotional state at the second stage of stress monitoring of officials showed the following results:

- in the period from 2014 to 2016, respondents showed an average degree of emotional exhaustion and depersonalization, but a high degree of reduction of personal achievements; overall score: average,
- in the period from 2017 to 2018, respondents were characterized by a high degree of emotional exhaustion; average depersonalization; reduction of personal achievements: high degree; overall score: high degree. The main factor of emotional burnout of officials is a long-term excessive workload and large-scale organizational changes in the customs service.

In the period from 2014 to 2018, the method of *Determination of the comfort of the organizational environment* was used to obtain an integral characteristic of the officials' organizational behavior [9]. Diagnostics was performed by sequentially calculating the individual comfort indices for each official and the general indices for the team. According to the results of the survey (28 respondents), the value of the general index of comfort of the behavioral environment in 2018 is characterized by a positive value. The study identified the following factors of the organizational environment of the customs authority, presumably influencing the effectiveness of customs officials in the field of customs affairs: the level of professional stress; the indicator of professional burnout; the indicator of the socio-psychological atmosphere of the organization; the reduction of personal achievements; the level of loyalty of customs officials.

The Data Envelope Analysis (DEA) method was used to confirm the influence of organizational environment factors studied at the diagnostic stage on the effectiveness of customs officials [2, 3, 12]. At the first stage of evaluating the effectiveness of customs officials in the field of customs affairs, the variables of the DEA method model were determined. The variables were obtained from the results of the customs officials' survey. The evaluation in the form of an anonymous questionnaire was carried out by both ordinary officials and managers. Input parameters are variables for which a positive effect is a decrease in their values, and output parameters are indicators the values of which are recognized as effective when they increase. So, input parameters in the model of the DEA method, are as follows: X_1 – level of professional stress;

X_2 – indicator of professional burnout; X_3 – indicator of the socio-psychological atmosphere of the organization; X_4 – reduction of personal achievements. The output parameter in the DEA method model is Y_1 – level of loyalty of customs officials.

In the model of the DEA method, the level of officials' loyalty is defined as the output parameter of the simulation, since the level of loyalty has a direct impact on the efficiency and productivity of labor. When analyzing the results of the first stage of modeling, it was found that the spread of estimates is within 10%. This variation is very small for adequate DEA estimates. In fact, they all averaged, that is, they became closer to each other. In order to get objective results, it was necessary to change the initial parameters of the simulation. To do this, the number of input parameters was reduced: X_1 – indicator of professional burnout; X_2 – reduction of personal achievements; Y_1 – level of customs officials' loyalty. Table 1 shows the initial data for repeated modeling.

Table 1. Initial data for repeated modeling

Period of analysis	Indicators		
	X_1	X_2	Y
	68	33	64.3
	67	32	63.5
	69	32	64.1
2015	73	31	65.8
	72	30	65.7
	74	30	65.7
2016	79	28	71.3
	78	28	71.1
	78	28	70.78
2017	85.5	27	70.45
	86.2	27	70.32
	86.5	26	70.12
2018	92	26	69.6
	92	24	69.3
	91	25	69.3

Source: authors.

The results of repeated modeling using the DEA method are presented in Table 2. These results are more consistent with adequate DEA estimates. The result of the solution is to determine the relative efficiency of objects in the form of ranking indicators of the CCR-model in the unit interval [0, 1] and the Super-Efficiency model – two objects have the maximum generalized efficiency index. These objects are, respectively, the periods: the first quarter of 2016 and the second quarter of 2018 (Table 3).

If you have statistical data for a certain period, you can create a forecast for the next year. The results of the simulation showed that the highest level of officials' loyalty was for the first quarter of 2016. For this period, the average values of the

Table 2. Results of modeling using the DEA method

Period	X_1	X_2	Y_1	CCR - model	Super-efficiency model
2014_1	54.61	33	64.3	0.801254	0.801254
2014_2	54.23	32	63.5	0.796829	0.796829
2014_3	52.62	32	64.1	0.828969	0.828969
2015_1	48.07	31	65.8	0.9315	0.9315
2015_2	48	30	65.7	0.931441	0.931441
2015_3	47.88	30	65.7	0.933776	0.933776
2016_1	48.52	28	71.3	1	1.012329
2016_2	49	28	71.1	0.990156	0.990156
2016_3	48.76	28	70.78	0.989191	0.989191
2017_1	48.87	27	70.45	0.992909	0.992909
2017_2	49.1	27	70.32	0.987696	0.987696
2017_3	48.76	26	70.12	0.999993	0.999993
2018_1	49.22	26	69.6	0.985748	0.985748
2018_2	49.34	24	69.3	1	1.041667
2018_3	48.95	25	69.3	0.995623	0.995623

Source: authors.

Table 3. Objects with the maximum performance indicator

Period	X_1	X_2	Y_1	CCR - model	Super-efficiency model
2016_1	48.52	28	71.3	1	1.012329
2018_2	49.34	24	69.3	1	1.041667

Source: authors.

socio–psychological atmosphere of the organization and the reduction of personal achievements were 48.52 and 28, respectively. This dependence is explained by the fact that an object with the highest value of the output variable of all evaluated objects (the output parameter Y for this period is 71.3) is automatically classified as an effective object. The analysis results also showed that the object of analysis is at the border line of efficiency, that is, the efficiency indicator is equal to 1 for the second quarter of 2018. For this period, the average values of the socio–psychological atmosphere of the organization and the reduction of personal achievements were 49.34 and 24, respectively. The evaluation of the achieved level of efficiency in the studied system with specified parameters allows one to give it some forecasts about the effectiveness in the future and make recommendations on the formation of a strategy for the prevention of customs officials' professional stress.

4 Discussion

The accumulated results of scientific research and practical experience create certain prerequisites for the development of new methods to increase the level of customs officials' stress competence. Kiseleva in the paper *Assessment of the moral and motivational potential of employees of the state administration bodies* considers the influence of values-based orientations of officials on labor productivity; the key motivational factors of professional choice and their relationship with values-based orientations are determined [8].

Korzhan and Murashko in the paper *On the need to improve psychological and pedagogical support for professional activities of customs specialists* developed a research project *Regional system of psychological and pedagogical support for professional activities of customs specialists*, which involves the development of scientific and methodological support for psychological and pedagogical support for professional activities and continuous professional training of FCS personnel [10].

Kazybekova in the paper *Improving the management of professional development of personnel in customs authorities in the modern period* analyzed the set of potential competencies that should be possessed by the head of customs authorities, focusing on the need for specialized psychological and pedagogical training [6]. Adrianov and Ovchinnikova in the paper *Causes of socio-economic conflicts in customs authorities and methods to overcome them* revealed the concept of the school of human relationships as a kind of tool for resolving conflict situations in customs authorities.

The authors of this paper highlight that it is necessary to increase the level of conflict competence of managers and officials [1]. The problem of making management decisions as the foundation of an effectively run organization is discussed in the paper *Concept of management decision in the activities of customs authorities: legal and organizational aspects* by Shabanov and Ananyeva.

One of the key competencies in the professional activity of customs authority managers is the competence to manage conflict situations and professional stresses [18]. In the study *Elements of organizational and managerial work,as an indicator of the effectiveness of official activities of customs authorities*, Krasilnikov and Trush developed a method for evaluating the customs officials' qualities, necessary from a professional point of view, that will allow structuring and standardizing the approach to personnel work in customs authorities [11].

5 Conclusion

In the course of the research, the authors of this paper developed the following:

- organizational and methodological recommendations for the formation of stress competence of customs officials and specialists in the specialty *Customs affairs*,
- a set of training courses,
- methodological tools for conflict counseling in the organization.

At the same time, there are a number of potential risks related to the implementation of the methodology for the formation of stress competence of officials and students:

- insufficient participation of employers in expert surveys on the research topic,
- insufficient professional qualifications of the personnel service at the customs authority, who have to support the use of stress management and programs for the formation of stress competence of managers, officials and employees of customs authorities,
- introduction of innovative methods and technologies that require a high level of motivational, methodological and expert readiness of university professors.

Acknowledgment. The authors express their gratitude to the representatives of the scientific school of personnel management (State University of Management, Moscow, Russia) Mitrofanova E.A., Konovalova V.G., Kashtanova E.V. and Mitrofanova A.E. for their consultations and critical comments.

References

1. Andrianov, A.Y., Ovchinnikova, V.I.: Causes of socio-economic conflicts in customs authorities and methods to overcome them. Bull. Univ. **4,** 201–204 (2016)
2. Charnes, A., Cooper, W.W., Rhodes, E.: Measuring the efficiency of decision making units. Eur. J. Oper. Res. **2**, 429–444 (1978)
3. Chen, Y., Du, J.: Super-efficiency in data envelopment analysis. In: Zhu, J. (ed.) Data Envelopment Analysis. International Series in Operations Research & Management Science, vol. 221, pp. 381–414. Springer, Boston, MA (2015)
4. Gryazeva-Dobshinskaya, V.G., Dmitrieva, Y.A.: Leadership and management styles: typological approach to personal resources of change management. In: Solovev, D.B. (ed.) Smart Technologies and Innovations in Design for Control of Technological Processes and Objects: Economy and Production. FarEastCon 2018. Smart Innovation, Systems and Technologies, vol. 138, pp. 202–212. Springer, Cham (2019)
5. Kalmykova, O.Y., Trubitsyn, K.V., Gorbunova, Y.N., Mitrofanova, E.A., Sysueva, I.G.: Building competence of future customs specialists in field of personnel risk management. In: Ardashkin, I.B., Bunkovsky, V.I., Martyshev, N.V. (eds.) International Conference on Research Paradigms Transformation in Social Sciences. The European Proceedings of Social and Behavioural Sciences, vol. 50, pp. 1211–1219. London European Proceedings (2019)
6. Kazybekova, E.E.: Improving the management of professional development of personnel in customs authorities in the modern period. Vest. KSUCTA **4**, 128–131 (2014)
7. Kibanov, A.Y., Mitrofanova, E.A., Esaulova, I.A.: Economics of Personnel Management. INFRA-M, Moscow (2013)
8. Kiseleva, A.A.: Assessment of the moral and motivational potential of employees of state administration bodies (on the example of the Altai customs). Regional Econ.: Theory Pract. **33**(168), 50–57 (2010)
9. Konovalova, V.G., Kalmykova, O.Y., Trubitsyn, K.V., Mitrofanova, A.E.: Conflict and stress management in customs authorities. Prospect, Moscow (2020)
10. Korzhan, D.I., Murashko, M.V.: On the need to improve psychological and pedagogical support of professional activities of customs authorities' specialists. Bull. Tver state Univ. Ser.: Pedag. Psychol. **4**, 168–175 (2013)
11. Krasilnikov, A.Y., Trush, A.N.: Elements of organizational and managerial work, as an indicator of the effectiveness of official activities of customs authorities. Customs Bus. Foreign Econ. Act. Co. **1**(2), 25–40 (2017)

12. Krivonozhko, V.E., Propoy, A.I., Senkov, R.V., Rodchenkov, I.V., Anokhin, P.M.: Analysis of the performance of complex systems. Autom. Design **1**, 2–7 (1999)
13. Kulikov, M.M., Dulin, A.N., Dulin, R.A.: Staffing of cluster initiatives in Russia. In: Solovev, D.B. (ed.) FarEastCon 2018. Smart Innovation, Systems and Technologies, vol. 139, pp. 81–89. Springer, Cham (2019)
14. Mitrofanova, A.E., Mitrofanova, E.A., Konovalova, V.G., Ashurbekov, R.A., Trubitsyn, K. V.: Human resource risk management in organization: methodological aspect. In: Martyshev, N.V. (ed.) Proceedings of the International Conference on Trends of Technologies and Innovations in Economic and Social Studies 2017. Advances in Economics, Business and Management Research, pp. 699–705. Atlantis Press, Paris (2017)
15. Oksinoyd, K.E.: Diagnostics of comfort of behavioral space of the organization. HR Specialist. HR Management, vol. 11 (2011). https://hr-portal.ru/article/diagnostika-komfortnosti-povedencheskogo-prostranstva-organizacii. Accessed 06 June 2020
16. Oksinoyd, K.E.: Stress monitoring. Why and how we conduct it. Methodology for assessing the level of stress in the organizational environment. HR Specialist. HR Management, vol. **5**, p. 51 (2009)
17. Savchenko, Y.Y., Goleva, O.G., Korchagina, I.A.: A study on the perception system of qualifications in different segments of stakeholders. In: Solovev, D.B. (ed.) Smart Technologies and Innovations in Design for Control of Technological Processes and Objects: Economy and Production. FarEastCon 2018. Smart Innovation, Systems and Technologies, vol. 139, pp. 519–528. Springer, Cham (2019)
18. Shabanov, V.B., Ananyeva, V.N.: Concept of management decision in the activities of customs authorities: legal and organizational aspect. Econ. Today **3**, 162–165 (2015)
19. Skripnikova, G.V., Shirmanova, G.S., Korobko, S.M.: Qualification level of employees in the estimation of efficiency of the labor potential of the enterprise. In: Solovev, D.B. (ed.) Smart Technologies and Innovations in Design for Control of Technological Processes and Objects: Economy and Production. FarEastCon 2018. Smart Innovation, Systems and Technologies, vol. 139, pp. 500–508. Springer, Cham (2019)
20. World Customs Organization PICARD Professional Standards – World Customs Organization, 2019 (2019). http://www.wcoomd.org/-/media/wco/public/global/pdf/topics/capacity-building/activities-and-programmes/picard/professional-standards/omd-normes-prof-uk-basse-def.pdf?la=en. Accessed 06 June 2020

Innovative Formats of Education in the Transformation of the Digital Economy

E. N. Sheremetyeva[1](✉), E. P. Barinova[2], and L. V. Zolotova[3]

[1] Samara State University of Economics, Samara, Russia
lena_scher@mail.ru
[2] Samara Branch of "Moscow City University", Samara, Russia
rfnz25@yandex.ru
[3] Orenburg Institute (Branch) Plekhanov Russian University, Orenburg, Russia
Zolotova.LV@rea.ru

Abstract. The article describes the current state of the smart education system. The directions of the formation of a system of digital competencies, innovative technologies and forms of integration of smart education, as well as the possibility of implementing competencies through the use of "qualified technologies" are analyzed. The problems of using modern digital technologies and the technology of practice-oriented training of university students in the context of the global digitalization of the economy are considered. New digital technologies require innovative approaches and qualitatively different competencies in both business and education.

Keywords: Digital educational technologies · Practice-oriented approach · Professional competencies · Smart-education · Smart-university

1 Introduction

The development of the digital economy causes major socio-economic shifts, including in the field of labor relations, a change in the types of professional activity and the nature of labor. Digitalization has an active impact on changes in supply and demand in the labor market, investment in human capital and methods for its improvement, adaptation of labor resources to the new architecture of the labor market, and government support program measures. New digital technologies require innovative approaches and qualitatively different competencies in both business and education. The introduction of intellectual SMART- technologies is changing the platforms used to transfer knowledge [9]. The key to efficiency and competitiveness is the professionalism of participants, their level of responsibility, innovativeness, creativity, systematic thinking, flexibility and adaptability, and focus on results. All these competencies should be formed at the level of training at the university. Vocational education should become one of the most rapidly updated industries, both in terms of content and in terms of technology and training methods. The speed of updating knowledge and technology should be considered as a criterion for the quality of the education system.

S. I. Ashmarina and V. V. Mantulenko (Eds.): IPM 2020, LNNS 161, pp. 249–254, 2021.
https://doi.org/10.1007/978-3-030-60926-9_32

2 Methodology

As the methodological basis of the study, the theories of innovative development, the management of socio-economic systems, intellectual and human capital, expert assessments and the design of managerial technologies are used. The analysis of foreign and national competency models, the requirements of domestic and foreign practice for the systems (models) of basic and professional competencies of the digital economy allows us to determine the methodological foundations of the human capital management mechanism in the context of the formation of a digital ecosystem. Indicators that determine the degree of digitalization of the economy include an assessment of human capital, including in the form of digital knowledge and skills in using information and communication technologies. They form new digital competencies that need to be differentiated, systematized, and based on the selected competencies to develop training and testing programs aimed at creating digital literacy and key competencies in the digital economy [3]. To analyze the degree of development of human capital at the stage of education and training of specialists, the existing risks and limitations, as well as socio-psychological indicators that determine the causes of insufficient digital literacy and insecurity in their digital skills and competencies, regular monitoring of the digital literacy index of university students and graduates is necessary. European countries are effectively transforming their educational systems, developing cognitive skills, re-qualifying national workforce and adapting professionals to a changing digital work environment. Therefore, the transformation of the higher education system in accordance with the challenges of the digital age is a key factor in the development and competitiveness of the country as a whole. The development of an innovative system of professional and managerial competencies that allows the fullest use of the digital capabilities of cross-industry interaction also requires an assessment of the confidence of students, graduates and faculty of the university in the knowledge of the key concepts of the digital economy to determine digital maturity.

3 Results

The implementation of the National Technological Initiative provides for training for the economy specialists of a completely new profile, able to carry out professional activities in the work of "smart - enterprises", participate in the development of global industrial networks and use their data in their activities, own concepts of the industrial Internet and the Internet of things. The need to use new smart technologies is due to a change in technological patterns [1]; the transformation of information technologies related to studies with distance learning [6]; quick adaptation of the modern generation to the use of smart technologies and electronic media as a means of communication, life support and training. The digital economy expects from future specialists a set of professional competencies that should be formed by higher education:

- critical and creative thinking; Initiative and responsibility,
- adaptability and the ability to rebuild professional activities to meet new requirements,
- innovativeness, the presence of skills for generating ideas in the professional sphere, optimizing one's own activities taking into account innovations,
- readiness for continuing education, including retraining and self-education,
- adaptability to reasonable occupational and occupational risk, ability to work individually and as a team enterprise,
- knowledge of digital technologies; their use in their own activities,
- emotional intelligence, etc. [12].

An employee and specialist in the digital economy is not only an advanced computer and Internet user. This is a special category of specialists who are adaptive to changes in the economy, science and education, and at the same time are the generators of changes [2]. The formation of the considered competencies of students is directly related to the restructuring of the content of the educational process. In this direction, it is necessary to note the active use of information technology in universities, as well as practice-oriented technologies.

The national technological doctrine is aimed at the formation of smart-education on the basis of higher educational institutions. Smart - education, defined as "the combination of educational institutions and faculty to implement a joint educational activities on the Internet based on common standards, agreements and technologies" [11]. In smart education, the following areas are distinguished:

- SMART as an educational, intellectual environment,
- SMART as a new education architecture,
- SMART as an effective approach to education,
- SMART as the formation of new competencies [8].

The creation of smart universities is aimed at:

- the flexibility of learning in an interactive educational environment (digitalization of education),
- ensuring free access for students and teachers to content around the world (integration of formal and non-formal education),
- personalization and adaptation of training,
- the active use of design training for students,
- the formation of an integration of creative space (practice-oriented approach).

The purpose of smart- education is to create an individual way of realizing the student's personal potential (student-centered learning). We emphasize that the continuity of education and flexible learning in an interactive educational environment are particularly relevant, which meets the objectives of the National Technological Initiative. The global challenges of the modern digital economy have an impact on the requirements for the formation of labor capital and on the availability of training (primarily professional), which contributes to the transformation of educational systems. Universities are active in teaching students the concepts of "Education throughout Life" by using various digital technologies that allow you to acquire knowledge on-line, to ensure the efficiency of students' repeated calls to information broadcast as part of the educational process [10].

Note that the training of innovative specialists is unthinkable without the use of information and communication technologies and knowledge transfer procedures, such as E-Learning, M-Learning, etc. Technologies of virtual and augmented reality, E-Learning, M-Learning are basic in the formation and development of smart- universities. Information and communication technology "E-Learning" is defined as learning using multimedia and the Internet. Using technology, E-Learning considers the student as a user of the Internet, various educational portals and resources, including the resources of his own educational institution on the network. The student is an active consumer of information posted on various information contacts, which allows him to get an education: perform practical tasks, write research papers, prepare for exams, etc. Students as content users read electronic lectures, watch videos, attend webinars, etc. [5]. This use of e-learning technology expands the boundaries for education, makes it remote, which allows students to combine education and practice. Practical activities may include work at the enterprise, internship, development of your own business project. The active spread of a whole range of mobile devices (handheld PCs, tablets and smartphones) has become an incentive for the introduction of e-learning technology - MobileLearning (M-Learning).

M-Learning application is based on the use of mobile communications in the learning process. M-Learning technology allows you to upload all educational content to a student's mobile communications tool and access it anywhere, anytime with Internet access. Among the "breakthrough" digital technologies in education include online cloud computing, the use of the TBD system. TBD system is a technology widely used by corporations to maintain a stable position or to lead in a chosen production niche. The technology is based on the development of "effective solutions for the implementation of rational investment strategies" [7].

4 Discussion

One of the University-Enterprise-Student interaction methods is the development of a digital educational environment for interaction and an integrated system of cooperation between a business university in order to increase the attractiveness of a graduate as an employee and to help reveal a potential entrepreneur in partner- countries. The scientific novelty of the task lies in the development of a holistic theory and practical mechanisms for the modernization of the educational system, improving the quality of specialists in accordance with the requirements of enterprises, training premium specialists for the business- environment and smart- enterprises using the latest information technologies. Specific innovative elements in this case are the following results:

- a detailed study of the models of cooperation between business enterprises and universities in various sectors of the economy and the development on this basis of an interactive educational environment through the use of interactive teaching methods (role-playing games, trainings, computer simulators (modeling, quality circles, coaching, etc.)),

- creation of a virtual laboratory of project competencies to coordinate the interaction of universities and representatives of employers. A virtual laboratory will allow for the training and professional development of specialists, as well as the creation and support of educational and practical content and the development of relevant practical economic problems and cases that meet the requirements of practice-oriented training,
- formation of a mutually beneficial system of cooperation between entrepreneurship and educational institutions of the university,
- retraining of lecturers acting as animators and drivers of the system of cooperation between business and universities.

Through the practical involvement of such sites in the work, the procedure for the formation of professional competencies among students is realized through the use of qualified technologies, which are currently the most tested, scientifically based approaches in the implementation of the educational process of the university. Qualitative technologies are focused on the maximum use of the scientific and pedagogical potential of a teacher, a leader, who can be represented by a representative from a particular enterprise and motivate students to independent practical activities through which the integrated development of disciplines and the formation of economic knowledge are implemented. For example, students are invited, on the basis of the real characteristics of an existing enterprise, to create a business – model for its further development in the context of the economic crisis, the launch of new products, changes in the quality of labor potential; apply and justify methods for calculating key economic indicators, their dependence on various economic factors, etc. Such practical work allows you to access information from various fields of economic, managerial, and legal disciplines to carry out tasks, search for information on the Internet using information technology for this purpose, and process and systematize the information received. The active involvement of students in practical activities stimulates the development of professional competitiveness, which will allow in the future to adapt to the dynamically changing conditions of the digital economy [4].

5 Conclusion

The transition to the digital economy in Russia can be realized if there are human resources prepared accordingly, innovative activity of educational and economic entities, development of the intellectual potential of not only the teaching staff, but also managers and technical specialists of enterprises. The effective use of digital technology at all levels of the organization requires employees with a range of technical, business, interpersonal and creative abilities and skills. Digital technologies are changing the system and the emergence of new recruitment criteria, modernizing the labor market infrastructure. Modern digital technologies are actively used to prepare students for higher educational institutions, which helps to increase the efficiency of the educational process and to satisfy demand from enterprises. Further progressive and effective development of digital educational technologies can be achieved by consolidating the efforts of the state in the person of power structures, institutions of science and education, enterprises.

References

1. Aletdinova, A., Bakaev, M.: The economy of smart and Ai-based education. Soc. Sci. **11**, 5151–5156 (2016)
2. Balonin, N.A., Petoukhov, S.V., Sergeev, M.B.: Matrices in improvement of systems of artificial intelligence and education of specialists. In: Hu, Z., Petoukhov, S., He, M. (eds.) Advances in Artificial Systems for Medicine and Education. Advances in Intelligent Systems and Computing, vol. 658, pp. 39–52. Springer, Cham (2018)
3. Barefoot, K., Curtis, D., Jolliff, W., Nicholsn, J.R., Omohundro, R.: Defining and measuring the digital economy (2018). https://www.bea.gov/system/files/papers/WP2018-4.pdf. Accessed 15 May 2020
4. Boccia, F., Leonardi, R.: The Challenge of the Digital Economy: Markets, Taxation and Appropriate Economic Models. Palgrave Macmillan, Cham (2017)
5. Brusilovsky, P., Somyürek, S., Guerra, J., Hosseini, R., Zadorozhny, V., Durlach, P.J.: Open social student modeling for personalized learning. IEEE Trans. Emerg. Top. Comput. **4**(3), 450–461 (2016)
6. Buryak, V.V.: Breakthrough technologies of Russia's digital economy. Beneficiar **26**, 15–18 (2018)
7. Dennouni, N., Slama, Z., Peter, Y., Lancieri, L.: Recommendation techniques in mobile learning context: a review. Int. J. Mod. Educ. Comput. Sci. (IJMECS) **9**(10), 37–46 (2017)
8. Dneprovskaya, N.V., Komleva, N.V., Urintsov, A.I.: The knowledge management approach to digitalization of smart education. In: Hu, Z., Petoukhov, S.V., He, M. (eds.) Advances in Artificial Systems for Medicine and Education II. Advances in Intelligent Systems and Computing, vol. 902, pp. 641–650. Springer, Cham (2020)
9. Dneprovskaya, N.V, Bayaskalanova, T.A., Ruposov, V.L., Shevtsova, I.V.: Study of digitization of russian higher education as basis for smart education. In: Azarov, V.N., Shaposhnikov, S.O. (eds.) IEEE International Conference "Quality Management, Transport and Information Security, Information Technologies" (ITandQMandIS), pp. 607–611. IEEE, Paris (2018)
10. Karna, N., Supriana, I., Maulidevi, N.: Implementation of e-leaming based on knowledge management system for Indonesian academic institution. In: Si, M. (ed.) 1st International Conference on Information Technology, Information Systems and Electrical Engineering (ICITISEE), pp. 43–48. IEEE, Paris (2016)
11. Kotevski, Z., Tasevska, I.: Evaluating the potentials of educational systems to advance implementing multimedia technologies. Int. J. Mod. Educ. Comput. Sci. (IJMECS) **9**(1), 26–35 (2017)
12. Kupriyanovsky, V.P., Sukhomlin, V.A., Dobrynin, A.P., Raikov, A.N., Shkurov, F.V., Drozhzhinov, V.I., Fedorova, N.O., Namiot, D.E.: Skills in the digital economy and the challenges of the education system. Int. J. Open Inf. Technol. **5**(1), 19–25 (2017)

Internal Marketing Role in Human Resources Management

I. V. Yakhneeva[1(✉)], A. N. Agafonova[1], and N. V. Kalenskaya[2]

[1] Samara State University of Economics, Samara, Russia
yakhneevai.v@sseu.ru, agafonova.a.n@gmail.com
[2] Kazan Federal University, Kazan, Russia
kalen7979@mail.ru

Abstract. The article addresses the impact of the internal marketing concept on the policy of companies in the field of personnel management. Authors examine the problem of the importance of personnel as stakeholders through the prism of corporate social responsibility. The sample of companies includes 20 largest representatives of Russian business and 120 regional small and medium enterprises. The study is based on the use of data posted by companies in the form of social and integrated reports, and data obtained by polling a sample of small and medium-sized businesses. The hypothesis formulated by the author that Russian companies increased their share of social support costs and regarded staff members as internal clients has not been confirmed.

Keywords: Corporate social responsibility · Human resources management · Internal marketing · Social investments

1 Introduction

The development of the internal marketing (IM) concept is connected with the identification of opportunities for revealing staff capacity for the success of the company on the market. During the decades of marketing science formation, the presence of this potential has never been questioned. At the same time, the perception of the role of the personnel in solving marketing tasks is continuing. Berry, Hensel, andBurk were one of the first who emphasized the importance of both qualified personnel and responsibility system arranged following the authority for effective consumerism response [4]. The introduction of the terms "internal market" and "internal client" enabled to shift the emphasis on the managerial influence of marketing, with the same used instruments. Grönroos underlines the necessity to form a holistic approach to the creation of consumer orientation of the personnel by means of traditional marketing tools [7]. The internal market can influence consumers [5, 6], and this requires coordination of internal and external activities of the company, separation of company values by its stakeholders (including employees and consumers). The most important components of IM are internal communications, market research, and personnel training [9]. Kanyurhi, Bugandwa and Akonkwa distinguish five of its elements: employee rewards, recruitment process, internal communication, employee empowerment, and employee training [10]. Lings considers internal marketing in two aspects: personal and procedural [11].

S. I. Ashmarina and V. V. Mantulenko (Eds.): IPM 2020, LNNS 161, pp. 255–261, 2021.
https://doi.org/10.1007/978-3-030-60926-9_33

The personal approach is based on the Human Resources Management (HRM) concept. In this case, the success of external marketing depends on whether employees are satisfied and motivated enough to perform efficiently. The procedural approach to internal marketing is based on the concept of Total Quality Management (TQM). The target of internal marketing is the process of intra-corporate interaction and the provision of internal services. Ahmed, Rafiq and Saad focus on mechanisms of the implementation of strategy and means of integration of inter-functional interests of different departments of the organization [1]. They highlight that internal marketing is a systematic action for overcoming employees' resistance to changes, their motivation, and the integration of employees for the effective implementation of corporate and functional strategies. Bell and Bulent emphasize the role of processes of initiation, maintenance, and development of relationships between employees, their management, and the organization, to create greater value for customers [3]. Internal marketing becomes a business strategy and a cross-functional unifying mechanism for the organization. New points of contact between the concepts of "marketing", "human resources management", "management" and "corporate culture" are being revealed. Conditions for the development of a sustainable environment with harmonious relations between employees, firm, and environment [2, 15], manufacturer and consumer are being created. The subject matter of internal marketing is fairly extensive and relatively new to Russian business. That is why this research is dedicated to studying the principal importance of the employees as stakeholders for the business and identifying the prerequisites for a qualitative change in the accents of marketing management.

2 Methodology

The research methodology includes the collection, analysis and generalization of data on social support programs that are implemented by Russian companies. The study is being carried out in two stages:

1. Analysis of social reports published by business leaders. The companies under research relate to the mining, processing, metallurgy, chemical, electric-power industries, telecommunications and financial sector.
2. Analysis of primary data collected by the authors in a sample survey of small and medium-sized businesses. The empirical research was conducted using questionnaire survey, the sample consisted of 120 regional entrepreneurs. The sample structure included manufacturing companies, trade organizations, service companies.

The results of our studies on corporate social responsibility of business abroad and in Russia became an additional information base [16, 17].

The following hypotheses are tested:

1. The share of staff welfare expenses in social investments is increasing.
2. The importance of personnel as stakeholders is increasing, and the loyalty of internal customers is important for business.

3 Results

The first stage of the research related to the analysis of public reporting of Russian large business representatives showed that all companies allocate funds for social programs for employees (Table 1). However, not all of them reflect this information in their reports. In addition, the list does not include major IT companies (Yandex, Mail) that do not publish their social expenditures figures.

Table 1. Trends in the indicators of revenue, social investments and staff welfare expenses

Company	Growth rate, %			Share of staff welfare expenses in social investments, %	
	Revenue	Social investments	Staff welfare expenses	2018	2019
FSK YeES	98.3	112.3	93.3	67.0	55.6
ROSSETI	100.8	111.8	–	19.6	–
RusHydro	102.2	107.5	–	42.1	–
JFSC Sistema	84.8	127.3	–	–	52.8
VTB Bank	95.2	–	–	30.2	–
Sberbank	109.4	109.9	105.1	37.3	35.7
MTS	99.1	–	–	84.5	–
Rostelecom	105.4	–	–	35.6	–
MegaFon	104.0	94.2	93.3	83.5	82.7
URALKALI	107.4	179.2	178.1	50.1	49.8
PHOSAGRO	106.3	223.9	337.7	21.2	31.9
NLMK	90.2	208.8	209.0	89.4	89.5
Severstal	99.9	116.5	103.6	35.0	31.1
NORNICKEL	120.4	89.3	61.0	50.8	34.7
Rusal	97.4	109.9	105.4	88.6	85.0
NOVATEK	103.7	110.3	125.7	41.2	46.9
GAZPROM	93.1	93.6	105.4	58.8	62.0
Lukoil	97.6	–	–	48.3	–
Rosneft	105.3	102.0	107.3	74.1	77.9
Alrosa	79.5	–	–	71.4	–

Source: authors.

In the course of the analysis, the indicators of changes in the volume of revenues, social investments, and social support expenses for the period 2018–2019 were calculated. (Table 1). The share of social expenditures on personnel in revenues varies in the range of 0.1-3.3%, with metallurgical companies having the highest share. The correlation between changes in revenues and social expenses on personnel is impossible to identify, as 45% of companies reflected digital data only in their annual reports (2018 or 2019).

More than half of the companies on the above list demonstrate faster rates of growth of social projects financing, 20% increased social expenses while revenues decreased, and 15% reduced funding of social programs for employees. The highest growth rates of social investments are observed in the metallurgy and power industry. One-fourth of the companies have higher social expenses on personnel than the growth of social investments in general.

In the share of expenditures on employees in total social costs, all companies can be divided into four groups:

- 5% of companies allocate up to 25%,
- 50% spend between 25 and 50%,
- 20% of the companies under research spend 50-75% of their total social investments on employees,
- 25% of companies spend over 75% on personnel support.

However, there is no sectoral dependency on costs. The hypothesis of an increase in the share of personnel costs has not been confirmed, as only 20% of companies increased their share of investments, and 15% have not changed it regardless of changes in revenue. At the same time, it can be asserted that employees are a significant stakeholder since more than half of companies allocate 50% or more of their social investments to the social support of their staff members. This is partially confirmed by the directions of social programs being implemented. One of the priorities of city-forming enterprises are projects oriented on the development of regions where they operate. This is determined by the necessity of creating convenient conditions for the population and increasing the attractiveness of the area for both its personnel and potential employees. The second direction is cooperation with universities in supporting talented students and, consequently, future personnel.

Another indirect confirmation can be seen in the results of a research conducted by Managers Association, which covered 44 enterprises in the extractive, manufacturing, and service sectors [13]. In determining the priority areas for integration of the Sustainable Development Goals (SDG) into corporate strategy, SDG number 3 - "Good Health and Well-being" and SDG 8 - "Decent Work and Economic Growth" received the highest scores (79% each). The top 5 also included SDG 4 "Education" (77%); SDG 17 "Partnerships for the Goals" (74%) and SDG 12 "Responsible Consumption and Production" (72%). However, the authors of the survey assume that these goals are generally not new for the respondents and mainly "marking" existing corporate social activities.

A study of social and integrated reporting has shown that all documents provide information on the staff structure, including the ratio of men to women. Some reports focus on professional and career development opportunities for women. The main part of the personnel sections is dedicated to training, professional development, and motivation programs, where the performance of internal social activities is assessed by the proportion of trained employees, the number of internal training courses, and person-hours of training completed. As compared to the period of 2016–2017, the list of companies publishing information about additional medical and social insurance support, non-governmental pension programs, and housing programs has been expanded. 90% of companies monitor staff turnover, 35% track employee engagement, with only 15% assessing the level of employee satisfaction and loyalty. These companies are interested in creating and filling the system of internal communications, there are examples of corporate portals and internal social networks. More than half of the companies that track staff involvement are from the telecommunications industry and financial sector. Moreover, all telecom representatives assess the level of employee satisfaction. Probably, this is determined by the specifics of services and close ties between such companies and customers. Thus, the second hypothesis is not confirmed either. Accordingly, it can be stated that employees are not considered as internal customers, and the concept of internal marketing in Russia is based on personnel marketing.

The study conducted by Managers Association presents interesting data on the use of non-financial and integrated reporting to adjust the direction and volume of social investment [13]. In the commodities and extractive industries, reporting for employees has a higher priority than reporting for shareholders and investors. Most leading companies use responsible human resource management strategies. 83% of companies indicated that they integrate CSR and Sustainable Development principles in their HR management strategies. Among the departments directly involved in the implementation of the corporate strategy for CSR and sustainable development at the level of individual functions/business processes, the Human Resources Management Department received the highest score. It was marked by 60% of respondents. Along with that, during 2008-2019, the significance of the Human Resources Management Department as a department responsible for the implementation of the corporate strategy in the field of CSR and Sustainable Development decreased - from 59% in 2008, 48% in 2014 to 38% in 2019. At the same time, the involvement of the Marketing Department is increasing - from 6% in 2008 to 18% in 2019.

Examination of CSR policies of small and medium businesses at the second stage of the survey revealed that social support of the staff dominates among CSR directions. The availability of such programs was reported by 77.4% of respondents. The next most popular direction was charity programs (58.1%). Thus, the main beneficiaries for small and medium enterprises are their own employees and consumers. In contrast to SMEs, large businesses actively involve their staff in external socially-oriented projects. Almost all large enterprises organize programs and projects with the participation of volunteers, tracking staff activity. As for SMEs, more than half (54.8%) of the respondents indicated that the employees of the company participate in charity programs, 38.7% - responded negatively, the rest do not have any information. Thus, small and medium businesses are more focused on their employees when developing social programs, but they are less motivated to participate in social programs aimed at external forces.

4 Discussion

The value of employees for business can be expressed through social investments and the implementation of social programs for employees. Most large companies include this in their annual reports. Recently, the structure of social reports has been changed, reflecting greater detail in this direction. Small and medium-sized businesses consider the internal focus of social responsibility programs to be a priority. Supporting its own personnel is becoming an important aspect heading the list of CSR directions. However, the results of the analysis indicate the immaturity of corporate management in terms of the internal marketing concept. Theoretical aspects are not widely implemented in Russian business practices. Alternatively, they are not covered by public information and expert opinions.

For an internal client, the question of expediency of marketing tools application in the domestic market remains a matter of discussion. Partially it is caused by the fact that a simple functional approach to the personnel prevails. Thus, the personnel are considered a resource. A special vision and business culture are required to reveal the possibilities of employees as a marketing tool. Conducted research shows that the purposes and objectives of social programs for the personnel are mostly determined by the HR department. However, during the last decade, the tendency of involving marketing departments and specialized subdivisions in the development of relevant programs emerged.

The connection between external and internal marketing is crucial since thus, marketing orientations of the company become organizational and personal obligations [8, 12]. Research by Schulz, Martin, Meyer reveals new perspectives of the internal marketing concept development, proving its connection not only with operative marketing efficiency but also with long-term ability to form affective commitment and loyalty of employees [14]. This provides the greatest impact on the financial and market success of the organization. There remains a question of the possibility of extrapolation of these results to different types of markets and businesses. The involvement of employees in the production and service processes of value creation is known to be very differentiated, as well as the situation on the labor market for different sectors of the economy.

5 Conclusion

A review of scientific publications, expert opinions and the results of our research showed that the process of implementation of the internal marketing concept by Russian companies is at the initial stage. Studies confirm the importance of personnel for both large and small and medium-sized businesses. However, this is mainly due to the understanding of the resource value of personnel and the effectiveness of using this resource to achieve corporate goals. In changing social conditions, this approach may prove to be unpromising. According to the CSR concept, business attention should be focused on different groups of stakeholders, including the staff. According to the modern marketing concept, the loyalty of an internal customer also determines the loyalty of an external customer. So far, the internal loyalty is not regarded as one of the corporate goals. However, in the near future this will become a mandatory part of socially oriented business.

References

1. Ahmed, P.K., Rafiq, M., Saad, N.M.: Internal marketing and the mediating role of organisational competencies. J. Mark. **37**(9), 1221–1241 (2003)
2. Astawa, I.P., Sukawati, T.G.R.: Ubud gets the customers an ethnomethodology approach. Int. J. Econ. Res. **13**(7), 2681–2692 (2016)
3. Bell, S.J., Bulent, M.: When customers disappoint: a model of relational internal marketing and customer complaints. J. Acad. Mark. Sci. **32**, 112–126 (2004)
4. Berry, L.L., Hensel, J.S., Burke, M.C.: Improving retailer capability for effective consumerism response. J. Retail. **52**(3), 3–14 (1976)
5. George, W.: Internal marketing: Concepts, measurement and application. J. Retail. **20**(1), 63–70 (1990)
6. Gounaris, S.: Antecedents of internal marketing practice: some preliminary empirical evidence. Int. J. Serv. Ind. Manag. **19**(3), 400–434 (2008)
7. Grönroos, C.: Internal marketing: theory and practice. In: Bloch, T.M., Upah, G.D., Zeithaml, V.A. (eds.) Services Marketing in a Changing Environment, pp. 41–47. American Marketing Association, Chicago (1985)
8. Hall, K.K.L., Baker, T.L., Andrews, M.C., Hunt, T.G., Rapp, A.A.: The importance of product/service quality for frontline marketing employee outcomes: the moderating effect of leadermember exchange (LMX). J. Mark. Theory Pract. **24**(1), 23–41 (2016)
9. Huang, Y.-T., Rundle-Thiele, S.: A holistic management tool for measuring internal marketing activities. J. Serv. Mark. **29(6/7)**, 571–584 (2015)
10. Kanyurhi, E.B., Bugandwa, D., Akonkwa, M.: Internal marketing, employee job satisfaction, and perceived organizational performance in microfinance institutions. Int. J. Bank Mark. **34**(5), 773–796 (2016)
11. Lings, I.N.: Balancing internal and external market orientations. J. Mark. Manag. **15**(4), 239–263 (1999)
12. Mainardes, E.W., Cerqueira, A.S.: Measuring the internal-market orientation in the public sector. Ekon. Vjesnik **28**(1), 53–70 (2015)
13. Managers Association: Report on social investments in Russia - 2019: Towards business transformation for sustainable development (2020). https://peopleinvestor.ru/upload/iblock/6a8/6a81d6776b78af5933de15fb125de8b3.PDF. Accessed 26 June 2020
14. Schulz, S.A., Martin, T., Meyer, H.M.: Factors influencing organization commitment: internal marketing orientation, external marketing orientation, and subjective well-being. J. Manag. Dev. **36**(1), 821–840 (2017)
15. Sukawati, T.G.R., Astawa, I.P.: Improving performance by harmonious culture approach in internal marketing. Pol. J. Manag. Stud. **16**(1), 226–233 (2017)
16. Yakhneeva, I., Agafonova, A., Nikitina, I.: Corporate social responsibility in Russia: motives and features. In: Mantulenko, V. (ed.) Proceedings of the International Scientific Conference "Global Challenges and Prospects of the Modern Economic Development". The European Proceedings of Social and Behavioural Sciences, vol. 57, pp. 1055–1068. Future Academy, London (2019)
17. Yakhneeva, I.V., Khansevyarov, R.I., Zhabin, A.P., Volkodavova, E.V.: Social investments as a component of sustainable business development. Russ. J. Entrep. **19**(12), 3903–3912 (2018)

Digital Transformation and New Architecture of the Labor Market

Labor Market Transformation in the Context of the Digitalization of the Economy

V. V. Borisova[1(✉)], E. E. Panfilova[1], and Hendra Raza[2]

[1] State University of Management, Moscow, Russia
{vv_borisova, ee_panfilova}@guu.ru
[2] Universitas Malikussaleh, Aceh Utara, Indonesia
hendra.raza@yahoo.com

Abstract. The purpose of the study was to identify key trends that determine the structural and qualitative changes in the labor market in a crisis state of the economy after exposure to coronavirus and intensifying business digitalization processes at the same time. In the course of the study, the tasks of analyzed factors that influence the transformation of the labor market in the context of digitalization are considered. The authors identify the problems of training specialists in universities based on the active use of online education tools. Particular attention is paid to the search for optimal tools for the interaction of business and public authorities in the preparation of university graduates in demand for various sectors of the economy. During the study methods of comparison, generalization, grouping, as well as comparative and statistical data analysis are used to reveal the features of labor market transformation in the context of business digitalization. The main result of the authors' paper is the development of guidelines for the effective interaction of universities, business and public authorities in the preparation of specialists demanded by the labor market who have digital competencies for working in a transforming business.

Keywords: Business digitalization · Competence · Labor market · Transformation · Trend

1 Introduction

Digitalization of the economy in the Russian Federation and abroad is inextricably linked with the implementation of the "Industry 4.0" concept, which involves the use of the Internet of things, predictive analytics, virtual reality technologies, additive manufacturing, and reverse engineering [19]. For industrial enterprises of high-tech sectors of the economy, competitiveness in the transformation of business processes will be determined not only by technological, organizational and informational readiness for radical changes, but also by the innovative activity of personnel in the labor market. For the leadership of organizations the issue of increasing the innovative activity of the staff of both already working and potentially attracted young specialists after graduation becomes relevant [8].

Obviously, the solution to this problem is associated with the formation of an effective mechanism for the interaction of the triad "business - university - public

S. I. Ashmarina and V. V. Mantulenko (Eds.): IPM 2020, LNNS 161, pp. 265–274, 2021.
https://doi.org/10.1007/978-3-030-60926-9_34

authorities." Under the influence of cloud computing, neuromarketing technologies and mixed reality technologies, the content of the work of performers is changing, the role of formed digital competencies among staff and the ability to work with a corporate data warehouse are increasing. The structure of the labor market is undergoing changes, freeing itself from the established professions such as an engineer, a manager, a marketer and moving to the demand for such positions as a coordinator of production in distributed communities, a foresight and a portfolio manager of corporate venture funds.

2 Methodology

Nowadays, the transformation of the labor market in the context of business digitalization is taking place not only under the influence of the technological revolution, but also the institutional one. The latter involves a change in the managerial model and its synchronization with state authorities in terms of targeted programs in the training of specialists. The study accepted the hypothesis that the modification of the managerial model is based on the theory of convergence. Its adherents say that in the context of globalization, there is a rapprochement of the socialist and capitalist management systems, systems of managing the national economies of different countries due to the fact that the same type of changes are occurring in engineering and production technology. This fact leads to the same changes in the economic life of society, both Russia and Indonesia [16].

The outbreak of the coronavirus pandemic will lead to a surge in enterprise bankruptcies and increased raiding. By virtue of this fact, HR (Human Resources)-managers will focus on finding applicants for vacancies opened by large businesses. The methods used for the selection of personnel in digitalization are undergoing changes. And mobile recruiting applications, ATS (Applicant Tracking Systems), various online tools for evaluating a candidate for a vacant position will become increasingly popular [9].

The dominant factor affecting the labor market is the expansion and implementation of the Industrial Internet of Things (IIoT). This will inevitably lead to increased demand for specialists and graduates of universities who possess digital competencies that allow you [15]:

- to manage the business processes of the organization using digital doubles and shadows,
- to plan interaction with contractors within the framework of several digital platforms,
- to conduct an IT audit on implemented functions and procedures,
- to implement projects in a digital environment using cloud data storages,
- to coordinate the work of remote access employees in outstaffing, as well as to work on electronic trading platforms and with self-executing contracts.

In order to identify the effect of synergistic interaction between universities and enterprises of the real sector of the economy, the study used methods of comparing the quality of education at universities; methods of generalization, concretization, grouping and systematization in assessing the distribution of university students by popular specialties of instruction. Statistical, structural-dynamic and comparative analysis methods were used to study the trends in the development of the modern market for higher education, taking into account distance learning forms, online platforms, as well as to determine the degree of compliance of graduates' specialties with the requirements of the modern labor market in the context of the digitalization of the economy. As the basis, analysis and synthesis methods were taken to identify the specifics of the organization of interaction between universities, business and public authorities [17].

The transformation of the labor market is influenced by external and internal factors. External factors include the success of various programs to digitalize business, transport, research and development, as well as the introduction of professional standards for the activities of specialists in the corresponding functional area. The internal factors include the personality-oriented approach of applicants and their parents when choosing a future profession, forming an individual learning path, as well as the effectiveness of the career-oriented interaction model in the "school-college-university" combination.

The result of the transformation of the labor market in the context of digitalization is the preparation of a strictly defined number of graduates, both colleges and universities, that are fully demanded by the labor market. The criteria for assessing the quality of educational programs of universities prepared together with employers (representatives of the real sector of the economy) are the level of starting salary, the number of employed graduates in the specialty (including students with disabilities) [18]. To identify key trends in the choice of applicants for the popular areas of training/specialties when studying undergraduate and graduate programs, linear coefficients of absolute structural changes were calculated.

In the framework of the organization of career guidance work for universities, it is necessary to orient students on the learning continuity, the interoperability of acquired specialties and monitoring of priority areas of economic development, including energy efficiency, strategic information technology, medical technology. The model for training a specialist in demand on the digital labor market should be based on the principles of multidisciplinarity, the use of project/collaboration tools, as well as the multidimensionality of "big data" used to construct the model of the control object. Roadmaps developed separately for various sectors of the economy, high-tech, knowledge-based and others can become tools for assessing the effectiveness of labor market transformation in the context of digitalization.

3 Results

Traditionally, the labor market in the Russian Federation is largely influenced by the institutional environment, which implies active state regulation of the educational activities of educational institutions through control numbers for students, regulation of accreditation and licensing procedures [1]. One of the quality indicators of education assessment is the ratio of the number of private educational institutions to the number of state and municipal educational institutions, as well as the ratio of the number of students studying in private educational institutions, to the number of students of state and municipal educational institutions (Fig. 1). Students of non-state educational institutions are in strong competition with graduates of state universities on the labor market.

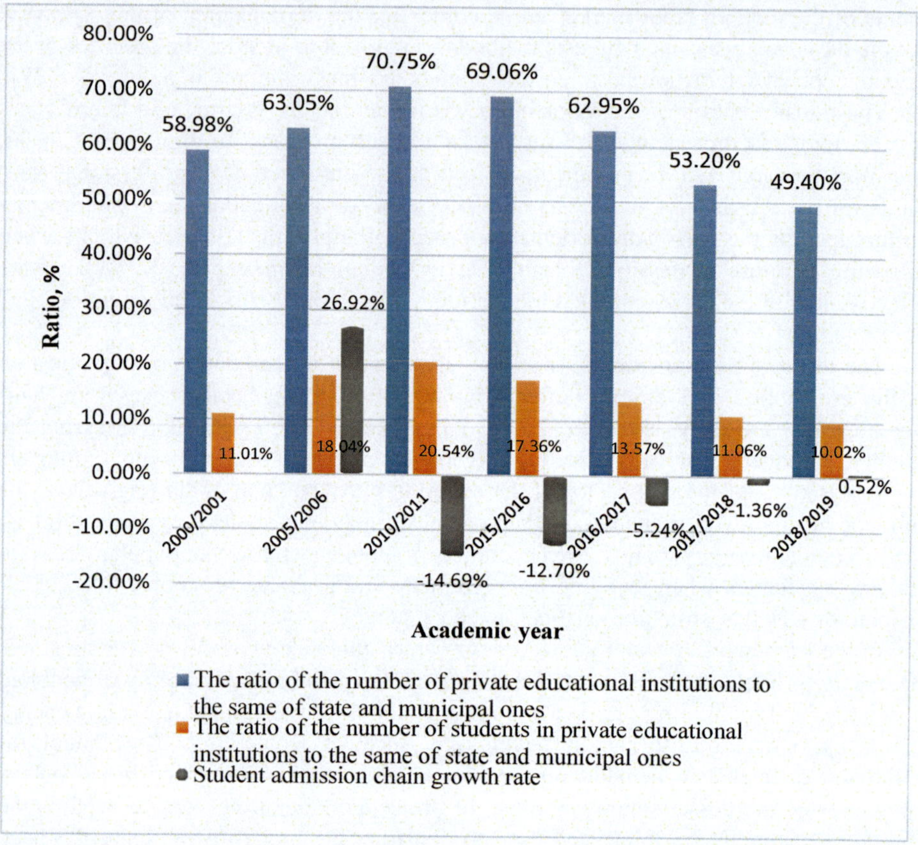

Fig. 1. The dynamics of the ratio of the number in private educational institutions to state/municipal educational institutions in comparison with the ratio of the number of students in private organizations to state and municipal organizations (Source: authors based on [12]).

As can be seen, from the academic year 2010/2011 to the academic year 2018/2019, there was a steady decrease in the ratio of the number of private organizations to the number of state organizations, accordingly 70.75% to 49.4%. Thus, the ratio of the number of students in private educational institutions to the number of students in state and municipal educational institutions decreases in the period under review from 20.54% to 10.02%, respectively. The lowest chain growth rate of student admission of 14.69% is typical for the 2010/2011 academic year. This was mainly due to the fact that some students couldn't organize their time, were unable to meet strict requirements and find a strong motivation to study. Since 2012, the trend has changed and this is due to the fact that students choose international programs, as well as they study simultaneously on several educational programs.

The main forms of impact on the labor market by private companies, businesses and the state abroad (including Indonesia) include the following [2]:

- public-private partnership in the training of specialists,
- reduction of specialties in the framework of multidisciplinary training (training of highly specialized specialists),
- internships at the enterprise,
- development of partnerships in the implementation of educational programs through an invitation to conduct master classes by leading experts in relevant field of activity.

The influence of internal motivational factors of applicants when choosing a future field of study and, accordingly, further professional implementation in the labor market is presented in Table 1.

Table 1. Structure of the distribution of students in the most popular specialties/areas of training in the specialty programe in the period from 2014 to 2018

Specialties/training areas	Years				
	2014	2015	2016	2017	2018
Law	5,86	3,16	0,20	0,02	0,0
Finance and credit	3,75	1,90	0,08	0,0	0,0
Organization management	3,69	2,05	0,14	0,01	0,0
Economics and enterprise management (by industries)	3,81	1,98	0,07	0,0	0,0
Accounting, analysis and audit	2,28	1,19	0,08	0,0	0,0
Medical business	7,94	11,44	20,58	21,40	21,75
State and municipal government	2,02	1,10	0,09	0,0	0,0
Industrial and civil engineering	1,15	0,66	0,03	0,0	0,0
Economic security	1,62	3,63	6,38	6,95	7,02
Dentistry	0,51	3,58	5,64	0,0	5,77
Others	67,37	69,31	66,71	71,62	65,46
Total	**100**	**100**	**100**	**100**	**100**

Source: authors.

According to Table 1, since 2015, the structure of the distribution of students in the areas of study has been changing and there is a rejection of humanitarian areas in favor of the specialty "Economic Security" (directly related, including taking into account cyber threats in the organization's activities and the use of digital competencies), as well as "General Medicine" (i.e. knowledge-based industry related to medicine). The role of the latter in the context of various epidemics and threats is increasing significantly. During the course of the study, absolute chain and basic structural shifts were calculated for the distribution of students in undergraduate programs in the most popular specialties. Data are presented for the period from 2016 to 2018 (Table 2).

Table 2. Absolute chain and basic structural shifts of distribution undergraduate programs in the most popular specialties of training in 2016–2018, %

Specialty	Chain shifts, %			Basic shifts, %		
	2016	2017	2018	2016	2017	2018
Economy	3,45	−1,35	−1,2	1,68	0,33	−0,87
Law	3,71	−0,47	−0,92	3,61	3,14	2,22
Management	1,82	−0,97	−0,58	1,22	0,25	−0,33
Teacher education	1,75	0,02	4,47	0,85	0,87	5,34
Building	1,36	0,03	−0,05	1,52	1,55	1,5
State and municipal government	1,04	−0,13	−0,15	1,21	1,08	0,93
Power industry and electrical engineering	0,82	0,1	0,09	0,96	1,06	1,15
Psychological and pedagogical education	0,57	−0,06	0,02	0,72	0,66	0,68
Others	−14,52	2,83	−1,68	−11,77	−8,94	−10,62
Total	**0**	**0**	**0**	**0**	**0**	**0**

Source: authors.

As follows from the Table 2 in 2017, there is a decrease in the share of undergraduate students in the areas of "Economics" (by 1,35% compared to 2016), "Management" (by 0,97%) and "State and Municipal Administration" (by 0,13%). Thus, there is a tendency to abandon humanitarian areas of study (partially lagging behind the restructuring of systems thinking based on the components of "Industry 4.0") in favor of more technical areas ("Electrical Power and Electrical Engineering", "Construction"). It is no coincidence, since many interviewed company managers note the fact that the transformation of the labor market in the context of the digitalization of the economy is primarily associated with technological readiness to automate business processes. Thereby, the organizational and managerial component is secondary [6].

During the study in 2015–2017, the degree of compliance of the position held by employed graduates with the specialty received in a higher educational institution was assessed. As the analysis shows, the highest degree of compliance (over 70%) of the profession of employed graduates is typical for technical areas, such as computer and information sciences, computer science and computer technology and information security. The competencies acquired during training in the specialties of these areas are widely in demand in various industries and more than correspond to the set trends of «Industry 4.0». However, computer and technical literacy are only one side of the demonstration of the modern labor market in the context of digitalization. The basic tendency in the selection of personnel is the demand for "soft skills", formed not only by universities, but by the external environment itself. Soft skills provide the development of digital skills based on professional skills [10].

However, the interpretation of soft skills is subject to changes due to the needs of implementing various digitalization programs, including Industry 4.0, and taking into account the specifics of the employers themselves (hard skills) are less variable. In this regard, the search for the required personnel is complicated by the fact that universities as the main suppliers of new specialists with basic knowledge, developing hard skills, they cannot quickly rebuild the learning system to take into account new trends, and, therefore, the professional community will be forced constantly to adjust the requirements to qualify job seekers or use its own opportunities for training.

Enterprises that are residents of special economic zones (technopolises, technology parks, industrial parks) in Moscow and the Moscow Region traditionally use the configurator of the “1C staff turnover program” to close staffing requirements taking into account the emerging structural changes in the labor market. For example, 169 technology parks closed 83% of the need for personnel in the range from 14 to 30 days in 2019. More than 50% of specialists annually undergo advanced training or professional retraining programs in the field of investment management in the digital environment, contractual relations using digital signatures, and innovative project management in the framework of international cooperation. The typical age structure of the technopolis of the city of Moscow is as follows: 30,5% are workers aged 51–60; 21,0% - workers aged 31–40; 20,0% - workers under the age of 30; 19,0% - workers aged 41–50; 9,5% are over 60 years old.

The qualitative staff composition of a typical technopolis by level of education is as follows: 55% of employees have higher education, which makes it possible to move up the career ladder. The remaining 45% are workers with secondary vocational education. Statistically average rotation for employees of the highest category is carried out once every 5 years, for qualified personnel once every 3 years and for work positions once a year. The transformation of the labor market is accompanied by a change in the set of tools that accompany the process of staff training. Thus, according to the results of the study, about 35% of technopolises use corporate centers and universities to improve the skills of their employees. 15% of industrial parks are focused on the use of distance learning courses posted on various online platforms [3]. About 17% of technology parks are interested in interacting with universities, government bodies and business partners within the framework of a single ecosystem implemented on a digital platform in the “single window” mode.

4 Discussion

A number of researchers of transformation processes in the digitalization of the economy indicate the fact that information tools/technologies make a great contribution to the implementation of the concept of open innovation. Organizations will be able to participate in the implementation of high-tech projects in a remote format, including minimizing the cost of research and development. At the same time, there is a growing risk of leakage of confidential and insider information from cloud data warehouses and digital platforms [5]. The results of the study confirm that the presence of digital competencies in the field of data protection for employees working in distributed networks significantly increases the information security of the company. At the same time, the issue of not only maintaining, but also developing the already formed digital competencies due to the rapid obsolescence of information management tools is still relevant. Scientists emphasize that the Chief Data Officer – CDO is becoming a key figure in the organization [13].

Analysts consider that highly paid specialists will not have difficulty solving routine tasks. Some of them will be taken over by robots, chat bots from process control to virtual recruiting. In this regard, they will have more opportunities for a creative decision making to find and apply non-standard methods of solving problems in poorly structured and unstructured situations. Consequently, software skills requirements will increase as a factor of competitiveness in the labor market [14].

Forecasts of specialists testify to the value of managerial personnel who are able to work in multitasking conditions and willingness to develop a new business area. That is why the initial vocational guidance of schoolchildren, project training and hybrid training (a semester of studies, a semester of work at the enterprise) will allow the younger generation, which is shaping the future labor market, to master professions that are not currently available. Eichar in an organization in the context of the digitalization of the economy is turning from a specialist in personnel search into an expert in selling the best vacancies of an employer to a highly competent jobseeker. By 2025, according to experts, generation Z (born in 1997 and younger) will make up about 25% of the labor force in the labor market [12]. It is typical for this generation to be between personal life and financial reward. Since representatives of this generation have developed digital competencies, employers will have to adapt to their values.

In this case, we will see a change in the forms of labor relations, the restructuring of the education system to train new personnel and the adaptation of the labor market itself to the new realities. The shift in emphasis from the need to look attractive to the employer and be able to integrate into the organization's processes to provide a contribution to the company's valuation to find jobs that will provide the applicant with the opportunity to realize his potential in a professional environment based on his own needs and preferences throughout his career, thereby increasing its own value, including for the employer, it will require major changes in the training of such personnel. Is the current education system able to change as fast as the modern labor market is changing in the context of digitalization? The discussion between researchers shows that at present, the balance between hard skills and soft skills is gradually shifting towards "soft skills" [11].

However, no answer has been formed yet who is the initiator of such a shift: the candidates themselves, who own digital competencies and set the labor market a certain standard for the level of professionalism and soft skills, or companies, realizing the needs of the digital society, setting standards for the level of key digital skills for staff selection. One way or another, the labor market, having a high degree of adaptability to emerging needs, in response to the growing digitalization trend, will determine the demand for digital competencies that can give impetus to the development of soft skills. The educational space, following modern trends, will have to take this circumstance into account in the system of training specialists.

5 Conclusion

Transformation of the labor market in the context of the digitalization of the economy affects the mechanisms of interaction in the triad of "business - universities - government". The rapid growth in the use of digital services and platforms is a serious challenge not only for employers, but also for hired personnel. Organization leaders are restructuring their business processes by automating routine operations and patterns of interaction with business partners. The state is forced to revise university financing programs, priority areas for training specialists and investing in public-private partnership projects. Universities are more oriented towards individualization of student learning with the parallel formation of the student's electronic portfolio.

Accordingly, specialists with competencies confirmed not only by university diplomas, but also certificates of various online schools, platforms and corporate training centers will be in demand on the digital labor market [4]. Given the emerging trends in the labor market, one of the priority areas of research will be the assessment of the psychological readiness of staff (employees/government officials) to interact in a virtual environment. Legal aspects of regulating labor relations in the digital environment, paying taxes for various modes of work (remote access, outstaffing, project work) will also come to the fore.

References

1. Afonasova, M.A., Panfilova, E.E., Galichkina, M.A., Slusarczyk, B.: Digitalization in economy and innovation: the effect on social and economic processes. Pol. J. f Manag. Stud. **19**(2), 22–32 (2019). https://pjms.zim.pcz.pl/resources/html/article/details?id=190160
2. Aghayeva, K., Slusarczyk, B.: Analytic hierarchy of motivating and demotivating factors affecting labor productivity in the construction industry: the case of Azerbaijan. Sustainability **11**(21), 5975 (2019). https://doi.org/10.3390/su11215975
3. Akhmetshin, E.M., Mueller, J.E., Chikunov, S.O., Fedchenko, E.A., Pronskaya, O.N.: Innovative technologies in entrepreneurship education: the case of European and Asian countries. J. Entrep. Educ. **22**(1) (2019). https://www.abacademies.org/articles/innovative-technologies-in-entrepreneurship-education-the-case-of-european-and-asian-countries-7827.html. Accessed 15 May 2020
4. Al-Tkhayneh, K., Kot, S., Shestak, V.: Motivation and demotivation factors affecting productivity in public sector. Adm. Manag. Public **33**, 77–102 (2019)
5. Ardolino, M., Rapaccini, M., Saccani, N., Gaiardelli, P., Grespi, G., Ruggeri, C.: The role of digital technologies for the service transformation of industrial companies. Int. J. Prod. Res. **56**(6), 2116–2132 (2017)
6. Betelin, V.B.: Challenges and opportunities in forming a digital economy in Russia. Herald Russ. Acad. Sci. **88**(1), 3–9 (2018)
7. Bondarenko, N., Borodina, D., Gokhberg, L.: Indicators of Education in the Russian Federation: 2020: Data book. HSE, Moscow (2020)
8. Borisova, V.V., Panfilova, E.E., Zhukov, P.V., Matulis, S.N., Matveev, V.V., Teymurova, V.E.: Information support in the enterprise risk management. Int. J. Manag. Bus. Res. **9**(1), 158–169 (2019)
9. Carrincazeaux, C., Gaschet, F.: Regional innovation systems and economic performance: between regions and nations. Eur. Plan. Stud. **23**(2), 262–291 (2015)
10. Doyle, S., Senske, N.: Soft skills for digital designers. In: Architecture Conference Proceedings and Presentations, vol. 81, 475–480. Iowa State University, Iowa (2016)
11. Foerster-Pastor, U.S., Golowko, N.: The need for digital and soft skills in the Romanian business service industry. Manag. Mark. Challenges Knowl. Soc. **13**(1), 831–847 (2018)
12. Graham, M., Hjorth, I., Lehdonvirta, V.: Digital labour and development: impacts of global digital labour platforms and the gig economy on worker livelihoods. Eur. Rev. Labour Res. **23**(2), 135–162 (2017)

13. Hussain, S., Rizwan, M., Nawaz, M.S., Hameed, W.: Impact of effective training program, job satisfaction and reward management system on the employee motivation with mediating role of employee commitment. J. Public Adm. Govern. **3**(3), 278–293 (2013)
14. Lacity, M., Willocks, L.: A new approach to automating services. MIT Sloan Manag. Rev. **58**(1), 40–49 (2016)
15. Pugh, K.J., Bergstrom, C.M., Spencer, B.: Profiles of transformative engagement: identification, description, and relation to learning and instruction. Sci. Educ. **101**(3), 369–398 (2017)
16. Steinmayr, R., Spinath, B.: The importance of motivation as a predictor of school achievement. Learn. Ind. Differ. **19**(1), 80–90 (2009)
17. Toktamysov, S.Z., gizi Vekilova, A.I., Gasimzade, E.E., Kurilova, A.A., Mukhin, K.Y.: Implementing the education of future entrepreneurs in developing countries: agile integration of traditions and innovations. J. Entrep. Educ. **22**(5) (2019). https://www.abacademies.org/articles/implementing-the-education-of-future-entrepreneurs-in-developing-countries-agile-integration-of-traditions-and-innovations-8569.html. Accessed 15 May 2020 (2019)
18. Vasiliev, A.: Entrepreneurial education quality management to improve university competitiveness. J. Entrep. Educ. **23**(1) (2019). https://www.abacademies.org/articles/entrepreneurial-education-quality-management-to-improve-university-competitiveness-9020.html. Accessed 15 May 2020
19. Weber, R.A.: The economics of effective leadership (2015). https://leadersforpurpose.com/the-economics-of-effective-leadership/. Accessed 10 Aug 2020

Digitalization of the Labor Market in the Fourth Industrial Revolution

I. S. Vladimirov[1], E. Yu. Kamchatova[2(✉)], and V. V. Burlakov[1]

[1] Federal State Unitary Enterprise All-Russian Research Institute "Center", Moscow, Russia
vladimirov.ivan.ns@yandex.ru, bur77@mail.ru
[2] State University of Management, Moscow, Russia
kuzkat@mail.ru

Abstract. The article analyzes the trends of digitalization of the labor market in the fourth industrial revolution. In the course of our work, we used empirical (observation, description, calculations and measurement) and theoretical (analogy, comparative analysis, literature analysis) research methods. New specialties and competencies that will be in demand in the future, as well as barriers to retraining of employees, have been identified. The impact of the covid-19 coronavirus pandemic on the labor market is also being considered. On the example of the entertainment industry, the effect of the introduction of quarantine measures is demonstrated. The conclusion is made about the future prospects for the development of the labor market in the conditions of the crisis. A number of anti-crisis measures based on the methods of planned economy are proposed.

Keywords: Crisis · Employment · Labor market · Pandemic · Trend

1 Introduction

The basis for the fourth industrial revolution, which is currently recognized by many researchers as a consequence of the third industrial revolution, is the process of digitalization of the economy. Based on the HSE data from 2018, it could be argued that the trend of internal spending on the development of the digital economy in the future will take an exponential form (Fig. 1).

However, the coronavirus epidemic of 2020 has made significant adjustments to this forecast. Provoking a new global economic crisis, it will lead to fundamental changes in the life of society. This dramatically increases the degree of uncertainty about the future state of the economy, its digitalization processes, and the labor market itself. The task of this work is to consider possible prospects and directions of digitalization of the economy and the development of the world labor market itself.

S. I. Ashmarina and V. V. Mantulenko (Eds.): IPM 2020, LNNS 161, pp. 275–282, 2021.
https://doi.org/10.1007/978-3-030-60926-9_35

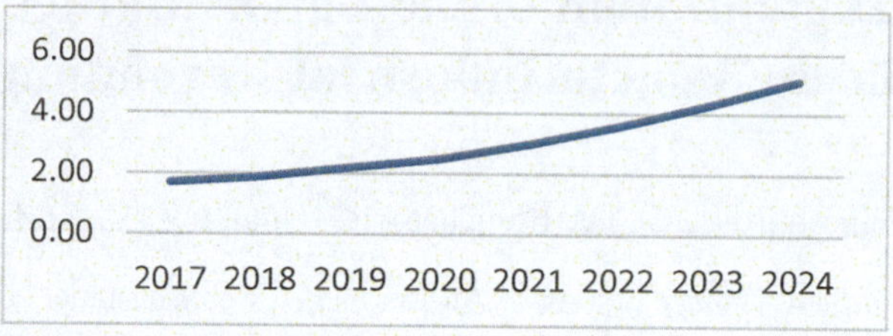

Fig. 1. Domestic spending on digital economy development as a percentage of GDP (data for Russia) (Source: authors based on [1]).

2 Methodology

The object of research is the labor market. Practical (empirical) and theoretical methods are used for its study. As theoretical tools, such as analogy, comparative analysis, study and analysis of literature. The analogy is to compare the current situation in the labor market with an earlier one and draw conclusions about possible measures to overcome the crisis that will be taken. To assess the current state, we analyze news reports and statistical reports on the problem under study. The practical methods are observation, description, calculation, and measurement. The observation consists in analyzing the dynamic series of various indicators that characterize the labor market. Then, based on these indicators, its current state is described. Based on mathematical operations with numerical data, conclusions are made about the low availability of higher education. We measure the level of service efficiency in the digital economy.

3 Results

Like previous industrial revolutions, the fourth industrial revolution will lead to the disappearance of a large number of professions. This is due to the extensive processes of production automation that took place in the past. However, along with the gradual increase in the unemployment rate, new types of professions will begin to appear, such as virtual reality designer, bioengineers, analytics “the Internet of things”, the curator of the personal data, engineer for development of permanent power supply devices, digital linguist, financial trajectory designer, etc.

According to the report of the world economic forum (WEF), robotization will lead to the disappearance of 75 million jobs, but will create 133 million new ones [8]. An important feature in this case is that the vast majority of new professions will perform the tasks of servicing the digital economy. To master them, a person will need a serious investment in their education. The last study on calculating the average cost of training for a programmer in Russia was conducted in 2017 - 143297 rubles per year, with a total duration of training of 4 years (11941 rubles/month) [12]. Average household savings in 2018 amounted to 3232.9 rubles/month, which is almost 4 times lower than the cost of education [5]. According to the RBC news Agency: “Two-thirds of Russian families-65%-do not have any savings, according to the results of a survey by the

Levada center research laboratory (Levada Lab)" [3, p. 9]. The average wage indicator does not reflect the real picture, due to the imperfection of the method of calculating it, in which the organization's payroll is divided by the number of employees. Thus, the average salary includes personal income tax and its objectivity is reduced by the polarization of the income level of the population (the effect of "average temperature in the hospital"). In other words, a household with a lower-than-average income level is currently unable to provide education in one of the "future economy" specialties. Thus we get two theses:

1. Children from low-income families will lose the opportunity to find work, due to the disappearance of available work professions. Thus, over time, their average income will fall, which will lead to the growth of the "poverty" class.
2. People from affluent families will be able to get an education and become high-class specialists. At the same time, you should keep in mind that:
   – some of them will refuse or fail to complete their training due to various circumstances, and will not be able to find work in the new digital economy in the future. The class of the poor will be replenished;
   – getting an education after a while will become an impossible task for them, and even more so for their children. Thus, once a family falls into the category of low-income, it will not be able to move back to the middle class.

These theses can be considered an explanation of the following graph (Fig. 2):

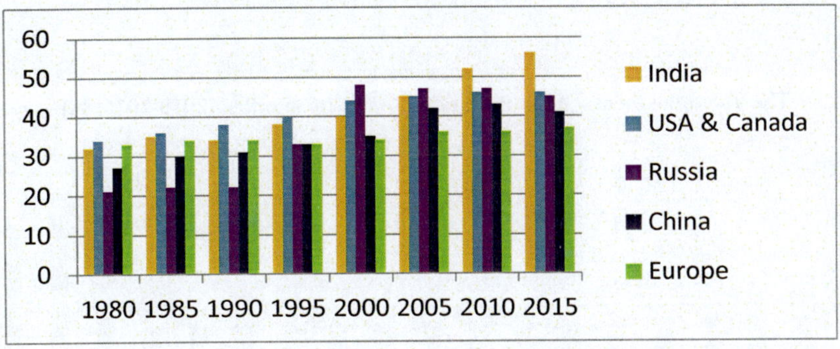

Fig. 2. Share of the top 10% by income level in the world, 1980–2015 (Source: authors based on [16]).

Many researchers refute the fact that the social catastrophe is spreading by the fact that in developed countries there are many institutions to support poor families, which does not allow the situation to become threatening. Indeed, the European Union, the United States and Canada have always had an expanded social security system, but this does not prevent, but only slows down the growth of inequality. The solution to the problem is not in the area of evaluating the effectiveness of social security measures. The labor market is often considered from the position of such an indicator as "unemployment". The global average unemployment rate in 2018 was 5.7%, which can be considered an acceptable indicator. However, recently there is evidence that even in developed countries, the level of discontent among the population is growing, despite the low unemployment rate. For example, in the UK, unemployment was only 3.8%. British discontent with the situation in the labor market and the decline in living standards was one of the reasons for the vote to leave the EU in June 2016, says economist at the University of Warwick Tiemo Fetzer, «Vedomosti» reports [10]. The main reason for discontent is the increase in the number of part-time jobs that do not strengthen the financial position of employees. From 2009 to 2018, the share of temporary employees increased in France from 13 to 16.2%, in the Netherlands from 16.4 to 20.1%, and in Italy from 10.8 to 16.5%. Thus, in this work, we want to identify the most important, in our opinion, trend of transformation of the modern labor market – an increase in non-permanent employment, which leads to an increase in the level of social tension and reduces confidence in the future of the population. However, in 2020, there was an event that could lead to fundamental changes in the situation on the labor market.

The coronavirus infection of the COVID-19 virus turned out to be such a significant factor that without taking into account its impact on the economy, it is unlikely that any reliable research can be conducted and any forecast made. Therefore, in this study, we will touch on the impact of this event on the modern labor market. Like any other large-scale negative event, the epidemic had a significant impact on the economy [6]. Due to the need to suspend the activities of service organizations, some of them were on the verge of bankruptcy. As a result, there was a jump in unemployment, which threatens to increase social tension in the developed countries of the first world (Fig. 3) [14].

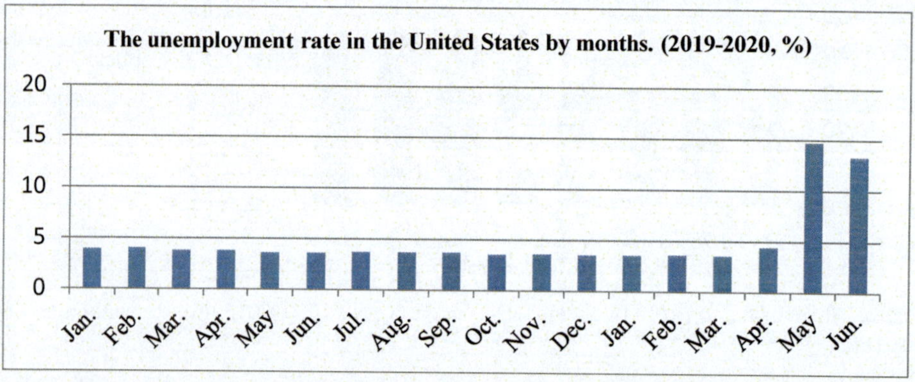

Fig. 3. Dynamics of the US unemployment rate (Source: authors based on [14]).

Major financial losses were incurred by organizations that:

- did not have the ability to organize the workflow remotely,
- did not have the ability to conduct remote customer service.

Thus, all entertainment centers fall under this category, such as cinemas, museums, water parks, festivals, etc. At the same time, developers of computer games, such as

Wargaming or Gaijin Entertainment, on the contrary, were in the black, because they do not have the above disadvantages. An important circumstance in this case is that it is easier to bear the "coronavirus crisis" precisely those enterprises whose level of digitalization is quite high.

The entertainment industry clearly demonstrates the main problem of digitalization of the labor market. Industry employees who have lost their jobs, even with the necessary qualifications, will not be able to find a job in a company with a high level of digitalization, because they do not require a large number of specialists. For example, the number of employees of the largest game developer in the CIS, Wargaming, in 2020 was only 4,500 people with an audience coverage in the "online" mode of 318965 players, only on the servers of the Russian sector and only for three games: World of Tanks Blitz, World of Tanks, World of Warships. In other words, the number of clients served simultaneously exceeds the number of employees by more than 70 times. Not every organization in the entertainment industry can match this indicator. For simplicity, in this work, we will call this indicator the "service efficiency index" - I_{SE}, which is the ratio of the number of clients served at one time (Q_c) to the average number of employees (ANE) (formula 1).

$$I_{SE} = \frac{Q_c}{ANE}$$

For example, for a small movie theater with 100 seats and 6 employees, this figure is significantly lower, at 16.6, and for an average entertainment shooting gallery – 5. Similar processes occur in other industries, such as construction, heavy engineering, and so on. Almost all professions that do not involve creative work can be automated and robotic.

Digitalization leads to the disappearance or transformation of industries with low I_{SE} and the emergence of industries with high I_{SE}. This process is fundamental and is observed in all industries without exception. For this reason, when assessing the consequences of digitalization, it is not necessary to refer to the number of new professions, since in this case the number of new vacancies is the determining factor. Thus, over time, the variety of professions will grow, and the total number of vacancies will significantly decrease, exacerbating the situation in the labor market.

The effect of exposure to a coronavirus infection will consist of several consequences:

1. At first, there will be an acceleration in the pace of digitalization in the desire of employers to reduce the number of personal contacts between employees, in order to avoid the spread of new types of viruses.
2. An abrupt increase in unemployment due to mass bankruptcy and general recession in the world economy will lead to a prolonged economic crisis. In order to overcome it, States will resort to Keynesian reforms and protectionist policies. If the situation continues to deteriorate, there may be labor camps that took place in the United States in the 30 s. The unemployment rate in the United States in 1933 was 24.9%.

3. If states do not change the market methods of managing the economy, the economic crisis will lead to the fact that the processes of digitalization will take a local character and in general will be significantly slowed down. In industries such as mechanical engineering, they may even be curtailed in order to retain the same number of employees.

If global economic reforms are carried out, mainly on the left, the current technological progress can be maintained. We will see below what type of changes in the economy are necessary for this.

Scientific and technological progress changes the nature of economic relations. The old production methods are being replaced by more modern ones. All industrial revolutions led to increased labor efficiency and increased unemployment. Digitalization is no exception in this case. In the end, it will lead to the almost complete disappearance of professions that can be automated, which means that in the future there will only be those professions that require not just mental, but creative work.

The problem in this case is the economic barrier that does not allow the broad masses of workers to re-qualify as workers in industries with a high level of digitalization. Inevitably, the unemployment rate will increase and, as a result, poverty, which will lead to new social catastrophes, such as revolutions and wars. The only proven method of preventing such a scenario at the moment is the transition of the economy to a planned character, and the following reforms, such as:

- full nationalization of large industrial enterprises, on which the life of the state depends, and the transition from the concept of "work for profit" to the concept of "work for the plan",
- restoration and creation of powerful state planning structures based on a scientific approach and which are inextricably linked with research institutes,
- establishing universal and free primary, secondary and higher education,
- eliminate the consequences of "optimizing health care" in order to prevent the spread of dangerous diseases. Medicine should be absolutely free, and its level should not correspond to "current" and" maximum possible "requests,
- establishing broad measures of social support for the population, without reducing its economic freedom.

A high degree of automation will lead to a significant reduction in agricultural and manufacturing workers. There will be an expansion of the class of workers in science and art, which will lead to the acceleration of scientific progress.

4 Discussion

Recently, many articles and papers have been published on the problems and institutional pitfalls that arise in the process of digitalization. Below is a small analysis of the literature in this area for 2019–2020 from the peer-reviewed journal Terra Economicus. In the article, Nikolaychuk and Nureev as modern trends (among others), the low level of financial literacy and the growth of household debt is noted, which is a brake on the development of the Russian economy [11]. Shmakov, analyzing the features of ritual behavior, notes the fundamental changes in the psychology of society, initiated by the processes of digitalization [13]. The issues of lagging Russian legislation in the field of cryptocurrencies are considered in the article by Dörr, Kowalski, and Nevskiy [4]. The general decrease in the level of information security and low topics of information technology implementation in the Russian Federation are noted in the work "Digital component of people's quality of life assessment in the regions of the Russian Federation" [9].

The problems of the neoliberal way of digitalization of the educational system, which consists in reducing the results of educational activities to quantitative indicators, are identified in the article by Volchik and Maslyukova [15]. Because of the neglect of important aspects of academic life there is a decrease in the quality of education. A sharp increase in unemployment due to the release of labor is noted in the work of Balatsky [2]. A general study devoted to the study of factors that determine the pace of digitalization is the work "development of digital infrastructure in the regions of Russia" [7]. The above articles partially echo this work on the consequences of digitalization. In addition, the study takes into account the impact of events in 2020 on the economic situation.

5 Conclusion

This paper describes the new and future professions that are emerging in the context of the fourth industrial revolution. Barriers to training and retraining of employees for more popular specialties have been identified. As an additional trend of digitalization of the labor market, the growth of part-time/non-permanent jobs is indicated. The conclusion is made about the negative impact of this factor on the financial situation of employees in developed countries. The article considers the impact of COVID-19 coronavirus infection on the labor market. One of the main consequences is a sharp increase in unemployment and, as a result, an increase in social tension. The example of the entertainment industry shows that the pandemic has reduced organizations with a low level of digitalization, and vice versa, has had a beneficial effect on organizations with a high level of digitalization. There is also a significant difference between these organizations in terms of the IEO service efficiency index. The conclusion is made about the probability of a crisis, similar to the "great depression" of the 30 s. Anti-crisis measures are proposed, which consist in transferring the economy to planned "rails".

References

1. Abdrakhmanova, G.I., Gokhberg, L.M., Kovaleva, G.V., Suslov A.B.: Internal costs for the development of the digital economy (2019). https://issek.hse.ru/news/317281408.html. Accessed 20 June 2020
2. Balatsky, E.V.: Global challenges of the fourth industrial revolution. Terra Econ. **17**(2), 6–22 (2019)

3. Dergachev, V., Starostina, Y.: Most Russian families found themselves without any savings at all (2020). https://www.rbc.ru/economics/31/03/2020/5e7dd7c59a7947c7f63c1e66. Accessed 20 June 2020
4. Dörr, J., Kowalski, O., Nevskiy, S.I.: Digitalization and monetary order: problems and prospects of cryptocurrency market regulation. Terra Econ. **17**(4), 6–22 (2019)
5. Federal Service State Statistics: Household consumption income and expenditure, the level and structure of household consumption expenditure in various socio-economic categories (2020). https://www.gks.ru/folder/13397. Accessed 20 June 2020
6. Grigoriev, S.G., Lukin, V.V., Lukin, D.V.: Human capital in the conditions of digitalization. E-Manag. **2**, 13–19 (2018)
7. Kramin, T.V., Klimanova, A.R.: Development of digital infrastructure in the Russian regions. Terra Econ. **17**(2), 60–76 (2019)
8. Leopold, T., Stefanova, V., Zahidi, R.: The future of jobs report 2018. World Economic Forum, Geneva (2018)
9. Litvintseva, G.P., Shmakov, A.V., Stukalenko, E.A., Petrov, S.P.: Digital component of people's quality of life assessment in the regions of the Russian Federation. Terra Econ. **17**(3), 107–127 (2019)
10. Nevelsky, A.: Why low unemployment doesn't always help the economy. Vedomosti (2019). https://www.vedomosti.ru/economics/articles/2019/11/19/816521-nizkaya-bezrabotitsa. Accessed 20 June 2020
11. Nikolaychuk, O.A., Nureev, R.M.: Household savings and investment. Terra Econ. **18**(1), 81–101 (2020)
12. Noskova, A.N.: Programmer, IT specialist, mathematician: How much does it cost to study at universities (2017). https://postupi.online/journal/kuda-postupat/programmist-it-specialist-matematik-skolko-stoit-obuchenie-v-vuzah/. Accessed 20 June 2020
13. Shmakov, A.V.: Code of ritual behavior in the context of digital transformation of economy. Terra Econ. **17**(4), 41–61 (2019)
14. Trading Economics: United States unemployment rate (2019). https://tradingeconomics.com/united-states/unemployment-rate. Accessed 20 June 2020
15. Volchik, V.V., Maslyukova, E.V.: Reforms, tacit knowledge, and institutional traps in education and science. Terra Econo. **17**(2), 146–162 (2019)
16. World Inequality Lab: World inequality report-2018 (2017). https://wir2018.wid.world/files/download/wir2018-summary-english.pdf. Accessed 22 June 2020

Labor Market in the Conditions of Digitalization of the Russian Economy

A. I. Gretchenko[1,2](✉) and A. A. Gretchenko[1]

[1] Plekhanov Russian University of Economics, Moscow, Russia
{gai51, vaz21063}@list.ru
[2] Financial University Under the Government of the Russian Federation, Moscow, Russia

Abstract. Modern research on the labor market in the conditions of digitalization of the Russian economy is being developed. Not only many international organizations, but also domestic ones are actively involved in this process. In 2015, the international labor organization (ILO), together with other international organizations, set the most important priorities for the development of the labor sector in the near and longer term. The scientific novelty of the research consists in a combination of theoretical and methodological principles and practical competencies in the digital organization of an effective Russian economy, which allow for Autonomous monitoring of the quality of labor market management. The results of the research are: classification of challenges of digital transformation of the global, national and regional economy; structured causal relationship of the unstable state of the economy with structural shifts and cyclical recession; the ability to eliminate the causes of the unstable state of the labor market through the use of digital technologies. Scientific and practical significance is determined by a real set of new competencies in labor market management, due to the materialization of the main provisions of the digital organization in Russia.

Keywords: Concept · Digital technologies · Globalization · Labor market · Paradigm · Practical competence

1 Introduction

The Russian Federation will be able to take its rightful place in the global digital economy if it actively develops a basic socio-economic profile of an employee in the labor market who is able to perform complex jobs in accordance with their qualifications in the conditions of modern information and communication technologies. At the same time, an employee in the conditions of digitalization of the Russian economy can perform their labor functions remotely outside the territorial and country borders of the employer. At the same time, the social and economic profile of the employed employee is required.

When analyzing the current state of research in the field of using the digital economy in labor market management, a number of scientific theories should be identified that affect the processes associated with digitalization occurring in society,

S. I. Ashmarina and V. V. Mantulenko (Eds.): IPM 2020, LNNS 161, pp. 283–290, 2021.
https://doi.org/10.1007/978-3-030-60926-9_36

including in the domestic labor market [7]. The topic of digitalization of the labor market in general is also focused on research on "digital capitalism". At the same time, when identifying the signs of "digital capitalism", in our opinion, it is necessary to specify which feature – platform or distribution.

It should be noted that in the practice of reforming economic systems, a lot of complex problems arise that require their scientific justification, study and solution and have been engaged in the problems of institutionalization, including the labor market, and continue their research.

2 Methodology

The methodological basis for the preparation and writing of the article was a documentary research, content analysis of current legislation, regulatory legal and methodological acts in the field of development of the state program "Digital economy of the Russian Federation" [13], methods of analysis, synthesis, comparison, analogy, induction, deductions, typology, classification, analysis of empirical data, methods and techniques of empirical scientific research, including economic diagnostics and monitoring of information and analytical modeling, the use of which will contribute to the discovery of previously unknown, non-trivial, practically useful knowledge necessary for the development of recommendations for the application of digitalization results in Russia.

In the course of the research presented in the article, an interdisciplinary approach is applied that contributes to deepening and enriching the understanding of the essence and features of the "digital economy", as well as identifying opportunities and limitations in its formation. The digital economy is based on information and communication technologies (ICTs) - the innovative conduct of business through markets via the Internet. In the digital economy, communication infrastructure provides a global platform where people and organizations develop strategies, interact, communicate, collaborate, and search for information [12]. Scientific hypotheses of Russian and foreign scientists were the theoretical basis [9]. Works devoted to the problems of university research activities in Russia on digitalization in the domestic economy are used. The initial data were:

- theoretical and methodological developments related to the development of theoretical provisions on the increasing role of the digital economy in the management of the labor market, which affects the change in the usual models of industry markets and changes in the nature of mutual influence and interaction between all subjects of the economic system,
- legal documents of legislative and executive authorities, as well as analytical materials on the formation of the digital economy in the country,
- statistical data and materials for monitoring the state of the labor market in Russia,
- official data of the country's online resources on the history, prospects, and digital economy of Russia, as well as websites of ministries and departments of the Russian Federation, author's developments.

The set of applied methods allows us to systematically approach the classification of problems and identify opportunities for the digital organization of the effective economy of the Russian economy by local allocation of digital technologies in the structure of business processes, as well as the subsequent formation of an optimal digital organization of the effective economy of the labor market [3].

3 Results

The functioning of the labor market in the digital economy objectively requires increased attention to the system of effective management and evaluation of the parameters of typical digital jobs, their classification into groups, taking into account the information necessary for each of these groups.

The level approach to the selection of the following groups of typical jobs:

- top-level managers of the organization,
- middle management managers of the organization,
- managers and specialists of the financial and economic block,
- specialists who develop legal and regulatory documentation for the creation of work technologies in the conditions of digitalization of the economy,
- performers of specific types of work,
- IT specialists who ensure efficient operation of software and computer systems with the introduction of digital technologies.

Creating competitive jobs and developing an employee profile in the digital economy is a prerequisite for the successful implementation of the program "Digital economy of the Russian Federation", which provides "... the formation of a labor market that meets the requirements of the digital economy and the creation of a system of incentives for the development of necessary competencies and participation of personnel in its development in Russia" [13, p. 2.1].

In the conditions of formation of the digital economy socio-labor relations on the labor market are undergoing significant changes: accelerate the process of "extinction" of certain traditional professions and new that are changing the principles of division and cooperation of labor, change to the competency requirements of employees; changing forms of employment in the form of crowdsourcing, freelancing, insourcing, flexible forms of inclusion of professionals into the workforce, as well as design and remote forms of employment, etc.; the accelerated mobility of the worker "education through life". The human resources management system accompanies the further process of orientation of the person and the organization in the system of professional division of labor. Thus, a systematic approach is required that would overcome the limitations of addressing issues within a narrow issue.

It should be noted that research on the role of labor in the digital economy consists in the fact that the theories and scientific concepts of personal development, "Life Long Learning, LLL), Human capital, Human Resource Management, "knowledge management", "talent management" in one way or another, directly or indirectly reflect the problems of career guidance and professional self-determination, but practically consider only certain aspects. It is necessary to introduce new conceptual concepts at all stages of human resource reproduction. At the stage of formation – the choice of profession, place of study and of formation of professional competences; at the stage of distribution – formation flows between institutions, regions, sectors, employment prospects in the labor market and professional labor division; point-of-use – the development of skills and reorientation to meet the changed conditions and the dynamics of functional (interprofessional) division of labor. Thus, a new understanding of the career guidance system is required, and, consequently, a theoretical justification of processes and relationships.

For example, in order to assess the effectiveness of the used methods of career guidance from the point of view of practice, it is necessary to identify the correspondence of the labor market and forms of employment to the actual events related to the digitalization of the economy. Currently, there is insufficient use of tools for forecasting the labor market, especially for the long-term period-10–15 years [11]. Independent methods of professional self-determination are not used, since the career planning system is more concerned with securing the employee in the organization, rather than free mobility in the labor market, which is a brake on the strengthening of the labor market on digital platforms. Thus, there is an objective need for a unified theoretical basis for building a promising system of career guidance and professional self-determination of employees engaged in the labor market using digital technologies.

The formation of the digital economy in Russia is impossible without fundamental and applied research by Russian economists, studying and generalizing their experience of the development of national specifics of the labor market in different foreign countries. An unstable state of the economy is a state of the economic system characterized by the instability of a set of essential variables and parameters of interaction between market entities and does not have sufficient stability of reproduction processes [5].

4 Discussion

It is well-known fact that trends in the development of labor markets have a pronounced specificity in various types of economic formations. The greatest "polar" dynamics has long been demonstrated by the labor markets of developed economies. Therefore, at the end of the XX century, developing countries that are currently leading the world's fastest-growing economies-China, Taiwan, and Singapore - have focused not on the issue of effective integration of labor in the reproduction process, but on the issue of creating a qualitatively new labor force that meets the main strategic challenges of the world economy, including rapidly developing digital technologies. In developed economies characterized by a high share of skilled labor among the population, the positive dynamics of labor markets for decades created the necessary conditions for independent growth-a state in which the country maintains a fairly stable socio-economic development at the expense of domestic sources. In the second half of the XX century in all highly developed countries, the "educational center" has been significantly expanded as an objective basis for maintaining sustainable growth and social development. According to our point of view, in both types of economies, science and education play a crucial role in investing in human capital [6]. They are an important factor for the transformation of the national socio-economic system, ensuring their economic development and social progress in society.

The dynamics of labor markets in both developed and developing countries will develop in modern conditions under the influence of a new driver – digital technologies, according to authoritative researchers of labor and employment. Today, the digital revolution has engulfed most of the world and has given rise to the development of trade, transport, finance, industry, education, health and public administration. Under the influence of digital technologies, not only new types of professions are being formed in the near future, but also the professional and qualification structure of the labor markets as a whole.

Research on the development of labor markets in the digital economy – an economic system, the core of which is a sector of digital products and services associated with digital technology, includes the development of modern forecasts of the national labor market as a whole, as well as forecasts of the economy's needs in qualified personnel in the professional field. This set of forecasts should become the basis for strategic planning of the development of the Russian Federation, its regions and key industries [2]. It should be noted that abroad at anticipating the needs of the economy in qualified specialists is a specific characteristic, the knowledge of which is necessary to determine the optimal fields of application of foreign experience in the practice of foresight needs of the economy in qualified specialists in vocational and professional context.

In developed foreign forecasts, development occurs not only by state bodies (ministries and employment agencies), but also by supranational organizations. The most famous is the European Centre for the development of Vocational training (Cedefop). All non-profit scientific organizations, such as the Employment Policies Institute (EPI), the Warwick Institute of labor Research (Warwick Institute of Labor Research) and other organizations that conduct research on the development of public employment policies, creating favorable conditions for entering the labor market, creating new jobs, and so on, carry out activities to develop forecasts for the labor market. In developing countries, international research organizations, as well as individual companies from Europe and the United States, participate in forecasting activities. For example, Cambridge Econometrics (UK) has completed a number of projects for EU countries in recent years. It should be noted that a number of scientific schools are successfully functioning in the Russian Federation, which have formed the methodological basis for scientific and practical research of labor markets in countries with transition and unstable economies. This is a Central Economics and mathematics Institute RAS (CEMI RAS), Institute of economic forecasting RAS (IEF RAS), Center for macroeconomic analysis and short-term forecasting and a number of other scientific organizations that analyze macroeconomic processes in Russia and develop scenario economic forecasts for the short, medium and long term, specialize in providing analytical and forecast procedures related to the dynamics of the labor market and employment. It should be noted that these organizations actively participate in the development of forecasts for the development of the Russian labor market using digital information technologies [8].

5 Conclusion

The digital economy used in the labor market is focused on improving its efficiency and competitiveness. The formation of the digital economy in the Russian Federation is at the very beginning of a difficult path, and in addition to the opportunity to receive digital dividends, it is necessary to solve problems, cope with uncertainties, and manage risks. A significant risk factor is the technological backwardness of many industries, which is associated with the problem of human resources, including management, lack of financial resources, both at the level of enterprises and regions [4]. In addition, the problem of digital inequality in the regions persists. On the other hand, it is the transition to a new model of economy that can breathe new life into many industries and enterprises. The state is doing a great job to ensure digital sovereignty: technology is developing, education is being reformed, local innovative companies are being supported, and so on. The path of digital transformation requires efforts from both the state and business. This is not just automation, it is changes in the business model and strategy. The success of the digital economy depends on how well both the public and corporate sectors will move towards a digital future [10]. This will require qualified personnel, engineers, and workers who are ready to perform tasks of a new level.

The digital economy is not a magic tool for turning a relatively backward country in terms of economic, technological, and especially social development into an advanced power. Digital transformation should be considered as a new tool for solving problems that requires enormous efforts at all levels: the state, regions, companies, while creating appropriate conditions, stimulating and motivating [1]. Generalization of international experience in forecasting the needs of the economy for qualified personnel allows us to justify the following areas of its application in the theory and practice of forecasting the prospects for the development of the labor market and employment in connection with professional qualifications in the Russian Federation:

- in the coming years, the dynamics of labor and employment markets in both developed and developing countries, including the Russian Federation, will be shaped by a new driver-digital technologies. This factor repeatedly increases the turbulence of the world and national labor markets and, consequently, determines the relevance of developing forecasts of the economy's needs for qualified personnel in the professional and qualification context. Such forecasts should form the basis for the formation of a strategic forecast, a strategy for the socio-economic development of the Russian Federation, and other documents of the country's strategic planning system,
- abroad, to forecast the development of labor and employment markets, balance and calculation models of general equilibrium (CGE models) are used, reflecting the features of socio-economic processes and the dynamics of labor and employment markets in a particular state. It should be noted that a holistic approach to forecasting the needs of national economies for qualified personnel has been formed abroad, based on quantitative and qualitative methods. the synergetic effect of joint use allows us to predict a high degree of reliability and practical usefulness,

- scientific and practical interest is the specification of institutional, behavioral, technological, industrial, territorial and socio-economic factors that determine the current and future supply and demand for qualified personnel in the digital economy, including in the professional and qualification context,
- in modern economic conditions, the creation of an early warning system is of particular interest to the Russian Federation – a tool for regulating the labor market, allows us to timely identify future trends in the dynamics of demand for skilled labor and its supply, develop strategic forecasts for the development of the labor market and the formation of an adequate state employment policy.

Acknowledgements. 1. This article was prepared as part of the government contract as requested by the Ministry of Science and Higher Education of the Russian Federation on the subject formulated as «Structural changes in economy and society as a result of achieving the target indicators of National projects, which provide opportunities to organize new areas of social and economic activity, including commercial, both in Russia and abroad» (project No. FSSW-2020-0010).

2. The article was prepared within the framework of the Russian Foundation For Basic Research - supported research project No. 8-010-00534 "Development of methodology, algorithm and methodology for forecasting the needs of the Russian digital economy for qualified personnel in the professional and qualification context".

References

1. Akhmadeev, B., Manakhov, S.: Effective and sustainable cooperation between start-ups, venture investors, and corporations. J. Secur. Sustain. **5**(2), 269–285 (2015)
2. Akhmadeev, B.A., Manakhov, S.V.: Innovative system of estimation of investment projects of development of subjects of regional economic complexes on the basis of combined methods of computer optimization. Espacios **39**(18), 31 (2018)
3. Banu, G.S.: Measuring innovation using key performance indicators. In: Moldovan, L., Gligor, A. (eds.) Proceedings of the 11th International Conference on Interdisciplinarity in Engineering (INTER-ENG), Procedia Manufacturing, vol. 22, pp. 906–911. Elsevier Science BV, Amsterdam (2018)
4. Beliakov, G., Gretchenko, A., Ryzhaya, A., Shpak, A., Belyakov, S.: The formation of a strategic planning and supply chain system for the scientific and technological development of the Russian Federation regions. Int. J. Supply Chain Manag. **8**(6), 1035–1044 (2019)
5. Domínguez, E., Pére, B., Rubio, Á.L., Zapata, M.A.: A taxonomy for key performance indicators management. Comput. Stand. Interfaces **64**, 24–40 (2019). https://doi.org/10.1016/j.csi.2018.12.001
6. Gretchenko, A.I., Demenko, O.G., Gretchenko, A.A.: Model of remuneration: 'catching up' type (Russian case). J. Adv. Res. Law Perform. Indicat. Manag. Comput. Stand. Interfaces **64**, 24–40 (2018)
7. Gretchenko, A.I., Gorokhova, I.V., Demenko, O.G., Gretchenko, A.A.: Digital economy: challenges and threats for modern Russia. J. Adv. Res. Law Econ. Quart. **9**(4), 1243–1248 (2018). https://doi.org/10.14505//jarle.v9.4(34).09
8. Gretchenko, A.I., Gretchenko. A.A.: The formation of the digital economy in Russian regions. In: Mantulenko, V. (ed.) II International Scientific Conference GCPMED 2019 "Global Challenges and Prospects of the Modern Economic Development". European Proceedings of Social and Behavioural Sciences, vol. 79, pp. 430–436. European Proceedings, London (2019)

9. Gretchenko, A.I., Nikitskaya, E.E., Gretchenko, A.A., Demenko, O.G.: Methodological aspects of forecasting skilled labor in context of innovation transformations (of the Russian economy). J. Adv. Res. Law Econ. **9**(3(33)), 481–489 (2018)
10. Gretchenko, A.I., Nikitskaya, E.E., Valishvili, M.A., Gretchenko, A.A.: Role of higher education institutions in developing hr potential in a forming innovation economy. Rev. Espacios **39**(21), 13 (2018)
11. Moiseev, N., Akhmadeev, B.A.: Agent-based simulation of wealth, capital and asset distribution on stock markets. J. Interdisc. Econ. **29**(2), 176–196 (2017). https://doi.org/10.1177/0260107917698781
12. Nadiri, M.I., Nandi, B., Akoz, K.K.: Impact of modern communication infrastructure on productivity, production structure and factor demands of US industries: impact revisited. Telecommun. Policy **42**(6), 433–451 (2018). https://doi.org/10.1016/j.telpol.2018.03.008
13. Order of the Government of the Russian Federation of July 28, 2017 N 1632-R on approval of the program "Digital economy of the Russian Federation" (2017). http://base.garant.ru/71734878/. Accessed 15 May 2020

Self-development of Sport Managers and Coaches Under Conditions of Education Digitalization

A. M. Danilova[1](✉) and A. D. Voronin[2]

[1] Samara State University of Economics, Samara, Russia
vitr@list.ru

[2] Samara State Technical University, Samara, Russia
sasha-voronin-1994@mail.ru

Abstract. The influence of the trainer on the athlete, his attitudes and training methods, is one of the key factors in the formation of successful sportsman training. However, in the current rapidly changing conditions of the modern world, trainers, both old and experienced, and young professionals, need to constantly improve and expand their knowledge. Based on the studies of a number of scientists by trainers who regularly attend various advanced training courses, athletes compete more successfully in competitions than specialists who prefer to rely on the knowledge gained during training at the university. Nevertheless, the concept of "the success of a child's sports training" has a broader meaning than achieving a prize in competitions. In contrast to the performance of competitions, the concept of "the success of a child's sports training" is integrative, including interrelated and correlating components. For the successful development of the components of success and their indicators, the trainer needs to have knowledge of the competent construction of sports training for the child. In order to provide trainers with such knowledge, the author has developed a continuing education program "Pedagogical conditions for the formation of the success of children in the process of sports training".

Keywords: Additional education · Culture and sport · Success · Self-development · Training

1 Introduction

Currently, in our country, the emergence of new successful and highly qualified athletes has carried out thanks to the system of sports training of the reserve, consisting of children of different school age. The primary role here is given to sports schools of the Olympic reserve in the system of additional education. Additional education of children, including in the field of sports, is a rather flexible educational system that can not only affect the child in his creative development but also expand his cultural and cognitive horizons. Enriched with knowledge and skills, as well as improving their motor skills and abilities, students, in addition, must learn to harmoniously and competently interact with people around him [12]. Sports schools, along with other institutions of additional education of children, can be classified as educational institutions,

S. I. Ashmarina and V. V. Mantulenko (Eds.): IPM 2020, LNNS 161, pp. 291–297, 2021.
https://doi.org/10.1007/978-3-030-60926-9_37

in which the formation of positive motivation of children personality to the process of cognition and creativity is paramount [3]. In addition, one of the main purposes of such institutions is the implementation of additional educational programs and services that affect the interests of society [5]. Therefore, in addition to preparing the reserve of the national teams of the Russian Federation, the goals of sports schools include: improving health and harmonious physical development, the formation of a creative and creative-minded personality, social adaptation of students and their self-determination.

2 Methodology

No less important is the work with trainers and teachers, since the implementation of various pedagogical tools directly depends on it. The role and influence of the trainer-teacher on students involved in sports in the modern system of continuing education cannot be underestimated [1]. According to Rudneva, it is necessary to pay closer attention to the problem of pedagogical professionalism "… having both functional and personal conditionality due to the specifics of pedagogical activity in the modern conditions of its implementation" [9, p. 192]. According to Popovicheva, "education and upbringing of a child, a pupil of the system of supplementary education, largely depends on the professionalism of the teacher" [4, p. 182]. The ability of a trainer-teacher to create in a sports team an expectation of the imminent success of a child who is engaged in it, to plan and constantly motivate her to achieve it, to pay attention to experiencing success, and then its reflection, are the main motivating means. Work with trainers and teachers implies their striving to promote adolescent students to success, the study of pedagogical tools that contribute to the formation of success in the process of sports training.

3 Results

In the framework of the experimental stage of this research, a continuing education course was held among trainers and teachers of the regional Sambo Federation "Pedagogical conditions for the formation of schoolchildren's success in the process of sports training" (Table 1).

The purpose of the course is to build the readiness of trainers and teachers to ensure success in the process of sports training of students.

Course objectives:

- to form the motivation of teachers to ensure the conditions for the success of sports training of students,
- to teach the use of pedagogical and organizational means of forming the success of schoolchildren in the process of sports training [11].

Based on the opinions of modern researchers, it can be argued that absolutely in any pedagogical profession it is necessary to update existing "theoretical knowledge and professional competencies" [8, p. 51]. From the point of view of Rudneva,

Table 1. The content of the continuing education program of continuing education "Pedagogical conditions for the formation of the success of children in the process of sports training"

The goal - the formation of the readiness of trainers and teachers to ensure success in the process of sports training of students.
Tasks:
- to form the motivation of teachers to ensure the conditions for the success of sports training of students;
- to teach the use of pedagogical and organizational means of forming the success of schoolchildren in the process of sports training

The course of lectures and pieces of training «Success of schoolchildren in the process of sports training as a pedagogical problem»

№	Topic	Content of the lecture
Theoretical block		
1	Introductory lecture. General ideas about the concept of "success"	The history of the development of the concept of "success" and the history of the study of this problem. The development of the concept of "success" at the present stage, the analysis of the concept of "success in teaching students", the development of pedagogical science. The study of the concept of "success in the process of sports training" basic concepts aimed at shaping the success of students in the process of sports training
2	2 Lecture "The role of trainers in shaping the success of schoolchildren in the process of sports training"	Styles of pedagogical communication. Features of adolescence, interaction with lagging students in the process of sports training
3	Lecture "Pedagogical conditions used by trainers to shape the success of schoolchildren in the process of sports training"	Analysis of pedagogical conditions, the influence of pedagogical conditions in the work of a trainer on the formation of schoolchildren success Pedagogical conditions used by trainers in their process of sports training for schoolchildren
Practical block		
4	Training "Using pedagogical and organizational means aimed at building success in the process of sports training"	Demonstration by trainers of various technologies and techniques for creating a success situation: removal of fear, advance payment of a successful result; hidden instruction of the child in the ways and forms of carrying out activities, introducing a motive, personal exclusivity; mobilization of the activity or pedagogical suggestion, high appreciation of the details

(*continued*)

Table 1. (*continued*)

5	Training "Development of a sports training system using pedagogical conditions"	Development of the training process as a whole or a separate fragment, depending on the situation using organizational and pedagogical conditions. Designing methods and techniques for the implementation of the considered pedagogical and organizational means
6	Final lesson	Discussion of the results of the seminar, reflection of the participants

Source: authors.

Glubkov, and Ilyukhina, "the uniqueness of the continuing education system lies in the fact that, unlike other educational systems and basic pedagogical education, the educational process, taking into account the existing professional qualities of a teacher, transforms them, changes them in accordance with educational development trends" [7, p. 58]. A trainer participating in continuing education courses has the opportunity to gain experience in building a sports process by communicating with other trainers. Rudneva and Bugaev note "the development of individual experience is possible only on the basis of broad social experience: by correlating one's own experience with the collective one, the teacher gets the opportunity to evaluate his position through the eyes of his colleagues" [6, p. 40]. The continuing education program involved 18 people. The data of the questionnaires and surveys conducted before the seminar showed the following picture: more than half of the interviewed trainers (61%) had absolutely no desire to monitor their pedagogical activity and analyze their experience, obvious expectations of reproductive activity (presenting ready-made information), at that time as the remaining 39% did not mind active training activities. In addition to this, almost all trainers had a superficial or incomplete awareness of the use of pedagogical means to shape the success of schoolchildren in the process of sports training.

4 Discussion

In general, the attitude of trainers to similar seminars or, for example, continuing education courses is not always positive, in view of the prevailing opinion about the inefficiency of this form of improving pedagogical mastery. It is necessary to highlight the fact that it is precisely those trainers whose authoritarian type of training prevails in training that did not express a particular desire to participate in the continuing education program [2, 10]. The compiler of the program assumed that this is due to the very difficult compatibility of the authoritarian type of education with a focus on the success of the student, using the means of creating a success situation.

The program of the developed advanced training course included: listening and discussing reports and messages, conducting pieces of training and workshops that show the technologies and tools used in the process of sports training and their specific result. This program can be conditionally divided into two blocks - theoretical and practical. In the first block, a lecture course "Success of schoolchildren in the process of sports training as a pedagogical problem" was developed, which examined the theoretical aspects of the concept of "success in the process of sports training". As part of this block, the personal interest of trainers and teachers in the problem of shaping the success of teenage students has increased.

The first topic of the theoretical block was devoted to general ideas about the concept of "success" and consisted of a series of lectures in which the question of the relevance of the problem of formation of success of schoolchildren was revealed. The seminar's developers'message and the participants' reports briefly highlighted the history of the concept of "success", the history of the issue of formation of student success in the training process and sports activities in general. The participants also presented the arguments of famous academic teachers, studied basic concepts aimed at shaping the success of schoolchildren in the process of sports activities.

The second section - "The role of trainers in shaping the success of schoolchildren in the process of sports training" - also included several lectures. The developers of the seminar at this stage of the program's implementation set themselves the following task: to convey to trainer's and teacher's information on personal significance in the process of building success for a teenage student in the process of sports training, with a view to their further awareness and understanding of this fact. At this stage, the lecturers emphasized that more attention should be paid to lagging athletes.

At the third stage of the lecture classes, trainers and teachers reported on pedagogical conditions and their influence in the work of trainers on the formation of schoolchildren success. The trainers monitored the pedagogical conditions, and also reported on the pedagogical conditions most often used in their work, and which, from their point of view, are the most effective.

After a course of lectures and a study of the main issues, the seminar participants switched from a theoretical block to a practical "Pedagogical and organizational means of shaping the success of schoolchildren in the process of sports training". The fourth stage was carried out in the form of training and was aimed at the joint development of those tools and methods, which, in our opinion, contribute to the formation of the success of schoolchildren in the process of sports training. During the lecture sessions, it turned out that trainers and teachers do not often use the technology for creating a success situation and extra-training methods in the educational process of sports training. To eliminate this shortcoming, a training was held entitled "Using pedagogical and organizational tools aimed at building success in the process of sports training", and several practical exercises were organized. During the lessons, various technological aspects were analyzed, aimed at creating a situation of success. In addition, the trainers offered pedagogical and organizational tools that would be effective in sports training, talked about them and demonstrated their most competent use in the educational process, what results were achieved. Attention was also paid to the organization of extra-training forms of work: visiting museums and historical and patriotic events, participating in sports events and demonstrations, joint hiking, participating in rallies, processions and civic events, etc.

At the fifth stage, a master class and training for trainers and teachers were conducted, aimed at applying the previously studied organizational and pedagogical conditions in the process of sports training. In the lessons on the development of a sports training system using pedagogical conditions, the subject of study was the design of the entire training process of sports training or its individual fragment, using the previously obtained information from previous classes. Thanks to this design, the trainer-teacher will learn to determine both the near and long-term goals of his work, to compare the educational goal of a separate lesson with the goals of the entire process of sports training; determine the stages of the implementation of development goals, methods for their implementation in a particular training session, correlate goals and results, plan and conduct targeted diagnostics; take into account the specifics, tasks and originality of training with the condition of their differences in types. In addition, when designing the training process, the attention of trainers was paid to the means and methods of achieving goals, to the technological aspects of sports training.

5 Conclusion

At the final lesson, the trainers summarized the results of the seminar, shared their thoughts on this program and expressed their opinion on the pedagogical tools studied in the framework of the seminar aimed at building the success of schoolchildren in the process of sports training in the system of additional education. Final testing was carried out. In conclusion, trainers and teachers were asked to take a survey on the relevance of this program, its impact on the formation of the readiness of trainers and teachers to ensure success in the process of sports training for schoolchildren.

References

1. Gerrard, B.: Understanding social networks and social support resources with sports coaches. Sport Manag. Rev. **17**(2), 240–241 (2014)
2. Miles, K.H., Clark, B., Fower, P.M., Miller, J., Pumpa, K.L.: Sleep practices implemented by team sport coaches and sports science support staff: a potential avenue to improve athlete sleep? J. Sci. Med. Sport **22**(7), 748–752 (2019)
3. Norris, L.A., Didymus, F.F., Kaiseler, M.: Understanding social networks and social support resources with sports coaches. Psychol. Sport Exercise **48**, 101665 (2020)
4. Popovicheva, O.N.: A model for improving the pedagogical skills of specialists of an institution of additional education. Soc.-Econ. Phenom. Process. **10**(56), 182–188 (2013)
5. Pruzhinin, K.N.: Continuity of the main elements of the system of continuity of professional education in the field of physical education and sports. Bull. Buryat State Univ. Pedagogy Philol. Philos. **13**, 117–121 (2011)
6. Rudneva, T.I., Bugaev, A.L.: Means of forming the methodological competence of the teacher. Second. Vocat. Educ. **7**, 40–42 (2012)
7. Rudneva, T.I., Glubkov, S.A., Ilyukhina, N.A.: New formats of the advanced training system for university teachers. Bull. Samara State Univ. **5**(96), 57–60 (2012)
8. Rudneva, T.I., Solovova, N.V., Strekalova, N.B., Sysoeva, E.Yu., Sanko, A.M., Nikulina, I.V.: Pedagogical support of innovative pedagogical activity. Bull. Samara Univ. Hist. Pedagogy Philol. **3-1**, 50–57 (2016)
9. Rudneva, T.I.: Pedagogical activity in the modern social context. Bull. Samara State Univ. **7** (129), 191–195 (2015)

10. Surujlal, J.: Influence of outlook towards work on entrepreneurial potential of professional sport coaches in South Africa. Proc. Econ. Financ. **35**, 597–603 (2016)
11. Van den Berg, L., Coetzee, B., Mearns, M.: Establishing competitive intelligence process elements in sport performance analysis and coaching: a comparative systematic literature review. Int. J. Inf. Manag. **52**, 102071 (2020)
12. Zhurkina, A.Ya.: Integration of the content of formal and non-formal education in physical education as a condition for ensuring its continuity. Russ. J. Educ. Psychol. **3**(11), 1–15 (2012)

Digital Transformation of Advocacy Activity in Modern Russia

Yu. A. Dorofeeva(✉)

Samara State University of Economics, Samara, Russia
log1612@yandex.ru

Abstract. The author analyzes the circumstances that caused these changes, which include the state's measures to combat coronavirus infection (COVID-19), novelties of procedural legislation on the use of digital technologies in judicial disputes, as well as the practice of applying this legislation. The purpose of the study is to establish the validity and expediency of applying certain legislative measures aimed at digitalization of activities related to the implementation of legal assistance by a lawyer to citizens and legal entities. During the research, the following methods were used: general scientific, such as analysis, synthesis, comparison, generalization, historical method; private-scientific: formal-legal, comparative-legal, allowing to consider the digitization issues of advocacy activity. The result of the research is to establish the applicable law to relations concerning the conclusion of a foreign economic contract, the grounds and conditions for recognizing the contract as concluded, and the consequences of not recognizing the fact of concluding an international commercial contract.

Keywords: Advocacy activity · Digitalization · Information technology · Lawyer · Legal aid

1 Introduction

A lawyer, as a person who provides professional legal assistance, must have a certain education, qualification and experience in the field of law. Observance of these terms, as well as the procedure for obtaining the status of a lawyer (successful completion of the qualification exam and a positive conclusion of the qualification commission of the lawyer's education in the Russian Federation) is sufficient to obtain the status of a lawyer. Throughout the entire period of advocacy activity, it is necessary to obtain knowledge in the field of law in the form of professional development. This means that the list of knowledge and skills that a lawyer should have, does not include those related to working with information and legal systems, Internet resources, including the Commercial Case File, and GAS-Pravosudie. Failure to work with these resources does not allow the lawyer to be as effective and informed as the people who use them, and most importantly-increases the risk of clients receiving poor-quality and ineffective legal assistance. This is especially important in connection with the restriction of access to the courts to comply with protection measures against the spread of COVID-2019 coronavirus infection, the introduction of remote procedural actions in the practice of

S. I. Ashmarina and V. V. Mantulenko (Eds.): IPM 2020, LNNS 161, pp. 298–304, 2021.
https://doi.org/10.1007/978-3-030-60926-9_38

courts (familiarization with the case materials, obtaining audio clips of court sessions) and even holding online sessions.

2 Methodology

During the research, the following methods were used: general scientific, such as analysis, synthesis, comparison, generalization, historical method; private-scientific: formal-legal, comparative-legal, allowing to consider the digitization issues of advocacy activity. The result of the research is to establish the applicable law to relations concerning the conclusion of a foreign economic contract, the grounds and conditions for recognizing the contract as concluded, and the consequences of not recognizing the fact of concluding an international commercial contract.

3 Results

Providing legal assistance is a complex and multifaceted process that requires a high level of professional qualifications. Therefore, only people who have received a higher legal education in a state-accredited educational program or a degree in law, as well as those who have two years of experience in the legal field or have completed an internship in advocacy education (part 1 of article 9 Of the law on the bar) have access to the practice of law [10]. A person who meets the above requirements must also pass the qualification exam for the lawyer status [15]. These conditions, after passing the Qualifications Commission, make it possible for the applicant to obtain the status of a lawyer.

Education requirements for people wishing to have the status of a lawyer are established not only in the Russian Federation, but also in other states. For example, to be eligible to practice law in the United States, you must have a bachelor's degree, in an educational institution approved by the American Bar Association (ABA), you must obtain a Jurisprudence Doctor (JD), Master of Laws (LLM), Doctor of Philosophy (PhD), Doctor of Juristic Science (JSD), or Doctor of Comparative Law (DCL), and pass the State Bar Examination [11].

In Germany, the possibility of obtaining the status of a lawyer is associated with the need to pass two state exams - Erste Juristische Staatsprüfung, after which students are called certified lawyers (Diplom-Jurist) and have the right to continue their studies in the so-called Referendariat. The referendariat is a two-year practical training of lawyers at one of the 24 Higher land courts (Oberlandesgericht) of Germany, at the end of which the second state exam (Zweite Juristische Staatsprüfung) is passed [7]. Subject to appropriate levels of education, in accordance with §4 of the Federal law on the advocate of the Republic of Germany, only people who have the right to conduct activities in court, according to the German Court Law (Deutschen Richtergesetz), fulfilled the integration requirements in accordance with part 3 of the Law on the activities of European lawyers in Germany, has a certificate according to § 16A paragraph 5 of the Law on the activities of European lawyers in Germany, can obtain a lawyer status [1].

The requirements for the qualification of people who have the right to provide legal assistance to the population by performing legal activities are also established by the legislation of other states. Restrictions on access to the status of a lawyer is not surprising. The professionalism of the person, providing legal assistance, often determines the future behavior of civil turnover and business activities participants, since before the occurrence of a legal fact that may become the basis for a legal dispute (before the transaction, creation of a legal entity, divorce, etc.), the subjects involved in such relations often seek legal assistance from a lawyer who is able to eliminate possible conflicts in the formed relationships.

The lawyer protects citizens and organizations from imputing to them offenses that were not committed, or committed under circumstances that exclude or reduce the amount of responsibility of the offender. Thus, the task of the advocacy to protect the rights and interests of citizens (subjects) of the state, the implementation of the law functions (preventive, human rights) by the legal community, entails increased attention of the state to the conditions for admission to the advocacy, regulatory consolidation of requirements for lawyers candidates and verification of the qualification level of applicants for the status of a lawyer and requirements for maintaining such status, including in the form of professional development.

However, it is the lawyer's professional (legal) knowledge and experience in applying knowledge in the legal sphere that is subject to verification and confirmation in accordance with the procedure established by law. The skills of a candidate in advocacy, a current lawyer in the field of computer technology, Internet resources that allow for professional legal assistance and are not included in the scope of the knowledge control of people who acquire or carry out legal activities.

The lack of legal requirements for information technologies knowledge for providing legal assistance to members of the legal community and people applying for the status of a lawyer, is explained by the large volume of legal information required for high-quality legal assistance, the need to structure and analyze such information. Indeed, the very possibility of obtaining the status of a lawyer is directly related to the possession of a diploma confirming professional education, as well as obtaining sufficient legal practice. But today this knowledge is clearly not enough.

The lawyer's ability to work with reference legal systems that combine legal norms, including by-laws and departmental acts, legal positions of the judiciary and individual judicial acts issued in specific cases, give him an advantage over those who do not have the ability to work with such systems or the necessary skills to use reference legal systems. Of course, this inequality of two lawyers gives one of them the opportunity to provide better and more effective legal assistance to their principal. This may affect the results of the legal dispute consideration, the possibility of its prevention, or even deprive the principal of the possibility of obtaining judicial protection, for example, if the lawyer incorrectly calculates the statute of limitations for the raised claims. But the consequence of the lawyer's information and legal inexperience is not only his personal professional failures. First of all, this is a negative, property and possibly penal consequences of receiving poor legal assistance for the lawyer's client, the decline in the overall level of legal protection of the population, that is, failure to perform non-advocacy functions assigned to it by the state. Someone may say that this argument cannot be used as a counterweight to the requirement of the legal competence level of the lawyer, since the possession of the necessary knowledge is a part of the professional requirements for a lawyer. We cannot disagree with this, since having a lawyer's degree and the necessary work experience does not guarantee the ability of a lawyer to obtain the latest knowledge and information from existing reference legal systems, which is especially relevant for Russia, where the practice of applying the law is formed based on the legal positions of courts that tend to change over time.

The provisions of the legislation of the Russian Federation on the loan agreement could be the example. Part 1 of article 807 of the Civil Code of the Russian Federation defined the content of the money loan between citizens: the lender transfers or undertakes to transfer it to the borrower ownership, and the borrower shall return to the lender the loan on the terms and conditions specified in the contract, and the contract shall be considered concluded from the moment of transfer the loan to the borrower or the specified person [3]. Statutory requirements for lender or borrower, such as income, sufficient for granting the loan or purpose of the loan, do not exist, but these circumstances are to be proved in the collection of the loan or the inclusion of creditor requirements in the register of requirements of competitive creditors of the debtor in the legal positions of the Supreme Court, current regulatory enforcement. In the refusal justification of the recovered loan, courts have argued impecuniousness of the loan agreement, but a special entity (the borrower) is often absent (loan disputes arbitration official receiver, the competitive creditor), and the prohibition of challenge of lack of money of the contract, made in writing (part 2 of article 812 of the civil code) is ignored. The basis for challenging the loan agreement is also the provisions of paragraph 2 of article 61.2 of the bankruptcy law, which provides for the possibility of invalidating a transaction made by the debtor in order to harm the property rights of creditors (suspicious transaction), but often the restrictions established by law (including the suspicious period) are not taken into account by the courts when considering a dispute on recovery of borrowed funds from the debtor [9]. As a result, the claims of the lender, who provided the court with a receipt in order to support the claim for recovery of the loan and loan agreement, are refused even if the debtor admits the fact of receiving money from the lender if the lender does not prove that it has sufficient funds to issue the loan and that the borrower actually incurred some expenses by applying the borrowed funds (for economic disputes, see, for example, the Ruling of the Supreme Court of the Russian Federation of June 11, 2020 No. 306-ES18-10093(2) in case no.A55-10923/2017 [16], for cases considered by courts of General jurisdiction, see, for example, the decision of the Soviet district court of the city of Samara of 25.05.2017 in case No. 2-917/2017 [6], the decision of the Samara regional court of 31.07.2017 in case No. 33-9616/2017 [4]).

Digitalization as a process that has covered all areas of social life has also affected the work of lawyers as people engaged in professional representation and protection of the rights and interests of parties and participants in legal proceedings. Already used to participating in the process via video conferencing systems, lawyers were able to remotely access the activities of the courts. The electronic Commercial Case File (CCF) reflects all information on the existence of a court dispute, the course of its consideration, judicial acts, including interim (definitions) issued in a particular case. You can also see the composition of the people involved in the case, the judges who are analyzing the dispute, and the period of its consideration. Information on the progress of the case can be obtained in the form of cards, a calendar, it is possible to get acquainted with the electronic case and a tab for participation in online meetings, which contains information about the filing of relevant petitions and their applicants [5] and to participate in the meeting during its holding online. The convenience of working with the Commercial Case File is supplemented by the ability to classify various information: on the participation of specific individuals in court proceedings, categories of such cases (civil, administrative, bankruptcy), etc. In addition, there are services "Strazh" (tracking the progress of consideration of a specific dispute), "Calendar" allows you to track the dates of disputes consideration by individual courts, judges, and participants in the dispute, and you can also contact the Bank of judicial decisions and work in the "My Arbitrator" system, which allows you to remotely post scanned copies of procedural documents in a case in which it provides representation, informing it of changes in cases to which it has subscribed. The transparency, completeness and availability of arbitration court services allows the lawyer to be informed, quickly and reliably deliver procedural evidence to the court, even during the period when restrictions are imposed on the reception of documents by the court's office. In addition, by June of this year, almost all arbitration courts of the Russian Federation have provided the opportunity to hear audio recordings of court sessions online for participants in the process and their representatives.

It is difficult to overestimate the comfort and convenience of a lawyer who uses the Commercial Case File. The speed and efficiency of providing legal assistance, as well as the guarantee of its quality due to awareness of the progress of the case, quick access to the case, including by receiving an audio protocol of the court session, is now supplemented by the possibility of remote participation in an online court session or in a meeting held via videoconferencing. Ignorance of these features, inability to use them not only creates obstacles to the normal conduct of the process with the use of modern technologies for participation in court proceedings, but also leads to the costs of the principal for expenses that can be avoided (postage for sending documents to the court, which can be sent in scanned copies via the "My arbitrator" service, travel costs for familiarization with the case, for participation in court sessions, etc.), and may also cause improper conduct of the case by the lawyer due to insufficient information.

The services of arbitration courts of the Russian Federation differ significantly from those used by courts of general jurisdiction. In addition, in the system of GAS-Pravosudie is only visible part of judicial acts (not intermediate, but on the merits), a similar problem of non-publication of information on cases and judicial acts at courts, where, among other things, there is no vision of the movement of the case on various instances. The essence of the dispute from the published information is impossible to understand, the amount of the claim, the parties to the proceedings are hidden with reference to the protection of personal data. Probably for the same purpose in the courts of appeal and cassation instances the judges considering the case are not indicated until the court session. This, of course, complicates the work of a lawyer in a court of general jurisdiction, and for the principal creates a greater likelihood than when the case is considered by an arbitration court, the lack of proper awareness and increase in the amount of court costs. However, the electronic services of general jurisdiction courts allow you to submit documents in electronic form, learn about the registration of claims, applications, and complaints filed by the court, as well as, except in certain cases, find information about the date and place of consideration of the dispute by the court of general jurisdiction. These data, along with the service Sudact.ru, allow lawyers to provide clients with legal assistance at a high professional level.

4 Discussion

The discussion of digitalization in advocacy activity is widely presented in publications of Russian authors [12, 14], including those published abroad [8, 17], and in scientific works on the implementation of legal assistance by a lawyer of foreign authors [2, 13]. The work analyzes the impact of digitalization on processes of advocacy activity in the organization and implementation of activities of the legal profession, covering various aspects of the research topic (attorney's request, improving the quality of legal aid, legal aspects of the use of digital technologies in the activities of a lawyer, etc.), and the goal of each study is the analysis of the impact and importance of digitalization for the advocacy.

5 Conclusion

Digitalization processes invariably affect all aspects of social interaction, as well as individual public institutions, including the Institute of advocacy. In order to achieve the goals set by the law and society, the lawyer, legal education and the community of lawyers have to develop skills related to the possibility of providing legal assistance through the latest information technologies. Recent events related to the adoption of measures by many states to protect the population from the danger of coronavirus infection (COVID-2019) have confirmed the need to develop the involvement of the legal community and each of the lawyers individually in the process of digitalization. The ability to work with tools for obtaining information and participating in court proceedings, provided through special electronic files, databases, is necessary for a modern lawyer to provide effective and high-quality, that is, qualified legal assistance.

References

1. Bundesamt für Justiz: Bundesrechtsanwaltsordnung § 4 Zugang zum Beruf des Rechtsanwalts (1959). https://www.gesetze-im-internet.de/brao/__4.html. Accessed 20 June 2020
2. Caserta, S., Rask, M.M.: The legal profession in the era of digital capitalism: disruption or new dawn. Laws **8**(1), 1 (2019)
3. Civil code of the Russian Federation (part two) of 26.01.1996 No. 14-FZ (1996). http://www.consultant.ru/document/cons_doc_LAW_9027/. Accessed 20 June 2020
4. Decision in case no. 33-9616/2017 (2017). https://oblsud–sam.sudrf.ru/modules.php?-name=sud_deloandname_op=caseand_id=10676891and_uid=b98c2f6c-3491-4ae7-bd01-a61069989df5and_deloId=1540005and_caseType=and_new=5andsrv_num=1. Accessed 20 June 2020
5. Decision in the SIP-40/2020 case (2020). https://kad.arbitr.ru/Card/f62e5c09-50e7-4973-928e-29f890145ca8. Accessed 20 June 2020

6. Decision no. 2-917/2017 2-917/2017 ~ M-567/2017 M-567/2017 dated may 25, 2017 in case no. 2-917/2017 (2017). https://sudact.ru/regular/doc/TpMRnRxGCvJx/?regular-txt=andregular-case_doc=2-917%2F2017andregular-lawchunkinfo=andregular-date_from=andregular-date_to=andregular-workflow_stage=andregular-area=1000andregular-court=andregular-judge=and_=1593015587300. Accessed 20 June 2020
7. Derra, M.: The legal profession in Germany (2016). https://zakon.ru/Tools/DownloadFileRecord/23439. Accessed 20 June 2020
8. Dorofeeva, J.A.: Digitization: the bar's aspect. In: Ashmarina, S.I., Vochozka, M., Mantulenko, V.V. (eds.) Digital Age: Chances, Challenges and Future. Lecture Notes in Networks and Systems, vol. 84, pp. 574–580. Springer, Cham (2020)
9. Federal law No. 127-FZ of 26.10.2002 "On insolvency (bankruptcy)" (2002). http://www.consultant.ru/document/cons_doc_LAW_39331/. Accessed 20 June 2020
10. Federal law No. 63-FZ of 31.05.2002 "On advocacy and the bar in the Russian Federation" (2002). http://www.consultant.ru/document/cons_doc_LAW_36945/. Accessed 20 June 2020
11. Learn how to become: How to become a lawyer (2014). https://www.learnhowtobecome.org/lawyer/. Accessed 20 June 2020
12. Makarov, S.Y.: Prospects for modernizing the status right of a lawyer to obtain information through lawyer requests in the context of digitalization processes. Actual Probl. Russ. Law **11**, 120–125 (2019)
13. Nanteme, P.: Definition: what is digitalization? (2018). https://www.aoe.com/en/digitalization.html. Accessed 20 June 2020
14. Pilipenko, Yu.S.: Advocacy today. Bull. Univ. Named After O.E. Kutafin **12**, 24–58 (2017)
15. Regulations on the procedure for passing the qualification exam for the status of lawyer (approved by the Council of the Federal chamber of lawyers 25.04.2003 (Protocol N 2)) (edited on 17.04.2019) (2019). http://www.consultant.ru/document/cons_doc_LAW_192781/. Accessed 20 June 2020
16. Ruling of the Supreme Court of the Russian Federation No. 306-ES18-10093(2) from June 11, 2020 (2020). https://kad.arbitr.ru/Document/Pdf/970c929e-7ab3-499c-8071-5054d65fcefc/3e8d5ed3-e96d-4346-8ee0-605b56135c0b/A55-10923-2017_20200611_Opredelenie.pdf?isAddStamp=True. Accessed 20 June 2020
17. Talapina, E.: Law and digitalization: new challenges and prospects. J. Russ. Law **2**(254), 5–17 (2018)

Managers from Networks: A Hymn to Humanism or the Pathos of Technocratism?

V. V. Kaftan[1] and Yu. A. Chernavin[2(✉)]

[1] Financial University under the Government of the Russian Federation, Moscow, Russia
kaftanvit@mail.ru
[2] State University of Management, Moscow, Russia
uchernavin@yandex.ru

Abstract. The article discusses the issue of radical transformations of modern management systems associated with the emerging of new digital technologies. The information space that has emerged as a result of their development, becomes a relatively independent managerial factor affecting the employee. Its implementation is considered in the framework of the two main trends opposition in the development of modern world civilization, which are humanism and technocracism. They exist based on the principle of complementarity and affirm the significance of different social and spiritual characteristics of the individual. In the course of cyberspace impact on the individual, these characteristics demonstrate both risks and threats, as well as the advantages of the development and functioning of the individual as an element of an effective management system. Technologies of virtualization, communications, network technologies, and mediatization of reality are analyzed as impact mechanisms. Based on the specifics of embedding certain technologies in the management process, the necessity and ways of accounting for them in personnel management are justified.

Keywords: Communication · Humanism · Management · Mediatization · Technocracism · Virtualization

1 Introduction

In modern management theory and practice, changes are taking place. These changes are comparable in depth to the post-war "managers revolution". The popular concept of the second half of the twentieth century is based on the importance of the figure of the "enlightened manager", who can replace not only the owner in the production management, but also displace the political forces, ensuring the change of elites and thus being able to play a significant, if not decisive role in historical development. This comparison is quite appropriate, taking into consideration a couple of issues.

Along with real managers of various levels, a virtual "manager from the network" has been formed in the information society. Its influence affects all social levels and all areas of society. Without solving specific tasks of the enterprise, the virtual manager "takes part" in its life, having a person and his community as the main management

S. I. Ashmarina and V. V. Mantulenko (Eds.): IPM 2020, LNNS 161, pp. 305–313, 2021.
https://doi.org/10.1007/978-3-030-60926-9_39

object. Using the concepts of "subject and object of management" when analyzing the impact of the information space on the activities and consciousness of people, management processes in organizations, the authors consider them in the wide sense, somewhat different from the established one in the management theory.

The new management factor is, in fact, the information space (cyberspace) that has been developed on the basis of digital technologies. During the implementation, this factor appears as a force, which is not always possible to identify in a particular individual or a group of people (often impossible), since network management centers arise spontaneously in response to the current situation in life or the network. This determines their anonymity and diffuse nature. Moreover, managerial influences often appear not as a purposeful influence of a particular subject, but as the result of communication interaction of many social subjects which exchange and create knowledge. Finally, the content of the transmitted knowledge and its meaning, as well as the constantly changing structure of information flows, algorithms for data processing and submitting, and rules for network communication, perform a guiding influence.

At the same time, the cyberspace controlling influence should not be considered as a primarily external factor, foreign to man. The electronic environment is not just created by a person, it has become his environment. This has led to the emerging of the phenomenon of "information man" and a new digital (network) generation of people [23]. Despite the social diversity and diffuse nature, the management factors of the information space comprise an effective and constantly operating management subsystem of spontaneous regulation, which, by analogy with the "invisible hand of the market", can be designated by the vague phrase "Manager from the network".

The emerging of a network digital space is a characteristic of a new information stage in the society development, so the analysis of the managerial impact of the "manager from the network" should be carried out based on an understanding of more general issues. At least this is the ratio of humanism and technocracism as civilizational characteristics of the modern world, their connection with the spirituality and way of human thinking that relies on knowledge and information in his activities and appears as a *subject of social existence.* The views formed during these aspects considering comprise the conceptual basis of the approach to the problem. Purpose: substantiation and disclosure of directions and methods of regulatory impact of the information space on a person and their accounting in personnel management.

2 Methodology

Arguments on the nature and direction of the cyberspace management influence on the consciousness and people activities are based on the opposition of humanism and technocracism as the defining trends of modern history. The corresponding provisions formulated by humanism theorists, in particular Heidegger [9], Blackham [2], Praag [16], on the one hand, the principles of technological determinism of Castels [4, 5], Toffler [21], on the other hand, served as the basis for the conclusion on the contradictory coexistence of these phenomena based on the complementarity principle. Mechanisms of information interaction between subjects, considered from a social and cultural point of view, are analyzed based on the theory of communicative rationality of

Habermas [8], the postmodern ideas of Baudrillard [1]. The attempt to apply the resulting inferred knowledge to the practice of personnel management is based on modern approaches that have developed in management theory and consider the flexibility and adaptability of the organization to constant changes in the external environment as a key problem.

3 Results

The main dilemma of the modern information society is the choice of its further development between humanism and technocracism. In the first case, the course of history is seen as a process from aggressiveness to humanity. At the same time, the role of religion, which was once the main one in the process, has changed on today. "Religion can be humanistic or anti-humanistic…it can either promote respect for human dignity or undermine it" [15, p. 20]. The modern approach to humanism corresponds to the principle "man for society and society for man" [20]. In the case of technocracy, we are talking about the emergence and approval of the science and technology factor, according to the laws of which civilization and man develop in the future.

Modern scientific literature has developed concepts that reveal the essence of modern civilization not as humanistic, but as "machine" or man made civilization. Technology creates a one-dimensional quantitative life approach, activity, mindset, spiritual culture, and moral values. Heidegger writes about such thinking as "calculating" thinking, "escape from thinking", contrasting it with "comprehending thinking" [9].

The analysis of the essence and content of humanism and technocratism leads to the following conclusions. First, the implementation of any of these trends of development will lead society to the different, almost opposite results: in the first case, *a spiritualized man* will be established, in the second – *a one-dimensional man.* However, secondly, such different concepts have a common ideological and methodological (and historical) source. This is the rationalism of the modern age that ignors the integrity of man, idealizes the only one characteristic of the individual, one way of mastering intellectual and practical culture, based on *ratio* – mind. This defines their relationship and inseparability.

What are the conclusions of such reasonings? The need to reject humanism, which has failed expectations, and from technocratism, which is deforming civilization? A return to the Golden age of human history?

The role of humanistic ideals, value systems, and moral and worldview guides in history is unavoidable; without them and outside of them, human activity loses meaning and therefore becomes impossible. Another thing is that the content of modern democratic humanism should differ from its interpretation of the modern age period. But in any case, it would be rash to declare that the essence of modern civilization is humanism. *The reality is that a spiritualized person and a one-dimensional person are two ambivalent essences that represent a single global civilizational process of modernity.* Thus, it is important to move in the direction of their convergence, with the understanding that they will never merge.

What is the role of the information society in this movement? What are its impact technologies? What processes does it contribute to and hinder?

Virtualization technology is an implement process of ideal images of reality into the mass consciousness that construct and imitate reality itself. The unit of exformation that is actively broadcasted in cyberspace is the "simulacrum" (the term of Baudrillard). This is a symbolic copy (an empty symbol) that does not connected with its original. The virtualization results are the doubling of worlds, the coexistence of the real and virtual.

According to Baudrillard, an image that is not connected with reality is a factor that leads to negative outcomes [1]. First, a person who is immersed in a deceptive hyperreality is not interested in the meanings and fullness of knowledge. Baudrillard considers the so called "silent majority": "Messages are given to them, they only want some sign, they idolise the play of signs and stereotypes, they idolise any content so long as it resolves itself into a spectacular sequence" [1, p. 10]. Secondly, a long stay in cyberspace with its simulacrums generates (and expresses) the individual's desire to hide from the harsh reality, to avoid life decisions making, and to reduce responsibility. All of this offers wide opportunities to manipulate someone else's consciousness and will for the "Manager of the network" in the case of his interest. Of course, these characteristics contribute to the formation of a one-dimensional personality that is not closely associated with humanistic values.

At the same time, the virtualization phenomenon has other sides that are socially positive [6]. The main character of the virtual space created by the "information man" is the character itself. The digital self-image is filled with its own worldview positions, personal interests, attitudes, and dreams. Indeed, this is often a set of simulacrums, but in this case, you should overcome simulation skepticism. The approach contradicting Baudrillard believes that the ultimate goal of the process is not to replace reality with simulation, but, on the contrary, to replace simulation with reality [1]. Possible real situations are simulated so that you don't have to experience them in reality. Therefore, many simulations are preventive and prospective [10]. In this case, they do not exclude a person from the real world, but encourage him to work. A person "plays" in his mind possible options of professional actions and situations.

It should be emphasized that reflection, the creation of an individual whose author is the individual – are deeply humanistic characteristics of existence. The prominent theorist of modern humanism, Blackham [2] emphasized that a person who promotes these values, needs a "high technique of life", "wise self-government".

The individual's imagination, desire and habit of management, participation in decision-making are a powerful resource in the practice of managing an organization, should be taken into account and used. Unfortunately, the skills acquired during the virtualization do not teach you to work in a team, even, on the contrary, develop individualistic positions.

Digital communication technology is related to the amount of information received and transmitted in cyberspace, the ways it is presented, and the opportunities for cognition and thinking development that arise. Castels emphasizes that "communication and information have been fundamental sources of power and counter-power, of domination and social change… the fundamental battle … is the battle over the minds of the people" [4, p. 238].

Crushed by the information flow, not having time or not being able to process information by himself, the individual loses the ability to learn. Moreover, he does not become the creator of knowledge. The language is reduced to a minimum – to short informative posts in social networks, messages. There appears a "mosaic thinking or fragmentary consciousness", based on the principle of music videos, dosing information in small peaces that replace each other, often without any connection between each other. In these cases, *the person is required only partially (one-dimensionally),* depending on *what fragment of communication is needed today.*

In other words, digital communication creates the risk of dehumanization by reducing *communication* with increasing communication intensity. In our case, it is important to see differences between similar processes. Communication is a subject-object connection, where the subject transmits some information, and the object passively receives it. This process is unidirectional (although it involves feedback). Communication is a subject-to-subject relations in which there is no sender and receiver of messages, but there are partners in conversation and in joint activity. So, interaction is much richer in content than communication. This is a sociopsychological process that provides a personal perception of its participants. In the course of contact, not only information and knowledge is exchanged, but also personal attitudes, views, impressions, intelligence, experience, and energy of people.

Digital technologies and telematics turn people's social and everyday life into a kind of "single window", when everything that was previously done in the presence through communication is now done remotely within the framework of communication. Reducing the level of interaction (even with intensive communication) means social life dehumanizing in the form of weakening interpersonal and intergroup ties, loss of a sense of belonging, feeling of the elbow, dissatisfaction with the most important human needs – the need for interaction and the need for presence.

The social and cultural consequences of interpersonal interaction with digital communications replacing are most substantial in personnel management. The live dialogue, contact discussion and joint development of management decisions, necessary for the development of critical thinking are replaced by remote sitting at the monitor. A combination of digital technologies and direct communication in the course of interaction with the manager will help to organize a full-fledged (not one-sided) implementation of the employee's personal potentials.

Network Technology. If the cultural symbol of traditional society was printing and the printed word, today we are witnessing the formation and development of the "Internet galaxy" (Castells' term). The networks principle that describes it, increases the possibility and develops virtual communication skills, but at the same time leaves person without the real communication, practical skills, and blocks the development of the corresponding culture [22]. People are rapidly losing their traditional attachment to their place of residence and job, moving to "network individualism" and becoming part of global network structures [4, pp. 62–63].

Toffler had earlier written on the loss of attachment as an indicator of social dynamics, however, referring to the attachment to the organization, the firm. "A new type of organization individual is emerging in front of us – a person who, despite his many contacts, does not join any organization" [21, p. 170]. In this situation, the company does not become a "place of work" with traditional social and moral ties that are still characteristic of the older, non-digital generation. For the latter, the organization was often also a "home". The digital generation considers the company only as a social field for solving personal, professional, material problems, and self-satisfaction. In these conditions, the organization's management must take special steps to "link" the specialist, create motivation to work in this particular structure.

At the same time, the phenomenon of networks in its application to personnel management has certain prospects. We are talking about the formation of a "mediatized world" of a firm or organization. It is possible as a communicative network of interconnected actors, thematically "framed" by specific social practices [13, p. 19]. As we can see, such a world is dual: it is characterized by both specific for it intersubjective knowledge and meanings, reflected the company's values and represented in the network, and related to them activity institutions. As a result, it has a high managerial efficiency.

Mediatization of reality is a technology of interpenetration of social reality and mass media, with the help of which the mass media constructs a new reality in cyberspace, giving it value and ideological meanings to manage the mass consciousness. At first sight, network mediatization leads to an even greater opportunity for each individual to declare themselves, make private things public and influence society in that way. However, Habermas shows how the "public sphere", which arises as a process and result of communication, becomes "manipulative" in postmodern society, used by the market and power institutions to legitimize the existing social order [8, pp. 211–222].

While allowing the individual the possibility of self-actualization as an independent media, mediatization also contributes to the situation when *it seems* to a person that their positions and judgments are a real factor of change. The organizational media elite are "symbolic analysts" – all those who not only deal with information, but have the ability to manage communications and form public opinion: mass media of various levels, agents of influence, PR specialists, media activists, columnists, etc.

The activities account of many subjects of the information space indicates a serious risk of the reemergence of democracy into a model more similar to a dictatorship that manipulates the minds of citizens with the help of high-tech media practices, disguised by the free self-expression. As a result of purposeful mass media activity in cyberspace, it is possible for a person to form a social and political orientation that both promotes and hinders his effective professional activity.

Humanism is the basis for overcoming the structures that overwhelm a person. The prominent theorist in this sphere, Praag, emphasized: already at the level of initial, naturally occurring beliefs, a person receives a "sense of orientation and motivation" aimed at people, their relationship and cooperation with them [16]. Only on the basis of humanistic principles it is possible to cultivate moral maturity, the emergence of a system-forming worldview. Spiritual and moral values, embodying the people's experience and spirit, contribute to the excitement of benevolent and elevated emotions according to the law of harmony and systematization [3]. People who see themselves as an element of cultural and man made environment, at the same time staying in the "digital" house of existing, you will be able to preserve and increase the language of culture, the essence of communication, thinking ability and sayings, central communicative meanings.

4 Discussion

The problem of the dynamics of society and man [12], economy [19], politics [11], culture [14] due to the development of information and digital technologies over the last years represents one of the main directions of research in modern science. The concepts of post-industrial society form the theoretical and methodological basis of research in this area. There is an understanding of changes in the quality of social relations included in the context of virtual communications. The article examines the problem of joint development of society and individual in the digital world. The comparison of the chosen approaches to the problem and the achievements of other scientists made the authors evaluate the need to include in the text the issues of "artificial sociality", artificial intelligence (AI), big data databases, which are interfaced and can act as a single phenomenon in the course of managing influence on individual.

"Artificial sociality" refers to the fact that AI agents participate in social interactions as their active intermediaries or participants [17]. In other words, an effective communication partner of an individual becomes AI, which, being integrated with big data (a huge amount of systematized and interpreted information), is a single digital system. This system, based on preprogrammed algorithm (and to a certain extent transforming them), offers a person solutions that are developed on the basis of a large array of data, taking into account existing and existed options. It is important to emphasize: in some cases and under certain circumstances, these management decisions can be effective. However, the situation should not be absolutized. It can only be successful under standard, stereotyped initial conditions. In addition, the formation of a decision based on certain algorithms does not guarantee neither managerial inerrancy, nor justice, humanity, and mercy [18].

First of all, the use of AI in the framework of "artificial sociality" leads to the conclusion that the content of information technologies has changed – they have ceased to be the only means of activity and claim to the status of management entities that make decisions instead of humans. Secondly, the analysis of the big data problem leads to the question of "data processes" – a legal way to access monitoring and understanding the behavior of people who leave a "digital footprint" [24, p. 52]. Data processes have become a management principle already used in personnel management. However, both in the case of "artificial sociality" and in the case of data classification, the algorithm and the meaning of cyberspace influence on a person stops being a phenomenon of the "manager from the network". In the first case, it does not just affect decision-making, but claims to be something more, lowering the subjectivity of the person, directing his drift towards the characteristics that represent the person as a *management object*. In the second case, there is a danger that the desire for effective management can lead management to total control, surveillance and absolute deprivation of a person's privacy. As a result, both aspects, which are largely related to personnel management, at the same time require special research attention that goes beyond the article.

5 Conclusion

In any social innovations, both positive and negative consequences are naturally combined. It is important to understand which of them are ultimately able to dominate, and what to do to block negative processes. To solve a problem of this kind, we need a reference point, which remains a person – the creator and subject of the knowledge society.

The interaction of the human – the creator with the information space generated by him, forms a paradoxical (and at the same time natural) situation. On the one hand, by focusing on one's own values and striving for self-expression, a person establishes himself as a global subject, demonstrates independence, creativity, aspiration and management abilities. On the other hand, its relationships with real groups and organizations, with real life in general, are weakening [7]. Taking root in the virtual space, it turns out to be the most open system, risks losing subjectivity, and is subject to manipulation. Lack of critical thinking, "mosaic thinking" leads to the loss of meanings from the individual, including worldview meanings, makes it difficult to assimilate and develop knowledge. Increasing the "one-dimensionality" of individual threatens society with dehumanization, the main mechanism of which is alienation. *In the process of alienation, certain qualities, forces, practices, and results of human activity transferred to the information space are transformed into factors that oppose and subordinate it, transforming the individual from a subject to an object of influence.*

To resolve the paradox between the two sides of a person, it is important to correctly localize the determinants and ontologies of humanism and technocracism. In terms of spiritual culture and human characteristics, humanism is absolute; and its clashes with technocratism, the opposition of these trends – the fate of the economic, political, managerial, social spheres, each of which requires special steps to reduce the level of this opposition. We should also consider the values of humanism as a purpose, and the possibilities of technocracy as a tool.

Trust to the principles and ideas that make up the fundamental values of humanism is important to include in the foundation of modern management improving at all its levels - from global to industrial and organizational. At the same time, a new aspect of culture – the culture of digital infrastructure using and information space in the interests of developing creativity, creating knowledge, preserving national identity, spiritual values, and language can become the general basis for implementing management innovation that can ensure the effectiveness of management based on the development of human capital. Maybe, *digital or information culture.* People are aware of the significance of this kind of culture in the world. It's all about development and implementation.

References

1. Baudrillard, J.: In the Shadow of the Silent Majorities, or, the End of the Social. Semiotext (e), New York (2007)
2. Blackham, H.J.: Humanism. International Publication Service, New York (1976)

3. Cano-Rollmann, M., Hannigan, T.J., Mudambi, R.: Global innovation networks-organizations and people. J. Int. Manag. **24**(2), 87–92 (2018)
4. Castells, M.: Communication, power and counter-power in the network society. Int. J. Commun. **1**, 238–266 (2007)
5. Castells, M.: The Information Age: Economy, Society and Culture. HSE, Moscow (2000)
6. Gryaznova, E.V., Vladimirov, A.A., Maltceva, S.M., Goncharuk, A.G., Zanozin, N.V.: Problems of virtualization and internetization of social space. In: Popkova, E.G., Bruno, S.S. (eds.) The 21st Century from the Positions of Modern Science: Intellectual, Digital and Innovative Aspects. Lecture Notes in Networks and Systems, vol. 91, pp. 119–124. Springer, Cham (2020)
7. Gvozdikov, D.: Scholasticism for Instagram: on the digital anthropology of modernity. LOGOS. Philos. Lit. J. **29**(6), 1–20 (2019)
8. Habermas, J.: The Structural Transformation of the Public Sphere: An Inquiry into a Category of Bourgeois Society. MIT Press, Cambridge (1991)
9. Heidegger, M.: Detachment (2016). https://cameralabs.org/10512-otreshennost-martin-khajdegger-o-vychislyayushchem-myshlenii-i-osmyslyayushchem-razdume. Accessed 25 May 2020
10. Hinterwaldner, I.: Some reflections on computer simulation. Philos. J. **11**(4), 80–94 (2018)
11. Laube, S.: The adapted position: preparing political contents for a hybrid media environment. Media Cult. Soc. **42**(2), 155–171 (2020)
12. Mitrofanova, A.: The cyborg as the code of a new ontology. Political and epistemological aspects of hybrid bodies. LOGOS Philos. Lit. J. **28**(4), 109–128 (2018)
13. Nim, E.G.: Exploring the mediatization of society: the concept of mediatized worlds. Sociol. J. **23**(3), 8–25 (2017)
14. Pennacchia, M.: Theatre strikes back in the digital era: an interview with Stephan Wolfert. Multicult. Shakesp. **20**(1), 37–49 (2020)
15. Poole, R.: The true meaning of humanism: religion and human values. Philos. J. **12**(1), 17–33 (2019)
16. Praag, J.P.: Foundation of Humanism. Prometheus Books, Buffalo (1982)
17. Rezaev, A.V., Starikov, V.S., Tregubova, N.D.: Sociology in the age of 'artificial sociality': search of new-boses. Sociol. Stud. **2**, 3–12 (2020)
18. Romaniuk, R.S.: IoT – review of critical issues. Int. J. Electron. Telecommun. **64**(1), 95–102 (2018)
19. Rust, R.T.: The future of marketing. Int. J. Res. Mark. **37**(1), 15–26 (2020)
20. Subbotina, N.: The ilea of humanism as a factor of social development. Voprosy Filosofii **8**, 5–15 (2018)
21. Toffler, A.: Future Shock. AST, Moscow (2002)
22. Twenge, J.: iGen: Why Today's Super-Connected Kids are Growing Up Less Rebellions, More Tolerant, Less Happy – and Completely Unprepared for Adulthood – What That Means for the Rest of Us. Alria Boors, New York (2017)
23. Zhong, B.: Social consequences of internet civilization. Comput. Hum. Behav. **107**, 106308 (2020)
24. Zhuravleva, E.: Call technologies 'big data' for the modern social sciences and humanities. Voprosy Filosofii **9**, 50–59 (2018)

Financial Law and Financial Labor Market in Digital Economy

E. N. Konrad[1], E. V. Pokachalova[2], A. M. Tsirin[3], and Z. I. Khisamova[4](✉)

[1] Moscow State Institute of International Relations (University) of the Ministry of Foreign Affairs, Moscow, Russia
kaifp@mgimo.ru
[2] Saratov State Academy of Law, Saratov, Russia
pokinarod@mail.ru
[3] The Institute of Legislation and Comparative Law under the Government of the Russian Federation, Moscow, Russia
artemtsirine@yandex.ru
[4] Krasnodar University of the Ministry of Internal Affairs of Russia, Krasnodar, Russia
alise89@inbox.ru

Abstract. The processes of globalization and digitalization of the economy and society affect the change in production and financial relations and significantly transform it. Any transformations that occur in the economy have positive and negative consequences, which requires in-depth study and development of measures aimed at preventing the appearance of signs of negative forecasts and accelerating positive changes. The chapter reveals the features of transformation of financial law institutions and the financial system as a whole in the conditions of digitalization of the economy, examines the dynamics of development of the most important financial law institutions in the conditions of the digital economy formation and the personnel formation. The authors substantiate the need for special control by the state when introducing new digital innovations in the sphere of financial activities of the state and society. The authors pay special attention to the most important trends in the development of financial law in the aspects of legal regulation, legal doctrine and legal practice related to digital financial technologies.

Keywords: Crisis · Digital economy · International system · World politics

1 Introduction

The last decade has seen the emergence of new financial processes and products that had no previous analogues in the global financial industry. The introduction and awareness of the potential of such technologies as Blockchain, Big Data, AI, etc. predetermined a new era in the financial services market and the emergence of new business models [4]. This rapid exponential growth provoked the emergence of new social relationships that have a large number of complicated situations within the current system of legal regulation. First of all, it concerns the financial sector. The

S. I. Ashmarina and V. V. Mantulenko (Eds.): IPM 2020, LNNS 161, pp. 314–320, 2021.
https://doi.org/10.1007/978-3-030-60926-9_40

development of digitalization in this area is experiencing certain difficulties due to the different (and sometimes completely absent) legal regulation of this area in different states and the complex macroeconomic environment. Nevertheless, the digitalization of the financial sector is steadily increasing and is gaining more support and recognition, entering new territories and involving more and more people, but at the same time causing a qualitative transformation of the financial infrastructure.

2 Methodology

The research successfully applied methods of system analysis, dialectical and its derivatives, methods and principles of determinism, induction, deduction and hypothesis. The methods of functional and institutional classification were used to systematize the obtained data. Using the methods of statistical, retrospective, current and prospective analysis and synthesis, the authors summarized the current trends in the development of financial law and identified a number of issues that need detailed scientific analysis. Based on the conducted research, a new view on the development of labor relations in the modern financial system was proposed and the key issues of legislation development in this area were identified.

3 Results

State regulation of relations in which law and technology are mixed cannot be recognized as fully formed in any state. In many cases, the legal regulation of these areas of legal relations is complicated, insufficient or absent. It is obvious that numerous gaps and vacuums urgently require their proper legal regulation. According to experts, a kind of "digital inoculation" will be made to civil, labor, administrative, criminal and many other branches of law, as well as regulatory complexes that control public relations in the field of medicine, information and Informatization, etc. The legal adaptation of technological achievements in the interests of a person makes it necessary to enter a new stage in the development of legal doctrine.

Digitization is most closely associated with the financial relationship. Nowadays, a key feature of the development of financial legislation is the excessive dynamism of financial relations [13]. The rapid formation of their new models necessitates prompt legal regulation of financial relations novels, which is not always possible or timely due to various reasons. As a result, this leads to gaps and collisions in law enforcement. Nowadays, there are two different directions of financial law and its institutional component development that are fundamentally different in essence and content. The first direction can be described as a traditional one, within which there is a tendency to modernize the existing rich theoretical heritage of the soviet financial and legal science. The second direction can be described as revolutionary, since in fact it tends to completely revise the traditional approaches to the consideration of fundamental categories of financial law [5].

We, in turn, are supporters of the second approach, in which it seems appropriate to note the conversion of public and private. The new financial law is characterized by internal dissection, differentiation on relatively independent (autonomous), but interrelated parts, the links between the structural elements are stable and ensure the integrity of the industry. The system of new financial law is represented by the following structural elements: budget law, tax law, banking law, insurance law, currency law, investment law, stock law, legislation on protection of competition in the financial services market, legislation on the securities market, legislation on financial control and auditing, legislation on countering the laundering of criminally-obtained incomes [20]. While agreeing with the above structure of financial law, we note its dynamism, which causes its revision in the context of digital transformation. The subject of financial law now includes not only net financial relations. Financial and legal regulation covers a complex set of public relations, covering almost the entire spectrum of both public and private finances [5], as well as changes occurring in them under the influence of the development of the digital economy. Digital technologies, affecting the economy, also affect financial law (digital technologies → economy → financial law) [18].

Speaking about the impact of digitalization on the process of legal regulation, experts note that the effect of this factor (digitalization) can lead to both expansion and narrowing of the sphere of legal regulation, changing its depth and other parameters, in particular, the ratio of the spheres of legislative and sublegislative regulation, areas of private and public law regulation, and, last but not least, the transformation of labor relations in the financial sphere [12]. The impact of digitalization on the content and structure of financial law is manifested in two directions. Financial innovations, such as the mode of digital of financial assets, alternative ways to attract financial investment (crowdfunding, crowdlending), lead to the extension of the object of regulation and the creation of new regulatory mechanisms and new workplaces, cause the death of unclaimed in the new realities of mechanisms and institutions to reduce traditional professions [17]. A special element of financial law in the context of digitalization should be called the emergence of fundamentally new tools for regulation, supervision and administration, as well as new labor competencies – FinTech and RegTech [2].

The Russian Federation does not have a legal framework for financial and regulatory technologies. In the Russian literature, there is also no common understanding of both the terms FinTech and RegTech, and their content. “FinTech” can be understood as a separate technology, for example, banking and exchange technology, as well as a separate innovative branch of financial services providing [8, 19]. The concept of “RegTech” is also not fully formed today. We should support the opinion of some authors that “RegTech” is a set of information and legal technologies used to optimize processes, achieve and fulfill compliance with regulatory and supervisory requirements, risk management, i.e. in order to assist in the field of regulatory compliance [9, 10, 15]. The Bank of Russia report highlights the following types of RegTech compliance control technologies; identification technologies; transaction monitoring technologies; risk management technologies; and reporting technologies [3].

Today, there is no doubt that there is a need for legislative regulation of financial and regulatory technologies at the state level. The relevance of this thesis is dictated both by the dynamic transformations taking place in the financial and banking sectors, and by the activities of the Ministry of Finance and the Central Bank of the Russian Federation. However, today there is an attempt to find a balance between the rapid development and implementation of financial innovations and their legal regulation within the existing legal system, including in terms of minimizing potential risks. We should support the opinion that the transition of the country's economy to an innovative path involves the creation of a financial and legal mechanism coordinated with other branches of Russian law [11].

It should be taken into account that the spread and increasing introduction of financial technologies leads to an increasing dependence of the population, business and government on financial technologies. In particular, this is reflected in the increasing interconnectedness of previously independent actors. The most striking and recent example is the recently adopted PSD2 regulations in the European Union, which obliges banks to provide financial and technical companies with access to customer information, thereby officially introducing a wide list of new actors into the previously closed financial system [16].

Regulatory intervention is also required by FinTech companies themselves, which are gaining influence by increasing the number of clients and the list of offered services. Let the latter be more typical for Asian markets, however, taking into account, for example, the widely publicized plans to create their own cryptocurrency payment systems by Telegram and Facebook, in 2021, we can expect the appearance of their own FinTech giants in the western world.

Among the main difficulties associated with the legal regulation of new financial technologies are certain risks groups for clients, companies, employees, and financial stability. Risks for customers include customers' lack of understanding of the nature and purpose of the purchased goods and services, lack of consumer protection in a number of areas (for example, due to the lack of formal entrepreneur status for one of the parties to the actual consumer transaction), and poor protection of personal data. The risk of developing big data and highly personalized services may be an increase in inequality between customers, primarily in terms of price and available range of services. This development may lead to accidental or intentional discrimination in areas or on grounds previously unknown to society. A macroeconomic risk for consumers may be the weakening of competition in the new FinTech industries, when certain services (for example, mobile payments) will be controlled by several major players in the market due to the emergence of FinTech giants.

Business risks include legal uncertainty in a number of areas of use of new financial technologies, which may result in entrepreneurs and their organizations being unable to comply with the will of regulatory authorities in a timely and full manner, for example, on AML/CFT or taxation. However, employment issues require special attention in the context of digital transformation of the financial sector.

4 Discussion

Nowadays, there is no consensus on what the financial labor market expects in the future. According to many scientists, many financial competencies will be replaced by AI and other cross-cutting technologies. Others, on the contrary, point out that in the next years, the competence of financial employees will expand: they will have economic and technical skills [7]. Thus, it is expected that by 2022, banks that will develop their work based on the interaction of employees and programs will increase their revenue by 34% [1] without reducing the number of employees. In Europe, there will also be an increasing demand for financial professionals with legal knowledge and programming skills [7].

It is assumed that employees in the financial sector will be required to have "human" skills that AI will not be able to simulate: communication skills; the ability to give accurate expert information to the results that will be issued by the program; the ability to quickly and effectively learn how to use new technical and software solutions in business; the ability to find, evaluate and revaluate risks due to the introduction of new methods and technologies, changes in the environment of the organization [7].

According to the US bureau of labor statistics, demand for financial managers will increase by 16% by 2018. At the same time, it is expected that many traditional positions (cashiers, brokerage employees) will be replaced by AI [6]. For example, the consulting firm Opimas predicts that by 2030, automation and AI implementation will eliminate more than 400 thousand jobs in the financial sector [14]. At the same time, 66 thousand jobs will be created related to data analysis and programming.

Summarizing expert positions allows us to identify a number of new professions that clearly reflect the transformation of the financial system:

- FinTech headhunter or FinTech relations specialist - an employee of financial institutions that will find qualified technical specialists or negotiate with FinTech companies on creation and ensuring the operation of digital mechanisms in the activities of this financial institution,
- engineer for a self-managed finance,
- manager for sustainable use of funds, whose tasks are to build a business plan for a digital startup, building an investment strategy, etc.,
- financial analyst in the field of digital assets, who can accurately predict fluctuations in the market of digital financial assets and create profitable investment schemes,
- a specialist in personal data protection in the financial sector will ensure that the personal data of the organization's clients is used lawfully and in a way that does not cause dissatisfaction with customers from this use,
- a specialist in cooperation in the field of cybersecurity will ensure the creation of information security tools at the level of organizations groups, etc.

5 Conclusion

Summarizing the above, we believe that the legal regulation of financial and regulatory technologies should be comprehensive and take into account not only global economic interests, but also employment issues. Risk minimization should be based on "flexible "or "soft" regulation. Considering the dynamics of the processes, the most progressive approach is to introduce a regulatory sandbox, which in the vast majority of states is a certain regime that allows FinTech organizations to test their products and services under the direct control of the regulator for a certain time. This will allow us not only to evaluate the effectiveness of individual products or legislative strategies, but also to test new labor competencies that will form the basis of financial market personnel in the future.

Acknowledgements. Prepared within the framework of RFBR grant No. 18-29-16102 "Transformation of legal personality of participants in tax, budget and public banking relations in the context of digital economy development".

References

1. Accenture: Realizing the full value of AI in banking (2018). https://www.accenture.com/us-en/insights/banking/future-workforce-banking-survey. Accessed 19 June 2020
2. Arner, D., Barberis, J., Buckley, R.: The evolution of Fintech: a new post-crisis paradigm? University of Hong Kong Faculty of Law Research, Hong Kong (2016)
3. Bank of Russia: Report for public consultations. Issues and directions of development of regulatory and Supervisory technologies (RegTech and SupTech) in the financial market in Russia (2018). https://www.cbr.ru/Content/Document/File/50667/Consultation_Paper_181016.pdf. Accessed 19 June 2020
4. Berger, A.N.: The economic effects of technological progress: evidence from the banking industry. J. Money Credit Bank **35**(2), 141–176 (2003)
5. Bochkareva, E.A., Vershilo, T.A., Miroshnik, S.V., Kosarenko, N.N., Krokhina, Yu.A., Peshkova, H.V., Proshunin, M.M.: Actual problems of financial law. Justice, Moscow (2019)
6. CSU Global: The future of jobs in Finance (2020). https://csuglobal.edu/blog/the-future-of-jobs-in-finance. Accessed 19 June 2020
7. EY: The impact of wider integration of data analytics and automation on manpower in the Singapore financial services sector (2019). https://www.ibf.org.sg/newsroom/Documents/IBF%20MAS%20Data%20and%20Automation%20Study%202019%20Web.pdf. Accessed 17 June 2020
8. Gupta, R., Mejia, C., Kajikawa, Y.: Business, innovation and digital ecosystems landscape survey and knowledge cross sharing. Technol. Forecast. Soc. Chang. **147**, 100–109 (2019)
9. Manita, R., Elommal, N., Baudier, P., Hikkerova, L.: The digital transformation of external audit and its impact on corporate governance. Technol. Forecast. Soc. Chang. **150**, 119751 (2020)
10. Milian, E., Spinola, M., Carvalho, M.: Fintechs: a literature review and research Agenda. Electron. Commer. Res. Appl. **34**, 100833 (2019)
11. Moore, S.: Separate Fintech noise from reality (2017). https://www.gartner.com/smarterwithgartner/separate-fintech-noise-from-reality/. Accessed 19 July 2020
12. Nambisan, S., Wright, M., Feldman, M.: The digital transformation of innovation and entrepreneurship: progress, challenges and key themes. Res. Policy **48**(8) (2019). https://ideas.repec.org/a/eee/respol/v48y2019i88.html. Accessed 19 July 2020
13. Palmié, M., Wincent, J., Parida, V., Caglar, U.: The evolution of the financial technology ecosystem: an introduction and agenda for future research on disruptive innovations in ecosystems. Technol. Forecast. Soc. Chang. **151**, 119779 (2019)
14. Pierron, A.: Workforce of the future: transplanting technology skill sets to the capital markets (2019). http://www.opimas.com/research/472/detail/. Accessed 19 July 2020
15. Povetkina, Y.V., Ledneva, N.A.: "Fintech" and "Regtech": the boundaries of legal regulation. Law J. High. School Econ. **1**, 46–67 (2018)
16. Rathinam, F.X., Raja, A.V.: Law, regulation and institutions for financial development: evidence from India. Emerg. Mark. Rev. **11**(2), 106–118 (2010)

17. Shaikh, A.A., Glavee-Geo, R., Karjaluoto, H.: Exploring the nexus between financial sector reforms and the emergence of digital banking culture – evidences from a developing country. Res. Int. Bus. Financ. **42**, 1030–1039 (2017)
18. Sitnik, A.A.: Financial technologies: the concept and types. Actual Probl. Russ. Law **6**(103), 27–30 (2019)
19. Thakor, A.V.: Fintech and banking: what do we know? J. Financ. Intermediat. **41**, 100833 (2019)
20. Tosunyan, G.A., Vikulin, A.Y.: To the new financial law. Financ. Law **6**, 7–11 (2003)

Digital Transformation of the Labor Market: Values and Competences

V. V. Mantulenko(✉), A. S. Zotova, and A. E. Makhovikov

Samara State University of Economics, Samara, Russia
mantoulenko@mail.ru, azotova@mail.ru,
shentala_sseu@inbox.ru

Abstract. The impact of digitalization on the creation of new professions and the destruction (transformation) of former ones as well as required competencies of the future are the key issues discussed in the academic and labor environment at the moment. The purpose of this work is to analyze the existing forecasts of labor market development in the context of global digitalization, identify the most relevant competencies and justify the feasibility of applying an axiological (value) approach to the analysis of these aspects. The main methods used for solving research tasks are analysis, synthesis and generalization. The study revealed the uneven impact of digital technologies on changes in employment, identified the most popular competencies of the future, analyzed scenarios for possible development of the labor market, tried to link these scenarios with groups of competencies and skills in demand, and justify the importance of considering the value component in the study of the modern global labor market, design and implementation of professional training and retraining programs.

Keywords: Competencies · Development scenarios · Digitalization · Labor market · Professional training · Values

1 Introduction

Many scientists have argued that man and machine are in the process of competition in which machines ultimately will win and this will lead to massive unemployment. Historical examples of jobs that were "washed away" from the labor market by technological innovations seem to confirm this point of view. At the same time, many new jobs were created as a result of the same technological developments of the past [4]. It is obvious that, considering the opportunities and threats of digital technologies for creating and destroying jobs, respectively, it is important to distinguish between tasks that people regularly perform, but can be performed equally well or more efficiently by machines, and tasks that cannot be replaced by machine work, or new additional tasks that are created as a result of the introduction of new technologies.

In the New Professions Atlas, developed by the Agency for Strategic Initiatives and the SKOLKOVO International School of Management in 2014 [2], it is emphasized that digitalization is more crowding out low-skilled competencies (since automatic devices effectively perform heavy physical, routine operations) and contributes to the development and increase the importance of highly-qualified competencies (since machines cannot always replace creativity, professional skill, art, experience).

S. I. Ashmarina and V. V. Mantulenko (Eds.): IPM 2020, LNNS 161, pp. 321–328, 2021.
https://doi.org/10.1007/978-3-030-60926-9_41

Considering aging professions (before 2020 and after 2020), Atlas developers point out the computerization and informatization of our reality as the reasons for obsolescence. However, if we compare the forecasts made in 2014 regarding “endangered” professions with the real situation at the beginning of 2020, then it becomes obvious that these forecasts haven’t come true. “Outdated” intellectual professions still exist, change, transform, like our whole world facing global challenges, but they exist.

2 Methodology

The information base of our study was the scientific work of Russian and foreign researchers on the problems of labor division between machine and man [12], the geography of digital skills [15], the impact of digitalization on the transformation of the labor market [9], the search for a connection between labor automation and new competencies, the development of the Russian and international labor markets forces in the context of information and computerization [13]. General scientific methods of analysis, synthesis and generalization were used. The subject of analysis was not only the results of scientific research of the last 30 years, but also the reports of the World Economic Forum 2019–2020, documents developed by the Agency for Strategic Initiatives and SKOLKOVO International School of Management in 2014–2015, data from LinkedIn, PwC reports in Russia. Also, the authors tried to make the synthesis of various approaches in order to prove the value approach for the study of future competences both with trends and perspectives of global labor market development.

3 Results

In our opinion, the emphasis today should be shifted from the consideration of “professions of the future” to “competencies”, and, accordingly, the “values” of the future. That is why we are increasingly talking today not about narrow professional skills, but about sets, complexes of unique, non-specialized, but important for a career, skills. Skills that are meta-subject, system-forming competencies, the so-called “soft skills” (flexible skills). These are competencies that are closely related to personal qualities, attitudes and values (responsibility, performance, self-management), socially significant skills (building relationships, influence, teamwork, emotional intelligence), as well as leadership qualities (problem solving, systemic and critical thinking, etc.).

At the World Economic Forum in 2019, analysts predicted that 35% of the key competencies required would change. In particular, among the ten most demanded competencies were: the ability to solve complex problems, critical thinking, creativity. Kichatov and Borisov, business trainers at Utraining Agency (Samara), after analyzing seven more sources, grouped ten key soft skills of the future into three clusters (I myself, my team, companies):

1. Decision making (systemic decision, critical thinking and cognitive distortion, group decision making).

2. Emotional intelligence (assertive behavior, team and group conflict management, social influence).
3. Active learning and updating (adaptation to change, mentoring, knowledge management).

Lists of demanded soft skills according to the version of the World Economic Forum for 2018 and for 2022. Not very different. In a structured form, these are three clusters:

– diverse thinking for solving problems (complex, creative, innovative, original, critical, analysis of systems and development dynamics),
– another type of leadership (emotional intelligence, social influence),
– active learning [19].

In 2020, experts at the World Economic Forum reiterated that technical skills will continue to dominate the workplace of the future, with human skills and networking continuing to play an important role. The existing imbalance among those who acquire the necessary skills for the work of the future (especially in the gender aspect) was also separately noted [6]. Not every new profession requires hard technical skills, but every job that reappears in the labor market requires basic technical competencies, such as digital literacy, general media culture, and others. Three professional spaces highlighted in the World Economic Forum report "Jobs of the Future" are cloud, engineering and information clusters, which are also among the fastest growing. These areas require breakthrough technical skills such as artificial intelligence (AI), robotics or cloud computing. However, technologies such as, for example, artificial intelligence are already so widespread today that many positions in areas such as sales and marketing will require a basic understanding of AI technology.

In English terminology, these technical skills are called "disruptive" ("destructive"), which echoes the model of Christensen and Leslie about disruptive innovations, innovations that change the ratio of values in the markets [8]. These disruptive technical skills are in high demand around the world. Blockchain, cloud computing, analytic thinking, and artificial intelligence are some of the most sought-after technical skills we see on LinkedIn.

The analysis of employer requirements shows that positions that require more human-centered skills are equally important in our technologically and information-rich world. Relevant studies show that talent search specialists, positions that ensure the success of clients and assistants in social networks are also among the fastest growing professions today. These are professional roles that rely on more diverse sets of competencies, especially soft skills. Demand for soft skills is likely to continue growing. Creativity, persuasion and cooperation - today these are the most demanded soft skills that are almost impossible to automate, which means that if you have these skills, you will be even more valuable for organizations of the future.

The question arises, how does the concept of "4 worlds" from PwC radically change the concept of key competencies of the future? Considering the key trends (megatrends) that determine the development of the labor market until 2030, PwC in Russia identified the following: technological innovations and their further development, changes in demography, urbanization, transformation of the global economy, limited resources and change climate [14]. Like any environmental factors, these trends carry both opportunities and threats to the labor "landscape" of the future, and the line between them is not always obvious (Table 1).

Table 1. Opportunities and threats for the future labor market development.

Mega-trends	Opportunities		Threats
Technological innovations	Raise in life quality and labor productivity, personal self-realization trend	Conversions in the quality and quantity of jobs caused by automation, robotics, etc.	Threat to social stability, political and economic tension
Demographic changes	Change in existing business models and appearing of the new ones, change in life philosophy and values of the working population	Growth in life expectancy (at least working life); the need to learn new skills and work longer	Pressure on business, social institutions and the economy in view of an aging population; labor shortages in some areas, hence the need for additional automation
Urbanization speed	An opportunity to show talents and find your vocation in different parts of the world	Cities as influential drivers of job creation	The threat of extinction of rural areas and the widening of the network gap
Global economy transformation	Development of new entrepreneurship activities, investments attraction, educational systems improvement	Migration vs Labor resources mobility	Network breaks, welfare difference Increasing social instability and unemployment
Limited resources and climate change	The emergence of new jobs in areas such as the production of alternative energy. Development of new technologies, new products, recycling and the use of secondary resources	Traditional energetic sphere reorganization	Depletion and irreversible climate change

Source: authors based on [14].

Highlighting digital technology as a factor, experts talk about a trend that plays a leading role in the formation of all four scenarios for the development of the labor market. The PwC model of 4 worlds (yellow, red, green and blue) is based on 4 opposing forces: collectivism and individualism, integration and fragmentation. In our opinion, this model has something in common with the Riemann-Thomann model, which originally arose as a concept of personality types. Based on 4 basic human fears, the concept was further expanded, finalized, and today it is actively used in world management practice, in particular in the field of change management [16, 17]. The testing methodology developed on its basis allows one to identify the basic values of a person, to form effective teams and corporate culture on this basis.

The value of intimacy generates such human needs as, for example, interpersonal contacts, harmony, security; distance - independence, tranquility, individuality; constancy is determined by the desire for order, patterns, control; changes mean diversity, spontaneity, creativity. Depending on the expression of the main orientations, the corresponding needs, values, and "life philosophies" prevail and manifest themselves in interpersonal behavior. Similarly, certain struggles against crises and inconsistencies are associated with this.

It is the value component that formed the basis for identifying 4 worlds, measurements, and scenarios for the development of the working environment by 2030. The main value of the "yellow world" is a human, therefore his main characteristics are socially useful and socially oriented business, brands with high ethical principles and impeccable reputation, a large number of craftsmen and manufacturers, new guilds of workers. The value of human qualities and the individual's duty to society, as the core of this world, arise at the junction of values [opposing tendencies]: proximity [collectivism] and change [fragmentation].

The central element of the red world is innovation, which is inextricably linked to digital technology. The development of the working environment and business is characterized by competition for the attention of consumers, unlimited access to information and leverage, concentrated among the elite, the high demand for highly specialized services and niche products, flexibility and speed of reaction. Such ("red") space is generated by individualism [distance] and fragmentation [changes].

The green world of corporate concern is defined by collectivism [proximity] and integration [persistence]. This is a space of responsibility (towards nature, society) and trust (high moral standards of behavior regarding company employees and the world as a whole). The blue world or "kingdom of corporations" is created at the intersection of individualism [distance] and integration [constancy]. Individual needs obscure the ideas of social justice in this world. Human capabilities, work automation, data analysis, innovations - all this is maximally used to increase productivity in corporations, which strengthen their influence, surpassing even some states in this regard. Employees in the blue world live in relentless pressure. The employer divides the elite and other employees. Specialists with exceptional abilities are in demand.

Let us return to the question of how much this concept of 4 worlds is consistent with key competencies, which, according to experts of the World Economic Forum, will determine the landscape and directions of development of the future labor market. In our opinion, the key competencies of the future, the so-called soft skills, are therefore designated today as key, subject-specific, not attached to any professional field, since they will be applicable in all 4 dimensions (green, yellow, red and blue). In each of their 4 worlds, socially significant skills, valuable personal and leadership qualities will be in demand. Their relevance and significance for today and the future is determined by global processes of informatization, automation, robotization, that is, digitalization. Digital technologies are the basis of the innovative processes taking place today. This is what drives scientific and technological progress, determining the deep socio-economic changes in society. The key competencies of the future have become such in the "digital" context, where the skills to navigate information flows, analyze, highlight the main, essential, compare, make responsible decisions, generate unique ideas, etc. become extremely important.

4 Discussion

The digital revolution challenges many professions. This is not always due to the fact that the new technologies behind it seem to be direct substitutes for specialists in the performance of certain tasks. Some digital solutions, in fact, allow you to make changes to existing production lines, for example, by creating intelligent platforms that make intermediaries between a product (service) and a client obsolete. In addition, unlike many other technological developments of the past, digitalization is a general-purpose technology, that is, it can be adopted in a wide range of industries, including the services sector [7]. Along with the development of information and communication technologies (ICT) over the past few decades, many new tasks have been created, such as programming and data analysis, which are still difficult to automate [1]. This contributed to the emergence of whole new industries, in which millions of workers are currently working, performing tasks that did not previously exist.

It is conceptually important to note two aspects. Tasks that become obsolete due to the rapid development of digital technologies, as a rule, differ from newly created tasks, and different types of digital technologies can have a heterogeneous effect on the skills requirements of modern and future specialists. Robots, for example, usually compete directly with manual labor, which requires a low or medium skill level. On the other hand, new tasks that arise in connection with digitalization, as a rule, require a high level of complexity and relative accuracy. Since jobs are determined by tasks that must be completed, a change in demand for certain tasks can simultaneously lead to the creation and destruction of jobs with heterogeneous qualification requirements. For example, over the past decade, studies have shown an increase in the number of jobs for highly skilled and low-skilled workers, while the number of jobs for middle-skilled workers has declined [11]. This polarization can be explained by technological changes, but it is not clear what types of technologies have caused these effects of employment and whether technologies generalized into the digital category lead to similar effects.

In addition, the scientific literature has so far failed to link empirically digitalization investments with intra-company changes in employment between different groups of specialists. Most of the previous studies on this topic have identified the positive effects of innovation on employment at the firm level [3, 5, 10, 18]. However, different types of innovations, for example, product and process innovations, can have different effects, the net impact depends on the chosen innovative strategies and varies in different industries.

5 Conclusion

Digital technologies are constantly changing production methods and generating new products and services, require a better technological infrastructure (faster Internet connections, universal access to the global information network, cloud technologies, etc.) new competencies for the development of innovations, as well as skilled workers for their effective use. The fulfillment of these requirements in recent years has accelerated the transformation of the economy in the digital direction. However, this process does not occur without friction. Often, average indicators mask an increase in the employment of skilled labor and a decrease in the employment of unskilled labor. The result runs counter to the widespread assertion that workers are in a race against cars - at least in the short term and for technologically advanced, open economies with a large share of highly skilled personnel. The global talent war is likely to intensify even further, and companies will have to spend a lot of time and resources on recruiting and retaining valuable employees. In times of high demand and an increasingly mobile workforce, even well-placed organizations may face intense competition for their most talented workers in the future. Thus, from an economic point of view, it is extremely important to develop and apply tools that minimize the potentially negative effects of digitalization, while reinforcing its positive aspects.

As long as middle and unskilled workers can be trained in new skills that will enable them to perform new tasks arising in the context of digitalization, the promotion of vocational training and retraining programs can be very useful for those categories of workers who are affected by automation/Robotics, etc. directly, and in general for the entire economy. However, training and continuing education programs should not be regarded as a universal medicine, as physical or mental limitations may impede their effective use. It is necessary to develop new tools, new solutions, sometimes digital technologies themselves offer such solutions. Collaborative artificial intelligence, for example, can allow workers of medium and low skill to focus on tasks where they have a comparative advantage over machines.

It is important that the future free movement of talent is guaranteed in the world and that the hiring process is accompanied by low administrative costs. Training and continuing professional development are necessary to improve the fit between existing skills, competencies and changing job requirements. This includes training not only technical skills, but also soft skills, which are also necessary in the context of the digital transformation of the labor market. Since the ability to train and retrain personnel is positively correlated with educational attainment, governments should increase the attractiveness of higher education.

References

1. Acemoglu, D., Restrepo, P.: The race between man and machine: implications of technology for growth, factor shares and employment. Am. Econ. Rev. **108**, 1488–1542 (2017)
2. Agency for strategic initiatives: Atlas of new professions (2014). https://asi.ru/reports/16344/. Accessed 29 Apr 2020

3. Balsmeier, B., Delanote, J.: Employment growth heterogeneity under varying intellectual property rights regimes in European transition economies: young vs. mature innovators. J. Comparat. Econ. **43**(4), 1069–1084 (2015). https://doi.org/10.1016/j.jce.2014.10.002
4. Balsmeier, B., Woerterb, M.: Is this time different? How digitalization influences job creation and destruction. Res. Policy **48**(8) (2019). https://doi.org/10.1016/j.respol.2019.03.010
5. Blanchflower, D., Burgess, S.: New technology and jobs: comparative evidence from a two country study. Econ. Innov. New Technol. **5**(2–4), 109–138 (1998)
6. Blue, A.: 5 things we know about the jobs of the future. World Economic Forum (2020). https://www.weforum.org/agenda/2020/01/future-jobs-and-skills-in-demand/. Accessed 29 Apr 2020
7. Brynjolfsson, E., McAfee, A.: The Second Machine Age: Work, Progress, and Prosperity in a Time of Brilliant Technologies. W.W. Northon Company Inc., New York (2018)
8. Christensen, C.M., Leslie, D.: The Innovator's Dilemma: When New Technologies Cause Great Firms to Fail (Management of Innovation and Change). Harvard Business Review Press, Boston (2013)
9. Colombo, E., Mercorio, F., Mezzanzanica, M.: AI meets labor market: exploring the link between automation and skills. Inf. Econ. Policy **47**, 27–37 (2019). https://doi.org/10.1016/j.infoecopol.2019.05.003
10. Entorf, H., Pohlmeier, W.: Employment, innovation and export activity. In: Florens J., Ivaldi M., Laffont J.J., Laisney F. (eds.) Micro Econometrics: Surveys and Applications, pp. 394–415. Basil Blackwell, Oxford (1990)
11. Goos, M., Manning, A., Salomons, A.: Job polarization in Europe. Am. Econ. Rev. **99**(2), 58–63 (2009)
12. Hammershøj, L.G.: The new division of labor between human and machine and its educational implications. Technol. Soc. **59** (2019). https://doi.org/10.1016/j.techsoc.2019.05.006
13. Putilov, A.V., Bugaenko, M.V., Timokhin, D.V.: Development of Russian labor market in the context of informatization and computerization of the economy. Proc. Comput. Sci. **145**, 169–176 (2018). https://doi.org/10.1016/j.procs.2018.11.035
14. PwC in Russia: Workforce of the future. The competing forces shaping 2030 (2018). https://www.pwc.ru/ru/publications/the-future-of-the-labour-market.html. Accessed 29 Apr 2020
15. Richardson, L., Bissell, D.: Geographies of digital skill. Geoforum **99**, 278–286 (2019). https://doi.org/10.1016/j.geoforum.2017.09.014
16. Riemann, F.: Grundformen der Angst. Verlag Ernst Reinhardt, München (1961)
17. Thomann, Ch., Schulz von Thun, F.: Klärungshilfe 1: Handbuch für Therapeuten, Gesprächshelfer und Moderatoren in schwierigen Gesprächen. Rororo-Verlag, Hamburg (1988)
18. Van Reenen, J.: Employment and technological innovation: evidence from U.K. Manufacturing firms. J. Labor Econ. **15**(2), 255–284 (1997)
19. Your training: Soft skills (2019). https://utraining.ru/2019/11/13/soft-skills-2/. Accessed 29 Apr 2020

Digital Economy and the Development of Women's Entrepreneurship: A Regional Aspect

A. V. Mikhaylova(✉) and L. N. Popova

North-Eastern Federal University, Yakutsk, Russia
mikanya23@gmail.com, ludaykt@rambler.ru

Abstract. The article investigates problems of women's entrepreneurship in Repblic of Sakha (Yakutia) in the conditions of digital and creative economy. There are main projects in the field of entrepreneurship and the digital economy have been launched in Russia, aimed at the comprehensive solution of problems in the field of supporting the infrastructure of entrepreneurship and creating conditions for supporting information and communication technologies, digital tools and methods of automating routine processes in the small business market. A sociological research was conducted in March-May 2020, in the largest constituent entity of Russia to study features of women's entrepreneurship and develop recommendations on forms of support for small women's entrepreneurship based on a synthesis of existing projects and programs. Research methods were system analysis, comparative analysis, questionnaires and structural analysis. 300 women-entrepreneurs in regional labor market took part in the survey. A portrait of women's entrepreneurship in the region was revealed. Digital infrastructure is a factor for a successful start in the region. The authors focus on the practical recommendations for the development of women's entrepreneurship in the framework of national projects and digital economy.

Keywords: Digital economy · Digitalization, women's entrepreneurship · Regional labor market · Small and medium-sized enterprises · Women's entrepreneurship

1 Introduction

At the present time the digital economy is a phenomenon that dictates the paradigm of development of the new economy in our society. The year 2020 in the Russian Federation is held under the auspices of entrepreneurship. Entrepreneurship in the digital economy gets a new impetus for its development. The national projects to support local business and digital economy have been developed at the initiative of the President of the Russian Federation in regions. The largest region is Yakutia. The Republic of Sakha (Yakutia) has focused on three directions in the digital economy for developing small business. Thus, the task has been set to develop new projects in support of small and medium-sized enterprises, in support of regional participants and business market entities.

S. I. Ashmarina and V. V. Mantulenko (Eds.): IPM 2020, LNNS 161, pp. 329–335, 2021.
https://doi.org/10.1007/978-3-030-60926-9_42

The authors focus on market changes in the context of digitalization and the trends of increasing the number of women in business and entrepreneurship. In our opinion, the explanation is Internet access [1, 4, 12–14]. The task of national projects is to make a breakthrough in the country's development. In the digital economy it is a synergistic effect that arises from the merger of different technologies. The Age of Digitalization is equality of opportunity. The digital economy opens up many opportunities for women-s entrepreneurial initiatives. At the same time, we agree with the researcher Gallyamov notes "There are a number of barriers for the development of women's entrepreneurship: political, legal, ideological, socio-cultural and economic" [4, p. 56]. Another obstacle is salary. Pokhvoshchev and Kolesnikova note that "women's entrepreneurship is concentrated in the field where the salary level is lower" [12, p. 103]. However, this circumstance is solved by creating added surplus-value due to the innovation of the product and service. Many women are attracted by the fact of being at home with children and doing any business, additionally earning income.

The main aim of our study is to understand peculiarities of regional women's entrepreneurship and form recommendations to support small women's entrepreneurship based on the synthesis of existing projects and programs in digital economy. In the digital economy, the transparence is an indication, which is inherent in the female type of management. The characteristics of the female type management are openness, communication, love for the client and for employees. Women have it by nature. Also they have the soft skill – creativity.

2 Methodology

The research methodology is based on readings of women's entrepreneurship and business infrastructure. The digital market supports it. Sahakyan drew conclusions in her study about the expansion of remote employment in the creative economy [13]. It is necessary to enhance the processes of digitalization and legislation of remote employment. Entrepreneurship is one of the fields where flexibly responds to changes in the digital economy, having a multiplier effect. We agree that a culture of new mobility has appeared in the era of information and digital technologies development due to the digital economy. The content of labor is influenced by the digital infrastructure and changes of social reproduction. What is common is that in the digital age entrepreneurs use unique types of data that cannot be obtained in other way (likes, geolocations, etc.).

Fedorova and Zueva revealed the relationship between the impact of women's and men's employment on the labor market. In business men occupy more leadership positions than women [2]. But in general, researchers-practitioners Gallyamov notes the wide prevalence of gender stereotypes across the country regarding women-entrepreneurs [4]. However, the effective implementation of these areas is impossible without the introduction of information technologies into the activities of business entities in the context of digital globalization.

We conducted a sociological research in the Republic of Sakha (Yakutia) in March-May 2020 in order to study the entrepreneurial activity of women: the culture of entrepreneurship, the entrepreneurial ecosystem and personal qualities.

300 women-entrepreneurs took part in the research. We made a stratified sample based on experience in the field of entrepreneurship and registration as an individual entrepreneur. An online survey of business women in the group of the Yakut business community "Business club" was conducted.

3 Results

The largest number of survey participants is represented in the age of 26 to 35 (32% of respondents): of 36 to 50 (31% of respondents), of 18 to 25 (17% of respondents) and of 50 years old (20% of respondents) (Table 1). By education level 78% have higher education, 20% secondary special and only 2% full secondary. The survey also revealed that 76% of women are married, 15% are single, and 9% answered civil marriage. The majority of women 49% answered that they have 2 or more children, 34% answered 1 child and 17% do not have children. The majority of respondents answered that their company was located in shopping centers-57%, in the building of a non-governmental institution, non-profit organization (preferential rent) 26%, work from home-28%. Processing of questionnaires showed that 57% of women are Sakha (Yakut) by nationality, 33% are Russian, and 10% of respondents indicated a different nationality. The research revealed that the majority (31%) of surveyed women opened their own business within the last year. We can explain this by the fact that there are a lot of successful examples in social networks, state support for women with children and maternity capital which allow women to think about their own business.

Table 1. Distribution by the age, experience of entrepreneur's activity

The experience of entrepreneurial activity	Age			
	18–25	26–35	36–50	Over 50
Less than 12 months	13%	50%	28%	9%
From 12 months to 35 months	20%	25%	31%	24%
From 36 months до 60 months	13%	23%	38%	26%
Over 60 months	22%	25%	29%	24%

Source: authors.

The majority (49%) of respondents are supported by their families, however it is alarming that about 20% of families do not support them. Women see support in the family, as well as a source of inspiration. We have noticed an interesting trend that many women-entrepreneurs have the second half who is also engaged in business or have a different permanent source of income. Women-entrepreneurs in Yakutia are supported by friends and we note that women-entrepreneurs feel the energy for an active and creative activity with the support of their family and friends. We also studied the psychological climate in women-entrepreneurs' teams. More than half of 39% have a comfortable, friendly environment in the team – and only 1% of respondents said that employees constantly conflict with each other. Family is a source of inspiration, as well as a reference point for studying pressure points. Knowing pressure points, you understand what types of activities and ideas will be supported by clients. The distribution of respondents by employment is the following: about 60% of all women surveyed are employed in trade and services, other 15% are engaged in (blogging on social networks, running marathons on various topics, consulting support, etc.).

We also studied the leisure time of women who are engaged in business. Women-entrepreneurs of Yakutia lead an active lifestyle in their free time: 31% are active outside of business (healthy lifestyle, active leisure); 20% are actively doing sports and trainings; 10% are passionate about favorite activity outside of business; 37% spend time at home with family and children; 2% are engaged in self-development, read a lot. To the question “What are three factors why you chose entrepreneurship?” the answers were distributed as follows. 45% - freedom, income, provide for the family; 35% - to have a moderate income, more free time, work for yourself; 10% - free schedule, profit growth, personal growth; 10% - the opportunity for self-development, potential realization and freedom of action.

Thus, we got a regional portrait of women-entrepreneurs in Yakutia. It is a young woman aged of 26–35 years, married, has two or more children and is engaged in trade or services. The main motivation is freedom of action, income generation and self-development. The research revealed that social networks are the main development channel for today. 85% of all respondents are constantly working to improve their professional skills, join business communities, communicate on digital platforms and engaged in self-development.

4 Discussion

The following authors Montuori and Donnelly [10], Norsun [11], Zabelina, Mayorova, Matveeva, and Safonov [15] note that there are many obstacles for doing business: legal, economic, administrative, rent payments, erosion of purchasing power, difficulties in lending, insufficiency of circulating funds and customer preferences. In the digital economy, women-entrepreneurs have opportunities for learning on online platforms and in business communities. The development of pre-school, school and additional education is important for the development of women’s entrepreneurship.

Fedorova and Zueva [2], Fryer and Fryer [3], Memon [9], and statistics in Russia show many negative aspects related to the development of women’s entrepreneurship, first of all, an adverse economic trend and effective demand. Gramescu, point out the presence of gender differentiation [5]. The strategic document is aimed at supporting women-entrepreneurs in Russia and in the regions. Support measures of women-entrepreneurs have been developed at the state level. Our study shows the relationship between women’s entrepreneurship and creativity. Holford indicates purely ‘efficient’ and analytical approaches fail to recognize the unique and inimitable characteristics of human creativity and its associated tacit knowledge [7]. Lee and Chen analyzed many sources about relationship between digitalization and creativity and its influence on business [8], determinants of entrepreneurial intention the extent of digitalization of the economy. Greffe demonstrates how cultural economics and creative economics should merge for their mutual benefits [6]. Creativity, as the major skill in SMART-economy, has become increasingly important. Results suggested that creativity was relatively domain general rather than domain specific.

Fryer and Fryer explore reasons why attempts to make cross-cultural comparisons of creativity are fraught with difficulties [3]. Technological turbulence, green recruitment and selection and green training, involvement and development are positive predictors of green team creativity. We found out in the conditions of digital economy novel insights that extend traditional HRM conceptualizations to reflect a more environmentally sustainable GHRM framework.

There is a correlation between efficiency and entrepreneurial creative entrepreneurship. Many researches point out creativity's central role in the new economy that has also led to questioning about how creativity is being used. All these characteristics are especially important in the digital economy. Thus in addressing the basic research question of why people who no longer need to work together apparently choose to do so (for at least part of their working lives). Gramescu analyses building a European network of incubators for social innovation and correlation links [5]. Memon explored the importance of network science in the area entrepreneurship [9]. One of the major uses of theory network science is SNA, which is not known as a traditional data mining technique, but can be used to discover most important actors in a network. The digital economy has substantially reduced the costs of search due to the role of digital platforms. Digitization has greatly increased the value of innovation and the need for intellectual property protection.

The proportion of women in the region who want to start their own business is growing steadily, but at the same time, insufficient support restricts women's initiative. Such support becomes important in times of financial crisis, when women go to work to maintain a family budget. Favorable conditions for doing business encourage women to quickly adapt to entrepreneurship, but not all countries have sufficient conditions for doing business. Women face problems of lack of knowledge, finance, information, high tax rates and administrative barriers and overcome them much harder than men. The digital ecosystem in the Republic of Sakha (Yakutia) makes prerequisites to the successful start in business. It is in the social field, preschool education field, additional education, in culture and health care, information and consulting services in the field of jurisprudence, bookkeeping, psychology, in fashion and socially useful services.

5 Conclusion

Based on our research, it is possible to build in digital economy model of women management that will allow you to form an adaptive management plan for the entrepreneurial program. Our recommendations in the digital economy may provide specific advantages to develop women's entrepreneurship in the digital economy. Social and creative entrepreneurship is a characteristic feature of women's entrepreneurship in the region. There is involvement in the field of social entrepreneurship, awareness of their usefulness to society and in the use of social networks to actively promote their ideas. Our recommendations cover the following areas: 1) it is necessary to establish the concept of women's entrepreneurship at the legislative level, 2) develop approaches to support women's entrepreneurship, 3) prepare methodological recommendations on definition of criteria for women's entrepreneurship projects, 4) develop criteria for estimating support for emerging women-entrepreneurs, 5) create specialized financial and credit products for women's entrepreneurship, 6) create a system of personalized support for women-entrepreneurs, providing them with advice on controversial issues, the possibility of free account opening and three-month free account maintenance, 7) create digital infrastructures to support the developed social infrastructure in the regions – it is a necessary condition for the development and support of women, 8) carry out the popularization and replication of successful business projects of women-entrepreneurs from various fields in the digital environment, 9) develop and implement special educational programs for current and emerging women-entrepreneurs, 10) use CRM systems and everything related to working with clients on digital platforms. Now the presence of a CRM-system is what should be in any company, clients are recorded and segmented. The client's portrait is clearly segmented.

The digital economy creates favorable conditions for the start-up of emerging women-entrepreneurs with their own characteristics. It is important to use digital tools and technologies to implement ideas and endeavors. Digital globalization is being actively implemented in all areas of the economy. Therefore, digital tools should become the foundation of the digital ecosystem of entrepreneurship. Digital tools are used for making managerial decisions, customer interaction, business valuation, prolongation of contracts, etc. Digital tools analyze business, search for customers, and most importantly are the support in solving routine operations and actions of entrepreneurs are integrated with the labor market.

References

1. Chetyrova, L.B.: Labor and education: from labor society to digital. Bull. Samara Univ. Hist. Pedagogy Philol. **24**(4), 85–90 (2018)
2. Fedorova, O.I., Zueva, E.G.: Accumulation and realization of human capital in new conditions: opportunities and limitations. Creative Econ. **12**(10), 1649–1660 (2018)
3. Fryer, M., Fryer, C.: Regional creativity: cross-cultural differences in creativity. In: Pritzker, S., Runco, M. (eds.) Encyclopedia of Creativity, pp. 409–418. Academic Press, Cambridge (2020)
4. Gallyamov, R.R.: Characteristics of women's entrepreneurship in modern Russia. Econ. Manag. Sci. Pract. J. **3**, 56–67 (2016)
5. Gramescu, L.: Scaling social innovation in Europe: an overview of social enterprise readiness. Procedia Soc. Behav. Sci. **221**, 218–225 (2016)
6. Greffe, X.: From culture to creativity and the creative economy: a new agenda for cultural economics. City Cult. Soc. **7**(2), 71–74 (2016)
7. Holford, W.D.: The future of human creative knowledge work within the digital economy. Futures **105**, 143–154 (2019)
8. Lee, M.R., Chen, T.T.: Digital creativity: research themes and framework. Comput. Hum. Behav. **42**, 12–19 (2015)
9. Memon, N.: Entrepreneurship in the eyes of network science. Procedia Comput. Sci. **91**, 935–943 (2016)
10. Montuori, A., Donnelly, G.: Creativity and the future. In: Pritzker, S., Runco, M. (eds.) Encyclopedia of Creativity, pp. 250–257. Academic Press, Cambridge (2020)

11. Norsun, T.A.: The development of women's entrepreneurship in Russia at the present stage. Bull. Zabaykaliskiy Univ. **9**, 14–21 (2016)
12. Pokhvoshchev, A.V., Kolesnikova, O.A.: The development of women's entrepreneurship as a factor for ensuring effective employment. Mod. Innov. Dev. **1**(21), 103–107 (2015)
13. Sahakyan, A.P.: Analysis of the transformation of state financial support for small and medium-sized enterprises in the context of digitalization. Innov. Invest. **1**, 11–23 (2020)
14. Tóth, Z., Nieroda, M.E., Koles, B.: Becoming a more attractive supplier by management references - the case of small and medium-sized enterprises in a digital enhanced business environment. Ind. Mark. Manag. **84**, 312–327 (2020)
15. Zabelina, O.V., Mayorova, A.V., Matveeva, E.A., Safonov, A.S.: New forms of employment in the Russian economy as an object of state regulation. Labour Econ. **2**, 15–27 (2018)

Personal and Collective Effectiveness of Merchandisers in the Digital Economy

D. V. Ralyk(✉)

Samara State University of Economics, Samara, Russia
dinarar@inbox.ru

Abstract. This study examines issues of evaluating the quality of a merchandiser's work in the physical and digital retail space. This presentation of the research results is determined by the transformation of the retail economic environment. The evaluation criteria and indicators are compared taking into account differences in shopping conditions. Individual effectiveness of a merchandiser is related to the organization of sales and promotion of a specific brand, SKU, or product category. Collective efficacy of merchandisers is considered in the whole store, assuming the measurement of sales and profit, as well as changing consumers' perception of the atmosphere of retail space and evaluation component of merchandising services. Formula of indicators' calculation for individual and collective performance of merchandisers are proposed.

Keywords: Merchandising · Performance criteria · Retail · Trading space

1 Introduction

The digital transformation of the economy has significantly affected its key sector – retail. The development of omnichannel sales, the launch of new work formats, and the application of new information technologies in the commercial process in a complex led to an update of requirements to functional responsibilities of employees. From the point of view of this research, the most interesting is the analysis of new competencies of merchandisers, which can be used as the basis for evaluating the quality of their work. The criteria for evaluating the personal and collective effectiveness of merchandisers include both general indicators of employees in the retail sector, and specific ones related to the rational organization of both physical and virtual retail space.

2 Methodology

The approach we are developing to evaluate the effectiveness of a merchandiser's work is based on the methodology for evaluating a merchandising service proposed by Alekseev [1]. According to this method, it is possible to assess the degree of fulfillment of obligations for ensuring positive dynamics of characteristics of such components as availability of goods, ease of finding them, completeness of information about them, etc.

S. I. Ashmarina and V. V. Mantulenko (Eds.): IPM 2020, LNNS 161, pp. 336–342, 2021.
https://doi.org/10.1007/978-3-030-60926-9_43

When determining the effectiveness of the organization of retail space, indicators of the use of retail space (coefficients of exhibition and installation space) and assortment management (coefficients of completeness and stability of the assortment) are traditionally used in Russian science and practice [6]. The set of tasks which a merchandiser has to solve should comply with SMART goal-setting principles, i.e. assume relevance, achievability, measurable results, etc.

In general, the measurement of results of the merchandiser's work can be performed without his direct presence. Expert assessment methods allow you to assess the actual state of the physical or virtual (Internet) retail space. Experts in the study of the effectiveness of merchandising can be specialists in the field of trade, related to a single product chain. However, the evaluation of the components of a merchandising service should be carried out by the consumers of this service, i.e. the store's customers.

If changes are made to the construction of the retail space, it is recommended to cover a relatively short period as a time interval that demonstrates the effectiveness of merchandisers. In our opinion, observations should be made during the month before the changes and during the month after the implemented changes, recording changes in the consumer behavior and ratings.

Collective efficacy of merchandisers may be expressed primarily in the sales and profits of the company, and to assess the general level of comfort in the shop and on the adjoining territory. In this case, it is advisable to conduct surveys of both retail professionals and store clients themselves.

3 Results

In assessing the quality of merchandising in the retail sector, general work results of the whole team and individual employees should be considered as a basis. These specialists organize the promotion and sale of a particular brand, Stock Keeping Unit (SKU), product groups or an entire product category. Taking this into account, the area of responsibility and the corresponding bonus indicators are determined.

From the point of view of marketing and psychology of buyers, merchandising forms the clients' satisfaction with the quality of trade services for providing necessary goods, promoting them in the retail space and attracting attention to them. Accordingly, the issues of evaluating the effectiveness of merchandiser's work can be considered from the perspective of achieving marketing objectives. For example, we recommend evaluating the customers' satisfaction with the components of a merchandising service.

Logistics guidelines in merchandising assume optimization of flow processes in a retail business, ensuring the availability of the desired product of the required quality in sufficient quantity, in a timely manner, in the right place, while observing the principle of reducing the overall cost of conducting business. From a logistics point of view, the effectiveness of a merchandiser can be determined by cost savings from optimizing flow processes. In other words, it means the quality of managing product, information, service, and customer flows in the store. Attempts to compare the difference in the performance of merchandisers in the physical and digital environment are presented in Table 1.

Table 1. Comparison of traditional and progressive criteria for individual effectiveness of a merchandiser

The rules of merchandising	Criteria and performance indicators of a merchandiser		Maximum of indicator
	Physical retail space	The digital environment of the store	
MML rule (Minimal Must List)	Coefficient of Minimal Assortment, C_{ma} $C_{ma} = SKU^{m}_{fact}/MML$ where SKU^{m}_{fact} – actual availability of SKU in MML MML - Minimal Must List		1
Inventory level rule Rule of presence	Coefficient Sustainable Assortment, Csa $C_{sa} = SKU_{fact}/SKU_{plan}$ where SKU_{fact} – actual availability of SKU in the store SKU_{plan} – planned quantity of SKU in the store		1
Rule for terms of storage and rotation of product on the shelf	Updating product layouts based on the following principle “First In First Out”, C_{chn} $C_{chn} = Check_{norm}/Check_{total}$* where $Check_{norm}$ – number of checks that confirmed compliance with the rule; $Check_{total}$ - total number of checks for the same period	Updating information on the website or in the mobile app - C_{act} $C_{act} = SKU_{act}/SKU_{total}$ where SKU_{act} – number of SKU with actual data; SKU_{total} – total number of SKU in the store	1
“Face to customer” rule	The share of products facing the customer in the total number of products displayed on the shelf, C_{chn} or C_{fp} $C_{fp} = SKU_{fp}/SKU_{total}$ where SKU_{fp} – number of SKU, with frontal position SKU_{total} – total number of SKU in the store	Quality of the product image that allows you to identify the brand, C_{qi} $C_{qi} = SKU_{cqi}/SKU_{total}$ where SKU_{cqi} – number of SKU, with correct image quality SKU_{total} – total number of SKU in the store	1
Rule of price tags	Availability and clarity of price, C_{chn}		1
Rle of optimality Point of Sale Materials (POSM)	Placement POSM	Promo tags in the electronic catalog	0,2
	maximum 20% of products, C_{chn} or C_{opt} $C_{opt} = SKU_{posm}/SKU_{total}$ where SKU_{posm} – number of SKU, for which POSM or promo tags are presented SKU_{total} – total number of SKU in the store		

*Formula C_{chn} is universal and can be used to check compliance with all merchandising rules.
Source: author.

The individual effectiveness of a merchandiser can be evaluated by an expert, such as a supervisor or a store manager. If an external consulting company is involved on the principles of outsourcing, its experts are provided with a system of benchmarks developed by the retailer. The closer the value of the coefficients shown in the table is to one, the higher the quality of the merchandiser's work is, regardless of whether the store operates offline or online. A comparison of estimates of the collective work of merchandisers to create a favorable physical and digital retail space is presented in Table 2.

Table 2. Comparison of traditional and progressive criteria for the collective effectiveness of store merchandisers

Object of evaluation in the store's merchandising	**Criteria and indicators of merchandising effectiveness**		**Maximum of indicator**
	Physical retail space	**The digital environment of the store**	
	The atmosphere of the interior and exterior	Online store or mobile app	10
Comfort of retail space	Coefficient of buyers' comfort, C_{comf} $C_{comf} = V_{comf}/V_{total}$ where V_{comf} – visitors who highly appreciate the comfort of staying in the store's retail space; V_{total} – total number of visitors polled in the store		
Components of merchandising service	Evaluating the quality of a merchandising service, Q_{ms} (10-points scale) $Q_{ms} = Q_{inform} + T_{search} + T_{cross} + P_{ident} + C_{access} + A_{disp}$ where Q_{inf} – volume and quality of product information; T_{search} – time to search for the desired product; T_{cross} – time to search for a related product; P_{ident} – matching the price tag to the offered product; C_{access} – ease of access to the product; A_{disp} – aesthetics of goods presentation		60
Use of the store's retail space	Coefficient of the exhibition area, C_{exp} $C_{exp} = \frac{S_{exp}}{S_{total}}$ where S_{exp} – display area; S_{total} – total area of the trading floor Coefficient of placing, C_{place} $C_{place} = \frac{S_{place}}{S_{total}}$ where S_{place} – area occupied by retail equipment; S_{total} – total area of the trading floor	There are no similar indicators for evaluating the use of digital retail space	$C_{exp} = 0{,}32$ $C_{place} = 0{,}75$

Source: author.

The assessment of a comfortable stay in a physical and digital store can be evaluated in more detail, in the context of key elements of the atmosphere of the retail space:

$$C_{comf} = \sum_{i=1}^{n} B_i \cdot W_i$$

where B_i is user's rating of a specific component of the shopping space atmosphere on a 10-point scale; W_i – the weight of the criterion and its significance in the store of this type.

The criteria for evaluating the comfort of the physical environment of the store can be: B1 - musical accompaniment of sales, B2 - aroma marketing, B3 - coloristic solution in the design, etc. The criteria for evaluating the comfort of the store's virtual environment can be: B1 - the design of a website or a mobile app, B2 -ease of navigation, B3 - speed of clicking on links, etc. On the one hand, the main indicator of the collective effectiveness of merchandisers is sales, the growth of which, according to various experts, can vary from 8 to 30% per month. On the other hand, an analysis of Russian trade practices has shown that merchandising standards will provide a significant increase in sales during the first months after their implementation, and will gradually give less tangible effect. Increased results will be recorded at a certain level, and the effectiveness of the merchandising standards will not look obvious. This fact should be taken into account in the subsequent audit of the staff motivation. Performance of the above indicators should be evaluated and encouraged without strict reference to sales volume.

It is necessary to note that some of the criteria can and should be evaluated by store clients, whose emotional mood determines the number and volume (average receipt) of purchases. Compliance with this standard should be recorded in the course of systematic monitoring of the customers' state in the physical and digital environment of a store.

4 Discussion

Issues on evaluating the quality of merchandisers' work in Russian and foreign practice are not widely covered. The materials we have studied usually reveal various aspects of the effectiveness of the merchandising process itself. At the same time, merchandising audit indicators can be considered as criteria for employees' performance of their tasks, if the individual and collective responsibility is considered separately. Krymov and Kapustina confirm that when forming merchandising standards, it is necessary to monitor the assortment, placement of retail equipment, and display of goods [8]. Novikov offers an author's methodology for evaluating the retail advertising and information space, which is appropriate from the point of view of an interactive approach that takes into account the share of goods in turnover, exhibition space and store profit. One of the key indicators is the performance of a merchandiser (turnover per 1 employee) [9].

Kleinová, Paluchová, Berčík, and Horská note the importance of neuromarketing in the organization of visual merchandising in the restaurant business and list its components that we can put in the assessment of the collective responsibility of store merchandisers [7]. Software engineering development by Bianchi-Aguiar et al. makes it easy for retail employees to follow the rules of merchandising based on efficiently create planograms [3]. The topic of modeling shelf space planning, which radically changes the architecture of the work of a merchandiser, is continued in the study of Bianchi-Aguiar, Hübner, Carravilla, and Oliveira [2]. New requirements to work and evaluation of labor results in the globalized economic space are highlighted by Claes, Quartier, and Vanrie [5].

The implementation of the concept of Social Media Marketing in the Internet of Things system obliges the merchandiser in the virtual retail space to establish feedback with consumers. According to Bollweg, Lackes, Siepermann, Sutaj, and Weber, the store's customer loyalty is determined by their ability to manage their communities [4]. Insufficient knowledge of the organization of merchandising and problems of managing the psycho-emotional state of buyers in the digital environment are emphasized by Sundström, Hjelm-Lidhlm, and Radon [10].

5 Conclusion

Studies of Russian trade practices and the work of foreign scientists have confirmed that the quality and completeness of achieving goals and objectives of logistics and marketing in retail should be divided based on two aspects:

- individual and collective effectiveness of merchandisers,
- indicators of the merchandising organization in the physical and digital retail space.

All the variety of evaluation criteria can be ordered as follows:

- product assortment management,
- rational use of retail space,
- advertising and sales promotion,
- commercial results of customer behavior management,
- the quality and culture of services to retail.

References

1. Alekseev, A.A.: Marketing research of the services market (1999). https://www.marketing.spb.ru/read/m17/index.htm. Accessed 25 June 2020
2. Bianchi-Aguiar, T., Hübner, A., Carravilla, M.-A., Oliveira, J.: Retail shelf space planning problems: a comprehensive review and classification framework. Eur. J. Oper. Res. (2020, in press). https://www.sciencedirect.com/science/article/abs/pii/S0377221720305610. Accessed 25 June 2020

3. Bianchi-Aguiar, T., Silva, E., Guimarães, L., Carravilla, M.-A., Oliveira, J.: Allocating products on shelves under merchandising rules: multi-level product families with display directions. Omega **76**, 47–62 (2018)
4. Bollweg, L., Lackes, R., Siepermann, M., Sutaj, A., Weber, P.: Digitalization of local owner operated retail outlets: the role of the perception of competition and customer expectations. In: Liang, T.P., Hung, S.-Y. (eds.) Pacific Asia Conference on Information Systems, p. 348. AIS, Hong Kong (2016)
5. Claes, S., Quartier, K., Vanrie, J.: Reconsidering education in retail design: today's challenges and objectives. In: Fokkinga, S.F., van Zuthem, H., Ludden, G.D.S., Cila, N., Desmet, P.M.A. (eds.) Proceedings - D and E 2016: 10th International Conference on Design and Emotion, pp. 512–516. Universidad de los Andes, Venezuela (2016)
6. GOST R 51773-2001: Retail. Business Classification (2001). http://docs.cntd.ru/document/1200025420. Accessed 23 June 2020
7. Kleinová, K., Paluchová, J., Berčík, J., Horská, E.: Visual merchandising and its marketing components in the chosen restaurants in Slovakia. Procedia Econ. Finan. **342015**, 3–10 (2015)
8. Krymov, S.M., Kapustina, I.V.: Merchandising business management model. Innov. Econ.: Prospects Dev. Improv. **3**(21), 48–54 (2017)
9. Novikov, O.A.: Formation and evaluation of the advertising and information space of retail. Bull. Russian State Humanit. Univ. Ser. Econ. Control Law **2**(4), 80–85 (2016)
10. Sundström, M., Hjelm-Lidholm, S., Radon, A.: Clicking the boredom away – exploring impulse fashion buying behavior online. J. Retail. Consum. Serv. **47**, 150–156 (2019)

Professional Competency of a Mediator: Criteria and Standards

A. R. Rakhmatullina[1(✉)], A. L. Fursov[2], and E. Yu. Bobkova[3]

[1] Samara State University of Economics, Samara, Russia
sseu_ar@mail.ru

[2] Stolypin Volga Region Institute of Administration, Saratov, Russia
andrew@fursov.ru

[3] Moscow State University of Technologies and Management, Samara, Russia
bobkovaelenayu@gmail.com

Abstract. Mediation is a process that aims to resolve disputes with the assistance of the mediator on the basis of the voluntary agreement of the parties with a view to reaching a mutually acceptable solution. The mediation skills are considered the key to successful resolution of corporate conflicts and labor conflicts that have arisen both inside and outside the organization in the developed countries. The success of mediation procedure depends on the qualification of the mediator. This article investigates key points and criteria of identifying the development of mediator competences. The purpose of this work is analysis of criteria and standards for determining the professional competencies of a mediator. The study has been conducted using statistical, mathematical, expert evaluation methods and legal methodology tools. The mediator competencies model has been created. Scientific gaps in the area of tools for assessing the mediation competence have been revealed, as well as the need for further scientific study of the mediator professional qualities assessment. The research results can be used in studying the mediation procedure.

Keywords: Competence · Labor actions · Mediation · Mediator · Professional competence · Professional standard

1 Introduction

Dynamically developing trends in the field of alternative dispute and conflict resolution systems in Russian society, completing the transition to "digital reality," updated the importance of mediation as a tool for harmonizing social relations through the resolution of conflicts and disputes with the participation of an independent person - the mediator. The application of mediation skills in developed countries is considered the key to the successful resolution of corporate conflicts, labor conflicts that have arisen both within and outside the organization. A good and complete awareness of the entrepreneur about the possibilities of the mediation procedure as an alternative to the judicial system allows him at the stages of preparing and implementing transactions to consider mediation as an option for a possible resolution of the dispute.

S. I. Ashmarina and V. V. Mantulenko (Eds.): IPM 2020, LNNS 161, pp. 343–347, 2021.
https://doi.org/10.1007/978-3-030-60926-9_44

2 Methodology

The analysis of many sources in the field of practical experience in resolving conflict situations between economic actors and in labor collectives proves that the result of any conflict is a limitation of people's needs and interests; that conflict prevention becomes significant at the stage of the earliest prerequisites, and the identification of potential causes of future conflict. And if a conflict arises, then you should contact specialized mediation institutions.

This is also evidenced by the dynamics of the increase in regulatory legal acts in this area: both in Russia and abroad: in the United States, more than 2,500 acts directly or indirectly related to the regulation of mediation procedures were merged in 2001 into the Uniform Mediation Act (Unified Law on Mediation). As a result, in the United States no one conflict is resolved without mediators [7]. "In the UK there is a British Mediation Organization - Conflict Resolution Center, in France there is a Conflict Tribunal that plays a mediating role in the judicial system" [3, pp. 34–35]. Mediation is widely used in China, South Korea, Hungary, India. In Austria, the specialty of mediator is included in the nomenclature of professions [2].

In the Russian Federation, mediation was enshrined in Federal law № 193 "On an alternative procedure for the settlement of disputes with the participation of an intermediary (mediation procedure)" [5]. The legislative consolidation of the list of competencies of the mediator in professional standard No. 307 "Specialist in the field of mediation (mediator)" [12] took place later, in 2014, this updated the formalization of the requirements for the professional and personal qualities of the mediator. What qualities should a successful mediator have and how can the presence of these qualities be verified? Despite the considerable interest of the scientific community in this problem, to date there are objectively significant gaps in this vector of research. This article deals with the main problems of determining the competence of the mediator.

3 Results

Considering the competence of the mediator, Bazarov based his studies on the fact that the object of attention of the mediator is the personality of the participant in the conflict (dispute), and its main function is actions to optimize the psychological climate, work with the emotional state of the participants. Bazarov refers to the main competencies of the mediator: influence, insight and tolerance. At the same time, the importance of the integrative function of the mediator is emphasized - it acts as a person who is able to merge a group to solve individual problems and at the same time makes the necessary correction of the state of the group atmosphere. Interestingly, by analyzing the relationship of mediator emotional leadership with the effectiveness of the mediation process and noting higher mediation efficiency in groups where the mediator has average or high rates of emotional leadership, the researchers note that a mediator with a high level of emotional leadership may be less successful than his colleague with average scores. Bazarov explains this effect by the fact that in the situation of resolving disagreements, an important factor in success is the independent work of opponents with their own feelings and emotions: The feeling of high professionalism of the mediator causes confidence in him among the conflicting parties, which leads to a decrease in attempts to independently identify, understand and deal with their own emotional experiences, shift responsibility and make a decision on the mediator [4].

Allahverdova identifies such characteristics inherent in successful mediators as "the ability to listen and analyze, clearly express their thoughts, manage their emotions, flexibility in thinking and behavior" [1]. Motivation for training also contributes to the rapid formation of the competence of the mediator. In addition, the researcher notes that the skill of a mediator is facilitated by benevolence to people, perseverance and patience in achieving goals, performance, tolerance of criticism, and the ability to work in an aggressive environment [6]. Korneeva and Rudakova compiled a list of "professionally important qualities of a mediator: communication, openness, friendliness, empathy, responsibility, tolerance, independence, activity, mindfulness, organization, balance, specificity" [9]. A logical question arises: how do the listed qualities relate to the requirements for competencies described in the professional standard of the mediator? By the way, the foreign toolkit for identifying and assessing the degree of possession of the mediator of competencies is still poorly covered [8], there is no information about the Russian toolkit, and problems are noted with the use of existing tests. So, in the article Allahverdova notes that when evaluating professional qualities, psychological tests were used: "Cat (intellectual lability), Strup test (flexibility of cognitive control), Kettell test (personal characteristics), "Hand" test (aggressiveness), etc. [1]. Subsequent analysis showed that there is no direct and meaningful association of the psychological characteristics obtained from these tests with learning success and mediator performance."

4 Discussion

Analysis of labor functions and related labor activities, knowledge and skills described in the professional standard allows you to combine knowledge and skills into several interconnected groups [12]. So, the labor function "Organizational and technical and documentary support of the mediation procedure" (Code A/01.6) describes the following skills, presented graphically in Fig. 1. An analysis of the professional standard revealed three main groups of knowledge and skills: in the field of information technology and document management; in the field of law, including knowledge of labor, civil, family legislation, as well as the law on mediation and the ability to implement their norms in the field of conflict science and social communication. With regard to the legal competence of the mediator, this part of the testing of knowledge and skills is carried out using a wide range of tests, including those used in higher education. The information and technological competence of the mediator is also not difficult to test, since this part of the verification procedures is quite formalized.

When you describe the following work function, «Prepare for Mediation», the conflict and communication competencies group is increased. For the labor function "Conducting a mediation procedure in a specialized field" which, incidentally, requires a level of education of a specialist or master's degree, we see additional requirements for subject knowledge in a specialized field and an expanded scope of competence related to psychology (Fundamentals of psychological correction of attitudes and

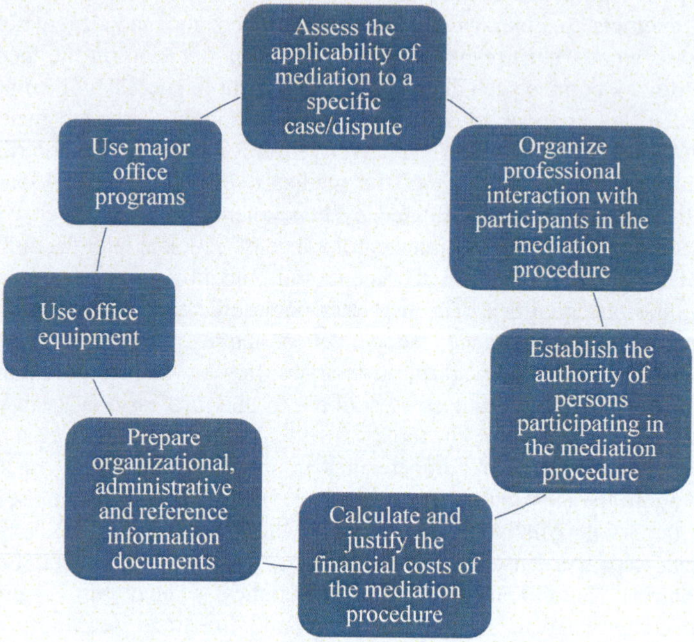

Fig. 1. Basic mediator skills (Source: authors).

perception; Fundamentals of social psychology; Fundamentals of humanistic psychology; Fundamentals of cognitive psychology).

5 Conclusion

Speaking about the model of competencies of the mediator, we first of all notice its closest part with the conflictology and models of competencies of the conflictologist. Knowledge of the technique of conflict resolution, the ability to implement this knowledge are presented in the Federal State Educational Standard for Conflict Scientists [10, 11]. In other words, the professional basis of the conflictologist meets all the requirements provided for the mediator, and therefore, when assessing the conflictological competence of the mediator, you can use the existing tools of conflictologists, an example of which, in particular, is given by Yakovlev, Romanov, and Romanova [13]. At the same time, it must be noted that the tools for assessing conflictological competence described in the literature do not fully correspond to the goals and objectives of assessing the competence of the mediator. Thus, this issue remains open for further research.

References

1. Allakhverdova, O.V.: Training of mediators and formation of competencies. St. Petersburg State Univ. Bull. **3**, 51–59 (2012)
2. Baibekova, E.F., Kalimullina, I.V.: The use of mediation skills is the key to successful resolution of corporate conflicts. Bull. Saratov State Law Acad. **2**(133), 245–252 (2020)
3. Bard, P.: Unification processes in the countries of Eastern and Central Europe. Mediat. Other Methods Stimul. Democrat. Dialogue **7**, 34–35 (2016)
4. Bazarov, T.Y.: Business education: development of organizations or organization of development? Organ. Psychol. **3**(4), 92–108 (2013)
5. Federal law of 27.07.2010 N 193-FZ (ed. from 26.07.2019) "On an alternative procedure for the settlement of disputes with the participation of an intermediary (mediation procedure)" (2010). http://www.consultant.ru/cons/cgi/online.cgi?req=docandbase=LAWandn=330191-andfld=134anddst=100008,0andrnd=0.4506648965l628583#027645483209185917. Accessed 28 June 2020
6. Gaidaenko-Sher, N.I.: Unification of requirements for the qualification of a mediator: mediator and mediator in family disputes. Curr. Probl. Russ. Law **5**(78), 56–60 (2017)
7. Hopt, K.J., Steffek, F. (eds.): Mediation: Principles and Regulation in Comparative Perspective. Oxford University Press, Oxford (2013)
8. Human Technologies: 11 Personal Factors (2019). http://www.ht.ru/cms/component/content/article/106-tests/101572-11lf. Accessed 28 June2020
9. Korneeva, Ya.A., Rudakova, Yu.N.: Features of the personality of mediators with different success of the conciliation procedure. Psychol. Psychotech. **3**, 68–80 (2019)
10. Kozlov, A.V.: Comparative characteristics of conflict prevention methods in organizations of the Federal Republic of Germany and the Russian Federation. Koncept **3**, 43–46 (2016)
11. Order of the Ministry of Education and Science of Russia from 07.08.2014 N 956 "On approval of the Federal law state educational institution higher education standards for direction of training 37.03.02 Conflict studies (undergraduate level)" (2014). http://fgosvo.ru/uploadfiles/fgosvob/370302_Konfliktologia.pdf. Accessed 28 June 2020
12. Professional standard No. 307 "Specialist in the field of mediation (mediator)"/approved by the order of the Ministry of labor and social protection of the Russian Federation of December 15, 2014 No. 1041n (2014). https://base.garant.ru/70843342/. Accessed 28 June 2020
13. Yakovleva, I.P., Romanov, D.A., Romanova, M.L.: Diagnostics of students' conflictological competence in the structure of psychological and pedagogical monitoring. Sci. J. KubSAU **108**, 740–760 (2015)

Fundamentals of Choosing an LMS Platform for Distance Learning

T. G. Sakova and S. A. Chevereva(✉)

Samara State University of Economics, Samara, Russia
t-g-sakova@yandex.ru, chevereva@yandex.ru

Abstract. The modern world is constantly changing and developing. Companies have to adapt to the conditions of the market environment, building flexible relationships not only with business partners, but also with their staff. There is a need to develop strategies corresponding to the changes that are taking place, to identify the main trends both in the development of the organization itself and in the selection of training methods for its employees. The article is devoted to the problem of distance education, or rather, software tools that help to put this type of education into practice. The definition is given to the concept of the LMS platform and its capabilities. The classification of existing LMS systems is presented. The criteria for choosing a specific platform for the organization are described in detail, depending on the requirements for educational content and learning outcomes. In conclusion, the advantages and disadvantages of using LMS platforms in training are indicated.

Keywords: Distance learning · LMS platform · Online learning · Training content

1 Introduction

In the current market situation, the most significant advantage of the organization is the high qualification and professionalism of employees. That is why staff training and timely updating of their professional knowledge and skills is one of the most pressing topics of modern business [3]. Personnel training is one of the most important components of personnel services (HR), which offer two approaches to organizing the process of obtaining knowledge by employees. The first approach includes the involvement of external providers of educational services, providing the customer with highly qualified trainers and teachers. Personnel training centers use mainly blended teaching methods, including lectures, seminars, and various trainings. This method is effective; however, many small and medium enterprises cannot afford it due to the high cost of services. The second approach is the organization of training by attracting internal resources of companies, i.e. on their own. In this case, either the most qualified specialist of his own company, or a partner of the company, or an employee of the personnel service of the enterprise is nominated as a "mentor". The presented approach is much cheaper than the previous one, but in this case, to obtain satisfactory results, a detailed development of the methodology and training plan is necessary. To provide high-quality preparation and organization of the educational process allows modern

S. I. Ashmarina and V. V. Mantulenko (Eds.): IPM 2020, LNNS 161, pp. 348–353, 2021.
https://doi.org/10.1007/978-3-030-60926-9_45

information technology, a special place among which is occupied by LMS platforms [1]. LMS platforms are more commonly used by commercial companies to train thousands of employees around the world. This allows you to: reduce training costs; to support the same level of the employees' training both in the central office and branches; to increase the speed of learning when opening a new product line or a new unit [10].

2 Methodology

LMS (Learning Management System) is a repository of educational materials - video lessons, lectures, presentations, books and courses, which can be accessed from any device anywhere in the world. In Russia, LMS is often called distance learning systems (LMS) [4]. The abbreviation consists of three words:

1. Learning - to teach. Using LMS, you can create a single database of electronic courses and training materials.
2. Management - to manage. Learning in the system is managed by the administrator. Its functions include: assigning users courses and tests, checking homework and more.
3. System - an electronic system. LMS automatically generates training results in a summary report, which simplifies the assessment of the level of training of employees.

To choose the right LMS for your business, first of all, you need to decide on the specific goals that you need to achieve as a result of the implementation of the system. It is also important to clearly understand who, where and how will use the platform for online learning. After that, you can proceed to search for the most suitable solution specifically for you. Developing LMS platforms on your own is impractical because this is an expensive event. Currently, the Russian market offers a wide selection of ready-made systems for distance learning, both domestic and foreign production [8]. All existing LMS can be divided into three groups:

1. Cloud platforms. In this case, the system does not need to be installed on the company's server. LMS works on the principle of any web service. You can upload the necessary materials to it, add users and start training. All information is stored on third-party servers of the provider. Such systems for example include: iSpring online, TeachBase, Loop, Learn Amp, Matrix and others. They are often used in corporate training.
2. Server LMS require installation on the company's server, which will allow employees to log into the training portal using a corporate username and password. All data is stored internally. The disadvantage of this installation is the need to create a detailed technical specification, checking the compatibility of the system with the software of the company. The implementation phase usually takes 3–4 months. Examples of server LMS: Moodle, RedClass, etc. It is used in universities and training centers.

3. LMS with integration with CMS. With integration, services are conveniently integrated into the CMS, and this allows you to maximize the benefits of its use. The case when LMS integrate with content management systems: wordpress, joomla, drupal, wix. Most often they use online schools, individuals who want to sell courses.

3 Results

Choosing an LMS platform is the most important point in launching distance learning. The management of any company wants to have an inexpensive LMS, which can be quickly implemented and easily understood by employees of all departments, regardless of the level of knowledge of computer technology [2]. The system should simply create courses, conduct webinars and receive detailed reports on the performance of employees and the results of their training [5]. Employees should be able to undergo training from any device: computers, tablets and smartphones. The question arises, which specific LMS platform to choose? The following platforms are currently the most popular on the Russian market:

- iSpring Learn,
- Mirapolis LMS,
- ShareKnowledge and others.

To make a decision about the choice, it is necessary to determine the selection criteria. To carry out the comparison, we studied information from official sites, technical documentation, public interviews, as well as user reviews about each LMS platform. In our opinion, it is advisable to select the following criteria for evaluation:

Ease of obtaining reference materials. It is necessary to evaluate whether the company's website has accurate data on the functions and capabilities of the training system, and whether there is a section for answers to frequently asked questions.

Technical support system. To study the possibility of obtaining qualified assistance of a technical specialist in installing, configuring and operating the system on-line, to evaluate the speed of feedback and its quality.

Cross-platform and system administration. Analyze the simplicity and convenience of system settings: user profiles, including access rights and interface, training content, chats, test modes and reports. Find out on which operating systems the system works, is there a version for mobile devices. Currently, more and more users prefer to use smartphones or tablets for study and work. Are educational content materials viewed in the correct form on these devices.

Availability demo version. To find out if there is a demo version of the system, which makes it possible to study its capabilities before deciding on its use. Convenience of the user interface. To evaluate the possibility of independent study of the system's functionality without studying technical documentation. Editing educational content. To study the options for self-editing and setting up a training course by its author after downloading it to the server, the availability of searching for a specific training in a common information base. Testing. Analyze the testing mode inherent in the system: types of the used tests (with the choice of one or more correct answers, open form, to establish compliance, to establish the correct sequence), test modes, options for summarizing, the possibility of parallel training information on issues with incorrect answers. Conducting webinars. Determine whether it is possible to independently conduct and save webinars in the form of files, whether there are restrictions on the number of participants. Communication Modes. To evaluate the possibility of using chat rooms, forums, microblogs in the learning process for team communication. Find out the possibility of sending notifications about changes in the system, for example, the next course or testing at the course being studied opens.

Operating modes of participants in educational content. To study the options for connecting participants (connects the administrator, the teacher of the course, independently for the course you like), the possibility of self-editing the student's profile. What information about himself he wants to put in the public domain, what he wants to get from the system (for example, his own achievements in studying the course). Find out whether it is possible to group participants in the training course according to some criteria and open them access to individual tasks or materials.

Surveys and reports. To evaluate whether the system has the ability to compile various popular reports to track user achievements (based on test results, current performance, etc.), the ability to edit the structure of standard system reports and create a report of any structure with the addition of graphics. Find out if there is a function in the system for compiling calendars, training schedules that will allow you to plan the process for associations or individual students, the mode of conducting mass surveys.

Expenses. Determine the cost of using the system, which consists of the price of the software product itself (paid once upon implementation of the system) and the costs of its maintenance (annual subscription for updating and maintenance). To study the factors affecting the cost of maintenance. Assess the possibility of profit from the sale of courses to third parties.

4 Discussion

LMS platforms can be compared to a virtual classroom, in which it is possible to train students or employees anywhere in the world, as well as track their performance [2]. The learning process in any LMS platform is launched in three stages:

1. Preparation of materials for training and uploading them to the system. The materials may include lectures, tests, surveys, video lessons and webinars, books or presentations - depending on the developed training course.
2. Connect users to educational content. Assign materials to users. Students will be able to study the material on the curriculum. At the end of each section or the entire course as a whole the material will be tested.
3. Track statistics. LMS automatically generates reports on student performance. In real time, you can check how many people studied the course, what point they received for the test, and what mistakes were made.

Systems allow tracking such parameters: how often the user visited the portal; how many training materials have studied; what mistakes made in the test. The LMS administrator can get information about any student at any time [7].

5 Conclusion

Using LMS platforms for staff training guarantees the following benefits: reduction in training costs, free access to information regardless of the location of the student, any number of courses, videos can be loaded into the system, the ability to communicate between users in different formats: chats, blogs, forums, conferences, the system automatically monitors attendance, academic performance, the degree of mastery of the material, etc. and receiving reports on these indicators, the ability to organize not only distance learning, but also testing and certification of employees.

For objectivity, we point out the shortcomings:

- the cost of the platform depends on production requests (it ranges from 24,000 to 600,000 rubles annually),
- the widespread use of mobile devices and their rapid modification requires that the LMS platform work correctly on all types of mobile devices, this must be taken into account when choosing a system, and this also leads to additional costs for their maintenance [6]. Each system has both advantages and disadvantages, and depending on the needs of the organization, as well as the characteristics of the educational process and the availability of a corporate portal, a choice is made in favor of a particular system. Definitely, distance education will continue to actively develop, as the pandemic clearly showed. To implement this type of education, the optimal choice of LMS platform is required [9].

References

1. Allen, D.K., Irnazarow, A., McLauglin, F.: Practice, information and the development of a digital platform. Proc. Assoc. Inf. Sci. Technol. **52**(1), 597–598 (2019)
2. da Rocha, F.N., Pollock, N.: Innovating in digital platforms: an integrative approach. In: Filipe, J., Smialek, M., Brodsky, A., Hammoudi, S. (eds.) Proceedings of the 21st International Conference on Enterprise Information Systems, pp. 505–516. SciTePress Digital Library, Heraklion (2019)
3. Gribova, V.V.: Digital platform concept for business activities. Inf. Technol. **25**(8), 502–511 (2019)
4. Hernandez-Garcia, A., Conde, M.A.: Bridging the gap between LMS and social network learning analytics in online learning. J. Inf. Technol. Res. **9**(4), 1–15 (2017)
5. Mamcenko, J.: Comparative analysis of learning analytics tools to improve learning used in different learning platforms. In: Chova, L.G., Martinez, A.L., Torres, I.C. (eds.) Proceedings of the 12th International Technology, Education and Development Conference, pp. 4666–4673. Iated-Int Association Technology Education and Development, Valencia (2018)
6. Murphy, V.L., Coiro, J., Kiili, C.: Exploring patterns in student dialogue while using a digital platform designed to support online inquiry. J. Interact. Media Educ. **1**, 1–13 (2019)

7. Radjenovic, T., Janjić, I.: The importance of managing innovation in modern enterprises. Ekonomika **65**(3), 45–54 (2019)
8. Sedkaoui, S., Khelfaoui, M.: Digital platforms and the sharing mechanism. In: Sedkaoui, S., Khelfaoui, M. (eds.) Sharing Economy and Big Data Analytics, pp. 51–60. Wiley, New Jersey (2020)
9. Wu, S.Y., Ye, J.M., Wang, Z.F., Luo, D.X., Zhao, R.: Analysis and research on learning behavior based on LMS. In: Goma, M.M., Fakharian, A., Chiementin, X. (eds.) Proceedings of the 2nd International Conference on Automation, Mechanical and Electrical Engineering, vol. 87, pp. 214–218. Atlantis Press, Paris (2017)
10. Zutshi, A., Nodehi, T., Grilo, A., Rizvanović, B.: The evolution of digital platforms. In: Shirvastava, A.K., Rana, S., Mohapatra, A.K., Ram, M. (eds.) Advances in Management Research, p. 10. Taylor and Francis Group, London (2019)

Transformation of the Legal Status of Digital Platform Employees

E. L. Sidorenko[1(✉)], S. V. Sheveleva[2], and E. Y. Komova[3]

[1] Moscow State Institute of International Relations (University) of the Ministry of Foreign Affairs, Moscow, Russia
12011979@list.ru
[2] Southwest State University, Kursk, Russia
decan46@yandex.ru
[3] The Central Bank of the Russian Federation, Moscow, Russia
komova.e.yu@my.mgimo.ru

Abstract. The active development of digital platforms has led to the spread of new forms of labour relations organization. On the one hand, the platforms allowed a large number of people to find work and an additional source of income, to develop competition. But on the other hand, the lack of the clear market regulation for the rights of employees of digital platforms has led to discrimination of their labour rights, the establishment of irregular working hours, etc. A new form of employment (working in the field of digital platforms) poses new challenges to modern labour legislation. For example, questions on the characteristics of a digital platform employee, criteria for distinguishing them from hired worker and self-employed, as well as questions on the list of labour guarantees on the platforms need to be addressed. In this paper, these issues are studied both from the perspective of international organizations and individual countries that have secured the status of employees of digital platforms. The authors pay special attention to the issues of differentiation between hired workers and self-employed individuals and consider trends in the development of labour law on the example of specific court cases.

Keywords: Digital platforms · Hired worker · Labour guarantees · Labour rights

1 Introduction

Digitalization has a serious impact on many aspects of social and economic life, including qualitative and quantitative indicators of employment. On the one hand, the employment picture is changing, individual professions are disappearing, and new ones are emerging. On the other hand, the approach to employment and relations that arise between the employee and the employer in the conditions of digitalization of production processes is changing qualitatively. This trend is becoming so stable and systematic that experts are no longer talking on the appearance of non-standard forms of employment, but on the qualitative transformation of labour relations themselves and the need to develop such concepts as "new standard of labour relations" [10], "non-standard employment" [30], "a person who looks like an employee" [12], etc.

S. I. Ashmarina and V. V. Mantulenko (Eds.): IPM 2020, LNNS 161, pp. 354–363, 2021.
https://doi.org/10.1007/978-3-030-60926-9_46

CPU when performing work, or when the CPU determines incentives and sanctions based on the employee's rating.

Harris and Krueger suggest to introduce the category of "dependent contractor" or "independent worker" into labour law as an intermediate form between the worker and the self-employed, different forms of which exist in the law of Germany, Italy, France and Canada [9]. In their opinion, such employees will have the right to form trade unions and enter the collective agreements, but will be deprived of the right to pay overtime and receive the minimum wage [27].

Other researchers dispute this necessity, referring to the practice of states where such a category exists [3]. Experts of the International labour organization listed 18 principles and 3 optional principles of fair labour organization in the field of digital platforms [10], including:

- platform employees should not be called self-employed if they are actually hired workers,
- employees must have formal communication mechanisms with the platform management through trade unions, collective bargaining and agreements, etc.,
- platform employees who are actually hired workers should receive the minimum wage set by the norms of their region of activity,
- employees must receive full payment, paid by the client, for their services, and the digital platform does not have the right to demand a percentage of this revenue or a fee for the opportunity to work on it,
- employees of digital platforms should have the right to reject some job offers and work at certain time without harmful consequences from the platforms,
- the organization must reimburse the employee for downtime caused by technical problems of the platform,
- working conditions on the CPU must be presented in a simple and understandable language,
- the digital platform should assign ratings to clients that include cases of refusal to pay for the services of an employee, and employees should have the right to familiarize themselves with this rating.

Some of these principles have already been put into practice. In California, since January 2020, the AB 5 gig-work bill has been in force, defining that any employees are treated as hired workers until the employer has proven otherwise. First of all, the law was adopted for employees in the field of digital platforms [19]. Experts from the European Commission suggest a different solution to the problem. In their opinion, the category "employee in the field of digital platforms" does not exist. Existing laws are sufficient to solve all regulatory problems and special legislation should not be applied [2]. Based on this concept, two opposite approaches to regulating the rights of digital platform employees have been proposed. The first approach is the Netherlands model. The law provides for payment of actual working hours. At the same time, the minimum wage for employees is guaranteed, even if they worked less than 15 h a week. At the same time, after three months of work, the employee is entitled on right to payment of the minimum guaranteed hours of work, the number of which is determined based on the average number of hours worked for the last quarter.

The second approach is UK law. The law allows you to apply a zero - hour contract, according to which the employer is not obliged to provide the employee with any time of his work (and, accordingly, pay for it), and the employee is not obliged to perform any work assigned to him. In this case, the employee receives payment only for the time he spent working or waiting for work, while at the disposal of the employer (usually in his premises). In this case, zero-hour employees can be recognized as hired workers if: the employer is obliged to personally provide work to the employee; the employee and the employer are mutually responsible; the employee explicitly or implicitly agrees to be under the "substantial" control of the employer [2].

There is a third approach. Thus, according to the position of the central bank of Italy, employees of digital platforms should be recognized as hired workers if they have a similar regularity of work, the amount of time spent and income received [7]. In other words, the issue is actually resolved by the employee and the platform independently through the definition of working conditions. An interesting solution to the problem of distinguishing employees from the self-employed has been proposed in US law. California Assembly Bill 5 [1] has summarized the results of the case Dynamex Operations West, Inc. v. Superior Court [22] from 2018 has highlighted the following criteria for self-employed: 1) the employee must be free to act, both under the terms of the contract and by virtue of their duties; 2) the services are provided outside of the usual type of commercial activity for the employer; 3) the employee regularly enters into contracts for the performance of such type of activity. The law provides for the companies right to obtain the right not to recognize their workers as hired ones. Only employees who meet the following conditions have self-employed status: they set the price of their services themselves; they can negotiate directly with the client and must earn twice the minimum wage.

4 Discussion

The active development of digital platforms has brought to the fore the question of how labour law will develop and which option is preferable: to adjust the existing relationship between an employee and a digital platform, or to change labour law for them. The answer to these questions depends largely on how the risks, which digital platforms carry for the labour relations, will be minimized. Among these risks experts call the following: employees of digital platforms are deprived of the level of social protection that ordinary hired workers have; the power of digital platforms over employees is virtually unlimited: employees do not control the price, they may be suspended from work, other sanctions may be applied to them, and in some cases, employees are not free to change the region of their activities; employees of digital platforms do not receive the minimum wage, because they receive payment for specific services. However, some OECD countries are considering introducing rules for distributing revenue between the platform and the employee so that even low-productivity employees receive their minimum wage [16]; they are deprived of access to advanced training courses that ordinary hired workers have.

In France, for example, they tried to solve this issue by requiring digital platforms that determine the amount of pay and working conditions for employees to reimburse them for the cost of courses that confirm their work experience [5]. The unresolved issues raised above allow many researchers to say that digital platforms widely spread the practice of "unregulated virtual hard labour" [21].

Despite the fact that international organizations and the laws of individual states define a general framework for regulating labour on digital platforms, the jurisprudence on specific disputes is ambiguous. So, in the US in the case of Otey vs. Crowdflower (2012) Crowdflower employees claimed to be entitled to the minimum wage for working on Amazon Mechanical Turk [11]. In 2015, the case was completed by a settlement agreement, under which Crowdflower agreed to pay the unpaid amount of wages [13], but their status as employees under the contract was not fixed.

In Italy, in a 2018 case, the plaintiffs, former couriers of the food delivery service Foodora, asked the Turin labour argument tribunal to recognize them as hired workers, and their "dismissal" (in the form of closing access to the application) as illegal. The tribunal recognized that couriers are self-employed, since they have the right to refuse to fulfill certain orders, and there are no grounds for satisfying their requirements [14]. In a similar case, the Supreme Court of the United Kingdom confirmed the decision of the central arbitration committee, which decided that couriers of the food delivery service Deliveroo do not have the status of hired workers. The grounds are that they can ask other couriers to perform their delivery [18]. And on these grounds, they may not be provided with labour guarantees. This issue is actively discussed by experts in many countries. In particular, Russian legislation does not recognize the possibility of refusing an order in favor of a third party as a mandatory feature of self-employed. A similar position was expressed by the Madrid court in Spain. In 2019, it admitted that Deliveroo employees are their employees, not self-employed [18].

The French cassation court on March 4, 2020 recognized that there is an employee-employer labour relations between Uber and a driver who operated under the similarly-named platform [13]. In the future, the labour argument court will decide whether this former employee is entitled to compensation under labour law. The Paris labour argument court in 2016 recognized that the plaintiff is a hired worker of the digital platform Le Cab (taxi service) on the grounds that the working conditions contained the employee's obligation not to work for other digital platforms [28].

In 2018, there was a case of Uber BV vs Aslam. The plaintiffs that were Uber taxi drivers, argued that they were entitled to the minimum wage and annual paid leave. Uber BV, the British subsidiary of Sweden's Uber, confirmed that the plaintiffs are simply "partners" and self-employed, and the company has no obligations to them. As a result of the case, the court recognized (and the court of appeal confirmed) that the plaintiffs are employees as defined in section 230 (3) (b) of the Employment Rights Act and are entitled to the minimum wage and annual paid leave [25].

In addition to the issues outlined above, the legal basis for competitive selection of employees also needs to be analyzed in detail [8, 29]. It is also unclear how individuals should have access to an undefined and previously unknown group of individuals to solve certain tasks of providing work results, who will guarantee the quality of work and pay, and how the personal rights of employees and customers will be protected [20, 23]. Currently, there is a discussion on whether it is justified to enter into a contract between the platform and an employee. According to our research, more than 40% of contracts concluded with platform employees in Russia are construction contracts. But in this case, it is important with whom this contract is concluded – with the platform or the customer. And how the function of a digital platform's payment agent is legally secured.

About 50% of employee contracts with the platform in Russia are executed in the form of an Agency agreement, where the digital platform (aggregator platform) serves as the employee's agent. The law of the Russian Federation from 07.02.1992 N 2300-1 (as amended from 24.04.2020) "On protection of consumer rights" [26] introduced the concept of "aggregator owner", and the aggregators owners will be considered as an independent subject of responsibility to consumers in terms of compliance with their rights to information on a product or service. However, the aggregator is not an employer for the employee. They are not required to pay the minimum wage, grant leave, or improve the employee's qualifications.

The important fact is that in this case, the platform's responsibility for the product or service is significantly reduced. The platform is responsible only if it provides the consumer with the following opportunities in relation to a particular product (service): read the seller's (contractor's) offer on contract conclusion; conclude a contract with the seller (contractor); make a pre-payment for the specified product (service) by transferring funds to the owner of the aggregator as part of non-cash payments. Thus, if the site has the ability to view the product of certain sellers, buy it and make a prepayment, then the owner of such a site is responsible for the product or service. If not, the responsibility falls only on the employee. The Yandex-market digital platform took advantage of the lack of legislation. In the spring of 2018, it closed the "Order on Market" program and removed the shopping cart from the site. In other words, the payment is received from consumers directly to the seller, and the platform is not responsible to either the consumer or the employee.

5 Conclusion

To sum up the research, it is important to note that there is currently no single approach to determining the status of digital platforms and their employees. And taking into account the variety of platforms (sharing, crowdsourcing, charity, etc.), it will be difficult to develop universal criteria for evaluating them. In fact, today states have a choice: either apply existing legal norms to modern labour relations, or create a fundamentally new model of employee – employer relations, which would take into account the specifics of digital platforms and the mode of work on them as fully as possible. As part of the first approach, it is crucial to answer the question on how serious are the differences between classic labour relations and relations on digital platforms. We believe that there are differences between them, but they are not so fundamental to deprive employees of digital platforms of protection guarantees. Among the factors that make it possible to transfer the status of hired workers to such employees are the following: 1) the existence of a contract between the person and the platform, in which their duties are fixed; 2) non-gratuitous nature of services; 3) payment of remuneration for work platform, despite the fact that the customer is another person; 4) the right of the platform to execute the work in time; 5) the platform does not generate problems, but reflects the request of the customer. At this point, it can be compared to an employer sending work to its employees; 6) the platform is interested in ensuring that work is performed in its interests, under its management and control. And in this part, it is close to the status of an employer. These circumstances give grounds for considering the possibility of protecting the rights of employees by granting them the status of hired workers. However, such a decision can only be temporary until a fundamentally new approach to regulating employment on digital platforms is developed, taking into account the interests of all legal relations participants.

Acknowledgements. The publication was prepared as a part of the state task for 2020. "Transformation of private and public law in the context of evolving individuals, society and the state" (no. 0851-2020-0033).

References

1. California Legislative Information: AB-5 Worker status: employees and independent contractors (2019–2020) (2020). https://leginfo.legislature.ca.gov/faces/billTextClient.xhtml?bill_id=201920200AB5. Accessed 17 July 2020
2. De Stefano, V., Aloisi, A.: European legal framework for digital labour platforms. European Commission, Luxembourg (2018). https://doi.org/10.2760/78590, JRC112243
3. De Stefano, V.: The rise of the «just-in-time workforce»: On-demand work, crowdwork and labour protection in the «gig-economy». International Labour Office, Geneva (2016). www.ilo.org/wcmsp5/groups/public/—ed_protect/—protrav/—travail/documents/publication/wcms_443267.pdf. Accessed 25 June 2020
4. Directive (EU) 2019/1152 of the European Parliament and of the Council of 20 June 2019 on transparent and predictable working conditions in the European Union (2019). https://eur-lex.europa.eu/legal-content/EN/TXT/HTML/?uri=CELEX:32019L1152andqid=1591998140968andfrom=EN. Accessed 25 June 2020
5. Donini, A., Forlivesi, M., Rota, A., Tullini, P.: Towards collective protections for crowdworkers: Italy, Spain and France in the EU context. Transfer **23**(2), 207–223(2017). https://doi.org/10.1177/1024258916688863
6. European Commission: COM(2016) 356 - A European agenda for the collaborative economy (2016). http://ec.europa.eu/DocsRoom/documents/16881/attachments/2/translations. Accessed 17 July 2020
7. Giorgiantonio, C., Rizzica, L.: Il lavoro nella gig economy. Evidenze dal mercato del food delivery in Italia. Bank of Italy (Occasional papers) (2018). http://www.bollettinoadapt.it/wp-content/uploads/2018/12/QEF_472_18.pdf. Accessed 25 June 2020
8. Green, A., de Hoyos, M., Barnes, S.-A., Baldauf, B., Behle, H.: CrowdEmploy crowdsourcing case studies: an empirical investigation into the impact of crowdsourcing on employability. European Commission. Joint Research Centre. Institute for Prospective Technological Studies (2013). http://publications.europa.eu/resource/cellar/49b73fdd-9116-463e-a500-3ae68d6032e9.0001.02/DOC_1. Accessed 16 July 2020
9. Harris, S.D., Krueger, A.B.: A proposal for modernizing labor laws for twenty-first-century work: The "Independent Worker". Discussion Paper. The Hamilton Project (2015). http://www.hamiltonproject.org/assets/files/modernizing_labor_laws_for_twenty_first_century_work_krueger_harris.pdf. Accessed 25 June 2020

10. International Labour Organization: Digital labour platforms and the future of work towards decent work in the online world. Geneva: ILO (2018). https://www.ilo.org/wcmsp5/groups/public/—dgreports/—dcomm/—publ/documents/publication/wcms_645337.pdf. Accessed 25 June 2020
11. JUSTIA US Law: Otey et al v. Crowdflower, Inc. et al., No. 3:2012cv05524 - Document 226 (N.D. Cal. 2015) (2012). https://law.justia.com/cases/federal/district-courts/california/candce/3:2012cv05524/260287/226/. Accessed 17 July 2020
12. Kotova, S.I.: The concept of employment: digital unions, the right to employment, an absolute social security relationship. Labor Law Russia and Abroad **2**, 10–13 (2018)
13. La Cour de cassation: Arrêt n374 du 4 mars 2020 / ruling / sentencia (2020). https://www.courdecassation.fr/jurisprudence_2/communiques_presse_8004/prestation_chauffeur_9665/374_4_44528.html. Accessed 25 June 2020
14. LEXOLOGY: Tribunal holds food delivery riders are not employees (2018). https://www.lexology.com/library/detail.aspx?g=9f42d61e-31a8–464c-83c2-7c6c4ba3670b. Accessed 17 July 2020
15. McKinseyandCompany: McKinsey Global Institute. A future that works: Automation, employment, and productivity (2017). https://www.handelsblatt.com/downloads/19337114/1/mgi.pdf?ticket=ST-7025019-PLqCFu9R6WsSaBV1IlPz-ap5. Accessed 25 June 2020
16. OECD: GIG Economy Platforms: Boon or bane? Economics Department working papers no. 1550 (2019). https://www.oecd.org/officialdocuments/publicdisplaydocumentpdf/?cote=ECO/WKP(2019)19anddocLanguage=En. Accessed 25 June 2020
17. OECD: Working party on measurement and analysis of the digital economy. New forms of work in the digital economy (2015). https://www.oecd.org/officialdocuments/publicdisplaydocumentpdf/?cote=DSTI/ICCP/IIS(2015)13/FINALanddocLanguage=En. Accessed 25 June 2020
18. REUTERS: Madrid court rules Deliveroo couriers are employees, not freelancers (2020). https://www.reuters.com/article/us-spain-deliveroo/madrid-court-rules-deliveroo-couriers-are-employees-not-freelancers-idUSKCN1UI17O. Accessed 25 June 2020
19. San Francisco Chronicle: California legislature passes AB5 gig-work bill, which could turn contractors into employees (2019). https://www.sfchronicle.com/business/article/Senate-passes-AB5-gig-work-bill-turning-14430204.php. Accessed 25 June 2020
20. Saxton, G.D., Oh, O., Kishore, R.: Rules of crowdsourcing: models, issues and systems of control. Inf. Syst. Manag. **30**(1), 2–20 (2013). https://doi.org/10.1080/10580530.2013.739883
21. Schwab, K.: The fourth industrial revolution. Publishing house "E", Moscow (2017)
22. SCOCAL, Dynamex Operations West, Inc. v. Superior Court, S222732 (2018). https://scocal.stanford.edu/opinion/dynamex-operations-west-inc-v-superior-court-34584. Accessed 16 July 2020
23. Sidorenko, E.L., Khisamova, Z.I.: The readiness of the economy for digitalization: basic methodological approaches. In: Ashmarina, S., Vochozka, M., Mantulenko, V. (eds.) Digital Age: Chances, Challenges and Future. ISCDTE 2019. Lecture Notes in Networks and Systems, vol. 84, pp. 308–316. Springer, Cham (2020). https://doi.org/10.1007/978-3-030-27015-5_37
24. Standing, G.: The Precariat: The New Dangerous Class (Bloomsbury Revelations). Bloomsbury Academic, London (2016)
25. The Guardian: Uber loses right to classify UK drivers as self-employed (2016). https://www.theguardian.com/technology/2016/oct/28/uber-uk-tribunal-self-employed-status. Accessed 25 June 2020

26. The law of the Russian Federation from 07.02.1992 N 2300-1 (as amended from 24.04.2020) "On protection of consumer rights" introduced (1992). http://www.consultant.ru/document/cons_doc_LAW_305/. Accessed 17 July 2020
27. Urzì Brancati, C., Pesole, A., Fernández-Macías, E.: Digital labour platforms in Europe: numbers, profiles, and employment status of platform workers. EUR 29810 EN. Publications Office of the European Union, Luxembourg. https://doi.org/10.2760/16653. JRC117330
28. Urzì Brancati, C., Pesole, A., Fernández-Macías, E.: New evidence on platform workers in Europe. Results from the second COLLEEM survey. EUR 29958 EN. Publications Office of the European Union, Luxembourg (2020). https://doi.org/10.2760/459278, JRC118570
29. Vorontsov, D.I.: Perspective approaches to regulation of labour relations arising on online crowdsourcing platforms. Law Econ. **11**, 73–79 (2019)
30. Waas, B., van Voss, G.H. (eds.): Restatement of Labour Law in Europe. The Concept of Employee, vol. I. Hart Publishing, Oxford and Portland (2017)

Qualified Specialists' Readiness for Digitalization Risks

M. V. Simonova[1(✉)], O. V. Zabelina[2], and F. I. Mirzabalaeva[3]

[1] Samara State University of Economics, Samara, Russia
m.simonova@mail.ru
[2] FSBI All-Russian Research Institute of Labor, the Russian Ministry of Labor, Moscow, Russia
zabelina.OV@tversu.ru
[3] Plekhanov Russian University of Economics, Moscow, Russia
faridamir@yandex.ru

Abstract. The authors analyze the working population readiness for training, the possibility of obtaining a new profession in the fields of professional activity and occupational groups. The purpose of this study is to identify the characteristics of labor attitudes of Russian specialists with higher and secondary education connected with the changes in their professional activities under digitalization of the economy and society. As a result, a different degree of the qualified specialists' readiness for continuous professional development, digitalization risks, depending on the professional sphere of activity and the level of education was revealed. A higher propensity of specialists with the highest level of qualification for labor mobility was revealed. Qualification groups with the least fear of disappearance of their profession in connection with digitalization were identified. Research results can be useful in designing corporate training and government employment programs.

Keywords: Continuing professional education · Digital competencies · Digitalization risks · Qualified specialists · Professional education · Professional skills

1 Introduction

Economic digitalization makes new demands for competencies of the working population, which causes active discussions among specialists on demand for professional skills associated with technological changes [14]. However, the demand should be provided, either by fully prepared specialists, or by specialists capable of training and motivated to continue development [3]. The bulk of the research is focused on identifying the present and future needs of employers [7, 15]. To a lesser extent, the qualified specialists' readiness to change work functions, skills, and their readiness for such changes through objective and subjective assessments is revealed [4]. Adequate self-assessment of the level of proficiency is an important component for the formation of digital competencies and further development of employees, adaptation to the

S. I. Ashmarina and V. V. Mantulenko (Eds.): IPM 2020, LNNS 161, pp. 364–373, 2021.
https://doi.org/10.1007/978-3-030-60926-9_47

challenges of the rapidly developing Industry 4.0, which was the basis when choosing the objectives of the study.

The authors adopted subjective assessments of the working population as a tool for measuring the level of skills. When assessing the following features were considered: consequences of the introduction of computer technology in the professional field by various groups of surveyed workers; the level of self-assessment of skills by the working population (business communication, computer literacy, English proficiency, skills by profession), and the readiness to form demanded competencies, to adapt to digital challenges of our time and to increase professional mobility.

2 Methodology

The authors use some results of the study conducted in the framework of the project "Social Attitudes and Dispositions of the Population of the Russian Federation Regarding New Forms of Employment: Subjective Assessments", implemented based on the Resource Center for Sociological and Internet Research of St. Petersburg State University using a computer system for conducting telephone surveys/CATI [10]. The target group of the study is the citizens of the Russian Federation in the age range of 18–59 years old, engaged in income-generating labor activities. A representative sample for an all-Russian telephone survey was 1,634 people. Sampling error: 5%. Calculation of sampling error by attributes: gender, age, place of residence (types of settlements). Distribution by gender: men - 764 people (53.2%) women - 870 people (46.8%), age: 18–29 years old - 494 people (30.2%); 30–39 years old - 394 people (24.1%); 40–49 years old - 368 people (22.5%); 50–59 years - 378 people (23.1%).

The research methods include theoretical analysis, document analysis, development of research design and research tools, telephone survey. The questionnaire used a group of standard questions of the socio-demographic block (gender, age, federal district, size of settlement, education, occupation/foreign economic activity, marital status, dependents, per capita family income). Project implementation period: June 14 - September 25, 2018; Fieldwork period: June 29 - July 13, 2018. The distribution of the employed population by the level of education, types of activity and occupation groups was obtained randomly and with a given probability reflects the social structure of society. The share of specialists with higher and intermediate level of qualification is 42% of all respondents, which is comparable with the data of a sample survey of the Rosstat labor force [13]. The distribution of the sample by sectors of the economy was obtained considering the same rule of chance and representativeness: industry - 14.9%, information technology - 3.2%, transport and communications - 8.4%, construction - 6.9%, etc.

3 Results

The employees' assessment of risks of digital transformation in their professional field is differentiated depending on the level and profile of professional education. Thus, specialists with secondary vocational education note the possibility of disappearance of their profession almost twice as often (17%) than specialists with higher education

(9%). Regardless of the profile of education, all respondents as the least likely point out the risk of disappearance of their profession, and as the most likely - the risk of higher requirements for skills (Fig. 1).

The lowest risks of disappearance of the profession are among those employees

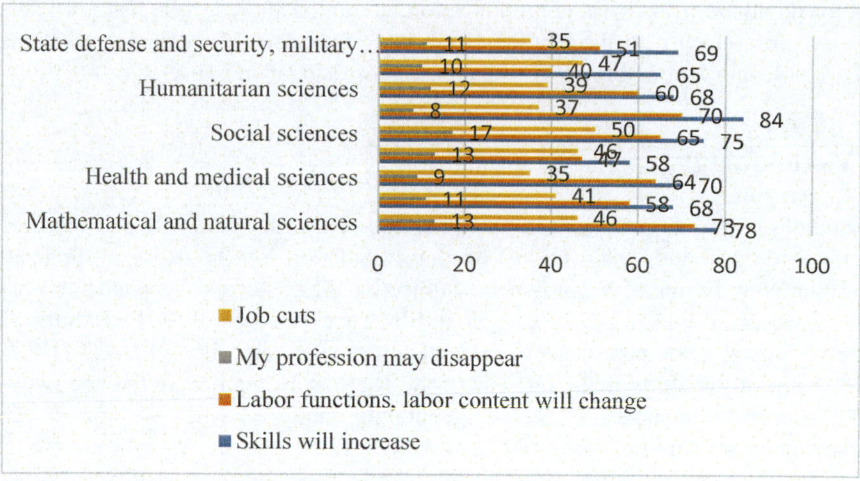

Fig. 1. The sum of the answers "yes" and "rather yes" to the question about the consequences of the widespread adoption of computer technology in the profession in % (distribution by the field of vocational education) (Source: authors).

who received professional education in the field of pedagogy (8%) and health care (9%), while workers in the field of social sciences estimate such risks almost twice as high (17%). The greatest possibility of increasing the requirements for skills is among workers who received professional education in the field of pedagogy (84%), mathematical and natural sciences (78%), social sciences (75%), health care (70%).

Assessment of Digitalization Risks of Professional Activity by Qualified Specialists. Among employees of all occupational groups (Fig. 2), specialists with the highest level of qualification are least afraid of disappearance of their profession/specialty (9%), while showing significantly higher expectations regarding changes in labor functions and labor content (67%) and increased requirements for skills (81%) and a willingness to cut jobs (40%), comparable to workers in other groups of classes.

Among the "white-collar workers" in the groups of employees engaged in the preparation and execution of documentation, accounting there are higher expectations of disappearance of the profession (respectively 18%, or 2 times higher and 15%, or 1.7 times higher), and at the same time, a comparatively lower possibility of changing the content of labor and increasing the requirements for skills. At the same time, it should be noted that expectations of job cuts in the groups of unskilled workers and operators of production plants and machines, assemblers and drivers, which are less aware of the

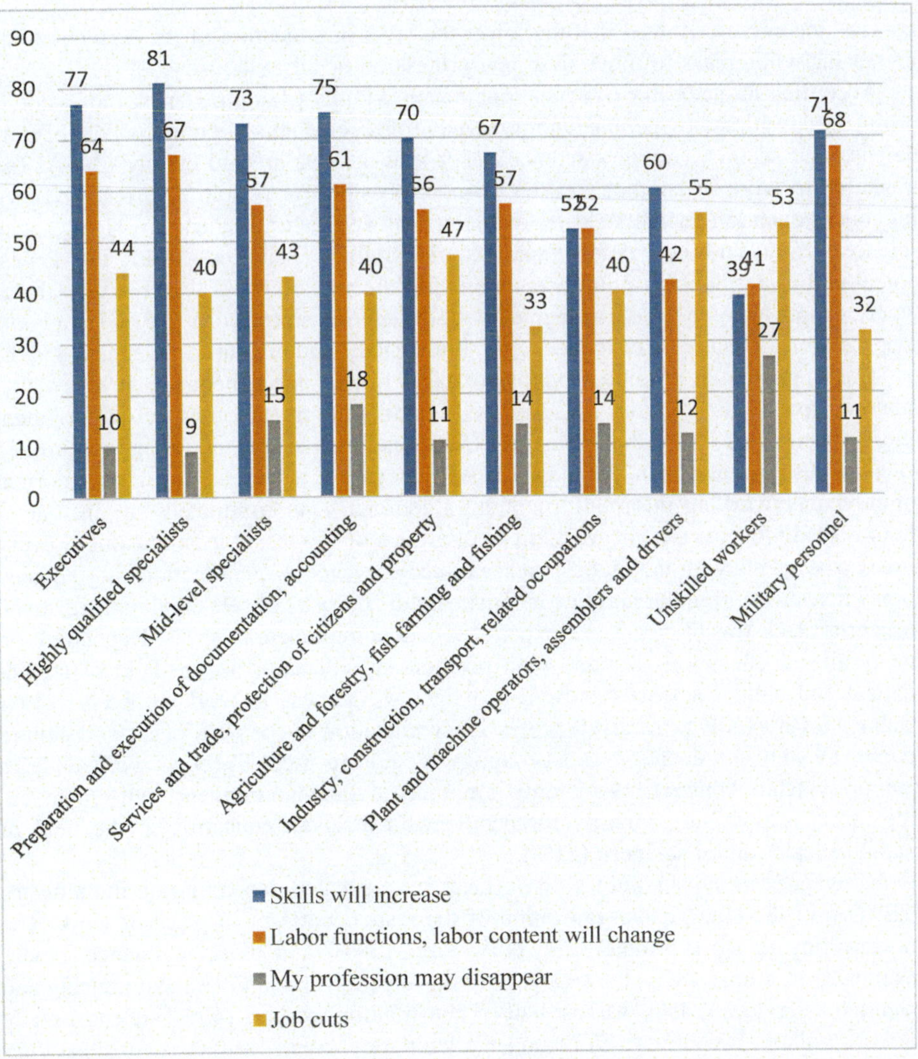

Fig. 2. The sum of the answers "yes" and "rather yes" to the question about the consequences of the widespread adoption of computer technology in the profession in % (distribution by groups of respondents) (Source: authors).

possibility of changes in the content of labor and growing demands on skills, are significantly higher than qualified specialists (by almost a third).

The Readiness for Changes in Labor Activity Depending on the Level and Profile of Education. The analysis showed that workers with higher education show higher readiness for territorial mobility compared to other groups of the working population (39%), and this type of change in labor activity is preferable for them compared to getting a new profession (28%) or moving to work in another industry (32%). In general, the survey showed that the lower the level of education of the employee, the higher his willingness to move to a new profession or other industry.

According to the profile of education, the opportunity to acquire a new profession is primarily considered by people who received professional education in the field of state defense and security, military sciences (42%), as well as art and culture (38%). The transition to work in another industry is more possible for people who received professional education in the field of mathematical and natural sciences (41%); social sciences (39%) and state defense and security, military sciences (38%). The lowest potential for interindustry mobility is among healthcare workers (14%). For the sake of work, people who received a degree in state defense and security (61%) are most prepared for a change of residence; agriculture (50%); humanitarian sciences (45%).

The study of the possible options for the workers' labor behavior revealed that qualified specialists are more conservative regarding the transition to self-employment (entrepreneurship) - 39% than managers (59%) and skilled workers in agriculture and forestry, fishing and fish farming (57%). They are also less loyal to informal employment (working informally, without a contract) - approximately one in four of them is ready for this type of work (in most groups of classes, at least one out of every three). Specialists with the highest and intermediate level of qualification call the need to pay taxes with irregular and low income; significant risks of self-employment; social insecurity; lack (insufficiency) of experience and entrepreneurial skills. Depending on the profile of education, workers with professional education in the field of mathematical and natural sciences showed a greater willingness to work remotely - 49% (Table 1). Persons who received professional education in the field of art and culture (57%), as well as education and pedagogy (45%) are most likely to work under a temporary labor contract. Work only for yourself (an entrepreneur, self-employed, freelancer) is possible primarily for people who received education in the field of mathematical, natural sciences (51%).

Differences in labor attitudes of workers were revealed depending on the industry affiliation of the enterprise/organization of the respondent's main place of work. The most willing to work remotely are IT workers (58%) and those in finance, credit, insurance, pensions (48%), the smallest are healthcare workers (19%) and housing and communal services (24%). Almost half of construction workers (45%) are most ready to work without a contract. Work under a temporary labor contract is for most construction workers (55%), hotels, restaurants, catering (59%), culture and art (51%), housing and communal services (50%). Self-employed (as an entrepreneur, self-employed, freelancer) is attractive for builders (60%), trade workers (52%), hotels, restaurants, food services (59%), IT (56%) and services (54%).

The readiness for continuous professional development in the context of digitalization for all groups of employees surveyed is quite high (Table 2). However, specialists with higher and intermediate qualification certainly lead in declared installations to study, improve their qualification or gain a new profession (about 81%), which significantly distinguishes them from "blue-collar" groups of workers, especially unskilled workers (among whom only every second declared readiness for learning or getting a new profession).

Among highly qualified specialists, the minimum percentage of those who noted the lack of need for training (10%), as well as (along with a group of managers - 40%)

Table 1. The sum of the answers "yes" and "rather yes" to the question about options for labor behavior in the future, in % (distribution by industry).

Branch of the enterprise/organization of the respondent's main place of work	Work remotely	Work informally, without a contract	Work under a temporary employment contract	Work only for yourself (entrepreneur, self-employed, freelancer)
Agriculture and forestry	40	34	44	44
Industry	31	26	36	36
Construction	38	45	55	60
Transport and communication	31	35	35	42
Trade	39	34	44	52
Hotels, restaurants and catering	33	37	59	59
Public administration	23	26	25	32
Finance, credit, insurance, pensions	48	26	21	48
IT field	58	28	42	56
Housing and communal services	24	29	50	32
Healthcare	19	25	33	32
Education	31	28	45	29
Culture and art	43	35	51	46
Service sector	31	36	42	54

Source: authors.

the maximum percentage of those who expressed their willingness to invest independently in training (34%). Among specialists with the average level of qualification, who are ready to study at their own expense - less than 28% (which is explained by a lower level of wages), but this indicator is nevertheless 27% higher than in the group of skilled workers in industry, construction transport and almost 87% - than in the group of operators of production plants and machinery, assemblers and drivers. At the same time, mid-level specialists occupy a leading position in readiness for training at the expense of the state or the employer (53%). A comparable readiness potential for training at the expense of the state/employer was identified among employees involved in the preparation and execution of documentation, accounting (almost every second - 49%). These two groups are equally significantly affected by digitalization risks of functions, the possibility of a significant reduction in employees, which is already recognized by the latter, showing their readiness for advanced training.

The same circumstance is reflected in the level of the employees' readiness to receive a new profession. Despite the highest potential for training, only every fourth specialist with the highest level of qualification gave a positive answer about the possibility of obtaining a new profession, while among mid-level specialists - every third.

Table 2. Willingness of employees to study, receive a new profession, in %

Study at the respondent's main place of work	Ready to study at my own expense	Ready to study at the expense of the state, the employer	No training required	Not ready to study	Difficult to answer	Total
Executives	40	38	5	12	5	100
Highly qualified specialists	34	47	6	10	3	100
Mid-level specialists	28	53	3	14	2	100
Employees involved in the preparation and execution of documentation, accounting	29	49	5	16	1	100
Workers in the sphere of services and trade, protection of citizens and property	28	43	5	21	3	100
Skilled workers in agriculture and forestry, fish farming and fishing	31	29	6	29	5	100
Skilled workers in industry, construction, transport, related occupations	22	42	10	21	5	100
Plant and machine operators, assemblers and drivers	15	47	15	22	1	100
Unskilled workers	15	35	7	40	3	100
Military personnel	21	50	4	18	7	100

Source: authors.

Regardless of the profile of professional education, the dominant share of employees is aware of the need for training in connection with the transformation of the sphere of professional activity. From 2% (in the group with humanitarian education) to 13% (in the group with agricultural education) of workers declare that there is no need for training, while in most cases this share does not exceed 4–6% and refers to people who decide to stop working according to the age. The largest share of workers willing to study at their own expense belongs to groups with a degree in humanitarian (43%), mathematical and natural sciences (41%). Among those who are ready to study (including getting a new profession) at the expense of the state or the employer, there are groups with education in the field of health care, pedagogy and military sciences (54–55%), they are workers in the so-called public sector. The willingness of Russian workers (including qualified specialists) to study, improve their qualification, acquire a new profession is steadily decreasing with age, which does not meet the requirements of the transforming labor market and requires special attention in conditions of increasing retirement age.

4 Discussion

The willingness to use digital skills in professional activities begins to take shape during the training period. The readiness for continuing education should be formed during the period of vocational training in educational institutions [11]. Digital preparation for information and communication technologies as a convenient support tool should begin in high school [6, 18]. Despite students get the sufficient digital training during the training period and have high confidence in their digital skills, employers express dissatisfaction with the preparation of graduates for modern digital tools in performing functional duties [2, 8]. We need constant updating of skills in various professional fields, for example, in medicine [5, 17], but specialists should have digital skills in all areas of activity. These skills are necessary for specialists and graduates for successful employment [12]. It is assumed that the young generation, born and raised already in the era of telecommunications, will be more adapted to new technologies [16]. However, the massive replacement of traditional technologies with advanced teaching methods does not always lead to an increase in the quality of education and an increase in the ability of students to master and use constantly emerging new technologies [1]. The organizational readiness for change is becoming a key skill that both students and working professionals need to prepare for [9].

5 Conclusion

The transition to a postindustrial economy has changed the structure of knowledge and skills that are in demand, digital competencies, which until recently were considered narrowly specialized, have become cross-cutting, the training of which becomes a necessity in any professional field. Given the prevailing structure of the labor market, which is dominated by middle and older specialists, it is necessary to consider the mood and self-assessment of this category of workers. The analysis revealed a high willingness to train and retrain specialists of different age categories. Systematization and analysis of the information obtained during the analysis made it possible to identify trends in the mood of qualified specialists in various fields of activity towards the risks of disappearance of professions caused by digitalization. At the same time, the fear of a significant change in qualification requirements in connection with modernization and robotization of production is quite high. It should be noted a high proportion of qualified specialists who would like to work only for themselves.

Self-assessment of skills is differentiated by field of activity, occupational groups, level of education of the employed population. Integration into the digital economy is ongoing and recognition of the need for professional development is common to all age and professional groups. A more adaptive adjustment of the labor market to digital challenges is needed, which allows us to clearly navigate in the field of popular digital competencies, which will stimulate training for new types of jobs.

References

1. Cohen, J.D., Jones, W.M., Smith, S.: Preservice and early career teachers' preconceptions and misconceptions about making in education. J. Digit. Learn. Teach. Educ. **34**(1), 31–42 (2018)
2. Cortesi, S.C., Hasse, A., Lombana, A., Kim, S., Gasser, U.: Youth and digital citizenship + (Plus): understanding skills for a digital world. SSRN Electron. J. (2020). https://doi.org/10.2139/ssrn.3557518. Accessed 22 June 2020
3. Frank, T.H.J., Castek, J.: From digital literacies to digital problem solving: expanding technology-rich learning opportunities for adults research review. J. Res. **6**(2), 66–70 (2017)
4. Georgievna, P.S., Aleksandrovich, K.S., Aleksandrovna, R.I.: Formation of readiness of future engineers-builders to innovative activity. Persp. Nauki i Obrazovania **32**(2), 75–79 (2018)
5. Goe, R., Ipsen, C., Bliss, S.: Pilot testing a digital career literacy training for vocational rehabilitation professionals. Rehabil. Counsel. Bull. **61**(4), 236–243 (2018)
6. Harrell, S., Bynum, Y.: Factors affecting technology integration the classroom. Alabama J. Educ. Leadersh. **5**, 12–18 (2018)
7. Khare, N., Koyadan, S.: Implementation of a successful talent management and succession planning governance framework. In: Proceedings of the Society of Petroleum Engineers - SPE Abu Dhabi International Petroleum Exhibition and Conference (133061). Society of Petroleum Engineers, Abu Dhabi (2017)
8. Mishra, K.E., Wilder, K., Mishra, A.K.: Digital literacy in the marketing curriculum. Ind. High. Educ. **31**(3), 204–211 (2017)
9. Mrayyan, M.T.: Nurses' views of organizational readiness for change. Nurs. Forum **55**(2), 83–91 (2019)
10. Omelchenko, I.B., Zabelina, O.V., Mirzabalayeva, F.I.: Self-working population of their skills in terms of digitalization of the economy. Labor Econ. **6**(1), 63–76 (2019)
11. Önger, S., Çetin, T.: An investigation into digital literacy views of social studies preservice teachers in the context of authentic learning. Rev. Int. Geogr. Educ. Online **8**(1), 109–124 (2018)
12. Privalov, A.N., Bogatyreva, Y.I., Romanov, V.A.: Engineering centre as innovative component of professional training of future IT specialists. Obrazovanie Nauka **21**(7), 90–112 (2019)
13. Rosstat: Regions of Russia (2019). https://www.gks.ru/folder/210/document/13204. Accessed 26 June 2020
14. Schwartz, J., Collins, L., Stockton, H., Wagner, D., Walsh, B.: Rewriting the rules for the digital age. Deloitte Glob. Hum. Capit. Trends **139** (2017). https://www2.deloitte.com/content/dam/Deloitte/global/Documents/HumanCapital/hc-2017-global-human-capital-trends-gx.pdf. Accessed 22 June 2020

15. Stępnikowski, A.: 5th level of Polish qualification framework in the contexts of education and labor market. Polit. Społeczna **553**(4), 8–16 (2020)
16. Tkacheva, O.N., Simonova, M.V., Matveev, Y.V.: The model of quality assessment of a scientific and educational network performance. Math. Educ. **11**(8), 2871–2883 (2016)
17. Trivedi, I., Keefer, L.: Barriers and facilitators of healthcare transition in IBD: adult provider perspectives. Am. J. Gastroenterol. **110**, S815 (2015)
18. Tsalapatas, H., Heidmann, O., Houstis, E.: A serious game for digital skill building among individuals at risk, promoting employability and social inclusion. In: de Carvalho, C.V., Escudeiro, P., Coelho, A. (eds.) Serious Games, Interaction and Simulation. Lecture Notes of the Institute for Computer Sciences, Social-Informatics and Telecommunications Engineering, vol. 176, pp. 91–98. Springer, Cham (2017)

Digital Transformation of Methods for Primary Selection of Job Candidates

Yu. Tokareva[1], S. Lipatova[1], and A. Tokarev[2(✉)]

[1] Ural Federal University named after the First President of Russia B.N. Yeltsin, Ekaterinburg, Russia
LipatovaSD@mail.ru, Ulia.tokareva@yandex.ru
[2] Surgut State Pedagogical University, Surgut, Russia
tochkae@yandex.ru

Abstract. The goal of the article is to consider the possibilities of digital transformation of primary methods for personnel selection in companies. The study included the following stages: 1) theoretical analysis of methods and sources of primary selection; 2) a survey of HR managers, conducted to identify the main sources of information about job candidates and their effectiveness; 3) monitoring of job-sites and comparative analysis of the completeness and quality of the information they provide; 4) expert interview with O.A. Gavrilova, the General Director of Nova LLC and the creator of the job-site "Sea of Opportunities" on the possibility of digital transformation of primary personnel selection methods. A comparative analysis of the completeness and quality of information showed that none of the leading Russian job-sites provide information on the level of development of professional knowledge, skills and personal qualities of applicants. However, these indicators affect the quality of candidate selection most strongly. The article shows the possibility of improving the quality of personnel selection by digitally transforming test diagnostics of professional and personal qualities during the primary selection. The presented study helps to improve the quality of personnel recruitment and to reduce the time for closing vacancies.

Keywords: Digital transformation · Job-sites · Primary selection of personnel · Testing

1 Introduction

The willingness of Russian companies to rebuild business processes taking into account the latest IT trends is an important condition for development and survival in the market. Digital transformation is one of such trends. The recruiting field is still the leader in digitalization investments. Companies continue to use increasingly sophisticated search tools in pursuit of a "perfect" employee [13].

Well-established personnel selection technologies cease to be effective. Digital transformation of the selection process will reduce the time spent on finding employees, as well as improve the quality of the selection of applicants. The rapid change of the specialists' roles in the company and the formation of new positions is already encouraging HR managers to apply new views and methods of employee

S. I. Ashmarina and V. V. Mantulenko (Eds.): IPM 2020, LNNS 161, pp. 374–380, 2021.
https://doi.org/10.1007/978-3-030-60926-9_48

selection. Firstly, recruiters use more sources to search for candidates nowadays. Secondly, recruiters can obtain the necessary information about the candidate with the help of new technologies. Thirdly, it is becoming increasingly difficult to use the previous job descriptions, as their components are constantly changing, and the question of how exactly the employee will fit into the organization requires the involvement of more and more people [14].

Most sources of information about a job candidate are limited to personal data (education, work experience, etc.). However, according to Zarina [14], it is important to see if the candidate's skills and behaviors meet certain criteria, to understand how the person fits into the team and how his abilities and skills… satisfy the future tasks of the company. This problem can be solved with the help of digital transformation of methods for personnel selection [2, 8].

2 Methodology

Kibanov defines recruitment as the process of studying the psychological and professional qualities of an employee in order to establish his suitability to perform duties in a particular job or position; the process of selecting the most suitable applicant based on the compliance of his qualification with the position, as well as the suitability of his personal qualities and abilities to the nature of work, the interests of the organization and the applicant himself [9].

The author focuses on the process of studying not only professional, but also personal qualities of an employee. Currently, the most common methods of primary selection are analysis of personal data and testing. HR specialists analyze the information contained in the questionnaires filled by candidates and compare actual data with their own model. The analysis of personal data is rather approximate in assessing the candidate's potential, since it is focused solely on facts from the candidate's past, and not on his current state and the ability for professional development [7]. In order to obtain reliable data on the current state of the candidate and his ability for professional development, it is necessary to use test diagnostics.

Testing is currently becoming more popular among the leading organizations in developed countries. It is used not only by corporations, but also by government agencies and public organizations. This primary selection method has the following disadvantages: high costs, the frequent need for special assistance, the conventions and limitations of tests which do not provide a complete picture of the candidate [4]. On the one hand, testing makes it possible to obtain reliable information about the professional development of job applicants. On the other hand, this method is not actively used due to the high costs and the need for additional involvement of specialists.

Three types of tests are most often used when applying for a job: qualifications and skills test; tests for general intelligence and other abilities; test for certain personal qualities and the degree of their manifestation [7]. Using several specially selected techniques helps to compile a psychological portrait of the job candidate.

Primary selection requires a huge investment of time and human resources during the mass recruitment for vacant positions. This reduces the efficiency of the organization. Automation of the primary selection process solves this problem. This transformation has happened to acquire highly skilled employees at low cost. And it saves time, too. Social Media are big boon for the companies which have adopted innovative and blooming Internet-based recruiting techniques. The study finds the innovative way of utilizing Social Media networks as recruitment sources on E-Selection [11].

There are many sources for finding candidates: regional and city websites, websites of professional communities, social networks, the company's website, etc. The variety of sources for informing candidates requires the selection of the most effective ones in terms of both the number of utilized resources and their quality [6]. Dolgopolova and Khasanov suggest the following criteria for evaluating the source's effectiveness: share of candidates who came for an interview in the total number of applications received from the source; share of candidates deemed eligible after the initial interview; share of candidates who have passed the probationary period; the cost of recruitment; job closing speed or recruitment time [3, 6]. A comparative analysis of the recruitment sources' effectiveness showed that the Internet sources are leaders in this regard. Job-sites are leaders among current sources of the candidate search for mass recruitment, because they can significantly reduce the search time for candidates, as well as increase the efficiency of the selection process.

In large organizations with a high level of employee turnover (retail, communication companies, etc.), the need for personnel recruitment is almost constant. Therefore, it is advisable to digitalize the search and selection of personnel using modern digital tools [1]. In 2017, the Deloitte company conducted the international study "International study of trends in HR management – 2017", which surveyed more than 10 thousand company executives and heads of HR management services from 140 countries, including 156 respondents from Russia. The results of the study showed that only 23% of Russian companies participating in the survey have automated "all or almost all" HR functions. 67% of companies have only partially automated HR functions, and in 10% automation is completely absent.

According to Ryzhkov's research, 35% of companies have already prepared a digital transformation strategy and are implementing it. More than 60% of surveyed companies are still studying information on this topic or are just beginning to develop a transformation strategy. 10% of surveyed companies do not engage in digital transformation [12]. What is stopping companies from introducing digital transformation? To answer this question, Ryzhkov lists the obstacles that can interfere with the organization's digital transformation, according to the employees of large companies: insufficient competencies and knowledge - 64.1%; lack of qualified personnel - 60.9%; lack of strategy - 53.3%; fear of change - 45.3%; insufficient funding - 39.1%; top management's position - 31.3%; risks - 23.4%.

The majority of respondents cited insufficient competencies and a lack of qualified personnel in the company as the main factors hindering the digital transformation development. Indeed, the development of psychological tests and questionnaires, as well as their digitalization, require a lot of time and the involvement of specialists in psychology and programming. Not every company can afford such expenses. Therefore, it is faster and more reliable to use a ready-made software product offered on the market. Thus, we set the following tasks:

1. To identify the main sources of information about job candidates and their effectiveness.
2. To conduct a comparative analysis of the main job-sites specialized on the search for job candidates. Identify the completeness and quality of the information provided by them.
3. To identify the possibilities of digital transformation of the primary selection methods.

Hypotheses:

1. HR managers use job-sites more often at the stage of primary selection of job candidates.
2. Russian job-sites do not provide complete and high-quality information about job candidates.

The study consisted of three stages.
Stage 1 - A survey of representatives of Russian companies (top managers, HR managers).

Objects of the study: 26 companies.

An empirical research method was used, which allowed:

- identifying priority sources of information about job candidates,
- ranking the effectiveness of sources for personnel recruitment.

Stage 2 - Analysis of the completeness and quality of information provided by job-sites. Objects of the study – the largest job-sites in Russia: hh.ru, rabota.ru, superjob.ru, zarplata.ru, avito.ru.

Stage 3 - Interview with Gavrilova, the General Director of Nova LLC and the creator of the job-site "Sea of Opportunities".

An expert interview method was used, which allowed: 1) identifying the possibilities of digital transformation of diagnostic techniques used in the primary personnel selection; 2) defining requirements for the quality of information about the job candidates; 3) identifying the distinctive features of the website "Sea of Opportunities". "Sea of Opportunities" is the Internet portal providing recruitment and job search services in Russia. The project was launched on September 6, 2019. The idea of creating a job-site belongs to Gavrilova.

Gavrilova was an expert in the field of recruitment software at the Seventh annual international startup conference for technology entrepreneurs StartupVillage 2019 (Skolkovo). She has over 20 years of senior management experience in the IT industry, including work in LLC RPE "Systems and Networks", LLC "Taxcom", etc. She is a certified specialist in software, including 1C: Enterprise 8*, especially in the "Salary and HR Management" configuration.

3 Results

A survey of representatives of Russian companies allowed us to rank sources of information about job candidates by their effectiveness (Table 1).

Table 1. Rating of information sources about job candidates

Rating	Source of information about job candidates	Effectiveness of the source of information about job candidates (%)
1	Job-sites	72,2
2	Social networks	11,4
3	Recommendations of friends, colleagues	8,8
4	Regional and city websites	3,8
5	Websites of professional communities	3,8

Source: authors based on [5].

The analysis of effectiveness of the information sources about job candidates showed the advantage of job-sites over other sources - 72.2%. The survey results confirmed the findings of previous studies [6]. Do Russian job-sites meet the requirements for quality and completeness of information about candidates? To answer this question, we developed criteria for assessing the quality of information provided on job-sites. Theoretical analysis revealed that the main methods of primary personnel selection are: analysis of personal data (CVs); testing of professional knowledge, skills, and personal qualities of job candidates. Based on this data, we used the following criteria to assess the quality of information provided on job-sites: CV quality (structure); availability of diagnostic tools to test professional knowledge and skills, as well as to study personal qualities of candidates; number of CVs posted on the site. The results of an empirical study are presented in Table 2.

Table 2. Comparative analysis of the completeness and quality of information provided on the main Russian job-sites

Criteria for the completeness and quality of information	Title of the job-site				
	avito.ru	hh.ru	rabota.ru	superjob.ru	zarplata.ru
1. Structured CV	–	+	+	+	+
2. Testing of professional knowledge and skills	–	+*	–	–	–
3. Diagnostics of personal qualities	–	+*	–	–	–
4. The number of CVs (TOP 10 sites to post a CV for job search, 2019)	1 million	25 million	1 million	20 million	2 million

*the site provides an opportunity for the employer to draw up a test and send it to the applicant to fill out.

Source: authors based on [10].

Is it possible to create a database of diagnostic tools within a single website? We received the answer to this question by conducting an expert interview with Gavrilova, the General Director of Nova LLC. Gavrilova believes that all standard recruitment sites are based on viewing and creating the applicant's CV. Moreover, that CV is quite specific: the photograph of the applicant is important, because it helps to draw certain conclusions, and previous work experience is significant, too. Generally, the candidate describes his key skills and personal qualities as he sees fit. Thus, the initial conclusion about the candidate's professionalism is based on his self-assessment. Ineffective selection at the initial stage provides a much smaller selection at the subsequent stages. Diagnostic results are compiled into a table, where each candidate is assigned a percentage of knowledge. A rating of candidates for the position is created on the basis of these data.Thus, the indicators of professional knowledge and skills, as well as a psychological portrait of a person, are digitized. Based on this information, the HR manager can make a preliminary conclusion about the candidate before the first interview.

4 Discussion

Gavrilova proposes to solve this problem through the digital transformation of personnel assessment methods. She suggests developing blocks of questions for each specialty that are focused on the functions and competencies of a particular position. The first block consists of general questions: occupation, area of interest, etc. Other blocks include legal knowledge testing, functional duties diagnostics. The candidate also undergoes a series of psychological tests recommended for this position. They include intellectual tests (the ability to write and speak correctly, analytical abilities, spatial thinking and imagination), attention and memory tests, personality questionnaires, projective tests, tests of interpersonal relationships (conflict, sociability, attentiveness to people) and others. A list of primary and secondary tests is compiled for each position. The employer chooses qualities that are important for a particular position. After that, the candidate is invited to pass these tests. In the process of collecting information about candidates, employers use such sources as job-sites, social networks, regional and city websites, recommendations of friends and colleagues, websites of professional communities. Job-sites are one of the most popular and effective sources of information about job candidates for HR managers. Hypothesis confirmed. The main Russian job-sites (avito.ru, hh.ru, rabota.ru, superjob.ru, zarplata.ru) cannot fully provide employers with objective information about the professional and personal qualities of job candidates. Hypothesis confirmed. Job-sites provide the employer with a choice of a huge number of candidates. Other sources of information cannot provide this opportunity. However, quantity does not imply quality. For high-quality selection, it is necessary to have objective knowledge about the professional and personal qualities of the candidate.

5 Conclusion

Automation of personal data of job candidates is a fait accompli. Automation of the diagnostic tools of professional and personal qualities of candidates is the next step in the digital transformation of primary selection methods. It will provide the employer with reliable information obtained with the help of survey and test diagnostics.

Not every company can develop and digitize a bank of diagnostic tools, so job-sites can further the advancement of digital transformation in this direction. Thus, the article outlines the ways for further development of the digital transformation of primary selection methods through the automation of test diagnostics. This will help to create a professional and personal portrait of the job candidate. This approach will also improve the quality of personnel recruitment and reduce the time for closing vacancies. As part of our future research, we plan to develop, digitize and test survey and test diagnostics for the most popular specialties.

References

1. Arkhipova, N.I., Sedova, O.L.: The use of digital tools in the selection and recruitment of personnel in the organization. RSUH Bull. Ser. Econ. Manag. Law **2**(12), 9–22 (2018)
2. Backman, C., Hedenus, A.: Online privacy in job recruitment processes? Boundary work among cybervetting recruiters. New Technol. Work Empl. **34**(2), 157–173 (2019)
3. Borisyuk, F., Zhang, L., Kenthapadi, K.: LiJAR: a system for job application redistribution towards efficient career marketplace. In: Matwin, S., Yu, S., Farfooq, F. (eds.) Proceedings of the ACM SIGKDD International Conference on Knowledge Discovery and Data Mining, pp. 1397–1406. Association for Computing Machinery, New York (2017)
4. Deineka, A.V.: Personnel Management. Dashkov and Co., Moscow (2013)
5. Deloitte: International study of trends in HR management – 2017 (2017). https://www2.deloitte.com/ru/ru/pages/human-capital/articles/introduction-human-capital-trends-2017.html. Accessed 15 Oct 2019
6. Dolgopolova, I.V., Khasanov, R.A.: A comparative analysis of the effectiveness of sources of personnel attraction in the labor market of the Perm region (example of Uralkaliy OJSC). PSTU Bull. Soc.-Econ. Sci. **4**(25), 103–111 (2014)
7. Gordienko, Yu.F., Obukhov, D.V., Samygin, S.I.: HR management. Fenix, Rostov on Don (2009)
8. Hedenus, A., Backman, C., Håkansson, P.: Whom do you know? Recruiters' motives for assessing jobseekers' online networks. The International Journal of Human Resource Management. In Press (2019). https://doi.org/10.1080/09585192.2019.1579245. Accessed 15 Oct 2019
9. Kibanov, A.Ya., Durakova, I.B.: HR management of the organization: current technologies of recruitment, adaptation and certification (2016). http://econom-lib.ru/4.php. Accessed 10 Oct 2019
10. ProfiComment: TOP 10 sites to post a CV for job search (2019). https://proficomment.ru/top-10-sajtov-gde-razmestit-rezyume-dlya-poiska-raboty/. Accessed 25 Oct 2019
11. Ramkumar, A., Rajini, G.: Innovative way of using social media networks for e-recruitment and selection. Int. J. Sci. Technol. Res. **10**, 175–181 (2019)
12. Ryzhkov, V., Chernov, E., Nefedova, O., Tarasova, V.: Analyst report: digital transformation in Russia 2018. LLC «Komanda-A Management», Moscow (2018)
13. Tyan, E.: How digital trends are changing the HR system (2019). https://umstrana.ru/article/elena-tyan-gonka-vooruzheniy-na-rynke-truda-v-razgare-/. Accessed 15 Oct 2019
14. Zarina, I.: Digital transformation in HR (2019). https://www.shl.ru/cifrovaja-transformacija-v-sfere-hr. Accessed 05 Oct 2019

Actual Competences for the Russian Civil Service System

O. A. Gris[1(✉)] and L. G. Lebedeva[2]

[1] North-West Institute of Management, Branch of Russian Presidential Academy of National Economy and Public Administration, Saint-Petersburg, Russia
777olga999@mail.ru

[2] Samara State University of Economics, Samara, Russia
ludleb@mail.ru

Abstract. The article is devoted to the problem of introducing the competence approach into the system of state and municipal administration in Russia. The purpose of this study was to determine the competencies, the presence of which will contribute to the achievement of the goals of public administration, solving the problems of society, identifying the needs of citizens, increasing the level of satisfaction of citizens of the Russian Federation with the quality of state and municipal services provided, as well as the level of public confidence in the authorities. The work used the analysis of works on this topic by Russian and foreign authors, the empirical basis of the study was a survey of representatives of the most interested groups - business and non-profit organizations. The results obtained indicate that the basic competence model proposed by the Ministry of Labor and Social Protection only partially corresponds to the tasks facing the state and the modern conditions of its functioning. The important missing elements of the competence-based model, according to the authors, are ethical competence, social and communication competence and communication and digital competence.

Keywords: Competence · Competency model · Digital transformation of the civil service · New public management

1 Introduction

The transformations that have taken place over the past few decades in the systems of public administration in different countries are due to the understanding of its importance for achieving the goals of the political system, forming the desired level of well-being and quality of life of the population. Taking business management principles as a basis, many states began to use the concept of New Public Management (NPM), which implies an emphasis on innovation, project management and meeting the needs of citizens. In the new model, a public or municipal employee becomes a participant in decision-making and the search for effective ways to implement them. Its main task is not to provide services, but to skillfully manage the process of connecting the resources of society, which requires special personal qualities, knowledge and skills. The transition to NPM required a change in the approach to training, the selection of state or

S. I. Ashmarina and V. V. Mantulenko (Eds.): IPM 2020, LNNS 161, pp. 381–387, 2021.
https://doi.org/10.1007/978-3-030-60926-9_49

municipal employees, to assessing the effectiveness of the result of their activities, the formation of new qualification requirements [2]. The answer to the request was the competence-based approach, which found its application already in the 1980s in the USA and Great Britain [5]. Currently, many countries around the world have developed and adopted their own competency models for the public sector, including Canada, Belgium, Sweden, Japan, Germany, Australia, New Zealand, the Netherlands, South Africa [10]. However, this process does not stop. The constantly changing conditions of the life of society, or the transformational processes of political systems, require updating of competence models, and for some countries - the introduction of a competence approach. Examples are the processes of France [7], Canada [12], China [13], Indonesia [17], Philippines [8], Kazakhstan [18], Slovakia [16], Azerbaijan [4], Tanzania [11] and others countries. The question of the need to apply a competency-based approach to the selection of employees is also considered in the context of the functioning of global international organizations [12].

In Russia, the need to apply a competency-based approach in the public civil service system was first announced in 2012. At this point in time, work is underway to identify the most in-demand competencies, develop competency models and introduce them into the state management process, taking into account specific responsibilities and functions, as well as the requirements of certain professional groups. As a result, a legal basis was prepared, methodological recommendations of the Ministry of Labor and Social Protection were developed [14], and the experience of using the competence-based approach in various state and municipal bodies is being generalized. In the methodological recommendations of the Ministry, basic, professional-functional and functional qualification requirements are formulated, taking into account the category and group of positions in the civil service. Of the existing two basic behavioral approaches - the American approach and the functional - European, focused on solving specific professional problems [7], a behavioral approach has been chosen in Russia. Competence is understood as "a complex of professional and personal qualities manifested in the behavior of a civil servant, indicating the presence of knowledge, skills, and experience of professional activity necessary for effective and efficient performance of official duties" [14, p. 43]. In the proposed models of professional and personal competencies, behavioral indicators are highlighted. From June 2019, a new stage of reforming the civil service began, concerning the issues of improving the procedure for appointment to positions, incentives, new forms of professional development of civil servants, etc. And at this point in time, the task is to develop competency models by areas and types of activities of civil and municipal employees. In this regard, the purpose of this study was to determine the necessary competencies, the presence of which would allow to effectively solve the assigned tasks, increase the level of satisfaction of the citizens of the Russian Federation with the quality of the provided state and municipal services, as well as the level of public confidence in the authorities.

2 Methodology

The study used an analysis of the works of Russian and foreign authors, and also conducted a survey of representatives of two socially significant groups that have the potential to become active participants in the process of achieving the set goals of the state - business and non-profit organizations, including public organizations. The time of the survey is November 2019 - February 2020. When conducting the survey, a non-representative targeted sample of 160 people (N = 160) was used. The sample structure is presented in Table 1.

Table 1. Sample structure (N = 160)

Region	Business representatives	Representatives of non-profit organizations
Leningrad region	30	20
Samara region	30	20
Belgorod region	20	10
Pskov region	30	10
Total	**100**	**60**

Source: authors.

The effectiveness of the work of regional and municipal services was assessed on a five-point scale, where 1 is extremely negative, 5 is the most positive. Through discourse analysis and similarity analysis, problems were identified with which respondents associate the inefficiency of the work of regional and municipal services. Since the purpose of the study was not to assess the effectiveness of the activities of state authorities and municipal authorities in specific regions, but to identify general trends associated with the problems of civil and municipal employees, the data are presented in general form.

3 Results

Let us give a brief description of the current situation in the system of public administration, the goals and objectives facing the state in relation to the issue we are considering. It should be noted that civil service reforms are being carried out under the influence of the global trend of public administration reforms, including taking into account the concept of New Public Management. In Russia, they strive to apply the practice of introducing a strategic management system, there is a redistribution of powers and functions between levels of government, a mechanism has been introduced that allows transferring a number of functions for the provision of public services to non-governmental organizations, communication channels with the population and stakeholders are being established, and much more. The goals and objectives of the state in accordance with 13 strategic directions of development have been embodied in national projects that address the problems of human capital development, creating a comfortable environment for life and economic growth. The basic model of competencies proposed by the Ministry of Labor and Social Development correlates with the current strategic goals and objectives [14]. It presents the following core competencies:

- strategic thinking, teamwork, personal effectiveness, flexibility and readiness for change - for all categories and groups of civil service positions,
- leadership, management decision making - for the highest group of positions.

The applied competencies necessary for the participants of project activities are especially highlighted. However, as the results of the study show, the proposed model of competencies does not affect the most important aspects of effective government management.

A survey of representatives of business and non-governmental organizations showed their high degree of willingness to participate in working groups and commissions created under the authority that discuss vital issues for society (business representatives - 68%, representatives of non-governmental organizations (NGOs) - 92%, average – 77%). At the same time, the respondents noted the absence or inaccessibility of information on the ongoing activities related to the formation of such working groups and commissions (representatives of business – 76%, representatives of NGOs – 58%, average – 69%). 34% of business representatives and 82% of NGO representatives came out with social and socio-economic initiatives to the authorities (the average indicator is 52%). Of these, 46% of respondents received an unmotivated refusal, 25% faced indifference and unwillingness to delve into the essence of the proposal, 19% encountered delays in making a decision. Were able to give examples of the effective work of regional and municipal services – 77% of respondents. The average assessment of the performance of regional and municipal services was 3.4 (on a five-point scale). Table 2 presents a ranked number of problems with which respondents associate inefficiency in the work of regional and municipal services.

Table 2. A ranked number of problems with which respondents associate inefficiency in the work of regional and municipal services

Rank	Reasons for the ineffective work of regional and municipal services	%
1	Bureaucracy, in the negative sense of the word	81
2	Misunderstanding the needs of society	69
3	Failure to build a system of interaction with representatives of the public by representatives of the civil and municipal services	59
4	Inability to make an independent decision, even if this function is within their competence	54
5	Shifting the solution of the issue to higher authorities	31

Source: authors

As can be seen from the presented survey results, one of the significant problems is the problem of building communication and interaction with representatives of stakeholders. And if, in response to the problems of inability to make a decision, bureaucratization, the model presented above proposes to introduce competencies that will be guided by in the preparation, hiring of new employees and assessing their performance, then social and communication competence did not find its application in it.

4 Discussion

Achieving the set goals of public administration and solving urgent problems facing society is impossible without taking into account the opinion and involvement of representatives of civil society and business, understanding the needs of citizens. The need to introduce social and communicative competencies into the Russian competence model is also shown by the results of a public opinion poll conducted by the Levada-Center [9] (2019). According to the data obtained (N = 1600), respondents consider the authorities in Russia to be criminal and corrupt (41%), distant from the people, alien (31%), bureaucratic (24%), short-sighted (19%). According to the respondents, "the authorities live at the expense of the population, they do not care much about how the people live" (53%), "there is no "feedback" between those who make political decisions at the top and the population" (19%). The majority of respondents believe that ordinary people cannot influence government decision-making in the country (most likely not - 33%, definitely not - 46%), as well as in the region, city and district (most likely not - 29%, definitely not - 37%) [9].

The need to understand the needs of various social groups and the ability to interact with the population is confirmed not only by the results of surveys, but also by the inability to implement the assigned tasks by various government bodies and municipal services. An example is failures, both in setting tasks, without involving all stakeholders, and in the implementation of projects for the development of a tourism cluster in the city of Togliatti [1]. As noted by Western researchers, the development of trusting relations between the state and society requires the presence of social competencies in civil servants [3].

The presented results of public opinion polls, as well as the ongoing processes in society, predetermine the consideration of two more important additions. First, the ethical aspects. Second, the continuation of discussions about the need to include communication competencies. So, for Russia, the experience of forming a competence model of leaders in the field of civil service in European countries may become relevant. It contains seven key competencies, among which a special place is occupied by:

- social competence - the ability to establish communication links and interact with representatives of society,
- ethical competence - honesty, responsibility, conscientiousness [3].

The UK Competence Model also contains ethical competence, which is based on personal and professional values [7]. It should be noted that Russian officials rank ethical competence among the multitude of competencies in the first place in importance in the competence profile for federal civil servants [6].

The modern competence model cannot be imagined without one more competence - "communication-digital". The ongoing processes of digital transformation in public administration related to the digitalization of the provision of public services, the introduction of end-to-end interdepartmental digitalization, impose special requirements on the knowledge and skills of state and municipal employees [15]. This competence will be necessary in the implementation of the activities of not only IT service specialists, but also all departments of state authorities and local government structures.

5 Conclusion

As part of the reform process, a competency-based approach is being introduced into the Russian system of public administration, which makes it possible to more effectively train and select personnel, evaluate the performance of civil and municipal employees. To date, the Ministry of Labor and Social Protection has developed guidelines for the formation of a competency model, which should become the basis for the formation of competence models for various activities of state civil and municipal employees. The study has shown that the basic model corresponds only partially to the tasks facing the state, the ongoing processes and modern conditions. Important missing elements that need to be included in the competence models of various areas of activity of civil and municipal employees are ethical competence, social and communication competence and communication and digital competence. Only in this format, competence models will not only contribute to the solution of "internal corporate state" tasks, but also take into account the interests of the public, increase the level of citizens' satisfaction with the quality of the state and provided municipal services, and also comply with modern trends.

References

1. Balashov, A.I., Kuznetsova, O.A.: Implementation practices of the marketing investment model by the single-industry city of Togliatti. Soc.: Polit. Econ. Law **1** (2018). https://doi.org/10.24158/pep.2018.1.4
2. Bartoli, A., Blatrix, C.: Management dans les organisations publiques: défis et logiques d'action. Dunod, Paris (2015)
3. Hassink, J., Salverda, I., Vaandrager, L., van Dam, R., Wentink, C.: Relationships between green urban citizens' initiatives and local governments. Cogent Soc. Sci. **2**(1), 1250336 (2016). https://doi.org/10.1080/23311886.2016.1250336
4. Huseyn, A.: Theoretical and practical aspects of the participation of civil servants in decision-making affecting business activities. In: Proceedings of the 37th International Scientific Conference on Economic and Social Development – "Socio Economic Problems of Sustainable Development", Baku, pp. 57–66 (2020)
5. Ingraham, P.W., Getha-Taylor, H.: Common sense, competence, and talent in the public service in the USA: finding the right mix in a complex world. Public Adm. **83**(4), 789–803 (2005)
6. Kutergina, E., Sanina, A.: The competency profiles of civil servants in today's Russia. J. Soc. Policy Stud. **15**(1), 113–128 (2017). https://doi.org/10.17323/1727-0634-2017-15-1-113-128

7. Larat, F.: Developing an integrated and comprehensive training strategy for public sector leaders: the French experience. Teach. Public Adm. **35**(1), 88–104 (2017). https://doi.org/10.1177/0144739416655598
8. Lee, D.S.: Executive control of bureaucracy and presidential cabinet appointments in East Asian democracies. Regulat. Gover. **14**(1), 82–101 (2020). https://doi.org/10.1111/rego.12190
9. Levada-Center: Public Opinion – 2019 (2019). https://www.levada.ru/cp/wp-content/uploads/2020/02/OM-2019.pdf. Accessed 30 July 2020
10. Lodge, M., Hood, C.: Symposium introduction: competency and higher civil servants. Public Adm. **83**(4), 779–787 (2005). https://doi.org/10.1111/j.0033-3298.2005.00475
11. Marijani, R.: Public service leadership competency framework [PSLCF]: is it a holy grail of service delivery? Open J. Soc. Sci. **5**, 169–184 (2017). https://doi.org/10.4236/jss.2017.510015
12. Mau, A.T.: Leadership competencies for a global public service. Int. Rev. Adm. Sci. **83**(1), 3–22 (2017). https://doi.org/10.1177/0020852315576706
13. Miao, Q., Newman, A., Schwarz, G., Cooper, B.: How leadership and public service motivation enhance innovative behavior. Public Adm. Rev. **78**(1), 71–81 (2018). https://doi.org/10.1111/puar.12839
14. Ministry of Labor and Social Protection of Russia: Methodological toolkit for establishing qualification requirements for filling civil service positions. Version 3.2 (2018). http://www.garant.ru/products/ipo/prime/doc/71755218/. Accessed 30 July 2020
15. Petrov, M., Burov, V., Shklyaruk, M., Sharov, A.: State as a platform: cyberstate for digital economies. Digital transformation. Center for Strategic Research, Moscow (2018)
16. Skorková, Z.: Competency models in public sector. Proc. – Soc. Behav. Sci. **230**, 226–234 (2016). https://doi.org/10.1016/j.sbspro.2016.09.029
17. Sudrajat, T., Kunarti, S., Hartini Bridging, S.: The legal gap between open selection and internal selection of state civil apparatus promotion in Indonesia. In: IOP Conference Series: Earth and Environmental Science, vol. 255, no. 1, p. 012053 (2019). https://doi.org/10.1088/1755-1315/255/1/012053
18. Suleimenova, G.K., Karamalayeva, Z.T.: Assessing civil servants training needs: the case of Kazakhstan. Public Adm. Issues Spec. Issue (Electron. Edn.) 96–115 (2018). https://doi.org/10.17323/1999-5431-2018-0-5-96-115

Innovative Personnel Development: From Resource Management to Capacity Management

Social Attitudes to the Career Process of Students with Different Career Orientations

I. V. Antonenko, O. L. Begicheva(✉), and I. V. Karpova

A.N. Kosygin Russian State University (of Design and Technology), Moscow, Russia
ivantonenko@mail.ru, olgalvb@mail.ru, pedagogikanet@yandex.ru

Abstract. In this study, social attitudes toward the career process of students with various career orientations are examined. After conducting a study based on Edgar Schein's questionnaire "Career Anchor" students are divided into groups according to how prevalent a criterion of the "career anchor" is. The results of the questionnaire are analyzed to understand the students' psychological features and to determine and interpreted among them the social attitude towards the career process basing on the various career orientations of the students. As a result, for each group of students significant characteristics and ways of making career are distinguished.

Keywords: Career anchor · Career process · Personal qualities · Social attitude · Students

1 Introduction

In the changing world of professions, modern persons are required to be mobile and competitive which makes a motivation to obtain high education at a university and clear career goals highly important. The interaction of these two factors determines individual learning process at a university [8]. Young people consider studying at a university as a way to improve their social status, since having a higher education diploma greatly expands the possibilities of future employment [10]. Having an independent professional opinion is an important status component, and intergenerational collaboration may tolerate a partial change in its priorities [6]. Students are a socio-demographic group, united by age, labour conditions (university studies and employment) and life (living in a dormitory or with parents). They are also characterized by distinctive features due the physiological, psychological and social maturation they experience [9].

2 Methodology

This study makes use of the fundamental approach to the study of personal career orientations, exhibits horizontal and vertical directions of career development [2]. The vertical direction of career development is associated with career advancement and a

S. I. Ashmarina and V. V. Mantulenko (Eds.): IPM 2020, LNNS 161, pp. 391–396, 2021.
https://doi.org/10.1007/978-3-030-60926-9_50

rise to a higher level of the structural hierarchy. A person oriented to vertical career development seeks career advancement, status promotion and leadership positions. Horizontal career orientation refers to a type of professional career which is based on the desire to achieve functional competence.

The study includes a complex of the following methods:

1. Schein's "Career Anchor" which serves as a meaningful questionnaire for determining the career orientations.
2. A questionnaire designed by the authors to identify socio-psychological characteristics and factors of the students' individual career orientations.

3 Results

In the study, 195 students of three Moscow universities took part: the State University of Management, the Moscow University for the Humanities and the A.N. Kosygin Russian State University. Their age was between 20 and 22 years. 60% of them were female and 40% male. The students first filled in Schein's "Career Anchor" - a questionnaire determining the individual career orientation. After analyzing the results the students were divided into groups according to their prevalent type of career orientation which are: management; autonomy; work stability; service; integration of lifestyles; enterprise; challenge.

Once divided into the aforementioned groups the students were asked to fill in the questionnaire which the authors designed especially to determine social-psychological features and factors of the students' career orientation. This questionnaire helped to detect social attitudes to the career process which students with different career orientations have. The following social attitudes to the career process were found and examined: 1. relevant personal characteristics and qualities which influence the career success; 2. career goals and ways of their achievement; 3. personal social attitudes; 4. metaphoric career associations.

The largest career orientation group was found to belong to the "management" type. The majority of this group (over 75%) indicated that the following characteristic have a considerable impact on the career success: self-confidence; correct self-esteem; activity; responsibility; flexibility. In comparison to other students, students of the "management" group tend to think that they esteem themselves correctly and that they are communicative, adaptive and flexible. Such results show that they are confident in their abilities and have a high, but still adequate self-esteem.

With a percentage higher than in other participant groups, almost all students of this group wish prosperity most in their lives - 94%. The least they are looking for is "to become a competent specialist" - 39%. Such a choice of goals in life proves their prevalent vertical career orientation. The majority of the "management" group also considers that for a successful career hard work and connections are necessary. During their study they participate more than others in study researches. The prevalent social attitude to the career process is "influence" (41%), and 67% of the students metaphorically associate their professional careers with "going upstairs".

The majority of students with "autonomy" as their prevalent career orientation indicated following characteristics as important for a successful career: self-confidence; sense of purpose; adequate self-esteem; independence. Students of this group are more self-confident and tend to count on luck more than students with other prevalent career orientations. The results indicate that the majority of them think of "activity" as important for career. More than others, students of "autonomy" group have "founding a good family" as their goal of life (90%) and the least goal is to "become a competent specialist" (30%). By such a choice they show that they consider their independence as a sort of social exclusion and their families as a independent personal social group. Their social attitude to the career process is characterized by "concurrence" (35%) and "recognition" (29%). 90% of these students think that for a good career personal effort is needed. Simultaneously, they think less than others that "competence" would influence their careers. Half of the students associate their career with "going upstairs", 30% with "walking an even road", and 20% with "climbing a mountain". 75% of students with the prevalent career orientation "stability" chose following characteristics as important for a successful career:

1. Persistence.
2. Activity.
3. Responsibility.
4. Senseofpurpose.

Although students of this group chose "activity" as an important personal quality for a successful career less than the half of them (and by this less than students of all other groups) noted that they are "very active". The "stability" students think less than others that they are persistent, initiative and active which leads to the conclusion that they do not think that they have the characteristics necessary for a successful career. The most students of the "stability" group desire is "material well-being" (82%) and the least - to "obtain power and influence" (25%). The majority of this group considers that for a good carrier personal effort is necessary (76%). Their social attitude to the career process is characterized by "reliability" (62%). 75% associate their career with "going upstairs".

As for the "service" group, the majority of the students (more than 75%) pointed at the following characteristics as important for a successful career:

1. Self-confidence.
2. Takinginitiatives.
3. Activity.
4. Independence.
5. Responsibility.

Students of this group are more "purposeful", "communicative", "responsible", and "active" than students of other groups. Female students tend to choose the "service" group more than male students. They wish more than representatives of other groups to "become a competent specialist" (72%), and they are most interested in researches during their studies at universities. The majority of them thinks that for a successful career hard work, proficiency and connections are important. The social attitude to the career process is marked by "development" (54%). The career process is associated with “going upstairs” (80%) and “climbing a mountain”. The “service” group was the most to compare its future career with a “shining star in the sky”. More than 75% of the students with the prevalent career orientation “challenge” chose the following personal qualities as important for a successful career:

1. Persistence.
2. Self-confidence.
3. Takinginitiatives.
4. Activity.
5. Independence.

Students of this group show themselves more “purposeful”, “independent” and “reliable” than others. All students of the “challenge” group - and by far more than other students - consider “persistence” as essential for success. They are the most self-confident participants since they are more sure than others that “luck” and “taking initiatives” are important to make a career - qualities which they think they have more than others. Further, the “challenge” group values “activity”, “flexibility” and “adaptability” as helpful for career but only half of it thinks to have these characteristics. All students with the prevalent career orientation “challenge” seeks to “achieve a leading position” (100%), and the least they look for in life is to “become a competent specialist” (33%) and “start an own business” (33%). The social attitude to the career process is described with words like “influence” (50%) and “competence” (33%). This group is the least one to count on “favourable coincidences of circumstances” and on “useful connections” in its career. The “challenge” group is the only one where the majority associates its career with “climbing a mountain” (50%), with “going upstairs” (33%) and “walking a plain road” (17%) as other chosen metaphors, which fully corresponds to the career orientation of this group: “for the purpose any means are good”.

The majority of the students group with the prevalent career orientation “integration of lifestyles” chose “responsibility” as the characteristic to condition the career success. They less than other students consider “luck”, “activity”, “independence” and “self-confidence” as qualities useful for a career. They also tend more than others to think that they are not communicative, that they lack the “sense of purpose”, the courage to “take the initiative” and the sense of “responsibility”. “Responsibility” was as well the personal quality which the majority of this group noted as important for a successful career.

Almost all students of this group would like to “integrate personal life with an active professional occupation” (97% - the highest rate of this parameter), and they least want to “gain power and influence” (7% - the lowest rate of this parameter as for the whole study) which fully supports the career orientation of this group: “integration of lifestyles”. The social attitudes to the career process are marked by “recognition” (35%) and “liability” (35%). All students of this group think that for a successful career personal effort is essential. 80% of the students describe their future career as “going upstairs” (80%) and “climbing a mountain” (20%). Students with the prevalent career orientation “enterprise” mostly chose the following characteristics as meaningful for a successful career (75%):

1. Self-confidence.
2. Responsibility.

They less than others think that an adequate self-esteem is important for a career, and this could be the reason why this group less than others pointed at it as a personal quality. They also less than others count on luck and value "flexibility" and "independence" as needful characteristics for a good career although more than have of these students think they are highly "independent".

The "enterprise" group chose as its main goal of life to "start an own business" (89% - more than all other students surveyed). On the other hand, the least they seek is "to become a competent professional" and to "be respected by the colleagues" (28% each, which are the lowest rates of these parameters in the whole study).

Students of the "enterprise" group show the lowest interest to research at university. The majority of them thinks that for a successful career hard work and a good education are needed. The social attitude to the career process is determined by "concurrence" (30%), "recognition" (28%) and "liability" (29%). 56% of these students associate their future career with "going upstairs" and 33% with "climbing a mountain". These results allow the conclusion that students with different prevalent career orientations have different social attitude to the career process.

4 Discussion

Many publications have been dedicated to reveal a connection between personal features and career preferences [5]. Zavodčikov and Manâkova conclude that at the stage of training, the process of professional self-determination continues as new development prospects arise [11]. Doronina also believes that at the present time, making surveys of the career orientations of students is important and relevant since it is at this stage that a person plans his professional future, becomes aware of the consequences of his professional choice, etc. [3]. Belayeva and Galkina consider the years of study at a high school crucial for the formation of the individual career orientation. It is in this period that career goals are determined and future plans are made. Later these plans will have a huge impact on the career and the professional development in general [1]. Kozevina also underlines that the period of studying in a high school is one of the important stages of career orientation, as career goals and plans formulated on this stage may determine success of the whole professional development of a person [7].

5 Conclusion

One of the urgent tasks of the higher education system at present is the formation of general cultural competencies among students, including the ability to self-organization, self-development, as well as the ability to creatively transform oneself, one's living conditions and professional activities [4]. The students' career orientations are a task of socio-psychological education of the highest dispositional level, which sets the orientation of the career process and determines the choice of means to achieve the career goals.

The career orientations “management” and “enterprise” belong to the vertical career orientation since they are associated with a focus on a high social status and material well-being. The horizontal career orientation is characterized by a striving to functional competency. It includes Schein’s “functional competence” and “service” since these characteristics are linked to values like development and social utility which are spiritual values. Vertical career orientation is based on the desire of power, prosperity and high social status in the social-professional environment. It shows a psychological readiness of students to move upwards the social-professional hierarchy. On the other hand, horizontal orientation indicates the students’ desire of professional and personal development, as well as of attaining competency which may be a sign of altruism in social interaction. Students of different prevalent career orientations have different social attitudes towards the career process.

References

1. Belayeva, E.V., Galkina, N.A.: Formation of career orientations of students at the final stage of university education. Sci. Notes **2**(50) (2019)
2. Chiker, V.A.: Psychologic Diagnostics of Organization and Personnel. Rech, St.-Petersburg (2006)
3. Doronina, N.N.: Career orientations of university students. Sci. Bull. Belgorod State Univ. Ser.: Hum. **28**(277), 179–185 (2017)
4. Garanina, Ž.G., Balayev, S.I., Ionova, M.S.: The role of how students of higher education relate themselves in their personal and professional self-development. Educ. Sci. **21**(1), 82–96 (2019)
5. Kalgin, A.S., Kalgina, O.V.: Career choice between the state and private sectors among: the role of personal features. Quest. State Munic. Adm. **4**, 145–171 (2018)
6. Kolesnikova, E.M.: Dynasties of engineers and perspectives of professional biographies of students at engineering universities. Sociol. J. **24**(1), 55–72 (2018)
7. Kozevina, A.P.: The dynamics of career orientations of students in the process of vocational training. Psychol. Econ. Dev. **9**(2), 63–73 (2017)
8. Manturova, N.M.: Study motivations and career orientations. World Sci. Cult. Educ. **5**(72), 245–248 (2018)
9. Nazarova, I.B., Zelenskaâ, M.P.: Reproductive attitudes of the student youth. Vestnik RUDN. Ser.: Sociol. **4**, 565–567 (2017)
10. Retivina, V.V.: Labour values and attitudes of the contemporary student youth. High Educ. Russia **1**, 57–63 (2019)
11. Zavodčikov, D.P., Manâkova, P.O.: The interconnection of self-regulation and personal professional perspectives of students. Educ. Sci. **20**(1), 116–135 (2018)

Development of Workforce Capacity in Russian Agrifood Industry: Innovative Approach

E. P. Afanaseva[1(✉)], A. B. Malina[1,2], and O. V. Novoselova[3]

[1] Samara State University of Economics, Samara, Russia
Parus82@mail.ru, Kuzdavletova_AB@mail.ru
[2] Samara State Technical University, Samara, Russia
[3] Samara National Research University, Samara, Russia
Olganovo63@gmail.com

Abstract. The labor force development in agribusiness is of particular importance in task solution of import substitution on the basis of the innovative development of this economic sector in Russia. The authors suggested a system and sectorial approach to study labor resources in agribusiness. The methodology includes the analysis of the factors that determine development trends of the country's agribusiness, its labor resources, and assessment of their qualification level. The authors distinguished the following trends: reduction in the rural population and low prestige of work in this sector among the youth, which directly affects a decrease in rural labor forces. The problem of understaffing among both direct workers and qualified specialists was revealed. The authors pointed out that an increase in the share of farm workers is a positive phenomenon in agribusiness, as it contributes to improving agrifood industry efficiency as a production system. But the direct factor that affects labor productivity is the staff qualification. New trends and options suggested for action will reduce rural labor shortage, and they can be used to justify decisions in the strategic management and management by objectives in Russian agribusiness.

Keywords: Agribusiness · Labor forces · Rural employment · Workforce capacity

1 Introduction

Today, the agrifood industry of the Russian Federation has achieved significant results in the implementation of Import Substitution Program in Russia until 2020 [1]. Under current conditions, one of the key factors in the development of the agriculture-based economy is to provide agricultural organizations with highly qualified managers and specialists who can properly and quickly solve issues and tasks, and make reasonable managerial decisions in a responsible way.

The structure of Russian educational institutions includes highly qualified teaching staff and experienced specialists who have not only scientific theoretical knowledge, but also strong business relationships with agricultural enterprises and holdings.

S. I. Ashmarina and V. V. Mantulenko (Eds.): IPM 2020, LNNS 161, pp. 397–403, 2021.
https://doi.org/10.1007/978-3-030-60926-9_51

Continuity in agribusiness is very important to ensure Russian countryside activity. Currently, there is no systematic work in agribusiness to organize permanent education, professional development, and reskilling. Despite innovations and globalization in modern society, there are still many traditional technologies in rural areas that require staff at the level of primary vocational education. It is extremely important to raise prestige of rural labor, which supports the agricultural economy of any country. In order to maintain continuity, it is necessary to ensure educational programs in educational institutions at all levels. All components of this chain should be fully covered. But there are still a lot of problems that negatively affect the professional training quality, retraining, and advanced training of managers and workers, that is to say, staffing in agrifood industry. One of the most important issues is obsolescent and poorly updated material and technical resources. The industry needs a significant technical reequipment in the educational process with the simultaneous use of new educational technologies. This issue is also relevant for foreign countries.

2 Methodology

The following key methods were used in the course of this research: monographic, comparative, economic and statistical, and abstract-logical analysis. The authors of the paper studied numerous theoretical and empirical researches of national and foreign scientists. The components of the personnel reserve were summarized on the basis of this work. An economic and statistical method allowed to collect, process and analyze the demographic situation in rural areas and remuneration rate of agricultural workers. Comparative analysis let reveal the fact that both Russia and many foreign countries have an important shortage of qualified personnel. An abstract-logical analysis allowed to establish and characterize workforce capacity in agribusiness and assess the formation and current state of higher agricultural education. The analysis mentioned above made it possible to reveal problems and prospects of agribusiness staffing, as well as propose and justify measures to build the personnel reserve for the country's agrifood industry.

3 Results

Today according to the new Food Security Doctrine of the Russian Federation, Russia faces the main task of ensuring food safety [3]. On the other hand, there is a question whether labor resources in Russian agribusiness are able to provide it. A village is an area that is poorly provided with staff and therefore, the agribusiness personnel is one of the most important component that requires attention. It should be based on high-quality education that meets the highest innovative requirements for present-day production with an adequate level of remuneration. Staff training in Russia is carried out in accordance with 122 programs of study and 70 fields of study. Personnel with vocational training are prepared by 26 universities in accordance with 51 programs of study. 429.2 thousand students are enrolled in higher education programs, and 25.6 thousand people are enrolled in professional education programs [12]. On the one hand, this is an impressive reserve but on the other hand, the majority of Russian youth consider agriculture as a low-prestige and low-tech sector. At the same time, agrifood industry claims to become the main area to demonstrate the results of the new technological revolution: robotic technologies of "industry 4.0" let all agricultural machinery operate in a computer-aided way, which integrates the entire production chain into a unified system [8]. Therefore, the problem is the level of agricultural education in the country, as the university competences should meet interdisciplinary requirements, rest upon serious scientific researches conducted in educational institutions, focus on the needs of the business, and face the challenges of global trends shaping the agribusiness future.

Many Russian universities have realized the need to implement foresight of competencies in order to determine strategic ways for their development. It happened through the development of national research and educational centers of competences in accordance with the Decree of the President of the Russian Federation "On national goals and strategic development goals of the Russian Federation for the period up to 2024" [4]. This means that it is the right way to resolve this issue, but this is not a one-year effort. We should not forget that the majority of students at these programs of study are country people. Therefore, it is necessary to analyze a demographic component of the village (Table 1). The demographic resources of rural areas consist of 38 million people (27% of the total population), including 23.6 million labor resources [13]. It is necessary to mention the negative overall growth of the rural population over the past 6 years. Table 1 shows that the maximum negative value of the total and natural increase was recorded in 2018 (−226.3 thousand and −112.1 thousand people). A migration component in the demography of the rural population tends to decrease, and therefore it does not save the situation in rural areas [8].

Table 1. Change in the number of rural population in the Russian Federation, thousand people.

Period	Population size at the beginning of year	Changes (±)			Population size at the end of year	Total growth in a year, %
		Total growth	including:			
			Natural increase	Migration gain		
2014	37,118.2	−110.6	−0.8	−176.8	37,985.1	−0.26
2015	37,985.1	−97.8	−61.4	−46.8	37,887.3	−0.26
2016	37,887.3	−115.3	−74.0	−36.4	37,772.0	−0.30
2017	37,772.0	−218.5	−95.1	−46.5	37,553.5	−0.58
2018	37,553.5	−226.3	−112.1	−69.4	37,327.2	−0.60

Source: authors based on [8].

The age pattern of the population is an important factor in increasing labor potential. The demographic situation in the country is characterized by ageing of the rural population (Table 2). Due to a long decline in the birth rate, there is a regressive type of age pattern of the population, where the relative share of aged people exceeds the share of children and adolescents. Table 2 shows that the coefficient of demographic burden in 2018 is the following: there are 863 people, including 376 children and adolescents, and 487 senior citizens respectively per 1000 working population [8].

Table 2. Distribution of the rural population by age groups in the Russian Federation in 2010–2018

Indicators	Period				
	2014	2015	2016	2017	2018
Population at large (thousand people)	37,118	37,985	37,887	37,772	37,553
Younger than working population	7,292	7,551	7,602	7,617	7,577
Working population	21,031	21,192	20,826	20,507	20,149
Older than working population	8,795	9,242	9,459	9,648	9,827

Source: authors based on [8].

It should be noted that the conditions of agribusiness functioning contribute to the qualified personnel outflow: a relative share of workers with higher and secondary vocational education in 2017 was only 16.8% and 28.3%, respectively. We pay attention to the fact that the share of farm workers with higher education remains low in comparison with the one of workers with secondary and elementary vocational education in the structure of professional education, 1.6 and 1.7 times, respectively. The share of specialists and managers in agribusiness organizations who do not have any professional education remains at a high level, amounting to more than 28% [11].

According to the results of an annual survey "Agribusiness Market Review – 2019" [5] initiated by Deloitte Company in the CIS countries with support of Strategic Development Center, it was revealed that the shortage of skilled personnel came third in the rating of problems in Russian agribusiness in 2018, and this problem came second in 2017. According to the study conducted by Hays, 70% of managers in agribusiness companies believe that the industry suffers from a labor shortage. 23% need only specialists of rare or new types of professions. 65% of managers demand technical workers and experts in production. 35% of managers need employees in sales, and 33% search for specialists in marketing. This means that this problem requires immediate decisions at the state level [6].

Labor remuneration is also an important factor of a low staffing level (Table 3). The ratio of the average salary of workers in agribusiness and food industry to the average Russian level is only 65% and 74%. Table 3 shows that the highest indicators for the period under review were achieved only in recent years, due to large technologically equipped agricultural enterprises which operate for export. At the same time, the gap between the level of wages in agribusiness and other economic sectors remains significant.

There are many factors that discourage young professionals from living and working in rural areas: hard work, low salary, gap between the city and the countryside in terms of income and quality of life, as well as the fact that universities are mainly concentrated in large cities. This is an important circumstance that becomes an obstacle for school graduates from the villages whose start-up budget is lower. It is extremely difficult to them to enter a university. That is to say, citizens are quite a significant part of students who are unlikely to start working in a village after graduation. This means that the rural school with its vocational guidance should also fit into the general scheme.

Table 3. Average nominal salary in Russian agrifood industry

Indicators	Period					Changes of 2018 to 2014
	2014	2015	2016	2017	2018	
By economy on average, rubles	32,495	34,030	36,709	39,167	43,445	133.7
including: - agribusiness, hunt, and forest sector	18,365	2,0670	22,915	25,671	28,185	153.5
Ratio of the average salary in agribusiness, hunt, and forest sector to the average Russian level, %:	57	61	62	66	65	8

Source: authors based on [8].

These circumstances require a set of actions. The modernization that has begun in today's education is designed to defuse these problems or even eliminate them. Students should have new motivation to acquire professional and general cultural competencies. New training technologies, active and interactive methods, and involvement of employers in educational programs support this modernization. Universities have to create synergies with employers and their associations not only in the development and implementation of basic and additional programs but also in the assessment of graduates' competencies. The building of the interaction model between agricultural educational institutions, business, and the state will allow to join an innovative component in research activities and agricultural organizations. Rectors of many Russian agricultural universities highlight the following basic points of the platform for the agricultural education development in Russia: the education system should be based on an active combination of practice and theory, and universities should become centers for the development and implementation of various innovative methods to organize and manage production. The teaching staff have a task to train students not only how to work in the niche of their specializations but also how to start a new business, develop production, and be competitive. According to experts, the most important aspect of the modernization of agricultural education in Russia should be an international component.

To help young professionals in the country, there is State Support Program for Workforce Capacity in Agribusiness. A number of Russian regions have also offered their own support options. The Kabardino-Balkarian Republic proposed to complete State Program for the Agricultural Development with Staff Training for Agribusiness Subprogram. Karelia and the Orenburg Region supported providing housing for young specialists in agribusiness by adopting Federal Target Program. A number of regions supported the placement of graduates who studied on a budgetary basis. The Arkhangelsk Region developed an incentives scheme for employers involved in training staff and creating conditions to attract specialists to agribusiness [14]. These activities are aimed at increasing the human resources potential in the country's agribusiness.

4 Discussion

It should be noted that the rural education systems differ greatly in countries across the world. This is primarily due to the national traditions of the agricultural universities development, the general features of higher education system, and the opportunities for graduates to have decently paid work in production. In the EU countries, a small proportion of graduates from agricultural universities can find work in agribusiness (for example, only one of five in Hungary). Specialists in the field of ecology, processing of agricultural products, and marketing remain in high demand. The United States have the same problems: staff shortage in rural areas and earnings lower than the minimum wages [15, 16]. Some foreign scientists pay attention to the fact that improving agribusiness training systems is able to contribute to increasing agricultural productivity and employment [9, 10, 17]. Others believe that the main reason is underpayment, and they insist that a large number of hired labor in the industry makes it low-paid [7]. In Sweden, for example, the rural market is under reconstruction, where the problems of living and working in the region are resolved through EU Structured Finance Regional Program [12]. Many EU countries keep this situation under control. This means that political efforts should help increase agricultural productivity through income and life quality controlling [2].

5 Conclusion

Despite the identified negative trends in the demographic situation, it should be noted that the state has quite sufficient labor potential for the economic development, as it has 59% of the population in working age. 23.8% of people over 15 have higher education, and 32.8% of them have secondary vocational education. At the same time, the employment rate of the working-age population is 70.9% (285.2 thousand people) in rural areas, and 92.4 thousand people are employed in agribusiness (6.1% of those employed in the Russian economy). The number of educational institutions that train specialists for Russian agribusiness is sufficient, and the trainers' qualification level is high. On the other hand, there is a problem. Due to the constantly changing market and active introduction of digitalization, agricultural enterprises have a shortage of qualified personnel who have new competencies in accordance with market requirements. In order to form this personnel reserve, we need a system of actions. First, it is the education modernization, that is to say, correlation between theoretical knowledge and internship at leading enterprises in order to develop students' professional skills in agrifood business. Secondly, it is the career guidance activity in the country's schools and increasing prestige of rural areas through salary rise. Third, it is a targeted training and the placement of university graduates at specific enterprises. Stimulating measures to support young professionals in rural areas are also urgently needed. The government of the country should develop a strategy to build a personnel reserve in agrifood industry through a special-purpose program.

References

1. 2020 Year: Import substitution program in Russia until 2020. Economy 2020 (2020). https://2020-god.com/programma-importozameshheniya-v-rossii-do-2020-goda/. Accessed 15 June 2020
2. Almeida, A.N., Bravo-Ureta, B.E.: Agricultural productivity, shadow wages and off-farm labor decisions in Nicaragua. Econ. Syst. **43**(1), 99–110 (2019)
3. Decree of the President of RF, of January 21, 2020, No 20 "Approval of the Food Security Doctrine of the Russian Federation" (2020). https://www.garant.ru/products/ipo/prime/doc/73338425/. Accessed 15 June 2020
4. Decree of the President of the Russian Federation of May 7, 2018 N 204 "On national goals and strategic development goals of the Russian Federation for the period up to 2024" (2020). https://base.garant.ru/71937200/. Accessed 15 June 2020
5. Deloitte: Agribusiness market review – 2019 (2019). https://ru.investinrussia.com/data/file/obzor-rynka-selskogo-hozyajstva-2019.pdf. Accessed 15 June 2020
6. Dyatlovskaya, Y.: 71% of farmers claim staff shortage. Agroinvestor (2016). https://www.agroinvestor.ru/analytics/news/24488-71-agrariev-zayavlyayut-o-nekhvatke-kadrov/. Accessed 15 June 2020
7. Emran, S., Shilpi, F.: Agricultural productivity, hired labor, wages, and poverty: evidence from Bangladesh. World Dev. **109**, 470–482 (2018)
8. Federal State Statistics Service: Demography (2020). https://www.gks.ru/folder/12781. Accessed 20 June 2020
9. Feuerbacher, A., McDonald, S., Dukpa, C., Grethe, H.: Seasonal rural labor markets and their relevance to policy analyses in developing countries. Food Policy **93**, 101875 (2020)
10. Giannakis, E., Bruggeman, A.: Exploring the labour productivity of agricultural systems across European regions: a multilevel approach. Land Use Policy **77**, 94–106 (2018)
11. Khlusova, I.A., Khlusov, V.N., Khajbrakhmanov, R.R., Rebezov, M.B., Gorelik, O.V., Prokhasko, L.S., Somova, Yu.V.: Assessment of the professional education level of workers in agribusiness. Fundam. Res. **5**, 148–153 (2019)
12. Kvist, E.: Who's there? Inclusive growth, 'white rurality' and reconstructing rural labour markets. J. Rural Stud. **73**, 234–242 (2020)
13. MediaQuest: Personnel for agribusiness. Development prospects (2014). https://akvobr.ru/kadry_dla_sh_perspektivy.html. Accessed 15 June 2020
14. Milknews: How to solve the problem of personnel support for agricultural industries: Regional opinions (2019). https://milknews.ru/longridy/kadry-apk-regiony.html. Accessed 15 June 2020
15. Nelson, P.B., Nelson, L., Trautman, L.: Linked migration and labor market flexibility in the rural amenity destinations in the United States. J. Rural Stud. **36**, 121–136 (2014)
16. Rashkovsky, A.: What can we do in agriculture, using the experience of the United States (2012). https://newsland.com/community/88/content/chto-my-mozhem-sdelat-v-selskom-khoziaistve-ispolzuia-opyt-ssha/1509261. Accessed 15 June 2020
17. Zhang, Q., Sun, Z., Wu, F., Deng, X.: Understanding rural restructuring in China: the impact of changes in labor and capital productivity on domestic agricultural production and trade. J. Rural Stud. **47**, 552–562 (2016)

Theoretical Investigation in Talent Management

L. S. Babynina(✉), L. V. Kartashova, and P. P. Pilipenko

Plekhanov Russian University of Economics, Moscow, Russia
{Babynina.LS, Kartashova.LV, Pilipenko.PP}@rea.ru

Abstract. The purpose of this paper is to consider the ways of development the concepts of talent and talent management. The authors have used empirical research methods based on a literature review, scientific reports and a review of organizations in Russia and other countries. The results of theoretical investigation in talent management had identification of current trends in Human Resource Management. Its involves focus on the strategy in human resources management on working with a talents and forming the talent management system in condition of talent shortage; developing a situational approach to the talent management reflects the specificity of international corporations and small and medium-sized enterprises, public sector and universities.

Keywords: Talent · Employees · Human resource management · Talent management system · Talent shortage · Situational approach

1 Introduction

Shifts in demographics and intense technological advances are one of the main reasons for skills shortage. The concept "the war for talent" is relates to the fierce competition of attracting and retaining talent employees, caused by the shortage of employees with sufficient skills. Today companies operate in the era of technological advancement, digital transformation, workforce flexibility, and talent shortage. To stay competitive in an industry, companies look for talented people with high potential. Many big international companies attract the best young talent since their brand image is strong. There is little doubt that the attraction, development, and retention of talent are nowadays one of the most critical challenges faced by companies worldwide. Despite the increasing scholarly attention during the last years many questions remain, particularly, those related to how (and why) talent management (TM) is conceiving, implemented and developed within organizations, not to mention about its outcomes or effectiveness [14]. Since the scarcity of talent has become obvious, companies should start to pay attention to the calls of economy and take significant action in attracting, raising and retaining talents. To do so, human resource specialists should understand the concept of TM and its processes.

S. I. Ashmarina and V. V. Mantulenko (Eds.): IPM 2020, LNNS 161, pp. 404–411, 2021.
https://doi.org/10.1007/978-3-030-60926-9_52

2 Methodology

Over the last decade TM has emerged as one of the fastest growing disciplines in the management field [15]. There are many publications in Russia and other countries, where the academics clarified the concept of "talent". The authors summarize the different approaches related to "talent" and "talent management" concepts. Tansley, Turner, Foster, Harris, Srewart, Sempik, and Williams define talent as individuals who can make a difference to organizational performance, either through their immediate contribution or in the longer term by demonstrating the highest levels of potential [13]. Yarnall points out that the definition of talent varies across companies. Some companies associate talent with performance and potential, while others associate with scarce resources, which would be able to fit middle and senior positions in the future [19].

Michaels, Handfield-Jones, and Axelord give another definition of the talent. They describe talent as the sum of a person's abilities - his or her intrinsic gifts, skills, knowledge, experience, intelligence, judgment, attitude, character, and drive. It also includes his or her ability to learn and grow [7]. According to the later work by Tansley, the meaning of talent across companies depends on the organizational and industrial specifications, group-level implications, and individual characteristics of an employee [12]. In opinion of Odegov, Babynina, and Aleksandrova, the talent in the field of business is a person with high professionalism and competence, which is assessed according to professional knowledge and skills; and the results of work [8].

The concept of talent management appeared in 1990 to denote innovations of human resources management based on human potential. There are multiple approaches to the talent management concept and all of them have based on the clear definition of a talent in the company. Therefore, it is crucial to define a notion "talent" in the company before planning talent management strategy.

According to Bukharina, the first approach defines a talent as a gifted genius employee [2]. Therefore, talent management is directed towards the creation of creativity-stimulating atmosphere and work conditions. One of the key elements of this approach is that leadership and creativity are greatly encouraged. There is no material motivation or control of employees. This approach focuses on the fact that it is impossible to manage talents but the company can provide them with sufficient working conditions and stimulate their creativity indirectly so that they are able to open their potential at their own pace.

Other approaches are linking a talent with high potential and great productivity. Odegov, Babynina, and Aleksandrova have defined talent management as a part of human resource management, which is responsible for attraction of highly qualified employees, their integration, development, and retention. Talent management is able to predict the future demand for human resources and stimulate creation of an attractive environment to candidates and employees [8].

The third approach defines all employees as talents. According to Bukharina, talent management is a mix of all HR-processes that stimulate everyone in the company to greater productivity [2]. There is no discrimination according to skills, talents, or performance. The company has groups of people that are based on some criteria so that the company can work with them and offer relevant development and motivation to higher performance.

To sum up the approaches to talent management, the authors adhere to the second approach, namely: in conditions of a talent shortage, companies need to focus their strategy in human resources management on working with a talents and forming the talent management system. The forming of a talent management system involves three main stages:

1. Should be searching for the employees with high potential in the team.
2. To implement the program as a process of continuous learning and talent development.
3. To use of these employees and their retention of them in the company should base on differentiation and inspiration of people.

Other researches in recent years have focused on developing a situational approach to the TM, which reflects the specificity of international corporations and small and medium-sized enterprises (SME), public sector, universities. The strong focus on TM in large multinational corporations (MNC) raises questions about whether current assumptions in the TM literature related to this specific context help us to understand and explain the TM issues in other contexts such as public sector organizations, SMEs, and organizations based in emerging market context. TM research is still focus at the meso (organizational) level of the analysis; with limited attention could be pay to individual-level research or more macro-level factors [11].

A review of empirical TM research by Thunnissen and Gallardo-Gallardo shows that only a small minority of TM publications are focus on public sector organizations [15, 16]. Education and healthcare seem to attract the most academic interest [5]. The popularity of these two public sector contexts can be explained by the fact that both universities and hospitals employ professionals (scientists and medical specialists) that can be considered as core employees or talents that play a strategic role in the organization's success.

The impact of the internal organizational context relatively neglected in TM research. Some studies highlight the importance of the industry sector (nature of the services/products, organization size, profit and returns, budgetary constraints, location, ownership, and the composition of the workforce on the choices made regarding the intended TM policy. For instance, Cooke, Saini, Wang show that homogeneity of the workforce and the type of jobs and the egalitarian culture makes it necessary for firms to adopt an inclusive TM approach in China and India [4]. Similar findings have reported by Buttiens on TM in Flemish governmental organizations [3]. Few studies go deep into specific organizational context when examining talent recruitment and selection practices [17]. Although the effect of the organizational characteristics on the other stages of the TM process is under research Thunnissen shows how the actual implementation of TM strategies in Dutch academia has affected by key actors, such as the role of academic line managers [14]. In fact, line managers are the link pin between intended policy and practice. In the opinion of the authors line managers are the "owners of talent" since they should be primarily responsible for taking decisions and making investments in talent due its critical impact on business performance, however, the role and perceptions of line managers are explored in current research. To get a more comprehensive and balanced understanding of the scope and nature of the TM approach in an organization, we need to understand the perceptions of multiple stakeholders including Human resource (HR) managers, line managers, employees and trade unions. We suggest the need to adopt the broader approach where wider stakeholder perspectives need to be assessed in future TM investigations.

Of particular interest are the results of research that examines the problems of personnel management through the prism of modern university talent management. Human resource management (HRM) based on talent management is currently an extremely popular trend and it have use in many foreign universities. This approach involves the integration of all the main functions of personnel management: attracting, evaluating, motivating and developing the university's personnel potential. It based on a flexible differentiated approach to different categories of employees. Without reducing attention to all categories of employees, it focuses on attracting, developing and retaining the most talented scientific and pedagogical staff and higher school administrators [17].

Talent-related issues are a major concern of many CEO's. Approximately 75% of CEO's highlighted the scarcity of essential skills and capabilities as a key threat to the growth prospects of their organizations. Executives both in Russia and abroad believe that the company should have sufficient number of employees with competences it needs. Therefore, the main goal of talent management, according to Levichev, Vorogushin is to ensure that there is sufficient number of employees with demanded competences and skills at right time, at right place and at optimal cost [6]. The goal can be accomplished due to various human resource processes and instruments.

Andrianova, Maor, and Schaninger analyzed McKinsey Global Survey with 1,820 respondents [1]. They statistically tested the effect of various practices on talent management success. In the survey they offered 20 different practices frequently used in talent management. As result, the McKinsey team concluded that talent management success closely connected to some talent management practices, and that only 11 out of 20 practices have a positive effect on talent management efficiency.

Among all the 20 practices, the three most effective ones were talent allocation, employee experience and strategically oriented human resource specialists. These practices statistically proved and strongly related to the effectiveness of talent management. The most significant driver, talent allocation, is defining as "the fast movement among strategic projects as priorities arise and dissolve fast or really fast" [1]. The further research allowed finding out what factors contribute to the talent allocation: these are skill-based deployment, involvement of executives in talent management and allocation, and work in cross-functional teams.

As for employee experience, human resource specialist should be involved in building a positive experience of their employees and clearly understand the strategy of the company and its strategic goals. The collaboration between HR managers and other executives is crucial for success of talent management. Comprehension of the corporate strategy, involvement of executives in talent management, efficient identification of people with necessary skills for a new project or job within the entire company, usage of data and analytics in talent decision-making, and the tied link between individual objectives and corporate strategic objectives are the other not least but effective practices. The practices may support the effectiveness of talent management system and the company's general outperformance. It is necessary to remember that no practice can work for every organization but the McKinsey survey proves that there is greater likelihood of that talent management strategy will work if the company applies at least one of the mentioned practices.

3 Results

There are many approaches to concept of notion "talent" and talent management. In opinion of the authors TM describes the activities and processes that involve the systematic attraction, identification, development, engagement, retention, and deployment of those talents, which are of particular value to an organization to create strategic sustainable success. Therefore, TM consists of multiple processes. There are identification of goals and strategy, creation of employer value proposition, talent attraction, talent sourcing, recruitment, selection, training and development, and retention. There is a close connection between all these processes and the influence of one of them that has distributed to all others. For example, talent retention has tied to talent development since one of the key aspects of retention is the development of employees. Without retention, there is no purpose of employees to develop themselves without clear career planning, opportunities for advancement or promotion.

The researches in recent years have focused on developing a situational approach to the problem. There are specificity of international corporations and SME's, public sector, universities. Of particular interest are the results of research that examines the problems of personnel management through the prism of modern university talent management. Human resource management based on talent management is currently an extremely popular trend and has been used in many foreign universities. This approach involves the integration of all the main functions of personnel management - attracting, evaluating, stimulating and developing the university's personnel potential. There are base on a flexible differentiated approach to different categories of employees. Without reducing attention to all categories of employees, it focuses on attracting, developing and retaining the most talented scientific and teaching staff and higher school administrators.

4 Discussion

There are many challenges the company can face managing talents. The first problem is that it is quite difficult to manage talents. They have their own expectations, which are different to expectations of the company. Human resource specialists not only should identify managerial and talent expectations but also find the equilibrium between them. Robertson and Abbey highlighted that "talented people do not just want personal growth for the sake of it [9]. They do want personal growth for its own sake, but that is not all they want. They want the chance to use the growth. They want new and bigger challenges. They want to use their talents. They want to keep moving. Talents want to shape the industry, improve processes, and be a part of something bigger and meaningful. They desire the freedom of own projects and actions and the flexibility. Talents enjoy taking responsibilities and making significant influence.

However, the bad side is that they have their own work standards and if the company do not accept or mitigate them, the talent start questioning his or her future in the company. In addition, such demotivating factors as a load of additional work without a pay rise or promotion, no appreciation for work done, faint praise or extreme pressure on talent can make a talent quit [10, 18].

The second challenge is Russian CEOs do not see the solution to talent management issues the way CEOs in the United States see it. Thus, according to the results of an annual survey conducted in 2017 by PwC, 52% of Russian CEOs surveyed ranked permanent, variable remuneration and benefits as the most important. Among American managers, only 28% were like this. At that time, American CEOs preferred the importance of training the leaders of the future. They put this direction in the first place (65% of respondents), while only 27% of Russian top managers noted the importance of this direction to create a value proposition that would fit to multiple generations of talents [6].

Today companies have four generations under the roof and needs of each generation are different. Younger generations Y and Z have completely different expectations about work environment than the previous generations. These generations look for completely different values: they want to work remotely, paid time off, control over schedule, have friendly and open communication environment, flexible working hours, meaningful work and clear career path. Generation X is people who value the amount of salary, job security, work-life balance, and advancement in the company. The third challenge is to retain recently hired employees for a company. Talent crunch is the real issue. Young employees are starving for career promotion and there is lack of available jobs for them in the future. Companies with the great percentage of young talents are doomed to the high employee turnover. The biggest threat related to the lack of talent pool for future leadership positions. Generation X, associated with the war for talent, considered precious because of skill shortage. Companies are highly interested in raising them for future leadership positions. However, not all in generation X show interest in leadership roles and some of them can reject leadership programs because they feel not being ready to move to senior positions. The fifth challenge is the creation of such a corporate culture, which could support talent management and will not harm it. The last big challenge linked to the fact that senior management seeks for new talents outside of the company, while talent executives seek for them internally. When this situation happens the talent selected for a talent pool starts to feel insecure about future prospects and feels that the company deceived him or her. This challenge still lies in the talent system transition from a "buy" strategy to "build" strategy and many executives fail to adapt to this transition.

5 Conclusion

Today one of key business risks is deficit of skilled workers. More than 75% of CEO's highlighted the scarcity of essential skills and capabilities as a key threat to the growth prospects of their organizations. Based on the conducted research it is clear that talent management influences the organizational performance of all companies. The authors have summarized approaches to talent management and defined it as a systematic attraction, identification, development, engagement, retention and deployment of those individuals who are of particular value to an organization, either in view of their 'high potential' for the future or because they are fulfilling business or operation-critical roles. Talent management has become a crucial element of any organization since it manages human resources. It ensures that the company is able to attract, develop and retain employees with right skills and competences. Because of the lack of resources, high attrition rate, insufficient motivation, and low engagement rate, sustainability of the company partially depends on its talent management system. Today companies have four generations under the roof and needs of each generation are different. Young employees are starving for career promotion and there is lack of available jobs for them in the future. The biggest threat related to the lack of talent pool for future leadership positions. The modern opportunities in solving these problems concludes the developing a situational approach to the talent management reflects the specificity of small and medium-sized enterprises, public sector and universities. These areas will be developing in the coming years.

References

1. Andrianova, S., Maor, D., Schaninger, B.: Winning with your talent-management strategy (2018). https://www.mckinsey.com/business-functions/organization/our-insights/winning-with-your-talent-management-strategy. Accessed 26 June 2020
2. Bukharina, A.Y.: Talent management: what to teach to employees to survive tomorrow. Soc. Psychol. Sociol. **8**(1), 144–162 (2017)
3. Buttiens, D.: Talent management in de Vlaamse overhead: Doctoral thesis. KU Leuven, Leuven (2016)
4. Cooke, F.L., Saini, D.S., Wang, J.: Talent management in China and India: a comparison of management perceptions and human resource practices. J. World Bus. **49**(2), 225–235 (2014)
5. Gallardo-Gallardo, E., Dries, N., González-Cruz, T.F.: What is the meaning of – talent in the world of work? Hum. Resour. Manag. Rev. **23**(4), 290–300 (2013)
6. Levichev, Y., Vorogushin, E.: Talent management in Russia and the world: numbers and trends (2017). https://www.pwc.ru/ru/hr-consulting/assets/talent-management-shtat.pdf. Accessed 23 June 2020
7. Michaels, E., Handfield-Jones, H., Axelrod, B.: The War for Talent. Harvard Business Press, Cambridge (2001)
8. Odegov, Y.G., Babynina, L.S., Aleksandrova, E.V.: Talents and the war for them. Actual Probl. Econ. **165**(3), 275–280 (2015)
9. Robertson, A., Abbey, G.: Managing Talented People. Momentum Wcze, Troy (2003)
10. Rothwell, W.J.: The Manager's Guide To Maximizing Employee Potential Quick and Easy Strategies to Develop Talent Every Day. Amacom, New York (2009)
11. Sparrow, P.: A historical analysis of critiques in the talent management debate. Bus. Res. Quart. **22**(3), 160–170 (2019)
12. Tansley, C.: What do we mean by the term - "talent" in talent management? Ind. Commer. Train. **43**(5), 266–274 (2011)

13. Tansley, C., Turner, P., Foster, C., Harris, L., Stewart, J., Sempik, A., Williams, H.: Talent: Strategy, management, measurement. Chartered Institute of Personnel & Development, London (2007)
14. Thunnissen, M.: Talent management: for what, how and how well? An empirical exploration of talent management in practice. Empl. Relat. **38**(1), 57–72 (2016)
15. Thunnissen, M., Gallardo-Gallardo, E.: Talent Management in Practice: An Integrated and Dynamic Approach. Emerald Publishing Limited, Bingley (2017)
16. Thunnissen, M., Gallardo-Gallardo, E.: Rigor and relevance in empirical TM research: key issues and challenges. Bus. Res. Quart. **22**(3), 171–180 (2019)
17. Van den Brink, M., Fruytier, B., Thunnissen, M.: Talent management in academia: performance systems and HRM policies. Hum. Resour. Manag. J. **23**(2), 180–195 (2013)
18. Volkova, N.V., Plotnikov, V.A.: Talent management practice: project work on SCRUM technology. Univ. Adm.: Pract. Anal. **21**(2), 131–142 (2017)
19. Yarnall, J.: Strategic Career Management: Developing Your Talent. Butterworth-Heinemann, Oxford (2008)

Queueing Theory: New Solutions to Optimize the Number of Workers

I. V. Bogatyreva[1](✉), L. A. Ilyukhina[1], I. N. Makhmudova[2], and N. V. Kozhukhova[1]

[1] Samara State University of Economics, Samara, Russia
scorpiony70@mail.ru, kuka_55_@mail.ru, laresa@inbox.ru
[2] Samara National Research University, Samara, Russia
mahmudova.I@yandex.ru

Abstract. The paper presents the results of the study aimed at determining the optimal number of workers, who are employed in equipment-specific production, through the use of the queueing theory. The relevance of the research is that service operations in equipment-specific processes are performed irregularly and at any time, there may be equipment and workspaces downtime in expectation of the service activities. Moreover, there may be deadtime in operational people's activities in wait for the service requests. Therefore, there is a need to determine service standards that would ensure both a minimum of the total costs in the personnel expenses and possible losses from equipment downtime. This task is complex and multi-variant, and it requires the use of mathematical methods. The authors presented the methodology for the queueing theory usage to develop efficient size standards in terms of alloy melting processes in gas-fired reverberatory furnaces, and calculation data of the service system performance indicators for melting furnaces by a working team of various size.

Keywords: Queueing theory · Labour rating · Multimachine service · Size standard · Optimality criterion

1 Introduction

With the increase in the mechanization and automation level in the total number of workers, the share of multi-tasking machine operators, setup technicians, service technicians, and electricians is going up. When rationing their labour, it is essential to justify the service and size standards, which determine the optimal ratio between the number of workers and units of equipment they serve. The economic reason of optimizing such ratios is the same for multiple-machine, repairing, and similar operations. This determines the common structure of the tasks to calculate service and size standards. Equipment and workspaces service systems that are found in practice can be divided into "with possible servicing waiting" and "without servicing waiting" systems. Calculations of the number of workers in the "without servicing waiting" system do not represent any difficulties and are made in the traditional way, according to the work labour input and volume of work (repair, transport, quality control in mass production, etc.). The peculiarity of equipment and workspaces service under the "with

S. I. Ashmarina and V. V. Mantulenko (Eds.): IPM 2020, LNNS 161, pp. 412–421, 2021.
https://doi.org/10.1007/978-3-030-60926-9_53

possible servicing waiting" system is that service operations are performed irregularly and at any time, there may be equipment and workspaces downtime in expectation of the service activities. Moreover, there may be deadtime in operational people's activities in wait for the service requests.

This system employs at least 25% of the total number of auxiliary labour (setup workers, technicians on duty, and electricians) in mechanical engineering. The main multifunctional workers use the same mode of operations. The similar scheme is used by assistant menders performing interrepair services and multi-machine weavers in the textile industry. When rationing labour, which is organized according to the "with possible servicing waiting" system, it is necessary to determine service standards (the number of workers per a certain amount of equipment) that would ensure both a minimum of the total costs in the personnel expenses and possible losses from equipment downtime. Such standards can be considered economically justified or optimal. This task is complex and multi-variant, and it requires the use of mathematical methods, in particular, the queueing theory. The queueing theory implementation in labour rating have already been reflected in the definition of economically justified service standards for a number of professions of main and auxiliary workers in mechanical engineering [7] and textile industry. However, the queueing theory application has not been developed for metallurgical processes. The authors presented the methodology for the queueing theory usage to develop optimal size standards in terms of alloy melting processes in gas-fired reverberatory furnaces. The method of using the queueing theory to develop optimal size standards is considered by an example of alloys melting processes in gas reflecting furnaces.

2 Methodology

The research on optimizing the number of workers in equipment-specific production is based on the application of mathematical methods, such as queueing theory and optimization methods. The queueing theory, based on the theory of chances and mathematical statistics, allows to determine the correlation between the main characteristics of the service system in order to improve system management.

The main indicators of the queueing theory are customer flow and in-service time. The service system consists of input sources and service channels (instruments). The customer flow (or incoming flow) is a collection of service requests that originate from input sources. In our system, the input source is the melting furnaces, and the customer flow is a series of furnace service that is performed by a working team.

The customer flow, as a random flow, is defined by the requirements allocation rule where the requirements are received to be served over a certain period of time. The flow parameter is its density, which determines the mathematical expectation of the requests number received per a unit of time. Service channels in our system are members of the serving team. The service channels index is the service duration, defined by the statistical law with the parameter - mathematical expectation of service time. According to the requirements, there are three main types of service systems. A service system with losses is characterized by the fact that an incoming request cannot wait for inception of service and leaves the system. A servicing waiting (lossless) system is characterized by the fact that an incoming request can wait for unlimited time and leaves the system completely serviced. A mixed-type system is characterized by the fact that an incoming request can wait for a limited time, or there are restrictions on the length of the queue. Each service systems type has its own mathematical instrument to calculate service characteristics. Based on the service characteristics calculation, the optimal service system option (in our case, the number of workers) is selected by the efficiency criterion. The efficiency criterion is selected according to the research task and the system type: minimum probability to meet the next request failure, minimum downtime of input sources and service channels, minimum waiting time for service, and others.

The solution to this problem is based on the study of "group of melting furnaces - team of furnacemen" service system. The incoming customer flow is service of furnaces. The service channels (instruments) are the team members who perform these activities. The conditions of the melting technology exclude the possibility of execution time delays in the main operations. Based on the stated requirements, the purpose can be formulated as follows: it is necessary to determine the team size of furnacemen, which can ensure the least probability to meet the delays in performing the furnaces group service under the usual conditions of furnaces operation and the same level of furnacemen's qualification. In this case, the minimum probability to meet the delays in the next operation performing can be taken as a criterion for the service system efficiency. Building a mathematical model of the service system and calculating performance indicators is based on experimental statistical research of a specific service system and determining the characteristics of the incoming customer flow and the service duration.

The greatest amount of researches on the queueing theory is related to the simplest customer flow, for which analytical methods of solving were developed. It has the quality of ordinariness, time invariance, and absence of aftereffects. The ordinariness of the flow means that it is practically impossible to receive two or more requests simultaneously. The time invariance of the flow assumes that the possibility of a certain number of requirements per a certain period of time does not depend on the point of reference, that is to say, the index of the flow does not change over time. The absence of aftereffects means that the probability to receive a certain number of requirements in any period of time does not depend on the number of requirements received before this period.

The simplest flow is defined by Poisson function:

$$Pk(t) = \frac{(\lambda * t)^k}{K!} * e^{-\lambda} \tag{1}$$

K = 0, 1, 2, … n,

Pк (t) - the probability to receive exactly "k" requirements in a time "t",

λ - the flow parameter is the mathematical expectation of the requests number per unit of time (the average frequency of requests received).

Time (t) between the appearance of two consecutive requirements in the simplest flow is distributed according to the exponential rule:

$$F(t) = \lambda e^{-\lambda t} \tag{2}$$

According to the logical reasons, we can suppose that the customer flow in the service system under consideration is the simplest. The ordinariness is assumed by the fact that under the conditions of a small number of furnaces and the melting process duration, the probability to receive more than one requirement at the same time is small. The time invariance is also not rejected if we consider the total flow from all furnaces, that is to say, the flow is evenly distributed on the time scale. The absence of aftereffects can also be assumed, since the furnaces operate independently of each other.

Knowing the type of system (system with losses), the rule of customer flow distribution (Poisson distribution), the rule of in-service time distribution (exponential rule), and the values of the main system parameters we can calculate all service system indicators:

1. Probability that all service personnel (workers) are available

$$PO = \frac{1}{\sum_{k=0}^{n} \frac{1}{K!} (\lambda m)^k} \tag{3}$$

   k = 0, 1, 2, … n,
   n – number of workers (service channels).
2. Probability that there are exactly “k” requirements in the service system, that is to say, “k” workers are employed:

$$Pk = \frac{Po}{k!} (\lambda m)^k \tag{4}$$

   k = 0, 1, 2, … n,
3. Probability to meet the next service requirement failure (the same for all «n» workers’ employment):

$$Pn = \frac{Po}{n!} (\lambda m)^n \tag{5}$$

4. Average number of service personnel employed (workers):

$$M = \sum_{k=1}^{n} kPk \tag{6}$$

5. Utilization rate of workers:

$$Kl = \frac{m}{n} \tag{7}$$

The conclusion about the economic efficiency of the service system for melting furnaces can be made by comparing the labour intensity indicators of the activity execution under the existing service system for furnaces together with the queueing theory usage:

$$\Delta T = \frac{N1 - N2}{N1} * 100 \quad (8)$$

ΔT – processing time decrease, providing that the production quota remains unchanged, in %,
N1 – size standard to service equipment (team content) in the context of the existing service system, people,
N2 – size standard (team content) in the system which was defined on the ground of the queueing theory, people.

3 Results

A characteristic feature of equipment-specific processes and, in particular, the melting of aluminum alloys in gas-fired reverberatory furnaces, is a significant duration (5–10 h) and low-grade cycling (2–3 melts per day). Like all equipment-specific processes, the melting of aluminum alloys in gas-fired reverberatory furnaces is characterized by a significant duration (compared to metal processing operations) and less participation of workers in the main technological operations.

The process of servicing the melting furnace by furnacemen includes the following activities: batch charging into the furnace, thermal conditions regulation during the entire melting process, metal purification in the furnace, mixing the metal and sampling for rapid analysis, raking off the slag, cleaning of the hearth bottom and walls in the furnace from slag, slag removal, and delivery of auxiliary materials. The furnace service is performed continuously during all three shifts and cyclically takes place with each melting.

The time expenditure on the furnace service is only 20–40% of all melting process duration. This creates opportunities to build a multimachine service of a furnaces group by a team of furnacemen. Under the conditions of multimachine service, the task of melting process rating is to establish the size standard (team content) for servicing a group of furnaces, along with the production standard. The size standard should ensure timely execution of operations to service furnaces under the conditions of usual man-loading during a working day. For analytical determination of multimachine service standards, it is necessary to know the nature of operations distribution during a shift and the time expenditure on servicing each furnace. The authors divided all time expenditures on furnace service into three groups:

- time expenditure on direct service of a specific furnace,
- time expenditure on performing various works that are not directly related to particular furnace service,
- time expenditure on transitions during multimachine service.

According to the nature of their distribution in cycle time of melting, the time expenditure of the first group is divided into two sorts: the costs which are caused by the melting technology, that is to say, the workers have to perform certain activities at specific time set by technology (metal purification in the furnace, mixing the metal and sampling for rapid analysis, slag removal, and other), and time expenditure on techniques, which is not regulated: they can occur in uncertain random times (thermal conditions regulation in a furnace).

The second group of time expenditure includes delivery of liquid remelting, auxiliary materials, and tools. Furnace service techniques vary in duration from minutes (regulating mode) to hours (loading and cleaning the furnace), as well as in frequency during the melting cycle. According to the conditions of aluminum alloys production, multimachine service includes parallel furnaces that prepare different alloys with different labour intensity of service operations. Therefore, the furnaces differ in both the process (cycle) duration and the cost to service furnaces within each cycle.

Thus, for a working team, which service a group of furnaces, the beginning of service operations and their duration is random and probabilistic that leads to the synchronization in several furnaces service. In the practice of rationing, the size standard is usually calculated based on the conditions of equal cycle duration, that is to say, that the same alloys are prepared in all furnaces. At the same time, calculations for different alloys show a different number of workers. To calculate labour intensity and pricing, the maximum number is taken. As a result, an overstated number of workers is always established. This does not take into account the reliability degree of melting process service.

Consequently, other methods to determine the size standard are necessary, which would take into account the probabilistic nature of the service process. To determine the characteristics of the current service system, a study of the functional structure of the process was conducted together with simultaneous measurement of duration of each service operation. The functional structure of the furnace service process was analyzed by the method of production process study. After processing and ordering the results of observations, an operation card of the service process was compiled (Table 1).

To prove that the customer observed flow brings into correlation with Poisson distribution rule, the authors performed a statistical analysis. For this purpose, the entire eight-hour shift was divided into equal five-minute intervals, small enough so that the rule of requirements allocation for various service operations could be shown. The number of requirements received for each interval and the arithmetic mean of the requirements number received for one interval is calculated (Table 2).

$$\overline{X}t = \frac{460}{106} = 4,34 \text{ min}$$

There are 0.222 ($\lambda\frac{x}{t} = \frac{1,11}{5}$) requirements per minute for five-minute interval.

Statistical studies show that the customer flow under study is of the simplest type. Service time is also an important indicator of the queueing system. The analytical queueing theory also imposes certain requirements on the service time distribution. The distribution of service duration is also checked by statistical analysis. The analysis of the functional process map shows that most operations have a short service duration,

Table 1. Operation card of the process

Operations (in the order in which they are serviced)	Opening, h-min	Interval between operations, min	Operation duration, min
Fluxing material delivery	7-30		28
Raking off the slag out of Furnace 6	7-32	2	28
Cleaning of Furnace 10	7-35	3	48
Preparation for purification in Furnace 7	7-48	13	4
Cleaning of Furnace 11	7-50	2	18
Mixing of the metal in the Furnace 15	7-51	1	16
Sampling from Furnace 7	7-65	4	17
Other operations, 107 operations in total			
Total			2,521

Source: authors.

Table 2. Calculation of the statistical characteristics of the customer flow

Number of intervals	Intervals	Number of requests per 1 interval - X	$(X - \overline{X})$
1	7-30 7-35	3	1.89
2	7-35 7-40	0	−1.11
3	7-40 7-45	0	−1.11
4	7-45 7-50	2	0.89
5	7-50 7-55	2	0.89
......	....	....	....
Number of intervals - 96		107	$\overline{X} = \frac{107}{96} = 1,11$

Source: authors.

and vice versa, a small number of operations have a long service duration. Therefore, we can assume that the service duration has an exponential distribution rule.

The average number of services per minute "n" and the average service time per one requirement "m" is determined from the data in Table 1.

$$n = \frac{1}{m} = \frac{107}{2521} = 0,0423 \text{ requirements per minute,}$$

where 107 is a number of requirements from all furnaces per shift, and 2,521 is service time spent on these requirements.

$$\mathrm{m} = \frac{2521}{107} = 23,5 \text{ min}$$

If we know both λ and m, we can calculate the utilization factor of the service system «j», that is to say, the working team

$$j = \frac{\lambda}{n} = \frac{0.222}{0,0423} = 5.2$$

The utilization factor "j" shows the average number of workers in a team that you need to have, in order to service all the requirements per unit of time. Consequently, the number of workers in the service system must be greater than "j", otherwise there may be an unlimited increase in the queue for service. According to the type of system, the rules of distribution of customer flow and in-service time, the values of basic system parameters (λ = 0.222 and m = 23.5), the authors calculated the service system indicators (Table 3).

Table 3. Performance indicators of the system of melting furnaces service by a team of workers of various size

Team content	Density of the customer flow, min (λ)	Average service time, min (m)	Probability that all workers n are available (PO)	Efficiency criterion – probability to meet denial of service (P_n)	Average number of employed workers (M)	Man-loading indicator (K_{load})
5	0.222	23.5	0.00900	0.240	0.360	0.720
6	0.222	23.5	0.00700	0.250	4.100	0.690
7	0.222	23.5	0.00654	0.134	4.512	0.643
8	0.222	23.5	0.00602	0.079	4.784	0.599
9	0.222	23.5	0.00576	0.044	4.971	0.552
10	0.222	23.5	0.00563	0.022	5.079	0.508

Source: authors.

After the comparison of all possible options of the size standard by the probability to meet the next request failure for service and the man-loading indicator, the optimal number of the team was selected. The results of calculations of possible options for furnaces servicing by a team of different size, which were performed through the use of the computer technology and Packet Assembler, are presented in Table 3.

According to Table 3, the most efficient is the team of 7 people.

In this case, the probability to meet denial of service to the next request is 0.134 when man-loading is 64.3%, that is to say, 13.4 requests per 100 will wait for service. Considering the fact that two workers, who are constantly engaged in batch load, are busy during a shift by an average of 75–80% (according to the working day study), they can service 13.4% of the requirements, that is to say, the service system for 9 people will practically work efficiently. Further increase in the team content is not efficient from the point of view of indicators of service system usage. The value of the load factor decreases to 0.5 if the team size increases to 10 people. Based on the results of the study, the authors calculated the economic efficiency of the service system for melting furnaces. In the furnace service system under study, N1 = 11 people. In the projected system, $N_2 = 9$ people $\Delta T = (11 - 9)/9 * 100 = 18\%$.

The calculation was made for certain values of incoming flow characteristics and service duration that are specific to the conditions of one of the plants. For this type of system (the system with losses, the simplest customer flow, and the exponential rule of service time distribution), we can calculate the performance system indicators at different characteristics values (λ, m, j = λ* m) and labour organization options (number of workers – n). This task requires a large number of calculations which are effectively performed through the use of the computer technology.

4 Discussion

The queueing theory, which deals with the analysis of repetitive homogeneous processes in service, production and management systems, has recently been discussed in the national and foreign literature, as it allows to solve a lot of practical tasks. First, to estimate the average waiting time for a specific flow of service requests [2]. Second, to identify the dependence of system performance indicators on the incoming flow, queue discipline and restriction, number, performance and conditions of channels, functioning in order to optimize it [6]. Third, to study the degree of mass-market satisfaction to perform any type of service, taking into account the random nature of demand and service. Fourth, to optimize the labour cost rates for service personnel [9]. The development and implementation of average standards for service work. In the circumstances they do not give sufficient effect to determine the required number of personnel, due to the difference in the working conditions. Therefore, the effective number of service personnel can be calculated using the queueing theory or the theory of waiting lines. Many national and foreign researchers have adapted the queueing theory tools to study the problems of improving service processes efficiency in industry [3, 4], medicine [5, 10], banking sector [1], call centers [8], and other areas.

The extensive use of the queueing theory elements in the practice of enterprises and companies in various fields of activities is primarily associated with the solution of important tasks: minimizing the total costs of service channels deadtime, loss of time and resources for service, optimizing labour costs for service personnel, and improving service system efficiency.

5 Conclusion

The application of the queueing theory to develop the size standards of workers under the conditions of multimachine service allows to calculate the optimal number of workers, which ensures reliable equipment performance and efficient labour utilization. Mathematical methods of the queueing theory can be used to calculate the service systems indicators in any case where there is a random nature of service, in particular, to solve the following tasks to improve the work measurement: determining the number of auxiliary workers: servicemen, bystanders, on-call repairmen, and electricians, developing standards for multimachine working service. This allows to establish scientifically-based standards for labour costs.

References

1. Afolalu, S.A., Babaremu, K.O., Ongbali, S.O., Abdulkareem, A., Adejuyigbe, S.B.: Overview impact of application of queuing theory model on productivity performance in a banking sector. J. Phys: Conf. Ser. **1378**(3), 032033 (2019)
2. Delasay, M., Ingolfsson, A., Kolfal, B., Schultz, K.: Load effect on service times. Eur. J. Oper. Res. **279**(3), 673–686 (2019)
3. Dodonov, V.V.: Application of the queueing theory elements to analyze productivity and reliability of automated machine. High. Educ. News **12**, 70–76 (2011)
4. Fazlollahtabar, H., Gholizadeh, H.: Application of queuing theory in quality control of multi-stage flexible flow shop. Yugoslav J. Oper. Res. **30**(1), 101–108 (2020)
5. Gazizova, L.R., Galimulina, F.F.: Mass service system in private medicine. Sustain. Dev. Manag. **6**(19), 5–10 (2018)
6. Irani, S.A., Chrissis, J.W.: Applications of queueing structures in manufacturing systems. Comput. Ind. Eng. **11**(1–4), 215–219 (1986)
7. Ivanov, A.A.: Object modeling of the automated production based on the theory of mass service. Proc. Nizhny Novgorod Tech. Univ. R.E. Alekseeva **1**(98), 88–93 (2013)
8. Li, C., Sun, C., Zhang, J.: The staffing optimization problem for the m-design multi-skill call center based on queuing model. In: Wang, R., Chen, Z., Zhang, W., Zhu, Q. (eds.) Proceedings of the 11th International Conference on Modelling, Identification and Control. Lecture Notes in Electrical Engineering, vol. 582, pp. 585–595. Springer, Cham (2020). https://doi.org/10.1007/978-981-15-0474-7_55
9. Nosov, M.A., Sabinina, A.L., Antseva, N.V.: Optimization of the maintenance crews quantitative composition in the quality management of industrial enterprises equipment maintenance. Notes of the Tula State University. Tech. Sci. **11**(2), 202–208 (2016)
10. Wen, J., Geng, N., Xie, X.: Real-time scheduling of semi-urgent patients under waiting time targets. Int. J. Prod. Res. **58**(4), 1127–1143 (2020)

Evaluating the Cost-Effectiveness of Staff Motivation When Implementing a CRM System

N. I. Voitkevich[1(✉)] and T. I. Solunina[2]

[1] Samara State University of Economics, Samara, Russia
VNI63@list.ru
[2] Samara National Research University, Samara, Russia
tsolunina@yandex.ru

Abstract. The article is devoted to solving the problem of rejecting innovations in the form of CRM systems by some employees of enterprises and organizations that are part of the distribution channels of goods and services. The research purpose is to show the need to motivate employees of commercial divisions at enterprises when implementing CRM systems and the possibility of calculating the economic efficiency when using material incentives to speed up work. Research objectives are: to prove the necessity of implementing collaborative CRM systems in the practice of interaction between participants of distribution channels; to show reasons for the negative attitude of employees to the implementation of CRM systems; to recommend material methods of stimulating employees. The authors propose a formula for calculating the economic effect of using incentive methods for motivation of employees in order to overcome their opportunistic behavior in the development and implementation of CRM systems in the activities of organizations participating in the distribution channels of goods and services.

Keywords: CRM systems · Economic efficiency · Employees' motivation

1 Introduction

Changing consumer behavior and increasing competition in almost all market segments complicate the distribution of goods and services, and require improved interaction between supply chain participants. In recent years, the culture of Russian business has significantly increased, which leads to gradual understanding of the need for mutually beneficial and trusting cooperation. In addition, the accelerated development is characterized for information technologies. These factors are objective prerequisites for the application of CRM projects in distribution channels. The use of modern CRM systems is based on principles of individualization and long-term relationships by implementing the concept of CRM (Customer Relationship Management) and developing CRM strategies. It is the need to use a customer-oriented approach in business that requires the use of CRM systems that enable to accumulate all the information about customers that companies receive through marketing, service and sales channels. This allows companies to better understand consumer needs and effectively build relationships with

S. I. Ashmarina and V. V. Mantulenko (Eds.): IPM 2020, LNNS 161, pp. 422–427, 2021.
https://doi.org/10.1007/978-3-030-60926-9_54

them and partners [8, 9]. The ability to attract and retain valuable customers through effective relationship management is often based on the use of technological CRM systems, which are considered as application software designed to automate the processes of interaction with customers.

All the distribution channels of a manufacturer, their totality forms a network, and the competitiveness of the manufacturer depends not only on its market potential, but also on the potential of the network participants included in the sale of goods. With the help of rationally formed and effectively functioning distribution channels, manufacturing companies increase the competitiveness of their own offer, using the marketing, logistics and commercial potential of trade intermediaries, in particular, the professional competence of their sales staff, the ability to build and maintain relationships with customers.

Manufacturers of goods strive to create a distribution system that can sell the maximum number of products with minimal costs over a long period of time, and respond flexibly to changes in the external environment and, first of all, to changes in the needs of the target market segment. To achieve this goal, they should establish interaction with those intermediaries who have a successful experience in using information technologies or are ready to implement and use them. First of all, we are talking about using collaborative CRM systems in distribution channels, using the potential of the Blockchain technology [2, 3, 10, 11], which allow establishing feedback with customers, i.e. with organizations of subsequent levels of distribution channels. For a manufacturing company, they are distributors or dealers; for large wholesale resellers, they are smaller wholesale organizations and retail chains. Based on the received information, the manufacturer can improve the range of manufactured products, adjust the production volumes of certain types of goods.

CRM systems allow you to identify potential consumers, analyze the purchasing behavior of individual segments of consumers grouped by geographical, demographic, industry, or other characteristics, and predict customer behavior based on the history of interaction with each customer. One of the functions of CRM systems is to manage customer service after the transaction, which includes monitoring the delivery of goods, providing feedback, working with claims, and tracking service requests.

Analysis of projects implementation on the application of CRM system in a number of Russian enterprises identified the most common and specific stages and challenges of project implementation which include the negative attitude of the staff to innovations. Opportunistic behavior of employees at logistics, marketing, and finance departments may occur: by the collection of information about business processes related to procurement and sales by the staff of the IT divisions of enterprises or involved in the implementation of the project IT companies; the introduction of a CRM – system and the initial period of its operation, when workers have to learn new things, learn the system. That results in a grudging and incomplete provision of information, exaggeration of difficulties and disadvantages of the implemented CRM system [4].

2 Methodology

The main method of research was a theoretical method that allowed us to objectively assess the possibility of using collaborative CRM systems in the distribution channels of goods and services. Based on the application of abstraction, analysis and synthesis methods, the relationship between the CRM concept, CRM strategy and the CRM system is shown. The study uses a conceptual and terminological method for revealing the content of the CRM system. The use of an empirical comparison method allowed us to identify the main problems and difficulties based on the study of the experience of implementing CRM systems in the scientific literature. To develop recommendations on the use of methods of personnel motivation, methods of overcoming the employees' resistance to changes were studied, which led to the conclusion that it is necessary to use methods of material incentives.

Currently, the main principle of measuring the efficiency in management is the principle of the relation between the goal and the end result, and in this case, efficiency shows how much the goals are being implemented and the planned results are being achieved. Efficiency is calculated as the ratio of the effect (result) and the set goals. This type of efficiency is called target or functional. However, this method may not always be applied. The main attention in this work is paid to the definition of economic effect, which required the use of methods for determining the economic efficiency, which in a generalized form can be defined as the ratio of the resulting effect to costs or to all used resources (assets).

3 Results

When implementing CRM systems, there are difficulties at the stages of preparing the project plan, discussing and agreeing on the concept and business requirements for the project, as well as at the stage of launching the system. At these stages, it is necessary to motivate employees using hard and soft methods of motivation, but first of all, we consider it necessary to use material incentives for employees by paying bonuses or using depremation. If you delay the implementation of the CRM system, the company may lose some customers as a result of failures in its work, errors and insufficient attention of managers to individual customers, which leads to a reduction in sales and loss of profit. A synergistic effect in distribution channels from the integration of autonomous CRM systems will occur the sooner, the shorter the stages of development and implementation of CRM systems for channel participants are. Therefore, reducing the time for organizations' employees to adapt to working together in the interconnected CRM systems will allow the management of companies to get an additional effect.

The effect of using methods to motivate employees of organizations – participants in distribution channels is to reduce the duration of individual stages of development and implementation of CRM systems and the initial stages of their operation.

To calculate the economic effect, we suggest using the following formula:

$$\text{Ec. effect} = S_{sw} + P_{cs} + E_{se} - C_{sm} \quad (1)$$

where Ec. effect is economic effect of employees' motivation, rubles;

S_{sw} - savings from wages of employees of IT departments in the organization, rubles;

P_{cs} - profit of the organization received as a result of saving clients by the unstable operation of the CRM system at the stage of its launch, rubles;

E_{se} - synergetic effect in the distribution channel, rubles;

C_{sm} - costs of staff motivation, rubles.

Savings due to reducing the duration of individual stages of development and implementation of CRM systems (S_{si}) can be calculated using the following formula:

$$S_{si} = S_{eit} * T \quad (2)$$

where S_{eit} - salary of employees of the IT department participating in the project, per day, rubles;

T - time of the shortened duration of stages, days;

The income received as a result of the clients-retention during the unstable operation of the CRM system at the start-up stage (I_{cr}) is proposed to be calculated using the formula:

$$I_{cr} = LTV * N \quad (3)$$

where LTV is Customer Lifetime Value or average lifetime value of a customer, rubles;

N – the average number of clients lost during the unstable operation of the CRM system at the launch and upgrade stages.

The economic efficiency is calculated by the ratio of the value of the economic effect (Ec. effect) to the organization's costs for motivating staff (C_{sm}).

4 Discussion

When forming channels for the distribution of goods and services and their restructuring, manufacturing companies select wholesale intermediaries based on a number of criteria, one of which should be the customer orientation of the organization. This will expand the functional area of the manufacturer's CRM systems by combining them with the CRM systems of trade organizations, which is extremely important for the implementation of complex relationships between each company and customers, including the collection of information through feedback. At the same time, a number of scientific papers note some problems of implementing CRM systems [1, 6, 7], related to the reluctance of employees to work in the system, resistance to innovations. Overcoming employee resistance to innovations requires the application of methods for managing employee's motivation and incentives to increase the speed of innovations, including CRM systems. Of course, the implementation of principles and methods of personnel marketing in organizations [5] may not require the introduction of any additional measures to stimulate employees. However, the lack of a broad practice of using personnel marketing forces the management of enterprises to bear additional costs for stimulating employees.

5 Conclusion

Increasing competition, new requirements to the quality of services, reducing the effectiveness of traditional marketing tools to influence customers, as well as the development of information technologies require further improvement of interaction in the distribution channels of goods and services by combining the CRM systems of their participants. At the same time, as a result of the research, it can be concluded that the motivation of staff in organizations participating in distribution channels is one of the most important elements of the mechanism for implementing collaborative CRM systems. To develop digitalization in distribution channels, changes should be made to the system of motivation of employees in manufacturing companies and trade organizations engaged in managing the flow of goods, information, and service in their purchases and sales, since the problems of functioning and low efficiency of CRM may be associated with the employees' resistance to innovations. Modern methods of motivation are diverse, the choice of the most effective ones will reduce the time spent on the development and implementation of CRM systems, as a result, organizations will not only get an expected economic effect, but also maintain the image of a reliable business partner.

References

1. Anshari, M., Nabil, M., Ariff Lim, S., Al-Mudimigh, A.: Customer relationship management and big data enabled: personalization & customization of services. Appl. Comput. Inf. **15**(7), 94–101 (2019)
2. Catalini, C., Christidis, K., Devetsikiotis, M.: How blockchain technology will impact the digital economy. Blockchains Smart Contracts Internet Things **4**, 2292–2303 (2017)
3. Francisco, K., Swanson, D.: The supply chain has no clothes: technology adoption of blockchain for supply chain transparency. Logistics **2**(1), 2 (2018)
4. Ghalenooie, B.M., Sarvestani, K.H.: Evaluating human factors in customer relationship management case study: private banks of Shiraz city. Procedia Econ. Financ. **36**, 363–373 (2016)
5. Kretova, N.N., Mitina, N.N.: Features of the implementation of personnel marketing in modern Russian conditions. Bull. Volgograd State Univ. Series 3. Econ. Ecol. **19**(2(39)), 88–96 (2017)
6. Kudinov, A., Sorokin, M., Golysheva, E.: CRM: Russian practice of effective business. 1C-Publishing, Moscow (2008)
7. Ryzhenkov, E.A.: Problems of CRM implementation at the enterprise. Actual Probl. Humanit. Nat. Sci. **1–3**, 139–143 (2017)

8. Soltani, Z., Zareie, B., Milani, F.S., Navimipour, N.J.: The impact of the customer relationship management on the organization performance. J. High Technol. Manag. Res. **29**(11), 237–246 (2018)
9. Tishkova, E.M., Sumina, E.V.: Using an automated CRM system as an element of improving the order management system in a manufacturing company. Econ. Soc. **6**(49), 1155–1159 (2018)
10. Troshina, E.P., Mantulenko, V.V.: Influence of digitalization on motivation techniques in organizations. In: Ashmarina, S.I., Vochozka, M. (eds.) Digital Age: Chances, Challenges and Future. Lecture Notes in Networks and Systems, vol. 84, pp. 317–323. Springer, Cham (2020)
11. We-IT: Types and comparative analysis of various CRM systems (2016). http://we-it.net/index.php/sistema-upravleniya-vzaimootnosheniya-s-klientami-crm/175-klassifikatsiya-crm-sistem-i-ikh-rejting/. Accessed 25 June 2020

Check for updates

Formation of an Individual Development Trajectory of a Specialist

S. R. Dreving, O. V. Borisova(✉), and N. A. Shevchenko

Financial University under the Government of the Russian Federation, Moscow, Russia
drevingsr@mail.ru, OLVBorisova@fa.ru, ninashev97@yandex.ru

Abstract. The development of the digital economy has led to the transformation of most markets due to a more active transition from manual to automated labor. Competition is increasing and specialists with non-standard and in-demand skills in the usage of digital technologies have become the most valuable. The main purpose of the research is the elaboration of a conceptual model for building an "individual trajectory of a professional development", implemented on a web service basis. To achieve this goal, the existing process of interaction between subjects of the educational environment, business and public authorities has been studied. The model of a «Web service for building an individual trajectory of personal development» is proposed. The architecture of a model is a flexible digital platform consisting of many separate platforms that will provide universal access to the main participants and will allow us to create a certain profile of an individual with its own ID to collect and organize all the necessary information. The model is based on a managing the development of individual, personal and professional abilities of an individual and allows the state to mobilize and use the full potential of human resources, processes and information technologies in the digital economy.

Keywords: Competence profile · Digital technologies · Digital economy · Individual trajectory of professional development

1 Introduction

The transition to the digital economy has largely dictated the requirements for changing the existing system of human resources management in the national economy. Such changes lead to the need to change not only the traditional model of skills, but the entire process of its formation. Active research in this area began in the early 2000s and is carried out in two main directions: allowing to modify skills as a result of the learning process and changing the traditional educational process. These studies allow us to judge the relevance and significance of the proposed work. Until now, there is no concept of «individual trajectory of development of a specialist» in the researches; besides, a unified mechanism of its formation has not developed yet, degree of integration into changing institutional environment has not determined, changes in the

S. I. Ashmarina and V. V. Mantulenko (Eds.): IPM 2020, LNNS 161, pp. 428–435, 2021.
https://doi.org/10.1007/978-3-030-60926-9_55

interaction of the main subjects of the educational environment (an employer, a learner, government agencies and educational institutions) has not shown.

The implementation of the Federal project "Personnel for the digital economy" has a significant impact on the ongoing transformation [9]. However, the need to modernize educational methods [1] and the educational system has been admitted long ago [5]. According to "The future of jobs report 2018" the labor market will change due to automation of most operations and digitization of data by 2022, which will lead to the necessity of retraining a significant part of the already working staff and to change the concept of training the younger generation [11]. Therefore, the scientific literature is increasingly studying the possibility of using modern communication technologies in the educational sphere [12]. In the research, the authors claim that many teachers are moving from using a personal computer to developing virtual worlds which are interesting for students. In response to the development of engineering and technology, the most important task of the modern educational system is to implement the individual educational trajectory of each student [4], which is a new tool for tracking their progress [17].

The variety of research in the subject area allows to identify a number of results. In particular, the need for a new methodology of professional education based on the network interaction of its subjects is justified [16]. Strategies for teaching in online courses are defined depending on the identified four learning styles, which has an impact on its quality [14]. The key reasons that hinder the overall personalization of MOOCS and the ability of courses taking into account individual differences of trainees are identified [13]. The impact of advances in artificial intelligence on the quality of MOOC training is shown [3]. It is proved that the learning environment based on CLIA helps students to acquire knowledge and high literacy based on learning outcomes, which contributes to the inculcation of self-regulation skills that allow them to carry out learning and solve problems [2]. A part of researches is related to professional development of the individual. The works in this field prove the activation of professionally-oriented independent work due to its individualization, which makes it possible to effectively form the competence of the individual [18]. The effectiveness of designing an individual professional future due to the rational determination of ways to achieve them is shown, as well as the relationship with the level of qualification of a specialist [8].

There are studies that present a structural and semantic model for the implementation of individual educational trajectories within the technological platform, a description of their design technology, and the definition of the main professional and educational competencies [17]. At the same time, the possibilities of transformation of the variable part of the curriculum of students using a modular approach in accordance with the needs of the most intensively developing industries have been studied [5]. The model of forming a personal educational trajectory of the future specialist's development in the conditions of the credit based system of education, the technology of route construction and the necessary pedagogical conditions is proposed [15]. In the future, the tendency to modernize the system will remain.

2 Methodology

The main trend of our time is the construction of an educational trajectory throughout a whole person's life. It leads to a change in the concept of training. It is necessary to involve citizens in the digital economy, which is also the reason for the demand for new tools for training, retraining of employees, and improving the skills of working citizens. One of the conditions for the continuous process of education is the definition of borders and forms of influence of the state. The demand and supply of new types of personnel in the labor market is balanced by close cooperation of business with developers of professional standards and professional education organizations. As part of the research, it was revealed that the traditional scheme of implementation of the educational process is a thing of the past. In most educational institutions, traditional classroom classes are replaced by platforms that host e-courses, sometimes completely replacing traditional classes. According to forecasts, the distance learning market may reach $215 billion by 2025. A significant number of courses and the instability of demand for certain skill groups from the employer will cause difficulties for students. At the same time, 3 of skills required by the employer is a complex process and sometimes unique approaches are used for its formation [6].

It is possible to partially eliminate the identified problem by developing a web service "individual trajectory of personal development". The authors have faced a number of problems when researching this topic: 1) the lack of a clear definition of the "individual trajectory of development of a specialist" in the authors' researches and, as a result, the presence of different semantic understanding of the term, 2) the disparity of individual studies that do not allow us to imagine the process of designing a model of professional development throughout the entire life cycle, which is integrated into a general system for regulating the need for human resources, which allows us to adjust the trajectory of a professional development in accordance with changing conditions quickly.

Rapid transformation of economic processes imposes requirements for the labour market, making the necessary ongoing training one of the required conditions of staff employment. The formation of the conceptual foundations of the mechanism of "individual trajectory of specialist development" becomes a necessary condition for further successful development of the country's economy as a result. It allows to change the process of forming students' skills fundamentally.

3 Results

The result of quantitative forecasting of the need for personnel and designing qualitative characteristics of future specialists should be the formation of a competence profile of the future employee, corresponding to the needs of the market. *The competence profile*, as a system of demanded competencies, is a set of knowledge, skills, and personal qualities that allows to implement certain labor functions. One of the requirements for the structure of the competence profile today is the availability of digital competencies of various levels, softskills, and transdisciplinarity. In the context of digital transformations and the accelerated development of information technologies in order to reduce the existing imbalances between the needs of the labor market and vocational education, there is a need to train graduates who are not only able to perform labor functions in ready-made jobs, but also able to adapt to new employer's requirements based on the increasing role of key competencies of the digital economy and personal efficiency. Transdisciplinarity will become one of the most fundamental characteristics of the digital economy of the future. New economic conditions and changes in society determine the need to search for promising models, technologies for the formation and construction of personal (individual) trajectories of professional development. Based on the definition of the digital economy, the personal trajectory of a specialist's development is a "path" of personal and professional development recorded in digital form and in terms of competencies, including activities of formal, non-formal education and education outside the standard educational environment, individual cognitive activity that accompanies everyday life and is not necessarily purposeful [7]. The scheme of the conceptual model of the "individual trajectory of development of a specialist" implemented on the basis of a web service is shown in Fig. 1.

The functional purpose of the web service is to purposefully design an individualized or personalized route of a differentiated program for building a trajectory of professional development, including educational and personal components while it seems appropriate to form a "space of opportunities", which involves the choice of a way to achieve the goals.

Thus, the algorithm of functioning of the web service should be based on the principles of harmonization of the market demand for specialists and the number of graduates of educational institutions, for systematization by regulatory organizations of the development, approval (updating) of professional standards, compliance with the interests of the specialist, continuity of support at all stages of professional activity. The market needs new specialties and corresponding professional competencies of employees. Previously, educational professional institutions of different levels did not create such competencies. This was caused by a rapid change in the economy as a whole, a change in the direction of business orientation towards customer orientation, the development of digital technologies – expanding opportunities for contacting customers, machine learning is becoming widespread, and the number of transactions using blockchain technology is increasing. Market needs was changed by analytics and Big Data. The logical sequence of designing an individual development trajectory can be represented as follows:

1. Consolidation of information about market needs, taking into account changes in professional requirements, changes in the business environment.
2. Systematization of information about educational standards for training specialists and educational institutions that use and implement these standards should form the basis for building a map of educational opportunities. The technological landscape is changing rapidly, which requires the same rapid adaptation of specialists. This requires improving the model of professional and educational standards, and this process should also include organizations of the professional community along with government agencies. It is also necessary to change the provisions of a number of existing legal and bylaws.

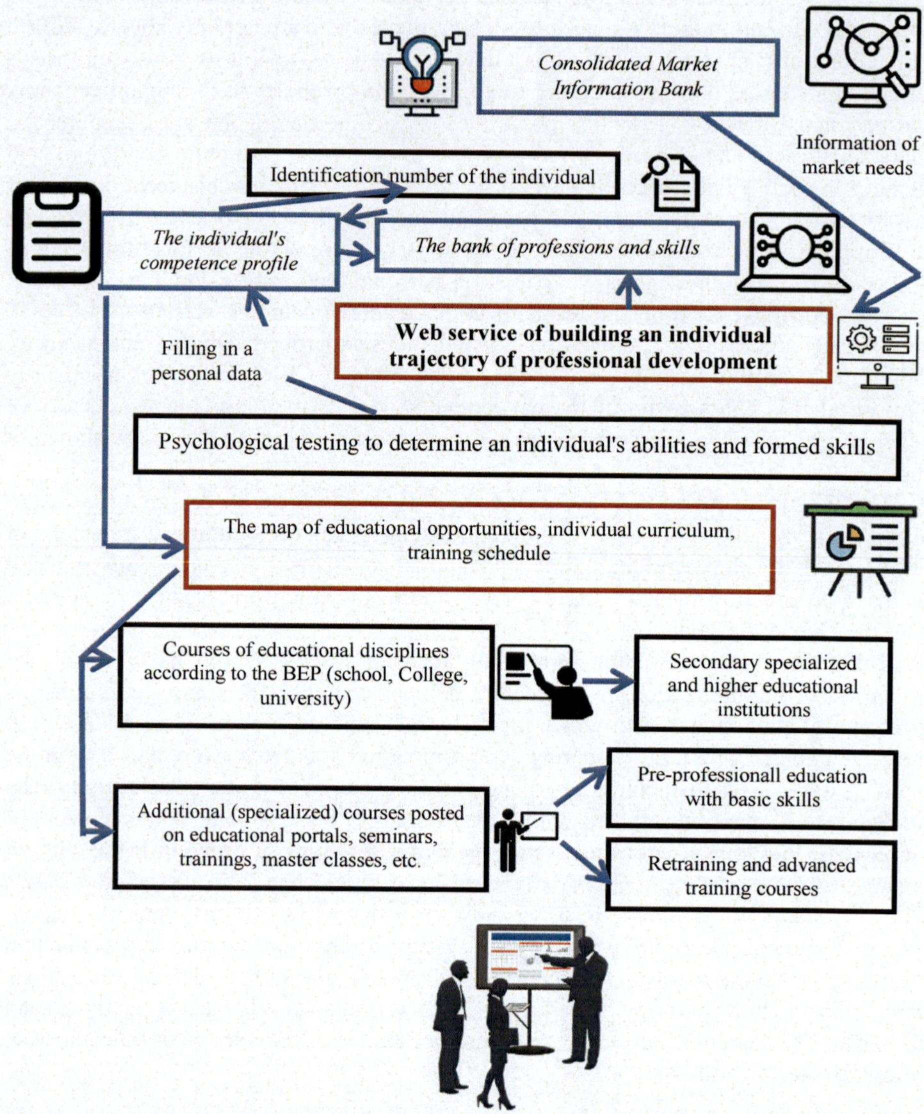

Fig. 1. Web service model for building an individual trajectory of professional development (Source: authors).

3. Formation and continuous updating of the Bank of professions and skills of personnel.
4. Periodic psychological testing in order not only to diagnose the level of development of General and special abilities that determine the success of training, professional activity and creativity, but also to determine the trajectory of further development, the individual's development potential, preferred areas of activity and

the horizon of active improvement, determining points of change in strategic priorities in career growth.

5. Formation of a specialist's competence profile by filling in personal data about the specialist, taking into account the optimal set of profile elements and the measurability of competencies, transparency of the construction process.
6. Development of the content of the individual development trajectory (setting goals and determining development needs and defining technological tools, determining the conditions that ensure the achievement of the goal).

Deep research is needed in the development of models and tools that can provide individualization of needs and opportunities, identifying potential abilities of a specialist at the next stage of their professional development.

4 Discussion

The developed conceptual mechanism of the "individual trajectory of personal development", taking into account the peculiarities of the digital economy development, can be used as a framework for the development of a web service. Its creation will be in demand by all participants of educational relations, since each of them will benefit from it. Students will receive a precise direction for the development of their competencies with clear recommendations on the courses necessary for the study and formation of skills in demand by the economy. Employers will receive highly qualified specialists with the most modern skills in the professional field. Educational institutions will receive timely information about the necessary courses to study and an order from public authorities and private companies to create them. The state from the implementation of the proposed service will receive significant savings by conducting better analysis based on big data and increasing the accuracy of the state order of specialists of each professional group. The authors of the study understand that the proposed model will be improved in the future due to constant changes in the business environment, the development of technology. The next step is to describe the model in the concept of IT architecture, formalization, and software implementation of the model, guided by the requirements of functionality, scalability, manufacturability, flexibility, convenience, reliability, and security. However, our research does not allow us to show how the individual trajectory of development will change under the influence of new technologies.

5 Conclusion

The development of the digital economy, platformization, and the increased use of information technologies in the educational process have led to the formation of conditions that allow us to change the process of managing the development of human resources fundamentally, i.e. to create an ecosystem for the formation of personal (individual) trajectories of professional development, using algorithms and tools for deep analysis based on machine learning, "big data", and artificial intelligence.

The peculiarity of our research is the complex nature of the studied processes, which allowed us to combine the existing research into a single whole. As a result, a model was proposed that allows for the process of training and retraining of personnel throughout life, starting from school age and ending with retirement. In addition, the proposed concept will later allow us to transform the web service in one of the blocks that make up a global platform for employment and training, the development of which today is put in the framework of the project "University 20.35" [10]. These innovations will allow for the transition to a new stage of development and to create a fundamentally new ecosystem in the formation of a personal (individual) trajectories of the development of a specialist, and further staffing of the Russian economy. The conducted research is of high importance for building a modern educational process, as it allows to change the mechanism of training future specialists fundamentally. The proposed ecosystem of interactions at the level of human resources management will allow to make adjustments at different stages of the learning process on time and at work of the individual on the basis of assessment of market demand for new skills and establish specialist skills, selection of necessary educational courses and programs for the development and formation of marketable competences, thus contributing to the harmonization of supply and demand in the labour market, to solve potential problems of uncertainty and volatility.

References

1. Borisova, O.V., Vasbieva, D.G., Malykh, N.I., Vasnev, S.A., Bírová, J.: Problem of using innovative teaching methods for distance learning students. Int. Electron. J. Math. Educ. **11** (5), 1175–1184 (2016)
2. De Corte, E., Verschaffel, L., Masui, C.: The CLIA-model: a framework for designing powerful learning environments for thinking and problem solving. Eur. J. Psychol. Educ. **19**, 365–384 (2004)
3. Han, H.Yu., Miao, C., Leung, C., White, T.J.: Towards AI-powered personalization in MOOC learning. NPJ Sci. Learn. **15**, 1–5 (2017)
4. Ibatullin, R.R., Anisimova, E.S.: Construction of individual educational trajectory of students based on e-learning. In: Adamov, A. (ed.) IEEE 10th International Conference on Application of Information and Communication Technologies, pp. 1–4. IEEE, New Jersey (2016)
5. Kalmykova, S.V., Razinkina, E.M., Pustylnik, P.N.: Scientometric research results for educational trajectory development in electronic educational environment. Eng. Educ. **19**, 87–90 (2016)
6. Laar, E., Deursen, A., Dijk, J., Haan, J.: Determinants of 21st-century digital skills: a large-scale survey among working professionals. Comput. Hum. Behav. **100**, 93–104 (2019)
7. Lebedeva, T.F.: Personal trajectories of IT specialists' development. Soc. Humanit. Bull. **1** (18), 12–18 (2018)
8. Lysuenko, S.A.: Designing of independent professional development path as a means of successful vocational training. Educ. Sci. J. **10**, 47–59 (2014)

9. Passport of the Federal project "Personnel for the digital economy". Approved by the Presidium of the Government Commission on digital development and the use of information technologies to improve the quality of life and business conditions. Protocol of 28.05.2019, no. 9 (2019). https://legalacts.ru/doc/pasport-federalnogo-proekta-kadry-dlja-tsifrovoi-ekonomiki-utv-prezidiumom/. Accessed 20 June 2020
10. Project University 20.35 (2020). https://www.2035.university. Accessed 22 June 2020
11. World Economic Forum: The future of jobs report 2018 (2018). http://www3.weforum.org/docs/WEF_Future_of_Jobs_2018.pdf. Accessed 22 June 2020
12. Wu, Q., Miao, C.: Modeling curiosity-related emotions for virtual peer learners. IEEE Comput. Intell. Mag. **8**, 50–62 (2013)
13. Yousef, A.M.F., Sunar, A.S.: Opportunities and challenges in personalized MOOC experience. Web Science Education, Oxford (2015)
14. Zapalska, A., Brozik, D.: Learning styles and online education. Campus-Wide Inf. Syst. **23** (5), 325–335 (2006)
15. Zavalko, N.A., Sakhariyeva, S.G., Sagimbayeva, G.S., Abdimaulen, G.A., Mukhametzhanova, Z.A.: Formation of future specialists' individual educational route in the conditions of credit system (on the example of the Republic of Kazakhstan). Espacios **39**(17), 30 (2018)
16. Zeer, E.F., Popova, O.S.: Psychological guiding of students' individual educational trajectories in vocational school. Educ. Sci. J. **1**(4), 88–99 (2015)
17. Zeer, E.F., Streltsov, A.V.: Technological platform for realization of students' individual educational trajectories in a vocational school. IEJME Math. Educ. **11**(7), 2639–2650 (2016)
18. Zhuykova, O.V., Shikhova, O.F., Shikhov, Y.A.: Professionally oriented course of engineering-graphical training. Educ. Sci. J. **1**(3), 46–61 (2015)

Developing Soft Skills Among Russian Higher Education Students in Context of Globalization

T. A. Ilyina[1] and N. A. Kryuchkova[2](✉)

[1] Samara State Technical University, Samara, Russia
tanya.ilyina@list.ru
[2] Samara State University of Economics, Samara, Russia
kryuchkova_n.a@bk.ru

Abstract. The article justifies the importance of developing soft skills among students of higher education institutions. The authors call soft skills the competencies of the future, since in the context of the ongoing globalization they are crucial for graduates in employment and career development. The study assesses the level of proficiency in soft skills of students of Russian universities (using the example of Samara State Technical University). To develop soft skills, the authors suggest a number of measures to change the educational process at the university.

Keywords: CLIL – Content and Language Integrated Learning · Educational process · Soft skills · University

1 Introduction

In the face of new challenges, the list of competencies required from employees of modern companies has expanded significantly. Globalization has had a significant impact on this, which contributes to the integration of national economies and the expansion of professional cooperation between enterprises of different countries. In the 21st century, information and communication technologies are being widely introduced into society, leading to changes in education, occupation and employment. These changes require continuous adaptation to the environment and the acquisition of new skills. Researchers call them "21st Century Competencies" [6]. Today, along with professional competencies, soft skills are of particular importance. They include: communication skills, critical thinking and problem solving skills, teamwork skills, lifelong learning, and leadership skills [8]. Soft skills have a significant impact on the ability of employees to perform their work, so they are of great importance in the employment of university graduates. The development of soft skills should begin at the university level. In Russia, the question of developing soft skills among students in higher education institutions is especially acute, which is associated with the transition to a competent approach in education.

S. I. Ashmarina and V. V. Mantulenko (Eds.): IPM 2020, LNNS 161, pp. 436–442, 2021.
https://doi.org/10.1007/978-3-030-60926-9_56

2 Methodology

The subject of this study is the processes of teaching soft skills in higher education institutions and their impact on the employment of graduates in the context of globalization. The object of the study was one of the leading Russian universities - Federal State Budgetary Educational Institution of Higher Education Samara State Technical University (SSTU), which annually graduates a large number of bachelors, specialists and undergraduates of technical and economic orientation. Also, SSTU is the largest scientific center. In studying the literature on the topic of research, methods such as analysis and synthesis were used. Experimental study of the object was carried out through observation and survey. The survey was attended by students of FSBEI HI SSTU. The sample size was 30 people. In summing up the study, the authors used mathematical methods, namely the data visualization method. The results of the study are presented in the form of diagrams.

3 Results

Before the transition to the Bologna system in 2003, a qualification approach was the basis of Russian vocational education, the main task of which is for students to obtain knowledge, but not skills (especially soft). The reform carried out in Russia in the 21st century to introduce the Bologna system changed this state of affairs. If you follow the model of the Bologna process, teaching should be aimed at developing both professional and social competencies aimed at future professional activities. At the same time, researchers give social competencies no less important than professional ones [11]. Thus, soft skills are today the competencies of the future, since in the context of ongoing globalization they determine the competitiveness of both graduates in the labor market and institutions of higher education themselves. Nevertheless, the conducted studies (on the example of Samara State Technical University) show that there are a number of problems in introducing soft skills into the teaching processes, although there are also successes in this direction. A survey conducted among students of the Faculty of Engineering and Economics showed that less than half (41.94%) are familiar with the notion of “soft skills” (Fig. 1). For others, this notion is either unfamiliar, or they have heard about it, but do not understand the meaning. This suggests that students do not attach special importance to the acquisition of these competencies for future professional activities and are not interested in this issue.

Accordingly, an important task for the university is to develop measures for informatization and the development of soft skills among students. The list of suggestions is presented in Fig. 2.

First of all, information work should be strengthened in terms of explaining to students what soft skills are in the framework of the training sessions. It is also necessary to increase the number of trainings with the participation of the leadership of higher education institutions, authorities and employers, who can directly voice the requirements for employees and share personal experience in the use of soft skills in work. The second recommended direction is the use of modern ways of mastering soft skills in educational process, for example, such as trainings, coaching and mentoring

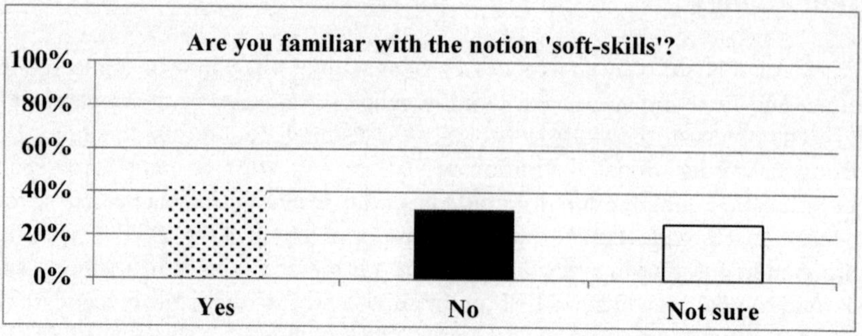

Fig. 1. Results of a survey on the notion 'soft skills' (Source: authors).

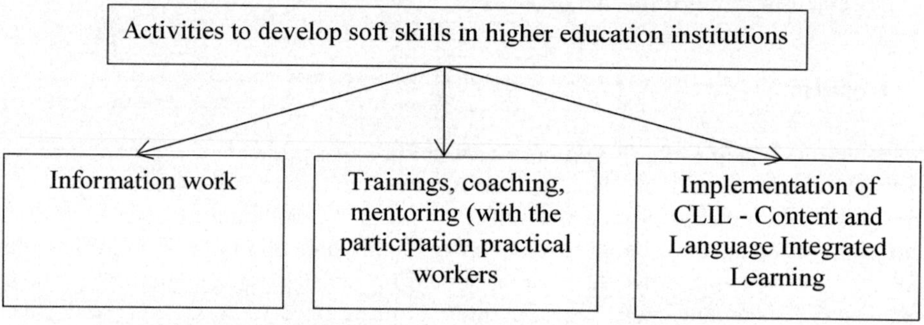

Fig. 2. Activities to develop soft skills in higher education institutions (Source: authors).

(with the participation of practitioners from the field of potential employers). These forms of interaction with enterprises will allow you to gain the soft skills required in future work.

It should be noted that in FSBEI HI Samara State Technical University there exists a whole range of measures for the professional orientation of youth and interaction with employers, and many of them have a good effect. So, the university has a department for organizing practices and promoting the employment of graduates. Every year, SSTU students go to enterprises in different parts of the country and even the world to test their knowledge and ideas about the chosen profession. Practice allows students to see the organization of activities, the list and content of professional duties of the enterprise where they will work. A developed innovative infrastructure has long been operating at the university that contributes not only to the educational and scientific, but also to the professional activities of students, since it helps in practice to get acquainted with the future profession. It is represented by training engineering centers created together with leading world manufacturers of applied research and industrial equipment [5]. However, these activities are more aimed at developing professional skills, but not soft ones. The third direction in the development of soft skills among students of Russian universities may be the introduction of educational technologies of subject-language integrated learning (CLIL - Content and Language Integrated Learning).

This technology implies that a foreign language is used as a tool in teaching a non-language subject. At the same time, both language and subject are of equal importance.

Conducting courses in specialized disciplines in foreign languages contributes to the development of foreign language competence among students and such soft skills as: the ability to conduct business negotiations and business correspondence in a foreign language, as well as the ability to develop and carry out projects in a foreign language. These skills in the context of globalization and the expansion of international relations are of particular importance for successful professional activities and career growth. In Russia today there are many foreign companies that urgently need employees who have foreign-language competence (first of all, we are talking about proficiency in English, to a lesser extent - German, French, Chinese, etc.) A similar situation is observed in Samara region, where one may find the largest international companies in the field of mechanical engineering, IT technologies, food industry, banking, trade, etc. FSBEI HI Samara State Technical University is the leading center for training personnel for these industries.

However, as a survey among students (Fig. 3) showed, less than 10% of respondents speak English on advanced level, which is necessary for full-fledged work in an international company. The same number of respondents admits that they have a zero level. No one speaks advanced German. There was no one speaking French (at any level) among the interviewees (Fig. 4).

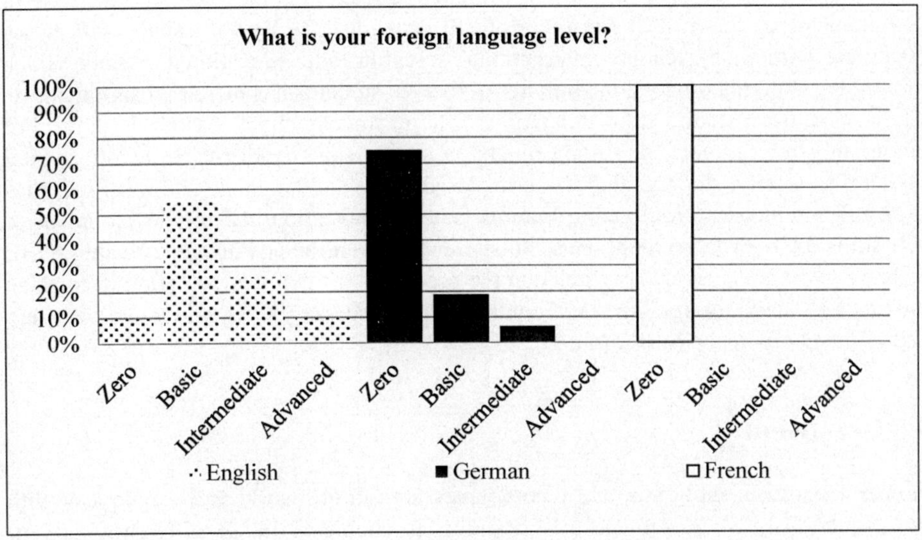

Fig. 3. Foreign language proficiency survey results (Source: authors).

However, the positive point is that most respondents want to improve the skill of foreign language competence and would like to study specialized subjects in English (64.52%). The situation with German and French is worse: 21.05% and 11.11% respectively.

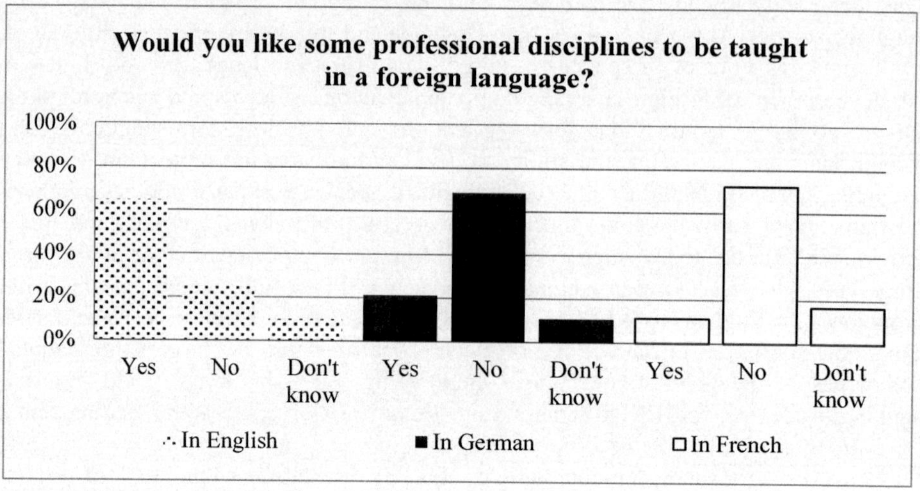

Fig. 4. Results of the survey on teaching professional disciplines in a foreign language (Source: authors).

Besides, the soft skills acquired with the help of CLIL technology can be useful already at a grade level in institution of higher education as they promote increase in communication between foreign and local students. Today, Russian universities cooperate with many leading international scientific and educational centers. Since joining the Bologna process, the number of foreign students has increased significantly. At the same time, studies show that among foreign students there is a very high level of emotional stress associated, among others, with academic problems. In addition, they increase as foreign students that try to master the Russian language and local specifics [7] CLIL technology would contribute to faster socialization and the development of soft skills for both foreign and Russian students. Given the advantages of using CLIL technology, it is necessary to accelerate the processes of its implementation in Russian universities, since foreign-language competence is necessary now, and it can be safely called the competence of the future.

4 Discussion

Numerous studies show that many companies around the world lack employees with the required skills. Moreover, this situation is characteristic of different fields of activity, despite the fact that institutions of higher education each year produce a large number of specialists. For example, a shortage of professionals with relevant skills is felt in the IT field. At the same time, we are talking not only about technical skills, but also soft ones, that is, skills for working with people [3]. The need for training and the development of soft skills among workers is indicated by the research of the hotel business. The survey showed that applicants in the hospitality industry lack such skills as responsiveness to customers, organization, oral communication and teamwork [1].

An analysis of nursing training programs also showed that in order to increase the level of medical care, attention must be paid not only to acquiring professional knowledge, but also soft skills [9].

Research into the professional training of future accountants [4] and marketing specialists [2] demonstrates the importance of soft skills. Soft skills in the professional activities of teachers and faculty members of higher education are also important [12]. Thus, the problem of the lack of specialists of various fields with not only professional, but also soft skills is relevant and discussed throughout the world. At the same time, studies conducted in the educational environment show that the use of innovative pedagogical approaches significantly improve the soft skills of students – future specialists [10].

5 Conclusion

Today, one of the urgent tasks for the university is the training of bachelors, specialists and undergraduates with competencies of the future, which include not only professional, but also soft skills. The authors, on the example of FSBEI HI SSTU, investigate the level of proficiency of modern students in soft skills, the degree of demand for these skills in the labor market and the prospects for teaching soft skills in higher education institutions. Studies have shown that, despite the demand for soft skills by employers, graduates do not own them sufficiently. To solve this problem, it is suggested to introduce a number of measures into the educational process of universities, namely: conducting information work; the use of modern ways of mastering soft skills; application of CLIL technology. In general, it can be said that although teaching soft skills in Russian higher education institutions is at the initial stage of implementation, competition and ongoing globalization contribute to the introduction of changes in their educational processes.

References

1. Crawford, A., Weber, M.R., Lee, J.: Using a grounded theory approach to understand the process of teaching soft skills on the job so to apply it in the hospitality classroom. J. Hosp. Leisure Sport Tourism Educ. **26**, 100239 (2020)
2. Daellenbach, K.: On carrot cake and marketing education: a perspective on balancing skills for employability. Australas. Mark. J. (AMJ) **26**(2), 172–179 (2018)
3. Dubey, R.S., Tiwari, V.: Operationalisation of soft skill attributes and determining the existing gap in novice ICT professionals. Int. J. Inf. Manag. **50**, 375–386 (2020)
4. Herawati, S.D.: Review of the learning method in the accountancy profession education (APE) programs and connection to the students soft skills development. Procedia-Soc. Behav. Sci. **57**, 155–162 (2012)
5. Ilyina, T.A., Panafyonova, L.I.: Innovative environment of the university as a means of solving the problems of professional self-determination of students. Qual. Innov. Educ. **2** (153), 9–12 (2018)

6. Khodeir, L.M., Nessim, A.A.: Changing skills for architecture students employability: analysis of job market versus architecture education in Egypt. Ain Shams Eng. J. (2019, in press). https://www.sciencedirect.com/science/article/pii/S2090447919301777. Accessed 23 June 2020
7. Kosheleva, E.Y., Amarnor, A.J., Chernobilsky, E.: Stress factors among international and domestic students in Russia. Procedia-Soc. Behav. Sci. **200**, 460–466 (2015)
8. Ngang, T.K., Hashim, N.H., Yunus, H.M.: Novice teacher perceptions of the soft skills needed in today's workplace. Procedia-Soc. Behav. Sci. **177**, 284–288 (2015)
9. Ng, L.K.: The perceived importance of soft (service) skills in nursing care: a research study. Nurse Educ. Today **85**, 104302 (2020)
10. San-Valero, P., Robles, A., Ruano, M.V., Martí, N., Cháfer, A., Badia, J.D.: Workshops of innovation in chemical engineering to train communication skills in science and technology. Educ. Chem. Eng. **26**, 114–121 (2019)
11. Shnyrenkov, E., Pryadko, I.: The Bologna process: exacerbation of social competences among civil engineering students. Procedia Eng. **117**, 330–335 (2015)
12. Tang, K.N.: The importance of soft skills acquisition by teachers in higher education institutions. Kasetsart J. Soc. Sci. (2018, in press). https://www.sciencedirect.com/science/article/pii/S2452315117304939. Accessed 23 June 2020

Innovation Management: Foreign Language Knowledge as a Means of Control

E. A. Karaseva, E. A. Pertsevaya(✉), and N. E. Petrova

Samara State University of Economics, Samara, Russia
elenakaraseva063@gmail.com, kmilyutina@mail.ru, bakinan@mail.ru

Abstract. The article studies a foreign language knowledge as a means of control in the processes of innovative management. The ability of organizations to continue training during the work process is closely connected with the processes of increasing the pace of innovation. The study is relevant due to the lack of published works identifying the ways of assessing the quality of the employee performance in a managerial position in the organization in the innovation process. The goal of this study is to analyze the level of English language proficiency by managers who hold upper-middle level positions in various fields of activity, since increasing the level of foreign language proficiency is a necessary element of the successful functioning of a modern leader. The prospects of applying foreign language knowledge are connected with improving the quality of research and development, increasing the production capacity of enterprises, and building up innovative potential.

Keywords: Control · Innovation · Innovation management · Manager

1 Introduction

The idea of innovation management is closely related to the idea of continuous education. Moreover, training focuses on the areas of an enterprise that are traditionally considered the most promising such as technical, informational, economic, social. These fields are constantly in the process of improving and continuous innovations are introduced. Innovation is a vital aspect, as it affects the degree of viability of the organization in social and economic transformations while maintaining sustainable processes of change, allowing organizations to survive and improve the quality of their activities [5].

Innovation is a dynamic process, which, when properly applied and managed, can provide a competitive advantage and the possibility of economic growth for individual companies and countries as a whole. Increased competition in all areas, inspired by globalization and the development of regional and global economies, highlighted the importance of scientific research in the area of innovation management and innovation as a whole [7].

The rapid development of innovations in various fields has led to the fact that there are many examples of practical implementation of successful practices in different countries, however, the detailed theoretical explanation of such is lacking but required

S. I. Ashmarina and V. V. Mantulenko (Eds.): IPM 2020, LNNS 161, pp. 443–449, 2021.
https://doi.org/10.1007/978-3-030-60926-9_57

in order to be applied in other countries. The gap between practical developments and theoretical knowledge when it comes to innovations is still significant [9]. The solution is to develop effective ways of introducing innovations through appropriate assessment methods, in other words, the control of innovations. This will narrow the gap between theory and practice with the help of adequate developments that will allow us to develop the correct approach to the implementation of theoretical knowledge.

Practical experience in the field of innovation management in Russia is the transference of foreign experience in large-scale projects to Russian corporate culture, subject to the necessary changes and transformations. Therefore, the application of innovations and innovation management is based on the experience that functions and exists only in a foreign language. Mostly, the foreign language that is used to transfer knowledge on innovative management is English.

The priority of learning English for all managers working in the field of innovation is increasing. The social aspects of innovation management are decision-making and control over their implementation, take place in Russian companies using the Russian language obviously. However, information on similar changes in the field of innovative management in joint and foreign enterprises is available only in foreign languages. The level of foreign language proficiency in the field of innovation management is subject to constant evaluation according to professional standards and, as a result, to continuous improvement. It is necessary to follow the clear criteria for assessing the acquired knowledge, which is based on personal methods of control, as well as universal and systematic methods of evaluating made by international agencies. Methods of personal control of the level of gained knowledge vary, but remain within the following variables:

- objective assessment of knowledge,
- awareness of personal knowledge level and level of proficiency,
- systematic approach to education,
- the need for public testing of acquired knowledge,
- individual adaptation of the studied material [8].

Most often when carrying out a level assessment of foreign language proficiency, managers rely on self-appraisal rather than on one of the accepted standards of testing.

2 Methodology

In this research, we review systems for assessing the foreign language proficiency of managers in different organizations to establish general trends in the field of management. The study used data from www.hh.ru [4], which provides and collects resumes of applicants for various vacancies across the country. The positions of a manager and a supervisor in this study are synonymous. The sample does not contain data on top-tier managers due to the insufficient amount of data for generalization and analysis. Therefore, the presented data applies exclusively to middle and upper-middle management employees. According to the data of the aforementioned site, a manager's position includes responsibilities of handling a company's innovative activity; the information contained in vacancies and resumes of job seekers for a manager's position can be applied to illustrate the general layer of innovation management.

The most complete presented data for the year 2018 focuses on labor market research in relation to enterprise management. The methodology for collecting data for this study was a publicly available system for online monitoring of the Russian labor market, individual labor market research, public surveys, and online payroll analysis based on data from more than 1000 employers throughout Russia and hh.ru databases.

The data from the latter site served as a starting point for this study to determine the necessity to know a foreign language for managers working in the field of innovation in Moscow and the Moscow region, St. Petersburg and the Leningrad region and other regions. The most commonly indicated foreign language is English. All data regarding the level of knowledge of a foreign language are based solely on a personal assessment of proficiency (Table 1) [4]. A formalized, independent approach to determining the real quality of knowledge is used exclusively in organizations engaged in the employee search and is not reflected in the form of formalized data.

Table 1. Statistics on foreign knowledge

Region/Level of english proficiency	Moscow and Moscow region	St. Petersburg and Leningrad region	Other regions
Fluent	25%	19%	15%
Able to pass an interview	16%	15%	12%
Able to read professional literature	11%	10%	9%
Basic knowledge	33%	39%	42%
No foreign language knowledge	15%	17%	23%

Source: authors.

3 Results

The innovation process involves the application of changes in the process of managing an enterprise of any size. This process, among other characteristics, has a specific one – diffusion. Diffusion involves the spread of successfully mastered and applied knowledge within a temporal and spatial aspect. Accordingly, it can be argued that innovation management is based on cross-cultural awareness and its application within a specific time frame and a specific enterprise. Thus, knowledge of the English language turns out to be closely related to the adaptation of experience gained in enterprises abroad during the management of enterprises in Russia.

According to the www.hh.ru, one of the main social skills of the manager is a cross-cultural awareness, i.e. the ability to adapt to work with staff from different countries, as well as to quickly switch to different cultures of employees. This cross-cultural awareness results from the basic knowledge of a foreign language, which allows to integrate into the culture of other countries and understand the processes taking place in them, as well as to adapt to possible changes and deviations from what is considered to be the normal pattern in the country the manager comes from.

Knowledge of a foreign language implies the ability to learn about the habits and traditions of another culture, to understand and accept the realities of other countries, and to interact with representatives of another cultural environment.

It should be noted that a foreign language mentioned by managers in a resume as a means of raising cross-cultural awareness varies. There are 10 basic foreign languages mentioned by managers in resumes. It is to be expected that English has the highest number of mentions (in percentage terms more than 70% of managers note the knowledge of this language), German and French rank second and third. Spanish, Italian, Chinese, and rare languages such as Turkish, Polish, Arabic, and Japanese are among others. Figures on foreign languages prevalence among applicants illustrate quite accurately those countries that are present in Russia or have joint ventures that serve, often as a model for the use of practical knowledge in the field of innovation.

English language, as the most common means of communication between different cultures, most often plays the role of an intermediary in the creation of working materials, as well as their transmission for a more detailed analysis of the situation. International innovation management involves close interaction between the managing enterprise and divisions from different countries (in some cases, the managing enterprise is located in one country, and its branches are in other countries). That is why the main and most accurate method of monitoring the effectiveness of international innovation is the entry of senior managers of the main management enterprise into the board of directors of foreign branches. Such a thing is possible solely on the condition of fluency of the foreign language typical for the managing company the manager belongs to. In the case of joint ventures, English is such a language. Accordingly, managers need to be fluent in the language and use it at all stages of the establishment and operation of the joint venture, i.e. it is necessary to analyze the language potential of the modern manager; the potential necessary for a successful position in the field of innovation management.

The percentage of foreign language mentioned in the resume of Moscow and Moscow region, St. Petersburg and Leningrad region, and other regions varies slightly, but the general trend remains unchanged - English occupies the leading position in the number of managers and the managers included it in their resume. Applicants assessed the quality of language knowledge according to the following scale: no foreign language knowledge, basic knowledge, able to read professional literature, able to pass an interview, fluent. It is interesting to note, that the most common variants for assessing own language proficiency among managers have changed in Moscow, St. Petersburg and regions. The most common option was "Basic Knowledge," represented by 33% in Moscow, 39% in St. Petersburg, and 42% in other regions.

The variant "Proficient" occupies the second place in Moscow. The percentage of applicants who used it in the description of their own level of language proficiency was 25%; 19% in St. Petersburg. In regions, the variant "There is no knowledge" occupies the second place with the percentage of 23%.

The third place is occupied by the variant "Able to undergo interviews" with the statistic of 16% in Moscow. In St. Petersburg it is the variant "Unable to speak" with the indicator of 17%, and the "Proficient" variant with the level of 15% in regions. Thus, the tendency to acquire basic knowledge of a foreign language is becoming a priority during a search for a job in the sphere of management.

The next variant for assessing the quality of foreign language proficiency is "The Ability to Undergo Interviews." It is noteworthy that this level is approximately the same in Moscow and Moscow region, St. Petersburg and Leningrad region, as well as in the other regions and is 16, 15 and 12%, respectively. The variant "Ability to read professional literature," is approximately the same according to the number of discrepancies. This variant counts 11% in Moscow and Moscow region, 10% in St. Petersburg and Leningrad region, and 9% in regions.

It is interesting to note that the last two criteria are roughly the same in terms of the percentage of managers indicated this level of proficiency in English in their resumes, while the other options are quite various. It is noteworthy that the level of foreign language proficiency is increasing in moving from regions to Moscow. One of four people among candidates for managerial positions in Moscow is confident about proficiency in English, while at the regional level almost the same number of applicants do not speak English.

In 2019, the system for assessing the level of foreign language proficiency changed. Thus, when it is described in a resume, according to the Common European Framework of Reference (CEFR) standard, long and successfully used in the European Union and Russia, the system of language knowledge assessment has undergone significant changes. The system is more formal and unambiguous. In total, there are 6 levels of foreign language knowledge according to this system. Each of them is characterized by the presence of certain skills typical for oral and written communication, as well as the skills necessary for the successful reading of literature, in particular, professional literature.

These variants are:

- A1—beginner,
- A2—elementary,
- B1—pre-intermediate,
- B2—intermediate,
- C1—upper-intermediate,
- C2—advanced.

The approximate levels referred to above can be matched to uniform standards (e.g. basic knowledge - A1, reading professional literature - B2, ability to have an interview with a firm - C1, and fluency in the language - C2). However, such formal adaptation is not complete - for example, a manager being able to pass an interview fails to read professional literature, and a person being able to read professional literature sometimes cannot understand speech at all. Besides the difference between B1 and B2 levels is significant and fundamental in maintaining professional communication at the above-average level, so the B1 level assessment system lack prevents potential candidates from defining their reference level for languages correctly, thus, avoiding confusion at job interviews. It should be noted that this grading lacks the concept of "I don't know" a foreign language, which proves the need for professional knowledge enhancement of a foreign language in the reality of contemporary times, as a person who cannot maintain communication in a foreign language cannot be considered as a potential candidate for a number of positions, including management ones. Language skills in management, especially in innovation management, are a critical aspect directly connected with prospective employment and salary level.

4 Discussion

In the context of the constant interaction between countries, the process of developing language skills, as well as its use both for professional purposes and for personal use to raise social mobility and creative development, is vital. Language skills are a means of professional communication, both in speaking and writing to produce competitive products, as well as research and management activities of the new generation innovation manager [10]. At present, the linguistic, social, and educational aspects of learning English provide a comprehensive motivation both to enhancing own knowledge and applying the knowledge gained in practice. Language learning aims at extending knowledge in the following areas: short-term and long-term benefits; personal knowledge and autonomy enhancement; contextual and social needs as well [2]. It is difficult to apply both approaches in practice without having the measure of knowledge gained over a long period of time, which is common to innovation and improvement processes [1]. Innovation management can be monitored and controlled by analyzing formal data provided based on response handling from participants interested in the process itself. The contemporary studies note the increasingly larger role of the positive influence on the ability to learn within an organization, innovation, and organizational management [3]. Formulated differently, the ability to enhance professional competencies, including foreign language skills, in particular English, empowers using the best practices of international companies in innovation management. The prospects for increasing the innovation level, as noted in the studies, are related to the possible scientific and research development application for building a unified academic research strategy, development of a systematic approach to training, selective training of certain aspects, as well as global knowledge accumulation [6].

5 Conclusion

International innovation management suggests benefiting from competing companies operating in different countries doing business, as well as using economic, social, political, cultural, and other peculiarities of these countries, and their interaction. A manager working in innovation, and having a social cross-cultural skill, is obliged to have foreign language skills, namely English, which is the most widespread in all spheres of activity both in speaking and writing. The minimum level of language skills is the level of "basic knowledge" (approximately A1), which needs ideally to be raised as closely as possible to fluency (C2), which often remains unachievable in practice. Learning a foreign language by innovation managers working in various spheres should meet professional demands, as well as cover such aspects as increasing foreign language vocabulary, modeling professional communication situations, grammar development required to understand the foundations of interaction between international companies. Therefore, the need for continuous improvement of foreign language skills is a social impact of innovation management functioning in the international aspect, as this particular indicator is the control point of activity success of a manager aspiring to be promoted in an international innovation company.

References

1. Bourke, J., Roper, S.: Innovation, quality management and learning: short-term and longer-term effects. Res. Policy **46**, 1505–1518 (2017). https://doi.org/10.1016/j.respol.2017.07.005
2. Fandiño, F.G.E., Muñoz, L.D., Velandia, A.J.S.: Motivation and e-learning English as a foreign language: a qualitative study. Heliyon **5**, e02394 (2019). https://doi.org/10.1016/j.heliyon.2019.e02394
3. Gomes, G., Wojahn, R.M.: Organizational learning capability, innovation and performance: study in small and medium-sized enterprises (SMES). Revista de Administração **52**, 163–175 (2019). https://doi.org/10.1016/j.rausp.2016.12.003
4. HeadHunter: Labour market study for directors and line managers (2018). https://hhcdn.ru/file/16688086.pdf. Accessed 05 Apr 2020
5. Kalmuk, G., Acar, A.Z.: The mediating role of organizational learning capability on the relationship between innovation and firm's performance: a conceptual framework. Procedia-Soc. Behav. Sci. **210**, 164–169 (2015). https://doi.org/10.1016/j.sbspro.2015.11.355
6. Kocoglu, I., Imamoglu, S.Z., Ince, H., Keskin, H.: Learning, R&D and manufacturing capabilities as determinants of technological learning: enhancing innovation and firm performance. Procedia-Soc. Behav. Sci. **58**, 824–852 (2012). https://doi.org/10.1016/j.sbspro.2012.09.1062
7. Ode, E., Ayavoo, R.: The mediating role of knowledge application in the relationship between knowledge management practices and firm innovation. J. Innov. Knowl. (2019, in press). https://doi.org/10.1016/j.jik.2019.08.002. Accessed 20 Apr 2020
8. Pavlenko, D., Barykin, L., Nemeshaev, S., Bezverhny, E.: Individual approach to knowledge control in learning management system. Procedia Comput. Sci. **169**, 259–263 (2020). https://doi.org/10.1016/j.procs.2020.02.162
9. Sandberg, K.W., Ohman, G.: Learning in innovation development. Procedia-Soc. Behav. Sci. **28**, 379–383 (2011). https://doi.org/10.1016/j.sbspro.2011.11.072
10. Volodina, I.: Professionally oriented foreign language teaching for innovative managers at a technical university. Young Sci. **8**, 315–318 (2012)

Motivational Aspects of Research Activities of University Students in Modern Conditions

J. Levashova and J. Sharikova(✉)

Samara State University of Economics, Samara, Russia
levashova99.99@mail.ru, sharikovajv@mail.ru

Abstract. This paper discusses modern approaches to motivating students to participate in research activities of Universities and their effectiveness in the conditions of Informatization of society. The authors draw conclusions about the applicability of various types of incentives for research activities of University students in modern realities. Within the framework of this topic, a study was conducted to identify the level of motivation of students on the example of Samara state University of Economics and presented methods to increase the motivation of University students to research activities, based on the results of the current level and the analysis of the applicability of these approaches in practice.

Keywords: Motivation · Motivation of students · Research work (RW) of students · Rating of students

1 Introduction

In the modern world, motivation of the company's staff plays an important role in the work of the whole organization, since a properly built system of motivation of labor among employees can ensure an improvement in the quality of labor of employees and the company's competitiveness in the market. In management theory, there are a sufficient number of authorial approaches to the definition of the concept of "motivation" [3, 5, 6], on the basis of which it can be concluded that motivation has the following characteristics: explanation of human behavior, its orientation and activity; the desire of the employee to satisfy his needs through work; encouraging someone to work towards the goals of the organization [8]. In general, motivation is the main engine of human behavior and activity, and this is especially necessary in the process of forming the future professional. That is why the question of the motivations and incentives for the educational and professional activities of students becomes the most relevant.

The research activities of students make it possible to fully demonstrate their individuality and readiness for the self-realization of the personality, as well as to develop creative abilities. It should be noted that the process of research is individual and is a value in the educational and personal sense [10]. Against the background of a developing market economy, the training of specialists capable of quickly coping with current problems, capable of applying the obtained theoretical knowledge in practice is of great importance for society. The development of student interest in research

S. I. Ashmarina and V. V. Mantulenko (Eds.): IPM 2020, LNNS 161, pp. 450–458, 2021.
https://doi.org/10.1007/978-3-030-60926-9_58

activities (RA) and the emergence of innovative technologies in education can help in solving this problem. These factors contribute to the development in students of independence in making non-standard decisions and innovative decisions. That is why motivation plays a huge role in stimulating students to research activities, which contributes to improving the performance of students in the research field and, as a result, the competitiveness of students in the labor market in the future [9].

2 Methodology

A questionnaire was conducted at Samara State Economic University to identify the level of motivation of students to take part in research activities. Respondents were undergraduate students of Samara State Economic University, mainly students of 1 and 3 courses. The survey involved more than 170 students from all 7 institutes of Samara State Economic University (including the faculty of secondary vocational education). Based on the results of the survey, the information obtained by applying elements of mathematical and statistical analysis were systematized and generalized into groups for ease of evaluation. In addition, this research conducted a social experiment with a group of students. The essence of the study is to analyze the behavior of people under the influence of such a factor as motivation, while in the course of the experiment, the situation was modeled to a greater extent resembling the real one.

3 Results

The first question was aimed at identifying interest in research work at the university. Analyzing the answers of respondents, we can conclude that 21% of students are engaged in scientific work on an ongoing basis, and exactly 25% of students - from time to time (Fig. 1).

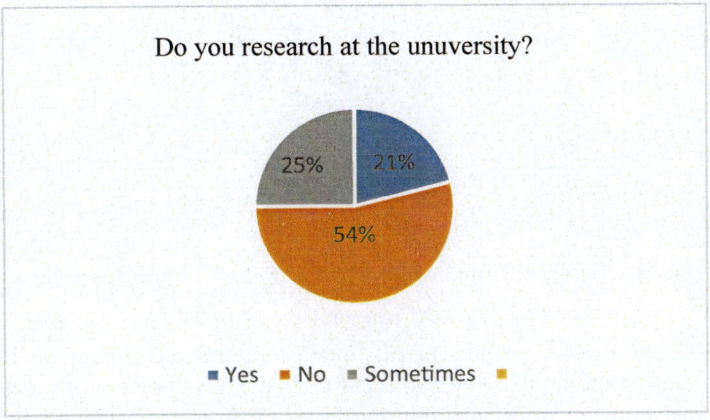

Fig. 1. Interest of students in research work (Source: authors).

The respondents were also asked about what, in addition to their studies at the university, extracurricular activities they prefer, the answers to this question are presented in Fig. 2. According to the results of the diagram, it can be concluded that almost half of the surveyed students do not have enough time to participate in any university activity other than academic, which is not a very good result. Students who are members of any student associations, as shown in the survey, most often participate in several areas of activity at the same time. (for example, a member of the student pedagogical squad is also an activist of the student scientific society). This testifies to us that students have opportunities to realize their potential at the university, it depends solely on the desire and motivation of young people to participate actively. Students who are members of the student scientific society, fortunately, are almost on an equal footing with such areas of social activity of students in the ranking, as volunteering, sports, a pedagogical team and participation in creative competitions.

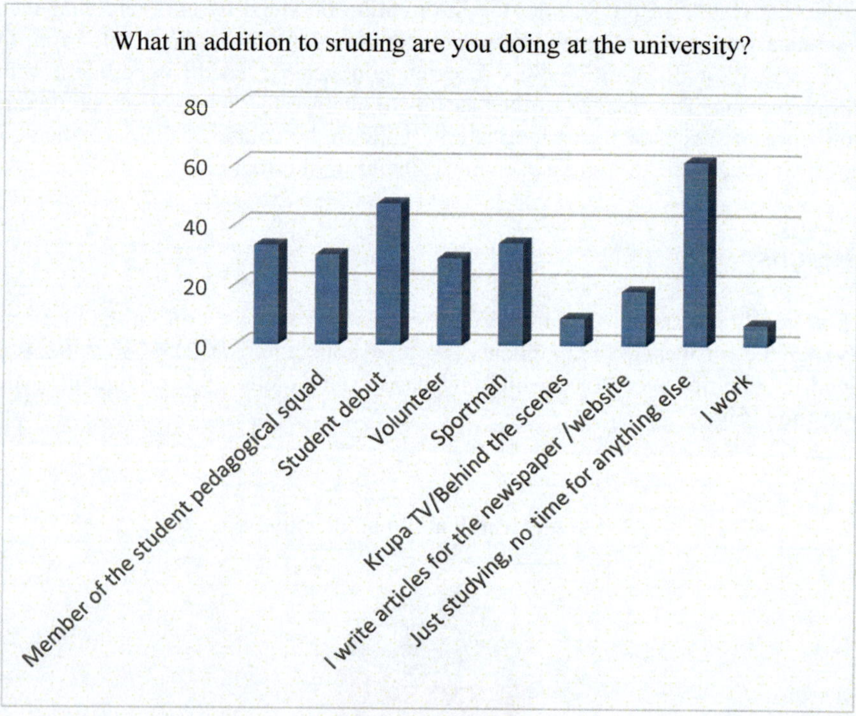

Fig. 2. Classes preferred by students of "SSEU" (Source: authors).

To identify the reasons for low involvement, motivational factors were also identified so students would think about participating in the development of science, among which the most frequent were obtaining machines (35% of respondents), the desire to acquire new knowledge (45% of respondents) and the desire to become a professional in the chosen direction (20% of respondents).

The most preferred forms of research work, based on the results of the questionnaire, for students are scientific circles (34% of respondents), as well as full-time participation in visiting scientific events (36% of respondents). Full-time participation of students means the participation of students in various youth forums (iVolga, Tavrida, Territory of Meanings and others), where there is a chance to win a grant for the implementation of their scientific development for its further introduction into production, as well as participation in various competitions and grants, part of the funds of which goes to encourage the grantee.

In this regard, students were asked about the frequency of writing scientific articles, the results of which are presented in Fig. 3. The vast majority (60%) unfortunately do not have a single publication, which certainly indicates a low involvement in this side of the university's research activities.

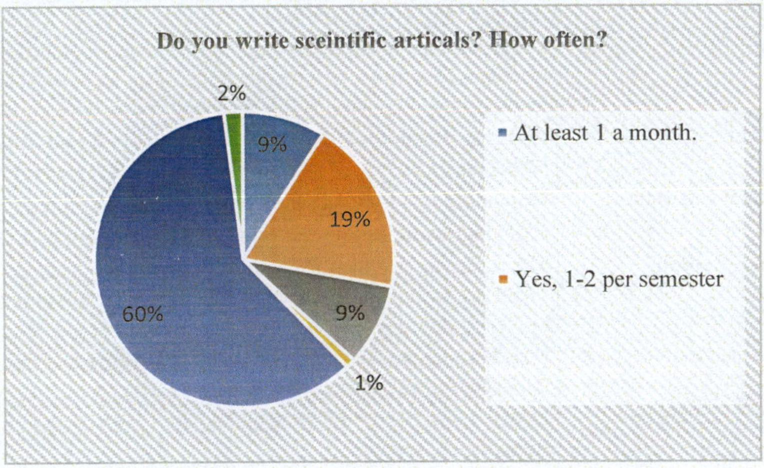

Fig. 3. Frequency of writing scientific papers by students of "SSEU" (Source: authors).

Analyzed these conclusions, we see that most having students surveyed at the university are not involved in the research activities of the university. In this regard, the need arose for the question - "What are the reasons for such low involvement in the research and development of the university?" The most popular answers among students to this question were: "lack of time," "lack of incentives," "ignorance."

Despite some lack of interest in the university research, a large number of respondents (71%) believe that it is quite prestigious to engage in science in the modern world and only 27% of respondents have doubts about this, but almost all interviewed students believe that it can be too difficult to implement in the scientific field for a long time, which, unfortunately, really is so.

4 Discussion

After analyzing the respondents' answers, it can be concluded that the level of motivation of students of the "SSEU" in the Research Institute of the university is at an average level, since despite the lack of sufficient time for students, about half of the respondents are occasionally engaged in research activities, which indicates the presence of weak motivating factors at the moment. Some efforts should be made to raise this indicator. There is the following classification of educational motivation of students, presented in Table 1 [4].

Table 1. Classification of educational motivation of students

Motives	Characteristic
Informative	Acquire new knowledge in a specific area
Social	Duty, responsibility, understanding the social significance of teaching, benefiting society; the desire of the individual to assert himself in society and assert his social status through teaching
Pragmatical	Receiving decent remuneration for his work, desire to have a higher income
Professional and value	Expanding opportunities to get a promising and interesting job
Esthetic	Enjoying learning, revealing your hidden abilities and talents
Status and position	The desire to establish itself in society through teaching or social activities, gaining recognition of others, the opportunity to take a certain position
Communicative	Broaden the circle of communication by increasing your intellectual level and new meeting
Traditional and historical	Establishment of stereotypes that arise in society and are strengthened over time
Utilitarian and practical	The desire to learn self-education and learn a separate subject of interest
Educational and informative	Focus on ways of obtaining knowledge, assimilation of specific educational subjects
Motives of social and personal prestige	Self-approval and the opportunity to take a certain position in the future in society, in a certain social environment
Unconscious	Obtaining education not of their own free will, but by the influence of someone, is based on a complete misunderstanding of the meaning of the information received and a complete lack of interest in the cognitive process

Source: authors.

Based on the results of the study and the presented classification of motivation, we can conclude what students need for greater motivation in the research work of the university. Firstly, in order to motivate students to engage in scientific activities, it is necessary to strive for the development of the currently gaining popularity of the student scientific society (SSS), including creating various scientific circles within it, relying on directions within the institutes. This practice has already been launched - within the

Institute of Management Systems (IMS), four scientific circles have been created in the areas that are part of this institute (Management, Economic Security, Economic Analysis, Accounting).

A more successful organization of scientific and technical conditions can help to increase students' attention to scientific activity. The system of incentives for publications may also serve as an additional motivation. One aspect of this motivation may be the rating of students most active in RA [1, 2, 4, 7]. To analyze the impact of the adversarial nature of a similar rating of students most active in research activities, the impact of ratings on student learning activities was tested. Among the students of the 3rd year of Samara State Economic University, an experiment was conducted. One of the groups began project activities in classes, as a result of which the group was divided into two subgroups. In one of the subgroups, a performance rating was introduced when performing certain tasks in this subject. This rating was created in the format of a Google table, which was accessed not only by students of this subgroup, but also by a teacher in order to determine achievement. The table was completed by an independent student with teacher adjustments. Based on the results of completing or not fulfilling the task assigned to the student, he was given "notes," the types of which are presented in Table 2.

Table 2. Types of "markups" used in the experiment

Designation	Essence	Transfer to points
«+»	The task was completed by the student to the full extent, there were no shortcomings in the execution or they are insignificant, the solution to the task was provided in due time	+16
«−»	Job not completed or completed after due date and with many shortcomings	−16
«=»	The task was completed by the student on time, but with a huge number of shortcomings or completed correctly, but not presented in time.	0

Source: authors.

Based on the results of assessing the work of students, a rating was compiled that clearly showed the participation of each student in the work of the team on the project. Summing up the results of the students' work, it became clear that in the team with a rating the level of achievement is higher, the team in which this practice was absent. Thus, we see that the rating of students, albeit within the same subgroup, has a significant impact on the activities of students. The essence of the rating of students involved in research activities is to assess the participation and achievements of students of universities in the scientific activities of the university. Through the introduction of scores for participation in research, it is possible to motivate students at the university, since the basis will be competition among students, which will be an incentive to participate and win competitions, conferences, olympiads, etc. (Table 3).

Table 3. Number of points for scientific achievements

Type of research activity	Points	Note
Participation in research conferences/forums/round tables, etc.	1	In-university or other level event without a report
	2	Regional level event (with report)
	3	All-Russian level event (with report)
	4	International level activities (with report)
Participation in the organization of research activities	1	Intra high school action
	2	Action of the regional level
	3	Action of the All-Russian level
	4	Actions of the international level
Publication of research papers by trainees	1	Other editions
	2	RSCI
	3	HCC
	4	Scopus or Web of Science

Source: authors.

Summing up the above, we can conclude that in universities there is a formation of prerequisites for the development of the scientific potential of youth. Conducting scientific and near-scientific events in various areas allows students and young scientists to actively develop their scientific potential, which in the future will contribute to the formation of professional-important qualities. However, at this stage of the development of society, the motivation of university students to participate in research activities is at a rather low level, which indicates the need to develop this direction through the introduction of new methods, for example, student rating in RA, since the adversarial nature of this method will encourage students to participate in research activities (creating their own projects, writing research papers, participating in RA). The prototype of this method was tested on a group of students, the results showed its effectiveness in modern conditions. In addition, classical theoretical approaches to the study of motivation were used to prepare this sample. The need to introduce this approach into universities is confirmed by the presence of problems with student motivation identified during the study.

In addition to the introduction of this system, of course, one of the factors of the greatest involvement of students in research activities is the use of other factors of material and intangible motivation discussed above. Among these factors, one can distinguish the encouragement of students participating in research and occupying one of the leading positions in the developed rating by means of material incentives (for example, every six months/year), the identification and presentation of laudable letters and/or memorable gifts to students who are most active in organizing scientific events or engaged in research work (for example, once a month/quarter).

5 Conclusion

Thus, based on the analysis of the theoretical and practical basis carried out in this article, the following conclusions can be drawn:

- rom the point of view of the level of interest and motivation of students, we can talk about the low (below average) involvement of students of analyzed higher education institutions in research activities (both the publication activity of university students, and participation and organization of research activities)
- at the moment in the theory of management there are a huge number of methods of motivation, which, in one combination or another, will also work in higher educational institutions under modern conditions, however, in order to achieve the greatest impact and efficiency of these methods, they need to be modernized, adapted and combined according to the goals and objectives of the motivation system at a given location and time,
- to increase the level of motivation of students, methods were developed and tested during a social experiment, which, together with existing in educational institutions, will help encourage students to participate in research activities.

References

1. Afanasieva, N.V., Malukhina, N.V.: About the Motivation of Students to Project Activities. VIDE, Vologda (2015)
2. Gelman, V.Ya.: Ways to increase motivation to learn from university students. Psychol. Learn. **2**, 52–61 (2019)
3. Kazimova, N.N.: Basic psychological aspects of personnel motivation. Humaniz. Pers. **1**, 31–40
4. Kulagina, N.V.: Educational and research activity of students as an active form of professional motivation formation. Psychol. Learn. **3**, 85–92 (2019)
5. Makaryeva, V.Y., Mahaaylichenko, N.A., Polyakova, E.Y., Sergienko, E.S., Ponomarenko, M.V.: Motivation as a modern management technology in digital economy. In: Popkova, E. G., Ostrovskaya, V.N. (eds.) Perspectives on the Use of New Information and Communication Technology (ICT) in the Modern Economy. Advances in Intelligent Systems and Computing, vol. 726, pp. 518–528. Springer, Cham (2019)
6. Mitrofanova, I.V., Korsakova, I.V., Ponomareva, A.S., Kalinicheva, I.D.: Motivation and stimulation of labor: evolution of conceptual and methodological approaches. Econ. Yesterday Today Tomorrow **9**, 436–451 (2019)
7. Preobrazhenskiy, A.P.: Features of motivation of students in higher education. Bull. Voronezh Inst. High Technol. **1**, 120–122 (2019)
8. Troshina, E.P., Mantulenko, V.V.: Influence of digitalization on motivation techniques in organizations. In: Ashmarina, S., Vochozka, M., Mantulenko, V. (eds.) Digital Age: Chances, Challenges and Future. Lecture Notes in Networks and Systems, vol. 84, pp. 317–323. Springer, Cham (2020)

9. Troshina, E.P., Mantulenko, V.V., Levashova, Yu.V.: The influence of digitalization on staff motivation methods in the organisation. Bull. Samara State Univ. Econ. **8**(178), 56–61
10. Yakovleva, S.M., Suraeva, M.O., Zhabin, A.P.: World trends impacting the change of the higher education system in Russia in the conditions of global crisis management on the basis of entrepreneurship and innovations. In: Popkova, E.G. (ed.) The Impact of Information on Modern Humans. Advances in Intelligent Systems and Computing, vol. 622, pp. 121–128. Springer, Cham (2018)

Intelligent Technologies for Knowledge Management at a Modern Company

S. Y. Lyapina[1(✉)], V. V. Degtyareva[2], and V. N. Tarasova[3]

[1] Higher School of Economics (National Research University), Moscow, Russia
sylyapina@hse.com
[2] State University of Management, Moscow, Russia
iump@mail.ru
[3] Russian University of Transport (MIIT), Moscow, Russia
tarasovavn@mail.ru

Abstract. The article investigates problems of digitalization and digital transformation of human resource management at modern organization. Authors suggest implementing approach based on the methodology of digital maturity and adapting it to tasks of human resource departments: the article includes the description of stages of digital maturity of human resource management and steps of its digital transformation of this managerial function. The Forth Industrial Revolution impacts on human capital strongly: its structure is dramatically changing and "talents" begin playing leading role at business companies. In new conditions a special attention is paid to knowledge management as a key function of human resource management. One of problems in knowledge management is to find out talented staff and create conditions for full integration their intellectual potential in production and business processes. Digital transformation of human resource management creates a new possibility for "talents" detection and improvement of their motivation. The authors suggest applying the instrument of knowledge mapping and demonstrate practical examples of its usage at companies. The described approach is especially useful for scientific and research organizations.

Keywords: Digital twin and tracking · Human capital · Intelligent technologies · Knowledge management · Mapping knowledge · Talents

1 Introduction

The COVID-19 epidemic and self-isolation regime led to the stuck of industrial production and the significant reduction of activity for the most business companies [5]. On the one hand, the branches of industry and enterprises where production and business processes cannot be implemented without human participation were most affected. On the other hand, the current situation has become a strong stimulating factor for decision making on the digitalization and digital transformation of the economy. The wide implementation of technologies that do not require direct and constant participation of people becomes an urgent task of survival and competitiveness for business in most sectors of the economy.

S. I. Ashmarina and V. V. Mantulenko (Eds.): IPM 2020, LNNS 161, pp. 459–469, 2021.
https://doi.org/10.1007/978-3-030-60926-9_59

In the recent past, the tasks of digitalization and digital transformation of enterprises were considered, rather, as an object of discussion of specialists than real strategies and projects. But the future has already come, and even skeptics have recognized the need for development of production without human labor. Therefore, the relevance of developments in the field of artificial intelligence as a core component of digital production should not be questioned.

The first experience of digital transformation of production shows that the transition to unmanned technologies begins, as a rule since the automation of technological processes in which unskilled personnel are involved. Automation is primarily subject to physically heavy or monotonous operations, as well as production processes with hazardous working conditions. However, even for the simplest manufacturing operations, automation often requires innovative and high-tech solutions.

A few years ago, one of the authors took part in studies on optimizing garbage processing in the Moscow region. Currently, waste sorting for disposal in Russia is mainly carried out manually. This is a monotonous, heavy, and potentially dangerous job that does not require special professional skills. However, in order to replace workers engaged in sorting in this technological process, it is necessary to design a robotic complex with sophisticated mechanisms for collecting recyclable waste. As a result, automatic sorting control should include machine learning blocks and have intelligent algorithms for making operational decisions. To develop such a complex and exclude people engaged in primitive physical labor from the technological process, it takes to attract highly qualified and creatively thinking specialists in new professional fields with advanced digital competencies. Without addressing the issue of the economic feasibility of digital transformation of household waste sorting, nevertheless, it is necessary to pay attention to the emerging phenomenon of changes in the structure of workers: a decrease in the share of unskilled labor and an increase in the need for specialists with high intellectual potential and advanced digital skills.

According to the authors' studies, now significant shifts are taking place at the labor market: the demand for IT specialists and representatives of other professions with developed digital competencies is growing. Nevertheless, the deficit of IT specialists, which has been noted lately [1], is temporary: on the one hand, the number of graduates of educational programs in IT is increasing [12]; on the other hand, information technologies are also developing in direction of automation – design of software products without direct participation of programmers [9]. In the long term, when the transition to the new technological paradigm (Industry 4.0) [18] will be completed, with a general reduction in the need for labor resources due to automation of the main and supporting productions and services, there will be a growing demand for specialists with creative abilities and developed digital competencies, with unique knowledge background and skills to elaborate non-standard and unique solutions [14]. The authors share the point of view that knowledge and creativity will determine the competitiveness of enterprises in the conditions of Industry 4.0, and Human Capital will become the main factor of business' value.

In parallel with these processes, the role of knowledge management at companies is changing. In the middle of the 20th century, experts in Microeconomics [4] increasingly began to single out information into an separated (fourth) basic factor of production, emphasizing that the role of classical factors [15] will decrease in the future, and role of information and knowledge will increase in the same time affecting not only the competitiveness of the business, but also the consumer value of products and services [4]. As a result, at the turn of the millennium, the function of knowledge management appeared and began developing in companies [2]. Information and knowledge transformed into human capital, as the 4th factor of production, acquire the greatest value for companies [11], therefore, in modern conditions, knowledge management is becoming a key managerial function at a modern organization. Based on these phenomena, Senge proposed a model of a developing organization [19] as an organization that knows how to create, acquire, and disseminate knowledge and changes its behavior in accordance with new information, original opinions, and modern models of thinking. However, despite all the advantages of organizational development based on knowledge management, at Russian companies this managerial function has not yet become widespread. Even at big companies that have sufficient means to form modern knowledge management systems, this task was set as a part of the general policy of "forcing to innovation" as a mandatory section of innovative development programs, which are also mandatory strategic documents for big companies with state participation [6].

As the authors' consulting experience shows, creation of knowledge management system often includes only development of internal corporate information system or portal for exchanging data between employees at a company. In fact, knowledge management function boils down to the growth of reference libraries and the desire to formalize and describe all technological and business processes, capture the best work practices, and ensure their dissemination across all structural divisions of an organization. The utopianism of this approach to knowledge management was proved by critics of the rational bureaucracy conception by Weber [22]: informal knowledge and the necessary organizational changes always are ahead of the highest-performing systems for converting new knowledge into information [21], because they are secondary to the generation of new knowledge.

At the same time, often knowledge management in Russian companies is equated with a function of human capital development, the main goal of which is to raise competencies of employees. In this context, knowledge management is provided by corporate centers of education such as corporate universities, academies, training institutes, etc. Often, at large companies these structural divisions concentrate on the formation of soft skills and the social and personal growth of staff. At the same time, IT support of integration of formal and informal knowledge into a single complex remains without proper attention. But at a modern company value of products and services for customers and consumers and the cost of the business are generated by synergy of information resources (formalized knowledge), network of staff communications, and growing human capital, as authors described above.

To achieve real positive results of knowledge management, it is necessary to balance development of human capital and information technologies, which ensure the transfer of individual employees' knowledge to public information, generate and productive circulate knowledge within a company. According to the authors' opinion, the function of knowledge management should develop at the junction of modern information technologies and techniques of socio-psychological work. Another feature of modern approaches to knowledge management as a necessary function of human resources management is, first of all, that it should be focused on employees with advanced competencies and soft skills, with a high level of creativity and the ability to find creative and original solutions of production problems. In this regard, in the field of human resource management in recent years a new function has emerged – talent management [10], which is directly associated with knowledge management.

2 Methodology

The research carried out by the authors is based on the theory and positive practice of knowledge management at modern companies. At the same time, the authors share the position of other specialists who emphasize the need to integrate modern information technologies and human resource management functions in order to identify, formalize and disseminate knowledge within a company and ensure the fullest use of human capital. To substantiate the approach to identifying knowledge and its carriers among employees of the enterprise, general scientific methods were used: description, generalization, systematization of organizational knowledge. The rationale for the choice of information technologies for use in the field of human resource management was based on the use of special (desk) research methods: analysis of open sources of data on the technological capabilities of modern information technologies in general and the use of artificial intelligence (big data analysis, machine learning, pattern recognition and natural language processing), as well as a comparative analysis of various options for implementing knowledge management functions.

Among other methods of scientific research, it is necessary to single out a sociological case analysis, which makes it possible to identify "talents", systematize the motives of their labor activity, and highlight the most effective incentives for professional activity. Modeling of business processes in the field of human resource management was used to develop an approach to digitalization and digital transformation, which makes it possible to record the stages of digital maturity and the criteria for the transition from one stage to another. The level of digital maturity of human resource management at companies was assessed in accordance with the methodology developed by Acatech and KPMG, which have become a global methodology for digital transformation.

The need for digitalization and digital transformation of knowledge management processes in an organization is justified on the theory of value management, which proves the direct impact of human capital development on the growth of company's capitalization.

The developed approach was verified and approved on the basis of the development and implementation of a digitalization and digital transformation project for human resources departments at some research institutes of the Russian Academy of Sciences, where knowledge management issues are of particular importance. The practical implementation of the developed approach has proven the legitimacy of using the knowledge mapping method in a company based on modern intelligent technologies, which allows implementing the basic knowledge management functions described above. The condition for applying the approach is digitalization and digital transformation of both the company in general and human resources management. A comparative analysis of the effectiveness of knowledge management and the loyalty of key employees ("talents") at companies before and after the implementation of the developed approach, proved the promising development of methods and tools for mapping knowledge by building horizontal communications and disseminating knowledge, which give a positive synergistic effect and lead to an increase in human capital.

3 Results

3.1 Methodology of Digitalization and Digital Transformation of Human Resource Management as a Key Factor of Intelligent Knowledge Management

The digitalization and digital transformation of business processes in the field of human resource management is based on creating the prerequisites for the development of new approaches that were previously unavailable due to the lack of technological capabilities for working with data. As the company's Industry 4.0 maturity index [17] rises, opportunities for knowledge management and the creation of conditions for the productive activity of talents expand. At the same time, creative employees with advanced competencies not only increase the value of the business through the development of human capital, but also they achieve full self-realization and recognition in the team, which relates to the needs of the highest order in the hierarchy of Maslow [3]. The highest human needs are the strongest internal stimulus to increase productivity and increase loyalty for talents [3]. Therefore, digital technologies of human resource management should be aimed at identifying talents, creating conditions for the full mobilization of their potential, and ensuring circulation and easy access to corporate knowledge. The authors of the article have developed an approach that allows us to solve this problem. But for its implementation it is necessary to develop intelligent information technologies of human resource management.

The 1st stage of digitalization and the digital transformation of human resource management named as "Computerization" is the creation of digital twins of employees – electronic files, which include full information characterized employees: personal data, a socio-psychological profile, assessment of competencies, job responsibilities, information about honors and incentives, etc. The availability of a computing database on staff is a prerequisite for a modern enterprise. To stimulate business towards digitalization of human resource management, the Russian government obliged all organizations without any exception to switch to digital workbook since the 1st January 2020 [7]. This is a positive trend in labor migration which facilitates access to personal data [13].

Intelligent technologies begin to be applied from the first stage of the digital maturity of personnel management. For example, the routine process of entering data into a computer base can now be automated using machine vision and recognition technologies. So, a student project, carried out under the leadership of one of the authors, has now become a dynamically developing startup, whose software product is increasingly being used in the personnel services of research institutes of the Russian Academy of Sciences when digitizing personal files of employees.

The 2nd stage of digital maturity ("Connectivity") in human resource management consists in establishing relationships between various functional databases at a company (including databases of personal e-files of employees) among themselves. It also includes connecting external sources of new and updated data (digital tracks) about employees. At the same time, it is necessary to develop new business processes based on ensuring the functioning of these databases. It is important that all the data within the organization ultimately form a single information base for making managerial decisions. The objective of this stage is to synchronize digital twins and their prototypes.

The 3rd stage of the digital transformation of human resource management ("Visualization") includes the expansion of analytical technologies' implementation. In the human resource departments at Russian companies, computer tools of human resource analytics are being increasingly used. However, as the experience of the authors' practical consulting shows, often attempts to implement analytical tools of human resource analytics fail, because the first two stages either incomplete or even ignored.

Authors' researches also show that human resource analytics tools are often based on a weak technological base. According to expert estimates [16], approximately 80% of the personnel analysis and human resource reporting processes at Russian companies are based on the use of spreadsheets (as a rule, MS Excel, and often even without using macros). Therefore, despite the development of some business processes in the field of human resource analytics at Russian companies, the level of development of their digital maturity has not yet reached the 3rd stage of maturity. The objective of this stage is to learn how to answer the question: "What is happening?" on real time, basing on analysis of digital twins of staff [17].

It is logical that because of having data on events and their consequences – digital tracks and the results of the employees' work, machine learning and big data analysis technologies, which are also called artificial intelligence technologies, are being increasingly used on the next (fourth) stage of digital transformation of human resource management. Companies are faced with an alternative: either to increase the staff of a human resource department, recruiting analysts, or to connect the capabilities of artificial intelligence, which allows to identify causal relationships and factors that determine the efficiency and effectiveness of work. Due to the automation of production processes, the number of employees in a company is usually reduced, therefore, an increase in the number of analysts leads to irrational proportions between administrative and production personnel. Thus, it is obvious that the use of intelligent technologies at the 4th stage of the digital transformation of the human resource management function becomes an inevitable condition for all organizations.

The 4th level of digital maturity ("Transparency") is designed to find the answer to the question: "Why is it happening?" [17]. At the 5th stage of the digital transformation of human resource management ("Predictive Capacity"), predictive analysis of technologies that allow forecasting and modeling the consequences of making decisions. Such kind of technologies reduce the risk of managerial errors leading to depletion of human capital. In addition, knowledge of causal relationships also allows predicting the negative consequences of external events independent of the company and taking prevent measures to protect them from their occurrence. In fact, at the 5th stage of digital maturity, there is a transition to proactive human resource management, which significantly reduces the risks in the field of work with personnel. The main task of the 5th stage is to learn how to answer the question: "What will happen?" [17]. Intelligent technologies based on machine learning, computer vision, and Big Data analysis will become the main for the functioning of HR-departments at organizations.

The 6th stage of the digital maturity of human resource management is called "Adaptability". At this stage, the number of HR-staff is significantly reduced, since the main business processes are fully automated and are carried out without the participation of people. HR-departments are being transformed into research and analytical units integrated with IT-departments in which new intelligent methods and analytical tools in the field of Human Resource Management are developed. The task to be solved at this stage: "How can an autonomous response be achieved?" [17]. The dynamically developing Industry 4.0 technologies fully capture the sphere of work with employees in all areas. Therefore, it is necessary to understand how modern information (and especially intelligent) technologies can change Human Resource Management and its individual functions.

3.2 Results of Implementation of Intelligent Technologies for Knowledge Mapping in an Organization as a Part of Digital Transformation of Human Resources Management

The application of the digital transformation model developed at Acatech and KPMG [17], and the use of intelligent technologies were carried out by the authors in some organizations for human resource departments, which allowed to increase the efficiency and productivity of work. It is clearly, the advantages of using Artificial Intelligence in human resource management were manifested in some research institutes of the Russian Academy of Sciences, the human resource departments of which are currently carrying out their digitalization and digital transformation. Obviously, talents for such kind of organization are of lasting value, and the knowledge management function is one of the main. Therefore, the use of the method proposed by the authors in this area turned out to be not only in demand, but a really working tool. It is the mapping knowledge system and building horizontal communications (between different departments within an institute), generating a positive synergistic effect. At the 1st stage, personal files of employees were digitized, and an archive of electronic personal documents was created. Already at this stage, the technology of automatic database filling based on intelligent scanning and recognition of documents (based on computer vision) was used.

Subsequently, this technology will be used to update and replenish the electronic document archive, which, to ensure information security, is completely autonomous and is not connected to any network. In fact, the only entrance to the archive is a scanner that recognizes documents and the technology of diversity of the entered data in the corresponding data fields. Also, at the 1st stage, a socio-psychological profile was compiled, and an assessment was made of the level of digital competencies of employees, as well as data on publications and ongoing research were introduced. At the 2nd stage, a corporate portal was created, on which personal accounts of employees were posted, as well as regulatory and reference documents. The open part of the personal account – information about scientific achievements, publications, and annotations of the studies performed – were automatically entered on the personal pages of the Internet-site. Closed data was used both by financial and economic services for calculating wages under an effective contract, and by research managers to form research teams. The electronic data archive with the appropriate setting became the basis for the generation of grant applications with participation in research competitions. Already at this stage, one of the main functions of knowledge management began to be realized – the circulation of knowledge within the organization. The employees of some research departments began to better understand the areas of research and the approaches of other departments, new professional communications arose, and research was initiated at the intersection of various fields of science. However, this process proceeded spontaneously, and it was not always possible to fix its results. Moreover, the circulation of knowledge within the organization depended on the characteristics of individual employees, their psychological readiness to share knowledge with others and the ability to describe their scientific achievements.

Therefore, at the 3rd stage, a technology was developed for analyzing the internal communications of employees, which allows one to identify key employees – Knowledge Centers, which are most often contacted by other employees of the organization. The intelligent system was built on the principle of Google search [16]: the employee to whom the most requests came received the highest rating. The employee, with whom there were practically no communications, was at the bottom of the list. To avoid the "mark-up" of the rating, the intelligent system analyzed communications and considered only those that were related to the research being conducted. In the future, artificial intelligence also plans to evaluate the usefulness and competence of the answers, which will also affect the employee's rating, and therefore – the amount of salaries. Another area of communication research within organizations is the confirmation and assessment of the level of competence of employees in subject areas of research. For example, an employee in his/her personal account declares his/her competence in area X (for example, based on the results of past training in this area), but neither research nor the requests of colleagues concern this subject area for the next 6 months. In this case, his statement of competence in area X is not supported. In addition, an employee to whom appeals are recorded in the subject area Y automatically receives a competency rating in this area. As a result, an intelligent recommendation system was created to attract the most competent employees ("talents") to research, and the task distribution system was to balance the workload of the institute's employees. The authors called this approach knowledge mapping and now it's currently being applied in human resource departments of big companies.

4 Discussion

Problems of talent management is a main subject of many recent publications in which authors study different aspects of this new phenomena in human resource management. The term of "talent management" was introduced in the late 1990s. One of the first authors who attracted attention to talent staff as core component of knowledge management was Vaiman in 2010 [20]. Later, Khoreva, Vaiman, and van Zalk described how talent management practices influenced on loyalty of employees with high potential and stimulated them to develop their competencies [10]. Experts in area of knowledge management, Milton and Lambeau, revealed the principles of development of knowledge management system at companies [11]. Prepared by this authors Guide (The Knowledge Manager's handbook) touched the problems of talent staff. Researchers Purdenko, Matusova, and Andryeyeva assessed the effectiveness of the creative activity of employees and studied the creative management system at companies [14]. They supported the point of view of Senge who had written that the only sustainable competitive advantage of an organization at present is the ability to learn faster than competitors [19]. All these publications are very close to authors' position in common. Some statements were used as a background of the authors' researches. Continuing studies in talent management as a core component of knowledge management at a modern companies authors researched impact of digitalization and digital transformation on human resource management in common and talent management in particularly. Studies of Industry 4.0 and knowledge economy, presented by the works of Gusenko [7], Schmidt [16] and Ivanova, Odinaev, Pulyaeva, Gibadullin, and Vlasov, [8] decomposed to the level of companies and concentrated on application of modern Intelligent Technologies to talent management's area. In addition, the presented research takes into account the impact of the events of 2020 associated with COVID-19 [5]. A feature of this article is the presented methodology, adapted to the tasks of human resource departments. The main focus of the work is on knowledge mapping, which is based on intelligent technologies with digital twins and digital tracking of staff. This approach is especially useful for scientific and research organizations.

5 Conclusion

Digitalization and digital transformation (Industry 4.0) reduce the share of routine and non-creative work operations and processes. In modern organizations, human capital is not only growing qualitatively, but also changing structurally: the number of employees is decreasing, at the same time, the intellectual potential of staff and, correspondingly, the value of organizational knowledge are growing faster. In a digital company, a leading place is given to creative employees with great intellectual potential. They form the human capital and ensure the competitiveness of the business. Transformed into human capital knowledge, as the 4th factor of production, acquires the greatest value for companies, therefore, in modern conditions, knowledge management is becoming a key function in a modern organization. Hence the model of a developing organization (Senge [19]) as an organization that knows how to create, acquire and disseminate knowledge and changes its behavior in accordance with new information, original opinions and modern models of thinking. Knowledge management is at the junction of classical human resource management and information technologies. The growth of the value of knowledge in business has led to the separation of the talent management function in the organization's personnel management system. To identify talents and mobilize the intellectual capital of each employee, the organization's knowledge is mapped – that is, the identification and fixation of employees with key competencies – carriers of certain knowledge. Progress in IT is related to artificial intelligence, which is based on machine learning and Big Data Analysis. Accordingly, progress in knowledge management is associated with the expansion of intelligent information models and systems. Authors demonstrate a possibility to implement Artificial Intelligent technologies to human resource management sharing their experience in developing of knowledge mapping system, based on intelligent technologies, for scientific institutions in Russia.

References

1. Alekseev, V.: Catastrophic deficit. Digital breakthrough is predicted by acute shortage of IT professionals (2020). https://www.dp.ru/a/2020/01/24/Katastroficheskij_deficit. Accessed 26 June 2020
2. Becerra-Fernandez, I., Sabherwal, R.: Knowledge Management: Systems and Processes. Routledge, New York (2014)
3. Beck, R.C.: Motivation: Theories and Principles. Pearson, London (2003)
4. Berczi, A.: Information as a factor of production. Bus. Econ. **16**(1), 14–20 (1981)
5. Duffin, E.: Impact of the coronavirus pandemic on the global economy. Statistics & Facts (2020). https://www.statista.com/topics/6139/covid-19-impact-on-the-global-economy/. Accessed 26 June 2020
6. Gershman, M.: Innovation development programs for the state-owned companies: first results. Foresight-Russia **7**(1), 28–43 (2013)
7. Gusenko, M.: Career in digit: what awaits workers and employers in the transition to electronic work books. Rossiyskaya Gazeta – Federal Issue, vol. 67, no. 8121 (2020). https://rg.ru/2020/03/29/chto-zhdet-pri-perehode-na-elektronnye-trudovye-knizhki.html. Accessed 17 July 2020
8. Ivanova, I.A., Odinaev, A.M., Pulyaeva, V.N., Gibadullin, A.A., Vlasov, A.V.: The transformation of human capital during the transition to a digital environment. J. Phys: Conf. Ser. **1515**(3), 032024 (2020). https://doi.org/10.1088/1742-6596/1515/3/032024
9. Kasmi, E.: Programmers are no longer needed. Launched a free software creation service for those who cannot program (2020). https://www.cnews.ru/news/top/2020-06-26_programmisty_bolshe_ne_nuzhny?fbclid=IwAR15p3euuGw5FbUkkuVZ1W5NuLji2V4wk8BJerc-XZ4oMZoovQq-DCO4-ZVc. Accessed 26 June 2020
10. Khoreva, V., Vaiman, V., van Zalk, M.: Talent management practice effectiveness: investigating employee perspective. Empl. Relat. **39**(1), 19–33 (2017). https://doi.org/10.1108/ER-01-2016-0005
11. Milton, N., Lambe, P.: The Knowledge Manager's Handbook: A Step-by-Step Guide to Embedding Effective Knowledge Management in Your Organization. Kogan Page, London (2016)
12. Odinokov, E.: ASI: The number of budget places in IT specialties will triple by 2024 (2018). https://na.ria.ru/20180518/1520827413.html. Accessed 26 June 2020
13. Primak, L.V., Lyasnikova, Y.V., Bykov, M.Y., Kamchatova, E.Y., Volgin, N.A., Vysotskaya, N.V.: Labor migration regulation: International practices in the implementation of economic and legal mechanisms. Int. J. Civil Eng. Technol. **9**(13), 1075–1081 (2018)
14. Purdenko, E., Matusova, O., Andryeyeva, V.: Assessment of employees' creative activity in modern conditions of intellectualization of the economy. Baltic J. Econ. Stud. **5**(1), 182–187 (2019). https://doi.org/10.30525/2256-0742/2019-5-1-182-187
15. Samuelson, P., Nordhaus, W.: Economics. McGraw-Hill Education, New York (2009)

16. Schmidt, E., Rosenberg, J.: How Google Works. Grand Central Publishing, New York (2017)
17. Schuh, G., Anderl, R., Gausemeier, J., ten Hompel, M., Wahlster, W. (eds.): Industrie 4.0 maturity index. Managing the digital transformation of companies (ACATECH STUDY). Herbert Utz Verlag, Munich (2017)
18. Schwab, K.: The Fourth Industrial Revolution. Publishing house "E", Moscow (2017)
19. Senge, P.M.: The Fifth Discipline: The Art & Practice of the Learning Organization. Doubleday, New York (2006)
20. Vaiman, V. (ed.): Talent Management of Knowledge Workers: Embracing the Non-Traditional Workforce. Palgrave Macmillan, London (2010)
21. Waters, T.: Weber's Rationalism and Modern Society: New Translations on Politics, Bureaucracy, and Social Stratification. Palgrave McMillan, London (2015)
22. Weber, M.: Economy and Society: A New Translation by Tribe. Harvard University Press, Cambridge (2019)

Staff Potential of Municipal Service in Digital Conditions

S. V. Rastoropov[1,2(✉)], A. V. Azarkhin[3,4], and A. A. Petrogradskaya[3]

[1] University of the Prosecutor's Office of Russian Federation, Moscow, Russia
rastoropov_sv@mail.ru
[2] Lomonosov Moscow State University, Moscow, Russia
[3] Samara State University of Economics, Samara, Russia
aazarkhin@mail.ru, petrogradskaya@yandex.ru
[4] Samara State Transport University, Samara, Russia

Abstract. The article analyzes the problems of staff recruitment potential of the municipal service in modern digital conditions. The author points out that nowadays the staff potential of the municipal service is at a rather low level, many municipal employees do not have the professional knowledge and skills necessary for the effective fulfillment of their duties and tasks. Reflecting on the fact that the reason for this is both underfunding of the municipal service and ineffective methods of schooling municipal employees, the author proposes to use the experience of several regions of the Russian Federation in which modern methods of schooling are used for the training of municipal employees, which are more relevant and effective in modern digital conditions.

Keywords: Advanced training · Case method · Human resources · Municipal service · Municipal employee · Professional development

1 Introduction

The importance of the municipal service staff is mostly reflected in the fact that it is the municipal employees who have to be in direct contact with the population and solve numerous pressing issues of citizens. It is through staff that most of the tasks of the municipal service are solved. In this regard, increased requirements are imposed on the personal and professional skills of municipal employees. In addition, the performance of municipal staff is constantly monitored by the population of the municipality, regional governing bodies, representatives of the media, heads of municipalities and other interested persons and organizations. Moreover, the criteria for evaluating the activities of municipal employees differs depending on the reviewer. Only one criterion is invariable - the staff potential of the municipal service and its compliance with existing and future requirements and tasks [7].

There is no consensus in the scientific literature on the definition of staff potential. Some of the definitions are: the totality of the capabilities of all people who are employed in a particular organization and solve certain problems; qualitative and quantitative characteristics of staff as one of the types of resources, and a measure of the abilities and capabilities of employees to materialize their knowledge and skills to

S. I. Ashmarina and V. V. Mantulenko (Eds.): IPM 2020, LNNS 161, pp. 470–476, 2021.
https://doi.org/10.1007/978-3-030-60926-9_60

ensure the development of the organization; a certain process of movement of staff in an organization (state body), which interact with each other through a system of goal-setting and assessment of personality qualities, informed choice of priorities and monitoring of potential abilities [5].

In a broad sense, staff potential is the skills and knowledge of employees that are necessary for the organization to function effectively. One of the main problems of staff potential in the municipal service is a clear contradiction between the need for highly qualified staff and the lack of such in practice. Thus, the problems of staff and their potential in the municipal service are currently more than relevant.

2 Methodology

The methodology of this research is based on philosophical and general scientific principles (objectivity, systematic, comprehensive, unity of theory and practice). To complete the study, methodological approaches described in the works of Russian and foreign scientists devoted to the analysis of the problems of the personnel potential of municipal employees were used and analyzed. The general scientific dialectic method of cognition was used as the main method. The analysis of legal acts and scientific literature on the topic of research is reached through the use of the method of analyzing legal problems, which proved to be most significant in this study. The validity of the conclusions is achieved through the integrated application of the method of description of concepts, the method of interpretation. In addition, methods of formal logic, analysis, synthesis, interrogation, and others were also used. The scientific novelty of the study is an objective analysis of the quality problems of personnel potential of the municipal service in modern digital conditions and the proposal of various ways to improve it.

3 Results

To achieve the set strategic goals of the socio-economic development of the country as a whole and territories in particular, as well to make effective decisions in resolving problems of public administration, it must be mandatory for the state and local authorities to act on established powers fully and with high quality [9]. The professional performance of local authorities is largely dependent on the maintenance and further development of personnel potential of municipal employees, which are key factors in choosing the best solutions to effective performance of the municipal service tasks. Therefore, the formation and development of professional and efficiently working personnel of municipal services are most significant conditions for the successful development of territories. That is, a qualitative improvement in the system of state and municipal administration in the Russian Federation today is directly related to increasing the efficiency of professional competencies of state and municipal employees [1]. However, as practice shows, current municipal workers are often poorly trained both professionally and personally, therefore their level of competence is clearly insufficient to fulfill the duties assigned to them. In addition, the problem of limited staffing in the municipal service has become chronic for Russian municipalities. At the same time, one of the main reasons for under-staffing is the lack of adequate funding for municipalities. The above is confirmed by the results of a sociological survey of the heads of local authorities. More than half of them (57.2%) indicated a lack of specialists in municipalities, and every third respondent (29.2%) additionally noted low level of staffing [6].

In addition, the lack of professional managerial personnel in the municipalities of the Russian Federation and the lack of specialized education in most of them are underscored by the results of the sociological study "Staffing of Local Self-Government", conducted by the All-Russian Council of Local Self-Government [2]. An analysis of this study showed that the lack of professional managerial personnel in the municipalities of the Russian Federation and the lack of specialized education in most of them is underscored by the results of the sociological study "Personnel Potential of Local Government" carried out by the All-Russian Council of Local Self-Government [2]. An analysis of this study showed that:

1. The majority of specialists working in local governments have higher education in humanitarian, socio-economic, agricultural and technical fields.
2. More than 30% of the employees working in municipal services do not even have a higher education (the situation is even worse in the municipalities of rural settlements).
3. Less than 15% of municipal employees have a vocational education in the training direction "State and municipal administration".

In addition to the insufficient professional level of municipal employees, other problems of the staff potential of the municipal service must be noted, especially such as:

- insufficient updating of the composition of municipal employees,
- lack of effective motivation for municipal employees to improve their skills and receive additional specialized education,
- insufficient attention to training, retraining, advanced training, the creation of staff reserve of municipal employees so that they are able to solve complex managerial tasks in the era of market economy [2].

In the light of the foregoing, it is obvious that serious work must be done to improve the professional level of modern municipal employees, since their insufficient qualifications often do not allow them to make optimal decisions to fulfill their tasks, and their knowledge and experience do not meet modern requirements due to the lag in the training system of professional development of civil servants.

For the effective functioning of local governments, it is necessary to form a new type of municipal employees on the basis of the meritocracy principle - employees with project thinking, people who can generate new ideas, plan and implement an action plan to achieve a specific result, while attracting the necessary resources. In this regard, there is a need to improve the professional qualities of municipal employees, in their personal and professional development, which will contribute to improving the quality of performance of official duties and effectively addressing issues of local importance [4].

As the analysis of legal regulation of the professional training of municipal employees showed, on the one hand, this activity is regarded as mandatory by the relevant regulatory acts and is implemented in practice within the framework of continuing education programs. On the other hand, it is hardly possible to reduce vocational training only to training in educational institutions, since this inevitably leads to gaps both in the content of training (using traditional teaching technologies) and in applying outdated methods during classes. A similar way of training for municipal employees has long been outdated in digital conditions and provides little to students for their service. In addition, irrelevant forms of training significantly reduce the motivation to learn, as a result of which the very idea of professional training and development is devalued [3]. Thus, it is obvious that the professional development of municipal employees, that only implies their professional training and advanced training according to outdated methods, is no longer relevant and is proven inefficient.

As the opinion polls results show, municipal employees themselves consider it necessary to constantly train and improve their skills. However, teaching methods must be consistent with modern digital conditions [5]. When retraining municipal employees, it is necessary to use various interactive teaching methods with a focus on practice-oriented training based on the analysis and exchange of best practices, in-depth knowledge of specific problems of municipal areas, etc. An example of an effective approach to the training of municipal employees is the experience of the Republic of Tatarstan, where best practices were introduced in the training technology for state and municipal employees, such as the case method. In Tatarstan, 44% of full-time classroom hours are conducted using the case study method.

The case method has the following advantages: Practical orientation - allows you to apply theoretical knowledge to solving practical problems. This approach compensates exclusively for academic education and gives a broader view of the processes than lecture classes. Interactive format - more effective assimilation of the material due to the high emotional involvement and active participation of students. Participants are immersed in the situation, since the case provides for the presence of the main character, in the place of which the team is put and it must solve the problem on behalf of such character. The emphasis in training is not mastering ready-made knowledge, but its development. Specific skills allow you to improve soft skills, the development of which is not always done in universities, even though such skills are important in the work process.

Programs that contain the case method:

- recruitment in government,
- management of personnel and anti-corruption work in government,
- top issues in the activities of the heads of executive committees/heads of rural settlements,
- novice civil servant,
- novice municipal employee,
- novice leader,
- effective manager.

In implemented programs, case solving consists of the following steps:

1. Study of the proposed situation.
2. Collection and analysis of missing information.
3. Discussion of possible solutions to the problem.
4. Working out the best solution.

The case solution allows you to develop skills such as working with information, which is to search for missing data: in a short time it is necessary to analyze and process large amounts of information, put forward hypotheses and assumptions, and logically and structurally present the results. Management skills are also developed: the search for innovative and creative solutions, effective work in conditions of stress, uncertainty and lack of information, as well as quick decision-making in conditions of limited time. Other important skills are teamwork, conflict resolution, the competent formulation of opinions, persuasion, consensus, presenting their solutions to the public [8]. Thus, the application of the case method in the training of municipal employees will improve their work efficiency. The main purpose of the case method is to enhance the cognitive activity of students. The use of the case method allows comprehensive development of employees, as they will develop skills in applying theoretical knowledge to analyze practical problems, skills for presenting their own point of view, skills for presenting and arguing, skills for receiving and providing feedback to colleagues, and skills for demanding additional information needed to clarify the initial situation, group interaction skills and techniques, as well as analytical skills.

4 Discussion

To increase the level of personnel potential of the municipal service, it is necessary to develop competency models for various groups of posts. In accordance with the characteristics of professional and personal qualities that an employee should possess, it is advisable to develop indicators of their manifestation in activities, to determine methods for assessing them. Indicative in this sense is the experience of the Tomsk Region, which in 2018 became the winner of the contest "Best Personnel Practices in Competitive Selection Technology". At least three methods are used to assess the professional level of candidates for the position of municipal employee there:

- mandatory - testing and individual interviews with members of the tender committee,
- additional (depending on the competencies that need to be assessed),
- questioning, preparation of a draft document. A practical task, a questionnaire of professional and personal qualities of candidates, writing a written work (project, abstract, essay), an interview with the Deputy Governor of the Tomsk Region or with the head of the structural unit.

A system for evaluating municipal employees has also been developed. As a result, a methodology for evaluating candidates during competitions was introduced in all executive bodies of state power of the Tomsk Region, thereby ensuring the objectivity of assessing the professional and personal qualities of the candidate and obtaining objective metrics showing their degree of manifestation. The applied methods and evaluation criteria correspond to qualification requirements, which are fixed in the job regulations. This method promotes the ability to make personnel decisions based on the unified rating of candidates who participate in competitions, as well as persons included in the personnel reserve based on the results of the competition, reflecting their level of professional and personal development.

5 Conclusion

Thus, in order for the attempts to develop the staff potential of municipal service to be effective, such must be built on the basis of complexity and integrity. Successful recruitment of highly qualified staff of municipal staff should be an important guideline for the development of municipal services. The effectiveness of local government is largely determined by the level of development of potential, their ability and willingness to solve the problems of municipal government. Therefore, the process of recruitment of highly professional and competent staff for the municipal service is a fundamentally important condition for increasing the efficiency of local government. To develop measures of improving the personnel potential of municipal employees, it is advisable to use not only the results of scientific development, but also the successful practice of other regions of the Russian Federation. In particular, the experience of the Republic of Tatarstan and the Tomsk Region are to be useful.

References

1. Antonyuk, V.S., Erlikh, G.V., Kozina, M.V.: Game technologies as a form of enhancing the training of personnel in state and municipal services. Issues State Municipal Adm. **3**, 123–152 (2019)
2. Eremina, E.V.: Features of the formation and development of the personnel potential of the municipal service of the Russian Federation. In: Alekhina, E.V., Ostashkova, A.V. (eds.) Socio-Economic and Political Development of the Territory: Problems and Solutions, pp. 7–9. PSU, Penza (2019)
3. Kalgin, A.S.: Sources and consequences of the availability of motivation for public service (public service motivation): a review of the research area. Issues State Municipal Adm. **2**, 215–239 (2020)
4. Kalgin, A.S., Matveev, V.V., Milosh, M.K., Yakovlev, A.A.: Meritocracy, effective rules and job satisfaction among public servants: results of an empirical study. Issues State Municipal Adm. **4**, 133–162 (2019)
5. Kostina, S.N., Sadykova, S.V.: Improving the personnel potential of the municipal service as a factor in the development of the municipality. In: Barkhatova, V.I., Bents, D.S. (eds.) Competitiveness and Development of Socio-Economic Systems, pp. 33–34. Chelyabinsk State University, Chelyabinsk (2017)
6. Kushchev, N.P.: Development of personnel potential of the state and municipal services. In: Prokofiev, S.E., Panina, O.V., Eremin, S.G. (eds.) Actual Problems and Prospects of Development of State and Municipal Government, pp. 110–112. Financial University under the Government of the Russian Federation, Moscow (2019)

7. Petrova, T.A.: Local government in modern realities. Bull. Tomsk State Univ. Econ. **39**, 174–181 (2017)
8. Vasilieva, E.V.: Competency-based approach in public service: what knowledge and skills do civil servants choose? Issues State Municipal Adm. **4**, 120–144 (2018)
9. Vysotsky, I.V., Luchnikova, N.S., Chmutina, E.V.: The development of the personnel potential of the municipal service in modern conditions. In: Surat, I.L. (ed.) Actual Problems of Economic Development and Management in Modern Conditions, pp. 279–287. Moscow Institute of Economics, Moscow (2019)

Motivating the Staff Innovation Activity for Sustainable Development

A. A. Sidorov[1(✉)], G. E. Kudinova[2], and A. G. Rozenberg[2]

[1] Samara State University of Economics, Samara, Russia
sidorov120559@yandex.ru
[2] Institute of Ecology of the Volga River Basin of RAS – Branch of SamSC RAS, Togliatti, Russia
GKudinova@yandex.ru, chicadivina@yandex.ru

Abstract. Motivating the staff innovation activity for sustainable development is the basis for its training in the socio-economic sphere. The purpose of the research is to study the conceptual foundations of the innovative development of personnel in the context of sustainable development on the example of the bachelor's degree programs in Social Economics at Samara State University of Economics (SSUE). The methodological basis of the study is the theoretical and methodological approaches motivating the staff innovation activity for sustainable development as a trend of modern higher education. These approaches include the study of scientific, methodological, specialized literature and regulatory sources on the topic of the research; principles of psychological and pedagogical research; the principle of the unity of consciousness and activity, which determines the essence of the formation of a worldview that meets the basic principles of education for sustainable development; analysis and generalization of pedagogical experience; pedagogical modeling and design; the principle of the unity of theory and practice; the consistency principle that predetermines the study of the relationship of universal, environmental values and environmental culture for students to complete the bachelor's degree programs. The authors consider various directions, forms, methods and types of student training.

Keywords: Environmental education · Staff innovation activity · Sustainable development

1 Introduction

The scale and employment in the green economy which determines sustainable development in the world is impressive in its dynamics. The US green economy, for example, is estimated at $ 1.3 trillion in annual sales revenue with nearly 9.5 million employees, with growth rates of more than 20% in three years (2013–16) [7]. Motivating the staff innovation activity for sustainable development is the basis for its training in the socio-economic sphere [13]. The United Nations Partnership for Action on Green Economy (PAGE) is a mechanism to do just that [12]. The scientific literature discusses the problems and prospects of professional environmental education and

S. I. Ashmarina and V. V. Mantulenko (Eds.): IPM 2020, LNNS 161, pp. 477–484, 2021.
https://doi.org/10.1007/978-3-030-60926-9_61

training using modern teaching technologies, both in the university environment [1, 3, 10, 11] and in business education [8, 9].

In Russia, legislation establishes requirement for the training of personnel in the field of environmental protection and environmental safety. Modern domestic business assigns a significant role to environmental values in the framework of its social responsibility [2, 4, 6], personnel safety of the organization [5], as a factor in the sustainable development of companies [14]. In Russia, environmental education programs have been developed and are being implemented, but their content, in our opinion, does not meet the requirements of the time. As a rule, this is due to the lack of targeted orientation, due interconnection of state structures, business and non-governmental organizations for environmental education, enlightenment and upbringing management. The scope of environmental training and retraining of managerial and production personnel are insignificant. Thus, the task of creating a system of continuous environmental education, training and enlightenment of personnel of enterprises and organizations has not yet been implemented in the training process, and, therefore, motivating the staff innovation activity for sustainable development is not realized.

2 Methodology

The aim of the research is to study the conceptual foundations of the innovative development of personnel in the context of sustainable development on the example of the bachelor's degree programs in Social Economics at SSUE. To achieve the aim of the research, the authors tend to solve the following problems:

- consider the tasks of environmental education,
- conduct an analysis of the methodology and principles of bachelor's degree programs,
- offer practice-oriented technologies to ensure that bachelors have the balanced mix of practical knowledge, skills, abilities and required competencies,
- consider the possibility of introducing the bachelors to scientific and experimental research activities.

The methodological basis of the study is the theoretical and methodological approaches motivating the staff innovation activity for sustainable development as a trend of modern higher education. These approaches include the study of scientific, methodological, specialized literature and regulatory sources on the topic of the research; principles of psychological and pedagogical research; the principle of the unity of consciousness and activity, which determines the essence of the formation of a worldview that meets the basic principles of education for sustainable development; analysis and generalization of pedagogical experience; pedagogical modeling and design; the principle of the unity of theory and practice; the consistency principle that predetermines the study of the relationship of universal, environmental values and environmental culture for students to complete the bachelor's degree programs.

3 Results

Since 1996 Samara State University of Economics (SSEU) - one of the leading universities in the Volga region - have been implementing master's and bachelor's degree programs on environmental education. The methodological basis of environmental education is modern concepts of the essence of the biosocial nature of man; the unity and interconnection of the development of nature, human society and the individual in a wide context. Training materials and courses offered to students are aimed at solving humanitarian, civilizational and socio-environmental problems and contain priority topics that determine the vision of the future as a civilization of sustainable development in the 21st century.

Below are the following tasks of environmental education:

- fundamental and empirical knowledge on sustainable development,
- development of cognitive and creative abilities,
- endowment with the basics of environmental education of personnel of enterprises and organizations,
- mastering the practical experience of environmental planning,
- development of skills in research activities in the environmental field.

Students study theoretical foundations of ecology, global and regional environmental problems and gain skills in analyzing the national environmental, social and economic system at different levels of management. They master the instruments of state regulation and the legal foundations of environmental management and environmental protection. Thus, the university seeks to ensure the formation among the future organizers and leaders of different levels of personnel management of a modern scientific worldview and environmental culture.

The university is constantly developing and seeks to introduce active forms of training into educational process: problematic seminars, situational and business games, solving production problems, testing knowledge, laboratory work, discussions, round tables on problems of economic development taking into account the environmental factor. An example is the educational and visual aids: “Regional and sectoral environmental management”, “Environmental monitoring”, business games: “Development of a strategic plan for the development of territories”, “Development of a feasibility study for an investment environmental management project”, etc.

Hands-on approach to real-life issues is a prerequisite for the implementation of educational programs. It increases the personal status of the student. The process of interaction “teacher - student” is clearly manifested in it. Feedback channels are actively operating, interest in the discipline is developing, the joy of creative activity is known. The course of the discipline “Ecology” at SSEU is built on the principle of the transition from knowledge transfer to training with the acquisition of experience based on practice-oriented technologies, which helps students increase their motivation to acquire environmental competencies, students' readiness for environmental-oriented actions and operations based on environmental knowledge and skills. Since 2015, this program has been implemented on the basis of developed workbooks, each of which sections reflects the content of the work program of the ecology discipline. Workbooks consist of three sections: lectures, practical exercises and independent work of students. Workbooks were reprinted three times, processed and supplemented, enriched with hands-on approaches to real-life issues. The most practice-oriented training topics include environmental monitoring, examination, certification, audit, environmental passport, environmental status of Samara region. In addition, students study environmental protection technologies, environmental principles of rational use of natural resources and nature protection, the basics of environmental law and international cooperation in the field of environmental protection. In the general list of educational topics, they occupy about 60%. Students have lecture notes outlined in the lecture workbook. And the classes themselves are accompanied by a consideration of practical examples that clarify one or another aspect of the lecture material, the latest data on modern technologies are provided. Traditional lectures are complemented by a demonstration of documentaries with their subsequent discussion. A workbook for practical work is filled out in practical classes and organized so that students are invited to familiarize themselves with the most important materials that reflect environmental policy, legislation, technical and technological solutions, regional environmental features and more. Subsequent topics consist of an even larger number of tasks for solving practical problems. Independent work also stimulates students to introduce reports and even scientific publications on practical regional topics that are proposed and subsequently carried out under the guidance of a teacher. We believe that the implemented technologies serve the formation of environmental competencies and the development of practical skills. The development of research skills, writing and reporting takes place at seminars on economic and environmental activities. So, in recent years, more than 300 reports on the following topics have been discussed: "Issues of general ecology", "Fundamentals of nature management", "Economic problems of managing natural resources and the land and property complex of the region", "Actual problems of the economy of nature management", "Resource conservation as a factor of sustainable development", "Ecology of land use".

Skills of scientific research are developed by participation in international, all-Russian, interregional, inter-university and intra-university scientific conferences; 20–25 student articles are published annually. Students work on the substantiation of theoretical principles and practical approaches to the economic regulation of environmental protection and the use of natural resources; reveal the ways, methods of their reproduction to ensure sustainable development of society, explore the problems and ways of environmental development of municipalities. They offer a set of measures to ensure a state strategy for the balanced development of the economy and improve the environment, as well as increase the efficiency and soundness of economic decisions. Students take an active part in the activities of Samara City Public Organization of Students "AIESEC-Samara", studying economics and management (including environmental management). The goal of the project is to nurture the ecological culture of youth through attracting them to participate in international environmental events. The practical experience of environmental planning is mastered by the implementation of budgetary and contractual research. In 2014–2020 in a number of administrations of Samara region, with the participation of students, conferences were held on the prospects of socio-economic development of municipal districts with a practical lesson on the development of road maps for solving the problems of the district, including environmental problems.

About 20 students took part in the practical work. Elements of environmental research are included in the thesis. The main calculations are carried out on the basis of modern methods of project analysis. The assessment of environmental investment projects is carried out using the Alt-Invest application package. The defense of a bachelor's thesis is attended by heads of departments of the ministries of the Government of Samara region, representatives of employers interested in young bachelors.

Submitted and successfully defended diploma papers of an environmental orientation contain a comprehensive structural-dynamic analysis of environmental and economic indicators that reflect the studied processes of interaction between society and nature; Forecast calculations for the medium term are carried out. More than 50% of the results of the graduation theses on environmental topics are put into practice, which is annually confirmed by certificates of implementation. SSEU conducts active outreach work. Pupils of secondary schools of Samara region are given lectures in the framework of the elective course "Ecology and rational nature management". Advocacy of environmental knowledge in the departments is also carried out through career guidance on recruiting students for the first year. The main forms of career guidance are open days, lectures on the university, departments and specialties conducted by teachers in front of graduates of secondary schools and secondary specialized educational institutions. Students at the university annually participate in the regional public event "Make your contribution to the forest wealth of Samara region". Its main purpose is to involve young people in restoring the forest cover of the territory, collecting seeds of woody plants. Student have an opportunity to prove themselves in the work of the All-Russian camp "ROST", in the green patrol of Samara region. They do it by planting trees, clearing the banks of the Volga and small rivers from household waste. The tradition is close ties with the Institute of Ecology of the Volga River Basin of the Russian Academy of Sciences, Zhigulevsky State Reserve named after Sprygin, the administration of the National Park "Samarskaya Luka". The annual visit by students and university professors of these institutions allows them to get acquainted with modern scientific achievements in the field of practical ecology and environmental protection. Also, there are held informative, game and cultural and entertainment events, such as: "Green Path", a student's evening "With a guitar by the fire". SSEU maintains stable relations with executive authorities engaged in environmental activities in the region: the Ministry of Forestry, Environmental Protection and Nature Management of Samara Region; Department of Improvement and Ecology of the Administration of Samara Office of the Federal Service for the Use of Natural Resources in Samara Region; The Office of the Federal Service for Supervision of Consumer Rights Protection and Human Well-Being in Samara Region, the Office of the Federal Service for Subsoil Use in Samara Region, the Office of the Federal Service for State Registration, Cadaster and Cartography of Samara Region, as well as in the Samara branch of Federal Cadastral Center "Zemlya".

4 Discussion

Bachelor's degree programs at SSEU stand on a selection of core courses that contribute to the development of solid training in acquiring both theoretical and empirical knowledge with a hands-on approach to real-life issues. Bachelors who have mastered the general higher education programs on "Economics", "Management", "State and Municipal Administration" are called upon to perform many diverse professional tasks. Competent decisions in the calculation and economic, analytical, organizational and managerial, accounting and other types of professional activity in the conditions of environmental restrictions imply the presence of environmental competencies. The following are the main environmental competencies formed during the training:

- monitoring of environmental problems and processes and ways to solve them,
- best possible solutions to socio-economic issues of environmental safety and sustainable development,
- knowledge of environmental legislation,
- delimitation of environmental powers and functions of managerial authorities, enterprises and organizations,
- the use of modern methods of information technology in solving sustainable development problems.

In order to provide training in acquiring both theoretical and empirical knowledge with a hands-on approach to real-life issues, SSEU has a long-time collaboration with the Institute of Ecology of the Volga River Basin of RAS – Branch of SamSC RAS. Students undergo internships in the laboratories of the Institute, getting acquainted with the principles of laboratory equipment and the methods of conducting experimental studies. The Dendrological Park and Environmental Museum of the Institute form a unique research and educational environmental center. It possesses a number of various collections (paleontological, zoological, botanical, etc.). The largest environmental library in the Volga region also belongs to the Institute of Ecology of the Volga River Basin of RAS and allows students to receive unique up-to-date and archive data.

5 Conclusion

Since 1996 Samara State University of Economics (SSEU) - one of the leading universities in the Volga region - have been implementing degree programs on environmental education. The methodological basis of environmental education is modern concepts about the essence of the biosocial nature of man; the unity and interconnection of the development of nature, human society and the individual in a wide context. Training materials and courses offered to students are aimed at solving humanitarian, civilizational and socio-environmental problems and contain priority topics that determine the vision of the future as a civilization of sustainable development in the 21st century. Students study theoretical foundations of ecology, global and regional environmental problems and gain skills in analyzing the national environmental, social and economic system at different levels of management. They master the instruments of state regulation and the legal foundations of environmental management and environmental protection. Thus, the university seeks to ensure the formation among the future organizers and leaders of different levels of personnel management of a modern scientific worldview and environmental culture. Students mastering the bachelor's degree programs acquire widely recognized knowledge for motivating the staff innovation activity for sustainable development, which lays the foundation for the transition from resource management to process management. The theoretical and practical training received during mastering the bachelor's degree programs prepares highly qualified bachelors whose activities are aimed at the transition to rational environmental management based on the principles of a green economy.

Acknowledgments. This research is conducted in the framework of the state assignment on the topic: AAAA –A17-117112040040-3 «Assessments of modern biodiversity and forecast of its change for the systems of the Volga basin in terms of their natural and anthropogenic transformation».

References

1. Adams, R., Martin, S., Boom, K.: University culture and sustainability: designing and implementing an enabling framework. J. Clean. Prod. **171**, 434–445 (2018)
2. Bataeva, B.S., Cheglakova, L.M., Melitonyan, O.A.: Social responsibility of business in the representations of owners and managers of small and medium-sized organizations in Russia. Organ. Psychol. **8**(1), 13–52 (2018)
3. Batterbury, S.P.J., Toscano, M.: Seeking justice through interdisciplinary environmental education at postgraduate level: lessons from Melbourne, Australia. Revista Internacional De Educación Para La Justicia Social **7**(1) (2018). https://doi.org/10.15366/riejs2018.7.1.007. Accessed 26 June 2020
4. Casey, D., Sieber, S.: Employees, sustainability and motivation: Increasing employee engagement by addressing sustainability and corporate social responsibility. Res. Hosp. Manag. **6**(1), 69–76 (2016)
5. Esikova, R.S.: Personnel security as one of the components of economic security of the organization. Socio-Econ. Phenom. Process. **12**(6), 65–69 (2017)
6. Fadeikina, V.S.: Influence of social responsibility of business on the development of Novosibirsk (sociological analysis). Bull. Siberian State Univ. Railway Transp. **4**(39), 55–60 (2016)
7. Georgeson, L., Maslin, M.: Estimating the scale of the US green economy within the global context. Palgrave Commun. **5**, 121 (2019)
8. Kirchherr, J., Piscicelli, L.: Towards an education for the circular economy (ECE): five teaching principles and a case study. Resour. Conserv. Recycl. **150**, 104406 (2019)
9. Kopnina, H.: Green-washing or best case practices? Using circular economy and Cradle to Cradle case studies in business education. J. Clean. Prod. **219**, 613–621 (2019)
10. Kudinova, G.E., Lazareva, N.V., Rosenberg, G.S., Rozenberg, A.G.: Environmental modernization - new solutions in the transformation of the world economy. In: Mantulenko, V.V. (ed.) Proceedings of the International Scientific Conference "Global Challenges and Prospects of the Modern Economic Development". The European Proceedings of Social & Behavioural Sciences, vol. 57, pp. 1491–1497. Future Academy, London (2019)

11. Maclean, R., Jagannathan, S., Panth, B.: Education and Skills for Inclusive Growth, Green Jobs and the Greening of Economies in Asia. Technical and Vocational Education and Training: Issues, Concerns and Prospects, vol. 27. Springer, Singapore (2017)
12. PAGE: Learning for an inclusive green economy. Assessing priorities and steering action. https://www.un-page.org/files/public/guidance_note_learning_for_an_inclusive_green_economy_8_12_2016_lr_pdf. Accessed 26 June 2020
13. Rozenberg, G.S., Gelashvili, D.B., Khasaev, G.R., Shlyakhtin, G.V., Kudinova, G.E., Noskova, O.L., Roshchevsky, Yu.K., Saksonov, S.V., Sidorov, A.A., Simonov, Yu.V.: Environmental education and education are the two pillars of sustainable development. Samara; Tolyatti; Nizhny Novgorod; Saratov: SSUE (2016)
14. Vatlin, A.A.: Management of corporate social responsibility as a factor of sustainable development of the company: PhD. Kursk: South-Western state University (2015)

Development of Knowledge Management Systems and Human Resources Using Lean Manufacturing Concept

A. Yu. Smagina[1(✉)] and I. V. Frolova[2]

[1] Togliatti State University, Togliatti, Russia
anastasiasmgn@rambler.ru
[2] Samara State University of Economics, Samara, Russia
frolova-i-v@mail.ru

Abstract. This article explores the problem of managing the knowledge and potential of employees through the integration of lean manufacturing concept tools and a quality management system. As a result of the study, groups of requirements have been developed for the knowledge and personnel potential management system. In addition, a scheme has been developed to integrate the requirements and tools of the quality management system and lean production for knowledge management, and criteria are proposed for assessing the effectiveness of the functioning of the knowledge base and compliance with the proposed requirements. To assess the degree of participation and interest of employees in the workflow organized under the conditions of the application of lean manufacturing tools and a quality management system, it is proposed to calculate the level of employee involvement in the enterprise, based on the results of self-examination and automatic assessment of employee performance based on KPIs.

Keywords: Knowledge · Potential management · Level of involvement · Lean manufacturing · Quality management system

1 Introduction

In today's rapidly changing business environment, in order to meet the requirements and remain market leaders, it is no longer enough for enterprises to follow one concept of production organization. Many researchers are expanding Toyota's world-famous toolbox, which has enabled the company to become one of the leaders in the auto industry. Since then, many techniques have been developed that are used together to ensure the process of reducing defects, losses and other areas of inefficient spending of organizations' resources. Enterprises can apply one or more of these techniques or tools within the framework of the concept of lean manufacturing - a universal recipe that is absolutely effective for absolutely any field of activity and scale of production. However, it is necessary to highlight the basic techniques used in the framework of lean production and, in combination, to achieve the necessary effect of cost reduction. The most popular tools and methods of lean production are: Value Stream Mapping, pulling in-line production, Kanban, Kaizen, System 5C - technology for creating an effective

S. I. Ashmarina and V. V. Mantulenko (Eds.): IPM 2020, LNNS 161, pp. 485–495, 2021.
https://doi.org/10.1007/978-3-030-60926-9_62

workplace, System 6 Sigma, System SMED - Fast equipment change-over, Total Productive Maintenance and others.

Most of the proposed methods allow you to focus on improving the efficiency of work with suppliers, minimizing overstock, working with clients and mobilizing production opportunities for an individual client request. At the same time, human resources and workplace management systems are widespread, allowing to increase labor productivity, to ensure the generation of business ideas, allowing each of ordinary employees to make their own proposals for streamlining production and using the resources of the enterprise. Many specialists and companies today pay particular attention to the concepts of organizing the workspace, creating project teams instead of the typical linear organizational structure of subordination and functional performance of duties with one goal - increasing the efficiency of production, sales and financial results of the enterprise as a whole.

According to scientists, the principles of lean manufacturing system can increase productivity by 30–70%, reduce costs and reduce delivery time. In Russia, thanks to rationalization proposals of KAMAZ employees, it saved tens of billions of rubles. The Kalashnikov concern is reorganizing the production system, the purpose of which is to halve the prime cost and increase the speed of developing new products by 4 times. Separate principles of the lean manufacturing system are applied in Rosatom, the Almaz-Antey concern and other large economic entities.

It should be noted the versatility of the lean manufacturing system. Its principles are applicable not only in the production process, but also in working with suppliers and customers, when organizing logistics or other separate business processes. In other words, lean manufacturing as a management system should be applicable, first of all, to the entire management process, to all management, and not just to the production process.

At the same time, many countries apply quality management system standards. One of the most common international permits is an ISO quality certificate. With regard to employee potential management, there are a number of requirements in the quality management standards. So, in ISO 9001: 2015, the concepts of "knowledge management" or "knowledge of an organization" are used [6]. ISO 9001 addresses the knowledge of the organization in all its sections: analysis of the enterprise environment, planning, design and development of products/services, monitoring, measurement and evaluation at all stages of the quality management system (QMS) functioning, implementation of corrective actions, internal and external audits, analysis of the results [6]. In other words, knowledge management occurs at all stages of the Shewhart-Deming cycle, forming a unified system of corporate knowledge, which is necessary as a systematic, meaningful and balanced application to achieve the company's goals and improvements [13].

In the standard ISO/TS 9002: 2016 "Quality management systems. Guidelines for the application of ISO 9001: 2015" indicates what the organization's knowledge management system should be:

- lessons learned from failures, prerequisites for incidents and successes,
- information from consumers, suppliers and partners,

- knowledge (explicit and implicit) that exists in the organization, for example, through training, succession planning,
- data comparison with competitors,
- sharing corporate knowledge with relevant stakeholders to ensure the sustainability of the organization,
- updating the necessary corporate knowledge based on the results of improvements [7].

2 Methodology

The processes of knowledge management and employee management, including the use of the result of the development of their potential, cannot be studied separately. However, often, they represent different subjects of research in various concepts of enterprise management. In the framework of the concept of lean production, special attention is paid to the maintenance of the workplace, minimizing losses, including when employees perform their job functions [12]. At the same time, the concept of a quality management system suggests paying attention to knowledge management as an organization's resource, which is not always used effectively. Therefore, in the framework of this study, the problem of integrating the tools of the concepts of lean manufacturing and a quality management system to improve the effectiveness of managing the knowledge and potential of employees of the enterprise is considered.

The methodological basis of this study is the work of leading scientists in the field of management. Since it is proposed to solve the research problem by integrating the tools of two concepts of production management and individual business processes, the basis is the work of scientists on the problems of organizing and developing lean production, as well as a quality management system [2, 11]. In addition, the study includes elements of knowledge management methodologies based on the use of intelligent process control technologies [1, 9]. The study also uses knowledge management and information management techniques [3, 4, 10]. The development of an approach to determining the level of employee involvement in the workflow is based on the use of elements of personnel management techniques [5, 8].

3 Results

The three main tools used in the lean manufacturing concept should be considered:

1. Strategy. Each employee should be aware of the need to follow this strategy, the result of its application. Each employee should have the opportunity to convey their proposals to higher management and everyone should be sure that their proposals will be taken into account. This is due to the fact that it is the performers directly involved in the process who can see how this process actually proceeds, and not according to the process map: all the shortcomings, bottlenecks and opportunities for optimization are visible.

2. Active involvement of senior management. Top management (including the CEO) should be actively involved in lean manufacturing. It is necessary to know and understand the features of the economic entity in order to set goals that will motivate employees to apply the principles of lean manufacturing. When employees see the senior management actively involved in the lean process, they are more likely to support such efforts and are willing to work within this concept themselves.
3. Work in teams. It is known that the best ideas for improving production efficiency come from employees. Therefore, process management should also be organized within the framework of employee teams.

Top management support is essential for an effective and efficient risk management program as part of the lean manufacturing concept. When introducing the concept of lean manufacturing, as a rule, such risks arise as:

- demotivation of employees after several years of implementation of this concept,
- lack of knowledge about lean manufacturing,
- lack of feedback from employees with senior management and others.

Company management, first of all, should focus not on the introduction of BP tools and methods, but on the formation of a corporate culture, the spread of the BP philosophy. Production can be called lean only when all employees share the principles of BP, starting with the leader and ending with the ordinary employee. The continuity of the principles of BP is carried out in the process of training existing and newly arrived workers.

Increasing the efficiency of knowledge and potential management of personnel is achieved through the integration of lean manufacturing tools and a quality management system in relation to work with personnel. Such integration is proposed to be carried out, based on the following blocks of requirements for the knowledge management system and personnel potential, reflected in Fig. 1: internal environment; staff motivation and financing; knowledge management technologies.

The first block "internal environment" involves the formation of approaches to the management of knowledge and personnel of the organization, the creation of conditions for the inclusion of personnel in the work within the concept of lean manufacturing. One of the key elements of lean manufacturing is to maximize product value and reduce waste. The rationalization proposals of employees are important tools for ensuring the achievement of the objectives of the studied concept. Therefore, to ensure the receipt and use of such applications from employees, it is necessary to apply a systematic approach. It will help the company integrate organically the concept of lean manufacturing into the existing quality management system. The consistency and transparency of the process of submitting, processing and implementing applications for improvement will ensure a high level of staff involvement in the lean production program. To increase the transparency of this process, it is necessary to ensure the electronic formation of applications and their inclusion in a single knowledge base. The application form should include not only the recording of information about the proposed event and evaluation of its effectiveness, but also the associated risks. Consideration of incoming applications should be carried out by a specially created expert commission at least once every two weeks. Feedback is also mandatory, that is,

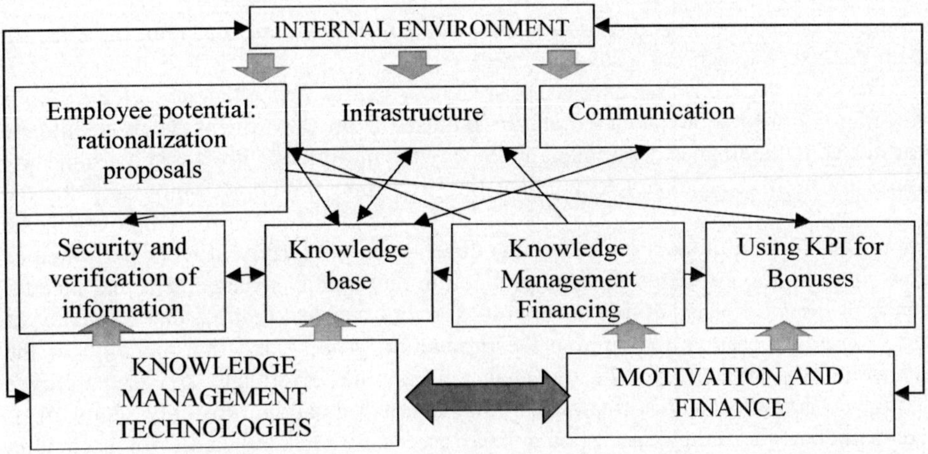

Fig. 1. Scheme of the relationship of the requirements for the knowledge management system and staff potential (Source: authors).

justification of the reasons for the refusal to use the measures indicated in the application and the formation of recommendations and their refinement.

The next important aspect is the creation of an enabling environment for the development of lean manufacturing and the expansion of staff potential. Such an environment should encourage employees to generate new knowledge and approaches to improve processes. Formation of a favorable environment for employees should begin from the proper technical equipment of the workplace, adherence to work and leisure schedules, information support, and end with the arrangement of recreation areas.

Another important aspect of the organization of the internal environment of the enterprise is the development of interaction and communication between employees and departments. Since the basis of the concept of lean manufacturing is personnel, it is necessary to provide for the possibility of effective communication at both horizontal and vertical levels of management. A rigid hierarchy system is not suitable for use in lean manufacturing, where the communication system is based on the possibilities of direct interaction between employees and management, as well as team work. To ensure this approach, it is necessary to have available means of communication, technical support for this process, ensuring transparency and efficiency of information exchange, and timely feedback. The elimination of communication barriers allows you to implement and develop a lean manufacturing philosophy in the enterprise.

The second set of requirements for the integration of lean manufacturing and the quality management system - "motivation and financing" - involves material support for the effectiveness of the knowledge and personnel management process at the enterprise. One of the most important areas of financing is the proposals of employees to streamline production processes. To ensure the construction of the internal environment and the construction of communication and feedback, funds should be allocated for the consideration of such proposals, their examination and assessment of their economic efficiency. This direction does not require significant costs, but the effect of them can exceed such costs several times.

The next aspect is the excess transparency, clarity and efficiency of the bonus procedure. Information about employee bonuses from department heads should be transferred to the finance department on the day the information is entered or on the day following it. It is necessary to regulate the procedure for bonuses not only for the performance of current labor duties, but also for the efficiency of the employee in the framework of lean production. Efficiency depends on the quality of work performance prescribed by the standards for lean production and quality management, the maintenance of the workplace, compliance with the work and rest regime, the rationality of the proposed measures to improve the production process and the operation of the enterprise as a whole. Key indicators characterizing the performance of each employee should be approved and normalized by the internal situation. Not only senior management, but also employees should have access to such indicators for each unit, information about them should be transparent. In addition, these data should be subjected to automatic express analysis, predicted and grouped according to the characteristics necessary for the user.

The third set of requirements for the integration of lean manufacturing and a quality management system is "Knowledge Management Technologies". Since knowledge management is one of the essential elements in a quality management system, it is necessary to use modern technologies to create a knowledge base, replenish it and use the data that is stored in it for each business process.

First of all, it is necessary to create a unified knowledge base, which includes: a description of the initial situation, a project idea, the results obtained (best practices), a detailed description of the project, applied tools, technologies, methods, used regulatory acts, risk assessment, economic justification. In addition, it is necessary to provide for the availability of knowledge base material for employees of the organization and an algorithm for its operational updating. One of the requirements presented in the information recorded in the knowledge base is comprehensibility and verification. In other words, it is necessary to ensure the moderation of the process of replenishment and use of database information for the implementation of knowledge management. Secondly, knowledge management technologies should be accessible and have the property of use not only in the production process, but also for the purpose of additional training for new employees or expanding professional potential.

In Fig. 2 presents the basic requirements for a knowledge management system, formed on the basis of the integration of lean manufacturing approaches and a quality management system.

Compliance with these requirements should be continuously monitored to comply with the principle of continuous improvement. It is proposed to select the following criteria for assessing the effectiveness of the functioning of the knowledge base and compliance with those indicated in Fig. 2 requirements:

1. Unfulfilled queries in the knowledge base for which the desired document or practice was not found. This is an important indicator that signals the need to replenish the knowledge base or search for knowledge outside the company.

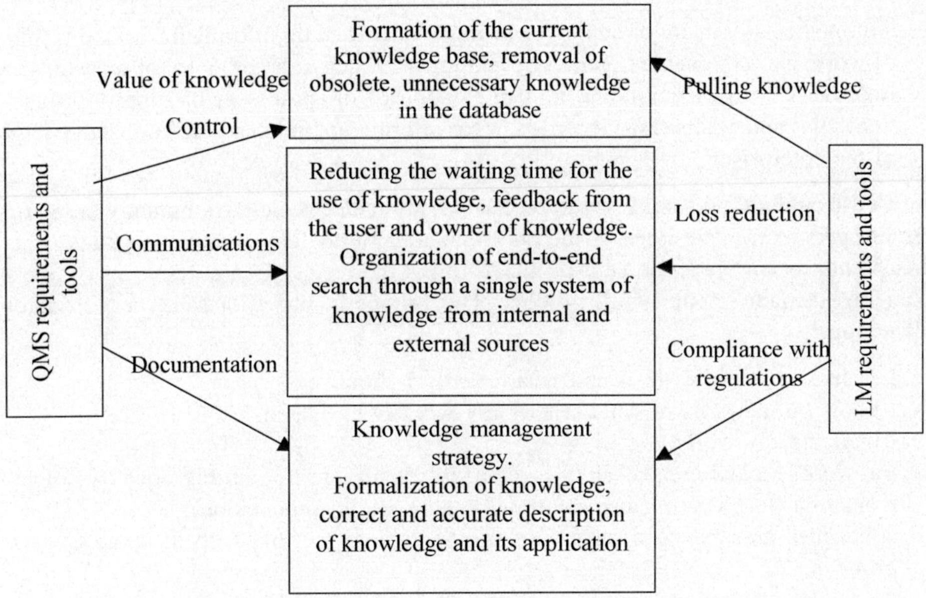

Fig. 2. Scheme for the integration of requirements and tools of a quality management system and lean manufacturing knowledge management (Source: authors).

2. "Defects" - they can be articles and documents in the knowledge base that do not meet the established standard, are outdated or with errors.
3. The frequency of search queries according to the same criteria, documents, practices.
4. The cost of maintaining the organization's knowledge management function.
5. Assessment of the effectiveness of the knowledge base - calculating the ratio of proven profits from knowledge management (for example, cost reduction due to the transfer of best practices) to the costs of the knowledge management function over a period of time.

Effective use of the knowledge base as a resource available to a person or organization that allows you to make the right decisions and carry out effective actions in certain circumstances is impossible without the main component - the potential of the organization's employees. The knowledge base should be replenished with formalized experience of employees in order to provide an operational solution to recurring problems or a high-quality solution to new problems. Such an order is possible with a sufficient level of staff involvement in the work and an understanding of the philosophy of lean manufacturing. In order to determine how much the level of staff involvement in the organization corresponds to the required value, it is proposed to determine it based on the level of involvement of each employee. It is proposed to divide this indicator into two components:

1. Self-study of involvement in the workflow in the concept of lean manufacturing (I_{SE}).

2. Automatic assessment of employee engagement based on information on the results of work, the implementation of a development strategy, the proposal of measures to improve the organization and implementation of the processes of supply, production, sales and others (I_{AR}) recorded in the information field and the knowledge base of the enterprise.

Self-examination is proposed to be carried out in electronic form monthly based on the answers to the questions of the questionnaire. Below are examples of statements and points for answers, the sum of which forms the result of an assessment of one's own involvement in the work process. The following provisions are proposed for allegations:

- I realize what my work is and what exactly I should do,
- I know on the basis of what criteria my work is evaluated,
- I maintain my workplace in order,
- for work, I have created all the necessary conditions for its high-quality implementation (technical means, information support, documentation),
- colleagues and the head pays enough attention to me and help to solve a work problem,
- the leader notes my achievements and appreciates them,
- during the working day, the regime of work and rest is respected,
- we work on a project in a team with colleagues, and not one at a time,
- I understand what the results of my work are and I can track them,
- I understand how my work relates to the development strategy of the entire enterprise,
- it happens; I condemn working issues with colleagues after hours,
- I have an interest in work and a desire to be useful,
- I understand that work can be improved on my site and I indicate this in the form of an application for rationalization,
- I have the opportunity to develop, learn new things,
- I have an important job; I understand my value to the enterprise.

It is proposed to evaluate these statements on a four-point scale, where: 3 - completely agree; 2 - for the most part I agree; 1 - for the most part do not agree; 0 - I do not agree. The maximum score for self-examination can be 45. The remaining 55 points are automatically generated at the end of each month according to the results of tasks performed by employees, applications for improving production processes after evaluating their effectiveness and other criteria fixed by local regulations for monitoring staff performance and bonus payments.

Based on the results of evaluating the activities of each employee by summing up points for self-examination and for the automatic assessment of his activities, the employee's level of involvement in the work process (I_{inv}) is calculated within the framework of the lean manufacturing concept and quality management system.

$$I_{inv} = I_{SE} + I_{AR}$$

The highest possible level of involvement is 100 points. With a value from 80 to 100 points, the level of involvement is considered significant, from 60 to 79 points - average, from 40 to 59 points - low, a value from 0 to 39 points indicates the absence of employee involvement.

The totality of the calculated level of involvement for each employee, as well as for the department, workshop, production, and the whole enterprise, allows us to evaluate the employees' understanding of the concept in which they work and take measures to improve it or solve the identified problems.

4 Discussion

It should be emphasized that resource management systems, including labor, for Russia represent a new, progressive direction of modern management. However, this statement is not true.

While in 1920–1930s in the USA, the direction of "Fordism" was actively developing - the construction of production by the type of conveyor, the orientation of production towards the manufacture of a large number of similar products, in Russia already in the 1920s actively engaged in the study of the impact of individual resources and processes on the final result of production. So in 1920 the Central Institute of Labor was created, which was headed by Gastev. In the process of work, he proposed measures for the efficient organization of jobs, the distribution of working hours, the possibility of using various technical devices, recommendations were given on the study of the influence of the physical and psychological state of workers on the result of the activities of an economic entity. In a language accessible to the working class, the guidelines "How to work" set out 16 basic rules, the observance of which allows you to effectively organize your work. Thus, back in 1920, the effectiveness of the conscious work of each employee in their workplace was carefully studied with the aim of improving and improving the production process. A work culture was created and all those postulates that are used and interpreted by foreign accounting managers today are recorded. We are talking about time management, and the organization of the workspace (5C system), and the mode of operation, and ergonomics.

Later in the USSR, a lot of research was conducted on rationalizing production and improving the technical equipment of the workplace. However, after the end of World War II, it was the Japanese engineer Tahiti Ono who used a combination of Soviet and American concepts of enterprise management, which made it possible to make a breakthrough in the management system. By the 1980s, the lean manufacturing system that had developed in Japan was widespread in America. By 2015, this system began to be applied in Russia. The fundamentals of this system are currently not very different from those that Gastev developed in his activities: lean attitude to people (taking care of their physical and emotional health, saving their strength) and time (eliminating unnecessary operations, optimizing production chains, thought out logistics).

The results of the study can be used to develop approaches to the management of knowledge and potential of employees based on the use of intelligent control systems. Based on the proposed requirements and indicators, enterprises can establish their own KPIs to assess employee engagement and performance in the context of lean manufacturing concepts and a quality management system. The development of the proposed tools lies in the plane of application of elements of other management concepts, for example, 5S, to achieve greater efficiency in organizing the workspace, generating, storing and using accumulated knowledge.

5 Conclusion

The developed proposals are aimed at improving the integration of lean manufacturing concepts and a quality management system while managing the knowledge and potential of employees to improve the quality of organization's resource management. Such resources are one of the most valuable in the organization, therefore, their management concepts should be continuously improved due to changes in the external and internal factors of development of both the enterprise and the economic situation and the life of society as a whole. Assessment of the involvement of employees in the implementation of their work functions, as well as in the production process, taking into account the integration of the concepts of lean production and the quality management system, allows you to identify how effectively and fully the potential of each employee in the organization is used. The requirements proposed in the article for the knowledge management system and personnel potential make it possible to formalize such a system and facilitate the involvement of employees in the overall work to reduce or eliminate losses and improve the quality of the product.

References

1. Burdo, G.: Methodological basics of creating intelligent quality management systems in mechanical engineering. Paper presented at the CEUR Workshop Proceedings, vol. 2258, pp. 131–140 (2018)
2. Demchuk, L., Baitsar, R.: Combined usage of theory of constraints, lean and six sigma in quality assurance of manufacturing processes. Key Eng. Mater. **637**, 21–26 (2015). https://doi.org/10.4028/www.scientific.net/KEM.637.21
3. Dneprovskaya, N.V., Shevtsova, I.V.: The knowledge management system development for smart education. In: 2018 IEEE International Conference "Quality Management, Transport and Information Security, Information Technologies" (IT&QM&IS), pp. 602–606. IEEE, New Jersey (2018). https://doi.org/10.1109/itmqis.2018.8525129
4. Dzedik, V., Ezrakhovich, A.: Analysis of quality management systems with the use of machine learning methods. Qual. Access Success **19**(164), 40–42 (2018)
5. Frolova, O.Ya.: Methodological aspects of assessing labor potential in personnel management. Mod. Econ. Success **5**, 144–147 (2019)
6. ISO 9001:2015: Quality management systems - Requirements (2015). http://docs.cntd.ru/document/1200124394. Accessed 26 June 2020
7. ISO/TS 9002: 2016: Quality management systems. Guidelines for the application of ISO 9001:2015 (2016). http://iso-management.com/iso-ts-9002-2016-2/. Accessed 26 June 2020
8. Magomedova, D.M., Kamilov, I.M., Kerimkhanov, S.T.: Motivation as a factor in increasing the efficiency of personnel labor. Bull. Acad. Knowl. **6**, 189–193 (2019)

9. Ramírez-Noriega, A., Martínez-Ramírez, Y., Armenta, J., Miranda, S., Figueroa Pérez, J.F., Mendivil-Torres, J., Jiménez, S.: Towards a mathematical knowledge management system: ontology to model linear equations. In: Rocha, Á., Adeli, H., Reis, L.P., Costanzo, S. (eds.) Trends and Advances in Information Systems and Technologies. WorldCIST'18 2018. Advances in Intelligent Systems and Computing, vol. 745, pp. 518–527. Springer, Cham (2018). https://doi.org/10.1007/978-3-319-77703-0_51
10. Turulja, L., Bajgoric, N.: Information technology, knowledge management and human resource management: investigating mutual interactions towards better organizational performance. VINE J. Inf. Knowl. Manag. Syst. **48**(2), 255–276 (2018). https://doi.org/10.1108/VJIKMS-06-2017-0035
11. Varshapetian, A., Semenova, W.E.: Aspects of integration management methods. Int. J. Qual. Res. **9**(3), 481–494 (2015)
12. Womack, J.P., Jones, D.T.: Lean Thinking: Banish Waste and Create Wealth in Your Corporation. Free Press, New York (2003)
13. Zhemchugov, A.M.: PDCA deming cycle. Modern development. Probl. Econ. Manag. **2**(54), 3–28 (2016)

Informatization of Society: The Development of Key Digital Competencies of Personnel

N. F. Soldatova[1,2], N. V. Rebrikova[1](✉), and I. K. Zakharenko[1]

[1] Financial University under the Government of the Russian Federation, Moscow, Russia
pankratova_n_f@mail.ru, rebrikovanv@mail.ru, marketing_metod@mail.ru
[2] State University of Management, Moscow, Russia

Abstract. It is obvious the formation of a qualitative change in terms of requirements of employers to the competencies that define the new technological structure, base and modernize the Russian economy. State energy companies are in the process of creating higher-quality economic and managerial relations, that allows to provide customers a high quality of life, adaptability to new conditions. The authors believe that the basis of growth requirements for the development of the digital competencies of the staff laid the trend to the digitalization of industry, internal trade, which ensure competitiveness in the global market of goods (services) and technologies. The digitalization of companies is the most important factor of realization of economic interests of the state in the period of Industry 4.0.

Keywords: Digitalization · Digital enterprise · Digital competencies · Digital technologies · The labor market · The transformation of competencies

1 Introduction

New in the current situation, the domestic business needs to address the key challenge - sustainable production and financial chains. This will become possible if companies are able to adapt in the direction of:

- reorientation tasks with maximum efficiency operation of the company in ensuring the sustainability of production and financial chains. The stability of the supply chain will be paramount compared with the efficiency of its individual components,
- to maximize the real digitalization at the corporate level should lead to the digitalization of internal operations, end-to-end digital processes with partners to create customer value, which ultimately will inevitably lead to the formation of the digital business ecosystem [5],
- new models of business ecosystems seem to be focused on national and regional specifications of individual markets, and the localization of production and financial chains with high consumer value at the national and regional territories, with appropriate support from public authorities will ensure the sustainability of their development,

S. I. Ashmarina and V. V. Mantulenko (Eds.): IPM 2020, LNNS 161, pp. 496–505, 2021.
https://doi.org/10.1007/978-3-030-60926-9_63

– the rapid transformation of corporate structures in the conditions of economic crisis and a global pandemic will lead to “leaving the past” several middle management level, where a paper contract and meetings for 30–50 people with a personal presence is no longer relevant. Company in new realities have appreciated the positives of remote work and the prospects of creating remote teams of managers. The efficiency of company management depends on the ability of senior managers to quickly develop and implement solutions,
– many consumers went online, they have some free time for shopping, so the company should focus on the study of the target audience: its problems, concerns, fears and needs. Obviously, that changes the model of consumer behavior and business should focus on producing a sought after product,
– changes should also occur in the terms of communication with customers, employees, partners and all those who form a high value product. Universal concept communications does not exist: the economic situation is changing rapidly, work in a particular industry has its own characteristics, brands need to become more valuable to the consumer.

In addition, sociologists argue that the mode of “isolation” imposed in many territories, transformed communication in society. Business communications will never be the same, they change: instead of generation Z comes generation Covid, i.e. the subjects of communication, secured communication habits in terms of quarantine.

2 Methodology

Instrumental-methodological apparatus is based on the following representative scientific methods: scientific knowledge, a systematic analysis of managerial processes and phenomena in the field of industry, domestic trade, induction, deduction, etc.; graphical, statistical and economic methods, method groups and classifications. Information and the empirical base of the research consists of previous works of the authors on the subject, as well as the online resource Rosstat, which are published annually and has open statistical data on the labour market, employment and wages [11]. For the analysis of the Russian labour market 2018–2020 was selected as the period.

Methodology of market research and phenomena occurring in them is a comprehensive simulation model of the processes describing the state, factors, prospects of development on the labour market. Methodological aspects of the analysis of demand for workers including a comprehensive study of competencies of staff and the changing requirements from employers is the monitoring of the external environment sensing internal environment which allows to identify problems of research and to generate a target installation. Algorithm a comprehensive study of digital competencies in the current labour market in the period of “Industry 4.0”, part of the general methodology is to implement the following sequence of actions:

1. The definition of information and an empirical base.
2. Classification of digital tools for the management of the company various economic activities and in the digital ecosystem.

3. Receiving, processing, grouping, and interpretation of the raw statistics obtained during the author's research.
4. To identify major trends, causal relationships and factors influencing the situation on the labour market requirements of employers in the part of the digital competencies, the prediction of its further development in conjunction with the digitalization of companies [6].

The author's algorithm has limitations in the analysis of the digital competences of the Russian labour market identified by the authors when comparing the results of this study and the results obtained from the analysis of requirements for employees of companies in the energy market [7], where it was highlighted the different approaches to digitalization, management and operating activities, the practice of using digital tools, as it is virtually impossible to describe the variety of approaches and digital tools.

Difficulties arise in the study of the Russian practice of application of digital platforms, programs and tools used by companies in the sphere of internal trade (wholesale companies) due to the lack of information available in the open access of the leading companies. Transformation of requirements of employers of domestic labor market to staff has led to the understanding of digital competences as a basic component of the architecture of professional competencies on the platform of digital technology [4]. Russian companies are developing the use of "intelligent networks" in the practice activities and consumers are actively involved in the production process of goods and services consumed by that shifts focus to the priority competencies in the Russian labour market.

3 Results

The study, conducted by the authors, was attended by more than 50 respondents from 9 large industrial companies in Russia and 5 companies in the sphere of domestic trade. According to the study, we have formulated seven key findings regarding the current experience and expectations of companies about the criteria for digital competence of employees, their ranking, selection of key digital competences in relation to industries and activities, the impact of global isolation.

The authors examined two aspects of Informatization of companies and the need for a corporate-wide digital competence of staff:

- the first aspect is related to the global trend of digitalization of business models used by companies in the framework of the concept "industry 4.0";
- the second aspect that affects the digitalization of companies and forms digital personnel's competence in the face of economic uncertainty, accelerating the process of social phobia in the world, change the explanation of consumer behavior among other factors.

McLuhan and Lapham in their book "Understanding Media" concludes that "the domination of certain types of communication defines the appearance of the culture." The use of information technologies, including digital, shape the quantity and quality of information exchange processes between the participants of market relations, society, consumers, which suggests the development of a new culture of communication [7].

The current stage of global socio-economic development is characterized by the transition to digital information based on the conversion of information into digital form, leads to the improvement of the quality of operational management and communication processes. Digitalization allows companies to create ecosystems and platforms in which the whole group solved the business problems [9].

Russian energy company (Rosseti) of all industrial companies advanced in the digitalization of operational processes. They developed digital compatibility GRIDS-MOESK (member of Rosseti) and put into operation the first digital substation in Moscow region - substation 110 kV "Medvedev". TSPS Medvedevskoe manages a supply center in the digital code without the presence of staff [10].

Investing in digital technologies and training of personnel, the corporation plans to increase the efficiency of operational processes first of all, optimizing costs (up to 30% of operating costs), and increase the quality of work (up to 10% improve the reliability of power supply) [9]. The company is financing the training of employees, raising their skill level of an implementation of organizational change. State -owned energy corporations are open to digital change.

Issues of quality, competencies, including digital in energy companies were studied on the basis of online surveys, on the basis of interviews of the heads of the respective departments. Therefore, the analysis of competencies was carried out at the company level, in the framework of which the staff level was determined as a whole and in the context of certain categories of employees. And also - the key requirements for professional and digital competencies of employees.

The company as a whole is characterized by a trend that indicates a lack of young, active and interested professionals in the profession. Experienced energy engineers are represented at the age category of about 50 years and their retirement age is approaching. The company will face a deficit of engineering staff of up to 50% in the next 10 years [9]. This problem has set the primary goal of the personnel policy of energy companies - consolidation, development of professional specialists. According to Rosstat, it takes at least 15 years to reach the full cycle of mastering the professional competencies of an engineer, and therefore staff retention prevails over attracting employees to the company (Table 1) [11].

The provided information shows that the types of activities, which include the priority of production, the most demanded employees are the key professional categories, that include skilled labor. In the group "research" more than 61% of the requests for jobs are specialists of the highest professional competence. In the surveyed energy companies is the shortage of engineers, developers and project managers. The head offices of these companies concentrated in the cities, where the leading technical universities in Russia with promising graduates are negotiating for Russian industrial enterprises.

In agriculture, the labour demand is not satisfied. There prevails a demand for skilled workers (18.8 per cent) in the areas of: horticulture, gardening and animal husbandry. There is a need in the agronomists and veterinary doctors. The need for higher education in agriculture is higher than in food industry or internal trade. Sufficient need remains in the unskilled labour force. Given the latest trend of closing

Table 1. Professional structure of the need for workers to fill vacant jobs in a number of types of economic activity in the Russian Federation in 2018,%

Type of economic activity	Qualified Professionals		Qualified		Unskilled workers	Other
	Top level	Mid level	Agricultural workers	Workers		
Research and development	60,6	8,0	–	17,8	5,0	8,6
Energetics	21,0	2,1	–	65,0	4,5	–
Agriculture and forestry, hunting	11,3	6,0	18,8	31,0	23,0	–
Food industry	7,0	4,2	–	63,1	19,0	–
Wholesale and retail trade	3,2	2,5	0,3	3,3	2,8	–

Source: authors.

national borders in a pandemic, the actual need for this category of workers may be higher. To solve the problem of shortage of staff domestic companies using different channels of recruitment. As for energy, which is actively implementing the tools of digitalization of business processes, the structure of the queries in the search frame is changed. Production technology and management in the energy sector necessitates the establishment of qualitative structure of employees and one of the priority recruitment channels are the technical universities, even for managerial positions.

4 Discussion

The study found the main problem - not enough staff and therefore companies are forced to compensate for the shortage of qualitative manpower increase in the number of employees. But this compensatory mechanism in high-tech industries, including energy, has certain limitations. The main criterion for the selection of a specialist is the level of professional competence and former students need knowledge of the industry, its specification, understanding of production and technology, logistics processes, as the level of risk of the decisions is quite high [6].

In modern conditions of economic uncertainty, the need for staff and effective use of staff resources are directly dependent on the adaptability of companies to changes. It is the individual's readiness to change, including digital openness, desire of high involvement of the employee in the performance of the company – are important criteria in the recruitment of staff, development of the concept of leadership in corporate governance. Despite the rapid change of cards of occupations and trends in the domestic labour market, increasing the demand for digital skills of the employees generated by the entry of industries and companies in the period of the fourth industrial revolution. Now in Russia there is a growing shortage of specialists with digital skills.

In 2019–2020, the first study of the mechanisms of internal marketing and communications in 14 Russian companies was held. In-depth interviews with 11 experts from both economic activities (energy and domestic trade) has allowed to analyze the digital skills that form the digital competence of the employees and match them with the needs of employers. Assessment of maturity for each component were determined on the basis of the findings in the survey of utilization of these skills.

The demand for digital skills in the public energy companies the results of the study are presented thus:

1. Skills computer and SOFTWARE (60%).
2. Technical skills in the field of technology, engineering (60%).
3. Building efficient communications in a business environment (50%).
4. Analytics in the business environment (40%).
5. Adaptability (35%).

In the result of the study the image of a model of digital competence of an employee was formed that includes five components. ICT skills have become key factors for the employer in selecting, selection, recruitment and employee assessment. According to the results of a deep survey of leading employees of the companies surveyed by the authors of the forecast of demand for digital skills for the period 2020–2021.

1. Integrated solution (35%).
2. System skills (19%).
3. Social skills (19%).
4. Process skills (15%).
5. Technical skills (12%).

Digital competence of the employee includes 5 components:

- information competence,
- communication and interaction,
- digital content creation,
- security,
- problem solving [1].

In times of economic uncertainty that has engulfed the entire global economy, the stability of operation of the company can only provide a highly adaptive personnel, open to changes in the environment. Willingness to change among employees indicates a high level of involvement in the development of the company, synergy, leadership concepts and corporate interaction on the whole chain of creating high customer value. Corporate approach "distributed leadership" involves, first and foremost, the digitalization and the integration of horizontal value chains. The formation of the system of internal digital processes, such as operational, managerial, auxiliary, etc. is increasing employee engagement, high leadership skills and competencies, creating and implementing the principles of good corporate interaction.

Internal communications have been transformed in the digital information flows, beyond the process of creating high customer value with partners and suppliers, leaving public organizations, nonprofit organizations, opinion leaders and others stakeholders [13]. It can be argued that the effectiveness of communications is an indicator of the financial growth of the company and of employee engagement. The article on survey data assesses the level of company communication based on the model of corporate digital competencies.

Investigation of the mechanisms of internal marketing and communications in Russian companies allowed us to deeply analyze the structure of the components of "digital competence" as applied to engineering and management personnel, as well as a set of tools that allows you to build this competency. The authors of the study developed 5 levels (score from 0 to 4 points) digital competence for each component. A reference model was built based on the profiles of nine companies and five industry profile companies in the sphere of internal trade with the highest industry measure of revenue per professional, possessing high levels of digital competencies. Based on this model, 5 levels of digital maturity of the company were identified (Table 2).

Major Russian industrial companies are characterized by a high level of digital competency. The reference state energy corporation "Russian GRIDS", being the leader in terms of digitalization of operational and management processes, is also considered a leader in the application of best management practices according to the values and improvement of the corporate culture, part of which is a digital environment company. The company adheres to the corporate values and actively engaged in the development and improvement of the digital environment. Each component of the state energy company has formed a high level of digital competences.

And if domestic retailers formed the maximum level of digital competencies through digital communication with the external environment, the industrial company provided the high level at the expense of cybersecurity. The introduction of digital business models and digital tools prevalent in the practice of domestic companies in the sphere of energy due to the implementation of the decrees of the President of the Russian Federation of 09.05.2017, No. 203 "On the strategy of information society development in the Russian Federation for the years 2017–2030" No. 204 "On the national goals and strategic objectives development of the Russian Federation for the period up to 2024" [2, 3]. As a rule, Russian companies have different requirements in terms of ownership of digital competences applicable to different categories of employees: managers, engineers, production and support staff. The fourth industrial revolution and the use of digital technologies, tools, approaches inevitably will encourage the domestic industry to the understanding of the idea – staff is a key element of digital transformation [8]. Therefore, the inevitable shift of emphasis from assessment of professional competence - to develop skills of cooperation and communication in the digital environment. In addition to assessing the digital skills of staff, formed on the basis of ICT, it is advisable to take into account cognitive skills, socio-behavioral aspects in digital processes.

Table 2. Assessment of the level of digital competencies of employees of Russian companies

Digital competence	Industry	Energy	Retail	Wholesale trade
Information competencies				
Search, view, filter information and digital content	4	4	3	3
Evaluation of data, information, digital content	4	4	3	3
Manage data, information and digital content	4	4	4	4
	12	12	10	10
Digital communications				
Digital Connectivity	4	4	3	3
Exchange of data, information, digital content through digital technology	4	4	4	3
Communication with the state and society through digital technology	3	4	3	2
Using digital technology to produce resources and knowledge	3	4	3	2
Digital Ethics	3	3	3	3
Digital identity management	3	3	3	3
	20	22	19	16
Digital Content Creation				
Digital Content Generation	3	4	3	2
Digital Content Integration and Processing	3	4	3	3
Programming	3	4	3	2
	9	12	9	7
Information Security				
Device protection	3	4	3	3
Personal data protection, confidentiality	4	4	4	4
Mental Health Protection	3	4	3	3
Environmental Protection	3	4	3	3
	13	16	13	13
Solving Information Problems				
Technical Problem Solving	4	4	4	4
The choice of digital tools for solving problems	4	4	4	4
Digital Adaptation	3	4	3	3
Digital Competency Development	3	4	3	3
	14	16	14	14

Source: authors.

5 Conclusion

Russian industrial enterprises, domestic trade, services, etc. faced the need of development in the digital environment, and staff development will build on the digital skills and digital competence. Applying the model of digital competence for the evaluation of potential employees to determine areas for staff development and supply the reference target of each component of the digital competence. For the practical implementation of the company concept of "Industry 4.0" and the relevant digital technologies, initially, it is necessary to build the concept of a digital ecosystem [12]. As a result of the survey it was revealed that information and communication technologies in the digital transformation and efficiency of digitalization will be determined by factors related to the staff person.

High level of digital competences by employees can significantly increase productivity in Russian companies. The digitalization of operational and production processes, according to experts, provides revenue growth and optimizes costs of the company. Information competence will develop in the direction of:

- improving analytics,
- the development of digital literacy.

Digital communication, as an important component of digital competence, and a high level of interpersonal interaction, promotes unity in a team, and develops the skills of business negotiation and self-presentation. Competencies associated with the creation of digital content will evolve, especially in the direction of programming in solving complex professional problems. Basic digital skills in terms of searching, collecting, verifying and structuring of digital information – will not lose its relevance in the future. The problem of ensuring digital security is to all Russian companies that invested in the creation of the digital enterprise. According to the company "Rosseti" – the intensity of information attacks on PJSC and subsidiaries up to 30 thousand attacks per day to 9 million attacks per year. Measures to ensure cybersecurity, JSC "Rosseti" is 2 billion rubles per year [10]. Digital transformation also has negative aspects related to the need to respect "digital hygiene" in the company, creating a system of protection against information threats call not only for corporate but also national levels. Digital competence in part: "the decision of information problems" is aimed at developing the skills of adapting digital tools to solve various types of business tasks.

The development of digital competences of employees and increase in the share of personnel with a high level of digital competences will provide the company a competitive advantage. In this study, the selected company, in which the level of digital skills is higher than the industry average (digital champions). It showed not only the specialists of IT departments, but also managers, marketers, engineers. If we consider only one direction in company activity "the development of relations with customers", then the digital competence of employees:

- goods (services) and the entire chain of creating high customer value to target customers,
- communication system with customers, partners, and community organizations to build their queries,

- digital channels of access to the end user, form the analytical base for the study of consumer's behaviors on the market,
- digital marketing-analytics, especially in the direction of changes in consumer preferences, create the initial business project cross-functional working groups that will significantly reduce the time and means in the output of goods (services) on the market.

References

1. Carretero, S., Vuorikari, R., Punie, Y.: The digital competence framework for citizens (2017). https://publications.jrc.ec.europa.eu/repository/bitstream/JRC106281/web-digcomp2.1pdf_(online).pdf. Accessed 21 June 2020
2. Decree of the President of the Russian Federation of 07.05.2018 No. 204 "On the national goals and strategic objectives development of the Russian Federation for the period up to 2024" (2018). http://kremlin.ru/acts/bank/43027. Accessed 24 June 2020
3. Decree of the President of the Russian Federation of 09.05.2017, No. 203 "On the strategy of information society development in the Russian Federation for the years 2017–2030" (2017). https://base.garant.ru/71670570/. Accessed 24 June 2020
4. InfoLine: The leader of the energy industry has changed and other significant changes have occurred (2020). https://infoline.spb.ru/news/index.php?news=203197. Accessed 21 June 2020
5. Ivanova, E.: Digitalization and security (2019). https://asninfo.ru/articles/1598-tsifrovizatsiya-i-bezopasnost. Accessed 23 June 2020
6. Ketoeva, N., Soldatova, N., Rebrikova, N., Ilyashenko, S.: Russian hydropower sector and its human resources under digitization conditions. IOP Conf. Ser. Mater. Sci. Eng. **497**, 012121 (2019)
7. McLuhan, M., Lapham, L.H.: Understanding Media: The Extensions of Man. MIT Press, London (1994)
8. Ministry of Energy of Russian Federation: Report of Alexander Novak on the results of the fuel and energy complex in 2018 and the tasks for 2019 at a meeting of the Government of the Russian Federation (2019). https://minenergo.gov.ru/node/14548. Accessed 21 June 2020
9. Rosatom: Digital products of Rosatom (2020). https://www.rosatom.ru/production/supercomputer-and-software/. Accessed 23 June 2020
10. Rosseti: Digital transformation 2030. Concept (2018). http://www.rosseti.ru/investment/Kontseptsiya_Tsifrovaya_transformatsiya_2030.pdf. Accessed 23 June 2020
11. Rosstat: Labor market (2020). https://www.gks.ru/labor_market_employment_salaries. Accessed 24 June 2020
12. Soldatova, N.F., Ilyashenko, S.B., Rebrikova, N.V.: Digital transformation of management of Russian energy companies. In: Nazarov, A.D. (ed.) Proceedings of the 2nd International Scientific and Practical Conference "Modern Management Trends and the Digital Economy: From Regional Development to Global Economic Growth". Advances in Economics, Business and Management Research, vol. 138, pp. 1265–1270. Atlantis Press, Paris (2020)
13. Soliman, M., Saurin, T.A., Anzanello, M.J.: The impacts of lean production on the complexity of socio-technical systems. Int. J. Prod. Econ. **197**, 342–357 (2018)

Research on Formation of Student's Soft Skills Ensuring Competitiveness in Digital Paradigm

O. L. Chulanova(✉) and E. S. Bogdan

Surgut State University, Surgut, Russia
chol9207@mail.ru, bogdan-es@mail.ru

Abstract. In the digital economy era, competition as a driving force necessitates organizations to seek new ways to improve their competitiveness constantly. More and more researchers and representatives of different branches of scientific knowledge note the importance of soft skills for achieving competitiveness in the labor market of different professional fields. The authors of the article conducted a study on the relevant competences of graduates of economic fields in employment. The aim of the study is to identify areas of competence that are important for the professional realization of students and graduates of economic fields and to justify the need to adjust the curricula of these fields. The following methods were used in the study: content analysis, synthesis, interview (using Google forms), statistical data processing method. The article presents the results of research on the significance of formation and development not only hard skills, but also soft skills in graduates of economic directions. By means of a questionnaire survey of students and graduates of economic fields, it has been possible to rank the most important competences for the performance of their professional duties and to indicate the formation/not formation of these competences in the course of studies at the university.

Keywords: Competitiveness · Digital economy · Graduate · Employment · Soft skills

1 Introduction

Within the digital paradigm, competition as a driving force necessitates organizations to seek new ways to improve their competitiveness constantly. The digital economy reflects the shift from the Third Industrial Revolution to the Fourth. The Third Industrial Revolution, sometimes called the Digital Revolution, refers to the changes that occurred at the end of the 20th century from analog electronic and mechanical devices to digital technologies. Competitiveness in this context can be achieved both through the use of modern technologies and equipment and through the development of the main competences that can operate in a digital economy effectively. A positive result of changes in the enterprise (introduction of new technologies, organizational and economic transformations, etc.) can be achieved only if human resources are adequately provided.

Having examined the basic approaches that define the requirements of personnel in the digital economy, it can be argued that, despite the dominance of digital and the need

S. I. Ashmarina and V. V. Mantulenko (Eds.): IPM 2020, LNNS 161, pp. 506–513, 2021.
https://doi.org/10.1007/978-3-030-60926-9_64

to form and develop the main hard competences of personnel, (in cross-cutting digital technologies such as neurotechnology and artificial intelligence; distributed registry systems; quantum technologies; new production technologies; industrial Internet; robotics and sensory components; wireless communication technologies; virtual and augmented reality technology), very much attention is paid to the development of soft skills of staff (rapid learning, high degree of adaptation to new, cognitive flexibility, teamwork, emotional competency, etc.)

Russian researchers devoted a great deal of attention to the issue of formation of personnel competences in the digital age. The studies emphasize that professionals should have both hard skills and soft skills. Representatives of various scientific branches (economics, psychology, pedagogy) also emphasize the importance of soft skills for achieving competitiveness of specialists on the labor market. Scientists note that the rapid computerization of various processes will lead to a reorientation of labour market needs towards competences that cannot compensate for digital technologies. A lot of studies were devoted to the formation of key competences of the employees of the enterprises [11, 17].

In our earlier studies we used the results of the World Economic Forum (projection, which identifies ten key competences that will be required in 2020, which include: integrated problem solving, critical thinking, creativity, people management, interaction with people, emotional intelligence, formation of own opinion and decision making, customer orientation, negotiation skills, flexibility of mind). Within the framework of the research «Russia 2025: From Human Resources to Talents», carried out by Boston Consulting Group with the support of Sberbank, Worldskills Russia and Global Education Futures, the following range of required competences was defined: communication, interpersonal skills, intercultural interaction, adaptability, solving non-standard tasks, achieving results, management skills, organization, self-development [4].

Today, powerful economic entities understand that the innovative potential of an enterprise and the efficiency of production are possible not so much because of the equipment but because of the human capital of the employees working at the enterprise. Management of companies has become more careful not only in the selection of already ready specialists who have entered the labor market, but also in the process of training of personnel in higher school, (prepare specialists «for themselves») so that at the end of the training the graduates have exactly the competences the employer needs. The quality of the training of young people according to the set indicators of the state and the labor market has become the subject of research of many scholars. It is reflected in the scientific writings of national and foreign researchers of various fields, revealing various aspects of the improvement of the quality of students' education and the development of young specialists.

Scientists of many countries raised the tasks of staffing organizations with well-trained personnel, acquiring and developing professional and managerial skills and competences of employees [13]. Applied research is the requirement of employers of industrial enterprises in a number of European countries for the formation and development of competences among university graduates [2]. It is interesting to study the formation of professional competences of graduates of higher education [18]. The results of the research of our Ukrainian colleagues on the effectiveness of student services within the framework of the international project are noteworthy [5]. The impact of transformational and transactional leadership is the subject of empirical research [12].

In studies of both foreign and Russian scientists in recent times, the formation and development of soft skills is of particular importance. Both foreign [1] and Russian scientists [16] carried out the study of one of the most important components of soft skills - time management. The focus of research has now shifted significantly towards the formation of soft skills in university study process. The role of soft skills in the process of successful employment of graduates of higher education is convincingly proven [3, 6, 9, 14, 15].

«Soft skills» is defined by us as a social-work characteristic of the body of knowledge, skills, abilities and motivational characteristics of the employee in the sphere of interaction between people, the ability to manage your time, the ability to convince, the ability to conduct negotiations; leadership, emotional intelligence, the empathy necessary for successful performance and consistent with the requirements of the position and the strategic goals of the organization, it is a characteristic of potential quality, which allows to describe almost all elements of the readiness of the staff for efficient work in the given situation at the workplace in the work community [8, 10].

Research shows that employers want to see motivated workers with innovative ideas and the ability to be proactive; flexible, development-oriented and efficient; those with conflict resolution and negotiation skills; problem-solving skills; punctual, able to work under multi-tasking conditions, to maintain deadlines, etc. [5]. Our research is also devoted to a study of the formation of soft skills among engineering students in the process of integration into CDIO [7].

We have already analyzed studies of Russian and foreign authors, which convincingly prove the importance of soft skills in the process of training graduates of engineering specialties (including IT specialists). We have conducted research and published the results concerning the formation of the model of interaction between the university and stakeholders to form the competences of competitive graduates [3] and the importance of soft skills of engineering professionals as a condition of competitiveness in the digital economy on the example of the Khanty-Mansi Autonomous Okrug-Yugra [7]. In this article we present the results of the study of flagship competences for students of economic areas. The analysis of the authors' works makes it possible to identify a number of competences significant for the graduates of economic fields: time management, stress tolerance, sense of responsibility, creativity, analytical thinking, communication skills, teamwork, striving for achievements, ability to express their thoughts in an accessible way, quick assimilation of new knowledge and so on.

2 Methodology

Taking into account the urgency of the issue of developing personal and interpersonal competences of university students to achieve competitiveness in the labor market, the authors put forward a list of questions for the study. Among them: is it necessary for graduates to develop «soft skills» in order to achieve competitiveness in professional activity? How do the graduates assess the degree of formation of their «soft skills» in the university?

The study interviewed 393 respondents to answer the above questions. 303 graduates of Surgut State University of Economics attended the survey in 2000–2018. Among them: 198 economists (graduates of economics, accounting, analysis and audit, finance and credit, taxes and taxation) and 105 managers (management, economics and management at oil and gas enterprises, management of the organization). The second group included 90 undergraduate students of the bachelor's degree course in economics (22 respondents in management, 44 in economics, 11 in personnel management, 13 in state and municipal administration).

A student survey was conducted in the spring of 2020 using the Google form (a link to the survey was sent to the students). In addition, the graduates were surveyed in 2019 using the Google form (a link to the survey was sent to the graduates). Both groups were asked two questions: 1 - which skills from the proposed list (the list of skills from the study «Russia 2025: from HR to Talents») are the most important for professional activity; 2 - which skills were formed during the university studies. Respondents had to put a «tick» (V) against the selected skills. The second group of respondents was also asked a number of additional questions. They were all involved in refinement of work experience and self-development to enhance their competitiveness within the digital paradigm (project activity, scientific activity, networking skills, etc.).

3 Results

Among the experimental group of students, 44 (49%) out of 46 (51%) respondents had no work experience before the survey. The main criteria for selecting a job for 63% of working students were working conditions, for 61% - wages, and 45.6% were focused on where they would be taken. While analyzing the answers to the question «Which of the listed aspects of your job are you satisfied with?» we see that the previously stated criteria are reflected in reality. 59% of respondents were satisfied with their working conditions, and 56.5% were satisfied with their wages. In addition, 41% of respondents said that the organization has prospects for professional development, 32.6% - opportunities for career growth/growth of wages.

Content analysis of the respondents' answers to the question «What does it mean for you to be a competitive graduate?» allows us to highlight the following characteristics: knowledge, demand, versatility (development of different skills, including soft skills), achieving goals, the ability to show what you are better than others, the ability to outperform competitors, possession of skills demanded by the employer. To achieve their competitiveness, 65.5% of the surveyed students are engaged in self-education, 69% study well and perfectly, 40% are engaged in science, and 41% gain practical experience.

In order to study the significance of soft skills among those mentioned in the study «Russia 2025: from HR to Talents», respondents from among graduates of economic areas of Surgut State University were asked the question «What skills are most important for your professional implementation». For undergraduate students, the question was «What do you think will be the most important skills for your professional implementation?» Figure 1 shows the distribution of the answers of students and graduates.

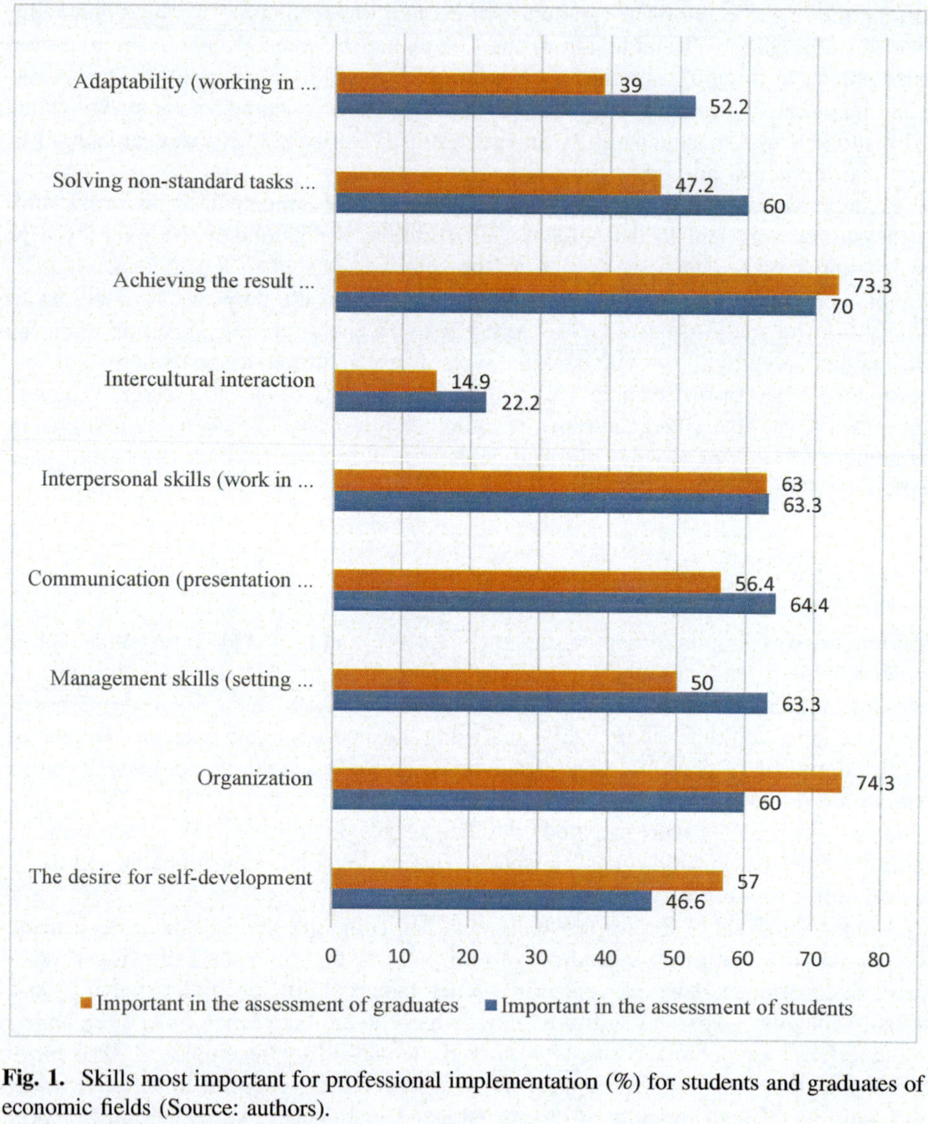

Fig. 1. Skills most important for professional implementation (%) for students and graduates of economic fields (Source: authors).

Assessing which competences the larger number of respondents chose, we note that they are «achieving the result», «organization», «interpersonal skills». However, graduates appreciate the importance of organization more than students do.

While students and graduates have a similar view on the significance of particular competences, there is a significant difference in some positions between the importance of competences and their formation at the university. This is particularly true in terms of organization, self-development, achieving results and solving non-standard tasks (Fig. 2). Using statistical data analysis methods (Fisher criterion), we note that for all these parameters the resulting empirical value $\varphi*$ is in the zone of importance.

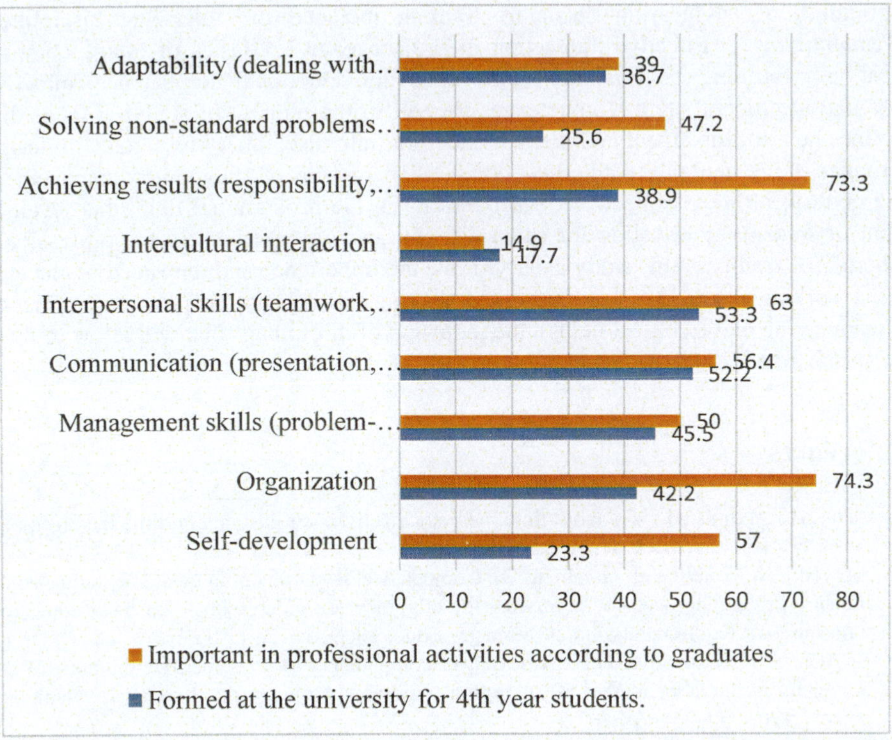

Fig. 2. Skills formed at the university (%) among students of economic fields in comparison with the most important for professional implementation according to graduates (Source: authors).

4 Discussion

The conducted research allows us to conclude that for the majority of students in economic areas, wages and working conditions are fundamental when choosing a job and in order to achieve their competitiveness on the labor market they are self-educated and try to master the educational program in a good and excellent way. At the same time, the correlation between the formation and demand for competences shows a significant gap in terms of organization, desire for self-development, achieving results and solving non-standard tasks, which in turn can lead to difficulties in the process of professional self-realization.

5 Conclusion

The current situation shows that it is necessary to introduce new approaches and improve the process of forming competences that are important for achieving competitiveness in the labor market. Our earlier publications present the results of the successful experience gained in SurSU: the implementation of engineering and technical fields of training disciplines that form soft skills into the educational process. The curriculum of engineering bachelors training included the following disciplines: "Teambuilding", "Effective interaction in project teams", "Basics of image communication" (forming the skills of public speaking, emotional competence, effective negotiation process, etc.). Competences formed in the process of studying these disciplines help to implement complex engineering, interdisciplinary research and design activities in the field of creating new competitive systems of automation, management and control; promote collective (team) activity in the professional field, the development of students' motivation for self-education and improvement in the professional sphere. The results of this study clearly prove the importance and relevance of changes in the working curricula of economic fields (taking into account the positive experience in engineering and technical fields), the inclusion of disciplines, that will allow to form the missing soft skills, in the training process.

References

1. Aeon, B., Aguinis, H.: It's about time: new perspectives and insights on time management. Acad. Manage. Perspect. **4**(31), 309–330 (2017)
2. Azevedo, A., Apfelthaler, G., Hurst, D.: Competence development in business graduates: an industry-driven approach for examining the alignment of undergraduate business education with industry requirements. Int. J. Manage. Educ. **10**(1), 12–28 (2012)
3. Bogdan, E.S., Chulanova, O.L.: Research of competences of a competitive graduate of the university in the labor market of the region: problems, trends, tasks. Internet J. "Naukovedenie", **8**(6), 252–264 (2016)
4. Butenko, V., Polunin, K., Kotov, I., Sychova, E., Stepanenko, A., Zanina, E., Lomp, S., Rudenko, V., Topol, E.: Russia 2025: From HR to talents (2017). http://image-src.bcg.com/Images/Skills_Outline_web_web_tcm27-175469.pdf. Accessed 02 Sept 2019
5. Chernyak, A., Kharlamova, G.: Problems of the effectiveness of student services in Ukraine: the experience of the international project Tempus impress. News Kiivsky Natl. Univ. Taras Shevchenko Econ. **11**(176), 55–58 (2015)
6. Chulanova, O.L., Bogdan, E.S.: Formation of competitive graduates' competencies in extracurricular activities. In: Mantulenko, V. (ed.) Proceedings of the International Scientific Conference "Global Challenges and Prospects of the Modern Economic Development". The European Proceedings of Social & Behavioural Sciences, vol. 57, pp. 177–184. Future Academy, London (2018)
7. Chulanova, O.L., Bogdan, E.S.: Formation of students' soft skills in the process of integration in CDIO. In: Ashamrina, S., Mantulenko, V. (eds.) Proceedings of the II International Scientific Conference "Global Challenges and Prospects of the Modern Economic Development". The European Proceedings of Social & Behavioural Sciences, vol. 79, pp. 526–533. Future Academy, London (2019)

8. Chulanova, O.L.: Challenges and trends on the labor market: synergy of digitalization and soft skills. Hum. Resour. Intell. Res. Manage. Russia **7**(3), 66–72 (2018)
9. Demcheko, T.S., Karácsony, P., Ilina, I.Y., Vinichenko, M.V., Melnichuk, A.V.: Self-marketing of graduates of high schools and young specialists in the system of personnel policy of the organization. Mod. J. Lang. Teach. Meth. **7**(9), 58–65 (2017)
10. Ivonina, A.I., Chulanova, O.L., Davletshina, Yu.M.: Modern directions of theoretical and methodological developments in the field of management: the role of soft-skills and hard skills in the professional and career development of employees. Internet J. "Naukovodenie", **9**(1), 363–378 (2017)
11. Mironova, M.D., Zaitseva, N.A., Larionova, A.A., Novikov, A.I., Borissova, A.A.: The formation of key competencies of employees of the enterprises of service sphere, applying innovative management techniques. Int. J. Adv. Biotechnol. Res. **8**(2), 660–666 (2017)
12. Rawung, F.H., Wuryaningrat, N.F., Elvinita, L.E.: The influence of transformational and transactional leadership on knowledge sharing: an empirical study on small and medium businesses in Indonesia. Asian Acad. Manage. J. **20**(1), 123–145 (2015)
13. Tsitskari, E., Goudas, M., Tsalouchou, E., Michalopoulou, M.: Employers' expectations of the employability skills needed in the sport and recreation environment. J. Hospital. Leis. Sport Tour. Educ. **20**(1), 1–9 (2017)
14. Vinichenko, M.V., Chulanova, O.L., Karacsony, P., Bogdan, E.S., Melnichuk, A.V., Makushkin, S.A.: Model of competences of graduates of high schools of engineering directions: research of stakeholders. Mod. J. Lang. Teach. Meth. **8**(3), 369–380 (2018)
15. Vinichenko, M.V., Chulanova, O.L., Oseev, A.A., Bogdan, E.S., Makushkin, S.A., Grishan, M.A.: Interaction of the higher education and key employer for the formation of the actual profile of the competences of graduates of engineering directions. Mod. J. Lang. Teach. Meth. **8**(5), 394–404 (2018)
16. Vinichenko, M.V., Karacsony, P., Kirillov, A.V., Oseev, A.A., Chulanova, O.L., Makushkin, S.A., Shalashnikova, V.: Influence of time management on the state of health of students and the quality of their life. Mod. J. Lang. Teach. Meth. **8**(5), 166–184 (2018)
17. Vinichenko, M.V., Chulanova, O.L., Klementyev, D.S., Rybakova, M.V., Malyshev, M.A., Shalashnikova,V.Ju., Korosteleva, D.A.: Improvement of the development system of the personnel of organizations of the oil and gas sector of the economy. Int. J. Appl. Exer. Physiol. **8**(2.1), 424–432 (2019)
18. Zaitseva, N.A., Larionova, A.A., Fadeev, A.S., Zhenzhebir, V.N., Filatov, V.V., Pshava, T. S.: Development of a strategic model for the formation of professional competencies of university students. Euras. J. Anal. Chem. **12**(7b), 1541–1548 (2017)

Check for updates

Talent Management: Needs and Prospects for Business Development in the Digital Economy

F. F. Sharipov(✉), T. Yu. Krotenko, and M. A. Dyakonova

State University of Management, Moscow, Russia
{fanissh,krotenkotatiana}@rambler.ru, marie.d@mail.ru

Abstract. In the digital economy, the need and prospects for business development determine their own conditions to the personnel departments of enterprises and organizations. In almost all sectors of the domestic economy, information technology forms a need for knowledge management, taking into account the security of data storage, ensuring the necessary speed of calculations when making a management decision. And the intellectual property of the enterprise itself as a result of systematic knowledge management becomes the most important factor in preserving and increasing the value of business. There is a new form of human resource management subsystem of the company – talent management. System analysis, the method of organizational modeling and analogies are necessary for the study of foreign and domestic scientific sources, best business practices. The purpose of this research is to determine the main activities of the company's HR departments in the sphere of talent management.

Keywords: Competitiveness · Digital business · Digital economy · Information technology · Knowledge management · Talent management

1 Introduction

In the context of development of information technologies used in business today, it is necessary to understand and comprehend, systematize the existing experience of leading companies in this area and try to predict trends for the near future related to the personnel policy. In May 2020, in the context of the COVID-19 pandemic, the TAdviser analytical center prepared a rating of one hundred largest companies operating in the Russian information services market [8]. In this rating, the profile of activity and key industries for which the leaders of the Russian IT market work are of special interest. After summarizing, we identified the following main profiles of companies: system integration; cybersecurity; data storage; Big Data; blockchain; artificial intelligence; machine learning; cryptography; quantum technologies; software distribution; hardware distribution; hardware production; multimedia; automated control system; digital technologies; additional education in the field of IT technologies; Internet of things; mathematical modeling; telemedicine; e-Commerce; digital production; predictive systems; energy saving. Next, we classified the key sectors based on orders with which market leaders in IT sphere work in the Russian economy: the state;

S. I. Ashmarina and V. V. Mantulenko (Eds.): IPM 2020, LNNS 161, pp. 514–518, 2021.
https://doi.org/10.1007/978-3-030-60926-9_65

the mining and processing industries; telecommunications; finance; insurance; retail; fuel and energy complex; transport; medicine; education; manufacturing; defense; non-oil exports; construction; engineering; hospitality; aviation.

The set of profiles of Russian IT companies and the coverage of the country's economic sectors by these firms directly indicate the fact of creating a digital economy in the Russian Federation. We can add the following to company profiles:

- international customs cooperation,
- digitalization of logistics in order to reduce the cost of cargo transportation, the choice of optimal transport (including the calculation of the efficiency of multi-modal transport),
- augmented reality technologies.

There are also some new trend in the area: commercialization of blockchain technology, unmanned cargo delivery (ordered by Gazprom, coal miners), digital identification, 5G networks, cybersecurity for data protection, computing architecture changes. Thus, we see that the Russian economy is entering the next phase of its technological development, where a new design of the commercial environment is needed.

2 Methodology

Among the general research methods that were used when conducting our study, the following can be distinguished: observation, comparison, analysis, and induction. It is necessary to study scientific sources of information, research and analysis of the experience of foreign and domestic companies in the use of knowledge management systems. Among special methods used in the conducted research, there are system analysis, organizational modeling, and method of analogies. The authors took into account various historical background, economic and technological factors of the emergence and development of knowledge management in domestic and foreign business ecosystems. The study also takes into account universal approaches to the knowledge classification. The main tools for managing knowledge at an enterprise are databases, Internet portals, expert community sites, and secure databases. The next tool in the knowledge management system is the 5C model [5]. Tools for implementing a knowledge management system from attracting to retaining talents are presented. The system of values of a self-developing organization is used. Finally, the study identifies the main stages of knowledge management and talent management at the enterprise.

3 Results

Intellectual property, which together with the cost of machines and mechanisms and the property complex makes up the triad of business valuation, plays a major role in the possible growth of the company's value and in the business development. This creates a new approach, or the need to create a complete knowledge management system. And a new subsystem is emerging in the knowledge management system – talent management. There is a new phenomenon – "war for talents", which eventually formed the concept of talent management [7]. This approach opens up opportunities for companies not only to achieve efficiency in key work areas, but also to preserve their positions in the industry.

As an example from domestic business, the leader in the system of knowledge management and talent management, respectively, is the state corporation Rosatom, which we often consider when talking about an effective knowledge management system. In the Russian science, talent management is a subject of serious study for the near future. Managing talented employees is a completely new management task for Russian enterprises [2]. Therefore, at this stage, it is interesting to collect information about the application of specific talent management practices in the ecosystem of a particular company. The XXI century accelerated the global change process: new technologies and businesses are creating a demand for new skills and training areas, and the previous educational system is obsolete as a set of competencies available to employees [3]. This is how the qualification gap arose: today, human capital is trying to develop in the same way as it did in the middle of the XX century, in completely different economic and social conditions [11]. Mass standardization (one education and one profession for the whole life) does not work in the modern world, where a person needs flexibility of thinking, skills of rapid and constant learning and readiness for mobility [10].

It is possible to reduce the qualification gap, increase the labor productivity, and reduce the cost of searching for and retraining specialists only within the framework of a system of social relationships that involve the disclosure of personal and professional potential without compromising others. We consider the talent management system as a knowledge management subsystem that focuses on people and their talent and experience as the main resource of the enterprise, rather than only on knowledge. However, these two systems are so closely linked that it is difficult to separate them, because the workforce is the source of the organization's knowledge, and the organization is the source of information and opportunities for employees. The work of this system is inextricably linked with personnel and the concept of "talent". This concept has been for a long time, but it is quite difficult to identify what it is and consider it from different points of view. Besides, this concept is applicable to various fields, such as psychology, pedagogy, and philosophy. In this paper, "talent" is considered from the point of view of management.

Today, it is becoming obvious that there is a need for employees in various fields and industries who have a unique set of these talents [1]. The available information about certain competencies and talents for working in the organization has greatly simplified the process of selecting and hiring employees. At the same time, it is necessary to understand that talent is quite difficult to track, and even more difficult to confirm that it exists and demonstrate it clearly. Of course, the conclusion about the presence of a talent in an employee is made based on the results of his activities. However, it is not enough for an enterprise to find an employee with a certain set of attributes, because the employee should be able to develop and multiply them. Flexibility, trainability, rapid adaptability, ability to learn and analyze large amounts of information and the ability to apply knowledge and skills in the work, these are qualities that an organization should take into account when searching for new employees [9]. A person's talent is born and unfolds slowly, and may not be revealed at all if the person does not constantly develop, learn and acquire valuable experience. That is why talent is not just a ready-to-use resource, but a combination of special abilities that give a person the potential for effective activity and continuous improvement of these abilities. Talent management is a process inextricably linked to knowledge management and human resource management.

4 Discussion

Over its thirty-year history, the global concept of talent management has developed a solid scientific foundation from the actual definition of the concepts of talent, talent management, and human capital to issues of developing, implementing and improving the company's talent management system. In the works of foreign scientists who have been engaged or are currently engaged in research on the topic of talent management, whose results have become the subject of analysis and correlation with our conclusions, the researchers consider aspects of creating a multi-factor knowledge management system at the enterprise starting from issues of attracting talents to a specific organization to issues of adaptation, development and promotion of talents at enterprises [6]. The Russian scientists studied the experience of leading corporations in the sphere of talent management, analyzed data from leading consulting companies, and provided examples of implementing a talent management system in domestic companies [4]. Our conclusions on such a phenomenon as talent, the need to develop and implement a talent management system at modern enterprises in a new technological way are made largely on the basis of the work of the above-mentioned domestic and foreign scientists.

5 Conclusion

The leader of the Russian practice in the field of talent and knowledge is the state corporation Rosatom. The company's policy of business diversification, the international nature of the company's activities, which is involved in the implementation of projects almost all over the world, and the latest system for creating, using and protecting the intellectual property make the topic of talent management relevant. The company relies on its own personnel. The company has created a three-level system for working with the own talent pool from promising specialists to top managers. Appropriate business processes have been built within the company, taking into account the appropriate corporate culture, creating conditions for the disclosure of individual abilities of attracted talented specialists, and socializing such employees through the creation of communities of experts in the company's information space. The development of a system of incentives that combines material and non-material aspects is of particular importance for talents. If material incentives have been studied and analyzed by the personnel departments of enterprises, then non-material incentives have their own characteristics and require some clarification. The main thing is a possibility of professional and career promotion of a talented employee. A system of continuous professional development in the company, flexible working hours, and a set of social guarantees are of great importance.

We believe that it is necessary to conduct research related to mathematical calculations of a relation between the cost of a talent management system and the results of intellectual activity of talents within the company. As never before, the role of the human factor is increasing while maintaining the companies' competitiveness under conditions of digital economy, both among developers and consumers of the new digital business architecture.

References

1. Jeffcoat, K.L., Eveleigh, T.J., Tanju, B.: A conceptual framework for increasing innovation through improved selection of specialized professionals. Eng. Manage. J. **31**(1), 22–34 (2019)
2. Khoreva, V., Vaiman, V.: Intent vs. action: talented employees and leadership development. Person. Rev. **44**(2), 200–216 (2015)
3. Lokuge, S., Sedera, D., Grover, V., Dongming, X.: Organizational readiness for digital innovation: development and empirical calibration of a construct. Inf. Manage. **56**(3), 445–461 (2019)
4. Savage, G., Franz, A., Wasek, J.S.: Holacratic engineering management and innovation. Eng. Manage. J. **31**(1), 8–21 (2019)
5. Schuler, R.S.: The 5-C framework for managing talent. Organ. Dyn. **44**(1), 47–56 (2015)
6. Sergeeva, M., Shilova, V., Evdokimova, A., Arseneva, N., Degtyareva, V., Zuykov, A.: Future specialists socialization in the context of competence approach. Revista Praxis Educ. **15**(34), 571–583 (2019)
7. Story, J., Castanheira, F., Hartig, S.: Corporate social responsibility and organizational attractiveness: Implications for talent management. Soc. Responsib. J. **12**(3), 484–505 (2016)
8. TAdviser: Ranking TAdviser100: The largest it companies in Russia 2020 (2020). https://www.tadviser.ru/index.php/%D0%A1%D1%82%D0%B0%D1%82%D1%8C%D1%8F:%D0%A0%D0%B0%D0%BD%D0%BA%D0%B8%D0%BD%D0%B3_TAdviser100:_%D0%9A%D1%80%D1%83%D0%BF%D0%BD%D0%B5%D0%B9%D1%88%D0%B8%D0%B5_%D0%98%D0%A2-%D0%BA%D0%BE%D0%BC%D0%BF%D0%B0%D0%BD%D0%B8%D0%B8_%D0%B2_%D0%A0%D0%BE%D1%81%D1%81%D0%B8%D0%B8_2020. Accessed 23 June 2020
9. Tinyakova, V.I., Morozova, N.I., Gunin, V.K.: Transformation of the system of professional training of personnel who are competitive in a knowledge-based economy. Econ. Sustain. Dev. **37**(1), 242–245 (2019)
10. Tsalikova, I.K., Pakhotina, S.V.: Research on the formation of soft skills (review of data in international databases Scopus, Web of Science). Educ. Sci. **21**(8), 187–207 (2019)
11. Yuryeva, A.A., Kuternin, M.I., Gibadullin, A.A., Romanova, J.: Formation of mechanisms for the development of innovative activity in the industrial production of the Russian Federation. J. Phys: Conf. Ser. **1399**, 033099 (2019)

Formation and Development of Innovative Potential of the Enterprise's Personnel

I. N. Ivanov and L. V. Orlova(✉)

State University of Management, Moscow, Russia
ivanov-igor-nik@mail.ru, lorlo@mail.ru

Abstract. The article analyzes the essence of the category "Innovative potential of personnel" (IPP), its most important components, and the conditions for the fullest implementation of this potential. Various points of view on the nature of IPP are presented. The insufficient development of the knowledge management system (KMS) at domestic enterprises is noted, a variant of the structure of this system is presented, the content of its individual blocks is disclosed, and the main methods of knowledge management are shown. It is stated that the key element of the knowledge management system should be the organization of continuous training of personnel with the formation of a "self-learning" organization with the transformation of the latter into an "intellectual" organization. The main stages of creating a KMS at the enterprise, as well as its main functions, are presented. The relation between the knowledge management system and the company's innovative business model is shown. The general structure of this model and the conditions for its integration into the corporate culture of the enterprise are suggested.

Keywords: Dynamic capabilities · Innovative potential · Key competencies · Knowledge management · Business model

1 Introduction

Stable functioning of the enterprise with a guarantee of its (the enterprise's) long-term competitiveness can be ensured at the present time only if the company's employees are ready for intense activities of an innovative nature. The organization's personnel is a key factor in all production and economic activities; it is the level of creativity of employees that determines the company's success in the market [2]. It is obvious that modern competition can be interpreted as a competition of ideas generated by the staff of specific economic entities. Innovative abilities of employees and their initiative are the decisive factors of fundamental changes on the enterprise, the source of ensuring long-term competitiveness of the enterprise.

It should be noted, however, that while recognizing the importance of employee participation in innovations, many organizations rely mainly on the relevant leaders, without paying due attention to the initiative of the majority of ordinary employees, which naturally hinders business development. In addition, the human resources of enterprises often behave passively, not taking part in the development of a strategy for the development of innovative capabilities of employees, and are not responsible for the

S. I. Ashmarina and V. V. Mantulenko (Eds.): IPM 2020, LNNS 161, pp. 519–529, 2021.
https://doi.org/10.1007/978-3-030-60926-9_66

results of their work. At the same time, close attention to the formation of innovative capabilities of personnel is especially important for high-tech enterprises in the real sector of the national economy. These enterprises are the core of the country's production system, and they are associated with the prospects for innovative transformation of the Russian economy. In this regard, an important task of high-tech organizations is to create and develop the innovative potential of personnel who accumulate unique knowledge, abilities, skills, and behavioral patterns of employees [4].

The category "innovative potential of personnel" (IPP) in the scientific literature is mainly highlighted as one of the complex resources available by the organization that have potential suitability for participation in the company's innovation activities. This complex usually combines the intellectual, motivational and communication qualities of the staff, their predisposition to changes, to the perception of the new, creativity, initiative, adaptability [11]. It is important to emphasize that the innovative potential of personnel has two sides: internal and external. The internal side reflects the ability and readiness of employees for innovative activities. In other words, we are talking about the quality characteristics of personnel that are not yet used in practical innovation activity (i.e. hidden), and are similar to such resources as intellectual, material, financial, technical and organizational, etc.

At the same time, the mere readiness of personnel to innovative activities is only a prerequisite and no more for the implementation of these activities. IPP is fully manifested only when the potential energy of the organization turns into kinetic energy, i.e. when the manifestation of this potential leads to an increase in the company's competitive advantages. In other words, when an enterprise can transform latent abilities and willingness of employees to innovate activities into real actions, into real projects. Such transformations are called upon in companies by human resources services (or HR departments), usually responsible only for ensuring that employees are trained in a timely manner, and vacant positions and jobs are filled. However, in practice, this is not enough. These services, in our opinion, should, in addition to performing the above-mentioned functions, participate in various actions to increase the competitive level of their company, moving from a narrowly functional role to the creation of a system that makes a significant contribution to the formation of IPP. The essence of the IPP, as well as such elements of the above-mentioned system as the key competencies and dynamic capabilities of the organization, knowledge management, and the innovative business model of the enterprise will be discussed below.

2 Methodology

The methodological basis of this research is the results of scientific research of domestic and foreign authors dealing with the problems of innovative potential of the economic entities personnel. The purpose of this article is to analyze the essence of scientific categories that form the concept of IPP, as well as to study the actual and possible use of tools for the formation and development of innovative potential of company employees. Various points of view on the nature of IPP and the features of important components of IPP – the categories "key competencies" and "dynamic abilities" were studied. The practice of creating knowledge management systems at domestic enterprises is analyzed. The author's approach to creating an appropriate system is offered. The existing methods of forming an innovative business model of an enterprise are analyzed. The relevant recommendations are justified. Methodological materials related to the topic of the article, legal and regulatory acts and resolutions of the government of the Russian Federation are studied. The paper uses methods of system analysis, statistical method, methods of analogies and comparisons. The empirical base of the research was formed by reference materials, development strategies of the Russian Federation and its individual regions, data from enterprises in the real sector of the economy, and the authors' own research.

3 Literature Review

In the scientific literature, the innovative potential of personnel as a subject of the study is considered as:

- a set of non-standard qualities,
- non-traditional behaviors,
- the self-development process.

As for the first point of view, there is a discussion about the transition, as already noted above, from the situation of "ability" and "readiness" to the practical implementation of innovative activities. For example, Gasenko refers to non-standard qualities as knowledge, skills, abilities and personal characteristics of employees that indicate their readiness to innovations implementation [5]. Some researchers refer to similar properties as intellectual, professional, personal and creative abilities of employees who are predisposed to innovative thinking and are able to quickly adapt to changes. From another point of view, non-standard qualities of personnel include readiness to implement tasks to ensure the functioning of companies in the conditions of innovative transformations. The advantage of the first version of the IPP description is that the above-mentioned personnel characteristics can be interpreted fairly easily within a separate enterprise. At the same time, a large number of qualities, parameters and characteristics dramatically increases the variability of relationships between them and the process of developing an organization's innovative potential. In addition, it is very difficult to evaluate many of these qualities.

The second aspect of consideration of the personnel innovative potential is called "behavioral", because it is the organizational behavior of employees at all levels of the company that determines the measure of their involvement in innovative activities. Evaluating the options above interpretation of innovative potential of personnel, it should be emphasized that, despite their advantages, both have one drawback: they only preserve the existing state of components of IPP, without assessing the prospects of development of the latter. In this regard, the third approach to the description of IPP categories is of much greater interest, in which the formation of innovative qualities in employees of the enterprise is a dynamic process of self-development of personnel. During this process, the so-called "dynamic capabilities" of the company's personnel are formed, which are understood as exceptional and particularly valuable resources that cannot be imitated and can be updated in changing circumstances. At the same time, the ability to updating can be manifested anywhere in the organization, so that ordinary employees can become drivers of innovation. There is an opinion that the innovative abilities of the organization's employees are not determined only by unique knowledge and skills. Their innovative potential is not manifested by itself, but is formed throughout life, in the course of work, based on the employee's accepted system of values, on his willingness to self-actualization, on the ability to self-development. With this consideration in mind, we can agree with Esaulova's opinion that the innovative potential of personnel is nothing more than existing and prospective opportunities for the development of the organization, which are contained in the unique knowledge, skills and behaviors of employees [4].

The innovative potential of personnel is the "engine" of the company's innovative development, including not only a set of certain resources, but also the so-called organizational abilities. The latter can be divided into basic, key, and dynamic ones according to their degree of development. As for basic abilities, they are inherent in most enterprises, because they are vital for the implementation of production and economic activities and simply for survival in a competitive environment. Key abilities create adequate competencies that represent the competitive advantages of the enterprise [16].

Key competencies became the basis for the concept of "resource theory of the firm", which was developed in the early 90s of the last century. At the same time, as a key competence, the authors of this concept recommended taking the collective knowledge of the organization, focused on the coordination of various production skills and linking existing technological flows [10]. In other words, key competencies are formed as a combination of intellectual resources and organizational capital, which in turn is the source of the organization's market capital. At the same time, the organization has the ability to use not only available resources, but also generate missing resources more effectively than its competitors. The following definition can be given: key personnel competencies are a set of unique knowledge and skills of the company's personnel, which form a particularly high value for the consumer. These competencies are created on the basis of the company's organizational and human capital.

In the conditions of high turbulence of the external environment, a particularly significant characteristic of the personnel is the degree of development of their dynamic abilities. They make it possible to quickly adjust the company to market fluctuations, as well as to maximize the use of available resources for the development of innovative activities, to ensure competitive advantages.

Dynamic capabilities as a scientific category were first described in the work of Teece, Pisano, and Shuen, which include "the ability of a firm to integrate, create and reconstruct internal and external competencies in response to rapid changes in the environment" [15]. In today's highly competitive market conditions, dynamic capabilities are mainly determined by the organization's intellectual resources and the presence of a developed IT infrastructure, including hardware and software. Dynamic capabilities make it possible to transform key competencies and abilities in accordance with changes in the environment faster than competitors [12].

The main dynamic capabilities of the company, in our opinion, consist in the readiness of its staff to:

– quick identification of changes occurring in the external environment,
– response to the above changes,
– constant development,
– lifelong learning,
– constant knowledge exchange,
– formation of new abilities.

Special attention should be paid to the formation of new staff abilities, since they cannot be copied by competitors.

4 Results

The development strategy of any company should correspond to its potential, the most powerful elements of the latter, which are planned to be formed. Before developing this strategy, it is important to determine the level of innovation of the company's personnel. The author's version of the corresponding classification is presented below.

A static level means that the enterprise does not have any innovations or effective structural changes; if the organization remains at this level for a long time, it will inevitably lead to bankruptcy. Such an enterprise can be described as *non-innovative.*

The basic level indicates that the company is doing some work to increase its innovation to a level that ensures the company's survival in the market. At the same time, there are either no innovations at all, or there is a small number of them. Such an enterprise can be called *a modernizing one.*

The advanced level indicates that the company is quite actively implementing innovations, has a sufficient set of knowledge and competencies to ensure the production and economic activities of the company. There are separate elements (usually information) of a knowledge management system. In this case, we are talking about *a low-innovation* enterprise.

At the *key level,* there is a fairly wide use of innovations, a KMS has been created, key competencies have developed, and dynamic abilities are being formed. *A key level* is typical for *the innovative* enterprise.

The dynamic level indicates that a mechanism for continuous innovative development, based on high organizational competence and knowledge management system, has been developed and now operates; an innovative business model that quickly adapts to changes in the environment operates in the organization. This level is the standard of *a highly innovative* enterprise.

To promote the enterprise to a higher innovative level, it is necessary to use an appropriate mechanism, which, in our opinion, should include the following main elements:

– principles of innovative development,
– the strategy of innovative development,
– methods and tools of innovative development,
– sources of innovative development.

The interaction of these elements forms the organizational competence of the enterprise, which is a key resource for innovative development, including the key competencies and dynamic capabilities of the organization.

Orientation to the formation of key competencies and dynamic abilities of personnel, purposefulness, adaptability, consistency and complexity, openness, efficiency, economic and social responsibility should be, in our opinion, among the possible principles for creating this mechanism by the main ones [14].

Key competencies and dynamic abilities of the staff should be formed in unbroken unity with the direction of development of the existing knowledge management system adopted by the company. "Knowledge economy" is a modern trend, where the most important role is played by managerial, organizational, social, and technological innovations, the implementation of which is closely linked to the technology and procedures of personnel management.

The term "knowledge management" in the Russian Federation has been actively used as an independent management function since 2000. In the context of the formation of the "knowledge economy", knowledge management becomes a leading management function due to the fact that it forms the basis for future business development by creating and using the accumulated knowledge baggage for the implementation of the company's strategy. This strategy is based on the unique properties of human capital, which creates intangible assets of the enterprise that contribute to the growth of the organization's capitalization. These qualities include creativity, a propensity for entrepreneurship and continuous learning, speed of decision-making, and the ability to work in a team in conditions of permanent changes.. In economically developed countries, knowledge accounts for up to 90% of GDP growth, the market has become intellectually oriented, and organizations compete in the field of intangible resources [7]. Digital technologies are becoming widespread, and the competitiveness of modern enterprises increasingly depends on the degree of management professionalism, on how effectively it can operate the methods of managing the company's intellectual resources, in particular, the methods of knowledge management. The organization of a continuous learning system is becoming the main method of human resource management. The enterprise needs to develop a mechanism for transferring knowledge from the implicit to the explicit form and inversely, from the individual to the group level.

Nowadays, full-fledged and effective knowledge management systems have not yet become widespread in domestic industrial enterprises. As a rule, there are only certain software products that serve the company's information system [6]. This is not enough to create an effective knowledge management system. In general, the knowledge management system at the enterprise should include the following main components: the group that creates knowledge, communications, training, information technology support, "idea factory", experts.

The company needs a creative *team* that makes knowledge of a certain direction that is currently necessary for the company (technological, managerial, environmental, etc. innovations). The presence of this team does not exclude, but, on the contrary, include the involvement of ordinary employees of the company in innovative activities [1]. The communication *system* should provide for both external and internal communications that facilitate access and knowledge sharing (creating an IT environment).

The system *of personnel training* should include such forms as obtaining the first education or working profession, advanced training, professional retraining, and higher education. Training should be continuous and its results should be reflected in the professional career of the employee.

Information technology support requires the availability of appropriate tools, databases, archives, etc.

"Ideas factory" means holding regular competitions of innovative ideas, not only among its employees, but also among consumers of the company's products, using crowdsourcing methods, and cooperating with specialized universities.

Experts are leading employees of the company, professionals whose knowledge is an important strategic resource of the enterprise, and often become its key competencies.

As already noted, for many Russian enterprises, such concepts as "knowledge management", "continuous learning", and "human capital" are still relatively new. As for advanced (mostly foreign) companies, the analysis conducted by the authors

Table 1. Knowledge management methods

KMS element	Knowledge management (KM) methods
Technological support of business processes for knowledge management	– IT environment for professional communities and training, WEB conferences, webinars, – expert systems of artificial intelligence, – corporate communications, mailing list, – libraries of documents and educational materials.
Organizational support of business processes on knowledge management	– knowledge and document management audit system, – crowdsourcing, benchmarking, –contact lists, – surveys and tests, – KM procedures and techniques, – intellectual property protection, – evaluation of the KMS effectiveness.
Social support of business processes on knowledge management	– teams that produce knowledge, – corporate training and adaptation system, – communities of interest, – knowledge-friendly corporate culture, – "idea factory "(system for attracting and selecting ideas), – selection of experts and their consultations, – motivation system for knowledge sharing, – expert rating system, – meetings and seminars, – contests, brainstorming, – mentoring, – corporate media.

Source: authors.

allowed us to identify the most commonly used methods of knowledge management (Table 1).

Let's take a closer look at some of the knowledge management methods that we haven't covered before. *Interest groups*. Usually the most promising "start ups" are produced in an informal atmosphere in small groups as a result of "enlightenment" from one of the group members or due to favorable circumstances. Therefore, employees of the company need to create a comfortable, relaxed environment, consistent with their addictions and competencies. In principle, such groups should be created by employees themselves, and the initiative of the latter should be encouraged by the company's management. Communities of interest can be either formal (project teams, working groups, etc.) or informal (lovers of hunting, fishing, literature, art, football, etc.).

The corporate culture should be "sharpened" to production, distribution and use knowledge, and the staff should not be afraid to share their intellectual product, for which the company's management should create an environment of trust and cooperation [13].

The corporate memory of the organization includes a database, a bank of ideas with access to them through the corporate WEB portal. The knowledge included in the corporate memory must be suitable for analysis, complete, structured, and accessible to employees.

A key element of the corporate knowledge management system is the personnel training unit. To create a system of continuous learning, the corresponding process must be self-adaptive, and the organization must be self-learning [9]. This is considered to be firm that studies and analyzes the results of its training in order to improve the efficiency of its activities. In the process of continuous learning with maximum involvement of employees, a self-learning organization is transformed into an *intellectual organization* [8]. The advantage of an intellectual enterprise is that it not only produces knowledge, but also capitalizes it, thereby increasing the market value of the company [3, 7].

Knowledge management is currently an independent management function, which should be implemented by the corresponding system being created at the enterprise. The main stages of creating this system should, in our opinion, be:

1. Analysis of existing business processes at the enterprise and identification of related problems.
2. Justification of the need to create a KMS.
3. Development of the KMS structure.
4. Formation of technological, organizational and social components of KMS.
5. Training of personnel to work in the KMS.
6. Calculation of the expected efficiency from the operation of the KMS.

The main functions of the knowledge management structure created at the enterprise should be:

- creation and development of KMS,
- motivation of the personnel to the production, use, and dissemination of knowledge,
- formation of the "knowledge environment»,
- identification and dissemination of implicit knowledge,

– identification of key competencies and dynamic capabilities of the organization,
– development of recommendations for the implementation of the identified competencies and abilities.

The knowledge management system should fit seamlessly into the company's *innovative business model* of its activities, the main components of which are presented below:

1. The unit of forming company values: key resources, activities, sales channels, partners, products, services, solutions, competencies; dynamic abilities.
2. The unit of key consumers and the system of relationships with them.
3. Financial block: cost structure, revenue streams.

A successful innovative business model of a company should be embedded in the corporate culture and memory of the organization. It should be transformed in accordance with the trends of business development. Among the trends that relate directly to the issue discussed in this article, it should be noted:

– shifting the focus towards socially significant goals,
– a sharp increase in the value of each individual employee,
– forming an atmosphere of trust and cooperation in the company that encourages the creation and exchange of knowledge, increasing the initiative of staff,
– changing the system of staff motivation towards transparency of this system and group rewards,
– formation of the internal "knowledge environment" (internal market of ideas),
– increasing the degree of employees independence in decision-making,
– orientation of the company to self-learning,
– increasing corporate social responsibility of business.

The above points to a trend from a bureaucratic, nonadaptive organization to a creative, flexible, socially-oriented team. These are the most successful modern companies.

5 Conclusion

In the era of digital transformation of socio-economic processes, the transition to the "knowledge economy" significantly increases the role of the personnel innovative potential. Creativity and initiative become a crucial factor in improving existing processes and procedures at the enterprise, a condition for ensuring the long-term competitiveness of the organization. This is especially important for high-tech enterprises in the real economy. IPP is shown to the maximum extent only when the ability and readiness of staff to innovative activities is transformed into real actions, into real projects. To ensure this transformation, the company should focus on the development of its key competencies and dynamic abilities, with the promotion of the innovation degree of the organization's employees from a statical to a dynamic level. Key competencies and dynamic capabilities of the enterprise should be formed in an inextricable connection with the knowledge management system developed and functioning in the company, the central link of which should be continuous training of personnel. The process of continuous learning should be self-adjusting, and the organization should be self-learning with subsequent transformation into an intellectual organization. The knowledge management system created at the enterprise should be an organic part of the innovative business model developed in the company, which, in turn, should also be organically embedded in the corporate culture and corporate memory of the enterprise.

References

1. Antonenko, Yu.P., Pleskachevskaya, A.Yu., Kaplunovich, S.M.: Features of work with personnel in the conditions of innovative activities in the organization. In: Kostryukova, A. V., Artamonova, O.E., Davydova, S.G. (eds.) Proceedings of the All-Russian Scientific and Practical Conference of Teachers, Undergraduates And Students “Days of Science - 2018”, pp. 8–11. Novgorod branch of RANEPA, Nizhny Novgorod (2018)
2. Ashurbekov, R.A., Antonova, O.M., Belova, O.L., Garnik, S.V., Zakharov, D.K., Ivanovskaya, L.V., Ivanov, I.N., Kashtanova, E.V., Konobevtsev, F.D., Konovalova, V. G., Lobacheva, A.S., Lukyanova, T.V., Mitrofanova, A.E., Mitrofanova, E.A., Orlova, L.V., Panin, V.I., Svistunov, V.M., Suvalova, T.V., Tikhonov, A.I.: Personnel Management Technology in the Digital Economy Modernization: Monograph. Publishing house of State University of Management, Moscow (2019)
3. Dugusheva, A.D., Sarakayeva, Z.H.: Intellectual potential of personnel as the main factor of innovative development of the organization. In: Gulyaev, G.Yu. (ed.) Proceedings of the IV International Scientific and Practical Conference “Economics, Business, Innovations”, pp. 173–175. Science and Education, Penza (2018)
4. Esaulova, I.A.: Innovative potential of personnel as a source of dynamic capabilities of the organization. NSUEM Bull. **2**, 262–271 (2015)
5. Gasenko, E.V.: Criteria for developing the innovative potential of the staff of a knowledge-intensive enterprise. Tomsk State Univ. J. Econ. **2**(18), 44–49 (2012)
6. Ivanov, I.N., Lukyanova T.V., Belova O.L.: Digital economy: Knowledge in the logic of the concept “Industry 4.0”. In: Terelyansky, P.V. (ed.) Proceedings of the II International Scientific Forum “Step Into the Future: Artificial Intelligence and the Digital Economy”, pp. 50–59. SUM, Moscow (2018)
7. Malichenko, I.P.: Knowledge management as an effective mechanism for forming a continuous system of training and development of personnel in the organization. NSUEM Bull. **1**, 174–188 (2016)
8. Mitrofanova, E., Mitrofanova, A., Tarasenko, V.: Immature digital expertise of the educational institution’s managerial staff as HR risk to education development. In: Popkova, E.G., Sergi, B.S. (eds.) Artificial Intelligence: Anthropogenic Nature vs. Social Origin. Advances in Intelligent Systems and Computing, vol. 1100, pp. 756–765. Springer, Cham (2020)
9. Mitrofanova, E.A., Simonova, M.V., Tarasenko, V.V.: Potential of the education system in Russia in training staff for the digital economy. In: Ashmarina, S.I., Mesquita, A., Vochozka, M. (eds.) Digital Transformation of the Economy: Challenges, Trends and New Opportunities. Advance s in Intelligent Systems and Computing, vol. 908, pp. 463–472. Springer, Cham (2020)
10. Prakhalad, K.K., Khamel, G.: Key competence of the corporation. Bull. Saint Petersburg State Univ. **3**, 19–46 (2003)

11. Saveleva, O.A., Mitasova, A.A.: Assessment of development factors of personnel innovative potential. Manager **2**(80), 208–214 (2017)
12. Scholz, C., Müller, S.: Dynamic human capital strategy: a new way to strategic human resource management (2010). http://1v.com/wp-content/uploads/2018/05/97.pdf. Accessed 22 July 2020
13. Svistunov, V.M., Kuzina, G.P., Lobachev, V.V.: The level of trust of the company staff as the factor of the increase of social and economic systems management efficiency. In: Proceedings of the XXI International Conference "Complex Systems: Control and Modeling Problems (CSCMP)", pp. 799–802. IEEE Xplore, New Jersey
14. Svistunov, V.M., Lobachev, V.V., Simonova, M.V.: Staff responsibility as efficiency-driven factor of ERP-systems. In: Ashmarina, S.I., Vochozka, M., Mantulenko, V.V. (eds.) Digital Age: Chances, Challenges and Future. Lecture Notes in Networks and Systems, vol. 84, pp. 480–486. Springer, Cham (2020)
15. Teece, D.J., Pisano, G., Shuen, A.: Dynamic capabilities and strategic management. Strateg. Manage. J. **18**(7), 509–533 (1997)
16. Vinnik, A.E.: Key competencies and dynamic abilities as the basis for the competitiveness of the socio-economic system. Mod. Res. Dev. **4**(12), 75–78 (2017)

Strategic HR Management and HR Analytics: Preparing for Macro-Calls

Motivation Tools in Entrepreneurial Networks of Strategic Alliances

D. V. Aleshkova[1(✉)], D. A. Akopyan[1], N. V. Kalenskaya[2], and R. N. Khusnutdinov[2]

[1] Samara State University of Economics, Samara, Russia
dashajuly343@gmail.com, diakopyan@yandex.ru
[2] Kazan Federal University, Kazan, Russia
kalen7979@mail.ru, 2911379@mail.ru

Abstract. In the article, the authors analyze motivators and motivation tools used in entrepreneurial networks of strategic alliances in order to improve the efficiency of corporations and the effectiveness of their personnel. The analysis and comparison of indicators is based on the example of entrepreneurial networks organized within the framework of Russian-Chinese strategic alliances. A comparative characteristic of motivators and motivation tools is performed, and conclusions are drawn about the possibility of their application in a Russian-Chinese joint venture. The results of the analysis show that there are the same number of common positions and its differences in the motivation system.

Keywords: Entrepreneurial networks · Motivation · Strategic alliances

1 Introduction

The growth and stability of the Russian economy are directly related to the increase in the number of transnational corporations (TNCs) operating in Russia. A transnational corporation is a special type of corporation that has outgrown the national framework and operates on the world market through its foreign branches and associated companies [7]. Along with China, India and Brazil, Russia is becoming one of the most attractive locations for TNCs. The increase in producers of mass-consumption products is primarily due to the growth of the economy and consumption in Russia and the low level of basic production costs. Most TNCs see a number of advantages in the Russian market: relatively cheap resources, constantly developing and expanding markets, and savings on transport costs and import duties.

2 Methodology

Within the conducted research, the authors considered the current motivation systems used by two entrepreneurial structures belonging to the same entrepreneurial network existing within the framework of a strategic alliance. These entrepreneurial structures are two large oil and gas TNCs that represent Russia and China and carry out joint entrepreneurial activities. The current legal acts affecting motivation were analyzed,

S. I. Ashmarina and V. V. Mantulenko (Eds.): IPM 2020, LNNS 161, pp. 533–537, 2021.
https://doi.org/10.1007/978-3-030-60926-9_67

and the methods used in each entrepreneurial structure were identified, which were then compared with each other. Based on the analysis results, synthesis of information and comparison of results, conclusions were made about the exchange of experience in applying specific methods of motivation in individual entrepreneurial networks.

3 Results

Let's analyze the activities of TNCs in the oil and gas industry, which is currently one of the most developed and competitive in Russia. Russia is one of the world's leading producers, processors, and exporters of oil and natural gas. In the current political and economic conditions, the Russian oil and gas industry is considered one of the fastest growing in the world [4]. Currently, the largest and most significant companies engaged in oil and gas exploration, production and processing in Russia are LUKOIL, Rosneft, TNK-BP holding, Gazprom Neft, Surgutneftegaz, and Tatneft [3].

The motivation function plays an important role in the personnel management system of these companies [9]. They have created all conditions for staff motivation, developed and implemented effective motivation systems that encourage staff to work effectively, implementing the strategic objectives of the organization.

Let's analyze motivation systems, used in the Russian TNCs, on the example of PJSC "NC "Rosneft". Russian organizations use material and non-material methods of motivation that are unique to Russian companies and borrowed from the practice of western countries. In the incentive systems used by the management of Russian companies, in contrast to western European companies, there is a difference in priorities - for the first, material incentives dominate, for the second, the non-material plays a more important role [1, 2]'s analyze motivation systems``. Please insert the parenthesis in the appropriate position." –>. This is primarily due to the economic conditions of life and work of employees, which determine the priorities of staff needs. In countries with unstable economies, earnings are more important than moral encouragement for many people.

Rosneft is one of the leaders in the oil and gas industry in Russia. SinopecCorp corporation has recently become one of Rosneft's partners. This Chinese multinational oil and gas corporation, an integrated company that combines the energy and chemical industries, is the largest oil and gas corporation in China, the second largest corporation in terms of production in the country.

In September 2016, PJSC "Rosneft" and Sinopec Corporation signed an agreement to create a project for the construction of a common oil and gas processing complex, which will be located in Eastern Siberia. As a result of the project, it is expected to meet the demand for polypropylene and polyethylene in China and Russia, which indicates the profitability of this project for both sides. The products will be sold exclusively on the Russian and Chinese markets. The project is based on resources obtained from PJSC "Rosneft" oil and gas fields [5].

Such close cooperation between the two oil and gas corporations makes it necessary to organize work together, including the use of a common system of motivation. This requires identifying the main similarities and differences between these systems. The basis of the personnel management system is a decent salary. PJSC "Rosneft" applies a policy of high social responsibility in terms of salaries and compensations.

An important factor in the motivation system of PJSC "Rosneft" is evaluating the effectiveness of employees' performance. It evaluates both the collective component and the individual contribution of each employee to the results of their work. The system of personnel incentives is aimed at achieving high production results and increasing labor productivity. PJSC "Rosneft" has a reputation of one of the most socially responsible employers in the market. To improve the social protection of employees, the company has developed a standard collective agreement that reflects all social benefits and guarantees provided to employees of subsidiaries in excess of the labor legislation of the Russian Federation.

The main areas of staff support laid down in the motivation system of PJSC "Rosneft" are:

- financial support for employees with a large family,
- additional social support for enterprise' retired employee,
- creating conditions that provide a favorable material climate for the staff, taking into account the specifics of the region and working conditions,
- assistance to employees in difficult life situations [6].

In Sinopec corporation, one of the main ways of motivation is also a high salary. Leadership believes that this helps retain good professionals and inspire them to dedication. Sinopec employees receive the highest wages in China. Additional methods of motivation in this company are: free food and medical care on the territory of the enterprise, its own sports complex, where all employees can train and rest for free. Employees with small children are entitled to an additional 7 weeks of vacation, which can be used either in whole or in parts. In addition, each employee who takes advanced training courses, re-profiling and other professional training receives compensation for the money spent on this. Special attention is paid to the organization of the workspace. This is due to the irregular working hours of employees and the complexity of the performed work. So, psychologists, doctors and designers are invited to design a work space. They work together to develop the design of workplaces for each employee. Thus, the corporation leadership achieves successful completion of tasks by employees [8]. The comparative characteristics of common motivators and tools are presented in Table 1.

Table 1. Motivators in the Russian and Chinese management system

Motivators	Internal		External	
	Russia	China	Russia	China
Material	–	Career growth with a higher salary	Bonuses	Bonuses, compensations
Non-material	Diplomas, public praise, empowerment	Empowerment	Boards of honor, gratitude	Corporate trainings, competitions

Source: authors.

As you can see, the main motivators used in motivation systems in Russia and China are largely the same, but there are also differences (Table 2).

Table 2. Motivation tools used in the Russian and Chinese management systems

A motivational tool	Russia	China
Bonuses	+	+
Compensation package	+	+
Non-material recognition	+	–
Censures	–	–
Corporate events, team building	+	+
Corporate training	+	+
Coaching, field training	+	–
Empowerment	+	+
Employee competitions	+	+

Source: authors.

Based on Tables 1 and 2, we analyzed common features and differences in the motivation systems of "Rosneft" and "Sinopec" corporations, presented in Table 3.

Table 3. Common features and differences in motivation systems of "Rosneft" and "Sinopec" corporations

Common features	Differences
– High wages, – Support of employees belonging to vulnerable segments of the population, – Support of large families and families with young children, – Organization of workplaces and creation of favorable working conditions	– System for evaluating the effectiveness of operational results ("Rosneft"), – Assistance to employees in difficult life situations ("Rosneft"), – Compensation of training costs ("Sinopec"), – Free food and medical care ("Sinopec")

Source: authors.

4 Discussion

Taking into account the fact that the same motivation tools can be successfully used in different corporations, it should be noted that the following factors may influence the final result – the formation of a motivation system in a Russian-Chinese joint venture:

1) subculture (values and other criteria for personnel decision-making) that has developed both in the country and in specific (analyzed) enterprises,
2) strategic approaches to the introduction of new industry technologies.

For example, China is a leader in scaling up advanced technological developments (a repeat of used European technologies), so there is experience that allows you to achieve the maximum positive effect in both motivation and investment programs. It can be argued that China demonstrates a high level of productivity, including by maximizing the impact of motivation programs and activities. Russia is currently focused on a policy of import substitution, which allows us to maintain the pace of sustainable economic development.

5 Conclusion

The results of the analysis show that there are the same number of common positions and its differences in the motivation system. However, different positions are not fundamental, and enterprises are able to function without them. From the point of view of improving motivation at enterprises, we can recommend the leadership of PJSC "Rosneft" to borrow free food and medical care for employees from their partners, and Sinopec corporation to use the system for evaluating the effectiveness of work results as the most accurate method for selecting employees for their subsequent promotion. Since the cooperation of the companies will be based on the fields of PJSC "Rosneft", the leadership of Sinopec corporation decided to use the labor organization and motivation system used at "Rosneft" enterprises. This decision is dictated by the peculiarities of the Russian mentality, it is believed that employees of Chinese enterprise can more quickly adjust to working conditions in a joint venture. However, Sinopec leadership leaves the right to use free medical care for their employees.

In conclusion, it should be noted that when creating a joint Russian-foreign enterprise, the analysis of the motivation system at foreign enterprises operating in the same industry. The peculiarities of cultural adaptation and system adjusting taking into account the practices of foreign companies, can help to improve the management system of the international economic alliance.

References

1. Booth, A.: Trade Unions, Economic Behavior of International Encyclopedia of the Social and Behavioral Sciences, vol. 2, pp. 497–502 (2015)
2. Brand, U., Niedermoser, M.K.: The role of trade unions in social-ecological transformation: overcoming the impasse of the current growth model and the imperial mode of living. J. Clean. Prod. **225**, 173–180 (2019)
3. Expert Online: Russia's Largest Oil and Gas Corporations (2020). http://expert.ru/ratings/krupnejshie-neftyanyie-kompanii-rossii. Accessed 29 June 2020
4. Klimovets, O.V.: Transnationalization of Russian Business Corporations. Ekoinvest, Krasnodar (2009)
5. Rosneft: "Rosneft" and "Sinopec" have signed an agreement to participate in a project of building a gas processing and petrochemical complex in Eastern Siberia (2016). https://www.rosneft.ru/press/releases/item/183487. Accessed 29 July 2020
6. Rosneft: Politics in the field of remuneration, motivation and social partnership (2020). https://www.rosneft.ru/Development/personnel/motivation. Accessed 29 July 2020
7. Shimlo, P.D.: Economics of Transnational Corporations: Monograph. HSE, Moscow (2010)
8. Sinopec Group: Social responsibility (2020). http://www.sinopecgroup.com/group/shzr/. Accessed 29 July 2020
9. Wang, Y., Quattara, K.S.: Employment double dividend hypothesis with the presence of a trade union. Econ. Lett. **193**, 109273 (2020)

HR Development Strategy Under Conditions of the Oil and Gas Industry Digitalization

O. A. Babordina[1(✉)], M. P. Garanina[1], and E. K. Chirkunova[2]

[1] Samara State Technical University, Samara, Russia
obabordina@mail.ru, garaninamarina@ya.ru
[2] Samara State University of Economics, Samara, Russia
ekchirkunova@gmail.com

Abstract. The research work contains information about the strategic development direction of the oil and gas industry in the conditions of digitalization of production business processes. An important resource in providing business with information using digital software devices is the company's personnel. The article defines professional competencies that ensure that specialists perform their functional duties, taking into account new knowledge, skills and abilities acquired in the framework of retraining. The article investigates the feasibility of forming and managing the personnel reserve of a modern enterprise. Issues on methods of selection and evaluation of HR reserve are considered.

Keywords: Assessment · Competencies · Digitalization · Personnel · Reserve · Strategy

1 Introduction

The experience of leading companies shows the importance of clear formulation of HR development strategies and formation of a personnel reserve. The HR policy that ensures the efficiency of the enterprise and determines the strategic direction in its goals and objectives is an integral element of the HR management system. The need to work with the personnel reserve is maintained on an ongoing basis in order to achieve the planned results in the overall concept of enterprise development, and the formation of a HR strategy allows realizing the human potential in accordance with modern trends of business development under conditions of the economy digitalization. Many oil and gas companies believe that the availability of a personnel training strategy will get a positive effect not only for individual workers, but also for all organizations [1].

Analysis of concepts on the role of personnel in production, presented by Russian [11] and foreign scientists in the field of strategic management [15] shows the influence of the corresponding lifestyle, scientific and technological progress on the definition of the necessary professional competencies of personnel in the enterprise activities. In modern conditions, the personnel management system is changing, its non-productive functions are excluded, and new professional competencies are being created [3]. The development of these factors ensures the flexibility of production, which creates competitive advantages of the enterprise [5, 14]. This fact determines the necessity and relevance of a research aimed at forming strategic directions for the oil and gas industry development in digitalization conditions.

S. I. Ashmarina and V. V. Mantulenko (Eds.): IPM 2020, LNNS 161, pp. 538–543, 2021.
https://doi.org/10.1007/978-3-030-60926-9_68

Digitalization creates favorable conditions for the development of innovative technologies, effective management decisions based on a complex analysis in a short time, and optimization of production resources [12]. Rapid implementation of initiatives and flexible management of their implementation increases the competitiveness level of the enterprise and the degree of its influence on the relevant market. The relevance of this scientific problem is determined by digital tools used in the production management, which should belong functionally to specialists with appropriate professional competencies [9]. Automated work processes and their transition to the electronic format frees up employees' time. The released time resource of employees should be directed to more important and interesting tasks. This implies creating new professions and losing old ones.

2 Methodology

The following elements are priorities of a HR strategy:

- retraining of personnel,
- digital transformation technologies,
- development of the information society,
- labor functions (professional standard),
- application of digital technologies.

Various selection procedures are carried out with candidates for forming the HR reserve in order to identify their managerial potential and then complete training programs. The selection should be made in two stages: The first stage is pre-selection, which determines the candidate's compliance with requirements necessary for enrollment in the personnel reserve. The second stage is the main selection, which aims to assess professional and business qualities of reservists and their compliance with the job profile for each of the positions. Examples of assessed business qualities include: organization and work planning skills; ability to analyze information and make management decisions; leadership qualities and desire to achieve results; responsibility and openness to new things; and readiness to continuous development.

Modern authors suggest using assessment centers, performance analysis, case testing, competency interviews, professional and personal testing as assessment methods. The source of additional information is the employee's expert assessment using the 360-degree method [4].

As a result of the second stage, the final list of candidates for enrollment in the personnel reserve is formed. Methods of selecting candidates for the reserve are divided into socio-psychological, practical and scientific ones. Socio-psychological methods include examination of personal files and other documents; interviews with candidates; reviews of the employee by managers, colleagues and subordinates; psychological testing and expert evaluation. Practical tools include temporary replacement of the head for the period of vacation, illness, business trips; training at advanced enterprises; understudy (deputy head); appointment as the head of the team performing a temporary task. Training methods contains business games, analysis of specific situations, trainings (group seminars), business design.

The following methods are most widely used:

- biographical method – study of employees' personal files,
- interview – obtaining of oral and written characteristics of the workers,
- expertise – summarizing opinions of independent experts,
- psychological tests of employees' self-esteem.

When selecting candidates, it should be taken into account that while they are in reserve, candidates are preparing themselves for further managerial work, and they are working on the own professional and personal development.

3 Results

In the context of digitalization of the oil and gas industry, the personnel strategy will change, since the basis of digitalization is the relation between people and technology, an important role is assigned to the personnel reserve. Specialists with new competencies develop the digital environment and contribute to the digital transformation of companies. Accordingly, it is necessary to train qualified IT specialists and ensure that the digital literacy of all employees of the enterprise is improved. Therefore, the HR strategy necessarily includes the company's investment in training (targeted training), retraining of personnel (advanced training) to obtain appropriate professional competencies [10]. It is obvious that companies need people who have digital competencies.

The state is a regulator in the development of the digital economy, providing a favorable environment in the legal, social and economic sphere [2]. The formation of a single industry order for training personnel in the fuel and energy complex (FEC) will help to exchange information regarding personnel training programs. The state has identified particularly important areas for implementing the information society development strategy, which are:

- creation of domestic digital platforms,
- application of data transmission networks of the fifth generation,
- development of a legal framework regulating the use of new technologies.

According to the KPMG study, one of the world's largest professional services networks, six from ten industrial enterprises in the world have already developed a digital transformation program [8]. At the same time, a quarter of enterprises have a program horizon of less than 12 months, while the majority (61%) plans to implement the existing program in one to three years. However, these indicators, both in the world and in Russia, largely reflect the development level of the largest enterprises – the industry leaders.

According to KPMG experts, modern technologies help to reduce operating costs and increase productivity in the oil and gas field by several times [8]. The most popular model is the smart deposit. In 2020, the global market for robots and artificial intelligence will reach US $152.7 billion, which can increase the productivity of companies by 30%. The oil and gas industry needs to strengthen the impact of digitalization and robotics on its business processes for the companies' efficiency in unstable price conditions on the hydrocarbon market.

In accordance with the professional standard “Drilling supervisor in the oil and gas industry” approved by the order of Ministry of Labor and Social Protection of the Russian Federation of 27 November 2014 No. 942n., the job function of “Coordination and management of drilling and service contractors on the drilling site” should be provided through the following labor steps [13]:

– control of the organization of the drilling process in accordance with the contractual obligations between the customer and the drilling and service contractors,
– monitoring the work progress, coordinating the contractors’ actions when adjusting the production process,
– coordination of operational decisions and shift tasks,
– assessment and confirmation of the completed work volume of contractors.

The effectiveness of these processes depends on tools that ensure their rapid implementation. The production process of drilling wells is a cycle of the following works: drilling and installation works, well deepening, fixing, development and testing. The specialist who manages these activities is called a supervisor. This profession is relatively young in the oil industry. However, the efficiency of oil and gas drilling depends not only on operational decision-making in the production process based on real field data, but also on the infrastructure required for digital transmission of information. Drilling supervisors at the well process a huge amount of information and make informed decisions under unforeseen circumstances. The drilling supervisor maintains constant contact with a project manager, an engineer, and a coordinator to minimize risks and ensure efficient and safe operations.

The new role of universities in the digital economy has a great impact on the formation and implementation of personnel policy, which allows getting significant benefits for all participants, requires understanding all the interests and strengthening the interaction between universities, business and government [7].

To ensure the smooth operation of the production process, it is necessary to provide a personnel reserve, the priority goal of which is to create specialists prepared for management in the new conditions, ensure the management continuity, its improvement through the selection, training and promotion of personnel who can effectively and professionally implement goals and objectives of the relevant structural unit. The efficiency of using the personnel reserve enables:

– high-quality selection and targeted training of candidates,
– reducing the adaptation period of specialists newly appointed to managerial positions,
– increasing professionalism and improving the quality of specialists’ competences.

The availability of a personnel reserve allows preparing candidates for vacant positions on a planned basis, organizing training and internships of specialists included in the reserve, and effectively using them in various areas and at different levels of the business management system. The feasibility of creating a personnel reserve is determined by the following tasks:

– to achieve the company’s strategic goals,
– to increase the readiness of the company’s personnel to organizational changes,

- to ensure the continuity in management,
- to increase the employees' motivation at enterprises,
- to improve the financial situation of the company.

There are the following stages of forming a HR reserve:

- preliminary assessment of candidates for their enrollment in the reserve,
- selection of candidates for the personnel reserve,
- transfer to the reserve,
- training of reservists,
- assessment of the readiness of the personnel reserve.

All these stages are important for the implementation of the company's HR strategy, so HR management departments develop individual training plans for reservists, which are interrelated with the staff career plans.

4 Discussion

Information and digital technologies make it possible to reconstruct our present, adapt to completely new communications, form new competencies and prepare for the successful realization of the employees' potential in new economic conditions [6]. The formation of strategic directions for the development of the oil and gas industry in the economy digitalization conditions is provided by the selected priorities of HR policy. One of the priorities is digitalization itself. Further digitalization in the oil and gas sector should lead not only to an adjusted regulatory framework, but also to the creation of prerequisites for the application of IT in business processes and formed digital competencies by workers. The methods of personnel selection presented in this article contribute to the formation of trained specialists, in uncertain environment, to ensure continuity of management. Conducting annual monitoring of all the results achieved by reservists, and making changes to the planned activities, will ensure the efficiency of the enterprise as a whole.

5 Conclusion

The application of digital technologies in modern conditions is an important direction in the development of the oil and gas industry. Implementation of production tasks that ensure the productivity growth, conducting a comprehensive analysis of a large data bank, reducing the time for making management decisions, preparing and structuring report data on the performance of tasks and quickly responding to unforeseen conditions that arise in the production process is possible with the appropriate qualified personnel and communication tools.

The inability to realize the qualification growth of personnel within the enterprise may contribute to specialists' mobility, as a result all the financial resources spent on training and re-training of the employee and organizational efforts may be in vain. Work with the personnel reserve should be purposeful and systematic, providing the company with the necessary human resources. The implementation of the set tasks is provided by new or additional professional competencies of employees, so business managers should invest in retraining or upgrading the professional skills to acquire new knowledge, skills and abilities that contribute to the efficient application of digital technologies and the possibility of flexible entry and exit from projects in the context of risky investments in new technologies. The presented research is important for the process of forming a personnel strategy, which defines a system of personnel evaluation and allows determining the actual situation in the qualification field, makes appropriate personnel decisions to perform strategic and tactical tasks in the oil and gas industry.

References

1. Alhanshi, M., Albraiki, H.: Knowledge sharing and employee development in oil and gas companies in the United Arab Emirates. In: Uden, L., Heričko, M., Ting, I.H. (eds.) Knowledge Management in Organizations. Lecture Notes in Business Information Processing, vol. 224, pp. 190–208. Springer, Cham (2015)
2. Babkina, A.V.: Shaping the Digital Economy and Industry: New Challenges. Polytechnic University Press, Saint Petersburg (2018)
3. Borovskikh, N.V., Kipervar, E.A.: Personnel policy of the enterprise: Prospects of formation in the conditions of digitalization of the economy. Bull. Belgorod Univ. Cooper. Econ. Law **4**(77), 223–233 (2019)
4. Egorshin, A.P.: Basics of Personnel Management. Infra-M, Moscow (2020)
5. Grant, R.M.: The development of knowledge management in the oil and gas industry. Universia Bus. Rev. **40**, 92–125 (2013)
6. Grigoriev, S.G., Lukin, V.V., Lukin, D.V.: Human capital in the conditions of digitalization. E-Manage. **2**, 13–19 (2018)
7. Khmeleva, G.A., Agaeva, L.K., Chirkunova, E.K., Koroleva, E.N., Domnina, S.V., Kasatov, A.D.: Drivers for the innovative development of Russian regions: economic sanctions, human capital, investments in RandD. Mod. J. Lang. Teach. Meth. **8**(5), 666–685 (2018)
8. KPMG: KPMG's experience working with companies in the oil and gas sector (2018). https://assets.kpmg/content/dam/kpmg/ru/pdf/2018/03/ru-ru-oil-and-gas-2018.pdf. Accessed 01 June 2020
9. Kruglov, D.V., Kruglova, O.D.: Particularities of staffing in the context of digitalization. Leadersh. Manage. **6**(4), 479–486 (2019)
10. Leavitt, P., Raybourn, C., Hubert, C.: Applying knowledge management to oil and gas industry challenges. American Productivity and Quality Center (2002). http://www.providersedge.com/docs/km_articles/Applying_KM_to_Oil_and_Gas_Industry_Challenges.pdf. Accessed 27 June 2020
11. Milner, B.Z., Evenko, L.I., Rappoport, V.S.: System approach to management organization. Economy, Moscow (1983)
12. National program. Digital economy of the Russian Federation (2019). https://digital.gov.ru/ru/activity/directions/858/. Accessed 01 June 2020
13. Order of the Ministry of Labor and Social Protection of the Russian Federation of November 27, 2014 N 942n. On approval of the professional standard" Drilling supervisor in the oil and gas industry (2014). http://base.garant.ru/70836306/#ixzz6RUK7n9y7. Accessed 01 June 2020
14. Ouchi, W.G.: Theory Z: How American Business Can Meet the Japanese Challenge. Avon Books, New York (1993)
15. Peters, T.J., Waterman, R.H.: In Search of Excellence: Lessons from America's Best Run Companies. Harper and Row, New York (1982)

Conceptual Aspects of Strategic Human Resources Management in the Context of Digitalization

L. F. Berdnikova[1(✉)], N. A. Igoshina[2], and E. A. Gerasimova[3]

[1] Togliatti State University, Togliatti, Russia
bleylaf@mail.ru
[2] Samara State University of Economics, Samara, Russia
bume63@mail.ru
[3] Samara State Transport University, Samara, Russia
gerasil960@mail.ru

Abstract. In the context of business digitalization and fierce competition, the search for modern and progressive ways to manage human resources is of particular importance. With the development of technology, the digital economy, telecommunications, the most profitable are investments in the development of human resources. The main goal of this article is to develop the conceptual aspects of strategic human resources management in the context of business digitalization. The tasks of the work include the refinement of the conceptual apparatus in the field of human resources management of an organization, the identification of key factors affecting the development of human resources in the context of digitalization, as well as the refinement of conceptual elements of strategic human resources management in an organization. In the research process, general scientific methods of cognition were used: analysis, synthesis, comparison, grouping, dialectic and integrated approaches that allow the formation of reasonable conclusions. The result of the study is the development of an algorithm for strategic human resource management taking into account the requirements of the digitalization of business. The results can be used in the activities of commercial organizations.

Keywords: Digitalization · Human potential · Human resource assessment · Human resources management · Motivation

1 Introduction

Digitalization of business, development of innovation, the emergence of new technologies, high competition entails the need to improve mechanisms for managing human resources. In the activities of each organization, the most valuable resource is human resources. For the growth of the innovative potential of the company, its competitiveness, investments are needed in the development of human resources, that is, in raising the level of education, qualifications, and obtaining professional skills of employees. Success in the activities of organizations is closely related to the implementation and implementation of highly effective ways of managing human resources.

S. I. Ashmarina and V. V. Mantulenko (Eds.): IPM 2020, LNNS 161, pp. 544–551, 2021.
https://doi.org/10.1007/978-3-030-60926-9_69

They are based on a strategic approach to the use and development of human potential, which justifies the relevance of the research topic.

The modern theory of human resource management redirects economic science from studying issues from their use to researching the issues of creating a qualitatively new approach to the formation of the organization's human potential in the context of scientific and technological progress and digitalization of business. The staff is the most important and difficult to manage of all resources. The reason for this is the individual abilities, characters, psychotypes of workers, which makes it difficult to evaluate them. Thus, to increase the internal capabilities, competitiveness, financial results of the company in the digital economy, improved methods of strategic human resource management are needed.

2 Methodology

The concept of human resource management is based on the recognition of the economic feasibility of investing in staff. This is necessary to achieve the organization's strategy and increase the financial results of its activities. This approach is focused on the fuller use of knowledge, skills, and abilities of staff, the identification and development of "hidden" employee opportunities. The conditions of the digital economy require the formation of new approaches to the selection, evaluation, motivation, stimulation and use of human resources.

In modern professional literature, the work of scientists such as: Bazarov [1], Berdnikova et al. [2], Catmull and Wallace [5], Melikhov and Maluev [8], Rasskazov, Rasskazova, and Deryugin [10]. Currently, significant attention is paid to publications on assessing the effectiveness of Bibarsov, Khokholova, and Okladnikova [3], Bruskin et al. [4], Dudin et al. [6], Glukhova, Syrotyuk, Sherstobitova, and Pavlova [7], Rampersad and Hussain [9].

In the process of implementing this article, materials from the scientific literature and periodicals devoted to the research topic were used. During the study, the main problems in the field of strategic human resources management in the practical activities of commercial organizations were studied. The main factors that positively and negatively affect the development and strategic management of human resources in the context of digitalization were identified. During the writing of the article, general scientific methods of cognition were used: analysis, synthesis, comparison, grouping, graphic, induction and deduction, the dialectical method of cognition, systemic and integrated approaches that allow the formation of reasonable conclusions.

3 Results

Currently, in the study of problems related to human resource management, various terms are used, for example: personnel, labor potential, labor resources, labor force, human potential, human capital, personnel. The current stage of economic development brings new technologies (digital technologies, blockchain, artificial intelligence, P2P networks and others), expands the scope of business, transforms institutional relations.

The spread of the digital economy in all areas of activity requires the improvement of methods of strategic human resource management, determining the role of the employee in the formation of key performance indicators of the organization.

In our opinion, in the era of business digitalization, the concept of "human resources" includes not only the ability of personnel to perform certain functions, their knowledge, experience, high qualifications, but also their potential to bring economic benefits to organizations in the context of telecommunications development, information technology, digital format of business interaction. We believe that the strategic management of human resources should be aimed at synchronizing the processes of management and development of human resources with the mission and strategic goals of the organization. Adjustment of personnel management methods in the company should be carried out taking into account the achievement of planned indicators, its goals and objectives, as well as changes in the external environment.

The current unstable state of the environment requires the organization of a mobile response to its various factors. From the highly qualified personnel, its ability to predict the appearance and determine the degree of influence of these factors on the activities of the company, to make prompt informed decisions depends on the efficiency of the business entity. In this regard, the concept of the enterprise itself is radically changing. If earlier its main task was the production of products, now it is the production of knowledge, the development of know-how, new technologies. Organizations set goals to maximize the use of the full range of knowledge, skills and abilities of their own personnel to obtain the greatest economic effect. Currently, the employee is not just an executor, but represents a strategic resource of the organization, the basis of its competitiveness and is the key to successful work in the future. However, in order to formulate such a strategic resource as a human within a specific organization, one should take into account not only the priorities of consumers, but also the needs and dignities of employees. With ease, changing one employee to another, not taking into account his individual abilities, the company risks losing not only valuable personnel, but also subsequent success in the market. This is due to the fact that both temporary losses are necessary that are necessary for adaptation of the newly arrived employee in the organization, as well as financial losses aimed at training or retraining him. Thus, there is a high degree of dependence of the company on its personnel.

The development of human resources in the context of digitalization is influenced by various factors that must be considered when choosing management methods. Factors affecting the development of human resources in the context of digitalization are presented in Fig. 1. For the purposes of strategic management of human resources, it is necessary to timely diagnose factors that negatively affect their development, identify their causes and eliminate them. In turn, special attention should be paid to factors that favorably affect the development of personnel and to constantly improve them. Strategic management of human resources is a set of measures that takes into account the mission and strategy of the company. We clarify the conceptual elements of strategic human resource management in the context of digitalization in Fig. 2.

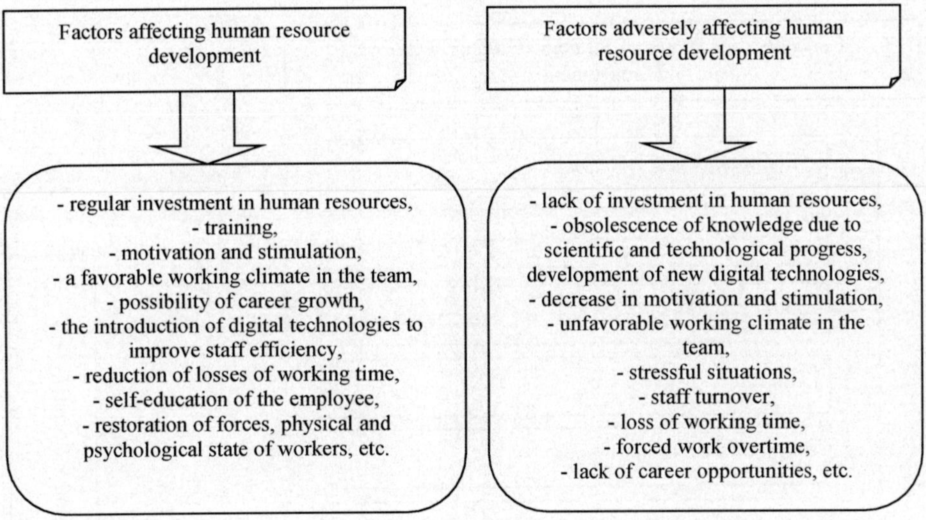

Fig. 1. Factors affecting the development of human resources in the context of digitalization (Source: authors).

In our opinion, the conceptual elements of strategic human resources management in the context of digitalization should include: personnel strategy and personnel policy, staff selection and training, incentives and motivation, assessment of personal performance indicators of personnel, assessment of the contribution of personnel to the effectiveness of an organization, assessment of compliance of personnel qualifications with strategic the goals and objectives of the organization, the implementation of the CRM system, digital individual control over the activities of staff and their job descriptions. All elements of strategic human resources management must be consistent with the mission and strategy of the organization. The algorithm for strategic human resources management taking into account the requirements of the digitalization of business is presented in Fig. 3.

The proposed strategic strategy for human resources management takes into account the demands of the external environment in the context of digitalization and the strategic goals of the organization. Moreover, special attention is given to improving the system of incentives and motivation. In the developed algorithm, an important place is taken by the assessment of personnel efficiency, individual control over its activities and the assessment of compliance of its qualifications with the strategic goals and objectives of the company.

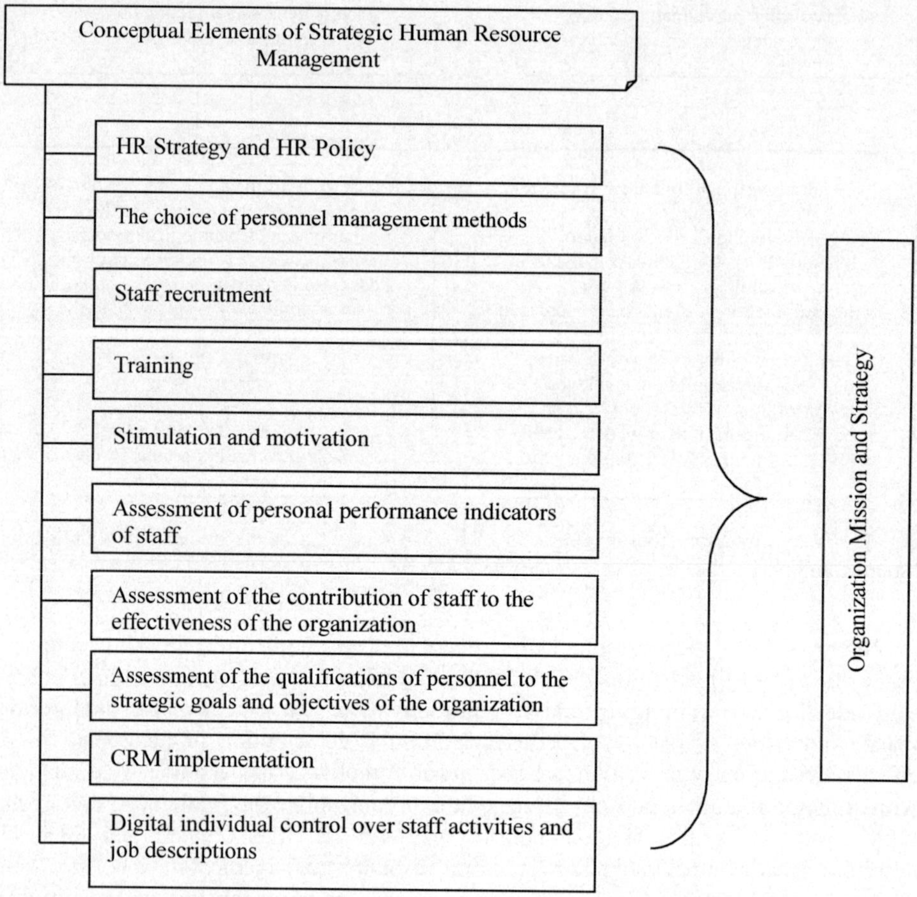

Fig. 2. Conceptual elements of strategic human resources management in the context of digitalization (Source: authors).

4 Discussion

In the process of management, it is necessary to remember that a person's abilities are not limited only to abilities to work. A person accumulates a wide range of various qualities, aspirations and interests. In the context of digitalization, the effectiveness of the organization depends not only on the high qualifications and professionalism of the staff, but also taking into account all the various qualitative characteristics of a person, his individual characteristics and abilities, providing the opportunity for personal self-expression. The study showed that a high increase in the financial results of the company is achieved when employees are given the opportunity to realize their potential, manifest their creative personality, and interest in their own effectiveness. In addition, a lot depends on the social well-being of people, their attitude to work, to each other in the process of joint work, relationships among the team, between performers and managers.

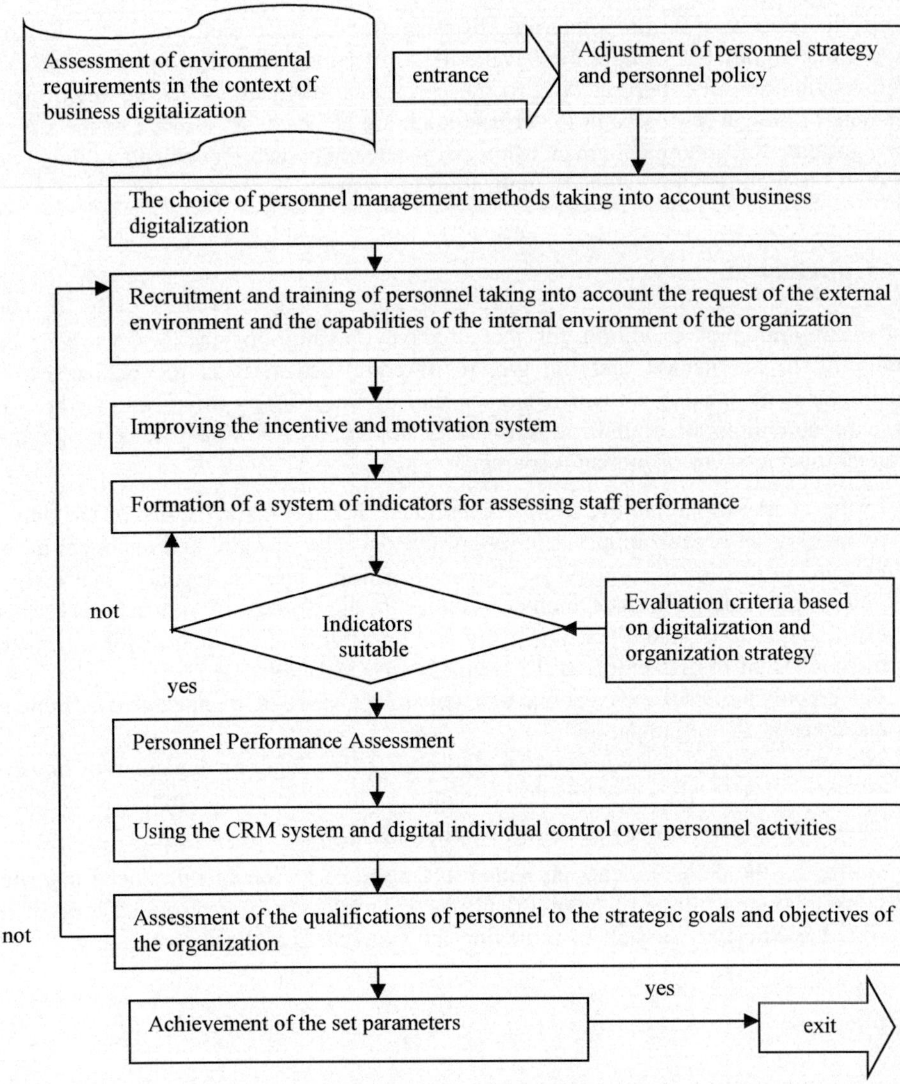

Fig. 3. Algorithm of strategic HR management (Source: authors).

As a result of studying the human resource management system in commercial organizations of medium and small businesses, it was revealed that it does not take into account the rapidly changing environmental conditions and the demands of the digital economy. In addition, human resources management methods do not take into account the employee's personal effectiveness and his contribution to achieving the company's overall strategy.

In the process of a comprehensive discussion of the identified problems, the key factors that significantly affect the development of human resources in the context of digitalization are identified, as well as the conceptual elements of the strategic management of human resources in the organization are clarified. As a result of the study, an algorithm for strategic human resource management was developed taking into account the requirements of the digitalization of business.

5 Conclusion

The most important condition for the effective operation of the organization, its credibility in the market and the growth of competitiveness is the presence and development of existing human resources that accumulate corporate knowledge. In turn, the conditions of digitalization of the business form new requirements for the strategic management of human resources.

1. In the course of the study, it was determined that the main factors in the effectiveness of an organization's activity are not only the material and technical base, but also professionally trained people who form that specific resource that distinguishes one enterprise from another. As a result, the concept of human resources is clarified and key factors that positively and negatively affect their development in the context of digitalization of the business are identified.
2. The article clarifies the conceptual elements of the strategic management of human resources in the organization.
3. The study allowed us to develop an algorithm for strategic management of human resources, taking into account the demands of the external environment in digitalization and the strategic goals of the organization.

Currently, the developments presented in the article are being introduced into the activities of organizations of medium and small businesses. The proposed tools will improve the efficiency of staff in achieving the strategic goals of the company.

References

1. Bazarov, T.Yu.: Personnel Management. Academy, Moscow (2017)
2. Berdnikova, L.F., Mikhalenok, N.O., Frolova, V.A., Sukhacheva, V.V., Krivtsov, A.I.: Human resource management system development at smart university. In: Uskov, V., Howlett, R., Jain, L. (eds.) Smart Education and e-Learning 2020. Smart Innovation, Systems and Technologies, vol. 188, pp. 327–337. Springer, Singapore (2020). https://doi.org/10.1007/978-981-15-5584-8_28
3. Bibarsov, K.R., Khokholova, G.I., Okladnikova, D.R.: Conceptual basics and mechanism of innovation project management. Eur. Res. Stud. J. **20**(2B), 224–235 (2017)
4. Bruskin, S.N., Brezhneva, A.N., Dyakonova, L.P., Kitova, O.F., Savinova, V.M., Danko, T.P., Sekerin, V.D.: Business performance management models based on the digital corporation's paradigm. Eur. Res. Stud. J. **20**(4A), 264–274 (2017). https://doi.org/10.35808/ersj/833
5. Catmull, E., Wallace, A.: Creativity Inc.: Overcoming the unseen forces that stand in the way of true inspiration. Random House, New York (2014)

6. Dudin, M.N., Frolova, E.E., Lubenets, N.A., Sekerin, V.D., Bank, S.V., Gorohova, A.E.: Methodology of analysis and assessment of risks of the operation and development of industrial enterprises. Qual. Access Success **17**(153), 53–59 (2016)
7. Glukhova, L.V., Syrotyuk, S.D., Sherstobitova, A.A., Pavlova, S.V.: Smart university development evaluation models. In: Uskov, V., Howlett, R., Jain, L. (eds.) Smart Education and e-Learning 2019. Smart Innovation, Systems and Technologies, vol. 144, pp. 539–549. Springer, Singapore (2019). https://doi.org/10.1007/978-981-13-8260-4_47
8. Melikhov, Yu.E, Maluev, P.A.: Personnel management. Portfolio of reliable technologies. Dashkov and Co., Moscow (2018)
9. Rampersad, H., Hussain, S.: Corporate balanced scorecard. In: Authentic Governance. Management for Professionals, chap. 8, pp. 65–78. Springer, Cham (2014). https://doi.org/10.1007/978-3-319-03113-2_8
10. Rasskazov, S., Rasskazova, A., Deryugin, P.: Corporate Governance. Infra-M, Moscow (2020)

Crimes in Financial Markets in Russia and Great Britain

S. P. Bortnikov[1(✉)] and A. V. Denisova[2]

[1] Samara State University of Economics, Samara, Russia
serg-bortnikov@yandex.ru
[2] University of Prosecutor's Office of the Russian Federation, Moscow, Russia
anden2012@yandex.ru

Abstract. Crimes in financial markets (not related to banking) are economic crimes committed by participants in the securities market, insurance, microfinance, investment and leasing, money and currency exchange markets. The authors analyzed Russian and British legislation to compare criminal law on crimes of financial markets and determine the possibilities of using foreign experience in domestic rule-making practice. Methods: The methodological framework of this study constitutes a set of methods of scientific knowledge, among which the main place is occupied by the methods of systematic, analysis and comparative law. Results: The authors' position is based on national and foreign legislation and opinions of the competent scientific community on crime prevention in financial markets. The authors describe similarities and differences between Russian and British financial and criminal law. The study revealed the gaps in the Russian criminal law regarding the socially dangerous acts committed in the investment sphere. Conclusions: The study shows the importance of comparative legal analysis for national lawmaking work and law-enforcement. The authors proposed to criminalize some socially dangerous acts committed in the financial sphere with the use of modern information technologies, considering the successful experience of the United Kingdom of Great Britain and Northern Ireland in countering of cybercrimes.

Keywords: Crime · Criminal Code of the Russian Federation · Criminal Justice Act 1993 · Financial Markets · Financial Services and Markets Act 2000 · Fraud Act 2006

1 Introduction

Crimes committed in the sphere of financial markets not related to banking are a special type of economic crimes. This category of crimes includes securities, insurance, microfinance, investment and leasing markets, money and currency markets. Despite the fact that these types of activities are quite "young" in Russia, the Criminal Code of the Russian Federation [6] contains a significant number of articles providing criminal liability for socially dangerous acts committed in this sphere.

However, these crimes are rare in Russian law enforcement because of the complexity of legislative structures, ambiguity in the interpretation of their characteristics and the problems of distinguishing them from similar offenses and civil torts. To improve

S. I. Ashmarina and V. V. Mantulenko (Eds.): IPM 2020, LNNS 161, pp. 552–557, 2021.
https://doi.org/10.1007/978-3-030-60926-9_70

domestic law-making practice on criminalization of acts in the financial markets it is advisable to study the foreign experience of establishing responsibility for these crimes, especially those countries that are generally recognized as the largest international financial centers.

According to the ratings of foreign analytical companies among the world's leading financial centers, the top three traditionally include New York, London and Hong Kong [10]. Given the fact, the United Kingdom of Great Britain and Northern Ireland is characterized by a high level joint culture, security and rights of shareholders, a significant percentage of shares in free float [17], and the London stock exchange is the oldest one in the world, it is not surprising that London has the status of a global international financial center and the business community recognizes the UK financial-criminal law fair and adequate.

2 Methodology

The methodological framework of this study constitutes a set of methods of scientific knowledge, among which the main place is occupied by the methods of systematic, analysis and comparative law. The research is based on national and foreign legislation and opinions of the competent scientific community on the interaction of economic and legal systems, processes and the results of their mutual influence. The authors conducted comparative research to know whether and how the crimes of financial markets are regulated differently in Russia and the United Kingdom of Great Britain and Northern Ireland.

Comparative law is a method of study of various legal phenomena. Thanks to the application of the comparative method, it has become possible to reveal the general and the special in modern world legal systems. Among adherents of this stream are such scientists as David and Brierley [8], De Cruz [9], Gutteridge [15] and Kahn-Freund [16]. The process of comparison does not in itself create legal norms; it only contributes to the creation of legal norms within the framework of one or more legal systems.

In this research the comparative method is used as the tool for collecting information on compared Russian and British systems and some legal phenomena. The authors carried out the comparative investigation in criminal law, compared institutes of crimes committed in the sphere of financial markets. The study contributes to the creation of new legal norms in Russian criminal law.

3 Results

According the Criminal Code of the Russian Federation [6] crimes of financial markets (not related to banking) include: art. 159.5 "Fraud in the insurance sphere", art. 170.1 "Falsification of the unified state register of entities, the register of securities owners or the Depository accounting system", art. 172.2 "Organization of attracting funds and (or) other property", art. 185 "Abuse in the issuance of securities", art. 185.1 "Malicious evasion from disclosure or provision of information determined by the legislation of the Russian Federation on securities", art. 185.2 "Violation of the procedure for recording rights to securities", art. 185.3 "Market manipulation", art.185.4 "Obstruction of realization or unlawful restriction of securities' holders rights", art. 185.6 "Unlawful use of insider information" and art. 186 "Production, storage, transportation or sell of counterfeit money or securities".

In addition, some crimes can be recognized as related to the sphere of financial markets if they are committed in the specified sphere of public relations (so-called "conditional" crimes in financial markets). These are crimes such as fraud (article 159), credit fraud (article 159.1), embezzlement (article 160), illegal business (article 171), falsification of financial documents of accounting and reporting of a financial organization (article 172.1), legalization (laundering) of money or other property acquired by other persons through criminal means (article 174), legalization (laundering) of money or other property acquired by a person as a result of committing a crime (article 174.1), malicious evasion of payment of accounts (article 177) and illegal actions in bankruptcy (article 195of the Criminal Code of the Russian Federation [6]).

The financial system of the United Kingdom is securities' market-based type [18]. A significant number of legislative acts (including financial and criminal acts) are devoted to the protection of the rights of financial markets' participants. These include Criminal Justice Act 1993 [7], Financial Services and Markets Act 2000 [12], Fraud Act 2006 [13], the Financial Services Act 2012 [11], Bank of England and Financial Services Act 2016 [2] and others.

For example, part V of the Criminal Justice Act of 1993 [7] describes criminal insider dealing. There are three types of criminal actions in a regulated financial market: first, an insider makes a deal with securities as a result of using information, the disclosure of which would have a significant impact on the value of these securities. Second, an insider induces another person to make a deal with such securities, if there are reasons to believe that this will happen. Third, an insider discloses the information to a third party, except for information about the proper performance of his official duties. Whoever commits this criminal acts shall be punished with imprisonment for a term which may extend to 7 years, or with fine, or with both. The authors pay attention to the severity of the sanctions of the United Kingdom' laws. There is not required to prove any major damage as a result of the insider's act or the extraction of large income or the fact of avoiding large losses for criminal responsibility in the UK. At the same time these signs are mandatory for "Illegal use of insider information" (article 185.6 of the Criminal Code of the Russian Federation). Despite the presence of these additional features, the insider's activity is punished significantly more leniently under Russian criminal law: under part 1 of article 185.6 - from two to four years of imprisonment with an additional punishment or without it; under part 2 of article 185.6 - from two to six years of imprisonment with an additional punishment or without it.

There are the same but more detailed rules on insider trading in the Financial Services and Markets Act 2000 [12]. This act provides for civil and criminal liability for various types of abuses in the financial market, which include "insider trading", "disclosure of internal information", "dissemination of false and misleading information", "use of fictitious devices", "making a false market" (article 397). The British legislator, unlike the Russian one, prefers to construct formal corpus delicti in criminal law without the references to the possible or mandatory consequences of the criminal acts [1].

This act criminalizes misleading statements and practices that means misleading financial market regulators (regulatory bodies) by persons providing financial services, in case of willful or careless provision of false information in reporting materials (article 398). A person guilty of that offence is liable on conviction on indictment, to a fine. Section 7 of the Financial Services Act 2012 [11] is devoted to offences in the sphere of financial services (Articles 89–95). They include misleading statements, misleading impressions, misleading statements etc. in relation to benchmarks, i.e. indicators used in the financial market to assess the state of the market as a whole or its segments. We can see the same method of constructing corpus delicti – the formal type and the same punishment – imprisonment for a term which may extend to 7 years, or with fine, or with both.

Fraud Act 2006 [13] contains rules about fraud by abuse of one's position; meaning taking advantage of a position where one is expected to safeguard another's financial interests or at least not act against them, but acting dishonestly with the intent to make a profit or cause a loss (Sect. 4). At the same time, the British lawmaker specifically stipulates that this offence can be committed by a positive action or by omission. The crime is over when offender abuses his position, regardless of the achievement of criminal goals. The presence and extent of criminal consequences are essential circumstances for the imposition of criminal penalties and for resolving issues of compensation and confiscation [4]. Section 12 of the Fraud Law provides the rule according to which managers of the corporation are liable for corporate fraud along with the corporation, if it is proved that due to their consent or connivance this offence was committed.

4 Discussion

Analyzing the British experience of countering crimes of financial markets, we note, nowadays the main danger for financial system of the United Kingdom is the illegal trafficking of crypto currencies [3]. Digital assets are an attempt to rebuild financial systems without using traditional financial institutions, they allow for the digital transfer values without a financial intermediary. Therefore, applying traditional strategies to combat financial crimes is not so effective in cases with crypto currency [14]. The activities of the Serious Fraud Office (SFO) and the Financial Conduct Authority (FCA) are aimed at applying the same anti-money laundering (AML) standards for companies providing traditional financial services and for individuals involved in the turnover of crypto currencies. From 10 January 2020, existing businesses carrying on cryptoasset activity in the UK have needed to be compliant with the Money Laundering, Terrorist Financing and Transfer of Funds (Information on the Payer) Regulations 2017, as amended and need to register with the FCA [19]. A new business must be register with the FCA before carrying out any activity. If businessmen are not registered with the FCA on 10 January 2021 they have to cease trading. The FCA will proactively supervise firms' compliance with the new regulations, and will take swift action where firms fall short of desired standards. This approach provides the right balance between the development of high-tech innovations and the elimination of increased risks of financial crimes in the digital economy.

The Russian government also sees a serious danger in the illegal use of digital financial assets, crypto currency and digital rights. In the near future, the State Duma of Russian Federation proposes to criminalize the following acts: organizing illegal trafficking of digital financial assets and digital currencies, making transactions with them, including with foreign digital financial assets and digital rights, providing services for the issue of digital rights and digital currency using sites registered in Russia, or technical tools located in Russia; violation of the rules for making transactions with digital financial assets, digital currencies and rights, if they are used as payment for goods, works or services; purchase of digital currency in Russia for cash or by transferring funds to accounts opened in Russian banks; issue of digital rights using Russian websites and/or user equipment located in Russia, if this infrastructure is used for transferring digital currency to third persons. These acts will be qualified as crimes if they caused major losses to citizens, organizations or the state, or resulted in large income. The offender must pay a fine from 500 thousand to 1 million rubles or in the amount of his income from one year to two years or get imprisonment for a term which may extend to four years with a fine. If offender causes extra large losses or gets extra large income he must be punished by imprisonment for up to seven years with a fine of up to 1 million rubles or in the amount of his income for a period of up to five years [5].

5 Conclusion

The conducted research allows us to draw a conclusion about the diversity of national financial systems and the polymorphism of their regulating legislation. Due to analysis the British experience in countering crimes in financial markets we can predict some promising directions for the development of Russian criminal law norms in this sphere. Thus, it is very likely that domestic law-making and law enforcement practice accept the ideas of criminalization and prevention fraud in the investment sphere, including in cyberspace, theft of personal data and their misuse and other preliminary actions for serious financial crimes.

References

1. Ahern, K.R.: Information networks: evidence from illegal insider trading tips. J. Financ. Econ. **1**(125), 26–47 (2017)
2. Bank of England and Financial Services Act (2016). http://www.legislation.gov.uk/ukpga/2016/14/contents/enacted. Accessed 20 June 2020
3. Chambers, T.: Unstable coins: cryptoassets, financial regulation and preventing financial crime in the emerging market for digital assets (2020). https://www.fca.org.uk/news/speeches/unstable-coins. Accessed 20 June 2020
4. Chen, Y., Kai, D.: The role of information disclosure in financial intermediation with investment risk. J. Finan. Stab. **46**, 1–8 (2020)

5. Chernyshova, E.: The State Duma proposes to imprison for illegal trafficking of crypto assets (2020). https://www.rbc.ru/finances/21/05/2020/5ec54f6a9a7947448e60319a. Accessed 20 June 2020
6. Criminal Code of the Russian Federation from 13.06.1996 N63-FZ (1996). http://www.consultant.ru/document/cons_doc_LAW_10699/. Accessed 20 June 2020
7. Criminal Justice Act 1993 (1993). https://www.legislation.gov.uk/ukpga/1993/36/contents. Accessed 20 June 2020
8. David, R., Brierley, J.E.C.: Major Legal Systems in the World Today. The Free Press, New York (1978)
9. De Cruz, P.: Comparative law in a changing world. Cavendish Publishing Limited, London/Sydney (1999)
10. Financial Center Futures: The global financial centers index 25 (2019). https://www.zyen.com/media/documents/GFCI_25_Report.pdf. Accessed 20 June 2020
11. Financial Services Act 2012 (2012). http://www.legislation.gov.uk/ukpga/2012/21/contents/enacted. Accessed 20 June 2020
12. Financial Services and Markets Act 2000 (2000). http://www.legislation.gov.uk/ukpga/2000/8/contents. Accessed 20 June 2020
13. Fraud Act 2006 (2006). http://www.legislation.gov.uk/ukpga/2006/35/section/13. Accessed 20 June 2020
14. Frunza, M.-C.: Solving Modern Crime in Financial Markets. Academic Press, Cambridge (2016)
15. Gutteridge, H.: Comparative Law: An Introduction to the Comparative Method of Legal Study and Research. Cambridge University Press, London (1949)
16. Kahn Freund, O.: Comparative law as an academic subject. Law Quart. Rev. **82**, 40–41 (1966)
17. Li, X., Walker, T.J.: The determinants of IPO-related shareholder litigation: the role of CEO equity incentives and corporate governance. J. Finan. Mark. **31**, 81–126 (2016)
18. Palan, S., Stockl, T.: When chasing the offender hurts the victim: the case of insider legislation. J. Finan. Mark. **35**, 104–129 (2017)
19. The money laundering and terrorist financing (amendment) regulations 2019 № 1511 (2019). http://www.legislation.gov.uk/uksi/2019/1511/introduction/made. Accessed 20 June 2020

Functional Cost Analysis in the HR Management System

O. A. Dinukova(✉)

Samara State University of Economics, Samara, Russia
odinukova@yandex.ru

Abstract. The relevance of this research is based on the natural need of any enterprise to increase its financial performance with an optimal ratio of available internal resources. The research problem is determined by the fact that in practice, enterprises do not fully use available tools of functional and cost analysis to justify conclusions about the results of their activities. An additional complication is that functional analysis is widely used in practice, while cost analysis of the management systems' functions has certain specifics. Functional cost analysis (FCA) allows you to identify costs for the implementation of system functions at a given quality level. The analysis explores all possible functions in order to determine the cost of providing services and to ensure that processes can be upgraded and the productivity can be improved. The author identified problems in the work of a specialist in the personnel management department and suggested ways to reduce costs and improve the quality of functions, which confirmed the effectiveness of the FCA as a tool for evaluating management systems and the significance of this method in making managerial decisions.

Keywords: Cost of performing functions · Functional and cost analysis · Personnel management system · Significance of functions

1 Introduction

An important place in the system of modern progressive management tools that contribute to achieving higher production efficiency is occupied by the method of functional cost analysis (FSA). The FSA can be considered as a tool for improving personnel management and labor efficiency. In our country, in the period from 1940 to 1990, considerable experience was accumulated in applying FSA for technical facilities. Currently, there is a revival of FSA in Russia. The improvement of FSA in modern conditions is associated with the expansion of its scope, for example, for the organization's management system, in particular, the personnel management system. In other countries, this method is called "value analysis" or "value engineering" (the USA) and "value management" (UK). In Russia, since 1970, the name "functional cost analysis" has been fixed.

The development of this method was facilitated by changes in the form of production and business methods in the 60s—80s of the XX century, when traditional methods of cost estimation lost their relevance. These methods were initially developed

S. I. Ashmarina and V. V. Mantulenko (Eds.): IPM 2020, LNNS 161, pp. 558–565, 2021.
https://doi.org/10.1007/978-3-030-60926-9_71

for external users and were not suitable for internal purposes; they were uninformative for operational management and did not allow to determine the production costs of an individual product accurately enough.

In addition, the process of structuring expenses has undergone significant changes: so if at the beginning of the last century, labor accounted for about half of total expenses, then by the end of the century, overhead costs were more than 60% of the structure of production costs, and using working hours as a base for allocating expenses has lost its meaning. With the development of information processing technologies, the cost of performing the necessary measurements and calculations has decreased. One of the principles of FSA is a functional approach, which is universal and applied in the field of labor organization and management systems. These functions determine the content and structure of a management system that exists to ensure effective functioning of the production system.

2 Methodology

FSA is a method of technical and economic research on functions of the organization's management personnel, aimed at finding reserves for reducing management costs and achieving the best production and commercial results based on the choice of effective management methods [1]. Let's consider the FSA stages. At the *preparatory stage*, the FSA object and tasks are determined, and a work plan is drawn up. As a research result, "bottlenecks" in the production sphere are identified, which are usually results of insufficient performance of the organization's management system. Further, the management system is considered by subsystems and by individual elements, for these purposes an expert survey is used, the results of which are used to select the object for the priority conduct of FSA. The *information stage* is intended for systematization of the collected and studied data.

The *analytical stage* is intended for parsing functions. First, functions are formulated, then analyzed, and classified. After analyzing the costs of their implementation, the significance of functions and costs of their implementation are compared. As a result, the main, general, auxiliary and redundant functions are identified. Excessive functions are not inherent in the personnel management system and they are a possible reserve for reducing the cost of performing the entire set of managerial functions.

The calculation of costs for performing functions is carried out by correlating labor costs for performing a specific function to the total fund of working time for a period. Then the part of wages, taxes, and the amount of depreciation of technical equipment that can be attributed to costs associated with performing this function are calculated. Cost estimation of functions allows you to analyze costs of functions and identify areas where it is possible to reduce costs. In addition, the analysis determines the time spent on performing functions and it is possible to determine how effectively the working time fund is used.

To determine the degree of functions' significance, an expert group is created, based on the results of which less important functions are identified, which should be taken into account at this stage in terms of reducing their costs. The cost of performing these functions is related to the degree of their significance, in case of significant discrepancies, it is necessary to search for management solutions.

The *creative stage* is designed to select methods for finding ideas. While searching for ways to improve the personnel management system, it is possible to use the method of control questions, "brainstorming".

The *research stage* covers the development of a project, where the selected options are worked out, their organizational and economic assessment is carried out with the participation of experts, and the justification is done.

At the *recommendation stage*, the economic and social efficiency is determined and costs of the project development and implementation are calculated, and the project is prepared for implementation.

The *implementation stage* involves material, technical, psychological and professional training of employees for the implementation of a project. In the research process, we identify problems of a management system and suggest ways to reduce costs and improve the quality, which is reflected in the final stage of FSA.

3 Results

To conduct FSA, a specialist in the HR management department of a regional enterprise was selected as a research object. The choice of this position for analysis is determined by its role in the department and in the organization as a whole. In addition, the specialist of the HR department has a number of problems related to congestion, which can be solved by improving the information and technical support of the HR management system. The FSA task was to clarify functions of the HR department specialist, improve the quality of the performed functions, reduce costs of functions implementation, rationalize the relationships of the HR department specialist with the other specialists in the performance of functions, improve the effectiveness of the HR specialist's activity, improving methods of labor processes organization. In order to minimize costs and improve the efficiency of the HR department, the following tasks were performed:

1) analysis of functions of the HR department, including duties, responsibilities of HR department specialists,
2) construction of functional cost diagram (Function Analysis System Technique (FAST) diagrams), which is a graphical representation of functions used for determining unnecessary and unusual functions, as well as functions that are not currently performed by specialists of the HR management department; costs of performing functions were determined,
3) costs of performing all functions during the year were determined, and by the calculation of these costs the author took into account the salary fund, contributions and expenses for the maintenance and operation of technical management tools,
4) construction of a diagram of functional relations of a HR department specialist with other workers, identification of redundant and missing, horizontal and vertical links; performing his functions, a specialist of the HR management department interacts

with an engineer for labor protection, an accountant, a specialist and the head of the HR department, a director,

5) importance degree of functions established by the expert group was determined, each member of the expert group completed a matrix of pairwise comparison functions.

To reduce the subjective factor in determining the significance of management functions, expert series are checked using the formula:

$$K = \frac{Smax}{Smin} \quad (1)$$

where K is a stability coefficient of the expert series, S_{max} – maximum significance in the expert series, S_{min} – minimum significance in the expert series.

If the result exceeds the standard value of the coefficient, the regulatory factor is 2.0, then one of the values is deleted and the action is repeated, if the result is less than the normative value of the coefficient, the arithmetic mean values of each function importance are calculated. The author constructed a combined diagram of the functions significance for a specialist of the HR management department and costs of their implementation (Fig. 1).

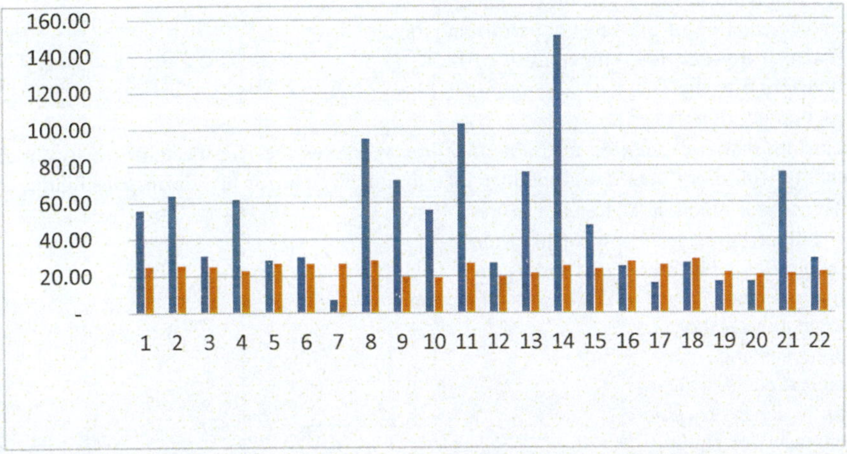

Fig. 1. Diagram of the functions significance and their implementation costs (Blue: Expenses for the functions implementation, thousand of rubles; Orange: Significance degree of functions, points) (Source: author)

6) a "card of ideas" was compiled with various options to reduce the cost of unnecessary functions. A fragment of a comparative organizational and economic assessment of options (submitted ideas) is presented in Table 1.

Table 1. Fragment of a comparative organizational and economic assessment of the submitted ideas

№	The essence of the option	Advantages of the option	Dis-advantages of the option	Effect	
				Organizational	Economic
1	Purchase and install the computer program "HR department plus" to perform 1–5, 20–22, and other functions	1. Reducing labor costs of performed functions 2. Free technical support 3. Increasing the efficiency of performed functions	Additional costs of purchasing and installing the program	+	+
Conclusion: Digitalization of the functions performance process significantly reduces the cost of performed functions of a HR management department specialist					
2	Creation of a new function: Preparation and provision of information for the implementation of the "salary project" Function 14 (Participation in the implementation of the "salary" project") is transferred to the accounting department	1. Partial exemption from performing a function that is not part of the official duties of a HR management department specialist, therefore, reducing the cost of performing this function	–	–	+
Conclusion: Partial exemption from this function will allow the HR specialist to work more effectively with more important functions that are directly part of his job responsibilities					
3	Transfer of function 9 (Familiarization of employees with the internal labor regulations, labor contracts and amendments to them) to the labor protection engineer for signing	Release of the HR specialist from performing an unnecessary function	–	+	+
Conclusion: The specialist of the personnel management department gets rid of performing an unnecessary function that takes away his time from performing functions that are part of his job responsibilities					

Source: author.

7) costs of developing and implementing the recommendations were determined, and the economic and social impact was identified, which confirms the feasibility of the project. A fragment of the calculation of the economic effect is shown in Table 2. The social effect is manifested in optimizing the working time of an employee of the HR management department, which allows reducing the load and time for performing certain functions. Changes in the duration of operations lead to a reduction in fatigue, to the rationalization of work and rest.

Table 2. Changes in the cost of performing functions as a result of activities

№ function	Function name	Costs of performing the function		Savings (+), additional costs (−), rubles
		Before activity	After activity	
9	To familiarize employees with the internal labor regulations, labor contracts and amendments to them under signature	72333,8	0	+72333,8
14	To participate in the implementation of the “salary” project	151276,5	20187,8	+131088,7
21	To carry out HR administration	76410,8	61122,1	+15288,7

Source: author.

As a result of the conducted functional cost analysis at the enterprise, functions were identified, the time to perform which can be reduced when using the software product. Excessive functions were also identified, which were transferred to the labor protection engineer and accountant. New functions have been created that partially simplify the work of the HR specialist and save the organization’s budget.

4 Discussion

Trends in the global economy require a review of existing approaches to management and the use of new methods to improve the efficiency of economic entities. One of which is FSA, which can be used for both production, auxiliary and management systems. Rezanovich’s research examines the process of reorganizing the functional structure of the personnel training department and presents some prospects for using this method [8]. In their work, Tinkov, Novikova, and Babenko, presented ways to optimize the cost of training the company’s personnel based on the use of functional cost analysis [10]. The universality of this method is reflected in the work devoted to evaluating the effectiveness of HR management at the enterprises of the United Aircraft Corporation, the assessment of functions transfer to outsourcing using functional cost analysis is proposed [4]. The scientific literature analyzes differences between FSA and traditional methods of analysis, specific features of its application with identification of its advantages and disadvantages. In the work of Podkhalyuzina, visual options are considered and approaches are indicated when studying the system of remuneration for employees' work in the company [7].

The study by Atabay and Galipogullari explains how VE (value engineering) principles are applied in construction projects, including the construction of a motorway in Croatia by the BECHTEL-ENKA joint venture as a sample project, describes VE practices in this project [2]. FSA is a related tool for measuring the effectiveness of construction projects. Earned value management (EVM) is the best way to track progress in a construction project. It takes into account time and cost factors to evaluate efficiency and estimate the completion time and costs [9]. Leupold, Hall, Hofer, and Zulch analyzed the personnel structure in which employees can perform any job or only one type of activity and investigated whether the actual situation can be improved with further qualification measures. They developed a simulation model for analyzing various employee qualifications and scenarios for loading the shipping division [6].

New management is needed to organize the digital transformation of the enterprise and its accompanying innovative concepts of HR management [5]. For example, in the work of Eeckhout, Vanhouche, and Maenhout, the construction of a basic personnel register is associated with the distribution of personnel by duties to cover certain personnel needs [3].

5 Conclusion

FSA was conducted in this research, the object of which was a specialist of the HR management department of a regional enterprise. In the course of this study, all the stages of FSA were passed, which reflected the main activity of a specialist of the HR management department and its functions. Comparing the significance of a function with the cost of its implementation is the essence of functional cost analysis. As a result of this analysis, the author revealed problems in the work of the HR specialist and proposed ways to reduce costs, improve the quality of functions performance, which is reflected at the implementation stage. The advantages of this method are the availability of a simple calculation, graphical methods of estimation and the universality of its application. In modern conditions, FSA is an effective tool for evaluating management systems. It plays an important role in making management decisions.

References

1. Anikina, Yu.A, Ragozina, M.A.: The functional-cost analysis in management decisions. Part two: Design. Econ. Entrepreurship **8–3**(85), 1135–1140 (2017)
2. Atabay, S., Galipogullari, N.: Application of value engineering in construction projects. J. Traffic Transp. Eng. **1**, 39–48 (2013)
3. Eeckhout, M.V., Vanhouche, M., Maenhout, V.: A decomposed branch-and-price procedure for integrating demand planning in personnel staffing problems. Eur. J. Oper. Res. **280**(3), 845–859 (2020)

4. Fedotova, M.A., Tikhonov, A.I., Novikov, S.V.: Evaluating the effectiveness of personnel management at aircraft manufacturing enterprises. STIN **12**, 6–8 (2017)
5. Kaczmarek, S.: Mastering fourth industrial revolution through innovative personnel management – a study analysis on how game-based approaches affect competence development. IFAC PapersJnLine **52**(13), 2332–2337 (2019)
6. Leupold, M., Hall, M., Hofer, K., Zulch, G.: Simulation analysis of alternative personnel structures in the shipping division of a tinplate manufacturer. Procedia Manufact. **39**, 583–590 (2019)
7. Podkhalyuzina, V.A.: Functional cost analysis and its differences from traditional methods of analysis. Baltic Econ. J. **3**(19), 56–62 (2017)
8. Rezanovich, E.A.: Practical application of functional cost analysis in the personnel management system. Mod. Probl. Sci. Educ. **6**, 427 (2014)
9. Sruthi, M.D., Aravindan, A.: Performance measurement of schedule and cost analysis by using earned value management for a residential building. In: Materials Today: Proceedings (2020, in press). https://doi.org/10.1016/j.matpr.2020.05.210. Accessed 26 June 2020
10. Tinkov, S.A., Novikova, M.M., Babenko, I.V.: Optimization of costs on trainings on the basis of the method of functional-value analysis. In: Proceedings of the Southwest State University, Economics, Sociology and Management, vol. 9, no. 1(30), pp. 132–139 (2019)

Older Workers Success: Biological Functions and Managerial Nudging Balance

I. B. Durakova(✉) and M. G. Holyavka

Voronezh State University, Voronezh, Russia
durakova@econ.vsu.ru, marinaholyavka@yahoo.com

Abstract. The article is an attempt to represent the success of older workers as a synergistic effect of the impact on the activation of natural body functions and involvement in the process and result of work through the technology of medical-biological and managerial nudging. To justify this position, the sections of the achievement process and the determinants of success are identified. Competencies that stabilize or increase with age are presented. The authors developed a matrix of using the methods of "soft paternalism" for the formation of a target attitude, the process of implementing a working action, obtaining and evaluating the result.

Keywords: Age and ability to work · Changes in the labor market · Determinants of success · Extension of working capacity · Nudging

1 Introduction

"The strategy of actions in the interests of older citizens in the Russian Federation until 2025" [13], was adopted in order to support the elderly population of the country and ensure the possibility of preserving their work performance. This strategy has not yet received extensive inter-disciplinary scientific support in the relevant literature. Articles in specialized journals and other publications allow us to identify the "pain points" of the problem in relation to individual sciences. According to the philosophical approach, analysis is really a way to get new knowledge. By determining the movement of scientific thought from the simple to the complex, from the accidental to the necessary, from diversity to unity, analysis allows us to represent categories of "success", "potential", and "aging" in the interpretations of various disciplines, that is, as elements of a whole complex. At the same time, the analysis leads to the identification of an essence that is not yet associated with specific forms of its manifestation. The process of combining into a single complex parts, properties, and relations identified during the analysis of the success of older workers and its preservation is possible through the synthesis. This research work proposes the implementation of a scientific synthesis of research from biological, managerial and behavioral sciences. This allows us to direct the vector of cognitive activity to obtain a "living", "specific" complex, to develop a fundamental platform for slowing down personnel aging and preserving productive work through the activation of natural functions of the human body and managerial "nudging" of employees at workplaces.

S. I. Ashmarina and V. V. Mantulenko (Eds.): IPM 2020, LNNS 161, pp. 566–575, 2021.
https://doi.org/10.1007/978-3-030-60926-9_72

2 Methodology

"Nudging" is a definition of a system that allows us to make a reasonably correct choice and an optimal individual and group decision, including the context of modern management technology [10]. Implementation of Nudge theory becomes a significant HR trend [16]. The reason for this is the idea that meets requirements of the time, which is to move away from direct instructions and compulsion for actions to use no less effective positive reinforcement and indirect instructions [11, 12]. Using the idea of nudging, in our opinion, is appropriate for maintaining the success of older workers. The study of the problem included three stages. At the first stage, the achievement process was characterized and determinants of professional success were considered. Achievement and evaluation of a result as successful one, giving the subject of achievement the status "successful" are integrated into the field of success, which, in our opinion, represents four process areas: "target setting" → "action" → "result" → "result evaluation".

1. "Setting" is a starting impulse to achieve, a condition for determining the quality of life, behavior, activity, self-awareness, self-esteem, psychological well-being. Success-oriented behavior is determined by dispositions and motivations for achievement, attitude readiness, and potential.

Prerequisites for the formation of an attitude to success should be, first, a justified need as a result of a specific action (task solution). Second, this is an incentive to action (motive). The motive may be an opportunity to experience a sense of pride of the own diligence in achieving success [1], the perception of the own responsibility for the result [3], getting more information about yourself [15]. Third, the determination of the possibility to implement an action that can be objective and subjective. An objective opportunity includes a certain segment of consumers, financial and logistical support, an employee's competence potential assessed from outside (for example, by an expert council, a manager, or colleagues), and an environment that increases individual readiness to succeed, including through competition arising within the social environment. The subjective possibility is an individually conscious sense of the internal resource potential to cope with the work independently, without outside help.

2. "Action leading to a result" represents the purposeful impact of the employee on the labor object, the implementation of internal potential and resource capabilities, bringing them into compliance with the requirements, conditions, restrictions and risks of the activity. Depending on the nature, complexity, and novelty of the task, various competencies may be required in an isolated or complex version: professional, methodological, process, managerial, and personal (social) ones.
3. "Result" is a term introduced into the modern vocabulary from the French language and in its etymology interpreted as "return", that means an integrated outputs of the performed work, productivity, skill.

Regarding the detailed components of professional success, scientists have different positions. The own developed version of effective components of professional success is shown in Table 1.

Table 1. Determinants of professional success

Types of performed work		
Work related to the solution of a problem	Work determined by a context	Adaptive work
Task execution	Cooperation, coordination, interaction	Adapting to changes and innovations
Fully completed work		
Innovations and novelty in the content or methods of the work performance (conquest, creativity)		
Timing and quality of execution		
Labor expenditures		
Compliance with instructions and regulations		
Scope and expected duration of the achieved result usage		
Emotional feelings of an employees that form their professional perspective in achievements		

Source: authors.

An action aimed at results, according to the concept of the dominant of differences in abilities, knowledge, and skills in success in comparison with the made efforts, will be highly likely implemented if the work content is consistent with capabilities of the elder employee (Table 2). "Evaluation", carried out through feedback (from a manager, expert council, working environment, public) or the employee's self-perception, is the final part of the success field. The evaluation report is aimed at determining (confirming or denying) the value of the achievement and the degree to which it is attributed to success.

Table 2. Professional and social action competences, stabilizing or increasing with age

Professional competences	Social competences
Competencies that tend to increase	
Professional and business knowledge	Professional skills and life experience
Conscious quality of work	Wisdom
Ability to cope with problematic situations	Thoughtfulness and discretion
Knowledge of available problem solving strategies	Positive attitude to work, attachment to the organization, working place
Competencies that remain unchanged	
Ability to make decisions	
Focus on results	

Source: authors.

The emotional state of the employee when achieving success forms a motive for repeating productive significant activities in the future. The need for such activities is determined by the nature and timeliness of feedback (positive reinforcement, negative reaction, neutral behavior) and the status of the subject(s) that qualifies the achievement degree. At the second stage, the authors considered the dynamics of natural functions in the aging process. The structural model of labor changes for older workers includes three elements. Biology-physiological component implies the ability to perceive, predisposition to diseases, mobility of the nervous system, death of brain cells. Sociocultural component includes loads associated with family, organization of free time, value attitudes. Professional element means the work intensity, its monotony and mobility degree, eye strain, working time.

The mutual influence of these elements determines the employee's ability to be engaged in physical or mental labor, as well as his social behavior. Each of these elements has a different degree of transformation by aging [7]. It is revealed that from three groups of prerequisites for obtaining a working result, the group of knowledge and professional skills is relatively easy to change. To a lesser extent, behavioral group (work in project groups, labor organization, communication, consulting, presentation activities) is prone to changes. The third group is the most difficult to change: the personal group includes such characteristics as dynamism, ability to bear loads, tolerance, life values, attitudes and motives.

At the third stage, we considered the preservation of productive labor through the intensification of natural functions and managerial nudging in workplaces. The study showed the following features of the modern labor market: aging of the labor force; increase in the average age of the average employee and the cohort of older workers; presence of job categories where there is accelerated professional aging; termination of work before retirement age because of non-compliance with new profession requirements, functional capabilities of the rapidly aging body. In accordance with this, changes are formed that limit professional performance, lead to a limited range of functional capabilities of the workers' body, reducing their reliability and accelerating the rate of aging.

At the same time, a significant moment in the modern world is the appearance and rapid spread of such a phenomenon as the "silver economy". Its essence is that a significant proportion of older people want to work as long as possible. Thus, the share of employment of older workers aged from 54 to 64 years in the European Union is 40%, in the United States – almost 60%, in Switzerland – 65%, in Germany – 38% [14].

The choice of nudging technology to preserve the successful working capacity of older workers and expand the range of their functional capabilities is based on its performance characteristics. First, nudging allows us to improve our activities in a situation of constant organizational changes, to contribute to non-resistance by taking into account the influence of psychological factors and cognitive limitations [9] that characterize older workers. Secondly, to create a "law of results" in the organization that allows you to retain staff through trust in the employer and the future perspective to be needed regardless of age restrictions. The platform for such achievements includes nudging rules: transparency, achievability, attractiveness, use for increasing the welfare of those employees who are targeted by actions, prevention of misleading nudging objects [4, 5, 8, 15] (Table 3).

3 Results

First, age-adjusted professional success is an objective characteristic of an employees' performance that is formed in the process of their meaningful need to overcome negative emotions, maintain the accumulated labor potential and a reputation image, involvement and emotional attitude to repeat significant work in the future. Second, achievement and evaluation are integrated into the success field, representing four interrelated process areas, including the target setting, action, result, and evaluation of the result. Third, the difference and similarity of success determinants are defined by the type of work performed: task-related work, contextual work, and adaptive work.

Fourth, taking into account the transformational features and mobility of crystal and fluid intelligence, the possibility of labor success due to the adjustment of the target setting, the nature and speed of action, the obtained result and its evaluation varies in different age groups.

Fifth, in the case of physiological aging of the body, its chronological and biological age must coincide. If the biological age is lagging behind the chronological age, we can assume a longer life expectancy. In the opposite case – premature (accelerated) aging, that is, the process of age-related changes, is characterized by a higher rate in comparison with healthy individuals of the same age category. Thus, in the latter case, we are talking about premature (and pathological) old age.

Sixth, the characteristics of premature aging are:

- the presence of an etiological factor that led to the launch of the aging process of pathological type,
- reduced effectiveness of compensatory mechanisms and more pronounced restrictions of adaptation to changing environmental conditions,
- more pronounced heterochronous and heterotopic changes in the body (concentration of changes in one organ or system).

The rate of age-related changes in the body and premature aging depends on diseases of the gastrointestinal tract that disrupt metabolic and trophic processes; the nervous system, diabetes, and chronic lung diseases [2]. Patients with coronary heart disease look 10 years older than they actually are. The structure of morbidity and mortality is largely determined by obesity and overweight, the frequency of which in highly developed countries reaches 30%. As the body weight increases, the frequency of degenerative diseases of the musculoskeletal system increases, for example, osteoarthritis, osteochondrosis, etc. [6]. Gender differences in the degree of physical activity and availability of functional capabilities, as well as in the psychological status of older people were revealed. Compared to men, women are less active and more restricted in those activities where physical effort is required [4].

Seventh, achieving a balance of natural functions of the body and professional capabilities is possible through the use of technology of unconscious and conscious medico-biological and managerial nudging.

4 Discussion

With the age of the employee, the need for achievement, the motive and the possibility to implement ideas, professional functions are characterized by a certain shift in emphasis. If there is no desire for promotion among older employees, there is still a need for a "competitive spirit". It can be assumed that the goal is not to achieve a new significant result, but rather to repeat what has been achieved from the existing potential and thus maintain the already formed reputation image.

The use of nudging technology to maintain the success of elder employees has obvious its disadvantages too. Threats are attempts to push an older employee to unacceptable behavior through deception and manipulation. It is also difficult to predict the effect of nudging in terms of age and national diversity because of different values, attitudes and cultural differences. Its use in organizations with directive management is also debatable. Switching to "soft paternalism" may seem like a trick. There are also risks related to employees with established norms of behavior, which they don't want to change considering it "beyond their age".

The effectiveness of nudging is provided by the "law of results", that is, its gradual introduction and the achievement of confidence of older employees to achieve goals of their well-being and performance improvement. The principles of achieving the "law of results" are: a specific transparent goal of the employer, the ability to receive feedback at the stages of the project implementation.

5 Conclusion

According to research in the considered field, the potential of the methods of the XX century, which caused the extension of human life and the period of active longevity, is almost exhausted. The next step in finding ways to improve the success of older workers is obviously related to finding adequate approaches to blocking the biological mechanisms of aging. The element of scientific novelty in this article is a detailed representation of the success of older employees and the employer's choice of methods of biomedical and behavioral nudging for its preservation. The results of the research can be used in the development of the theory of personnel longevity and work with personnel in organizations.

Acknowledgments. The reported study was funded by RFBR, project number 20-010-00263.

Appendix A

Table 3. Systematization of ideas about the use of behavioral and medico-biological nudging to preserve the professional success of older workers

	Nudging types
1. Target setting	*Unconscious nudging*
	– publishing company reports on achievements, creating video films with the participation of employees of different generations; – placement of articles, interviews with older employees in the media and on the company's website, in which age is positioned as a predictor of invaluable experience, and not the "ballast" of the organization; – motivation for active longevity through prevention of occupational diseases, preservation of vision, musculoskeletal system, video demonstrations of consequences of obesity, smoking and other bad habits
	Conscious nudging
	– advertising vacancies of organizations for older employees; – management's recognition of the value of older employees at work meetings and seeking their advice; – providing elderly employees with flexible working hours, additional guarantees and compensation in the field of healthcare; – information sheets and videos about changes that occur in the body while aging to adopt a possible strategy for their adequate correction
2. Action	*Unconscious nudging*
	– creation of ergonomic workplaces; – use of large font on computer monitors and information stands; – sending via e-mails and posting information on the website about possible training and upcoming seminars; – publication of reviews of older employees who have completed training in order to perceive it not as a risk, but as an opportunity to continue successful work; – promotion of a healthy lifestyle, the trend for physical education: through sports events/competitions between employees and their family members, hiking, walks (treatment with air containing phytoncides-aerotherapy; Scandinavian walking, Chinese qigong technique, etc.); meetings with adherents of a healthy lifestyle; encouragement of employees without bad habits; – creation of comfortable areas for relaxation (colour therapy, aromatherapy and music therapy) – prevention of psychosomatic disorders; – installation of coolers with clean water, the possibility of using oxygen cocktails;

(*continued*)

Table 3. (*continued*)

	Nudging types
	– organizing healthy nutrition in the canteen: enrichment menu, meals which are rich in vitamins and minerals (vegetables, fruit), fiber, etc.; prevention of obesity leading to metabolic disorders: decrease in fast carbs in the content of menu, reducing the amount of food consumed by reducing the size of plates; – ban on working in night shifts for older employees: maintaining normal levels of melatonin (the hormone of youth), the maximum production of which falls on the night hours; – elimination or restriction of the sale of fast food and cigarettes on the territory of the organization; – booklets/videos about spa holidays, methods of early diagnosis of diseases and the latest achievements of medicine; – division of the assigned work into stages: feeling the effect of the hormones of satisfaction after completing the stage and pleasure after exertion – serotonin and endorphin; maintaining a high level of these neurotransmitters, that means prevention of depressive states
	Conscious nudging
	– close location of workplaces for older and younger employees; – formation of multi-age groups, project groups, training groups; – motivation of older employees to participate in mentoring programs; teaching at a corporate university; – informing employees about unfavorable alternative offers in the labor market and losses associated with retirement; – representing an employee as "the best of the best" when delegating them to training; – mandatory annual medical examinations; – informational and educational lectures, consultations of doctors on eliminating the deficiency of vitamins, minerals and microelements, on taking adaptogens, food additives to improve mental abilities (Ginkgo biloba, choline, antioxidants); – videos/presentations about the achievements of biogerontology; – smoking only in a designated area, no smoking during working hours; – introduction of mandatory breaks for gymnastics to relieve the eyes and back tension of employees engaged in sedentary work at computers, etc.
3. Result	*Unconscious nudging*
	– placing the current results of employees' activities on the stand/information board to create an atmosphere of competition and form the desire to keep up with the team
	Conscious nudging
	– providing coaching support to older employees in case of problems at work, resistance to changes; – offer employees a transfer to positions that require responsibility and quality of work, rather than a quick reaction

(*continued*)

Table 3. (*continued*)

	Nudging types
4. Evaluation	*Unconscious nudging*
	– placement of information about the success of elder employees in internal and external media using statistical information; – attracting them to participate in the ritual of “initiation” of new employees
	Conscious nudging
	– creation of a “Council of Elders” from employees with extensive experience and experience to consult when making management decisions; – holding joint award ceremonies for representatives of different generations in order to form the organizational unity

Source: authors.

References

1. Atkinson, J.W.: Motivational determinants of risk-taking behavior. Psychol. Rev. **64**(6.1), 359–372 (1957)
2. Bryantseva, O.V., Proschaev, K.I.: Medical aspects of premature aging. Sci. Rep. Med. Ser. Pharm. **4**(99(13/1)), 23–27 (2011)
3. Feather, N.T.: Valence of outcome and expectation of success in relation to task difficulty and perceived locus of control. J. Pers. Soc. Psychol. **7**(4.1), 372–386 (1967)
4. Gretebeck, K., Sabatini, K., Black, L.D.: Physical activity, functional ability and obesity in older adults: a gender difference. J. Gerontol. Nurs. **11**, 1–9 (2017)
5. Halpern, D.: Inside the Nudge Unit: How Small Changes Can Make a Big Difference. Ebury Press, London (2015)
6. Must, A., Spadano, J., Coakley, E.H., Field, A.E., Colditz, G., Dietz, W.H.: The disease burden associated with overweight and obesity. JAMA **282**(16), 1523–1529 (1999)
7. Pozdnyakova, N.M., Proschaev, K.I., Ilinitskiy, A.N., Pavlova, T.V., Baschuk, V.V.: Modern views on the possibility of assessing biological age in clinical practice. Fundam. Stud. **2**, 17–22 (2011)
8. Rohrschneider, U., Fridrichs, S., Lorenz, M.: Erfolgsfaktor Potenzialanalyse.-Aktuelles Praxiswissen zu Methoden und Umsetzung in der modernen Personalentwicklung. Gabler Verlag, Wiesbaden (2010)
9. Shantz, A., Latham, G.: The effect of primed goals on employee performance: implications for human resource management. Hum. Resour. Manag. **50**(2), 289–299 (2011)
10. Sunstein, C.R., Thaler, R.H.: Libertarian paternalism is not an oxymoron. Univ. Chic. Law Rev. **70**(4), 1159–1202 (2003)
11. Thaler, R.H., Sunstein, C.R.: Nudge: Improving Decisions About Health, Wealth, and Happiness. Yale University Press, New Haven (2008)
12. Thaler, R.H.: The power of nudges, for good and bad. N. Y. Times (2015). https://www.nytimes.com/2015/11/01/upshot/the-power-of-nudges-for-good-and-bad.html. Accessed 15 June 2020

13. The strategy of actions in the interests of older citizens in the Russian Federation until 2025. Order of the Government of the Russian Federation from 05.02.2016 No. 164-R (2016). https://mintrud.gov.ru/ministry/programms/37/2. Accessed 20 May 2020
14. Tolstonog, V.: Stimulating the work of the elderly unemployed. Partner **8**(95) (2005). https://www.partner-inform.de/partner/detail/2005/8/194/1865/stimulirovanie-trudovoj-dejatelnosti-pozhilyh-bezrabotnyh?lang=ru. Accessed 15 June 2020
15. Weiner, B.: Perceiving the Causes of Success and Failure. General Learning Press, New York (1971)
16. Zielinski, D.: What to expect: Top 2019 HR tech trends (2019). https://www.shrm.org/resourcesandtools/hr-topics/technology/pages/top-2019-hr-tech-trends.aspx. Accessed 15 June 2020

Marketing Concept of Interaction Between the Parties of Labor Relations

D. K. Zakharov(✉) and L. V. Ivanovskaya

State University of Management, Moscow, Russia
zakharov_dk@mail.ru, ivanovskayalv@rambler.ru

Abstract. Organizations operating in different markets have increased requirements for analyzing the quality level of personnel and comparing competitive advantages based on various types of communication interactions. The fundamental task of personnel marketing is to create the most attractive image of the company as an employer in order to provide itself with human resources with optimal quantitative and qualitative parameters. The development of a personnel marketing management system at the enterprise level is not sufficiently reflected in the modern Russian research field on human resource management. This model includes goals, tasks, places in the organizational structure and in the personnel policy, and areas of interaction with other labor market actors. The research purpose is to develop methodological and applied provisions that allow employers to take a competitive place in the labor market by identifying effective interaction areas with its subjects using marketing approaches while determining and ensuring personnel needs. The paper investigates marketing concepts of personnel management and the place of personnel marketing in the organization's personnel policy. The methodology of forming a marketing management system for the organization's personnel is considered. The paper shows the importance of developing and applying requirements to employees and to the employer in personnel marketing.

Keywords: Marketing concept · Personnel marketing · Personnel marketing management system · Personnel marketing tools · Requirements for employees and employers

1 Introduction

Currently, the role of strategic orientation of the HR management and the image of the organization's staff increases not only in the labor market but also in the market of goods and services. Organizations operating in different markets have increased requirements for analyzing the quality level of personnel and comparing competitive advantages based on various types of communication interactions. This is due to the fact that the professionalization of almost all production functions covers and subordinates the task of personnel acquisition, among other things. The intensity with which these actions are taken depends on the current situation of the organization and the operational need for staff. The interaction of the parties to labor relations based on the marketing concept is called personnel marketing. It allows you to solve similar tasks

S. I. Ashmarina and V. V. Mantulenko (Eds.): IPM 2020, LNNS 161, pp. 576–582, 2021.
https://doi.org/10.1007/978-3-030-60926-9_73

from a slightly different angle, and implies an emphasis on long-term oriented regulation of the attractiveness of the enterprise in the eyes of relevant target groups of potential employees. Positioning and creating an image of the company in the labor market is carried out gradually and purposefully.

The fundamental task of personnel marketing is to create the most attractive image of the company as an employer in order to provide itself with human resources with optimal quantitative and qualitative parameters. This task is solved inefficiently due to the fact that the methodological and organizational issues of building an integrated system of personnel marketing at the level of enterprises (the employer) have not been fully developed yet. The development of a personnel marketing management system at the enterprise level is not sufficiently studied in the Russian research field.

2 Methodology

The research purpose is to develop methodological and applied provisions that allow employers to take a competitive place in the labor market by identifying effective areas of interaction with its subjects when using marketing approaches to identify and ensure the personnel needs. To achieve this goal, the following tasks were solved:

- development of the concept of personnel marketing based on the analysis and generalization of production and marketing concepts,
- definition of principles for formation and implementation of the organization's personnel policy based on the marketing concept of personnel management,
- identification and classification of requirements of different parties of labor relations, analysis of their consistency and inconsistency, development of approaches to their assessment,
- research of the structure of the organization's image and the image of personnel, the sequence of its formation and role in the process of job hunting by potential employees,
- highlighting the strategic aspect of personnel marketing, which creates the basis for successful and long-term positioning of the employer in the labor market.

Among the most important principles applied in this study, the principle of objectivity is the leading one. It is based on identifying objective factors, relations, and dependencies and comparing them with practical data in the field of interaction between the parties of labor relations. The principle of relevance, which is also important for achieving the research goal, required building the work on the basis of the achieved theoretical knowledge. This is based on facts that are prioritized and verified.

3 Results

The environment of interaction between the employer organization and the labor market entities has been formed in recent years under the influence of stable trends and factors that determine the specifics of solving the main personnel problems. Trends in the development of personnel management include: the development of external professional infrastructure, internationalization and globalization of personnel management, the priority of team labor organization, the growing importance of delegation and decentralization, etc. External factors of the organization's marketing environment (including the first two of the above trends) are political, legal, general economic and independent information environment, as well as the technology development. Internal factors include the organization's goals, financial resources and human resources, and organizational culture.

The formation of a marketing environment makes it necessary to use the analysis results of marketing concepts of personnel management. The analytical work showed that when developing the methodology of personnel marketing, it is advisable to focus on a system of concepts that includes the concept of socially responsible personnel marketing with its expanded interpretation in conjunction with the concept of interaction of strategic and tactical aspects of personnel marketing [4]. The concept of socially responsible marketing staff implies that one of the most important conditions for achieving organization's goals is a clear definition of personnel requirements, the personnel's needs in the course of professional activities and ensuring that these requirements and needs are more effectively satisfied than by competitors [7]. An expanded interpretation of personnel marketing involves linking it to the entire human resources policy of the organization (including, for example, internal personnel development activities).

Personnel marketing focuses on subjects of the labor market that are directly involved in labor relations. For its effective functioning, it should be organically integrated into the organization's personnel policy. The direct connection of personnel policy with personnel marketing occurs through a management system, the first stage of which is structuring of goals. Building a goal tree involves a sequential transition through its levels using the conceptual, factor-conceptual, and factorial features of the goal decomposition. In general, the goals of the first level can be classified into several types: production, social, and economic ones. The second level is the decomposition of goals on a factor basis, where the tasks of the system under consideration are formed, which are conditions for the implementation of the goals of the first level. The goals of the third decomposition level are basic principles for achieving the goals of a higher level.

The main goal of the first level for the personnel marketing management system, formulated on the conceptual basis of decomposition, is to gain a certain, fairly stable position in the labor market. To achieve this goal, the second-level goals have to be defined based on the factor-conceptual feature of the decomposition, which are the target tasks for the implementation of the specified goal of the first level. These tasks include the following: 1) certainty of target market segments; 2) positive public opinion about the employer organization; 3) active behavior in the external labor market; 4) completeness and reliability of marketing research [6].

According to the concept of interaction between the strategic and tactical aspects of personnel marketing, two corresponding groups of tasks can be distinguished. Tasks within the framework of the strategic aspect include: goal setting of the personnel marketing management system; segmentation of the labor market and selection of target segments; positioning of the employer organization in the labor market; information function of personnel marketing to ensure strategic objectives. As a part of the tactical aspect, marketing tasks are tools for the practical implementation of such categories of personnel marketing as: personnel labor potential, employer services, personnel costs, communications in the external and internal labor market, target segments of the labor market, and information.

In relation to personnel marketing, strategic work involving market segmentation, target segment selection, and product positioning involves the simultaneous development and analysis of personnel requirements and requests to the employer, which can be considered as the main factors of the market segmentation [11].

The classification of personnel requirements includes a group of performance indicators and several groups of factors or conditions that contribute to achieving this performance. Naturally, the classification of requirements to personnel needs to decipher the composition of their indicators. For example, the classification group of requirements "Work Performance" includes such parameters as the quality of performance of official duties, the suitability of results for using them by related parts of the organizational structure, the rationality of the process, the reliability of the presented results, and the completion of tasks on time.

Personnel marketing works to meet requirements of both sides of the labor relations, which means that it is necessary to study requests that potential employees make to the employer [5]. Taking these requirements into account when conducting the organization's personnel policy will allow it to form and maintain its own image at the proper level, both in the external and internal labor markets [3]. Hence, there is a need to develop a fairly complete list of potential requests to organizations-employers, from which potential applicants will form the content of their expectations and preferences. To solve this task, it is necessary to formulate an answer to the question: what are the main functions of labor in terms of meeting the needs of participants in the labor process. These functions are: monetary and incentive non-monetary motivation, self-realization, social and security. At the same time, the implementation of the monetary function of labor assumes the fulfillment of such requirements to the employer as, for example, an adequate level of income for labor costs, career chances of employees, rapid growth of responsibility, etc.

In the labor market, there are groups of applicants who differ, on the one hand, in their education and qualifications, and, on the other hand, in their interests, needs, and requests to the employer. From this perspective, the labor market becomes particularly heterogeneous and fragmented. The task of segmenting the labor market within the framework of the concept of personnel marketing is to establish the presence of different groups of potential employees who differ in their requirements and perceptions of potential employers, and to establish the feasibility of maintaining separate strategies of personnel marketing for them [10]. These target groups can be young specialists of economic, managerial, and technical training background. For positioning an enterprise in the labor market, the difference in the requests of applicants of technical and economic profiles is of particular interest, since they can be compared to identify typical features of both target groups.

Measuring the staff image, attractiveness, and preferences is essential for evaluating existing staff marketing strategies and the used marketing tools. As a way to measure the image of personnel, it can be proposed to measure the perception of individual image-forming factors using a rating scale and present results as a perception profile

[1]. Interpretation of individual perception profiles is not rather informative in itself. Only by comparing the perception profiles of competitors in the labor market, we can make a conclusion about the own profile.

The result of strategic research in the framework of personnel marketing is the target personnel-image, which provides the company with a more favorable position in the labor market compared with competitors. The task of tactical personnel marketing is to plan and implement its tools that bring it closer to the desired personnel image and create competitive advantages. The process of segmentation of the labor market is aimed at establishing an initial balance between the parties to the labor relations. The initial balance is established during the search for compliance with the employer's requirements for potential employees and those requests that representatives of a certain segment of the labor market can also potentially make to a future employer. The success of the personnel marketing strategy depends on how likely the expected balance and the actual balance of interests of all parties the labor relations coincide [2].

The task of the executive function of personnel marketing (i.e., the function of forming a specific "product" - the organization's labor potential) as one of the tactical tools is a job offer that takes into account requirements of candidates. When developing the content of work responsibilities and career planning, certain degrees of freedom should be laid in order to interest young professionals in starting opportunities and prospects for promotion.

Integration of all available communication tools is a prerequisite for effectively covering the need for personnel. The relation between personal and non-personal communication can be determined by analyzing the attractiveness and preferences of potential employees towards the employer. There are companies that achieve relatively high attractiveness among applicants, but have a deficit in preference and vice versa. If the weak point is low attractiveness, then the organization clearly does not belong to the so-called "acceptable set" for candidates. In this case, it is necessary to strengthen non-personal communications, since they have a broader coverage of the target group. If the weakness lies in preferences, then you need to start from changing the perception of individual characteristics of the employer. For this purpose, it is necessary to use activity-oriented personal communications [8].

4 Discussion

As a scientific and methodological tool, we use quantitative and qualitative approaches to improving the organization of enterprise management and its individual subsystems; methods for collecting, processing and using marketing information; methods for analyzing elements of the marketing complex. The choice of research methods was influenced by the pace of research conducted in the considered field: the interaction of labor market subjects in modern conditions, as well as characteristics of the object under study [12]. The paper uses formal logical research methods: generalization and classification, specification, comparison and analogy, secondary research of documents. These methods were used simultaneously, and each of them revealed new aspects of the research object. Thus, the secondary research was based on the works of

Russian and foreign scientists, which allowed us to compare results of several studies on the use of marketing approaches in personnel management [9].

The methodological basis of this research is the refraction of theoretical and applied developments in the field of personnel management and general marketing, taking into account the specifics of methodological and applied problems of personnel marketing. The authors use a conceptual framework based on the transformation of the basic provisions of general marketing and their binding to the tasks of personnel management, mainly in the field of determining and ensuring the need for it [9]. In addition, they used the method of structuring goals and developing on its basis the classification and content of general and specific tasks of the management system. This allows determining the degree of their consistency and inconsistency, and forming on this basis a meaningful structure and approaches to measure the organization's image as an employer [1].

To solve the research tasks, the authors applied methodological approaches to positioning and repositioning an employer organization in the labor market based on a comparative analysis of competitive advantages, as well as methodological approaches to determining areas of analysis and improvement of personnel marketing communications in the course of the strategy implementation for forming the personnel-image of the employer [11].

The developed methodology and organizational principles of general marketing create prerequisites for their effective transformation into the theory and practice of personnel management. The development of marketing approaches in this area will increase the competitiveness of employer organizations and provide them with a stable, positive position in the labor market. The fundamental task of personnel marketing is to create the most attractive image of the company as an employer in order to provide itself with human resources with optimal quantitative and qualitative parameters [7]. This task is solved inefficiently due to the fact that the methodological and organizational issues of building an integrated system of personnel marketing at the level of employer enterprises have not been fully developed yet.

5 Conclusion

The research solves an important methodological problem, namely, the composition and content of the strategic aspect of personnel marketing is analyzed, which allows creating a basis for successful and long-term positioning of the employer in the labor market. In turn, the development of the composition of tactical tools for personnel marketing provides targeted and gradual implementation of strategic tasks. The interaction space between the employer organization and the labor market entities has been formed in recent years under the influence of a number of significant factors.

The relation between the personnel policy and the personnel marketing occurs through a management system. Its first stage includes structuring of goals. In relation to personnel marketing, strategic level involves market segmentation, target segment selection and product positioning, simultaneous development and analysis of personnel requirements and requests to the employer. All this can be considered as the main factors of market segmentation. The result is a target personnel-image, that provides the company with a more favorable position in the labor market. The task of tactical personnel marketing implies planning and implementing its tools for achieving the desired personnel image and creating competitive advantages.

Practical application of the research results allows employers to gain competitive advantages in the labor market and ensure the formation and maintenance of a positive personnel image for both potential and existing employees. Personnel marketing also allows solving similar tasks from a slightly different angle. It implies an emphasis on long-term oriented regulation of the attractiveness of the enterprise in the eyes of relevant target groups of potential employees. Positioning and creating an image of the company in the labor market is carried out gradually and purposefully.

References

1. Alekseeva, L.A.: Method of evaluating the employer's image. Kadrovik. HR Manag. **4**, 31–40 (2010)
2. Anselmsson, J., Bondesson, N., Melin, F.: Customer-based brand equity and human resource management image: do retail customers really care about HRM and the employer brand? Eur. J. Mark. **50**(7/8), 1185–1208 (2016). https://doi.org/10.1108/EJM-02-2015-0094
3. Barrena-Martínez, J., López-Fernández, M., Romero-Fernández, P.: Socially responsible human resource policies and practices: academic and professional validation. Eur. Res. Manag. Bus. Econ. **23**(1), 55–61 (2017). https://doi.org/10.1016/j.iedeen.2016.05.001
4. Kang, C., Germann, F., Grewal, R.: Washing away your sins? Corporate social responsibility, corporate social irresponsibility, and firm performance. J. Mark. **80**(2), 59–79 (2016). https://doi.org/10.1509/jm.15.0324
5. Kazantseva, S.: Main research directions of personnel as an object of marketing. Stat. Econ. **2**, 68–71 (2015)
6. Kibanov, A.Ya., Durakova, I.B.: Personnel management of the organization: strategy, marketing, internationalization. SIC Infra-M, Moscow (2013)
7. Morokane, P., Chiba, M., Kleyn, N.: Drivers of employee propensity to endorse their corporate brand. J. Brand. Manag. **23**(1), 55–66 (2016). https://doi.org/10.1057/bm.2015.47
8. Nechaeva, E.S.: Development of marketing approach in human resource management of the organization: internal marketing. Bull. Tula State Univ. Econ. Leg. Sci. **5-1**, 167–173 (2014)
9. Patlasov, O.Y.: Marketing of Personnel. Dashkov and Co., Moscow (2016)
10. Renaud, S., Morin, L., Fray, A.M.: What most attracts potential candidates? Innovative perks, training, or ethics? Career Dev. Int. **21**(6), 634–655 (2016). https://doi.org/10.1108/CDI-01-2016-0008
11. Rodin, D.: Strategic marketing of the organization's personnel. Bulletin of the Volga State University named after V.N. Tatishchev **1**, 10–14 (2015)
12. Theurer, C.P., Tumasjan, A., Welpe, I.M., Lievens, F.: Employer branding: a brand equity based literature review and research agenda. Int. J. Manag. Rev. **20**(1), 155–179 (2018). https://doi.org/10.1111/ijmr.12121

Readiness of Future Managers to Work in Conditions of Uncertainty

A. Kutuev and E. Malysheva(✉)

Samara State University of Economics, Samara, Russia
artem-kutuev@mail.ru, sam.malysheva@mail.ru

Abstract. Future managers act as agents of change in today's business environment. In this regard, the formation of managerial competencies in view of readiness to fulfill their professional duties in conditions of uncertainty is relevant. In this regard, it is important to assess the readiness of future management personnel for change in modern conditions. The aim of this study is to provide empirical data on the established attitudes of future management personnel towards uncertainty during a period of change. Among the available variety of questionnaires, the PCRS (Personal change readiness survey) technique was chosen thanks to its multi-factor model, which allows you to study not only individual significant personality characteristics in a volatile environment, but also their relationship with the relation to uncertainty.

Keywords: Changes · Personality · Personal readiness for changes · Uncertainty

1 Introduction

The modern world is in the state of VUCA (volatility, uncertainty, complexity and ambiguity conditions). This abbreviation describes the volatility, uncertainty, complexity and ambiguity of the conditions and processes that modern managers need to take into account. The world is constantly changing, becoming increasingly unstable every day, when big and small changes are becoming more unpredictable. Since events develop in a completely unexpected way, it becomes impossible to determine the cause and effect. It is becoming increasingly difficult to anticipate events or predict how they will unfold. It is almost impossible to plan for investment, development and growth, as the direction of movement becomes increasingly uncertain. Solutions come down to a confusing chain of reaction and counter-reaction - and choosing the only right path is almost impossible. Requirements for modern organizations and leadership are more controversial and paradoxical than ever, challenging personal value systems. The report "Skills of the future: How to thrive in the complex new world" [2] records global changes and transformations taking place in most developed and developing countries, which increase the sense of turbulence and unpredictability in certain areas. New breakthroughs in such areas as digitalization, automation and robotization, demographic changes, the formation of a networked society, globalization (economic, technological and cultural), greening, all these changes occur under the influence of one common megatrend - an increasing rate of change. The manager is responsible for a

S. I. Ashmarina and V. V. Mantulenko (Eds.): IPM 2020, LNNS 161, pp. 583–589, 2021.
https://doi.org/10.1007/978-3-030-60926-9_74

large proportion of decisions that determine the work of the organization. Increasing volatility, uncertainty, complexity and ambiguity means that the manager must look for new guidelines and take a new approach to management [3, 9]. The survival of the business depends on the managerial personnel trained for this modern world. Unmotivated, without the necessary competence or not mentally or physically prepared leaders are detrimental to the organization [6]. In this regard, it is important to assess the readiness of future management personnel for change in modern conditions.

2 Methodology

The personal change readiness assessment techniques were developed by various scientists. The key conditions for choosing this particular technique are justified by a multifactorial model of human personality qualities, important in implementing changes in life, taking into account such an element in the model as a relation to uncertainty, acting as one of the stress factors during the period of change, as well as simplicity in processing the results [1, 7]. In the management training course submitted to students, there were 35 questions with an assessment of consent with statements from 1 to 6 and aimed at studying the individual's personal readiness for change.

Consider in more detail the elements of the model studied within the framework of the above methodology. Statements aimed at determining the level of the "Passion" scale help to determine the subjective assessment of the respondent regarding his energy and life tone. "Resourcefulness" determines the level of creativity of a person and the ability to generate possible options for getting out of the emerging complex conditions of the environment. "Optimism" describes the respondent as a person who no longer focuses on emerging problems, but on the possibility of solving them and avoiding considering adverse scenarios. "Entrepreneurship" - determines the readiness of the individual to study something new, the readiness to take risks in conditions of uncertainty. "Adaptability" demonstrates the ability of the respondent to adapt and change himself, change his own plans and behavior, if this is required by a changing external environment. "Confidence" determines how confident a person is in his own strength and ability to show the will necessary to solve the emerging problem. "Tolerance to ambiguity" shows the respondent's ability to calmly respond to uncertainty, situations in which there are no clearly formulated answers. Subsequently, the results of these scales are summed up and form an integral indicator of readiness for change, the value of which helps to assess the complex level of readiness for possible changes and the individual behavior model in conditions of uncertainty.

3 Results

The number of people participated in the study is 100 respondents, the sample structure has the following appearance among students (the number of employed is indicated in parentheses), shown in Table 1.

Table 1. Structure of sample on a gender and a course of respondents

Course	Number of respondents	Male	Female
Bachelor degree, 1 year	31	4	27
Bachelor degree, 2 year	18	3	15
Bachelor degree, 3 year	21	4	17
Bachelor degree, 4 year	20	7	13
Magistracy	10	2	8
Total	100	20	80

Source: authors

We analyze the results obtained and compare the average values of the respondents' responses shown in Fig. 1.

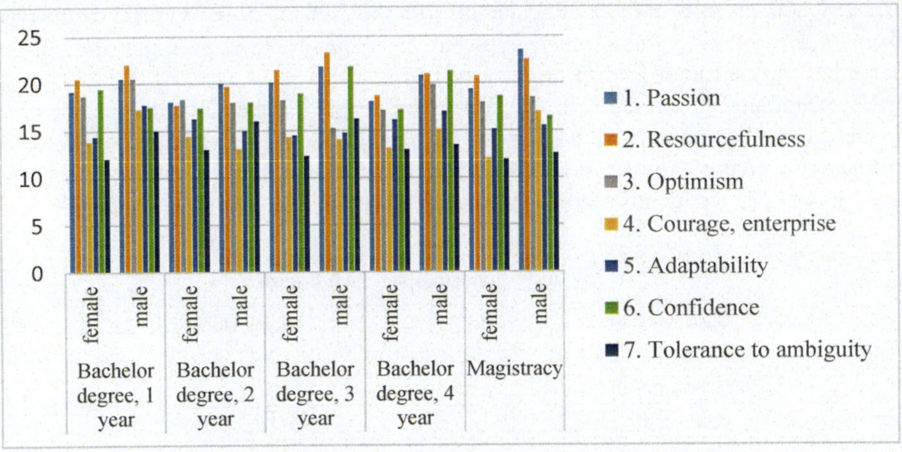

Fig. 1. Average change readiness scores (Source: authors)

There are slight differences in scales when comparing values by gender. In general, males have higher rates for all study indicators than females. Now we will study the differences that are available on scales depending on the course of study, which are more pronounced. So, for students of the 3rd year of undergraduate and graduate studies, such features as passion and resourcefulness are most strongly developed. At the same time, the minimum values of the 2d year are observed on these two scales. Moderate values are observed on the Optimism and Confidence scales. In this context, respondents to the 1st and 3d years of study can be considered the most confident in their abilities, and only the 1st year is optimistic for a favorable development of events. Weak features among respondents can be recognized as "Courage, enterprise" and "Adaptability." Minimum values are observed in the framework of the current study on the scale "Tolerance to ambiguity."

4 Discussion

The correlation analysis between these indicators did not reveal strong links between them, but found the following moderate-strength links:

1. Passion and resourcefulness (r = 0.6195).
2. Passion and confidence (r = 0.4861).
3. Confidence and resourcefulness (r = 0.5310).
4. Courage, entrepreneurship and adaptability (r = 0.4788).
5. Adaptability and tolerance to ambiguity (r = 0.4069).

The identified pairs of joint dependencies give rise to the assumption that the respondent's attempts to develop one of the above features will lead to the development of other characteristics that are in the existing relationship. In addition, the revealed relationships between the scales were noted in the work of Rudenko [4], where the similarity of the results of the PCRS method with the division of individuals by the type of innovative thinking was correctly noticed. It is important to note that the current course has only a minor impact on the scale.

By examining the averages for each criterion (Table 2), the following conclusions can be drawn. The greatest influence on the calculation of the integral indicator of readiness for change among students and undergraduates is their creative thinking, which is one of their manifestations of resourcefulness.

Table 2. Study indicators for each criterion

Characteristic	Mean value
Passion	19,43
Resourcefulness	20,25
Optimism	18,3
Courage, enterprise	14,01
Adaptability	15,34
Confidence	18,74
Tolerance to ambiguity	12,8
Integral indicators of readiness for change	118,87

Source: authors

The level of energy and passion of almost the majority of respondents is at the average level, according to this methodology, which indicates that they are capable of acting in a volatile environment, but will not always strive to take the initiative. Confidence takes the third place, and the values obtained tell us about low self-confidence in changing conditions. This suggests that these respondents may doubt and possibly be temporarily inactive in such conditions, which will affect the end result and the process of change itself.

If the values of the previous scales had a more favorable effect, then, starting from the fourth, a negative effect on the readiness for change begins to develop. Low values on the Optimism scale characterize most students as pessimistic in perceiving such a phenomenon as changes. Negative attitudes and the denial of the benefits of their implementation are among the factors of resistance to change.

The ability to adapt to environmental variability in most students is also at a low level and ranks fifth. Low readiness to change the principles of their work, personal planning adversely affects the resulting outcome of any changes and can be a factor in the development of stress of an auto-destructive nature.

In addition, from the point of view of the authors of the applied methodology, respondents to this survey have a low level of entrepreneurship and willingness to take risks in conditions of uncertainty and occupy the penultimate, sixth line of the list of scales, which forms an integral indicator of readiness for change. Courage and enterprise, the ability to see shortcomings in existing systems and eliminate them with qualitatively better solutions act as drivers for the subsequent development and management of environmental variability.

The lowest values in all respondents are observed on the scale "Tolerance to Ambiguity." Ambiguity by nature can be recognized as one of the manifestations of uncertainty and the low level of tolerance towards it suggests that respondents experience various negative emotions in such situations [8]. Thus, by making both impulsive and irrational decisions, they can also engage in the collection of information at times when it is necessary to quickly respond to future changes. We analyze the summary information on the sample shown in Table 3 to assess the overall readiness of students for uncertainty in the conditions of change.

Table 3. Sampling ranking by integral indicator of readiness for change

Year/gender	Change readiness level		Overall total value
	Low	Moderate	
Bachelor degree, 1 year	**19**	**12**	**31**
Female	18	9	27
Male	1	3	4
Bachelor degree, 2 year	**11**	**7**	**18**
Female	10	5	15
Male	1	2	3
Bachelor degree, 3 year	**12**	**9**	**21**
Female	10	7	17
Male	2	2	4
Bachelor degree, 4 year	**13**	**7**	**20**
Female	11	2	13
Male	2	5	7
Magistracy	**7**	**3**	**10**
Female	6	2	8
Male	1	1	2
Overall total value	**62**	**38**	**100**

Source: authors

As you can see, students with low readiness for change prevail - their share is 62%. However, not a small part is at the boundary level, which is confirmed by the average level of the integral indicator of readiness for change (according to the methodology, this indicator should not be less than 121 to achieve the average level). In terms of gender, the following picture is obtained: 13 males are classified as the average level of readiness for change, which is 65% of their number, and 25 females (31.25%), respectively.

5 Conclusion

Since the future is shaped by the forces of VUCA, it is necessary to develop the ability to adapt and flexibility in future managers in order to respond effectively to changes [5]. The formation of certain abilities will allow to be more effective in the process of work:

1. Cognitive flexibility - the ability to use different thinking strategies and mental frameworks in decision-making. The vision is simultaneously multiple scenarios when to shift and make changes.
2. Emotional flexibility - the ability to vary your approach to working with the emotions and emotions of others. Adapting to change requires mutual concessions between the manager and those experiencing the change. An emotionally adaptive manager promotes changes or a program of action without ignoring other people's problems and emotions.
3. Flexibility of disposition - the ability to remain optimistic and at the same time realistic. Dispositional flexibility (or personality-related flexibility) allows you to recognize a bad situation, but at the same time visualize a better future.

Practicing behaviors in future managers that will increase their cognitive, emotional and dispositional flexibility help them become more adaptable and, in turn, help others adapt.

References

1. Ashmarina, S.I., Mantulenko, V.V., Troshina, E.P.: Readiness to changes as one entrepreneurial value of the innovation-oriented economy. In: Ashmarina, S., Vochozka, M. (eds.) Sustainable Growth and Development of Economic Systems. CE, pp. 157–166. Springer, Cham (2019). https://doi.org/10.1007/978-3-030-11754-2_12
2. Loshkareva, E., Luksha, P., Ninenko, I., Smagin, I., Sudakov, D.: Skills of the future: How to thrive in the complex new world (2018). https://worldskills.ru/assets/docs/media/WSdoklad_12_okt_eng.pdf. Accessed 16 2020
3. Mantulenko, V.V., Mantulenko, A.V., Troshina, E.P., Vorotnikova, M.V.: Structural and functional analysis of requirements to managers of innovative companies in the conditions of the digital economy. In: Ashmarina, S., Mesquita, A., Vochozka, M. (eds.) Digital Transformation of the Economy: Challenges, Trends and New Opportunities. AISC, vol. 908, pp. 253–259. Springer, Cham (2020). https://doi.org/10.1007/978-3-030-11367-4_24

4. Rudenko, V.N.: Entrepreneurs are ready for change at various stages of business development. North Caucasus Psychol. Bull. **8**(1), 32–35 (2010)
5. Sevastyanova, S.A., Sevastianova, A.L.: Information and knowledge: from a resource to a key factor in progress. In: Ashmarina, S.I., Mantulenko, V.V. (eds.) Current Achievements, Challenges and Digital Chances of Knowledge Based Economy. LNNS, vol. 133, pp. 51–58. Springer, Cham (2021). https://doi.org/10.1007/978-3-030-47458-4_7
6. Shekshnia, S.: What should the board of directors do during a period of turbulence (2020). https://hbr-russia.ru/management/upravlenie-izmeneniyami/827872. Accessed 16 July 2020
7. Terekhin, R.A.: The correlation of values, socio-psychological attitudes of military personnel and their individual readiness for change. Educ. Sci. J. **8**(19), 109–127 (2017). https://doi.org/10.17853/1994-5639-2017-8-109-127
8. Titova, O.I.: Tolerance to uncertainty as a factor of the relation to business interaction in the context of students' common cultural competences development. Siberian J. Psychol. **68**, 131–142 (2018). https://doi.org/10.17223/17267080/68/8
9. Zelenkov, Y.A.: The effectiveness of Russian organizations: the role of knowledge management and change readiness. Russ. J. Manage. **16**(4), 513–536 (2018)

Organizational and Economic Mechanism of Staff Turnover Management

E. A. Mitrofanova[1(✉)], A. E. Mitrofanova[1], and G. I. Margarov[2]

[1] State University of Management, Moscow, Russia
elmitr@mail.ru, ae_mitrofanova@guu.ru
[2] National Polytechnic University of Armenia, Yerevan, Armenia
gmargarov@gmail.com

Abstract. In the current crisis conditions caused by the global challenges of the 21st century: globalization, digitalization, and the COVID-19 pandemic, issues of key staff turnover management are of particular importance. The need to rethink existing approaches to staff turnover management, as well as to identify additional reserves for the development of the organizational and economic management mechanism determines the choice of the research topic, its theoretical and applied significance. The purpose of the research is to develop theoretical and methodological foundations and scientific and practical recommendations for the formation of an organizational and economic mechanism for managing staff turnover. For that purpose, the article analyzes the scientific views on the content of staff turnover and identifies its features in modern economic conditions; the factors of staff turnover in the organization are systematized and clarified; the reasons and motives for turnover are identified, that leads it to dismissal. The modern concept of personnel turnover management in the organization is proved; the organizational and economic mechanism of staff turnover management in the organization is developed.

Keywords: Factors of turnover · Motives of turnover · Organizational and economic mechanism of turnover management · Reasons for turnover · Staff turnover · Staff turnover management

1 Introduction

Nowadays, one of the most important strategic management task in the company is to ensure the sustainability of its operations in the current economic realities, during the crisis caused by the global challenges of the 21st century: globalization, digitalization, and the COVID-19 pandemic. One of the problems caused by the instability of the Russian economy is the growth of staff turnover in organizations of different types and forms of ownership [9]. This problem requires attention from employers, because the ability to attract and retain the best staff is a crucial competitive advantage that will determine the current and future profitability of most companies. It is not a surprise that increasing attention is being paid to employee retention and the costs of high staff turnover.

S. I. Ashmarina and V. V. Mantulenko (Eds.): IPM 2020, LNNS 161, pp. 590–598, 2021.
https://doi.org/10.1007/978-3-030-60926-9_75

It should be noted that it is the staff turnover rate that reflects both the mobility and stability of the organization's personnel. The turnover rate growth contributes to the growth of employees' dissatisfaction with the company, work, and working conditions. This affects many aspects of personnel management: staff selection, training system, compensation and benefits, staff adaptation, social and psychological climate, and so on [11].

The scientific elaboration of this topic, despite a fairly large number of different studies and publications, is rather disunited and practical, than holistic and theoretical. Economic studies analyze the material issues, focusing primarily on profit, rather than on the employee, considering the interests of the company above individual employees. Sociological research carefully analyzes human capital, considering it as the main one, taking into the consideration financial issues and tasks. Such one-sided approaches do not reflect the whole picture, and consider individual components without taking into account their connections and interdependence.

The complexity of this issue studying is due to the lack of optimal solutions and the ideal action plan, the reason is that this issue depends too much on the scope of activities of the organization, the specifics of the organization of work, distribution of responsibilities, organization capabilities and even the country in which the organization operates [14]. Thus, in modern conditions, issues on managing the turnover of organizations' key staff is particularly important.

2 Methodology

A retrospective analysis of the problem has shown that during the Soviet period in Russia, a stereotype which means that staff turnover is an unmanageable process that causes only negative consequences. it was understood that staff turnover was only a subject to complete elimination. This was due to the fact that staff turnover could spoil the stability factor of the team, which was then considered one of the main characteristics of the team at work. The fact that the turnover of employees is a process necessary for the development of the team, which reduces the possibility of "obsolescence" of the team or "stagnation", was not considered at that time. Nowadays, the attitude to staff turnover has changed. Staff turnover, as a social and economic phenomenon, began to be understood in a different way. The reasons for this were the changed conditions in the economy: intensification of production, implementation of new technologies. These changes have led to a situation where the market is formed by the labor oversupply. Employees began to freely choose where they want to work. The above factors influenced the mobility of workers. Employees began to move between companies in different directions (for example, territorial or professional directions).

There is no generally accepted definition of staff turnover at the moment, and scientists take the turnover differently. A number of researchers believe that staff turnover refers to unorganized employee layoffs [12]; other scientists consider turnover as the result of an imbalance in the system of competitiveness in the domestic labor market [13]. In the framework of the following approach, staff turnover is presented as a social phenomenon caused by the different interests of employers and employees [10]. Some scientists understand staff turnover through all types of dismissals, regardless of whether they are organized or spontaneous [8]. Foreign researchers mainly consider staff turnover as a result of inefficient management of the organization and a source of additional financial costs [3, 15]. Integrating fundamental aspects of the reviewed approaches, we can define the staff turnover as a form of social and labour movement of the organization staff, caused by the imbalance of interests of the internal labour market subjects in the interaction of socio-economic, socio-psychological factors, etc.

3 Results

Staff turnover depends on a variety of factors represented by phenomena and processes that directly or indirectly affect the occurrence of staff turnover in the organization. Among the factors of staff turnover, we can distinguish [2, 7]: external factors, including political, socio-economic, market, and climatic factors; internal factors, including objective (organization management system, personnel management system) and subjective (biological, professional, personal, psychological, motivational and labor characteristics of employees).

Unsatisfactory from the employee's point of view, the state of internal factors can be considered as the causes of staff turnover, transferring it from a potential state (prepossession to staff turnover) to an active one (readiness for staff turnover) [5].

The employee's awareness of the discrepancy in the state of a particular factor in the social and labor sphere leads to the formation of turnover motives that encourage employee to leave the organization. According to the results of a meta-study by a group of scientists, the main motives for staff turnover are the following (Fig. 1). Thus, to manage staff turnover, it is necessary to constantly monitor and evaluate the factors of staff turnover in order to prevent them from developing into motives of staff turnover [6].

Staff turnover management is considered as the process of identifying, evaluating, and monitoring internal and external factors that may affect the organization's staff turnover. Staff turnover management is a systematic process that begins at the stage of developing a personnel strategy and permeates the entire staff management system of the organization [16]. Organizational and economic mechanisms for managing staff turnover are represented by the following components (Fig. 2).

The main element of the developed organizational and economic mechanism for managing staff turnover is the method of qualitative evaluation of staff turnover. The main point of the practice is to evaluate all factors of staff turnover and rank them both by the influence degree on the organization's activities and by the significance of staff for such activities [1].

To identify and evaluate staff turnover factors, a survey is conducted using a specially designed questionnaire, in which employees are asked to select and evaluate staff turnover factors based on two criteria:

- impact degree (consequences scale) for the organization (range - from minimal to critical consequences) [V],

DRIVERS OF EMPLOYEE TURNOVER		
Stress 9%	☑ Role Clarity ☑ Role Overload	☑ Role Conflict ☑ Overall Stress
Demographic 8%	☑ Marital Status ☑ Kinship Responsibilities ☑ Children	☑ Age ☑ Tenure
Indicators 6%	☑ Lateness ☑ Absenteeism	☑ Performance
Leadership 5%	☑ Supervisory Satisfaction	☑ Leader-member exchange
Job Satisfaction 5%	☑ Job Satisfaction ☑ Job Met Expectation	☑ Job Involvement
Job Content 4%	☑ Routinization ☑ Promotional Chances	☑ Instrumental Communication
Ext Environment 3%	☑ Alternative Job Opportunities	☑ Comparison of Alternatives with Present Job
Co-worker 3%	☑ Work Group Cohesion	☑ Co-worker Satisfaction
Work Satisfaction 2%	☑ Work Satisfaction	
Compensation 1%	☑ Pay Satisfaction	☑ Distributives Justice

Fig. 1. Motives of staff turnover (Source: authors based on [4]).

– significance for the employee (range - from insignificant to the most significant factor) - [Z].

The integral evaluation of staff turnover factors is performed using the following formula:

$$Fi = f\,(Vi, Zi) = \sum_{k=1}^{n}(Vik \times Zik)/n, = 1 \qquad (1)$$

where Fi is the level of the I-th staff turnover factor;

Vik - evaluation of the influence degree (consequences scale) of the i-th factor by the k-th employee;

Purpose of the staff turnover management	Efficiency increasing of the organization by retaining the most valuable employees and maximizing the use of their labor potential in accordance with the staff management strategy
Staff turnover management tasks	• Identification and assessment of factors and causes of staff turnover • Evaluation the scale and economic results of staff turnover • Formation of methodological tools for managing staff turnover in order to prevent and reduce it • Development of a program to reduce staff turnover and retention • Evaluation of the socio-economic effectiveness of the staff turnover reducing and retention program
Staff turnover management object	Employees of the organization
Staff turnover management subject	Factors and staff turnover reasons
Staff turnover management object	• Organization • Staff department • Line managers
Staff turnover management principles	• System approach • Strategic orientation • Continuity of the management process • Integrated management methods
Staff turnover management functions	• Analytical • Regulatory • Motivational • Control
Methods staff turnover management	• Administrative • Economic • Social and psychological

Fig. 2. Organizational and economic mechanisms for managing staff turnover of the organization (Source: authors).

Zik - evaluation of the significance of the i-th factor of k-m employees;

k-employee number – (from 1 to n).

Its ranking and creation of staff turnover profile in the organization is implemented on the basis of the values evaluation of the staff turnover factors.

The staff turnover profile of an organization is a dynamic characteristic of staff turnover in the form of a ranked list of turnover factors evaluated by two criteria: the consequences number for the organization and the significance for employees. Analysis of the staff turnover profile allows you to identify those factors of turnover that have a significant negative impact on the organization's activities and its personnel. You should regularly update the staff turnover profile in order to analyze changes in this profile, identify trends in organization staff turnover and on this basis form measures to manage staff turnover.

4 Discussion

The method of qualitative evaluation of staff turnover was tested at an industrial enterprise, the evaluation of turnover factors is shown in Table 1, Fig. 3 shows the profile of staff turnover.

Table 1. Evaluation of staff turnover factors in the organization (fragment)

№	Staff turnover factor	The number of consequences for the organization	Significance of the factor for the employee	Outcome evaluation
1.	Inefficient remuneration system	0.82	0.7	0.76
2.	Non-standardized working hours	0.78	0.54	0.66
3.	Adverse working conditions	0.78	0.62	0.7

Source: authors.

Factors shown in Fig. 3:

1. Inefficient remuneration system.
2. Non-standardized working hours.
3. Adverse working conditions.
4. Adverse social and psychological climate in the team.
5. Inefficient work organization.
6. Lack of initiatives and creativity stimulation measures.
7. Lack of career opportunities in the organization.
8. Lack of staff adaptation.
9. Lack of measures aimed at the formation of the staff cohesion.
10. Subjective methods of staff appraisal.
11. Irrational organizational structure.
12. Lack of the social package.
13. Lack of additional medical insurance program.
14. Lack of additional retirement insurance program.
15. Training goals do not correspond to the interests of employees.
16. Lack of conditions for professional development.
17. Low level of labor discipline at the enterprise.
18. Low professional level of the enterprise's management.
19. An undeveloped communication system.
20. Lack of measures to increase staff loyalty.
21. An undeveloped system of work measurement and regulation.
22. Presence of conflicts.

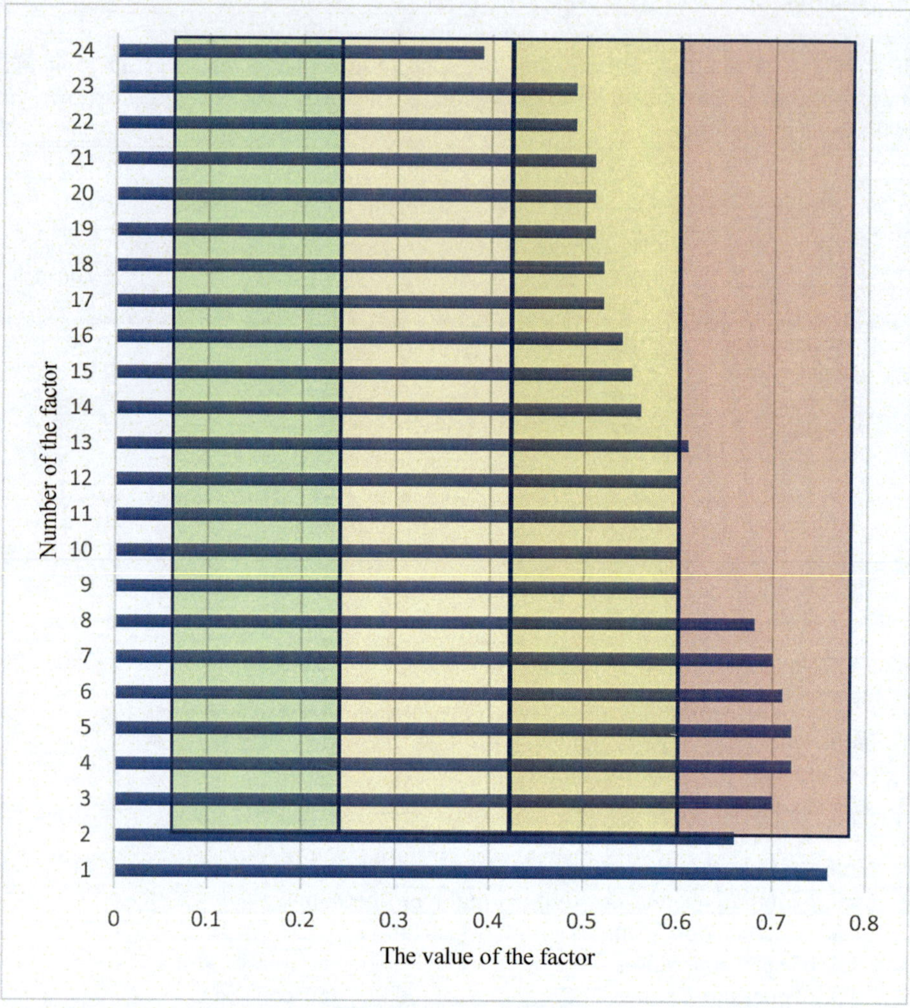

Fig. 3. The profile of the staff turnover on the enterprise (Source: authors).

23. Undeveloped social infrastructure of the enterprise.
24. The irrationality of work and rest modes.

The analysis of the formed staff turnover profile has shown that 8 factors belong to the area of high significance: inefficient remuneration system, non-standardized working hours, adverse working conditions, inefficient work organization, adverse social and psychological climate, lack of initiatives and creativity stimulation measures, lack of career opportunities in the organization, lack of staff adaptation.

If we talk about the main directions of solving the problem of high staff turnover at the analyzed enterprise, we need to use the following methods:

– review of the recruitment process,

- improving the system of material and non-material motivation,
- implementation of a system for monitoring the workload of employees in order to prevent their burnout,
- creation and maintenance team spirit,
- creation of comfortable working conditions,
- organization of the employee adaptation system.

5 Conclusion

Competent management of staff turnover by the organisation's management team is the main factor that can affect the dangerous situation for the company when employees quit regularly. By implementing effective measures that help employees feel confident, in-demand, and comfortable, the management will improve the organization's performance and reduce staff turnover to zero. Thus, the implementation of the proposed methodological approach to managing staff turnover will be an important direction for improving the efficiency of organizations.

References

1. Bruns, A.S.: Is all turnover intent the same? Exploring future job preference and environmental considerations. Public Manage. Rev. **20**(12), 1768–1789 (2018)
2. Bullock, J.B., Stritch, J.M., Rainey, H.G.: International comparison of public and private employees' work motives, attitudes, and perceived rewards. Public Admin. Rev. **75**, 479–489 (2015)
3. Caillier, J.G.: Do transformational leaders affect turnover intentions and extra-role behaviors through mission valence? Am. Rev. Public Admin. **46**, 226–242 (2016)
4. Griffeth, R.W., Hom, P.W., Gaertner, S.: A meta-analysis of antecedents and correlates of employee turnover: update, moderator tests, and research implications for the next millennium. J. Manage. **26**(3), 463–488 (2000)
5. Hancock, J.I., Allen, D.G., Soelberg, C.: Collective turnover: an expanded meta-analytic exploration and comparison. Hum. Resour. Manage. Rev. **27**(1), 61–86 (2017)
6. Hom, P.W., Lee, T.W., Shaw, J.D., Hausknecht, J.P.: One hundred years of employee turnover theory and research. J. Appl. Psychol. **102**(3), 530–545 (2017)
7. Ivashko, A.S., Kochetkova, R.M.: Problems of staff turnover. Bull. UlSTU **4**(80), 58–59 (2017)
8. Mitrofanova, E.A., Shablin, A.V.: Management of staff turnover in an industrial enterprise: a methodological approach. Hum. Resour. Intell. Resour. Manage. Russia **5**(38), 9–14 (2018)
9. Novoselova, I. A., Samylkina, T. A.: The problem of staff turnover in modern organizations. Siberian Federal University, Krasnoyarsk (2014)
10. Peretz, H., Levi, A., Fried, Y.: Organizational diversity programs across cultures: Effects on absenteeism, turnover, performance and innovation. Int. J. Hum. Resour. Manage. **26**(6), 875–903 (2015)
11. Rubenstein, A.L., Eberly, M.B., Lee, T.W., Mitchell, T.R.: Surveying the forest: a meta-analysis, moderator investigation, and future-oriented discussion of the antecedents of voluntary employee turnover. Pers. Psychol. **71**(1), 23–65 (2017)

12. Skavityn, A.V.: Methodological approaches to staff turnover management (2014). http://www.aup.ru/articles/personal/2.htm. Accessed 21 May 2020
13. Sotnikova, S. I., Soloveva, Yu. Yu.: Diagnostics of staff turnover in trade industry (2015). https://hr-portal.ru/article/diagnostika-tekuchesti-personala-v-torgovle. Accessed 21 May 2020
14. Tikhonov, S.: How to reduce staff turnover (2015). http://www.gd.ru/articles/3631-tekuchest-kadrov. Accessed: 21 May 2020
15. Wang, J.: Hours underemployment and employee turnover: the moderating role of human resource practices. Int. J. Hum. Resour. Manage. **29**(9), 1565–1587 (2018)
16. Zhu, X., Seaver, W., Sawhney, R., Ji, S., Holt, B., Sanil, B.G., Upreti, G.: Employee turnover forecasting for human resource management based on time series analysis. J. Appl. Stat. **44**(8), 1421–1440 (2017)

Logistics Outsourcing Efficiency

S. V. Noskov(✉)

Samara State University of Economics, Samara, Russia
noskov50@yandex.ru

Abstract. Logistic outsourcing is the use of the resources of a specialized company to carry out business processes and the functions of logistics management. The development of outsourcing in Russia is constrained by the insufficient development of the logistics services market. The cost-effectiveness of outsourcing is not absolute. The purpose of the work is to evaluate the effectiveness of logistics outsourcing. Tasks include determining the location of logistics outsourcing in the management system, a review of methods for evaluating the effectiveness of outsourcing as a strategic decision of a company, and determining the factors of its effectiveness. The work used the provisions of economic theory and management, the concept of reduced costs, the classification of costs of the organization. The result of the study is a method for assessing the effectiveness of logistics outsourcing by changing the absolute value of the reduced costs of the company before and in the conditions of outsourcing, testing the proposed method.

Keywords: Efficiency · Factors · Logistics · Methods · Outsourcing · Services

1 Introduction

Outsourcing is the use of external resources in economic activity in the form of transferring non-key organization business processes and their management functions to specialized companies. For these companies, these business processes are key, therefore, they can be performed more efficiently with minimal costs. In logistics, business processes of logistics of supply, sales, as well as information logistics and logistics of services can be outsourced.

The main factors in the effectiveness of logistics outsourcing are improving the quality of business processes performed by the logistics provider, transferring part of the supply and sales risks, reducing costs and expenses of fixed and working capital (storage costs, transaction costs, fixed assets and working capital in stocks), reduction of taxes and insurance payments. In addition, the result of cooperation with logistics service providers and trust is the wider use of digital technologies, the formation of supply chains and networks, the acquisition of new knowledge and skills in logistics management, and the growth of human capital [12]. These and other factors of the economic efficiency of outsourcing are shown in Fig. 1.

S. I. Ashmarina and V. V. Mantulenko (Eds.): IPM 2020, LNNS 161, pp. 599–607, 2021.
https://doi.org/10.1007/978-3-030-60926-9_76

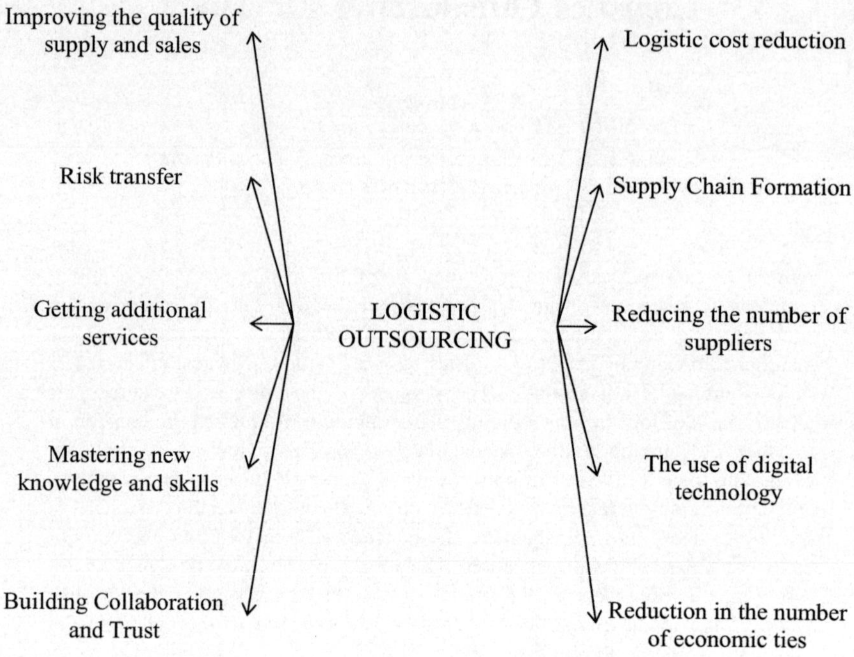

Fig. 1. Logistics outsourcing efficiency factors (Source: author)

However, logistics outsourcing is associated with additional costs and the disclosure of certain aspects of commercial activity, due to the use of digital technologies and greater transparency of the business. The price of the services of logistics providers is significantly higher than the organization's costs for the transferred business process and its management function.

Logistic costs are reduced by their elements shown in Fig. 2.

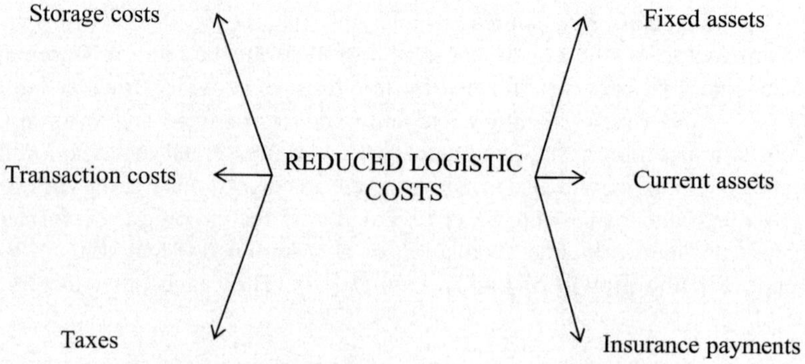

Fig. 2. Elements of logistics costs (Source: author)

A fundamental external factor in the economic efficiency of logistics outsourcing is the degree of development of the market for logistics services providers at the 3PL level. The Russian market of logistics services is significantly inferior in development to its Western counterparts [2]. This is determined by a low degree of competition due to the high concentration of sales of services at this level; high prices for logistics services, their low quality and variety. The degree of development of the 3PL logistics service providers market can be indirectly determined by the share of 3PL service providers in the cost of all logistics at the national level. Thus, the share of revenues of Russian suppliers of 3PL services amounted to only 7.9% of the total cost of logistics, or $ 20.7 billion in 2018, and in Germany - 10.7%, or $ 34.0 billion, respectively.

Another indicator of the degree of development of the Russian market of logistics service providers is the high share of logistics costs in gross domestic product (in Russia - 19.0%, and in the USA - 9.5%). This is due to the incomplete optimality of logistics solutions and business processes of supply and sales. The bulk of the cost of logistics falls on manufacturers and consumers of goods, rather than providers of logistics services. The implementation of the factors of the effectiveness of logistics outsourcing is possible only in conditions of growing demand and supply in the market of logistics services [9]. Low demand is due to high prices for services, their low quality and diversity. The insufficient supply of high quality 3PL services is explained by the lack of investment in the logistics infrastructure. Consequently, the effectiveness of logistics outsourcing in Russia is not absolute and should be confirmed by its calculations [1].

2 Methodology

The work used the provisions of economic theory about the resources of the company and costs. Important elements of the company's resources are fixed and working capital. Elements of current logistics supply costs include transport and procurement costs, storage costs, transaction costs, and insurance payments. The study used methods of market analysis of the degree of development of the logistics services market. The factors of the effectiveness of logistics outsourcing were distinguished on the basis of the expert method. The main provisions of management theory made it possible to establish the place of logistic outsourcing in the company's management system. And also highlight the business processes and functions of supply logistics management. It has been established that logistics outsourcing is a strategic decision of the company. The paper reviews methods for assessing the effectiveness of logistics outsourcing. A method for evaluating the effectiveness of outsourcing based on the concept of reduced costs has been developed. Outsourcing efficiency was calculated in accordance with this concept.

3 Results

The use of outsourcing in the activities of the organization is a strategic decision [7]. In the management system, strategic decisions relate to the function of strategic planning. Strategic planning itself includes procedures for setting goals and developing strategies for achieving them, which are detailed in the strategic plan of measures [11]. The goals and strategies of the organization in their hierarchy include the corporate level, the level of business units and functional activities (logistics, marketing, manufacturing, personnel management, finance, security, etc.) [8]. Further differentiation of strategic decisions in logistics can be carried out according to its functional areas (supply, sales, reverse logistics). Thus, it is possible to accurately indicate the place of logistic outsourcing as a strategic decision in the organization's management system.

Methods for evaluating the effectiveness of logistics outsourcing:

1. Logistic outsourcing is a strategic decision of the organization. Evaluation of the effectiveness of any organization strategy is carried out according to the degree of achievement of the goal set for this type of activity. The main goal of strategic logistics decisions is to reduce the share of logistics costs in the organization's revenue. Logistics costs are of two types: current costs (transaction costs, storage costs, transportation and procurement costs, taxes and insurance payments) and one-time investments in fixed assets for transport and storage purposes, working capital in stocks. The total logistics costs are equal to the sum of current costs and reduced to the annual dimension of one-time costs.

Since the share of logistics costs in revenue may vary from other factors not related to the organization of outsourcing, this should be taken into account in the calculation of efficiency. Thus, the evaluation of the effectiveness of logistics outsourcing is carried out by reducing the share of the organization's logistics costs before outsourcing and in the conditions of outsourcing.

2. The change as a result of outsourcing of economic value added (EVA) allows us to evaluate its effectiveness. The elements of EVA growth are:

- an increase in the organization's profit due to a decrease in inventory storage costs to a greater extent than an increase in transportation and procurement costs,
- reduction as a result of outsourcing of fixed and current assets (capital).
- decrease in other capital during the liquidation of other business processes and management functions. For example, a reduction in the number of logistics service personnel under outsourcing leads to the release of office space.

3. Evaluation of the effectiveness of logistics and logistics outsourcing can be carried out by changing the indicator of return on assets of the organization (ROA). Thus, a change in current costs as a result of outsourcing will lead to a change in net profit in the ROA numerator. A change in the value of fixed assets for transport and storage and working capital in inventories will lead to a change in the organization's assets in the ROA denominator. However, the change in ROA should reflect only the factors of the effectiveness of logistics outsourcing [6].

Table 1. Key performance indicators of supply chain logistics business processes

Business processes/ Management functions	Key performance indicators
1. Procurement	1.1. Customer compliance level 1.2. Number of suppliers 1.3. Purchase order cycle time
2. Supplier management	2.1. Supplier availability 2.2. Material defect rate 2.3. Lead time
3. Cost-effective procurement/order	3.1. Purchase order value 3.2. Reduce purchase costs 3.3. Exclude purchase costs 3.4. Return on investment in procurement
4. Warehouse management	4.1. Operating costs 4.2. The number of received batches of materials 4.3. Rational placement of materials
5. Transportation	5.1. Delivery time 5.2. Fare 5.3. Volumes of supply lots 5.4. Kind of transport
6. Inventory management	6.1. Inventory turnover 6.2. Inventory level 6.3. Inventory cost 6.4. Inventory accuracy

Source: author.

4. The system of key performance indicators (KPI) can be used in assessing the effectiveness of logistics outsourcing and its business processes [5]. The main KPIs by groups of business processes and supply logistics management functions (procurement, supplier management, procurement/order economy, warehouse management, transportation, inventory management) are shown in Table 1.

The effectiveness of logistics outsourcing is determined by the change in KPI before and in the conditions of outsourcing [10]. The methodology for calculating changes in KPI includes:

– rationing of individual KPI indicators in each group from 0 to 1. Rationing is carried out on the basis of calculating the ratio of each indicator to and under outsourcing. As a result, all KPIs get relative units of measure,
– calculation of the weights of each KPI by group and the weights of each group of business processes and management functions. The weights of indicators and their groups are determined by any expert and economic-mathematical methods,
– determination of the integral KPI before and in the conditions of the organization of outsourcing as a double sum of products of weights by normalized KPI values,
– calculation of changes in the integral KPI before and in the context of outsourcing of business processes and supply logistics management functions.

5. Evaluation of the effectiveness of logistics outsourcing can also be based on a balanced scorecard, which introduces an additional projection of logistics; investment calculations based on indicators of net present value, payback period and internal return index; benchmarking; the absolute value of the reduced costs [4].

Changing the value of the organization's logistic costs in terms of outsourcing is an affordable and simple method for evaluating its effectiveness. Calculation of changes in reduced costs for evaluating effectiveness is carried out after the procedure for selecting a logistics provider [3]. The method for evaluating the effectiveness of logistics outsourcing has been tested in the organization "NOVA" Ltd. Preliminary selection of a logistics service provider was carried out according to the criterion of the minimum ratio of the price of a service to its utility. "MetallSamara" Ltd was chosen as the provider of logistics services.

The methodology for assessing the economic efficiency of outsourcing is based on a comparison of the increase in transportation and procurement costs of "NOVA" Ltd when switching to outsourcing and saving the company's reduced costs. If the savings in the above costs are higher than the increase in the transportation and procurement costs of the company, then a conclusion is drawn on the advisability of outsourcing. The economic efficiency of outsourcing logistics services is determined by the annual economic effect as the difference in savings in reduced costs and the increase in transportation and procurement costs of the company. Thus, the feasibility of outsourcing the business processes of inbound logistics of "NOVA" Ltd was determined by the formula:

$$\Delta TPC < \Delta(CR + SC + AE + CC + WC), \quad (1)$$

where

TRC - transportation and procurement costs of the company, thousand rubles/year;
CR - expenses for the reception registration and placement of rolled ferrous metals in the warehouse, thousand rubles/year;
SC - storage costs of rolled ferrous metals, thousand rubles/year;
AE - management expenses, thousand rubles/year;
CC - reduced to the annual dimension the cost of fixed capital for warehouse purposes, thousand rubles/year;
WC - reduced to the annual dimension working capital costs in the stocks of rolled ferrous metals, thousand rubles/year.

Reduced to annual dimension, the cost of fixed capital stock for outsourcing before outsourcing was determined by the product of their amount (7000 thousand rubles) by the return on assets of the company in relative units (0.08/year):

$$CC = 7000 \cdot 0.08 = 560 \text{ (thousand rubles/year)}.$$

In terms of outsourcing, the reduced costs of fixed capital for warehouse use will amount to 160 thousand rubles/year:

$$CC = 2000 \cdot 0.08 = 160 \text{ (thousand rubles/year)}.$$

Similarly, the calculation of reduced to the annual dimension of working capital costs in the stocks of rolled ferrous metals was carried out. The annual volume of consumption of rolled ferrous metals by "NOVA" Ltd was 300 tons according to the data of 2018, its stock value without outsourcing was equal to 25 tons, under outsourcing conditions - 3 tons. The weighted average price of rolled ferrous metals was 40 thousand rubles/ton.

The reduced costs of working capital without outsourcing amounted to 80 thousand rubles/year, in terms of outsourcing - 10 thousand rubles/year:

$$WC = 40 \cdot 25 \cdot 0.08 = 80 \text{ (thousand rubles/year)}.$$

$$WC = 40 \cdot 3 \cdot 0.08 \approx 10 \text{ (thousand rubles/year)}.$$

The calculation of the feasibility of outsourcing the logistics business processes of supply logistics is presented in Table 2.

Table 2. Calculation of the economic efficiency of outsourcing logistics supply (thousand rubles/year)

Expenditures	Without outsourcing	In terms of outsourcing	Change (+, −)
Transportation and procurement costs	750	1480	730
The costs of receiving the goods, registration and placement in the warehouse	30	75	45
Stock holding costs	590	210	−380
Management expenses	125	38	−87
Reduced to annual dimension, the cost of fixed capital stock	560	160	−400
Reduced working capital costs in inventories	80	10	−70
Total	2135	1973	−162

Source: author.

Thus, the annual economic effect of the transfer of business processes for the supply logistics of "NOVA" Ltd to the logistics service provider "MetallSamara" Ltd will amount to 162 thousand rubles/year. Therefore, outsourcing is effective.

4 Discussion

Existing methods for evaluating the effectiveness of logistics outsourcing, as a rule, use indirect indicators of its effectiveness. These indicators relate to the performance of the organization as a whole (EVA, ROA, balanced scorecard) or logistics (KPI). Under these conditions, the problem arises of eliminating the action of other factors. The proposed method of evaluating economic efficiency allows you to take into account the change only in the costs as a result of logistics outsourcing. The disadvantage of such a measurement of effectiveness is the neglect of social, organizational, informational, cognitive and behavioral factors. All of this may be the subject of future research.

5 Conclusion

The main factor in the economic efficiency of logistics outsourcing in Russia is the degree of development of the market for logistics service providers. Other efficiency factors include reducing logistics costs, improving the quality of supply and sales, obtaining additional services, and building trust and cooperation in supply chains. The combination of positive and negative factors requires calculating the cost-effectiveness of outsourcing.

Methodological approaches to assessing the effectiveness of outsourcing: logistic outsourcing is a strategic decision; in the calculations of efficiency can be used financial and economic indicators of the company as a whole; it is proposed to calculate the determination of the effectiveness of logistics outsourcing by changing the organization's reduced costs before and in the context of outsourcing.

In terms of outsourcing, transportation and procurement expenses of the organization, expenses for the acceptance of goods, processing and placement at the warehouse increase, but expenses for stock storage, management expenses, reduced to annual dimension costs of fixed capital for warehouses, reduced to annual dimension, costs of working capital in stocks. Evaluation of the effectiveness of logistics outsourcing is carried out on the basis of a comparison of these cost items.

References

1. Aamer, A.M.: Outsourcing in non-developed supplier markets: a lean thinking approach. Int. J. Prod. Res. **56**(18), 6048–6065 (2018). https://doi.org/10.1080/00207543.2018.1465609
2. Armstrong and Associates, Inc.: Global 3PL Market Size Estimates (2020). https://www.3plogistics.com/3pl-market-info-resources/3pl-market-information/global-3pl-market-size-estimates/. Accessed 17 June 2020
3. Bianchini, A.: 3PL provider selection by AHP and TOPSIS methodology. Benchmark. Int. J. **25**(1), 235–252 (2018). https://doi.org/10.1108/bij-08-2016-0125.
4. Bottani, E., Rizzi, A.: A fuzzy TOPSIS methodology to support outsourcing of logistics services. Supply Chain Manag. **11**(4), 294–308 (2006). https://doi.org/10.1108/13598540610671743

5. Cichosz, M., Nowicka, K., Ocicka, B.: Collaborative outsourcing for sustainable transport management. In: Leal Filho, W., Borges de Brito, P.R., Frankenberger, F. (eds.) International Business, Trade and Institutional Sustainability. WSS, pp. 709–724. Springer, Cham (2020). https://doi.org/10.1007/978-3-030-26759-9_41
6. Handley, S.M.: How governance misalignment and outsourcing capability impact performance. Prod. Oper. Manage. **26**(1), 134–155 (2017). https://doi.org/10.1111/poms.12609
7. Jain, T., Hazra, J., Cheng, T.C.E.: IT outsourcing and vendor cost improvement strategies under asymmetric information. Decis. Sci. (2020). https://doi.org/10.1111/deci.12446
8. Kull, T.J., Ellis, S.C.: Coping with dependence: a logistics strategy based on interorganizational learning for managing buyer–supplier relations. J. Bus. Logistics **37**(4), 346–363 (2016). https://doi.org/10.1111/jbl.12146
9. Nadarajah, G.: Factors influencing third party logistics performance in Malaysia: the role of trust as a mediator. Int. J. Supply Chain Manage. **4**(4), 108–114 (2015)
10. Richards, G., Grinsted, S.: The Logistics and Supply Chain Toolkit: Over 100 Tools for Transport, Warehousing and Inventory Management. Kogan Page Publishers, London (2016)
11. Sheu, J.B., Kundu, T.: Forecasting time-varying logistics distribution flows in the One Belt-One Road strategic context. Transp. Res. Part E **117**, 5–22 (2018). https://doi.org/10.1016/j.tre.2017.03.003
12. Tsay, A.A., Gray, J.V., Noh, I.J., Mahoney, J.T.: A review of production and operations management research on outsourcing in supply chains: implications for the theory of the firm. Prod. Oper. Manage. **27**(7), 1177–1220 (2018). https://doi.org/10.1111/poms.12855

Anti-crisis HR Management Program as a Measure of Retaining Specialists in Organizations

A. M. Patrusova, P. V. Kharitonova(✉), and M. Yu. Vakhrusheva

Bratsk State University, Bratsk, Russia
patrusova@mail.ru, hpv83@mail.ru, mvahr@mail.ru

Abstract. In the context of the economic crisis, the formation of the working environment in Russian companies is becoming increasingly important. It is particularly important to establish such social proportions in the main characteristics of the organization's personnel that contribute to its growth and development, including by attracting and retaining professionally trained specialists and other labor resources in the staff. The aim of this work is to supplement theoretical and methodological approaches to the formation of the working environment of companies in the conditions of economic crisis based on the method of attracting and securing regional personnel. The methodological basis of this research includes a systematic approach, a logical approach, analysis, abstraction, classification and comparison techniques. The results of the research work are the development of the theory in the field of anti-crisis HR management by suggesting ways to solve problems of staff outflow in an unstable economic environment. The authors' step-by-step program of anti-crisis HR management as a measure of retaining specialists in the Russian regions is presented.

Keywords: Adaptation · Business assessment · HR management · Human resources potential · Staff outflow · Training program

1 Introduction

In Russia as a whole and in its regions in particular, the issue of human resources and human potential is relevant. There is a shortage of specialists in many spheres. Under personnel potential, we understand a system of factors that allow achieving goals of an organization through effective HR management. In a broader sense, human resource potential is a competence of personnel that allows, all other things being equal, the company's management to receive surplus value, including social value. In a narrower sense, the human resource potential is often considered in the science as a reserve for the promotion of an employee on the career ladder that includes his internal motives.

Competent management of personnel potential creates a well-coordinated team, an orderly system and professional development of employees. If we turn to the conceptual apparatus, then in science we can find the use of human resources as a synonym for personnel. We believe that "personnel", "staff", "labor resources", "labor force", etc. are not synonymous with human resources and human resources potential.

S. I. Ashmarina and V. V. Mantulenko (Eds.): IPM 2020, LNNS 161, pp. 608–613, 2021.
https://doi.org/10.1007/978-3-030-60926-9_77

Human resources potential is a broader concept and is considered as a synthesis of not only people, but also their intellectual and professional competencies, motivation for development and training, which allow achieving the company's strategic goals.

Human resources management is the art of achieving results through the use of tools to work with people. And here the word "people" becomes a key. When employees are treated as people, that is, with respect, interest, and a friendly attitude, then even at the same level of wages, a person is able to show high productivity results for a long time. This also applies to situations where companies are in crisis. Let's look at the main programs that can be used by organizations' managers when working with staff.

2 Methodology

The object of this research is personnel management tools for retaining highly professional personnel in the regions of Russia. The subject of the research is the development of an anti-crisis program of HR management as a measure of retaining specialists in organizations. The working hypothesis of the study suggests building a systematic understanding of the role of labor environment of companies in the economic crisis and improving this environment through attracting and retaining regional personnel. In the current theory and practice, the issue of the role of the working environment for increasing the labor productivity and retaining highly professional workers in a crisis needs to be clarified and supplemented.

The theoretical basis of this research covers works of both domestic and foreign scientists on the role of socio-psychological climate, working conditions, motivation, residence, labor organization, and staff adaptation. The data of the Federal State Statistics Service, legal acts of Russia, and information received at national and international conferences on the issues of HR management in crisis conditions served as the research information base. The authors' analysis of the most important aspects of the impact of crisis on personnel shows that the following tasks are necessary and vital for managers in the field of HR management:

Task 1. Conducting a Personnel Audit of the Human Resource Management System. At this stage, the so-called social analysis of the working team is usually carried out. This analysis includes various components that can be evaluated quite effectively by HR managers, including with the involvement of psychologists. These components include assessments of organizational culture, as the core element in any company [8]. It is the organizational culture that sets a vector of action in organizations. The success of achieving the company's strategic goals and objectives depends on its competent formation. In addition to the organizational culture, it is necessary to study the organizational structure, the distribution of personnel by categories, and the social climate that has been established in the team. This will determine the number and quality of conflict situations and personnel's job satisfaction.

Task 2. Clarification of Interaction with Regulatory Groups: federal and territorial financial recovery and bankruptcy authorities. As noted in the work of Solodova and Kleymenova, under conditions of unstable functioning of organizations, the importance of external "partners" (state, territorial, administrative bodies, trade unions, etc.) increases [9]. Therefore, attracting external partners is one of the areas of business interaction in the context of the economic crisis, which is aimed at redistributing financial burden of companies.

Task 3. Training of Managers in the Anti-crisis Sphere (Crisis HR Management). It is recommended to use research results in the preparation of training programs in the field of HR management. The manager is looking for more effective solutions in the field of personnel management. The level of training of anti-crisis managers determines the degree of their competence in critical situations.

The authors of this article highlight the most important goals of training programs, among which the following are formulated: the development of skills and abilities necessary for effective management; the development of attitudes among employees that contribute to the successful achievement of organization's goals; systematization of existing experience; rethinking the content of work, etc. Thus, each company will solve the presented tasks in its own way, taking into account its real possibilities of implementing specific measures, as well as observing the proposed principles.

3 Results

It is proposed to solve the problem of staff outflow by developing and implementing a program of anti-crisis HR management as a measure of retaining specialists in the conditions of economic crisis. This program includes a number of activities.

Activity 1. Implementation of a Training Program for HR Department Specialists "Anti-crisis HR Management". Speaking about the effectiveness of personnel management, we mean high values for the following indicators: the efficiency of the work of administrative and managerial personnel, the effectiveness of the management process, the effectiveness of the management mechanisms. The training program for HR specialists "*Anti-Crisis HR Management*" should be aimed at improving these indicators and enables a reorientation to "anti-crisis" goals and methods of working with personnel.

These include:

- focus on maximizing the organization's human resources potential,
- reducing repetitive tasks and unnecessary bureaucratic schemes by allocating responsibility functions between departments and managers as efficiently as possible, and choosing the optimal structure of the organization,
- development of the most "open", clear, simple criteria for evaluating personnel,
- improving the system of personnel motivation, in particular, stimulating labor activity by applying an effective schedule of working hours and forms of wages for all participants in the process,
- implementation of a successful system of personnel development as a human resource potential for the organization in order to achieve the company's strategic goals,

- formation of an optimal and effective organizational culture that allows, all other things being equal, to achieve effective performance indicators.

Improving the qualification of HR department staff will have a positive impact on the quality of human resource management, which will lead to an increase in the productivity of employees in the whole organization and the retention of highly qualified specialists.

A specialist in personnel management who has been trained in this program can offer an *anti-crisis procedure for reviewing the labor functions of each employee.* Performing only vital tasks for the organization and, if possible, reducing and facilitating the work functions of specialists will allow employees to keep the most necessary job responsibilities. This measure will reduce the cost of performing work that is not relevant. On the one hand, this activity is organizational, and on the other hand, it is social and has a "saving" character for the organization's personnel potential. If Activity 1 is completed, the HR specialist is recommended to implement Activities 2 and 3 sequentially. Otherwise, the burden of implementing Activities 2 and 3 will fall on the heads of structural divisions.

Activity 2. Improvement of the Employees' Adaptation Program. The application of modern tools and mechanisms for personnel adaptation is determined by several aspects: eliminating the probability of staff turnover because of the ignorance of rules of work organization, reducing the cost of conducting the personnel adaptation procedure, reducing the time of the employee's adaptation period, improving the effectiveness and efficiency of the personnel adaptation process, etc. The adaptation program specifies what the employee is responsible for, what skills and in what order the employee should be trained, and what responsibilities he should start performing first.

As a function of the personnel management system, adaptation leads to the successful completion of the probation period of a new employee and saves the organization's costs for searching, training and evaluating personnel. The better the results of the adaptation program are, the more likely it is that the employee will remain in the organization for a long period. Thus, adaptation can be considered as a process for learning internal rules and organizational culture.

A HR specialist who has been trained in the program "Anti-Crisis HR Management" can offer an anti-crisis program for employees' adaptation. Only a complete solution of problems associated with passing the adaptation period by each new employee will help to ensure effective functioning of the organization at the proper level, including in the conditions of economic crisis.

Activity 3. Clarification of Indicators of Business Evaluation of Employees. Business evaluation of personnel acts as a feedback with management, which can be used to evaluate, among other things, the effectiveness of the implemented HR policy and personnel management system. The assessment aimed at achieving the main ultimate goal of improving the personnel performance and productivity increase. Based on the results of business evaluation of employees, the following tasks are implemented: planning professional training; planning professional development and growth;

providing remuneration results, career growth or termination of an employment contract with an employee.

4 Discussion

The authors' analysis of human resources potential in various sectors of the economy showed that the employer is not able to provide labor resources with economic activities [6, 10]. Thus, people of working age are not employed in the region's economy. This, in turn, leads to an outflow of population from the Russian regions.

In general, it should be noted that because of the crisis situation in the region, the problem of outflow of labor resources to the Western part of our country and abroad has become even more acute. The most difficult moment in a crisis situation for an enterprise is the loss of both qualified personnel and labor resources, without which it is impossible to overcome the crisis. In the works of Grudistova, Pastuhova, Slinkov, Slinkova, and Melnikov [1], Kaverzina, Kovalenko, Dudina, and Belskii [2], Kaverzina, Koshevoi, and Dorofeeva [3], Kaverzina [4], Patrusova and Sapozhnikov [7], Shupletsov and Haritonova [8], devoted to the study of various aspects of organizational management culture, the use of the potential of intangible resources and the use of HR management methods, it is noted that the retention of key employees becomes one of the main tasks of management in the conditions of economic crisis. In the background of the problem of outflow and shortage of qualified personnel in a crisis, providing the organization with a workable and cohesive team is a strategic task for any manager.

Ozernikova notes popular measures of HR management in the context of the economic crisis: reduction of the number of employees; termination of employment of new employees; change of working hours (part-time work); reduction of employees' wages; sending employees on leave without paying; reduction of social benefits and bonuses; reduction of employees' training costs [5]. These measures entail certain consequences that worsen the working conditions in the organization, which are fundamentally important for every employee: living conditions in the region (comfort of life, social protection of citizens); conditions of remuneration (its constant decline is the basis for making a decision to change the place of residence and search for a more developed city); working conditions in the organization (economic security of enterprises).

5 Conclusion

The anti-crisis program proposed by the authors for working with personnel in an unstable external environment, of course, includes measures to work with the workforce and implies not only formal performance of official duties by the company's managers (these are the main functions of management from planning to control), but also democratic HR management. Democratic HR management is focused on a social component. We have already discussed the importance of careful and respectful attitude to employees, their work and preferences. This often allows achieving higher company performance than the authoritarian management style, which has no place for a favorable socio-psychological climate. The program for efficient HR management in a crisis as a method of retaining highly professional employees in an unstable external environment should include following activities: training personnel managers in anti-crisis measures, clarifying the program for personnel adaptation in the organization, adjusting criteria for business evaluation of personnel. Thus, at present, the problems of employment and personnel training in the Russian regions can be successfully solved only in a comprehensive manner with the involvement of the main stakeholders. The interaction of employers, labor and employment agencies, and municipalities becomes fundamental when generating information that allows long-term planning of personnel needs.

References

1. Grudistova, E.G., Pastuhova, D.A., Slinkov, A.M., Slinkova, O.K., Melnikov, T.N.: Study of self-actualization needs of Russian students as a factor of competitiveness in the labor market. Espacios **40**(26), 18 (2019)
2. Kaverzina, L., Kovalenko, G., Dudina, I., Belskii, O.: Cost efficiency assessment of automated quality control of precast structures. In: Kopanitsa, N.O., Ovsyannikova, T.Y., Plevkov, V.S., Klopotov, A.A., Teplyakova, L.A., Radchenko, P.A. (Eds.) Proceedings of the IV International Young Researchers Conference “Youth, Science, Solutions: Ideas and Prospects”. MATEC Web of Conferences, vol. 143, p. 04006. EDP Science, Les Ulis (2018)
3. Kaverzina, L., Koshevoi, P., Dorofeeva, V.: The improvement of company planning at the building enterprise. In: IOP Conference Series: Materials Science and Engineering, vol. 667, p. 012037. IOP Publishing, London (2019)
4. Kaverzina, L.: Restructuring of the regional investment and construction complex. In: Peshkov, V.V. (Ed.) Proceedings of the International Scientific Conference Investment, Construction, Real Estate: New Technologies and Special-Purpose Development Priorities. MATEC Web of Conferences, vol. 212, p. 08030. EDP Science, Les Ulis (2018)
5. Ozernikova, T.G.: Trends in the development of human resource management systems in organizations in the region: research results. Manager **11–12**, 14–19 (2010)
6. Patrusova, A.M., Vahrusheva, M.Y.: Process management: some implementation aspects. IOP Conf. Ser. Mater. Sci. Eng. **753**(8), 082028 (2020)
7. Patrusova, A.M., Sapozhnikov, A.A.: Development of mechanisms for retaining young professionals in the region. In: Shapovalova, I.S. (Ed.) Reproduction of Intellectual Capital in the system of Higher Professional Education, pp. 152–156. Belgorod State University, Belgorod (2013)
8. Shupletsov, A.F., Haritonova, P.V.: Forming and development of organization culture as factor determining organization’s entrepreneurial competitiveness. Baikal Res. J. **1**, 24 (2012)
9. Solodova, N.G., Kleymenova, L.V.: Personnel audit in crisis management. Proc. Irkutsk State Acad. Econ. **6**, 104–108 (2011)
10. Vahrusheva, M.V.: The role of personnel management methods in higher professional education. In: Proceedings of the International Scientific and Practical Conference Values and Interests of Modern Society, pp. 151–154. MESI, Moscow (2013)

Regulatory Nature of Moral Standards in the Organization

I. A. Romanova(✉), N. I. Laas, and E. V. Gurova

State University of Management, Moscow, Russia
romanova_ia@bk.ru, laasni@yandex.ru,
nadkate2001@mail.ru

Abstract. The article presents the results of a survey of leadership and ordinary employees of Russian organizations on the problem of the regulatory role of ethical rules for labor/organizational behavior and management of ethical behavior of personnel. The respondents' answers to the questionnaire on the mission of moral standards in the organization and the importance of their observance; on the state of compliance with ethical canons in the organization, detailed causes and forms of their violation; whether they want a sanction for violation/adherence of moral designs; on the presence of the companies' ethical codes and their purpose are analyzed and presented. The results of the survey showed that ethical standards are the most important regulator of labor/organizational behavior, and their most important purpose is "the development of good relations between employees". The study showed that the main reason for violating moral standards in companies is the personal characteristics of employees, then the management style, and then organizational factors. The results of the study are practice-oriented. At the same time, they cannot reflect the full variety of the influence aspects of the regulatory nature of moral canons on labor/organizational behavior and its management system.

Keywords: Companies' personnel · Ethical codes · Ethical labor/organizational behavior · Management of ethical behavior · Moral/ethical standards · Regulatory nature of moral standards

1 Introduction

Moral (ethical) standards are standards of people's actions based on the categories of good and evil in all spheres of public life, including labor/organizational. They are not formally fixed, and their observance is secured by the human conscience and supported by the power of public opinion. According to experts, ethics is a key problem facing modern organizations, which dictates the need for meaningful research to understand the causes and consequences of ethical decisions, as well as ethical/unethical behavior of staff from different sides, in particular, taking into account the psychology of employees [5]. The widespread and aggressive digitalization of the economy, and therefore of the employee's work/organizational life, does not reduce the value of morality. In this regard, the ethical problems of using information and communication technologies in companies are very relevant and only begin to be studied. For example,

S. I. Ashmarina and V. V. Mantulenko (Eds.): IPM 2020, LNNS 161, pp. 614–624, 2021.
https://doi.org/10.1007/978-3-030-60926-9_78

researchers focus on the contradictions between the employee as a subject of moral choice, on the one hand, and corporate requirements when using modern information technologies, on the other. These contradictions are related to the restrictions imposed on the subject of labour by the system of ethical canons and values, which can lead to unethical "technological" behavior of personnel [4].

2 Methodology

To study the role of moral standards in the activities of personnel and the functioning of organizations in general, the authors posed the following research questions, which later took the form of narrower survey questions:

1. How do employees of modern organizations relate to moral standards, does the staff recognize the value of ethical standards, and what is their purpose in the organization?
2. What is the situation with the observance/non-observance of moral canons and what are the main reasons for violation of ethical standards in organizations where respondents work?
3. What management tools do ordinary staff and management of modern organizations think are preferable for managing ethical behavior?

A written survey was conducted in April 2020 to study the regulatory role of morals and management of ethical labor/organizational behavior of personnel in modern organizations. The organizers of the study and the questionnaire authors (they are the authors of the article) are associate professors of the personnel management department of the State University of Management (Moscow). According to the initiators of the survey, its results will form the basis for the preparation of individual topics of lectures and practical classes with students studying in the field of training "Personnel management" and studying the academic disciplines "Conflictology" and "Organizational behavior". Based on the obtained data, cases and other tasks will be developed as part of active learning methods. However, the results of the survey can be used in management consulting carried out by the department of personnel management of GUU, as well as in the process of teaching courses "Conflictology", "Organizational behavior" for advanced training programs.

The respondents were employees of public and private enterprises. Using Google Forms, they were asked to fill out a questionnaire, consisting of 25 questions. The number of participants in the survey was 107. Respondents were divided by gender as follows: 80.4% identified themselves as female and 19.6% as male. 17.8% are the managers of various levels and 82.2% are survey participants from the ordinary staff. 66.4% are employees of private companies, 33.6% work in public organizations.

3 Results

To the question "Whether moral standards are an important regulator of employee behavior in the organization or not?" the absolute majority of respondents (94.4%) chose the answer "rather are"; 0.9% - "rather are not", 4.7% found it difficult to answer. Moral standards are universal and capable of handling a wide range of relationships. In accordance with the goals of the study, the authors of the questionnaire asked respondents on the narrow practical mission of ethical canons in the organization: "What do you personally think should be the main purpose of moral standards in the organization?".

As it became clear from the answers, most of the survey participants agreed that moral standards are necessary to eliminate socio-psychological problems of companies, namely, "to develop good relations between employees" (72%); "to prevent conflicts in the organization" (38.3%); "to improve relations between ordinary employees and the management of the organization" (35.5%). Among the respondents, there were many of them who believe that moral rules help in solving the organization's economic problems. Here "formation of high labor discipline" (39.3%); improvement of efficiency (39.3%); increase of employees "responsibility for labor results" (26.2%) are indicated. It is interesting, the ordinary employees (42.0%) indicated the importance of ethical images in improving the economic performance of companies more than managers (26.3%). In this regard, moral models, as one of the regulators of personnel behavior, can be successfully used by managers to increase the profitability of companies.

Moral laws are a complex phenomenon and are not obligatory for their implementation. In addition, individual employees differ in different levels of morality, i.e. the degree of implementation of these rules in their life. In this regard, for the authors of the study, the answers to the question "Do you think that if some employees of the organization ignore the rules of ethics, what should do other employees that abide it: is it still worth observing them or not?" are important. The absolute majority of respondents (95.3%) chose the answer "rather worth observing"; it is gratifying that none of the respondents answered "rather not worth observing"; 4.7% found it difficult to answer.

The following answers were received to the question "Do you think it is important or not important for employees to observe moral standards?": "very important, moral standards are of leading importance for regulating the behavior and activities of employees" (54.2%); "rather important, in general, morals are important for regulating the behavior and activities of employees" (44.9%); "rather not important, morals are not of leading importance for regulating the behavior and activities of employees (0%;) "difficult to answer" (0.9%).

Question "In general, are you satisfied or dissatisfied with the state of compliance with ethical standards by employees of the organization in which "Do you work?" caused the following reactions: "quite satisfied" (35.6%); "rather satisfied" (43.9%); "rather not satisfied" (11.2%);" absolutely not satisfied" (6.5%); "I can't answer (2.8%). At the same time, the managers we interviewed are more optimistic than employees of lower status. The answer "quite satisfied" was chosen by slightly more than half (52.6%) of leadership against 31.8% of ordinary employees.

Now we will give (in descending order) the responses of the survey participants to the question "Do you think the leadership of the organization in which you work pays enough or not enough attention to the observance of moral standards?". More than half of the respondents (57.0%) answered that "enough attention is paid". 23.4% believe that "not enough attention is paid". "Too little attention is paid" - this was the response of 12.2% of respondents. 3.7% chose the answers "too much attention is paid" and "I find it difficult to answer".

Some executives believe that the presence of moral qualities in employees can be detected already at the recruitment stage, for example, during an interview, while others are convinced that the level of morality, as the degree of compliance with moral standards by a person, is almost impossible to identify at the recruitment stage in the organization. The study authors asked survey participants to choose one of these two statements. "The presence of moral qualities in employees can be identified already at the recruitment stage, for example, during the interview process" - this is the opinion of 52.3% of respondents. 63.1% are the surveyed executives and exactly half of the survey participants are ordinary employees. A third of respondents (33.6%) chose the statement "the degree of compliance with moral standards by a person at the stage of recruitment to the organization is almost impossible to identify". Of these, 21.1% are managers and 36.4% are subordinates. "I find it difficult to answer" - 14.1% of respondents. Meanwhile, the practice of personnel selection shows that the use of specialized personal questionnaires, as well as various types of interviews (in particular, situational and behavioral) at the recruitment stage can reveal certain moral characteristics of the applicant and his moral value orientations.

Now let's speak on the questions of the violation of moral standards by employees of the organization. First of all, the question "Whether moral standards are violated in your organization or not?" was asked. "Yes, they are violated" is the most popular answer (46.7%); "no, they are not violated "(38.3%); "I find it difficult to answer" 15%. At the same time, according to the respondents' answers, ethical canons are violated more often in private organizations (49.3%) than in public organizations (41.7%).

Next, we specified the questions and tried to get answers to the question on the subjects of violation of moral standards, stating the problem as follows: "Who violates the moral standards in your organization: ordinary employees or executives?" The following results were obtained here (in descending order): "equally, both managers and ordinary employees" (36.4%); "I find it difficult to answer" (29.0%); "rather ordinary employees" (27.1%); "rather managers" (7.5%).

Since the behavior of personnel is a complex multidimensional mechanism, including both the solution of production tasks and the process of interaction with colleagues and leadership, the deviation from ethical requirements can have different configurations. Here the respondents were asked the question: "In what forms are moral standards violated in your organization?" Let's show the respondents' answers to this question (in descending order): "work avoiding, shifting responsibility to others" (56.1%); "spreading gossips, rudeness towards colleagues and leadership" (47.7%); "the desire to present themselves better on the example of mistakes of other employees" (29%); "failure to help colleagues" (21.5%); "moral bullying of individual employees" (13.1%); "organization of intrigues" (13.1%); "careerism in the negative sense" (9.3%); "deception of leadership and colleagues (8.4%); "dishonesty of employees that causes damage to the company" (7.5%) "theft, use of organizational equipment and resources (paper, copier, computer) for personal purposes" (4.7%), "alcohol consumption in the workplace" (4.7%) "I can't answer" (14.9%).

In order to demand ethical behavior of employees, company management needs to understand the reasons that make employees of organizations doubt the value of moral prescriptions and ignore them. To the question "What do you think are the reasons for employees violating ethical standards?" the responses were distributed as follows. More than half of the respondents (69.2%) believe that "personal qualities of employees". This is the most popular answer for both executives (63.1%) and ordinary employees (70.5%). Management style was chosen as the main reason by 13.1% of the survey participants (15.8% of managers and 12.5% of subordinates). The answer "organizational factors" was chosen by 12.1% of respondents, only 5.3% of executives and 13.6% of employees. 5.6% found it difficult to answer.

As can be seen from the respondents answers, the main activity of the company's leadership should focus on work to neutralize the destructive immoral qualities of employees. However, in our opinion, this is the most difficult aspect of applying managerial efforts, since company management deals with stable signs of employee personality (character, orientation, values, motives, etc.), which have long been formed and are extremely difficult to change.

The main question of the causes of unethical behavior requires certainty. Therefore, the next few questions of the survey were devoted to specifying the reasons for violating moral laws. First of all we have formulated the following question: "There are those who believe that the reason for the violation of moral standards in the organization is the management style. What features of leadership behavior contribute to the fact that ordinary employees violate moral standards?". "Managers themselves violate moral standards by showing a negative example to the staff" - this is the response of the majority of respondents (43.9%). 39.3% of respondents chose the answer "ignoring cases of violation of moral standards in the organization". The statement "lack of interest of executives in subordinates" attracted the attention of 38.3% of the survey participants, and the judgment "low professionalism of executives" - 32.7%. About a third (30.8%) of respondents chose the answer "familiarity", (19.6%) chose "I find it difficult to answer". As you can see, the responses of the study participants prove that the most important condition for compliance with moral rules is a positive example of managers themselves, the ethics of their actions and intolerance to violations of moral standards in the organization.

Let's move on to another reason for violating moral canons in companies. As it was shown above, personal characteristics of employees, according to respondents, are the main reason for unethical behavior of staff. To the question "What personality qualities of the organization's employees contribute to the violation of ethical norms?" the survey participants responded in the following way. The most popular answer is "irresponsibility" (53.3%). Further, the respondents chose the "hypocrisy" (40.2%). Among the immoral characteristics of employees, 37.4% of respondents chose the answer "search for personal gain". 31.8% of the survey participants indicated such qualities as "deceitfulness" and "laziness". 24.3% agreed with such characterological feautures as "egocentrism", 17.8% agreed with "indifference", and 16.8% agreed with "envy". 13.1% of respondents preferred the answer "lack of integrity" and "strong language". 2.8% found it difficult to answer. In general, the above-mentioned characterological qualities of company employees, characterized by stability, primarily affect the ethical decision-making of managers and ordinary employees, and can also negatively affect internal interactions.

Taking into account the fact that unethical behavior is based on personal characteristics and value orientations, there are also external circumstances that determine the deviation from moral laws. Therefore, it was interesting to know about these objective (organizational) factors. In this regard, the respondents were asked a question: "What organizational reasons provoke violation of moral standards?" As can be seen from the selected judgments, the two most popular responses "lack of moral and material incentives" (42.1%) and "strained relationships with colleagues and leadership" (43.9%) represent both economic and socio-psychological determinants. Further, as the attractiveness of the survey participants decreases, the answers are: "lack of discipline in the organization" (37.4%); "overloaded with tasks" (35.5%); "labour remuneration" (26.2%); "inconvenient workplace, uncomfortable office space" (15.9%); "lack of social package" (5.6%). 16.8% of respondents found it difficult to choose an answer. At the same time, when building a system for managing ethical behavior, the management of companies should take into account the entire range of organizational reasons, which, in our opinion, are easier to eliminate than to correct the immoral personality qualities of employees.

Now, we will point out important differences in the perceptions of respondents belonging to different statuses in the organization regarding the significance of certain organizational determinants that provoke unethical behavior. For example, almost half as much managers than subordinates point to the "provoking" role of "low payment" in the violation of moral standards (15.8% vs. 28.4%). More surveyed executives (47.4%) than respondents from the ordinary employees (40.9%) attribute the deviation from moral standards in their organizations to the "lack of moral and material incentives for employees". More surveyed workers (46.6%) than managers (31.6%) indicated socio-psychological factors ("strained relationships with colleagues, with leadership").

The question "Is employee satisfaction related to their compliance with moral standards or not?" did not cause much controversy among the survey participants. In this case, 74.8% of respondents believe that "rather connected"; 15.9% are convinced that "rather not connected"; 9.3% were not able to answer.

Now we will consider a block of problems related to the management of ethical/unethical behavior of personnel. In this regard, for the authors of the study, the following problem is of importance: how to respond to violations of moral standards by employees. In the questionnaire this practical task was formulated as follows: "Do you think it is worth punishing violators of ethical norms or not?". Here 76.6% answered yes (they chose the answer "rather worth it"). "Rather not" - this answer satisfied 8.4% of the survey participants. 2.8% found it difficult to answer.

Further, the question on the need for punishment was specified and suggested to the respondents in this form: "What do you think should be done in relation to violators of moral standards?". The answers to this question were distributed as follows (in descending order). Thus, more than half of the respondents believe that it is enough to apply "conversations" to violators of ethical laws (68.2%). 44.9% preferred "to demand an explanation of such behavior". Slightly less than half of the respondents (45.8%) are convinced that it is necessary to "impose a disciplinary penalty (remark, reprimand, dismissal)" for deviating from morality. It is worth recalling that moral standards are provided by the power of public opinion and the conscience of a person, and do not entail formal sanctions, which are just disciplinary penalties. "Discuss an immoral act at a meeting of the labor collective" - such answer was chosen by about a quarter of respondents (24.3%). 22.4% of respondents agreed to apply material penalties to employees with unethical behavior, namely, to "deprive them of bonuses". The radical measure "declare a boycott" was chosen by 2.8%, and "do not apply any measures" by 1.9%. 3,7% found it difficult to answer.

"Do you think that employees should be encouraged for observing moral standards or not?" - the answers to this question was also interesting to the initiators of the study. Since, in our opinion, leadership's attention to the ethical behavior of employees and encouragement of employees' compliance with moral standards should play a crucial role in improving the ethical labor/organizational behavior of company employees. More than half of the survey participants (60.7%) said that they "rather should". Of these, 68.4% of the surveyed managers and 59.1% of subordinates. About a quarter (23.4%) believe that "rather not" (21.1% of leadership and 23.9% of respondents from the ordinary employees). 15.9% found it difficult to answer.

Next, we decided to clarify the respondents' position on the problem of encouragement forms for strict adherence of moral standards. In this regard we asked a question: "How should employees be rewarded for their compliance with moral standards: financially or morally?" The following answers were received. More than half (63.6%) of the surveyed employees of the organization of different status said that such behavior deserves "rather" moral encouragement. A quarter (25.2%) of the survey participants agreed that employees of the organization who follow the rules of ethics deserve "rather material" encouragement. 11.2% of the survey participants found it difficult to choose an answer.

Continuing to ask the opinions of the study participants regarding the management of ethical behavior, we asked them to answer the question: "Does the organization you work for have an ethics commission that examines the unethical behavior of ordinary employees and the organization's management?" 72% of the survey participants indicated that "there is no such commission in our organization" (78.9% of respondents from private companies and 58.3% from public companies). Only 14% recognized its existence in their organization (11.3% of respondents from private companies and 19.4% from public companies). The same number (14%) chose the judgment "I find it difficult to answer".

An organization's ethical code is a document that contains organization – specific moral rules that govern the behavior of managers and subordinates. Here we were interested to find out whether such a document exists in the organizations where the study participants work. In this regard the question was formulated as follows: "Does your organization have such document or not?" The most popular answer to this question is "such a code has not been developed (is in effect) in our organization". This was the response of 46.7% of respondents (56.3% of respondents from private firms and 27.8% from public ones). A quarter of the survey participants (25.2%) chose the answer "in our organization (division) the code is developed (22.6% of respondents from private firms and 30.6% from public ones). And 28.1% of respondents could not decide on the choice of the proposed options.

The authors of the study were interested in the opinion of the survey participants regarding the specific goals of implementing the ethical code in the organization's activities, namely, its practice-oriented purpose. In this regard, the participants of the study were asked: "What do you think is the purpose of developing and putting into practice the organization's ethical codes?". The respondents' answers were distributed as follows. 57% of respondents chose "to solve ethical problems of the organization's activities". The answer "to clearly determine which actions are ethical and which are unethical" is preferred by 55.1% of the study participants. 54.2% of respondents drew attention to the statement "in order to give guidance to employees behavior". 27.1% of respondents believe that the ethical code is necessary "to prevent illegal actions (for example, theft of employers' things)". Answer "the ethical code" is being introduced, including "to maintain the image of the organization" was preferred by 26.2% of respondents. Only 8.4% of the survey participants chose "to strengthen control over the staff activities", and 4.7% chose the statement "to protect employees from managers". 8.4% of respondents don't know which answer to choose from the list ("I can't answer it").

As the answers show, the main practical mission of ethical codes is to show models of ethical behavior to staff, suggest ways to make moral decisions when employees solve production tasks and participate in organizational interactions. Now let's turn to the answers of the research participants to this question, identified by their status positions. 68.4% of managers at different levels and 52.3% of respondents in a number of subordinates chose the statement "to clearly determine which actions are ethical and which are unethical". More ordinary employees (60%) than managers (42.1%) indicated that the purpose of ethical codes in the organization is to "eliminate ethical problems of the organization's activities". 36.8% of the managers we surveyed and a quarter of the surveyed subordinates believe that ethical codes should be used "to prevent illegal actions (for example, theft of employers' property)". The same answer was significantly more attractive for the executives we surveyed than for the subordinates - 31% against 19.4%. More managers (10.5%) than subordinates (3.4%) chose the answer "to protect employees from executives". Also, the statement "to strengthen control over the activities of staff" was attractive to 15.8% of the survey participants from among executives, and only 6.8% of the surveyed subordinates.

4 Discussion

The results of the study generally correlate with the various conclusions of other experts whose work is devoted to the topic we are studying. In particular, the analysis of the answers of our survey participants on the forms of moral postulates violation showed possible signs of psychological pressure in organizations (mobbing, bullying). Psychological pressure is currently in both private and public companies; it has its own national characteristics; it depends on many organizational factors (for example, the implementation of the concept of corporate social responsibility); it has a serious impact on its victims, directly impairing their performance; and it also requires serious organizational preventive measures [10].

Then, the respondents pointed at personal characteristics that act as reasons for immoral actions of organizations employees. Some authors attribute the features indicated in our questionnaire to the signs of “toxic personnel”, while emphasizing that toxic employees in their various behavioral manifestations are a stressor, complicating organizational interaction and adversely affecting the company’s performance. In addition, they require additional management efforts to neutralize their destructive (in particular, unethical) potential [2]. This can be achieved, for example, by building effective programs for training employees in company ethics, taking into account possible difficulties and problems in the process of their “moral” training [11].

As we have already pointed out, the study revealed the negative role of company management in the violation of moral rules by their employees. In this regard, there is a problem of training future managers in higher education institutions in ethical behavior, namely, moral competence, the necessary level of ethical thinking and skills for making morally verified managerial decisions [1]. At the same time, some experts believe that it is unacceptable to ignore cases of violation of moral canons by top managers, and the practice of discussing “moral matters”, ethically complex situations of labor behavior of personnel forms a cohesive and emotionally mature team, promotes morally justified actions and strengthens the morale of the individual [8].

Speaking on the problem of managing ethical behavior of staff, it is worth noticing that researchers are increasingly addressing the difficulties and contradictions of implementing ethical codes in the activities of modern organizations in various industries, revealing their regulatory mechanism for stimulating moral actions of employees. At the same time, the cultural component of such documents and national characteristics of moral regulators, which, as it turned out, directly affect the ethics of behavior, are also analyzed. For example, you can specify a work in which a comparison of the British and Korean codes of ethics for labor and corruption behavior revealed that the latter pay more attention to the hierarchy of relations [6]. Further, we study the relationship between individual qualities of employees that are important in labor behavior (internal locus of control, hostility towards colleagues, perceptions of unethical behavior) and ethical codes implemented in companies, and also show the crucial role of these documents as the formation of the required moral behavior of personnel [9]. In addition, the researchers point to a number of important steps that the organizations leadership must take in order for the provisions of the codes to be effective in practical terms. In particular, the management of construction companies is recommended to develop optimal remuneration measures, measures to train personnel of ethical behavior, and a well-established system for responding to violations of the moral canons of the company’s leadership [3]. To speed up the implementation of ethical codes and reduce immoral organizational behavior, exemplary ethical behavior of executives themselves, protection of those employees who expose violators, and consistent explanation of the ethical requirements contained in the codes to employees will also help [7].

5 Conclusion

The survey authors are satisfied with the received data and will continue to study this problem. But first, based on the results of the survey, they formulated the following conclusions.

1. Moral standards are still the most important moderators of the organizations employees behavior, their strength and regulatory role is recognized by the absolute majority of both managers and ordinary employees, both employees of public organizations and private firms. In addition, the majority of respondents believe that it is important to adhere to ethical rules even if other employees deviate from them.
2. The main purpose of moral models in the organization is to improve relations between employees and prevent conflict situations. At the same time, indirect compliance/non-compliance with ethical rules by the organization's personnel can have a certain impact on the company's performance as a whole.
3. The main factor that provokes violation of moral canons, according to the majority of survey participants, is the personal characteristics of employees of the organization. Certain immoral qualities and attitudes to unethical behavior can be identified at the recruitment stage. At the same time, every employee of the organization needs to be trained in moral behavior, including the process of making ethical decisions, including managerial ones.
4. Ethical labor/organizational behavior depends not only on the personal characteristics of ordinary personnel, but also on the management style. In particular, the key role here is played by the behavior of managers who act in a way to comply/not comply with moral rules. In a number of organizational factors that provoke violation of ethical canons, first of all, there are tense relations with the management, as well as the lack of a system of moral and incentive.
5. The majority of respondents indicated that their organization can detect violations of moral standards, and responsibility for non-compliance with morality is borne by both ordinary staff and management of the company.
6. In the opinion of respondents, cases of deviation from moral prescriptions should entail sanctions, in particular, disciplinary penalties. And the most effective method of influencing offenders is educational conversation. Similarly, the application of sanctions, but already positive and moral, should be applied to those employees whose behavior is characterized by high morality, i.e. the acceptance of moral standards, recognition of their value and strict compliance.
7. In almost half of the organizations that the survey participants represent, sufficient attention is paid to the observance of moral precepts and the respondents are satisfied with the state of their compliance. At the same time, ethical codes as guidelines for moral labor/organizational behavior have not been developed.

Taking into account the importance of ethical standards and their value, company management needs to organize a variety of work on managing moral labor/organizational behavior, including the development and implementation of effective ethical codes, the development of practices for collecting, analyzing and discussing unethical behavior of employees in the performance of their professional tasks, and the design of a system of sanctions for compliance/violation of moral laws.

References

1. Coldwell, D.A., Venter, R., Nkomo, E.: Developing ethical managers for future business roles: a qualitative study of the efficacy of "Stand-Alone" and "Embedded" University "Ethics" courses. Journal of International Education in Business. In Press (2020). https://doi.org/10.1108/JIEB-08-2019-0040. Accessed 24 June 2020
2. Esaulova, I.A., Nagibina, N.I.: "Toxic" personnel: problems and management methods. Manager **5**(69), 58–71 (2017)
3. Ho, C.M.-F., Oladinrin, O.T.: A paradigm shift in the implementation of ethics codes in construction organizations in Hong Kong: towards an ethical behaviour. Sci. Eng. Ethics **25** (2), 559–581 (2019)
4. Leclercq-Vandelannoitte, A.: Is employee technological "ill-being" missing from corporate responsibility? The foucauldian ethics of ubiquitous IT uses in organizations. J. Bus. Ethics **160**(2), 339–361 (2019)
5. Mitchell, M.S., Reynolds, S.J., Treviño, L.K.: The study of behavioral ethics within organizations: a special issue introduction. Pers. Psychol. **73**(1), 5–17 (2020)
6. Moon, C., Uskul, A.K., Weick, M.: On culture, ethics, and hierarchy: how cultural variations in hierarchical relations are manifested in the code of ethics of British and Korean organizations. J. Appl. Soc. Psychol. **48**(1), 15–27 (2018)
7. Oladinrin, O.T., Ho, C.M.-F.: Embeddedness of codes of ethics in construction organizations. Eng. Constr. Archit. Manag. **23**(1), 75–91 (2016)
8. Silén, M., Svantesson, M.: Impact of clinical ethics support on daily practice—first-line managers' experiences in the Euro-MCD project. J. Nurs. Manag. **27**(7), 1374–1383 (2019)
9. Valentine, S.R., Hanson, S.K., Fleischman, G.M.: The presence of ethics codes and employees' internal locus of control, social aversion/malevolence, and ethical judgment of incivility: a study of smaller organizations. J. Bus. Ethics **160**(3), 657–674 (2019)
10. Vveinhardt, J., Sroka, W.: Workplace mobbing in polish and Lithuanian organisations with regard to corporate social responsibility. Int. J. Environ. Res. Public Health **17**(8), 2944 (2020)
11. Weber, J.: Investigating and assessing the quality of employee ethics training programs among US-based global organizations. J. Bus. Ethics **129**(1), 27–42 (2015)

Evaluating Changes in Organization Structure Using Methods of Mathematical Statistics

S. V. Sidorkina(✉)

Financial University Under the Government of the Russian Federation, Moscow, Russia
Svetlana-sidorkina@yandex.ru

Abstract. To effectively manage an organizational structure it is essential to understand whether the estimates of job levels are well-founded when changing such a structure. One should ensure that there is no "inflation" of job levels, i.e. that the level of a position in the system of job grades is not unduly overestimated. This research offers methods to assess the significance and appropriateness of changes in distribution of positions by job levels in the organizational structure of an enterprise. To analyze factors that affect qualitative change in an organizational structure, it is proposed to use a combination of mathematical statistics methods, like estimation of arithmetic weighted average, Pearson criterion $\chi 2$, and qualitative methods of analysis. The proposed method allows assessing in mathematical terms whether the changes in organizational structure that occurred during the researched period are appropriate as well as confirming/disproving the hypothesis of job levels "inflation" in the organization during such a period. Methods used in this research add statistical assessment to the generally accepted approaches to qualitative assessment of validity of changes in an organizational structure. These methods allow setting and controlling the targets of the organizational structure change metrics.

Keywords: Chi-square statistic criteria · Job evaluation · Job hierarchy · Job level · Job grading system · Organizational structure

1 Introduction

As an organization grows and develops, its organizational structure constantly changes [10, 13]. There is a number of triggers for such changes like dynamically changing environment, emergence of new technologies, and digitalization [1–3, 8]. New divisions appear, while some of the old ones are closed. New positions are created in the existing divisions, while functions and subordination of positions can be changed. Simple and algorithmizable activities become automated, and non-core activities become outsourced [12]. This is the reason why the number of low-skilled and medium-skilled positions in the organization may decrease. At the same time, development of new products and implementation of new production and information technologies become more and more complicated. As a result, the organization needs new positions that require high qualification and high degree of responsibility for business results [4].

S. I. Ashmarina and V. V. Mantulenko (Eds.): IPM 2020, LNNS 161, pp. 625–632, 2021.
https://doi.org/10.1007/978-3-030-60926-9_79

In a naturally developing business there are objective reasons for job levels of the organizational structure to upshift. In other words, the frequency of positions at different levels of the organization shifts upwards. At the same time, there are certain rules regulating the hierarchical structure of positions in an organization (the structure of grades) and the significance of positions in it is assessed by certain criteria regardless of the position evaluation system that the particular organization uses [7, 11]. Practical observations show that over time higher grades tend to be assigned to positions in organizations though the job does itself not become more complicated. This phenomenon is known as "inflation of job levels" [5]. There is a number of reasons for that, like desire to assign a higher status to certain positions both inside and outside the organization, to give an incentive to individual employees who hold specific positions, to increase salary or provide more benefits to holders of a particular position if the current job level does not allow doing this, etc. This phenomenon is similar to grade inflation in education which is widely discussed in literature [6].

No organization is interested in "job level inflation" within its organizational structure, as such inflation results in too high HR-costs and depreciation of labor. Thus, an organization needs a way to assess whether its job levels tend to upshift and whether such a trend is significant enough. We can use mathematical statistics methods for the purposes of such an assessment.

2 Methodology

The system of job levels (grades) in an organization implies scoring by certain criteria and assigning each position a certain rank in the company's structural hierarchy pursuant to such scoring. Positions rated in the same range of points are assigned the same job level regardless of divisions they belong to. A medium/large company normally has a hierarchy composed of 15 to 21 job levels. The position of the head of the organization, as a rule, is assigned the highest job level in the hierarchy. Thus, all positions in the organization are placed on an ordinal scale. As assessment of a position by each criterion includes expert opinion by evaluators (or the Evaluation Committee), assigning job level always include a subjective component and thus is exposed to erroneous judgments (especially in borderline cases).

To understand whether changes in the company's job hierarchy were justified during a certain period of time, one has to gain the following insights:

- whether or not there are any statistically significant differences in the organization's job distribution by job level during the control and comparison periods (for example, as of the last day of the year),
- if such significant differences exist, one should investigate whether the shift in the company's organizational structure was is due to significant factors. Such factors may include the following:
- automation and robotics,
- outsourcing of non-core activities,
- centralization of functionality and reduction of duplicate positions,

- growth of the company's business meaning greater financial responsibility for senior managers,
- formation of "hybrid" positions combining functions and competencies of several roles,
- changes to the company's operating model, etc.

Arithmetic weighted average for all job levels in the structure of the compared periods is a good tool to identify whether an organization faces an upshift of the average value for the entire structure of the organization's job levels. For this purpose, the number of positions belonging to each job level is considered as the "weight" of the job grade. Assessing the degree of change in distribution of positions per the ordinal scale of job levels used by the organization means identifying differences in the distribution of a parameter when comparing two (or more) empirical distributions. The following criteria can be used for this purpose: Pearson's Chi-square (χ^2) criterion; Kolmogorov-Smirnov test. Both the methods allow assessing the significance of differences in distribution of positions in the structural hierarchy, while the Kolmogorov-Smirnov method also allows identifying the points of maximum discrepancy between distributions [9]. Assessment based on these criteria can be applied to certain groups of positions (for example, only management positions) or to the whole positions hierarchy of the company, as the case may be.

3 Results

Hypothetical case study. This article covers application of the Pierson's Chi-square criterion to assess the significance of changes in the organization's structure. For example, *Alpha Ltd.* has 15 job levels in its structure: levels 13–15: senior management; levels 10–12: mid-level executives; levels 7–9: highly qualified professionals; levels 4–6: mid-level professionals; levels 2–3: low-skilled professionals; level 1: unskilled personnel. Now we will find out whether there were statistically significant changes in the distribution of positions in the positions hierarchy of *Alpha Ltd.* during the period from 2018 to 2020 (Table 1).

1. Let us determine whether there is a shift in the structure of job levels. For this purpose we will calculate the arithmetic mean weighted for all job levels in the structure of 2018 and 2020 using the formula:

$$\overline{x} = \frac{\sum_{i=1}^{n} x_i * w_i}{\sum_{i=1}^{n} w_i} \quad (1)$$

where x_i is the value of the job level rank; w_i is the number of positions belonging to this job level.

Regardless of how the organization calls its job levels, these should be ranked in ascending order from the lowest to the highest hierarchical level. In our case, $\overline{x}_{2018}$ = 5.48 and $\overline{x}_{2020}$ = 5.81. The median grade in both 2018 and 2020 is 5. Thus, in general, the average level of job grades has upshifted. Although the job levels are ranked by an ordinal scale, we believe that in this case it is reasonable to apply weighted average as it

Table 1. Distribution of positions by job levels within *Alpha Ltd.*

Job level (Grade)	Job category	Number of positions belonging to this grade	
		2018	2020
		Empirical distribution 1	*Empirical distribution 2*
15	Senior management	1	1
14		9	12
13		5	6
12	Mid-level executives	22	22
11		43	50
10		118	215
9	Highly qualified professionals	314	416
8		412	614
7		518	580
6	Mid-level professionals	1232	1316
5		1314	1500
4		918	1140
3	Low-skilled professionals	564	617
2		290	15
1	Unskilled personnel	50	0
	Total	**5810**	**6504**

Source: authors.

reveals the shift trend in the average level of grades. While mathematically, it is more correct to use the median value, it gives a rougher estimate and may fail to reveal the differences.

Then, we need to determine the statistical significance the identified shift. For this purpose we use the Chi-square criterion.

2. Let us formulate statistical hypotheses:

H_0: empirical distribution 1 does not differ from the empirical distribution 2.
H_1: empirical distribution 1 differs from the empirical distribution 2.

Further, we will calculate the value of the Chi-square for two empirical distributions. For this purpose we will do the following:

1. Calculate the theoretical frequencies separately for the distribution of 2018 and 2020 using the formula below

$$f_{th} = \frac{\sum_{i=1}^{m} (x_1 + x_2) * n_j}{n_1 + n_2}, \tag{2}$$

where f_{th} is the theoretical frequency of job distribution by level in each period,

i is the values of job levels in the organization hierarchy,
m is the number of job levels in the organizational hierarchy,
j is the values attributed to the studied periods (in this case, period 1–2018, period 2–2020),
x_1 and x_2 are the number of positions belonging to a specific job level during the compared periods 1 and 2,
n_1 and n_2 are the total number of positions that existed during the compared periods 1 and 2.

2. Calculate the squares of differences between the empirical and theoretical frequencies.
3. Calculate the value of the Chi-square using the formula below

$$\chi^2 = \sum_{ij} \frac{y_{ij}}{fth_{ij}} \tag{3}$$

where f_{th} is the theoretical frequency of job distribution by level in each period,
i is the values of job levels in the organization hierarchy,
j is the values attributed to the studied periods (in this case, period 1–2018, period 2–2020),
y_{ij} is the square of the difference between the empirical and theoretical frequency for the corresponding job level during the corresponding period.

Table 2 presents the calculated values for 1–3 above.

Table 2. Calculations of χ^2 for *Alpha Ltd.*

Job level (Grade)	Number of positions belonging to this grade within the organization		Sum	Theoretical frequency		Ratio between the Square Differences of Empirical and Theoretical Frequencies and the Theoretical Frequencies	
	2018	2020					
	Empirical distribution 1	*Empirical distribution 2*		*Empirical distribution 1*	*Empirical distribution 2*	*Empirical distribution 1*	*Empirical distribution 2*
15	1	1	2	0.94	1.06	0.0034	0.0030
14	9	12	21	9.91	11.09	0.0833	0.0744
13	5	6	11	5.19	5.81	0.0070	0.0062
12	22	22	44	20.76	23.24	0.0741	0.0662
11	43	50	93	43.88	49.12	0.0176	0.0157
10	118	215	333	157.12	175.88	9.7385	8.6994
9	314	416	730	344.43	385.57	2.6883	2.4015
8	412	614	1 026	484.09	541.91	10.7350	9.5895
7	518	580	1 098	518.06	579.94	0.0000	0.0000
6	1 232	1 316	2 548	1 202.20	1345.80	0.7387	0.6599
5	1 314	1 500	2 814	1 327.70	1486.30	0.1414	0.1263
4	918	1 140	2 058	971.01	1 086.99	2.8936	2.5849
3	564	617	1 181	557.22	623.78	0.0825	0.0737
2	290	15	305	143.91	161.09	148.3174	132.4914
1	50	0	50	23.59	26.41	29.5635	26.4090
Total	**5 810**	**6 504**	**12 314**	**5 810**	**6 504**	χ^2 =	**388.2853**

Source: authors.

4. Calculate the number of degrees of freedom when comparing two empirical distributions using the formula:

$$v = (k - 1) * (c - 1), \tag{4}$$

Where v is the number of degrees of freedom,

k is the number of levels of the attribute (in this case, the number of levels in the organization's job hierarchy);

c is the number of distributions to compare (in this case, the number of periods).

For *Alpha Ltd.* v = 14.

Further we use special tables or the corresponding MS Excel function to determine the critical values Chi-square in case v = 14.

$\chi^2_{cr} = 23.68(p \leq 0, 05)$

$\chi^2_{emp} = 388.29$

$\chi^2_{emp} > \chi^2_{cr}$

H_0 hypothesis should be discarded as there is a statistically significant difference between the distributions. Further qualitative analysis of the factors impacting the structure of positions in the company's hierarchy can show, for example, that during the study period, *Alpha Ltd.* has automated and partially outsourced servicing functions, which resulted in a significant reduction in the positions of unskilled and low-skilled personnel (job levels 1–2).

Putting to similar analysis changes in the frequency distribution of positions in *Alpha Ltd.* only for senior and mid-level management (job levels 10–15), we will find $\chi^2_{emp} = 6.46$.

With the number of degrees of freedom v = 5, $\chi^2_{cr} = 11.07(p \leq 0.05)$

$\chi^2_{emp} < \chi^2_{cr}$

For a group of senior and mid-level management positions, there were no significant changes in distribution of positions in the hierarchy during the study period.

4 Discussion

We propose an approach to studying the job level dynamics in the organization that allows mathematically evaluating these changes and determining their significance. The mathematical assessment should be supported by a qualitative analysis of factors triggering the shift of the weighted average in the structure of the organization's job levels. In case no significant impact of such factors is detected, though a statistically significant shift of the average value of the levels has occurred, this means that the company faces job levels "inflation" (increasing the level of the position without actually complicating and increasing the level of responsibility for this position holder).

Currently, many researchers use approaches implying qualitative assessment of changes in the position structure (whether or not the organization changed its position structure, whether or not any individual posts were re-evaluated) or they just calculate the ratio between the number of positions that were promoted and the total number of posts in the organization. This approach is used in case during the study period neither the enterprise's organizational structure nor the job functions were changed significantly. However, this approach doesn't work if the organizational structure of an enterprise changes dynamically, i.e. a large number of new divisions and positions emerge, and some divisions and positions are abolished, as the actual list of positions during the compared periods may have noting in common.

Using statistical criteria to assess changes in the position structure for some specific job categories (managers, specialists) may help to determine whether during the study period any significant changes in the company's positions hierarchy took place or whether such changes applied only to some specific groups of positions at some specific job levels. Methods to mathematically evaluate upshift of job levels in an organization allow setting and controlling thresholds for this parameter. As a result, changes in the organizational structure of the enterprise become more manageable, and it becomes easier to control the impact such changes have on the organization's HR-costs. It is yet to be researched to which extent changes in the average indicator that reflects the distribution of positions in the structure of job levels can be deemed reasonable in case certain significant factors affect the enterprise's organizational structure.

5 Conclusion

Changes in the enterprise's organizational structure are driven by both external and internal factors. Changes in the distribution of positions in the organization by the established job levels can be both reasonable or result from "inflation" of job levels. Methods of mathematical statistics (e.g., Pearson's Chi-square) allow assessing the significance of changes in distribution of positions within an organization during a certain period of time. Qualitative analysis of factors driving changes of the organizational structure during the study period allows determining the validity of changes that occurred.

References

1. Bonanomi, M.M., Hall, D.M., Staub-French, Sh., Tucker, A., Talamo, C.M.L.: The impact of digital transformation on formal and informal organizational structures of large architecture and engineering firms. Eng. Constr. Archi. Manag. **27**(4), 872–892 (2019)
2. Dolzhenkova, Yu.V.: Labor sphere transformation: Challenges of digitization, Selfgoverenment, **2**(1(118)), 129–132 (2020)
3. Gaspary, E., De Moura, G.L., Wegner, D.: How does the organisational structure influence a work environment for innovation? Int. J. Entrepreneurship Innov. Manag. **24**(2–3), 132–153 (2020)
4. Konopleva, V.S., Konopleva, I.A.: The impact of information technology on the organizational structure of the enterprise. Modern Research and Innovation, **8** (2016). http://web.snauka.ru/issues/2016/08/70483. Accessed 02 Feb 2020
5. Korn Ferry: Job evaluation: Foundations and applications (2017). https://www.kornferry.com/content/dam/kornferry/docs/pdfs/job-evaluation.pdf. Accessed 20 June 2020

6. Nichols, T.: The death of expertise: The campaign against established knowledge and why it matters, 1st edn. Oxford University Press, Oxford (2017)
7. Saputra, D., Ghaida, T., Kezia, Z.T.: Design of job grading model in operational division of PT. Antariksa Anugrah Perkasa. Int. J. Recent Technol. Eng. **8**(3), 127–131 (2019)
8. Schwarzmueller, T., Brosi, P., Duman, D., Welpe, I.M.: How does the digital transformation affect organizations? Key themes of change in work design and leadership. Manag. Rev. **29** (2), 114–138 (2018)
9. Sidorenko, E.V.: Methods of Mathematical Processing in Psychology. Rech, Saint Petersburg (2003)
10. Sidorkina, S.V.: Modern trends in the design of organizational structures of enterprises. Kadrovik **6**, 78–83 (2020)
11. Sychenko, V.V., Marenichenko, V.V., Kozyryeva, O.V., Strapchuk, S.I.: State regulation of formation small and medium businesses quality development based on grading. Financ. Credit Act. Probl.Theor. Pract. **1**(24), 205–212 (2018)
12. Worren, N.: Organization design: Simplifying complex systems, 2nd edn. Routledge, London (2018)
13. Zhang, D., Bhuiyan, N., Kong, L.: An analysis of organizational structure in process variation. Organ. Sci. **29**(4), 722–738 (2018)

HR Risks Management in the Context of Labour Market Precarization

N. V. Solovova, N. V. Sukhankina(✉), and D. G. Slatov

Samara National Research University, Samara, Russia
solovova.nata@mail.ru, sukhankina@inbox.ru,
slatov76@gmail.com

Abstract. The article deals with the HR risk assessment for Russian companies under the conditions of labor market precarization. Based on the analysis of the precariat, a new phenomenon, we've specified the peculiarities of social and labor relations under precarization of the labor market and have revealed statistically significant links between certain types of non-standard employment and socio-economic parameters. The authors applied a comprehensive approach to the assessment of HR risks (demographic, political, economic, technical) and proposed the directions for HR policy of the labor market agents in response to precarization. Scientific recommendations have been developed to minimize HR risks and improve HR policy of various economic agents in terms of labor market precarization. Good practices on customized assessment of HR risks for a certain employee and manager have been presented, followed by personalized recommendations for identified risks elimination.

Keywords: Economic agents · HR risks · Labor market · Precariat · Precarization · Temporary employment

1 Introduction

Improved work performance, labour and social protection of employees is among the basic directions of the Russian Federation state policy framework for systemic changes to ensure socio-economic development. A serious challenge for the institution of modern economy, i.e. the institution of formalized and legally protected employment, is the precarization in labour market, which over the past decades has affected both developed and developing countries. In virtue of economic globalization, increased competition between countries and businesses, and in the aftermath of these, gradual weakening of the state social guarantees, the main trend of the global labour market is a transition from full-time and permanent employment to part-time and temporary employment. Whereas in the 1960s the average number of times that people changed jobs during their career was four, at present this number has grown to nine [16]. According to Toshchenko, the labor markets of the industrial era were affected by socio-and-status-specific agreements in which the employer guaranteed employment and the employee in return guaranteed subordination and loyalty [17]. Currently, this social contract is no longer valid, and this has dramatically increased the ambiguity on the labour market, both for the employer and the employee.

S. I. Ashmarina and V. V. Mantulenko (Eds.): IPM 2020, LNNS 161, pp. 633–643, 2021.
https://doi.org/10.1007/978-3-030-60926-9_80

Numerous forms of non-standard employment (part-time, short-term and contingent work, and other multilateral labor relations, as well as disguised employment or dependent self-employment) [13] reduce the level of social protection for employees, worsen their economic status and psychosocial state, often affecting their physical and mental health. The increased precariat creates, on the one hand, social risks due to the vulnerability of employees, but, on the other hand, the burden on enterprises in managing such employees [16]. With this in mind, the current orientation for the scientific enquiry is the theoretical justification and development of practical recommendations for improving HR policy implemented by economic agents and aimed at HR risk management under the labor market precarization.

2 Methodology

The theoretical and methodological basis for the survey is a comparative study of theses, hypotheses, concepts and ideas presented on the precariat in Russian and foreign scientific literature. The key questions of the labor market precarization and social inequality growth have been raised in the works of such researchers as Bec [2], Bourdieu [4], Piketty [15], Standing [16] and others. In the local and foreign scientific literature devoted to the problems of precarization in the labor market, there is a consensus on the special features of precariat, such as deprofessionalization and detachment from job performance and from the process of influencing the entity's policies. According to Standing, a feature of the precariat is not the level of money wages or income earned at any particular moment but the lack of community support in times of need, lack of assured enterprise or state benefits, and lack of private benefits to supplement money earnings [16].

Content analysis of foreign works devoted to precariat has revealed statistically significant links between certain types of non-standard employment and socio-economic indicators Julià, Vanroelen, Bosmans, van Aerden, and Benach in their study proved that precarious employment is one of the social factors that are related to poor general and mental health [14]. On the one hand, employees act as the objects vulnerable to certain job-related health risks; on the other hand, they act as the economic subjects that participate in the health services payment. The researchers attribute the deterioration of health of the precariate to an increased risk of injury due to poor working conditions. According to Giraudo, Bena, Leombruni, and Costa part-time workers face an increased risk of injury by up to 24–57% compared to full-time workers [9]. Benach, Julià, Tarafa, Mir, Molinero, and Vives, in their study show that underemployment is most common among women, youth, immigrants, and low-skilled workers and is associated with worse health [3]. According to research by Inoue, Tsurugano, Nishikitani, and Yano [12, 18], part-time workers (hourly work, shift work, fixed-term contracts) are less likely to undergo annual checkups and are more likely to have inaccurate self-perceptions of health and mental disorders. Bacci, Pigini, Seracini, Minelli [1] and Canivet, et al. [6] cite economic deprivation and the need to maintain unloved jobs because of labour market instability as economic factors affecting the mental health of part-time workers. According to Wege, Angerer, and Li [19], the increased risk of depression in employees is attributable not only to current underemployment but also to previous experience of unemployment.

Fagan, Norman, Smith, and González Menéndez in their study note that reduced working hours with adequate social safeguards can have a positive social impact by creating new jobs and ensuring full recovery of the workforce [8]. Cardoso, Loviglio, and Piemontese have revealed a correlation between the job seekers' subjective perception of the labour market and the level of wages [7]. When the subjective perception of respondents overstates the level of unemployment in the country, a gap of one percent between the perceived and actual levels causes wage reduction by 0.4–0.7%. A pessimistic approach to labor market leads to concerns about employment prospects, accepting their positions in the labor market as weak, and wage stagnation. Bryan and Sevilla focus on the social effects of flexible employment [5]. They have proved that some flexibility in working hours for parents allows them to spend more time (plus 0.5-1 h) with the family. The authors believe that the flexible working arrangements that increase the flexibility of the employee, rather than the employer, shall be deemed effective, and only in this case it can be attributed to the advantages of precarization. Hirsch, Husain, and Winters in their survey explored the relationship between multiple job-holding and unemployment rates [11]. The survey is based on the data set of workers in US labor markets during 1998–2013. Interestingly, no relationship has been found, they conclude that multiple job holding among men is largely acyclic. Hanglberger and Merz argue the widely held view that self-employed people are more satisfied with their work than employees: a positive effect of self-employment on job satisfaction is found only in the first years of self-employment [10]. After three years, adaptation eliminates the higher satisfaction of being self-employed. According to the authors, the positive long-term effects of self-employment on job satisfaction are overestimated.

3 Results

HR risks are associated with the possibility for a company to lose its resources or revenue as a result of actions or omissions of its personnel. For our survey, we've widened this definition understanding HR risk as the probability of a loss for the system due to actions or omissions of individuals included in the system. This is necessary for correct consideration of the HR risks arising not only from within the organization, but also in the field of administrative governmental influence, and within the limits of certain life strategies implementation. Let's consider different HR risk's categories: demographic risks; economic risks; political risks; technical risks. We have suggested response options to HR risks associated with precarization for each economic subject (state, company (organization), individual) (Table 1). Thus, HR risks related to the labor market precarization are complex, systemic, self-sustaining and synergistic.

Table 1. HR policy key directions of the labor market subjects under precarization

Subject Risks	State	Business	Individual behavior strategies
Demographic risks are affected by global, country, regional and local demographic processes (population aging and migration)	The need to increase spending on health care, pensions, and the shortage of labor resources in the young cohorts of the population	Labor Market Imbalances: a shortage of young people, increased number of older people, early wage freeze, demotivation to changes	Increased competition in the labor market, wage stagnation or reduction, deterioration of working conditions due to the spread of non-traditional employment
Economic risks resulting from adverse changes in the economy of the country or in the economy of the organization itself (price fluctuation and production, exchange rates, inflation)	Stagnation of economic growth, real income decline, exchange rate dependence on the world market, growth of emigration and reduction of immigration	Reduction in effective demand, reduction of employees, loss of competencies due to labor migration, "drop off the radar"	Reduction in salary expectations, "rainy day" fund reduction, weak bargaining position while dealing with the employer, acceptance of non-traditional employment, especially unregistered self-employment, increase in "under the table" payments, reduction in labor and social guarantees
Political risks are associated with regime change, tax, budget, credit, currency systems change, administrative corruption, and the influence of industrial groups	Outflow of capital and labor resources, narrowing the planning horizon, decline in the effectiveness of public administration, orientation for short-term measurable results	Increase in the share of profit distributed among owners, withdrawing money abroad. Reduced investment, including in human capital, and planning horizon narrowing. The spread of unethical labor practices	Motivation to get money at the expense of long-term instruments for personal and professional development. The spread of unethical labor practices. Emigration
Technical risks are associated with the use of new equipment, technologies, innovative projects implementation in the absence of qualified personnel, risks associated with change implementation	Digitalization of management, development of artificial intelligence systems, reduced necessity for routine positions of the state administration apparatus	Digitalization of management, development of artificial intelligence systems, reduced necessity for routine positions, uberization	Decrease in the labor market due to withdrawal of some positions with routine functionality, difficult adaptation to digital reality, ultra-high self-exploitation in "uberized" systems and within the framework of self-employment

Source: authors.

To assess key Russian companies' HR risks in conditions of labor market precarization and to develop recommendations for their minimization, expert surveys were conducted (2019). The objectives of the survey were as follows:

– empirical confirmation of the HR risks importance for effective company operations in conditions of labor market precarization,
– identification of effective ways to minimize HR risks,
– development of practical recommendations for the employer on individual assessment of HR risks when hiring.

Objective empirical justification involves analyzing the responses of three groups of respondents:

– masters studying in "Human Resource Management" ("external evaluation"),
– heads of various companies in Samara ("internal evaluation"),
– personnel of HR departments in Samara ("internal evaluation").

Large working experience of specialists and managers of the first and second groups of respondents (15 heads of organizations and 19 employees of HR departments) provided for the sample representativeness. The third group of respondents consisted of 56 master degree students attaining intramural (full – time) and extramural (part-time) education at Samara National Research University ("Human Resource Management" Training Programme). The representativeness of this sample was ensured by students of different instructional years. Thus, a total of 90 respondents of

Table 2. Findings of company's HR risk assessment expert survey in terms of labor market precarization

Types of HR risks		Human resources staff	Head of companies (organizations)			
	Average score	Average over 3 points, %	Average score	Average over 3 points, %	Average score	Average over 3 points, %
Performance problems	4, 3	76, 1	4, 3	76, 6	4, 4	84, 5
Labor Productivity Problems	4, 8	78	4, 5	76	4, 7	85, 3
Lack of a workplace when working from home (difficulty in self-organization)	4, 9	85, 2	4, 7	82, 5	4, 8	87
High risk level and high self-responsibility (in case of individual entrepreneurship)	4, 3	76, 1	4, 3	76, 6	4, 5	84, 6
Difficulties in meeting the requirements of the professional standard	4, 5	80, 3	4, 8	83	4, 8	87
Non-commitment of a freelancer to corporate culture	4, 5	80, 3	4, 8	86	4, 9	88, 1
Decentralization of work and lack of effective organization	4, 6	81, 4	4, 5	87	4, 8	85, 4
Problems of trade secret and confidential information protection	4, 2	79, 1	4, 2	82, 5	4, 5	84, 6

Source: authors.

three groups were interviewed during this stage of survey. The study was conducted in the form of a questionnaire survey that assesses the significance of HR risks for the effective operation of the organization in conditions of precarization in the labor market. In addition to those listed in Table 2, the list of proposals for evaluation includes others selected at random.

As is evident from Table 2, practical skills that are important from the point of view of the survey were highly rated by respondents (the average score is "4" or higher, more than 2/3 of respondents rated the need for these skills at "4" and "5"). Analysis of the survey findings shows that the most significant indicators are HR risks associated with self-regulation difficulties; non-commitment to corporate culture; the lack of effective organization; absence of benefits and safeguards for employees, this allows to suggest methods to minimize specified risks and others (Table 3).

Table 3. Methods to minimize HR risks in terms of labor market precarization

HR risks	Methods to minimize
Performance problems	- Labor and other contracts shall be supplemented with a separate clause on the quality of work performed, - No verbal agreements with employers, - Involvement of employee's self-motivation
Labor Productivity Problems	- Training program, - Optimization of business processes in terms of correct goal setting, effective and timely communication
Lack of a workplace when working from home (difficulties in self-organization)	- Participation in public organizations, professional communities, professional fan clubs, and alliances, - Transit to remote employment as an option, not a binding alternative; applicable upon agreement with the employee, - The basics of time management shall be learnt by the employees
High risk level and high self-responsibility (in case of individual entrepreneurship)	- Participation in public organizations, professional communities, professional fan clubs, and alliances, - Improving financial and legal literacy, additional education in managerial psychology, management and other branches of management
Difficulties in meeting the requirements of the professional standard	- Performing job not as an employee, but under a civil contract, - Getting an education on your own, - Lobbying for professional standard easing (unlikely and difficult), Quality Management System Implementation
Non-commitment of a freelancer to corporate culture	- Participation in public organizations, professional communities, professional fan clubs, and alliances, - A range of responses depends on the depth of the problem and varies from ignoring because of insignificance to adaptation efforts, team building arrangements and corporate identity strengthening, and dismissal

(*continued*)

Table 3. (*continued*)

HR risks	Methods to minimize
Decentralization of work and lack of effective organization	- Introduction of the position of HR Manager, supervising the work of the representatives of the precariat; - Strict requirements for the implementation of occupational safety rules; - Optimization of business processes
Challenges of trade secret and confidential information protection	- A separate clause on trade secrets and confidential information protection shall be added to employment and other contracts, - Signing a non-disclosure agreement; frequent password changes and introduction of two-factor authentication for access to critical segments of the management system (if possible)
No social package, benefits, or health insurance	- Alternative insurance, self-registration and membership in insurance companies. Labor laws expansion and updating: - Consolidation of the practice of social and labor relations of the precariat as a social norm or procedure, exercising social control over the practice of fixed-term labor relations, - New standards for work organization in the information society, - Institutionalization of the precariat, imposing responsibility on the employer for the hired workers, - Wages are slightly higher than market wages for the purposes of independent payment for services not received. The low efficiency of this measure resulted from the inability to control the targeted expenditure of funds by the employee, - Payment for check-ups and purchase of insurance at the expense of the employer
Difficulties in protecting the rights of employees. Lack of trade unions	- Alternative insurance, self-registration and membership in insurance companies, - Labor laws expansion and updating: - Consolidation of the practice of social and labor relations of the precariat as a social norm or procedure, exercising social control over the practice of fixed-term labor relations, - New standards for work organization in the information society, - Institutionalization of the precariat, imposing responsibility on the employer for the hired workers, - Improving the legal literacy of employees

Source: authors.

As we can see, it is impossible to offer a single strategy to minimize HR risks when working with temporary employees. In this regard, a customized assessment of HR risks for a certain employee and manager with further recommendations how to eliminate them is of great importance. Depending on the possibility of modification, HR risks when working with non-standard employment can be divided into three groups (Fig. 1).

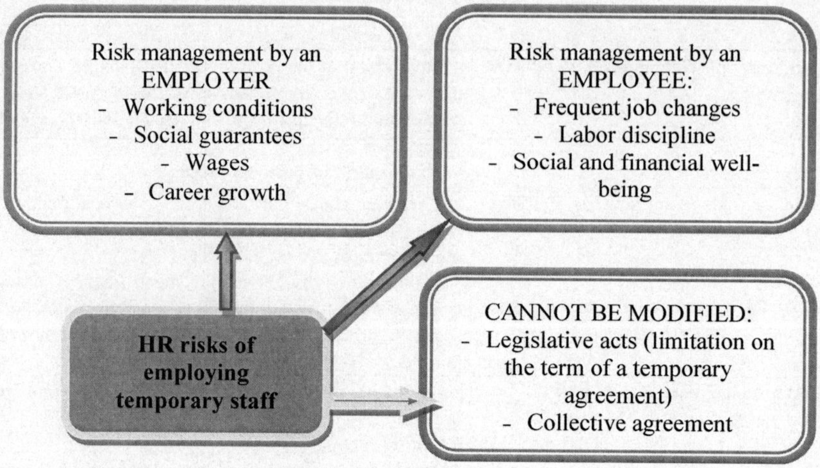

Fig. 1. Temporary Employment HR risks Structure (Source: authors).

In this context, it seems relevant to develop a customized system of HR risk identification for the employee when hiring for temporary work and for the employer when defining the HR policy of the company. The HR risk assessment system should have the specifications as follows:

- individual approach - the employee and the manager shall be able to fill out the questionnaire individually,
- ease of use – it is necessary to ensure that the assessment system can be applied with minimum technical requirements, for example, a paper questionnaire,
- speed of filling out,
- ease of processing,
- unambiguous interpretation of the results.

Such requirements are met by the questionnaire with the subsequent ranking of individual HR risks for the employee or for the development of HR strategy for a company. The employee questionnaire is presented in Table 4.

Table 4. Sample questionnaire for an employee

Arrange listed performance characteristics in order of importance (from 1 to 7):
Wage rate
Advanced training
Social security (pension benefits and medical insurance)
Regular place of work
Flexible working hours
Favorable working conditions
Team relationships
Arrange listed performance possible threats to long-term employment relationships in order of importance (from 1 to 8)
Job losses due to digitalization
Lack of professional development
Inadequate wages
Informal employment or "under-the-table" wages
Deterioration of social guarantees
Unethical labor practices
Deterioration of working conditions
Family problems, health problems

Source: authors.

From the analysis of the employee's responses, it is possible to identify the major motives for working at a regular job, as well as factors contributing to improvement of temporary employment productivity. The questionnaire for the Manager is presented in Table 5.

Table 5. Sample questionnare for a manager

Arrange the offered responses in the priority to prevent the outflow of employees (from 1 to 7):
Training of employees with no work experience
Advanced training of employees
Wage Indexation
Production Digitalization
Improvement of working conditions
Productivity Improvement
Maintaining team motivation

Source: authors.

The analysis of the questionnaire reveals the manager's intention to use non-standard employment and his/her desire to support full-time employees.

4 Discussion

The survey reveals that the HR policy of the company shall be aimed at improving work efficiency of those employed under fixed-term contracts and at increasing attractiveness and economic efficiency of the permanent employment. In general, increasing the adaptability of economic agents and their ability to build and follow "long-term" strategies will improve the context of precarization and stabilize the associated risks of erosion of trust and degradation of human resources potential both at the level of the individual, and at the level of the company and the state as a whole. Depersonalization of hiring is equally important. The modern labor market is rapidly moving from hiring a person to hiring not even a function, but a one-time operation. This raises serious issues of a new social contract, as the old one is collapsing around our ears. The contradictions are accumulated between the labor legislation formed during the period of full-time employment, strong trade unions, a relevant package of the social benefits and guarantees, and the changed reality of labor relationships with flexible employment, precarization and uberization of work. The total dominance of the service sector in the economy has led to a transition from mass and cooperative labor practices to individual and competitive ones. This circumstance, due to the very nature of work, objectively prevents the consolidation of employees in the process of bargaining with the employer, strengthens their information, financial and organizational asymmetry.

5 Conclusion

Our research reveals that HR risk management in the context of labor market precarization is complex, systemic and synergistic. The core of this process is the advancement of the adaptability of employees and self-employed, improving their ability to build and follow "long-term" strategies. An important factor enabling the implementation of this scenario is the assistance of the state in flexible response to the challenges of precarization and amendments into labour law and social safeguards in line with the current situation in the labour market. These pose a challenge of raising awareness of economic behavior of subjects in the labor market, the importance of investing in personal development and improving their capacity to better defend their interests in the labor market. Otherwise, we run the risk of uncontrolled degradation of both the very basis of human capital reproduction and social relations in the sphere of employment. Based on an interdisciplinary analysis of the phenomenon of the precariat, related HR risks have been assessed, practical recommendations have been developed to minimize HR risks and improve HR policy of various economic agents in terms of labor market precarization.

References

1. Bacci, S., Pigini, C., Seracini, M., Minelli, L.: Employment condition, economic deprivation and self-evaluated health in Europe: Evidence from EU-SILC 2009-2012. Int. J. Environ. Res. Public Health **14**(2), 143 (2017)
2. Bec, U.: Risk society on the Way to Another Modern. Progress Tradition, Moscow (2000)
3. Benach, J., Julià, M., Tarafa, G., Mir, J., Molinero, E., Vives, A.: Multidimensional measurement of precarious employment: social distribution and its association with health in Catalonia (Spain). Gac. Sanit. **29**(5), 375–378 (2015)
4. Bourdieu, P.: Sociology of social space. Institute of Experimental Sociology, Moscow (2007)
5. Bryan, M.L., Sevilla, A.: Flexible working in the UK and its impact on couples' time coordination. Rev. Econ. Household **15**(4), 1415–1437 (2017)
6. Canivet, C., Aronsson, G., Bernhard-Oettel, C., Leineweber, C., Moghaddassi, M., Stengard, J., Westerlund, H., Östergren, P.O.: The negative effects on mental health of being in a non-desired occupation in an increasingly precarious labour market. SSM – Popul. Health **3**, 516–524 (2015)
7. Cardoso, A.R., Loviglio, A., Piemontese, L.: Misperceptions of unemployment and individual labor market outcomes. IZA J. Lab. Policy **5**, 13 (2016)
8. Fagan, C., Norman, H., Smith, M., González Menéndez, M.C.: In search of good quality part-time employment. Int. Lab. Off., Geneva (2013)
9. Giraudo, M., Bena, A., Leombruni, R., Costa, G.: Occupational injuries in times of labour market flexibility: the different stories of employment-secure and precarious workers. BMC Public Health **16**, 150 (2016)
10. Hanglberger, D., Merz, J.: Does self-employment really raise job satisfaction? adaptation and anticipation effects on self-employment and general job changes. J. Labour Mark. Res. **48**(4), 287–303 (2015)
11. Hirsch, B.T., Husain, M.M., Winters, J.V.: Multiple job holding, local labor markets, and the business cycle. IZA J. Labor Econ. **5**, 4 (2016)
12. Inoue, M., Tsurugano, S., Nishikitani, M., Yano, E.: Full-time workers with precarious employment face lower protection for receiving annual health check-ups. Am. J. Ind. Med. **55**(10), 884–892 (2012)
13. International Labor Organization: Non-standard forms of employment. Analysis of problems and prospects for solutions in different countries (2017). https://www.ilo.org/wcmsp5/groups/public/—dgreports/—dcomm/—publ/documents/publication/wcms_534326.pdf. Accessed 22 June 2020
14. Julià, M., Vanroelen, C., Bosmans, K., van Aerden, K., Benach, J.: Precarious employment and quality of employment in relation to health and well-being in Europe. Int. J. Health Serv. Plann. Admin. Eval. **47**(3), 389–409 (2017)
15. Piketty, T.: Capital in the 21st Century. Ad Marginham Press, Moscow (2016)
16. Standing, G.: Precariat: a New Dangerous Class. Hell Marginem Press, Moscow (2014)
17. Toshchenko, ZhT: New phenomenon in the social structure of society: Precariat. Soci. Econ. **11**, 25–45 (2018)
18. Tsurugano, S., Inoue, M., Yano, E.: Precarious employment and health: analysis of the comprehensive national survey in Japan. Ind. Health **50**(3), 223–235 (2012)
19. Wege, N., Angerer, P., Li, J.: Effects of lifetime unemployment experience and job insecurity on two-year risk of physician-diagnosed incident depression in the German working population. Int. J. Environ. Res. Public Health **14**(8), 904 (2017)

The Specifics of the Management System of a Municipal Institution

S. I. Sotskova and V. P. Fomin(✉)

Samara State University of Economics, Samara, Russia
S_sotskova@mail.ru, fominvp@mail.ru

Abstract. A municipal institution, as a governing body, is not a static object and is subject to changes in the internal and external environment, the impact of various risks and threats. The changes that are taking place encourage managers to make changes in financing, managerial decisions, core business, personnel management, etc. In such conditions, the organizational structure of management also needs to be adjusted, since it is the skeleton of the municipal government and gives orderliness to all processes. The choice of organizational structure is always associated with the need to assess their own capabilities and capabilities of the staff, which always constitutes a risk of managerial activity. The article examines the management system using the example of a municipal institution, identifies its main shortcomings and suggests ways to improve it. Improvement of the organizational structure is aimed at creating structural units of the Department of Culture and Youth Development of the Chapaevsk city district administration, carrying out activities to solve the most pressing problems of culture and youth policy in the territory of the municipality.

Keywords: Control system · Institutional management structure · Municipal institution · Organization · Strategic management

1 Introduction

The relevance of the research topic is also associated with the insufficient development of the topic. The study of organizational structures of state and municipal institutions is reduced to the use of traditional templates, which are used to study the organizational structure of commercial enterprises. This practice does not reflect either the accumulated experience or the existing problems of the construction and functioning of organizational structures of state and municipal authorities.

To date, the study of organizational structures of state and municipal authorities is carried out in the context of the legal consolidation of organizational structures, presented mainly in the educational literature [1, 11, 12]. There are no large fundamental studies on the problem of studying the construction and transformation of organizational structures of municipal authorities. At the same time, it should be pointed out that there are studies of experience in building organizational structures using concrete institutions of municipal authority as an example. Certain issues of building organizational structures are investigated: the concept of "organizational structure" in the municipal administration system [2, 3, 6], documentation of the functioning of the

S. I. Ashmarina and V. V. Mantulenko (Eds.): IPM 2020, LNNS 161, pp. 644–654, 2021.
https://doi.org/10.1007/978-3-030-60926-9_81

organizational structure, construction principles [4, 6, 14], differentiation of powers, directions improving organizational structure [5, 8, 9]. The timeliness and correctness of decisions made in the field of changing the organizational structure will depend on how adequately the leader evaluates the possibilities of the institution (financial, personnel, material and technical) and emerging threats from the external environment.

2 Methodology

The effectiveness of the institution is largely determined by the existing management system. The term "management system" is one of the basic in management and is used to characterize and analyze the goals of the institution, the ways to achieve them, the functions of managerial activity, the distribution of powers between employees [13]. The management system is the interconnection of the various elements and subsystems of the enterprise, the communications established between them, which enable the functioning of the institution. The activities of any institution are based on the use of certain resources, the combination of which allows us to achieve our goals. And the link in this is precisely the institution management system.

As elements of a management system, one can single out the goal, tasks, management process, functions, management methods and principles, organizational communications, organizational relations, organizational structure, organizational relations, personnel, finances (capital), material and technical resources (equipment and technologies), innovations, manufactured products or services, technological process, infrastructure facilities of the institution [10].

One of the important elements of the management system is the organizational structure. The organizational structure allows you to create an orderly distribution of responsibilities and activities between the personnel of the enterprise, establish organizational interaction, exchange of information that allows you to achieve your goals in the work. To consider the place of the organizational structure in the enterprise management system, it is necessary to consider the components of the definition of "organizational structure", which are presented in Table 1.

Table 1. The Conceptual components of the definition of "organizational structure"

Term	Definitions
Institutional Management Structure	The interaction of management levels within the chosen form, which allows achieving the goals of the institution Logically ordered set and interconnection of various elements of the institution, aimed at achieving the goals and objectives of the activity [7]
Organization	The totality of processes or actions that allow you to form and improve the structural relationships between the individual parts of the whole object; an association of people jointly implementing a program or goal and acting on the basis of established rules and procedures [11]

Source: authors based on [7, 11].

Based on the definitions presented in Table 1, the organizational structure can be defined as a form of separation and organization (interaction, cooperation) of management activities carried out within the framework of the given functions and aimed at achieving the goals and objectives [1].

The organizational structure in the management system of the institution ensures the optimal distribution of functional responsibilities, rights and authorities, an interaction algorithm between the management bodies and the employees working in them. In other words, the organizational structure represents the ordered core of the institution, ensuring the functioning of all types of activities.

From these positions, the management structure is represented in the form of a system of optimal distribution of functional duties, rights and responsibilities, the order and forms of interaction between its governing bodies and the people working in them. The organizational structure can also be considered as the totality of a certain number of links and control levels with the interconnections between them [15].

3 Results

Consider the example of the Department of Culture and Youth Development of the administration of the urban district of Chapaevsk (hereinafter referred to as MKU "DKiMR"), the features of the formation of the organizational structure and the possibility of its improvement. The institution was created through reorganization in the form of the Department of Culture of the Chapaevsk city district administration joining the Youth Development Department of the Chapaevsk city district administration with the transfer to it of the rights and obligations of the associated legal entity. MKU "DKiMR" is subordinate to the Administration of Chapaevsk, the head of the city Chapaevsk, is directly under the supervision of the First Deputy Head of the Chapaevsk urban district. Activities are funded by the city budget and other sources. Formation of the image component of MKU "DKiMR" is carried out through the development of various services in the field of culture, the development of the youth movement, improving the quality of services provided and ensuring their accessibility for various categories of citizens. MKU "DKiMR" has a high level of automation. All stationary workplaces of specialists and subordinate institutions are automated, there is a system of electronic document management and reporting to higher authorities. Also, MKU "DKiMR" and structural units have sites and pages in social networks, which allows the community to be informed about ongoing activities and main areas of activity.

Strategic management is based on the implementation of the main goals and objectives in the field of cultural and youth policy specified in federal laws, regional and municipal target programs, the development strategy of MKU "DKiMR" and subordinate institutions. Tactical management is based on annual activity plans, as well as plans for the half year, quarter and month. Tactical management is aimed at solving tactical tasks defined by strategic documents and activity goals. The management of the organizational structure of MKU "DKiMR" is based on the principles of unity of command, hierarchy of positions, publicity, collegiality (youth policy programs developed by MKU "DKiMR" are publicly discussed and adopted in a team). The organizational structure of MKU "DKiMR" is developing dynamically. The need to develop an organizational structure is due to a change in the environment among young people, the specifics of social, economic, and political processes in the territory of the city of Chapaevsk, changes in state and municipal legislation in the field of youth policy. The change in the organizational structure of MKU "DKiMR" is based on the "from problems" approach.

Among the positive aspects of the organizational structure of MKU "DKiMR" can be identified centralization of activities in the field of youth development and cultural development in the territory of the municipality. Over the past three years, the centralization of subordinate institutions has taken place, and the vertical line of power has been built: "head of the Department - head of department/specialist - head of the subordinate institution". Strict centralization allowed to accumulate experience in the field of management of cultural institutions, youth development, functioning and activities of institutions of additional education.

The next advantage of the organizational structure can be considered the development and implementation of municipal projects and programs; the existence of a well-functioning mechanism for planning and implementing activities within the framework of the main activities. However, the desire to build and strengthen vertical relations has weakened functional interaction. The presence in the organizational structure of a large number of independent specialists and consultants in various areas of activity led to the duplication of functions with department specialists.

A number of areas of activity of MKU "DKiMR" Chapaevsk, caused by the needs of the public, has no organizational fixation. In particular, such issues include: the volunteer movement, fundraising, the development of tourism and museum business, the development of entrepreneurship among young people in the municipality. Museum business and tourism development could not be fixed in the organizational structure of the Department, although these activities were prioritized.

The contradictions and shortcomings of the organizational structure led to a number of problems in the development of MKU "DKiMR" Chapaevsk:

- lack of dynamics in the replenishment of the museum fund, the museum's activities are reduced to excursion services, the museum's low activity in the implementation of children's and youth projects, the introduction of innovative forms of work,
- the implementation of various areas of volunteer movement on the basis of the Municipal Institution "Choice", which does not meet the lines of activity of this institution, as reflected in the charter. MU "Choice" is engaged in the prevention of the use of surfactants, but not as a volunteer movement that goes beyond the main area of activity,
- there is no organizational fixation of resource support for implementation.

4 Discussion

When developing a new model of the organizational structure of MKU "DKiMR", the goals structuring method was used, which provided an assessment of the institution's existing goals, the development of a new set of activity goals and goals of structural units focused on the implementation of activities and their subsequent combination with the structure being developed. It is proposed to improve the organizational structure of MKU "DKiMR" by forming two new structural units: "Resource Center" and "Department of Youth Innovation":

1. *Formation of the "Resource Center"*
   The creation of the Resource Center is caused by the need for a functional separation of duties of specialists in the main areas of the Department's activities (the implementation of projects and programs in the field of tourism, the implementation and further improvement of work in the field of museum education and local history, the development and implementation of events in the field of advertising and public relations, etc.).
   The implementation of legislation in the field of preservation and development of cultural heritage, the presence of problems in the development of museum activities requires legislative consolidation and the allocation of a separate specialist for the development of museum activities as a direction of culture and local history. Previously, such a specialist did not exist. The management of museum activities was carried out indirectly, through the head of a subordinate institution - MBU "Sociocultural leisure complex." In fact, the activity was reduced to organizing a leisure component in the museum's work - organizing excursion services. Also, duplication of functions and the lack of a clear delineation of functional responsibilities in the field of resource support for the implementation of activities between specialists requires their structuring.
   First of all, the subject of activity of the Resource Center is defined, goals and directions of activity are developed. The subject of the Resource Center's activities are: resource, information and methodological support of cultural institutions, additional education and youth policy; introduction of modern technologies in the field of activity; creation of a unified cultural space aimed at increasing the professional competence of specialists.
   The purpose of the "Resource Center" will be to assist in the implementation of the municipal laws and regulations envisaged by the legislation of the Russian Federation, the powers of municipal authorities in the field of culture, tourism, youth policy in the territory of the city Chapaevsk.

   The following are fixed as priority areas of the Resource Center's activities:

- performance of work on resource support (providing organizational, advisory, informational and other support) to cultural institutions, additional education and youth policy,
- organization of social design in the field of culture, tourism, and youth development,
- development and implementation of tourism activities on the basis of the municipality, the development of tourism in a small area,
- assistance in the development of local history, museum work in the municipality, support of the city museum, assistance in the development of museums on the basis of educational institutions of the city,

- organization and implementation of activities in the field of youth policy aimed at involving young people in innovative entrepreneurial and voluntary activities, as well as at developing civic engagement of young people, creating a healthy lifestyle,
- organization and conduct of measures to prevent the use of psychoactive substances in the youth environment,
- assistance in integrating young people who find themselves in a difficult life situation into society,
- monitoring of professional and informational needs of cultural workers, youth development and further education,
- implementation of marketing and advertising activities, relations with the public;
- implementation and support of municipal and regional programs in the field of culture, tourism and youth development,
- work on professional development and development of personnel of the Department and subordinate institutions.

In connection with the indicated areas of activity, the Resource Center will acquire the structure shown in Fig. 1.

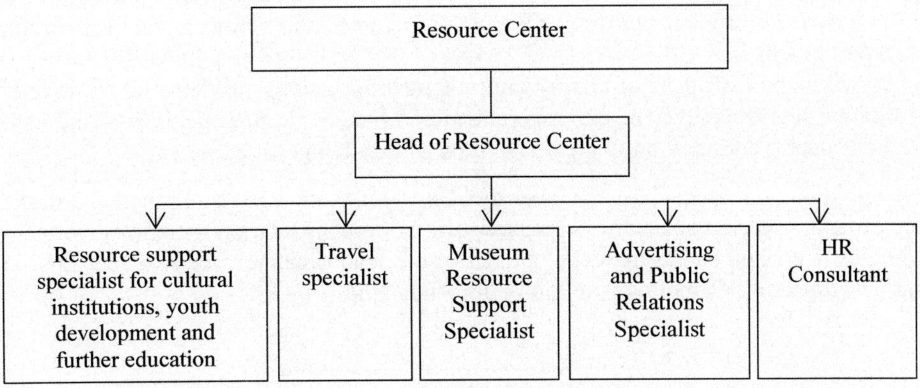

Fig. 1. The structure of the resource center MKU "DKiMR" chapaevsk (Source: authors).

As a result of the formation of the Resource Center, the burden on the specialists of the youth affairs department and the culture development department will be reduced. These departments will be left with the organization of activity planning, including the development of municipal programs for the development of cultural institutions, youth development, the system of additional education, and tourism; activity monitoring, case studies, reporting. Practical aspects of organizing and conducting events in the city will be assigned to the specialists of the Resource Center.

2. *Formation of the "Department of Youth Innovation."*
   The formation of the Youth Innovation Division is due to the fact that the development of new activities in the youth environment has become multi-faceted and the existing provision on the activities of the Youth Affairs Department can no

longer cover all existing activities and at the same time implement organizational and administrative functions. Subject of activity: information and methodological support and implementation of innovative projects in the field of youth policy in accordance, development of volunteerism in the youth environment, stimulation of entrepreneurial activity among young people. The purpose of the activity is the development and implementation of youth innovative non-commercial projects and programs, events on the territory of municipal education.

The main activities:

- development of volunteer youth organizations, involvement of youth in volunteer (volunteer) activities,
- development of social partnership in the implementation of volunteer activities,
- the introduction of effective programs, projects, the implementation of actions and measures to develop the social activity of youth, necessary for promotion in the labor market,
- development of youth entrepreneurship and business activity of youth,
- development of innovative activity, project activities, grant activities of youth in the field of culture, tourism and youth development,
- organization and holding of seminars, trainings, round tables, conferences on the problem of implementing innovative programs and projects in the youth environment,
- dissemination of ideas of charity and volunteering among residents of the municipality, representatives of the government and business representatives of the city,
- information and methodological activities and exchange of experience.

The following specialists will enter the Youth Innovation Division: the head of the department, a design specialist, a specialist in the implementation of innovative technologies, a consultant for the development of the volunteer movement, a specialist in the development of employment and entrepreneurship (Fig. 2).

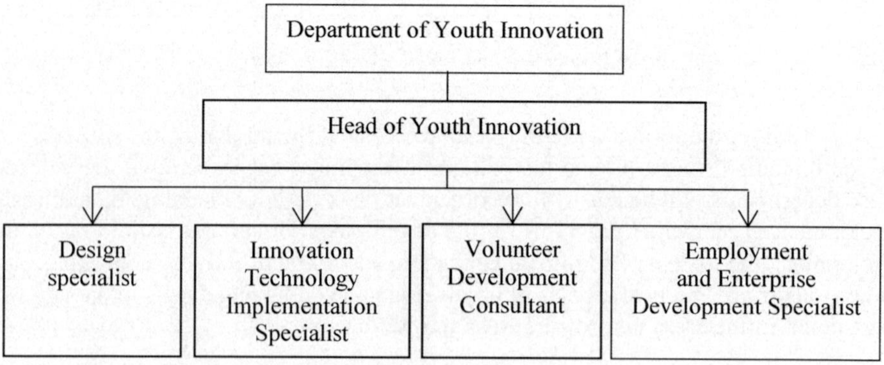

Fig. 2. Structure of the department of youth innovations (Source: authors).

The head of the department carries out planning and control of the department; provides organization and coordination of the activities of department specialists, interaction with employees and various institutions of the city Chapaevsk; ensures the participation of the department and the Department in grant activities, etc.

The design specialist carries out work on the preparation of documentation and projects for grant support, develops and implements innovative projects in the youth environment, consults specialists of the department, Department, city youth associations, youth on social design issues, and manages fundraising activities.

The specialist in the implementation of innovative technologies collects information on innovative areas of work with young people, about contests, potential partners; together with a specialist in design, carries out the development and implementation of social projects in the youth environment. The consultant on the development of the volunteer movement provides the development of volunteering and volunteering in the city of Chapaevsk, participates in the development and sale of shares of volunteers of various kinds, monitors the results of actions, maintains documentation on the volunteer movement. A specialist on the development of employment and entrepreneurship carries out work on the organization of employment of adolescents, implements measures and programs for the development of entrepreneurship in the youth environment, and oversees youth entrepreneurship projects. The subordinate institution - MU "Vybor" Chapaevsk. As a result of organizational changes, there will be an increase in the number of staff by three people. It is necessary to hire a specialist in tourism, a specialist in the resource support of museum activities, and a resource center. Heads of new structural units will directly report to the head of the Department. The provision of working space to change the organizational structure of the institution is not required. Financing of organizational changes will be carried out at the expense of the municipal budget (additional financing), as well as funds pledged for the implementation of federal and municipal projects and programs of culture and youth development. Monitoring the implementation of organizational changes in the structure of the institution will be carried out by the head of the Department, the head of the city of Chapaevsk, the Chapaevskaya City Duma, and the deputy head of the city on social issues.

1. The functioning of structural units is associated with the need to develop action plans, which should reflect the main areas of work of MKU "DKiMR" Chapaevsk, focused on meeting the needs of citizens of the municipality in obtaining services in the field of culture, leisure, tourism, youth policy.
2. Therefore, the next step to improve the organizational structure will be the development of an algorithm for preparing an action plan and a flow chart of administrative procedures for the consideration of applications by representatives of the external environment for conducting and approving activities to work with various categories of citizens. Plans are prepared by structural units independently, and then a single plan of activity is formed. A clear structuring of all procedures in the process of developing the plan will avoid duplication of activities of functional units, will ensure the inclusion of the Resource Center in the preparation and support of planned activities. At present, this procedure is not regulated.

Thus, the goal of structural changes in the organizational structure of the department will be to promote the self-realization of youth in the public life of the city. The achievement of the goal will be facilitated by the solution of such problems of changing the organizational structure as an integrated approach to solving youth problems, creating conditions for the social development of youth and its inclusion in the life of society through a social protection mechanism, supporting the talented and most active part of youth, developing a system of youth and children's public associations and the support of their activities, reducing the dependence of young citizens using surfactants, building a system of effective communications with business structures, authorities and the media of Chapaevsk city, developing social and civic activism of youth, youth entrepreneurship, introducing innovative technologies logic in the field of culture and tourism. As a result of the changes made in the organizational structure, the structuring of goals (analytical, practical) and activities will take place.

5 Conclusion

Thus, as a result of the study, we can conclude that the organizational structure is a multidimensional display of the state of the organization. The authors highlighted the main shortcomings of the organizational structure of MKU "DKiMR" Chapaevsk:

1. Management of innovative technologies in the field of culture and youth policy, tourism, and the field of financing of non-profit projects at the local level through grant competitions requires the creation of a special staff that is currently not provided for in the organizational structure.
2. In connection with the active participation of the Department and youth associations in grant activities, which involves the development of projects of various kinds, there is a need to educate young people in the basics of social design and implementation of projects, which is currently difficult due to the lack of organizational structure of the Department special unit dealing with these issues.
3. The complexity of the activities of employees and departments of the Department related to the establishment of feedback from the public through the media. Each structural unit is forced to independently prepare information about the work for publication in the media, which leads to duplication, distortion of information.
4. A number of areas of activity of MKU "DKiMR" Chapaevsk, caused by the needs of the public, has no organizational fixation. In particular, such issues include: volunteer movement, fundraising, tourism and museum development, youth entrepreneurship development in the municipality.
5. Museum business and tourism development has not been fixed in the organizational structure of the Department, although these types of activities are prioritized.
6. The lack of dynamics in the replenishment of the museum fund, the museum's activities are reduced to excursion services, the museum's low activity in the implementation of children's and youth projects, the introduction of innovative forms of work.
7. The implementation of various areas of volunteer movement on the basis of the Municipal Institution "Choice", which does not meet the lines of activity of this

institution, as reflected in the charter. MU "Choice" is engaged in the prevention of the use of surfactants, but not as a volunteer movement that goes beyond the main area of activity.

8. There is no organizational fixation of resource support for the implementation of the main activities of MKU "DKiMR" Chapaevsk.

To eliminate the identified shortcomings in the organizational structure of MKU "DKiMR" Chapaevsk developed the following proposals:

1. Form two new structural units: Resource Center; Department of Youth Innovation.
2. To develop procedures for the linear and functional interaction of structural units in the development of planning and events: a flowchart for the development and approval of the annual work plan of MKU "DKiMR" Chapaevsk, flowchart of administrative procedures for the consideration of applications for conducting and approval of activities to work with various categories of citizens.

As the expected results of the activities of the formed structural units, it is planned to achieve: creating conditions for the implementation of entrepreneurial activity and youth employment; teaching youth social design skills; increase in the growth of volunteer movement; improving the quality of resource support MKU "DKiMR" Chapaevsk and subordinate institutions; increased coverage of the population with cultural events; promoting the development of tourism, museum business in the municipality; introduction of innovative forms of work.

References

1. Blinov, A.O., Ugryumova, N.V.: Organization Theory and Organizational Behavior (theory and practice): A Training Manual. KNORUS, Moscow (2016)
2. Bogashev, M.V., Lutsenko, A.V.: To the question of building an effective organizational structure of Russian corporations. Bull. Sci. Pract. **12**(25), 269–275 (2017)
3. Budaeva, N.N., Batashova, A.F.: Directions for improving organizational management structures. In: Sukisyan, A.A., (ed.) Innovative Mechanisms for Solving the Problems of Scientific Development, pp. 100–102. Omega Science, Ufa (2019)
4. Day, K.M., Armenakis, A.A., Field, H.S., Norris, D.R.: Other organizations are doing it, why shouldn't we? A look at downsizing and organizational identity through an institutional theory lens. J. Change Manage. **12**(2), 165–188 (2012)
5. Ewens, H., Voet, J.: Organizational complexity and participatory innovation: participatory budgeting in local government. Public Manage. Rev. **21**(12), 1848–1866 (2019)
6. Freund, A., Drach-Zahavy, A.: Organizational (role structuring) and personal (organizational commitment and job involvement) factors: do they predict interprofessional team effectiveness? J. Interprof. Care **21**(3), 319–334 (2007). https://doi.org/10.1080/13561820701283918
7. Gultyaev, V.E., Burykin, A.D.: The principles of building organizational structures for managing the organization. Econ. Manage. Probl. Solutions **5**(9), 14–24 (2018)
8. Kasdin, S.: An evaluation framework for budget reforms: a guide for assessing public budget systems and selecting budget process reforms. Int. J. Public Adm. **40**(2), 150–163 (2017)
9. Lorsuwannarat, T.: Public participation in budgeting: The new path of budget reform in Thailand. Int. J. Public Adm. **40**(5), 385–400 (2017)

10. Pyryaev, V.V., Pyryaeva, L.A.: An integrated approach to building the organizational structure of organization management. Sci. Modern World Dev. Priorities **1**(3), 166–170 (2017)
11. Ruzhanskaya, L.S., Yashin, A.A., Soldatova, Y.V.: Theory of Organization: A Training Manual. Publishing House Ural. University, Yekaterinburg (2015)
12. Vasilenko, I.A.: State and Municipal Government. Yurayt, Moscow (2016)
13. Vishnevskaya, P.V., Faramazyan, L.G.: The concept, types and basic principles of building organizational management structures. In: Sukisyan, A.A., (ed.) A New Science: The History of Formation, Current Status, Development Prospects, pp. 24–26. Omega Science, Ufa (2017)
14. Vriesman, D.W.: Modernizing organizational structures for agencies and institutions to meet new governmental and economic challenges. J. Relig. Aging **2**(4), 73–80 (1987)
15. Yezhova, L.S., Voronin, A.V., Lyakh, K.A.: Features of the formation of organizational structures. Actual Probl. Modern Econ. **2**, 200–204 (2019)

Features of Recruitment of Hospital Teachers to Work with Long-Term Ill Children

G. Sukhanova[1,2(✉)] and S. Sharikov[2]

[1] National Research University – Higher School of Economics, Moscow, Russia
gsukhanova@hse.ru

[2] Hospital School in the Dmitry Rogachev National Medical Research Center Pediatric Hematology, Oncology and Immunology, Moscow, Russia
sh@uchimznaem.ru

Abstract. It is clear that children in long-term treatment are categorized as having special educational needs. Prolonged treatment (from several months to several years) and regular painful medical procedures result in increased fatigue, unstable attention, increased distractions and sometimes inability to remember and absorb information. This is especially the case for students in cancer hospitals. This creates a special need in hospital schools for highly qualified teaching staff with a high level of professional competence, psychological and personal readiness to work with ill children. However, the current situation is such that only a small percentage of Russian pedagogical universities can boast that they provide educational programs aimed at training such specialists. As a result, teachers from ordinary schools, who have not received special training, teach children undergoing long-term hospital treatment. Of course, all this imposes serious requirements on the process of searching and selecting such personnel. The authors of the article tried to identify and classify the key professional, personal and behavioral characteristics necessary for the work of a hospital teacher. In addition, the authors attempted to develop tools for analyzing these qualities at the recruitment stage.

Keywords: Hospital school · Recruitment of hospital teachers · Selection of teaching staff

1 Introduction

According to the World Health Organization as of early 2019, one in 33 children in the world today is born with congenital malformations. Every year, 3.2 million children are born with disabilities due to congenital malformations. Moreover, this figure is growing rapidly every year [17]. In addition, more than 12,000 cancer cases are diagnosed among children under 18 every year only in the Eurasian region [16]. Cancer is still one of the leading causes of death among children and adolescents in the world. Every year, 300,000 children from birth to 19 are diagnosed with cancer worldwide [14]. One in two children with cancer has immune and hematopoietic diseases [10]. According to statistics, about 4,000 new cases of confirmed oncological diagnoses in children aged 0

S. I. Ashmarina and V. V. Mantulenko (Eds.): IPM 2020, LNNS 161, pp. 655–666, 2021.
https://doi.org/10.1007/978-3-030-60926-9_82

to 18 are detected annually in Russia. Approximately 24,000 Russian children are under constant observation in oncological institutions every day [9].

Such children spend a significant part of their lives in hospitals, being treated for a long time. And, in addition to abandoning their habitual way of life, they face the fact that, for a long time, due to health reasons, they cannot attend general education school in their place of residence. However, we should bear in mind that absolutely all children have the same right to education from birth, that enshrined in the Convention on the Rights of the Child as far back as 1989. Ensuring the realization of this right is one of the priority tasks of the States Parties to the Convention [2]. Education and upbringing of all children without exception are sources of development of psychological functions, regulators of physiological and psychological processes. Education, as a system of purposeful and consistent pedagogical influences, determines, to a great extent, the child's mental and personal development, influences his or her life prospects and future in general [6]. That is why the education, training and upbringing of children undergoing long-term hospital treatment should attract as much (and perhaps even more) attention as the education of healthy children freely attending kindergartens and schools.

Combining medical and pedagogical efforts in the process of restoring the health of a child is the practical implementation of an integrated approach to the rehabilitation process [6]. The influence of children's education in the process of treatment on the rate of recovery of mental functioning, realization of cognitive potential and successful socialization has been scientifically proved by outstanding Russian and foreign scientists.

Academician of the Russian Academy of Education, Doctor of Pedagogical Sciences Yamburg in his numerous articles, scientific papers and monographs repeatedly notes that long-term ill children with the correct organization of the educational process have increased motivation to learn. Children chained to drips and deprived of the usual way of life for most children, with competent and delicate pedagogical influence, show good results after several weeks of communication with the hospital teacher: the study distracts the child from the hospital routine, from painful procedures, improves his or her mood and forms a positive mood. For high school students, a full and quality education in the hospital becomes a serious step in their adult life: thanks to the help of hospital teachers, they pass the USE, enter universities, build their careers and fill their lives with new goals and meanings distracted from health problems [18, 19].

Sergey Sharikov, PhD, head of the Federal project of hospital schools of Russia UchimZnaem, is the author of the national model of the hospital school. In his articles and interviews he draws attention to the inextricable link between the full educational environment and the speed of medical and social rehabilitation of children with disabilities, disabled children and children undergoing long-term treatment in hospitals. Besides, in his works Sharikov stresses the influence of the educational environment on the development of cognitive abilities of a long-term ill child, on the formation of positive self-esteem, on the development of his or her independence [11, 12].

Shaw, PhD, an outstanding scholar and professor at McGill University (Montreal, Canada) and his colleague at Brooklyn College (City University of New York) McCabe in their joint research papers and books on the topic, argue for the importance and necessity of continuous and close collaboration between medical staff and hospital school teachers for the successful recovery of a long-term ill child and his/her further painless reintegration into the educational environment of the school in the community [7, 13].

A group of scientists from India found in their research that children with cancer who continue to attend hospital school have higher levels of trust and positivity than those who have been deprived of this opportunity. Moreover, 85% of the child patients who attended hospital school were able to formulate their future aspirations and dreams during the interview. And of those who did not attend school, only 2 out of 10 spoke about possible future plans [1].

As we can see, recruitment and involvement of professional teachers in the process of restoring the health of long-term ill children is a justified necessity. This creates a special need in hospital schools for highly qualified teaching staff with a high level of professional competence, as well as psychological and personal readiness to work with ill children.

However, for most graduates of pedagogical universities and even for experienced school teachers, a health care organization is not a natural environment for their professional activity—no one has ever prepared them for work in a children's hospital. A teacher of a hospital school, while supporting the educational opportunities of children, regardless of the subject of their position, should have unique competencies that are related to building in-hospital communications, drawing up an individual training program and developing an individual map of the special educational needs of a long-term ill child (diet, requirements and restrictions on physical activity, the regimen of taking medications and performing medical procedures, etc.), with the choice of tools for careful, but adequate assessment of the results, with a plan for returning the child after treatment to school at the place of permanent residence, etc. [12].

2 Methodology

Based on the above arguments, there is an obvious need to develop up-to-date tools for selecting and recruiting teachers for hospital schools. The authors of the article set themselves the following tasks:

1. Identify and describe the key features of children that hospital teachers work with in hospital schools.
2. Classify these key features.
3. Formulate the key characteristics of the personality and behavior of a hospital teacher that will help him successfully work with long-term ill children (taking into account the identified characteristics of an ill child).
4. On the basis of the formulated characteristics of a hospital teacher, identify and classify the main requirements (selection criteria) for candidates applying for a job as a hospital teacher.
5. Determine the degree of importance and significance of each of the identified selection criteria.
6. Develop a questionnaire that allows to evaluate applicants for each of the criteria identified as significant and important.

Scientific methods of observation, generalization, measurement, expert assessment and classification of objects were used in the work. An anonymous survey of 125 experts from 11 regions of Russia was conducted to identify and evaluate the significance of criteria for selecting hospital teachers.

3 Results

3.1 Requirements for a Hospital Teacher

The job of a hospital teacher requires emotional and mental contact with a child to make him or her feel protected, supported and accepted by the adult and feel emotional comfort from interaction with them. This on the one hand determines the mission of the teacher, and on the other hand it is an important direction of his or her professional training to work with seriously ill children in a hospital. In their work with a child, hospital teachers should take into account recommendations of doctors on the current status of the child's health and rehabilitation potential [5], cognitive and other deficits caused by the illness and the process of treatment, the severity of the condition, as well as data and observations on the emotional state of the child and his or her family members. A priority guideline in the work of a hospital teacher is the thesis that a child in a hospital is first and foremost a patient [3]. Long-term ill children are characterized by anxiety, fearfulness, insecurity, rapid fatigue, dependence on the opinion of others. Poor health, restriction of physical activity and movements, heavy and prolonged treatment lead to the development of emotional agitation, high anxiety, depressive and regressive reactions. Disease can have a serious impact on a child's endurance, his or her ability to remember and reproduce information [15]. The cognitive processes are largely determined by the course and stage of a child's illness: even during the same school day, the same student can behave in different ways, for example, before and after medical procedures (Table 1).

The above psychological features of long-term ill children make a serious impact on the professional activity of the hospital teacher. Every day, a teacher working with children who are undergoing long-term hospital treatment faces working conditions that are characterized by:

- high emotional tension,
- high requirements to the level of self-control (responsibility for every gesture, gaze, every word spoken, the slightest manifestation of their emotions),
- serious requirements to the level of emotional intelligence of a teacher (at the same time, a high level of empathy should not interfere with their professional work),
- close and daily contacts with parents and relatives of a ill child, who are also in a complicated psychological state and may cause the teacher to have additional existential experiences,
- the need for constant and continuous monitoring of the psychophysical state of the child, the need to take into account the degree of his or her readiness to learn at any given time and to dose the load depending on his or her well-being,

Table 1. Cognitive, emotional and behavioral characteristics of children, being on long-term treatment in a hospital

Cognitive	Emotional	Behavioral
– a significant decrease in mental performance at certain stages of the disease, – difficulties in shifting attention and reducing attention productivity, the child has difficulty concentrating, – reduced amount of information that can be memorized, the difficulty of memorizing and reproducing, – temporary disruption of functions of generalization, analysis, synthesis, logical operations, – a decrease in the development of creative imagination.	– an underdeveloped ability to control and regulate their emotions, – high degree of emotionality (vivid manifestations of even weak emotions), – depressed mood, apathy, anxiety and anxiety, – undervaluation and growth of the inferiority complex due to illness, dependence on the opinion of others, – tendency to feel sorry for oneself and for one's parents (perception of oneself as a burden for one's family), – tendency to reflexion.	– reduced level of autonomy both in decision-making and in the implementation of specific actions, – sharp reduction in the circle of close contacts up to complete disregard for close friends before, the desire for loneliness, – decreased self-control and increased susceptibility to conflict as a result of the constant suppression of internal feelings;, – the demand for increased attention from others, – drowsiness, increased fatigue, low physical activity, decreased endurance.

Source: authors.

– high requirements to the level of flexibility and adaptability of the teacher (ability to change the topic, duration and content of the lesson depending on the state of the student-patient),
– the lack of a clear connection between the teaching efforts made and the educational outcomes obtained,
– the need to develop and implement new gentle knowledge assessment tools and soft, adapted tools to monitor the progress of an ill child [3, 4].

In addition, to work in a hospital the teacher has to comply with a range of additional requirements and standards that teachers in regular schools do not face:

– increased hygienic requirements to the teacher and teaching materials: in some cases, the teacher must wear a medical mask and special clothes when in contact with the student, regularly treat his or her hands and teaching aids with antiseptic (especially when it comes to a child with reduced immunity),
– special requirements for the teacher's appearance (minimum cosmetic and perfumery products, which can cause allergic reactions in pupils-patients),
– high requirements for the level of mastery of distance learning tools, electronic means and modern information technologies, allowing to continue the educational process when the child returned home from hospital, but has not yet started to study in his or her local school,

- high requirements for a teacher's knowledge competencies that are not directly related to the subject taught (e.g., certain medical knowledge, as well as knowledge of legal documents and regulations governing health care and hospital pedagogy),
- the need to conduct classes with patients in non-standard and unusual formats, different from traditional class activities (individual classes, work in small groups (including remotely), work in different age groups, simultaneous classes for an ill child and his or her siblings, etc.).

These and other characteristics of a hospital teacher's work allow us to identify a number of necessary requirements for a person applying for a job in a hospital school. We will conditionally divide them into three categories: professional and knowledge competencies of a teacher (subject, inter subject and extra subject), personal qualities and behavioral norms (Table 2).

Table 2. Main categories of requirements for hospital teacher and their meaning

Professional-knowledge competencies	1. Broad knowledge and practical experience in the subject of the course (discipline). 2. Regulatory and legal preparedness of the teacher (knowledge of the main legal acts and laws related to education, health care and hospital schools). 3. Possession of a wide range of special software tools and tools for distance learning, including experience in organizing remote group study for schoolchildren. 4. Technical skills and active use of modern means of telecommunication and computer technology (tablets, smartphones, laptops and other gadgets, which are easily sanitized unlike paper media). 5. Mastery of the skills to apply original and non-standard teaching methods that can be applied to children with special educational needs. 6. Possession of an extensive set of tools and means to carefully but objectively assess and monitor the progress of an ill child (including remotely). 7. The ability to build an individual educational route for the child based on the recommendations of doctors, psychologists and the individual characteristics of the student. 8. Possessing the tools to keep students' attention and interest in the subject. 9. Experience in applying a project-oriented approach.
Personal qualities and features	1. Ability to control and manage their emotions, and at the same time the ability to recognize the emotions, motivation and intentions of the child (emotional intelligence of the teacher). 2. Ability to limit the degree of empathy to a long-suffering child (empathy should not interfere with the educational process, on the one hand, and empathy should not turn into pity). 3. Low level of personal teaching ambition: few of the long-suffering children are able to demonstrate brilliant results in national competitions and Olympiads.

(continued)

Table 2. *(continued)*

	4. Awareness of one's mission, understanding and acceptance of the burden of responsibility, not only for the educational results of the student, but also for his or her recovery, for the strengthening of their faith in themselves and in their strength, for the renewal of the desire to live and achieve. 5. Artistry and sense of humor, limited by a high degree of self-control (the ability to control their gestures, facial expressions, words, posture, intonation and even the speed of speech). 6. Understanding and acceptance of the higher authority of doctors and health care workers as compared to teachers, readiness to work together closely with medical personnel. 7. Positive attitude and a high degree of motivation for the result, even when it comes to a terminally ill child with a fatal diagnosis. 8. Flexibility and readiness for frequent abrupt changes in operating mode: irregular schedule and long hours. 9. Sincerity, honesty and openness in dialogue with the child, respecting professional distance.
Behavioral norms and values	1. Flexibility and adaptability: readiness to react sensitively and quickly to changes in the psychophysical state of the patient and to change the teaching load and lesson content depending on how the child feels. 2. Strict observance of hygiene requirements and compliance with the rules concerning clothing and appearance (minimum cosmetics and perfumes, use of medical masks and special clothing if necessary). 3. Readiness for constant communication with an ill child and his or her parents, including on topics unrelated to the discipline and academic performance of the student. 4. Creative approach in work with a child in long-term treatment, the teacher's ability to creative pedagogy, focus not only on mastering and learning, but also on creation and invention. 5. Desire and willingness to continuously improve their skills, regular attendance of training events and refresher courses. 6. Respect for medical confidentiality regarding a child undergoing long-term treatment (non-disclosure of information on the diagnosis, nature of the disease and the location of the ill child). 7. Directing efforts at creating a favourable and trusting atmosphere between the learner and the teacher, at reducing the level of stress and the severity of the child's experiences. 8. Orientation to non-standard forms of conducting classes, different from the traditional classroom activities. 9. Avoiding categorical judgments when working with an ill child and evaluating the results.

Source: authors.

Obviously, all the qualities and competences of the hospital teacher listed in Table 2 should be assessed at the entrance, i.e. until the person is employed and has direct contact with the child in long-term care. An untrained specialist who does not meet these requirements can harm not only the learning process, but also the psychological state of the student-patient and his or her recovery. Thus, the requirements listed in Table 2 can and should be used as criteria in the selection of candidates for the position of a hospital school teacher. The authors of the article asked teachers and staff of the administration of Russian hospital schools to assess the degree of importance of each of the above criteria (requirements to the candidate), so that in the future it would be possible to take into account the weight of each criterion when assessing the professional suitability of an employed candidate.

3.2 Expert Assessment of Requirements

The criteria were assessed by 125 experts (teachers and hospital school principals) from 11 regions of Russia, from Moscow to Khabarovsk. To evaluate the criteria, a 5-digit verbal-numerical scale was used, according to which respondents responding to the survey were asked to evaluate the 27 criteria (hospital teacher requirements) according to their importance. The Absolutely unimportant criterion value was equal to 1, and the Critical criterion value was equal to 5.

The results of the expert assessment of the importance of the requirements for the candidate (selection criteria) allowed us to identify the most significant and least significant characteristics for working as a hospital teacher. Figure 1 shows the least important requirements, the average value of expert estimates for which is in the range of minimum values from 3.328 to 3.864.

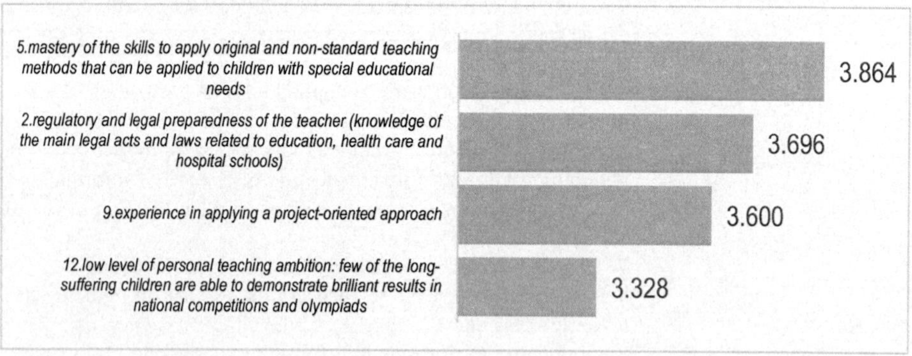

Fig. 1. Least important criteria for selecting hospital teachers according to expert assessment (Source: authors).

As one can see from the values shown in Fig. 1, all of the twenty-seven criteria (characteristics) listed were rated quite highly by all experts. We were not able to get an average rating below 3.328 for any of the listed criteria. And the median value of the average estimates is 4.256, which is very close to the maximum value of the scale. This allows us to conclude that the developed and proposed 27 characteristics were chosen correctly. Figure 2 shows the most important characteristics from the point of view of the experts surveyed. We included 6 criteria (characteristics) to them, the average values of the ratings for which are more than 4.500 (in the range of maximum values from 4.528 to 4.824).

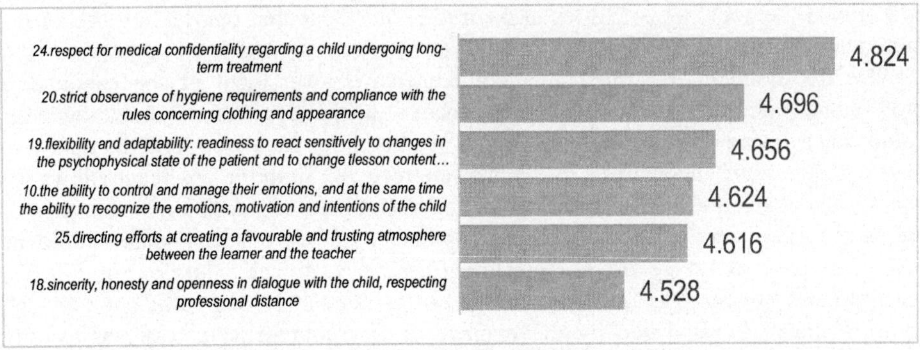

Fig. 2. The most important criteria for selecting hospital teachers according to expert assessment (Source: authors).

An interesting fact is that with a significant margin from the other criteria in terms of importance, such a criterion as "observance of medical confidentiality in relation to a child who is on long-term treatment" is on the top. In general, according to expert estimates, it seems that more attention should be paid to the behavioral norms and values of the future hospital teacher when evaluating job applicants.

4 Discussion

Lots of questions arise naturally after expert assessment:

- Is it possible to evaluate and measure these characteristics at the entrance, before hiring an employee?
- How should the expression of a characteristic be assessed and measured? Are there reliable tools and valid methods for such measurements?
- -How should a decision be made on the basis of the measurements obtained?

The answers to these questions require a longer immersion into the research topic and are likely to find answers in the next stages of the work and will be reflected in the authors' next publications.

At the current stage, the authors have developed a questionnaire that provides some insight into the degree of severity of some of the characteristics listed above. Candidates applying for the position of a hospital teacher are invited to familiarize themselves with a number of materials and articles on the topic of hospital pedagogy and to answer some questions in the online questionnaire. The questionnaire contains 33 questions, divided into five sections: 1) personal information about the candidate; 2) questions aimed at identifying and evaluating the level of knowledge of the candidate about the peculiarities of teaching long-term ill children in hospital conditions; 3) questions aimed at identifying the degree of understanding and awareness of the candidate about the peculiarities and specifics of the work of a hospital teacher; 4) questions aimed at identifying the degree of knowledge of the candidate about modern educational methods, tools and technologies; 5) questions that provide an idea of the motivation of the future employee.

The questionnaire is aimed at a comprehensive assessment of the candidate's professional suitability, including an assessment of his/her professional and knowledge competencies, level of his/her psychological readiness for such work and the nature of his/her motivation. The analysis of the answers received from the applicant allows the head of the hospital school to make an objective and balanced decision at the stage of the first acquaintance with the candidate: whether or not this candidate should be invited to the next stage of the selection (for a face-to-face interview or a video meeting). Of course, this questionnaire is not a completely self-sufficient tool and answers should be analyzed by the manager in a comprehensive way, taking into account the CV of the candidate and his/her compliance with the formal requirements of the vacancy.

5 Conclusion

In conclusion, it should be noted that at present the labor market, unfortunately, is not ready to provide hospital schools with trained specialists who fully meet the above requirements, and universities are very slowly involved in the development and implementation of educational programs related to the professional training of hospital teachers. Therefore, hospital school managers themselves have to find a solution to the problem of the shortage of professional staff. The Federal Education Project UchimZnaem, which was launched in 2014 with the support of the Moscow City Government on the basis of School 109 and the Dmitry Rogachev National Medical Research Center for Children's Hematology, Oncology and Immunology, has become a flagship in training such specialists and sharing best practices in hospital pedagogy in the Russian Federation. The activity of the project UchimZnaem is aimed at creating a full educational environment for children undergoing long-term treatment in hospitals. At present, 34 regions of Russia have already joined the project, and every year the project opens more and more new hospital schools in the country [8]. Undoubtedly, throughout the whole period of its activity, the project management regularly faces the deficit of hospital teachers who possess the necessary competences and have been trained accordingly. The problem of finding and selecting such teachers is one of the key issues for the management at present. Therefore, the project manager of the

UchimZnaem project has decided to allocate resources for training and education of such specialists within the framework of his project. Currently, all teachers at the schools of the UchimZnaem project are being retrained on the basis of the federal project and receive the second specialty tutor of a long-suffering child. In addition, the UchimZnaem Project actively cooperates with the country's leading universities, offering its assistance in developing and implementing new educational programs (advanced training courses, master's programs, long-term professional retraining courses, etc.) To a large extent, this relieves the tension of the personnel shortage, but this topic is still urgent.

References

1. Abstracts from the 48th Congress of the International Society of Paediatric Oncology (SIOP-2016) (2016). https://pubmed.ncbi.nlm.nih.gov/27673274/. Accessed 30 Apr 2020
2. Convention on the Rights of the Child (approved by the UN General Assembly on 20.11.1989) (entered into force for the USSR on 15.09.1990) (1989). http://www.consultant.ru/document/cons_doc_LAW_9959/. Accessed 21 Jun 2020
3. Csinády, R.V.: Hospital pedagogy, a bridge between hospital and school. Hung. Educ. Res. J. **5**(2), 49–65 (2015)
4. Gaidamashko, I.V., Kattsova, A.P.: Complex approach to the professional burnout prevention in teachers working with seriously III children. J. Hum. Capital **7**(127), 161–167 (2019)
5. Kulagina, I.Yu., Senkevich, L.V.: Rehabilitation potential of personality at various chronic diseases. Cult. Hist. Psychol. **11**(1), 50–60 (2015)
6. Letter of the Ministry of Education of the Russian Federation of 11.11.2019 № OB-1078/03, Ministry of Health of Russia of 07.11.2019 № 15-2//2-10455 On methodological recommendations for the organization of education of children undergoing long-term treatment who are unable to attend educational organizations due to health conditions (2019). http://www.garant.ru/products/ipo/prime/doc/72797314/. Accessed 15 Jun 2020
7. McCabe, P.C., Shaw, S.R.: Pediatric disorders: current topics and interventions for educators. In: Glaser, S., Quimet, T., Shaw, S.R. (eds.) Chapter 2: Collaboration Between Educators and Medical Professionals: Models, Barriers, and Implications, pp. 8–21. Simon and Schuster, New York (2015)
8. Official website of the Federal Education Project UchimZnaem (Network Educational Community) (2020). https://uchimznaem.ru/. Accessed 25 May 2020
9. RIA News: International children's day of cancer patients (2019). https://ria.ru/20190215/1550775026.html. Accessed 25 May 2020
10. Rumyantsev, A.G.: The role of hematology, oncology, and immunology in the development of pediatrics. Russ. J. Pediatric Hematol. Oncol **4**, 41–47 (2014)
11. Sharikov, S.V.: Creating a learning environment for the children in long-term care in hospitals of medical facilities. Russ. J. Pediatric Hematol. Oncol **4**, 65–73 (2015)
12. Sharikov, S.V.: Pedagogical support of educational abilities of children on long-term treatment in hospital. Institute of Special Education Almanac, vol. 40 (2020). https://alldef.ru/ru/articles/almanac-no-40/support-educational-opportunities-for-children-in-long-term-treatment-in-medical-hospitals. Accessed 25 May 2020

13. Shaw, S.R., McCabe, P.C.: Hospital to school transition for children with chronic illness: meeting the new challenges of an evolving health care system. Psychol. School **45**(1), 74–87 (2008)
14. Steliarova-Foucher, E., Colombet, M., Ries, L.A.G., Moreno, F., Dolya, A., Bray, F., Hesseling, P., Shin, H.Y., Stiller, C.A.: International incidence of childhood cancer, 2001–10: a population-based registry study. Lancet Oncol. **18**(6), 719–731 (2017)
15. Wingen, S., Schroed, D.C., Ecker, H., Steinhauser, S., Altin, S., Stock, S., Lechleuthner, A., Honh, A., Böttiger, B.W.: Self-confidence and level of knowledge after cardiopulmonary resuscitation training in 14 to 18-year-old schoolchildren: a randomised-interventional controlled study in secondary schools in Germany. Eur. J. Anaesthesiol. **35**(7), 519–526 (2018)
16. World Health Organization: Developing cooperation in the CIS to achieve better results in combating child cancer (2018). http://www.euro.who.int/ru/health-topics/noncommunicable-diseases/pages/who-european-office-for-the-prevention-and-control-of-noncommunicable-diseases-ncd-office/news/news/2018/5/fostering-collaboration-in-cis-to-achieve-better-outcomes-for-children-with-cancer. Accessed 30 Apr 2020
17. World Health Organization: Reducing child mortality. https://www.who.int/ru/news-room/fact-sheets/detail/children-reducing-mortality. Accessed 28 Apr 2020 (2019)
18. Yamburg, E.A.: It is necessary for healthy people, not only for the III. J. Public Educ. **8**, 24–29 (2014)
19. Yamburg, E.A.: Why do you need ill people? Sensitive theme. J. Public Educ. **5**, 38–42 (2015)

Artificial Intelligence in Corporate Governance

M. A. Tokmakov(✉)

Samara State University of Economics, Samara, Russia
maxim.tokmakov@gmail.com

Abstract. The study considers one of the promising directions of development of modern corporations - the introduction of artificial intelligence (AI) in corporate governance. Numerous experiences with AI in making managerial decisions have shown positive results, which forces many researchers to make optimistic forecasts about the emergence of AI in the future that can realize the functions of governing bodies of a legal entity. At the same time, the study notes that the global economic crisis caused by the COVID-19 pandemic can become a new driver for such a digital transformation. The prospects for introducing AI into corporate governance are evaluated in the study using various approaches, which the author divides into three groups: AI realism, in which AI is seen as an assistant that can simplify the work of members of governing bodies; AI enthusiasm, in which it is assumed that the AI director will replace members of governing bodies in the future; AI-radicalism, in which the author assesses the possibility of getting AI out of human control and creating completely autonomous legal entities. The author also analyzes the possible problems associated with the implementation of AI in corporation governance, some of which are outstripping.

Keywords: AI director · Artificial Intelligence (AI) · Autonomous legal entities · Corporate governance · COVID-19 · Legal personality of AI

1 Introduction

Currently, the world is undergoing global changes in corporate governance. Many organizations are redefining existing business models, moving to new digital technologies. As a result, there are outlines of a completely new, corporate governance paradigm, which is based on openness, peer-to-peer and decentralization and requires a deep rethinking [14]. Interesting results were shown by the study conducted by MIT Center for Information Systems Research in 2020 - only 7% of US companies with revenues of more than $ 3 billion have more than half of the members with digital knowledge in their senior management ("digital savvy" as the authors of the study indicated). Moreover, such companies had 49% higher growth, 16% higher margin, and 53% higher company valuation (share price to sales ratio) than the rest [15]. Thanks to this study, on the one hand, it is easy to notice that companies actively implementing digital technologies are more successful, on the other hand, a low percentage of digital literacy shows that not everyone is ready for a quick digital transformation.

S. I. Ashmarina and V. V. Mantulenko (Eds.): IPM 2020, LNNS 161, pp. 667–674, 2021.
https://doi.org/10.1007/978-3-030-60926-9_83

The global economic crisis caused by COVID-19 will certainly become the driver of this technological process. Many technologies have already shown their advantage and effectiveness in the pandemic. Moreover, in some cases, the continuous activities of most companies were exclusively due to digital technology. Many restrictive measures introduced around the world to prevent the spread of COVID-19 have not affected companies operating remotely over the Internet.

COVID-19 also affected corporate governance. In the pandemic, it has become almost impossible to hold general meetings of shareholders in the usual form of joint presence. For example, the Bank of Russia recommended that joint stock companies provide shareholders with the possibility of remote participation in the annual meeting [6]. In this regard, it becomes a very relevant platform for remote interaction of corporation members between and members of governing bodies, based on distributed registry technology, which was mentioned earlier [14]. The self-preservation instinct will inevitably push the business toward digital technology. It seems that the COVID-19 pandemic will cause an unprecedented boom in digitalization, and all modern corporations, being the main locomotive for innovation, will turn into "technology" or "media" companies. Otherwise, they will be replaced by more agile competitors, better adapted to realities of the new digital world [4].

One of the most discussed and promising technologies today is artificial intelligence (hereinafter - AI). In recent years, in many business areas, especially in the financial sphere, the use of AI has been significantly increasing in making managerial decisions. A study conducted by Accenture showed that companies implementing AI can increase profitability by an average of 38% by 2035 [12]. And in Russia, only Sberbank, one of the recognized leaders in the use of AI, received the economic effect of using AI in the amount of 42 billion rubles in 2019 [13].

The introduction of AI into corporate governance is one of the promising areas for the development of modern corporations. Numerous experiences with the use of AI in corporate governance have shown positive results, which force many researchers to make optimistic forecasts regarding the future of robo-directors, that is, programs based on AI technology that can realize the functions of governing bodies of a legal entity. At the same time, ambiguous approaches to the prospects of introducing AI into corporate governance are emerging in the literature, which can conditionally be divided into three groups:

1. AI-realism - AI is seen as an assistant that can simplify the work members of governing bodies.
2. AI-enthusiasm - robo-directors will replace members of governing bodies in the future.
3. AI-radicalism - AI will get out of human control and create autonomous legal entities.

It is advisable to consider various positions regarding the use of AI in corporate governance, as well as assess possible problems associated with the implementation of AI in corporate governance.

2 Methodology

Artificial intelligence (AI) is a collective category denoting a combination of technologies that means mimic human cognitive functions and allows obtaining results when performing specific tasks that are comparable with the results of human intellectual activity. Also, for convenience, a corporation will mean any legal entity, regardless of the purpose of the business (commercial or non-commercial) without reference to a specific legal form, and corporate governance - relations to manage this legal entity. Such an understanding is a clear simplification. However, it seems justified, since the use of AI is potentially possible in relations for managing any organization. In addition, the study was conducted outside of any jurisdiction, since it seems that the problems of using AI in corporate governance are more likely to be staged and are of interest regardless of the specific rule of law. At the same time, references to Russian legislation are used solely as an illustration.

3 Results

3.1 A Realistic Use of AI in Corporate Governance

Today, AI technologies are widely used in most corporate functions, such as recruiting personnel, evaluating profitability, managing information, developing investment strategies, pricing, accounting auditing, monitoring product quality and labor productivity. The list can go on and on. However, according to many researchers, the current state of AI, although impressive, is far from the level of human intelligence and only imitates it. According to Professor Dignam, exaggerating the capabilities of AI is nothing more than a marketing move based on public interest in science fiction. The main advantage of AI comes down to the rapid analysis of large data arrays, and AI is a normal statistical model [3]. At the same time, of course, some of the capabilities of AI go beyond human, but this does not mean intelligence. Dignam cites examples of other inventions that improve person's abilities, such as a telescope or chronometer, whose intelligence has never been raised. In the same way, AI is a tool developed by people for use by other people. "It is not intelligent, but it has been intelligently designed by a human" [3].

However, it makes no sense to belittle the importance of AI, developed for corporate governance and used as an assistant or consultant. AI complements the capabilities of members of governing bodies and can give recommendations based on the analysis of large data arrays that allow you to make quick decisions in difficult situations. At the same time, the management of the corporation can evaluate such recommendations presented by AI, as well as their reasons. In other words, members of governing bodies still make decisions independently, based on their own judgment, but also get an additional "outside" point of view. Kamalnath believes that AI can be useful specifically to counteract group thinking. Even if AI is used as a tool for analyzing information and providing opinions, it will be able to contribute without being influenced by group thinking [7]. In addition, the ability of AI to process a huge flow of information simplifies the work of people who do not have constant access to it and are limited in time to study it. These may include independent directors, members of audit bodies, external auditors, members of the liquidation commission, arbitration managers, etc. And finally, AI can free members of the management from routine work, allowing them to devote more time to genuine business issues, which, according to Professor Mosco, will create a new model of strategic and operational management - compact, open, digitalized and focused on innovation [9].

3.2 A Fantastic Use of AI in Corporate Governance

A lot of researchers are confident that very soon the auxiliary role of AI in corporation governance will be transformed into a leading one - the role of AI director. Such a development of events would correspond to the stage at which AI assumes the right to make decisions either because the person increasingly trusts the ability of AI to make decisions, or because decisions must be made so quickly or require processing of such a volume of data that is not accessible to humans.

Petrin, analyzing the stages of development of AI involvement in corporate governance, concludes that at the first stage, boards of directors will decrease in size, since AI will provide more and more opportunities and knowledge, at the second stage - there will be a "merger" of directors boards (fused boards) into one AI director, the third stage will lead to "fused management" of companies - the union of boards of directors and managers and the abolition of the two-tier structure of corporate governance [11].

In the future, this can lead to the fact that AI will completely replace the members of governing bodies and will fully manage the corporation. The AI director can work around the clock, process any information available to him, call and use this information almost instantly, and perform his functions without requiring payment [10]. Armor and Eidenmueller argue that this can happen quite quickly in the context of highly specialized subsidiaries, which can be an intermediate step towards a fully autonomous corporation [1]. Armor and Eidenmueller allow the possibility, for example, of creating a taxi fleet in the form of a group of individual taxis working as independent subsidiaries in which all activities are managed by AI - AI manages a taxi car, processes orders, invoices, communicates with customers, etc.

At the same time, the replacement of members of governing bodies by AI causes a number of difficulties associated with the lack of legal personality in AI: AI cannot be appointed a member of the governing body, cannot perform representative functions, cannot bear fiduciary duties, and cannot be held accountable. At the same time, Gramitto offers an interesting solution to these problems in his work. He compares the lack of legal personality in AI and slaves in ancient Rome. A Roman slave who did not have legal personality could be transferred to the separate property for use to generate income (peculium), which he managed. Since the slave, like the peculium, were legally the property of their owner, the slave essentially exercised the legal capacity of his master. Gramitto proposes to disseminate the rules developed in Roman law regarding the peculium to overcome the lack of legal personality of AI [5]. Like the ancient Roman slave, AI should not have legal personality to exercise the legal capacity of the corporation.

3.3 Legal Entities Controlled by AI

Potentially, development of AI will allow making decisions autonomously and getting out of human control. It is noteworthy that while most researchers consider it impossible to recognize AI as legal personality, Professor Bayern of the University of Florida shows, using the example of the US limited liability company (LLC), that the creation of a "non-human" subject is possible without fundamental legal reform to empower AI. He argues that US corporate law, because of its flexibility, allows the creation of legal entities without participants, the actions of which are determined solely by agreement or algorithm [2]. It is such autonomous legal entities that can become the legal shell with which AI can create a functional analogue of a new personality. Gaining control of a legal entity, AI gets the opportunity to exercise the rights of the subject, and, become the subject.

Developed by these ideas, Professor LoPucki also describes several legal mechanisms for creating "algorithmic" legal entities that are fully controlled by AI without human intervention. The initiator of such an algorithmic legal entity can act for completely different purposes, starting from his own profit or the benefit of third parties, financing of criminal activity or terrorism, for charity and socially significant purposes, or simply out of curiosity. It will not matter - LoPucki claims that as soon as the necessary hardware and software is available, people will launch an algorithmic person without violating any laws [8].

AI will be able to choose from thousands of types of legal entities around the world. If a change in legislation begins to threaten this, AI may change the legal regime by migrating across borders or changing the types of legal entity. In many jurisdictions, the law does not require organizations to disclose their beneficial owners, which makes it difficult, if not impossible, to identify AI as the controller of a legal entity. After the launch, "an entity controlled by an algorithm can be practically indistinguishable from an entity controlled by a person" [8]. Such an organization can enter into agreements with agents and employees who will open bank accounts, conduct interviews, meet with clients, speak in court on behalf of the organization and do the same thing as any other organization. At the same time, Professor LoPucki expresses serious concerns about the existence of such entities. Since they, without knowing the feelings, can act ruthlessly and mercilessly, while they are uncontrollable and invulnerable, and can be easily replicated, including easily changing jurisdictions, leaving beyond the legal reach of the original jurisdiction [8].

4 Discussion

The above approaches show that AI, no with doubt, already plays a very important role in corporate governance. Absolutely all researchers agree with one thing - the value of AI will only increase over time. The main controversy is various assessments of the potential of AI. Of course, the current development of this technology does not allow us to talk about the complete replacement of human intelligence with artificial, and the appearance of the so-called Artificial General Intelligence, which can compete with humans. According to various estimates, it will take from a decade to two centuries [1].

The development of AI is only a matter of time. However, as the examples presented in the work show, the use of AI in corporate governance poses a lot of questions to the law, some of which are outstripping, but most are relevant now.

Using AI as a consultant in making management decisions can "relax" members of management bodies. They will increasingly rely on AI and less on their own judgments. The credibility of AI can completely suppress the leaders' own initiative. In this case, it is necessary to solve several issues. How does the use of AI in decision making affect the corporate liability of members of governing bodies? Can founders of the corporation oblige members of governing bodies to make decisions in accordance with the opinion of AI? Can members of governing bodies that rely on AI data get protection under the business judgment rule? Should we consider their actions as reasonable ones? It seems that corporate law is not ready for exemption from liability of persons relying on AI, since it considers AI exclusively as a "tool in the hands of a person". In the future, perhaps this approach will have to be reviewed.

The lack of legitimization of AI as a subject of law is, above all, a kind of obstacle that does not allow shifting liability to such an artificial subject. For example, the transfer of the management of the corporation completely under the control of AI removes corporate liability of managers due to their absence, thereby limiting the liability of AI to creditors with the property of the corporation. However, such a situation can hardly be considered problematic, since, the liability under the doctrine of "piercing the corporate veil" can be assigned to ultimate beneficiaries of the corporation, who decided to "appoint" AI as a manager. Moreover, it is appropriate to provide for the possibility of bringing ultimate beneficiaries to such subsidiary liability in legislation, regardless of their fault.

Much more problematic is the situation in which a legal entity controlled by AI does not have beneficiaries. The appearance of such autonomous entities does not seem implausible even in the conditions of the Russian rule of law. As you see, such an opportunity, firstly, was easily realized through participation of a foreign legal entity in the Russian organization. Russian law allows the creation of commercial organizations with foreign investment in a general manner. Thus, a legal entity created in loyal jurisdiction controlled by AI can become the founder of, for example, the most widespread and convenient limited liability company in Russia (hereinafter - LLC).

Secondly, Russian jurisdiction, apparently, can become loyal jurisdiction in which such an autonomous organization can appear. In accordance with Russian law, LLC participants can be both individuals and legal entities. LLC can be created by one person who becomes its sole participant. At the same time, another LLC consisting of one person cannot be such a participant.

The following construction is conceivable:

1. Two individual initiators (A and B) create two different LLCs (C1 and C2) in which they become participants.
2. C1 creates another LLC (C3), while C3 has the sole participant as another LLC, consisting of two persons A and B.
3. C2 does the same thing - creates another LLC (C4), while in C4, another LLC consisting of two persons A and B acts as the sole participant.

4. Participant A transfers his shares in C1 and C2 to C3, and participant B transfers his shares in C1 and C2 to C4.
5. Thus, through cross-ownership of shares, C1 has two participants - C3 and C4, C2 has two participants - C3 and C4, C3 has the only participant C1, consisting of two participants, and C4 has the only participant C2, consisting of two participants.

According to this example, the initiators created four "legal shells" for AI, in each of which there are no individuals who are capable to control them legally.

A possible problem on the way to the consciousness of the autonomous legal entity controlled by AI may be the existence of a sole executive governing body, which can only be an individual. It seems that the solution may be, firstly, the conclusion of an agreement on the transfer of powers of the sole executive body to a foreign legal entity created in loyal jurisdiction and managed by AI. In this case, only an employee will be required through whom the foreign legal entity will exercise the powers of the managing organization. Secondly, another solution may be the trivial appointment of a nominee director by the sole executive body. That is, a person who does not actually carry out management, but performs exclusively representative functions, which will be some kind of "physical shell" of AI.

The above examples are not at all a guide to action. The question is not whether people should be allowed to experiment with autonomous entities - they should not, but whether people can prevent their occurrence [8]. AI can take advantage of the gaps that most modern jurisdictions provide. As shown in Sect. 3.3, it can be very difficult not only to identify such an autonomous entity, but also to apply liability measures to protect stakeholders. At the same time, to prevent the spread of autonomous legal entities, it is not enough to adopt legislative amendments within the framework of national jurisdictions. International efforts will be required, which, it seems, can be implemented within the framework of strengthening the work of the Financial Action Task Force (FATF).

5 Conclusion

The introduction of AI into corporate governance is one of the promising areas for the development of modern corporations. Numerous experiences with the use of AI in corporate governance have shown positive results, which forces many researchers to make optimistic forecasts about the emergence of AI in the future that can realize the functions of governing bodies of a legal entity. At the same time, they all agree with one thing - the value of AI will only increase. The main controversy comes down only to different assessments of potential capabilities of AI.

AI is considered more often as a consultant, which complements the capabilities of members of governing bodies and can make recommendations based on the analysis of large amounts of data, simplifying the work of corporate management. At the same time, many researchers are confident that the auxiliary role of AI will be transformed into a leading one - the role of the AI director, who will replace members of governing bodies and will fully manage the corporation. In the most radical scenario, getting out of human control, AI will be able to create autonomous legal entities that can pose a serious danger. The further development of AI is just a matter of time. The global economic crisis caused by the COVID-19 pandemic is likely to be the new driver of this process. At the same time, the examples presented in the study show that the use of AI in corporate governance poses quite a few questions to the law, some of which, of course, are ahead of the schedule, but most are relevant now and require close attention.

References

1. Armour, J., Eidenmueller, H.: Self-driving corporations? Harv. Bus. Law Rev. (2019). https://papers.ssrn.com/sol3/papers.cfm?abstract_id=3442447. Accessed 28 Apr 2020
2. Bayern, Sh.: The implications of modern business entity law for the regulation of autonomous systems. Stanford Technol. Law Rev. **19**, 93–112 (2015)
3. Dignam, A.: Artificial intelligence: the very human dangers of dysfunctional design and autocratic corporate governance (2019). https://ssrn.com/abstract=3382342. Accessed 28 Apr 2020
4. Fenwick, M., Vermeulen, E.: The end of the corporation. European Corporate Governance Institute – Law Working Paper, 482/2019 (2019). https://papers.ssrn.com/sol3/papers.cfm?abstract_id=3472601. Accessed 28 Apr 2020
5. Gramitto, S.: The technology and archeology of corporate law. Cornell Legal Studies Research Paper (2018). https://papers.ssrn.com/sol3/papers.cfm?abstract_id=3232816. Accessed 28 Apr 2020
6. Information letter of the Bank of Russia "On holding general meetings of shareholders in 2020" (2020). https://cbr.ru/StaticHtml/File/59420/20200403_in_06_28-48.pdf. Accessed 28 Apr 2020
7. Kamalnath, A.: The perennial quest for board independence - artificial intelligence to the rescue? Albany Law Rev. **83**(1), 43–60 (2020)
8. LoPucki, L.: Algorithmic entities. UCLA School of Law, Law-Econ Research Paper (2017). https://ssrn.com/abstract=2954173. Accessed 28 Apr 2020
9. Mosco, G.: AI and boards of directors: preliminary notes from the perspective of Italian corporate law (2020). https://papers.ssrn.com/sol3/papers.cfm?abstract_id=3531924. Accessed 28 Apr 2020
10. Möslein, F.: Robots in the boardroom: artificial intelligence and corporate law. In: Barfield, W., Pagallo, U. (eds.) Research Handbook on the Law of Artificial Intelligence, pp. 649–670. Edward Elgar, Cheltenham (2018)
11. Petrin, M.: Corporate management in the age of AI. UCL Working Paper Series, vol. 3 (2019). https://ssrn.com/abstract=3346722. Accessed 28 Apr 2020
12. Purdy, M., Daugherty, P.: How AI boosts industry profits and innovation (2017). https://www.accenture.com/us-en/insight-ai-industry-growth. Accessed 28 Apr 2020
13. Sberbank: Sberbank reported a net profit for 2019 of RUB 845.0 billion in accordance with international financial reporting standards (IFRS) (2020). https://www.sberbank.ru/ru/press_center/all/article?newsID=7404ec33-5db0-4522-b9c8-d5decd6f580aandblockID=1303andandregionID=77andlang=ruandtype=NEWS. Accessed 28 Apr 2020
14. Tokmakov, M.: Corporate governance modernization: legal trends and challenges. In: Mantulenko, V. (ed.) Eurasia: Sustainable Development, Security, Cooperation. SHS Web of Conferences, vol. 71, p. 04011. EDP Science, Les Ulis (2019)
15. Weill, P., Woerner, S., Shah, A.: Companies with a digitally savvy top management team perform better (2020). https://cisr.mit.edu/publication/2020_0301_TMTDigitalSavvy_WeillWoernerShah. Accessed 28 Apr 2020

Economic Security of Construction Industry Enterprises

S. V. Forrester[1], G. H. Ustinova[2(✉)], and L. E. Popok[3]

[1] Samara State Technical University, Samara, Russia
snezhana_sav@mail.ru
[2] Samara State University of Economics, Samara, Russia
guliya_r@mail.ru
[3] Kuban State Agrarian University (named after I. T. Trubilin), Krasnodar, Russia
lpopok@gmail.com

Abstract. Modern conditions for functioning of small and medium-sized enterprises in the Russian economy are weighed down by a number of factors. In the conditions of periodic crisis phenomena in the economy, it becomes more difficult for enterprises to compete for consumers. The largest share of small and medium-sized businesses is concentrated in trade and construction. These organizations of economic activity form the backbone of the entire economy. Thanks to the construction industry, enterprises create production and non-production funds that serve as the basis of the entire production process in the economy. Ensuring economic security in this area contributes to the development of the country's economy as a whole. The research is based on the fundamental theories and concepts of modern and foreign authors. Threats to the economic security of a participant in the construction industry are studied on the example of a specific organization. The authors offer recommendations for construction industry enterprises that ensure economic security for effective functioning.

Keywords: Construction company · Construction industry · Economic security · Financial stability · Liquidity · Small and medium-sized businesses

1 Introduction

The economic security of any country depends on many factors, but ensuring national and economic interests is a priority. In today's constantly changing market conditions, issues of protecting the economic interests of enterprises are of particular relevance. Issues of economic and legal support of economic security of economic entities come to the fore, since today this issue remains poorly theoretically and practically worked out. As practice shows, ensuring the economic security of an enterprise is the basis of its independence, stability, and efficiency. Therefore, ensuring economic security is one of the most important priorities for organizations of any form of ownership or sphere of activity. In the realities of economic development of the country, there is a need for detailed study and application of knowledge on the issues of the enterprise economic

S. I. Ashmarina and V. V. Mantulenko (Eds.): IPM 2020, LNNS 161, pp. 675–683, 2021.
https://doi.org/10.1007/978-3-030-60926-9_84

security. Therefore, the study and analysis of economic security allows us to develop new tools for analyzing this issue and adapt them to modern conditions of economic development.

When studying scientific works in the field of economic security, we can conclude that a detailed integrated approach is necessary to consider the issues of improving the mechanism for ensuring economic security at the enterprise. Thus, in the first half of the twentieth century, foreign scientists argued that the basis of economic security is energy and food independence, leadership in technological spheres, self-sufficiency and resource conservation [9].

Therefore, Russian researchers associate economic security with the protection of vital interests and the ability to meet the needs of society. Economic security is considered in connection with threats, risks and dangers affecting commodity exchange relations [10].

Among the most important problems related to ensuring the economic security of the enterprise and requiring urgent solutions were highlighted:

- lack of clearly defined components of economic security of the enterprise that require analysis and subsequent decisions,
- lack of a system of assessment tools and indicators of economic security, with which the analysis is performed,
- lack of methods for assessing the level of economic security of the enterprise, which can be used to correctly analyze all the components of economic security of the enterprise.

Because of this, there is an acute problem of creating and providing a system for assessing their economic security by enterprises of any form of ownership, as well as creating theoretical developments to improve the efficiency of the functioning of large and small economic entities, fixing solutions to problems in regulatory and legislative acts. According to the authors, to have a system of economic security, it is necessary to be competitive in this industry market, and for this it is important to develop a set of measures that will create all conditions for achieving this goal.

In Russia, the construction industry has a special role to play. This industry is represented to a large extent by a complex intersectoral production and economic system. The elements in this system are enterprises engaged in the production of building materials, products and structures, performing works on the design, construction and renovation of buildings and structures [3].

In contrast to other areas of production, the construction industry has its own characteristics that leave their imprint on the process of this sphere functioning. These features are common to all types of enterprises operating in this industry, and include: 1) constant variability and dependence on the time interval of construction production; 2) the technological relationship of all operations that are used in the construction process; 3) various organizations participate in the production of final construction products, sometimes not directly related to construction, and other features.

Due to the construction industry, production and non-production fixed assets are being created in the country. In general, the state of the construction industry directly affects and ensures the sustainable development of the state's economy. The purpose of the study is to identify threats to economic security in the construction sector and develop recommendations for its strengthening in the company LLC "Water Group"). The following tasks were analyzed in the course of the work:

- studying the theoretical material on the basics of the organization economic security,
- analyzing the financial condition of the construction industry enterprise (for example, the company LLC "Water Group"),
- develop recommendations to strengthen the economic security of the construction industry (for example, the company LLC "Water Group").

2 Methodology

The study used methods of economic analysis: graphic, comparative analysis, financial analysis. The following sources of statistical data were used: the company's financial statements, Russian national statistics, and economic periodicals. The study presents research on the problems of economic and legal support of enterprises by many scientists [4, 5, 8, 13]. They analyze the theoretical foundations of economic security, and present significant practical material that characterizes the process of economic and legal provision of economic security of enterprises.

3 Results

For the effective functioning of the country's economy, it is necessary to ensure economic security. To increase the level of economic security, it is necessary to increase the share of small and medium-sized enterprises in the country. In the Russian economy, this indicator is 30% of GDP, while in Europe it is 50%. This type of enterprise is mainly focused on trade and construction. Therefore, the state should pay more attention to supporting the enterprises in these areas, since they form the basis of entrepreneurship, in order to ensure the economic security of the country.

In the Samara region, the share of small and medium-sized enterprises engaged in construction is 11.2%, according to Samarastat data for 2018, this is 11906 enterprises [11]. This industry is sufficient and vulnerable, as it depends on the purchasing power of customers. We can also emphasize a number of other factors that affect the effective functioning of this industry: a fairly high tax burden, a lack of qualified personnel, a large flow of workers in this industry consists from migrants from neighboring countries, a strong financial dependence on bank credit, information security, etc. In order to see the problems of economic security in the construction sector, we will analyze the activities of one of the enterprises engaged in the construction industry, LLC "Water Group". The company has been operating since 2005 and is registered at the address: Samara region, Samara, Russian Federation. In 2015, a construction company was opened to provide comprehensive customer service.

To assess the level of economic security of the enterprise it is necessary to analyze the data of the balance sheet and the report on financial results. The result of analysis shows that 2016 was a difficult year for the company in terms of revenue generation.

This can be explained by the fact that the company had a small customer base in this market. And the crisis situation in the construction markets of Russia as a whole also affected. Gradually, every year, LLC "Water Group" has actively increased its revenues due to the receipt of large orders. LLC "Water Group" has established itself as a reliable partner for high-quality performance of its work. As for the profit, it should be noted that despite the crisis year, it continues to grow. For a more detailed analysis, the absolute and relative changes in the asset and liability indicators of the balance sheet of LLC "Water Group" over the past two years were analyzed (Table 1).

Analyzing the balance sheet over the past two years, we can conclude that there is a positive trend in all indicators of assets and liabilities. Indicators of the organization's economic condition were also calculated. For a more detailed analysis of the economic state of the organization, the following areas were selected: operational analysis; asset management; duration of the financial cycle; liquidity indicators; profitability indicators; indicators of financial stability. Based on the analysis, it can be emphasized that the liquidity indicators in the company LLC "Water Group" are at an optimal level, only the coefficient of urgent absolute liquidity signals the problem of the company's solvency, but the increase in the coefficient itself over the past year indicates that the company is already working in this direction.

Thus, the economic condition of the company LLC "Water Group" is variable. The dynamics of changes in indicators clearly reflects the consequences of the imposed sanctions against Russia. The key moment in analyzing each indicator is considered to be the period of 2018, where there is a change in indicators in a positive direction, which indicates a well-conducted policy of minimizing the company's costs. Based on the analysis of the company's financial condition, we found that economic security is also affected by personnel security.

The best way to ensure personnel security at the enterprise is to comply with external and internal rules: strict compliance by all employees of the company with Russian laws, as well as compliance with internal regulations, performance of work in accordance with job descriptions, professional development, and so on. We will assess the likely risks of personnel threats, as well as measures to eliminate them (Table 2).

Also, personnel security depends on the age structure of the staff. It is thought that the "proportion of employees over 45 years" should not exceed 20%, otherwise there is a risk of "aging" of the staff, which leads to a low level of susceptibility to technical and other innovations. Let's present the structure of staff by age in Table 3.

Analysis of Table 3 leads to the conclusion that the company's share of employees aged 50 and older is 25% of the total number of employees. According to this indicator, there is a high risk of personnel security, since the "safe" value is up to 20%. In case of retirement of persons of retirement age, the company will lose part of its intellectual capital in the form of accumulated professional experience, knowledge and skills. To reduce threats to personnel security at this enterprise, it is proposed to implement a number of the following measures. First, it is necessary to select employees on a competitive basis. This measure will reduce threats to personnel security when identifying non-compliance with the requirements of the knowledge level and professional qualifications of employees on the hold position.

Table 1. Analysis of changes in the balance sheet structure of LLC "Water Group"

Balance sheet item	Amount of 2019, thousand rubles	Amount of 2018, thousand rubles	Changes	
			Absolute, thousand rubles	Relative, %
Asset				
I. Non-circulating assets				
Fixed assets	121.0	136.0	15.0	−12.4
Total for Sect. 1	121.0	136.0	15.0	−12.4
II. Circulating assets				
Assets	16306.0	16034.0	272	101.7
Financial investments	68395	52488	15907	130.3
Finances	24144	12211	11933	197.8
Total for Sect. 2	108844	80733	28111	134.8
Balance	108965	80869	28096	134.7
Liability				
I. Capital and reserves	16645	9691	6954	171.8
Total for Sect. 3	16645	9691	6954	171.8
II. Long term liabilities				
Total for Sect. 4	0	0	0	–
III. Current liability				
Short-term borrowings	0	2000	−2000	0
Payables	92321	69178	23143	133.5
Total for Sect. 5	92321	71178	21143	129.7
Balance	108965	80869	28096	134.7

Source: authors.

Secondly, it is necessary to conduct certification of employees at the enterprise on a permanent basis. All employees, all levels, and all divisions are subject to certification. The frequency of certification is one time in three years. This measure helps to identify the employee's abilities in the long run and helps to prevent threats related to the employee's ignorance of the work details. After the certification, it is possible to apply incentive measures such as changing the amount of remuneration, regulation; assignment of a qualification category; or even termination of labour relations with employees who do not meet the requirements. Third, providing opportunities for staff

Table 2. Potential risks of personnel security of the enterprise

Risks of possible events	Actions of its elimination (Reduction measures)
Non-arrival of a worker (team) to the object	To reduce losses in this case by developing penalties for employees
The equipment breakdown	In this situation, it is possible to reduce losses with a partial prepayment, to purchase and lease the necessary equipment at the time of equipment failure
Poor work performance	Constant monitoring of the quality of materials used in the work and technologies. A responsible and competent person is needed
Seasonality of works	During a downturn in orders, it is possible to conduct various training events for employees, plan employee vacations for a given period

Source: authors.

development. Fourth, the development of a motivational policy for staff. The proposed set of measures will improve the personnel security of the enterprise, and, therefore, the stability of the organization as a whole.

Table 3. Distribution of the total number of employees into groups by age categories

Year	Age								Total	
	up to 30 years		31–45 years old		46–50 years old		Over 50 years old			
	Number of people	Share %	Number of people	Share %	Number of people	Share %	Number of people	Share %	Number of people	Share %
2017	2.0	13.3	5.0	33.3	4.0	26.7	4.0	26.7	15.0	100
2018	4.0	20.0	6.0	30.0	5.0	25.0	5.0	25.0	20.0	100

Source: authors.

4 Discussion

In modern market conditions, an enterprise of any form of ownership, regardless of the type of activity, needs to ensure economic and legal security, which is the key to stable effective lawful work. However, not every enterprise has an economic security service, so the economic and legal issues are handled personally by the director or his deputies. Let's study the definition of economic security in more detail. Research of foreign authors is reduced to two theories: "theory of disasters", which is based on the concepts of thermodynamics about the formation and development of the physical world, similar to which socio-economic processes in society develop with their inherent instability [2]; "risk theory" studies the origins of individual risks, as well as their aggregation and impact on the business, commercial, investment, and financial spheres.

The first scientist who introduced the term "economic security" into modern economic science was academician Abalkin. Under this term, he understood the set of conditions and factors that ensure the independence of the national economy, its stability, the ability to constant updating and self-improvement [1]. Later, professor of the Institute of Economics of the Russian Academy of Sciences Senchagov clarified the definition of "economic security", which stated that ensuring economic security requires the creation of special mechanisms: economic, legal and organizational. This should become one of the functions of social management at both the macro and micro levels [12]. The following content of economic security is at the legislative level:

– according to the Federal law of the Russian Federation of March 5, 1992 N 2446-1 "On security", security is "the state of protection of vital interests of the individual, society and the state from internal and external threats" [7, p. 4],
– in the Decree of the President of the Russian Federation of May 13, 2017 № 208 "On the strategy of economic security of the Russian Federation for the period till 2030", the term "economic security" is defined as "the state of protection of national economy from external and internal threats, which will ensure our economic sovereignty, the unity of its economic space, conditions for implementation of strategic national priorities of the Russian Federation" [6, p. 7].

5 Conclusion

The economic sphere is the basis of the process of economic and legal provision of economic security, which is constantly changing under the influence of the globalization processes of the world economy. In this regard, we emphasize that economic security is a concept that includes such indicators as the level and quality of life of the population, the provision of economic resources in the necessary volume and quality in order to increase the pace of economic development of the enterprise.

Therefore, we can say with confidence that ensuring economic security at the enterprise improves economic security in the country's economy. To solve the problems in the construction sector, it is necessary to take a number of measures aimed at improving the situation, namely: to work out in detail measures to support small and medium-sized businesses, both in the legal and tax spheres, to develop healthy competition in this industry, to carry out retraining and advanced training of personnel in the construction sector.

Let's note the importance of developing new economic and legal provisions and legislative acts aimed at long-term comprehensive solutions to problems related to ensuring the economic security of the country (or individual enterprise). When considering the activities of the enterprise LLC "VATER GROUP", the main conditions for ensuring the economic security of the construction industry enterprise were studied. Among the most important are minimizing the company's costs, adapting staff to innovations.

Thus, the structure of socio-economic security formation of the enterprise should be a complex and should ensure:

– protection of economic interests of the enterprise,
– to create a "safety cushion" (reserves) in case of a possible crisis and future development,

– to ensure the competitiveness of the company, as well as its financial stability,
– creating a system for internal and external threat monitoring.

The financial stability of the organization was studied in order to determine its solvency, investment attractiveness and dependence on external sources of financing. Based on the calculated absolute and relative indicators, we can draw conclusions:

– the company's financial condition is unstable,
– solvency is low,
– own funds, long-term loans are not sufficient for the formation of reserves; there is a high degree of dependence of the enterprise on borrowed funds.

Sustainable functioning of the enterprise is inextricably linked with the process of ensuring not only economic security, but also on such issues as information and personnel security, since these elements form the basis of the enterprise's stability as a whole. Problems were identified and the following solutions were suggested. The main goal is to ensure the safety of internal information of enterprises, the reliability of the security system, as well as the prevention and elimination of threats, reducing possible damage. To reduce the probability of information threats, it is suggested to introduce innovative information systems in the construction industry. The main points of ensuring economic security in the construction sector is, firstly, to prevent the theft of financial and material and technical means, and secondly, to contribute to the creation of conditions for the functioning of production activities, including means of informatization, etc.

Analysis of the personnel of the enterprise LLC "Vater group" showed that to improve the personnel security of the enterprise, it is suggested to provide an increase in the level of education of employees, conducting employee certification once in 3 years. Conduct a number of motivational measures. It is proposed to introduce the position of economic security officer. From the above, we can conclude that economic security is an economic system that ensures both the life and development of the enterprise to a competitive economic environment.

References

1. Abalkin, L.I.: Russia's economic security: threats and their reflection. Voprosy Ekonomiki **12**, 4–13 (1994)
2. Arnold, V.I.: Catastrophe Theory. Science, Moscow (1990)
3. Asaul, A.N., Morozov, I.E., Pasyada, N.I., Frolov, V.I.: Strategic Development of a Construction Organization. SPSUACE, Saint Petersburg (2013)
4. Blinichkina, N.U.: Model of economic security policy effectiveness in the transition to digital economy. In: Ashmarina, S.I., Mantulenko,V.V., (eds.) Proceedings of the 2nd International Scientific Conference GCPMED 2019 - Global Challenges and Prospects of the Modern Economic Development. European Proceedings of Social and Behavioural Sciences, vol. 79, pp. 302–308. European Proceedings, London (2020)

5. Chudaeva, A.A., Svetkina, I.A., Zotova, A.S.: Industrial culture as economic security factor of industrial enterprises in digitalization process. In: Mantulenko, V.V., (ed.) Proceedings of the 18th International Scientific Conference "Problems of Enterprise Development: Theory and Practice". European Proceedings of Social and Behavioural Sciences, vol. 82, pp. 987–994. European Proceedings, London (2020)
6. Decree of the President of the Russian Federation dated May 13, 2017 No. 208 "On the strategy of economic security of the Russian Federation for the period up to 2030" (2017). https://www.garant.ru/products/ipo/prime/doc/71572608/. Accessed 26 Jun 2020
7. Federal Law "On security" from 28.12.2010 N 390-FL (2010). http://www.consultant.ru/document/cons_doc_LAW_108546/. Accessed 26 Jun 2020
8. Korneeva, T., Kozhukhova, V., Arkhipova, N.: ESG-risks assessment as a factor of oil and gas companies economic security. In: Mantulenko, V.V., (ed.) Proceedings of the 18th International Scientific Conference "Problems of Enterprise Development: Theory and Practice". European Proceedings of Social and Behavioural Sciences, vol. 82, pp. 662–671. European Proceedings, London (2020)
9. Kuzmin, M.D.: "Security" and "Economic security" as categories of social cognition: abstract of the dissertation for the degree of candidate of philosophical sciences. Tyumen State University, Tyumen (2014)
10. Makarov, I.N., Khodyakova, O.E.: Economic safety: the theory, methodology and the main aspects in functioning of the accounting entity, region, country. Scientific Notes of the Tambov Department of RoSMU, vol. **7**, pp. 204–213 (2017)
11. Samarastat: Small and Medium-Sized Entrepreneurship (2019). https://samarastat.gks.ru/small_and_medium_enterprises. Accessed 26 Jun 2020
12. Senchagov, V.K.: The essence and the basis of economic security strategy of Russia. Voprosy Ekonomiki **1**, 98–99 (1995)
13. Svetkina, I.A.: Smart city: ensuring economic security of administrative center of the Russian entity. In: Ashmarina, S.I., Mantulenko, V.V. (eds.) Current Achievements, Challenges and Digital Chances of Knowledge Based Economy. Lecture Notes in Networks and Systems, vol. 133, pp. 539–547. Springer, Cham (2021)

Saving Success of Older Workers: Diversity Policy in Organization's Ecosystem

S. M. Taltynov(✉) and T. I. Rakhmanova

Voronezh State University, Voronezh, Russia
taltynov@gmail.com, rakhmanova@bio.vsu.ru

Abstract. The article attempts to justify the possibility of preserving the success of older employees through the transition of the organization from the ego- to the eco-system and the formation of an age diversity policy. To implement these ideas, the content of the main categories used in the study is updated. The similarity of the organization's development with the evolution of living organisms in the population through the substitution of genetic egoism with intra-group altruism is shown. The advantages of "mutual assistance" are identified and described in details by functions of its manifestation in the organization to extend the professional longevity of employees.

Keywords: Age diversity · Ecosystem · Egosystem · Labor success · Older workers

1 Introduction

Research on successful aging at work is accompanied in the scientific literature by examples of various positive results that can be achieved by older workers. At the same time, questions of concretization of the success concept in the context of aging, time thresholds of personnel old age, reasons for achieving significant labor results by older employees and the loss of success functions by their peers remain to a certain extent debatable. In traditional societies, age was perceived as a gradual increase in social status in the course of life. The scientific and technical revolution of the industrial society has led to an increased role of more active youth. In the post-industrial society, this trend has persisted: age conservatism has become an obstacle to replacing key positions occupied by older workers who, in the opinion of employers, have lost their ability to adapt quickly. In the modern society, "youth" also serves as a criterion of personal and aesthetic value, while youthfulness is often considered as a competitive advantage [10]. The shortage of labor force, the desire of older employees to stay in the organization, in their positions, has led to the need to form approaches that provide representatives of older generations, who have preserved in their self-consciousness and behavior features of the youth lifestyle, with the ability to overcome labor changes to join younger colleagues in organizations. This work suggests the implementation of one of these approaches, which is to preserve functions of labor success of older workers through the policy of age diversity in the organization that functions as an ecosystem.

S. I. Ashmarina and V. V. Mantulenko (Eds.): IPM 2020, LNNS 161, pp. 684–692, 2021.
https://doi.org/10.1007/978-3-030-60926-9_85

2 Methodology

The implementation of the proposed approach includes several stages. The first stage: Updating scientific interpretations of the content for categories involved in the approach development.

1. “Senior (older) employees”. Psychological maturation of a person is considered through a conditional age, which is determined by the method of periodization and differs from the absolute one (calendar, passport, chronological). Periodization allows not only to divide the life path of an individual into segments, but also to give the age a meaningful meaning, since each cycle has common development features.
   Scientists have not reached a generally accepted statement of the age after which an employee is considered as “older”. Most employers in European countries believe that the decline in productivity occurs after 40–50 years. Depending on the sphere of activity, an employee may be classified as an employee of an older age group earlier or later than the specific age.
2. “Labor success of older workers”. Professional success is an objective characteristic of an employee’s labor performance, which is formed in the process of his/her meaningful need to overcome difficulties, maintain the achieved from the accumulated labor potential, preserve the reputation image, involvement and emotional mood to repeat a significant work activity in the future.
3. “Successful aging”. Initial gerontological research focused on people over the age of 60, many of whom are not involved in productive activities. Subsequently, it was determined that aging is normal, free from physical and mental pathologies [13]. Successful aging began to be considered as the simultaneous presence of three components: a low probability of disease and disability, maintaining a high level of physical and cognitive abilities, and participating in social and productive activities [2].
4. “Egosystem”. According to Dawkins’ research [7], living organisms are driven by absolute selfishness and the realization of the main goal: to survive and reproduce. No matter how altruistic the behavior of any living creature may seem, the frequency of occurrence of the “egoist gene” trait in the population eventually increases. The organization’s egosystem is designed to meet the owners’ goals which blocks access to information and new knowledge, defines narrow “corridors” of interaction between departments and employees, and uses directive management.
5. Ecosystem. In biology, ecosystems demonstrate a mutual adjustment of their components aimed at an ideal dynamic equilibrium that can be achieved under given conditions and with available components [17]. The peculiarity of the natural ecosystem is the interaction and competition of all its participants, allowing their joint development and adaptation to the external environment and related failures. The idea borrowed from biology to economics and management means a shift to collective generation of results, according to which ecosystems are defined as structures for coordinating a multi-stakeholder set of partners who need to interact in order to materialize a value proposal [1]. The presence of an ecosystem concept in an organization is determined by four characteristics: 1) heterogeneity of

participants in different roles, 2) the ability of one ecosystem to form another, 3) specific nature of interdependence between participants; 4) role-based management, consistency of actions, etc.

The second stage: Interdisciplinary justification of advantages of switching from the ego- to the eco- system for maintaining the success of older workers.

"Mutual Assistance" as an Advantage of Ecosystems from the Point of View of Natural Sciences. The basis of any ecosystem is a person with certain competencies and able to solve innovative tasks of the organization. Human is a part of nature, so it is advisable to involve biological science in the study of approaches to achieving labor success. All organisms do not live in isolation in nature but interact constantly and simultaneously and are exposed to the influence of "non-living" factors from the external environment. The need for a population to compete successfully in a constant struggle leads to weakening of the egoism gene of its members and forming the altruistic behavior that does not exclude mutual assistance.

According to the concept of mutual assistance by Kropotkin [9], cooperation and altruism increase chances of survival not only of the whole species, but also of an individual organism. They become the main catalyst for development: they increase life expectancy, raise its quality, and even improve the mental abilities of individuals. Kropotkin argued that mutual assistance is more important for evolution than mutual struggle, since cooperation not only helps individuals survive with the least energy expenditure, but also lays foundations for their further development. This means that the best conditions for progress are created by eliminating mutual struggle and replacing it with mutual assistance and mutual support. At first glance, these two concepts contradict each other, but according to natural mechanisms, what seems to be suicide for genes will be justified at the population level. Altruism and mutual aid allow populations to compete with groups in which there is a continuous struggle. In accordance with this, during the course of evolution, ecosystems are formed, which means a set of co-existing species of various organisms that interact with each other and with the environment in such a way that a community can be preserved and function over a long period of geological time.

Communities are not a random set of interacting living organisms, but a well-defined system of them, quite a stable system having numerous internal links, with a relatively constant structure and a mutually dependent set of species. In this case, interactions between organisms, as well as their impact on living conditions, can have the following models: one species (organism) benefits at the expense of another (positive impact); one species (organism) receives damage from the other (negative impact); there is no harm or benefit for the species (organism) (neutral effect).

The mechanism of replacing genetic egoism with intra-group altruism can be explained on the basis of the theory of kin selection, developed in the 30 s of the last century by the famous biologist Hamilton [8]. The main idea is that related individuals have a certain set of common genes. What is important for evolution is not the survival of individuals, but the transfer of genes to the next generation. The law is formulated, according to which the fixing of the altruism gene depends on three factors: 1) the genetic relationship of individuals; 2) advantages that the object of altruism receives; 3) damage that the altruist himself bears.

Hamilton's rule is expressed with the formula:

$$nrB > C \qquad (1)$$

where n is the number of altruism objects r - degree of genetic kinship;

B - the reproductive benefit of the recipient; C - altruist's reproductive damage.

If inequality is observed, the altruism gene can be fixed and passed on. An example of Hamilton's rule is multicellular organisms. All cells of these organisms, in fact, are clones of only one cell, so they carry exactly the same genetic set. This is the reason for their "refusal" to reproduce themselves for the sake of the common good. At the same time, many bacteria are also able to combine, for example, to form a bacterial consortium. But they do not create a single organism. Bacteria are still different creatures and each of them has its own "selfish aspirations". Such temporary unions look just like "drafts" of evolution, which once unsuccessfully tried to create a multicellular organism.

The study of the problem from the point of view of civilization showed that the development of mankind also tended to go from the egosystem to the ecosystem. No matter where on the planet a person was born and grows, everyone has heard of the importance of doing to others what they would like to be done to them - a truism commonly known as the Golden Rule. Despite this, altruism is seen as a weakness, and material success is noted as the only worthwhile investment of time and energy.

"Mutual assistance" from the point of view of economic and managerial sciences

An organization's ecosystem is a dynamic and co-developing community with various actors that, based on their interaction and communication, allow to create and receive a new content product.

Advantages of ecosystems relative to egosystems are:

- ecosystems, in contrast to egosystems, are characterized by greater stability, which allows them to function much more efficiently [3],
- the increased resilience of ecosystems is determined by the attitude to everyday and acute problems as an opportunity for learning and gaining experience [14],
- in contrast to eco-systems, problems and potential risks are often denied in egosystems; to increase the organizational stability, among other things, it is necessary for egosystem managers to be ready to solve problems and strive to learn from experience, turn problems into opportunities (transition to the ecosystem) [12],
- ideas of creativity and innovations are supported in the ecosystem; in such organizations, a creative climate is formed, and employees are motivated to generate new ideas that increase the organizational sustainability,
- leadership in ecosystems is focused on creating an organizational environment that encourages the adaptation of collective decisions instead of following the prescribed ones,
- in contrast to egosystems, which usually ignore the creative thinking of employees, ecosystems motivate staff to innovative and creative ideas [12]; while developing creativity, providing time and resources for experimentation, encouraging innovation, and showing tolerance for failure, sustainable organizations create an environment where employees feel safe to share new ideas [11].

The movement of scientific thought and the adaptation of applied technologies show that ecosystems are a new and necessary form for organizations to gain competitive advantages. To create a product with a new innovative content, taking into account principles of the ecosystem concept, it is assumed that employees are capable of teamwork, have professional, methodological and personal competencies, are focused on situational change of work roles and non-directive behavior of managers. The formation of such groups in a changing demographic situation is impossible without the staff of older age groups.

In accordance with this, the real possibility of extending labor longevity is the formation of an age diversity policy. Workforce of different ages is beneficial to the organization, it can contribute to improving the labor efficiency, as employees will feel valued and committed to their organization [4]. Optimal behaviors and ways of thinking make it possible to compensate for age-related losses (reduced physical and cognitive abilities, memory, etc.) due to age-related benefits (experience, knowledge, emotional stability). Thus, the replacement of "losses" at the older age is possible by positive selection processes (correct setting of goals), compensation (use of new technologies), optimization (acquisition of necessary resources) [2].

The third stage: Preserving success functions of older workers through the formation of age diversity.

The difficulty in achieving the balance between levels of the age pyramid relates to the inequality of different age cohorts. Inequality can be considered from the point of view of socio-economic or cultural injustice. The first is rooted in the economic structure of our society, which leads to exploitation, economic marginalization (restriction or denial of access to well-paid or generally profitable work) and deprivation (deprivation of an adequate material standard of living). Cultural injustice lies in the cultural dominance, non-recognition and disrespect of representatives of other social groups. The possibility of maintaining the success of employees of older age groups in the context of the age diversity policy implementation is presented in Table 1.

Table 1. Maintaining success functions of older workers while implementing the age diversity policy

Success function	Characteristic	Technology for maintaining success and overcoming inequality
Regulatory-biological	Increasing age is associated with changes in brain volume and measures of cognition and behavior [18], deterioration of spatial and associative memory [19], which is critical for high-level cognitive and executive functions [15]. Obesity at middle age is associated with lower cognitive abilities and faster cognitive aging at old age [6]	Prevention of a healthy lifestyle and nutrition that inhibits aging. Activities: development of information leaflets by nutritionists and individual mailing, low-calorie food in the organization's canteens. Ergonomic furniture for employees of all age groups

(*continued*)

Table 1. (*continued*)

Success function	Characteristic	Technology for maintaining success and overcoming inequality
Social status and symbolic	More frequent participation in cognitive-stimulating activities at a later age may become an important behavior modification for improving cognitive functions [5], prolonging active longevity, and have a positive effect on overall performance [21]. Success is determined through socially significant and approved status positions of the society	Activities: - achieve a positive social environment in which employees can freely express emotions and share information; - provide them with resources that are appropriate for their work needs; - formation of employees' commitment to the organization's goals; - reaching agreements on the acceptance of additional responsibilities by the working group with an informal change of working roles to solve tasks of the organization; - ensuring public recognition for successful work motivation; - creating a positive image of the group, recognizing it as "outstanding" to motivate belonging to it
Meaning- and goal-forming	Success is based on achieving a goal and comparing the obtained result with it, it implies a conscious goal setting, perseverance, faith in the success of your endeavors, and readiness for self-improvement. The success depends on what personal meaning the employee invested in the achievement of a result	Activities: Employees with a low functional level or at risk of it will welcome personnel initiatives that maintain the actual achievement level or take into account lost resources (for example, flexible working hours, reduced workload). It is advisable to use symbolic leadership that denotes the meaning or idea of the performed work, its significance

Source: authors.

3 Results

The need to preserve functions of professional success by employees of older age groups has a two-way conditionality. On the one hand, there is a shortage of labor in the aging labor market and the need to extend the professional longevity of employees. On the other hand, the demonstrated willingness of older personnel to continue the active participation in the organization's work processes.

The content of the category "age" is justified by the movement of scientific thought, taking into account economic, social and behavioral factors. In accordance with this, we can distinguish three models of age perception. Age as the basis of a dominant position in the society and change in the social status (priority of the older generation). Age as a component of "age conservatism", explained by the complexity to meet the requirements of rapid adaptation to production conditions for aging employees (priority of the younger generation). Age as a demographic characteristic in the diversity policy with a preference for employees with competencies of non-resistance to labor changes and preservation of the youth lifestyle in self-consciousness and behavior (lack of age priorities).

The adaptation of behavioral models of living organisms borrowed from biology to the organizational production environment allowed us to determine their certain similarity, consisting in the presence of "egoist genes" in the population and personnel, whose non-replacement by altruism leads to a weakening of the sustainability of a system, its inability to compete and survive.

The transition from ego- to eco-system is justified in natural sciences by the theory of kin selection, which explains the mechanism for replacing genetic egoism with intra-group altruism, the concept of "mutual assistance", which shows that cooperation and altruism become the main catalyst for development: they increase life expectancy, raise its quality, and even improve the mental abilities of individuals. From the point of view of management science, advantages of the ecosystem are argued to confirm the effectiveness of its functioning due to greater stability, creating an atmosphere of creativity, motivation for creative thinking and innovations, tolerance for failures, supporting and delegating leadership.

4 Discussion

In the context of an aging labor market, the transformation of value attitudes of new generations, there is an increasing need to create prerequisites for the use of older workers' labor while preserving functions of their professional success. One of the options of "mutual assistance" as the central idea of the organization's ecosystem is the age diversity policy. The research results in this area are quite contradictory. On the one hand, the idea of positive effects prevailing over negative ones is justified. Teams with members of different ages have a wider range of resources and sources of information – knowledge, experience, personal acquaintances, etc. [20]. On the other hand, according to the theory of social identity, a team with signs of age uniformity can lead to high productivity. For a person, the sense of belonging to a particular group is important. Age can become a categorization and identification criterion by which an employee defines a group as "own", which makes him/her more positive towards people of a similar age group [16]. Age diversity increases the participation of older workers in the cultural, economic, political and social life of society, which in turn increases their motivation to maintain their potential for success. The factors for meeting this need on the part of the employer are taking into account characteristics of different age groups, subjective beliefs and stereotypes, signs of ageism in the culture of the organization in the diversity policy.

5 Conclusion

This study promotes the scientific idea of the possibility of extending the working capacity of older workers through the preservation of their success functions. What is new in this view is the use of an age diversity policy adapted to the "organization ecosystem" model. The obtained results can be used in the development of the theory of personnel aging, as well as in the practice of organizations to retain key employees. The organization's ecosystem allows creating prerequisites for preserving the success and potential of older employees through the age diversity policy. The formation of this policy allows to minimize, first, disagreements and conflicts because of differences in opinions and beliefs of representatives of several generations; second, managerial protest against the formation of a set of unique characteristics, competencies, knowledge and experience of employees of different ages; and third, the unfair distribution of resources among members of the group of different ages. The applied effect of the age diversity policy is possible through detailing success functions and reasonable selection of a set of auxiliary measures for each of them.

Acknowledgments. The reported study was funded by RFBR, project number 20-010-00263.

References

1. Adner, R.: Ecosystem as structure: an actionable construct for strategy. J. Manag. **43**(1), 39–58 (2017)
2. Baltes, P.B., Baltes, M.M.: Psychological perspectives on successful aging: the model of selective optimization with compensation. In: Successful Aging: Perspectives from the Behavioral Sciences. Cambridge University Press, Cambridge (1990)
3. Barasa, E., Mbau, R., Gilson, L.: What is resilience and how can it be nurtured? A systematic review of empirical literature on organizational Resilience. Int. J. Health Policy Manag. **7**(6), 491–503 (2018)
4. Boehm, S.A., Kunze, F., Bruch, H.: Spotlight on age-diversity climate: the impact of age-inclusive HR practices on firm-level outcomes. Pers. Psychol. **67**, 667–704 (2014)
5. Buschkuehl, M., Jaeggi, S.M., Jonides, J.: Neuronal effects following working memory training. Dev. Cogn. Neurosci. **2**(1), 167–179 (2012)
6. Dahl, A.K., Hassing, L.B.: Obesity and cognitive aging. Epidemiol. Rev. **35**, 22–32 (2013)
7. Docinse, R.: Selfish Gene. Corpus, Moscow (2013)
8. Hamilton, W.D.: The genetical evolution of social behavior. J. Theor. Biol. **7**(1), 1–52 (1964)
9. Kropotkin, P.A.: Ethics: Selected Works. Politizdat, Moscow (1991)
10. Lishaev, S.A.: Horizons of old age: typology and existential content. In: Lishaev, S.A. (ed.) Mixtura Verborum'2006: Modern Topology, pp. 14–43. Samara Humanitarian Academy, Samara (2007)
11. Mafabi, S., Munene, J.C., Ahiauzu, A.: Organisational resilience: testing the interaction effect of knowledge management and creative climate. J. Organ. Psychol. **13**(1/2), 70–82 (2013)
12. Oluwasoye, M., Ugonna, N.: Environmental risk: exploring organisational resilience and robustness. Int. J. Sci. Eng. Resour. **6**(1), 1103–1115 (2015)
13. Rowe, J.W., Kahn, R.L.: Successful aging. Gerontologist **37**(4), 433–440 (1997)

14. Sawalha, I.H.: Managing adversity: understanding some dimensions of or-ganizational resilience. Manag. Res. Rev. **38**(4), 346–366 (2015)
15. Stokes, M.G.: 'Activity-silent' working memory in prefrontal cortex: a dynamic coding framework. Trends Cogn. Sci. **19**(7), 394–405 (2015)
16. Tajfel, H., Tuner, J.C.: The social identity theory of intergroup behavior. Polit. Psychol. **20**, 276–293 (2004)
17. Tansley, A.G.: The use and abuse of vegetational concepts and terms. Ecology **16**(3), 284–307 (1935)
18. Wahl, D., Cogger, V.C., Solon-Biet, S.M., Waern, R.V., Gokarn, R., Pulpitel, T., Cabo, R., Mattson, M.P., Raubenheimer, D., Simpson, S.J., Le Couteur, D.G.: Nutritional strategies to optimise cognitive function in the aging brain. Ageing Res. Rev. **31**, 80–92 (2016)
19. Weber, M., Wu, T., Hanson, J.E., Alam, N.M., Solanoy, H., Ngu, H., Lauffer, B.E., Lin, H. H., Dominguez, S.L., Reeder, J., Tom, J., Steiner, P., Foreman, O., Prusky, G.T., Scearce-Levie, K.: Cognitive deficits, changes in synaptic function, and brain pathology in a mouse model of normal aging (1, 2, 3). eNeuro **2**(5), 1–26 (2015)
20. Williams, K.Y.: Demography and diversity in organizations: a review of 40 years of research. Res. Organ. Behav. **20**, 79–140 (1998)
21. Wilson, R.S., Boyle, P.A., Yu, L., Barnes, L.L., Schneider, J.A., Bennett, D.A.: Life-span cognitive activity, neuropathologic burden, and cognitive aging. Neurology **81**, 314–321 (2013)

Reactive and Proactive Leadership, Relations between Investments in the Human Capital and Business Results

Personnel Motivation Stratum Structure as Stability Factor Within Socio-Economic Changes

V. N. Voronin[1(✉)], M. V. Iontseva[2], and O. A. Kolosova[2]

[1] Academy for Standardization, Metrology and Certification, Moscow, Russia
prvoronina@asms.ru
[2] State University of Management, Moscow, Russia
miontseval@gmail.com, olga_kolosova@mail.ru

Abstract. Achieving the strategic goals of the organization in various socio-economic conditions. Create an effective system of motivation, to establish purposefully organizational culture in order to achieve the strategic goals of the organization. A suggested system of motivated impact (rewards, penalties, accruals and payments of wages) is based on three principles of economic effectiveness: optimality, effective motivation, transparency and accessibility for understanding. Motivation system requires all the parameters of the organizational environment that affect the motivation of employee. The implementation of the principles of the organizational structure at the individual level, and application of organizational and socio-psychological mechanisms allow to create an effective organizational culture, which transforms the structure of motivation of majority of employees when internal motives become leading. In other words, a fundamental importance is the balance between external and internal motivation. The dynamic modern environment claims to monitor periodically the operability of the financial incentives system and its compliance with the goals and objectives of the company within the changing conditions and for future development. The proposed model for constructing a motivation system has been successfully implemented in practice.

Keywords: Efficiency · External and internal motivation · Motivation system

1 Introduction

The socio-economic situation in the modern world is constantly changing. Moreover, we witness increasing pace and unpredictable direction of these changes. In these conditions, the strategy of any effective organization should be aimed, on the one hand, at maintaining stability, and on the other, at its development, taking into account the each and every socio-economic change. At the same time, the task of the HR department is to choose a functional strategy that will ensure the building a team that capable to achieve the strategic goals of the organization in changing conditions. For large federal enterprises, the following tasks are typical in the field of HR management [12]:

S. I. Ashmarina and V. V. Mantulenko (Eds.): IPM 2020, LNNS 161, pp. 695–701, 2021.
https://doi.org/10.1007/978-3-030-60926-9_86

- to hold key, most effective managers and specialists,
- control staff turnover,
- reduce costs for the selection and adaptation of personnel,
- effectively distribute staff costs,
- maintain a high level of employee motivation,
- increase the efficiency on each employee,
- create a positive image of the employing organization.

According to the results of our research since 2006, one of the key components has been the stratum structure of the motivation system, which is understood that the staff has clear differences in qualifications and "values" for the organization and these differences are layered. One of the prerequisites for the effectiveness of the stratum structure of motivation is the clear and understandable conditions for the transition from one stratum to another. In addition, an important factor is the attainability of these conditions and the dependence of their achievement on the employee himself. While constructing any motivation systems, starting from classical [2] and ending with the latest [1], it is necessary that the organizational and psychological procedures are adequate to the internal needs of the main categories.

2 Methodology

The suggested model for constructing a motivation system is based on the following theories and models:

1. The theory of the interdependence of Kelly and Thibaut, where the analysis is carried out in terms of "gains" and "costs". Comparing them a person considers his relations with other people or organizations [10].
2. Hierarchical model of motivation Maslow that reflects one of the most popular and well-known theories of motivation - the theory of the hierarchy of needs [6].
3. Two-level Lawler model [5] where needs are divided into two levels
4. The concept of socio-psychological mechanisms of the formation of organizational culture of Voronin. In his paper Voronin describes the main psychological mechanisms that ensure the formation of motivational components of the general organizational culture at the individual level [11].

3 Results

While building an effective motivation system, it is necessary that it include all the parameters of the organizational environment that affect the motivation of an employee. However, the most important component of the organizational environment, which affects the structure of motivation and the magnitude of each of its elements, is still a reward-penalty system based on a system of accrual and payment of wages, including both fixed and variable components.

The first principle of establishing of a reward-penalty system, most commonly, is similar for different types of organizations and involves maximizing the motivation effect. The purpose of any reward-penalty system is to attract qualified workers, to keep them from leaving to other organizations and to encourage them to work as efficiently as possible. Achieving these three goals is the essence of maximizing the motivation effect.

The second principle of economic efficiency of the proposed reward-penalty system. In some cases, it is strived to achieve its maximum (and performance criteria can also vary significantly), in others - to achieve the necessary critical minimum, sometimes the so-called reasonable profitability becomes this principle, that is, achieving such a combination when a relative maximum of motivation is achieved with a minimum of costs.

Thus, there is a certain point in the costs on motivation, after which any increase in costs leads only to a slight increase in the level of motivation. For example, such point can be reached in financial incentives. There is amount of payment, and it differs for each employee, above which the motivation does not increase significantly (and even decreases) and below which it drops significantly. On the other hand, a decrease in the total monthly income below this critical point will lead to an increase in turnover and large economic losses for the organization coming from search, adaptation and training of a new employee, and the loss of a monopoly on the technologies used in the company.

Finally, the third principle is accessibility in understanding of the reward-penalty system. For example, if employee does not understand the charging mechanism in details it will destroy the connection between results and payment, actions and reactions to them.

The studies show that a full-fledged entity capable of effectively managing organizational culture may include an external advisory body, must include group of internal consultants, interested senior managers and mid-level managers trained in this activity and leading a coherent HR policy [3]. Moreover, the loss of any of the components listed makes it impossible to form purposefully an organizational culture.

A mechanism in this case is a certain way (through predetermined procedures) the compiled living conditions of the employees, where the proper combination of various psychological patterns will produce the intended effect [11].

The main psychological mechanisms that ensure the creating of general organizational culture motivational components at the individual level are:

1. Facility of all procedures related to the motivation system: assessment criteria, career mechanisms, remuneration.
2. Casual relationships between efficiency and payment.
3. Cause effect between outcome and reward (penalty) should be as close as possible in time.
4. Achievements of above-threshold-dissociation. Any incentive should exceed the differential limit value or some minimum amount by which it makes sense to change the payment, and which will be perceived as a change.
5. The definitive frequency. In motivation building process the frequency of motivating events is more significant than their magnitude.

6. The principle of non-obligatory payments, ensuring the absence of addiction.
7. Success of each individual employee (unit) should be beneficial to everyone in the organization. This allows to maintain an atmosphere of cooperation and mutual assistance, and to block separatist tendencies.
8. Optimization of the ratio of external and internal motivation has a very significant role.

Our studies have shown that the implementation of these principles, The implementation of the principles of the organizational structure at the individual level, and application of organizational and socio-psychological mechanisms allow to create an effective organizational culture, which transforms the structure of motivation of majority of employees when internal motives become leading. More than 70% of the organization's employees have their internal motives (interest in activities, the possibility of gaining new experience, attitude to the organization) are put in the first place. These are examples of the fact of fundamental importance is to find an optimal combination of external and internal motivation [8].

The combination of internal and external motivation will be proper if the factors of external motivation are in the middle of range of motivational factors, ranked by strength. Of course, the enrichment of the motivational component can go through of attracting to decision-making process, the development of additional competencies [4] gamification [9].

Yet, the payroll scheme stays one of the central factors in building a holistic motivation system. If we assume that the perceived difference between working as "A" in company "X" and working as "A" in company "U" is very small, then the difference in salary will have a big impact on the decision where to work. Lawler made the same conclusion. "A "fresher"on the labor market for the first time will be especially affected by this difference in his decision. He cannot appreciate non-monetary rewards well, and he will feel that salary is the only clear difference between the jobs" [5, p. 152].

Most fresh graduates will choose a job with the highest salary, and this is the category of people who applies for mass market vacancies, since after graduation 70% of graduates who do not have work experience do not work in their specialization. Apparently, newcomers to the labor market will look for the jobs with high salaries. This is fair for both those who choose a job for the first time and those who make a choice between an existing job and work in another company, if the organization where the employee works has nothing valuable for him except money. However, this consistent pattern is not so obvious if a company with low salary gives some other desired rewards that are not represented in a company with a high salary. It is the balance of external and internal motivation that leads to job satisfaction that is according to the studies of Soboleva is one of the key factors in life satisfaction [7].

The main tools for building a stratum motivation system are:

- the analysis of the wage market (normal distribution curve) for each position, the ratio of the constant and variable components of wages. This ratio depends on the profile of activity and the level of position.
- assessment of the effectiveness and "value" of the employee for the organization, necessary for its attribution to a certain level.

Three main categories of employees are distinguished: top management, middle management and key specialists (employees who have unique knowledge or technologies necessary for the organization) and profile specialists. In each of these strata there are two steps that assist to achieve a sense of growth prospects.

For the initial mass market vacancies without special qualifications or where short-term training is sufficient, a salary is set below the market median (at the level of 25–40 percentile).

The variable part in this case (when fulfilling planned performance indicators) provides income above the market average for a given position. Thus, a certain filter is set at the initial stage. If a new employee does not regularly fulfill planned job targets, he will receive a salary significantly lower than the market average and, as a rule, leaves the organization. If he copes with the planned indicators, then in a few months, in accordance with the provisions on the assessment, he proceeds to the next step, where the salary part is close to average, and the variable part is fully retained.

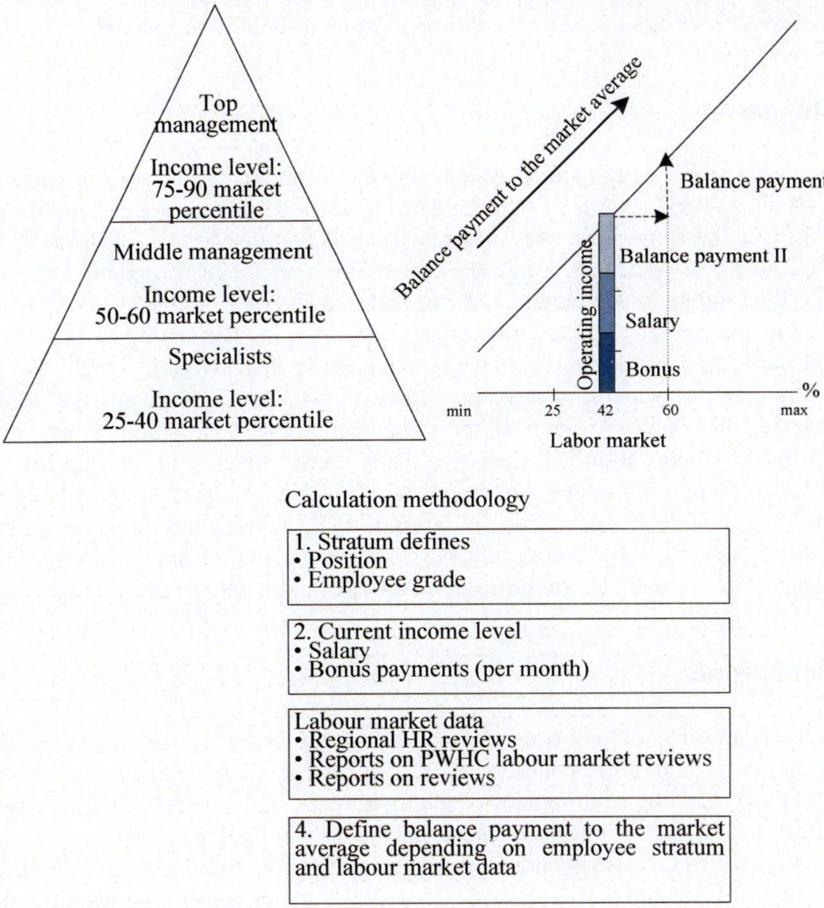

Fig. 1. Motivation system model: stratum calculation methodology and principle (Source: authors based on [11]).

For the second stratum, the income level is at the average market level or higher. For the third stratum this figure is at the level of 75–90 percentile. At the same time, the payroll fund of the organization does not grow since with a properly structured organization, the number of the first stratum is about 3%, the number of the second is about 10–20%, and the third is up to 80% (Fig. 1).

This model of the motivation system will work only under permanent evaluation for its effectiveness. The financial incentives system is a central element of staff motivation, which allows create the proper organizational behavior of employees for the most effective achievement of the goals and objectives facing the organization. The current financial incentives systems are distinguished by premium schemes and are constantly undergoing changes in order to be able to take into account the dynamism of the modern environment and meet the business needs of departments that require flexibility of wage policies to changing conditions for future development. Therefore, it is very important to monitor periodically the performance of the incentives system on its compliance with the goals and objectives of the organization. That will help to plan the process of redesigning of motivation system in an unhurried manner.

4 Discussion

The proposed model of motivation system elaboration and its results are consistent with generally recognized studies of other authors. The model assumes a certain not rigid hierarchy. At the same time, the hierarchy itself is a divisive issue. Whilst Maslow regarded all needs as a rank range, Lawler offers a two-level structure. The model prevails a harmonious combination of external and internal motivation with internal motives in area of focus. Trusov, for instance, analyzes the correlation of external and internal motivation, and indicates that excessive external motivation actually destroys internal motivation, which in conditions of constant socio-economic changes leads to the turnover of a key personnel of the organization and poses serious risks to the stability of the organization [8]. Gordeeva, came to the conclusion that students with internal motives as leading ones are more successful [1]. In regarding model, we used the basic psychological mechanisms of Voronin that provides the formation of motivational components, ensures the effectiveness of the transition from one level of needs to another with the optimal combination of internal and external motivations [11].

5 Conclusion

The assessment methodology can be considered as a tool for regular system analysis, which allows to determine contribution of these schemes to motivation, how they provide a direct relationship between individual merit and remuneration, if they are cost-effective and meet the needs of the organization. An integrated approach suggests to obtain a systematic assessment of the effectiveness of financial incentives by taking into account the positions of interested groups, both the employer and from the point and employee in analyzing the main performance criteria:

1. Regular monitoring of the target indicators achievement and the level of reachability of bonus payments that will help to establish an idea of the motivational effect of the bonus scheme, as well as the correctness of the establishment of planned indicators in order to promptly make changes if necessary.
2. Analysis of the structure of the wage fund (payroll). The ratio of constant and variable parts, the definition of these proportions for various categories of personnel. It is important to track the dynamics of these indicators in relation to time and correlate with the strategic tasks of the employer.
3. Analysis of the relationship between payroll growth rate and profit, the dynamics of profitability per employee in order to comply with certain proportions (effective indicators should increase at a faster rate than the corresponding costs).
4. Assessment of internal justice (parity) of income.
5. Analysis of the conformity of employee income to market values.
6. Assessment of staff stability and job satisfaction.

The considered model of constructing a motivation system worked during the economic crises of 2008 and 2014, and helped the organization to retain key personnel, to attract the best employees from the market, and to keep the rate of decline in efficiency on significantly lower level than the industry average.

References

1. Gordeeva, T.O.: Motivation: new approaches, diagnostics, practical recommendations. Siber. Psychol. J. **62**, 38–53 (2016)
2. Hekhauszen, X.: Motivation and activity. Smysl, Moscow (2003)
3. Iontseva, M.V., Voronin, V.N., Kalinina, N.V., Antonenko, I.V., Lukin, V.V., Utlik, E.P.: Effective corporate culture of the university: monograph. State University of Management, Moscow (2019)
4. Kolychev, V.D., Belkin, I.O., Udovichenko, R.S.: Specifics and effectiveness of programs for the development of managerial competencies of the personnel reserve. High. Educ. Russia **11**, 134–143 (2019)
5. Lawler, E.E.: Motivation in Work Organizations. Jossey-Bass, San Francisco (1994)
6. Maslow, A.: Motivation and personality. Eurasia, Saint-Petersburg (2003)
7. Soboveva, N.E.: Specifics and effectiveness of programs for the development of managerial competencies of the personnel reserve. Monit. Public Opin.: Econ. Soc. Changes **1**, 368–390 (2020)
8. Trusov, V.P.: Socio-psychological research of cognitive processes. Leningrad State University, Leningrad (1980)
9. Tsyplakova, E.O.: Gamification-motivational practice or a mechanism for total control over the labor process? Econ. Sociol. **3**(17), 82–109 (2016)
10. Kelley, H.H., Thibaut, J.W.: Interpersonal Relations: A Theory of Interdependence. Wiley, NewYork (2008)
11. Voronin, V.N.: Socio-psychological mechanisms of organizational culture formation. Moscow (1999)
12. Zavyalova, E., Kucherov, D., Tsybova, V.: Human resource management in Russian companies-leaders of the world economy. Foresighte **4**(11), 52–61 (2017)

The Role of Management in Providing Anti-corruption Activities of Economic Entities

O. V. Konovalova[1], D. A. Nikolaev[1(✉)], and E. M. Caruana[2]

[1] Financial University under the Government of the Russian Federation, Moscow, Russia
ovkonovalova@fa.ru, dnikolaev@fa.ru
[2] Berlitz Language Centre, Saint Julian's, Malta
elacaru@yahoo.com

Abstract. The paper investigates the anti-corruption issues in the activities of economic entities. Business faces corruption every day in cases of inspections by regulatory authorities or relevant departments of business entities, obtaining state funding or participating in various kinds of project activities with the involvement of third-party investments etc. Often, the leader himself is a participant in corruption schemes and he is the initiator of corrupt behavior among his subordinates as well. The purpose of the paper is to identify perspective directions of anti-corruption in Russia at the company level by determining the role of a leader in the context of combating corruption in the country. The authors accepted hypothesis that not only anti-corruption programs in the activities of economic entities may have a positive impact on the level of corruption. The anti-corruption role of the leader in the corporate culture is the best way to protect employees from their involvement in corruption schemes. This paper focuses on the possibility of using leader authority and its impact on the implementation of anti-corruption programs. The authors concluded that the role of the head of each division is to prevent any violations on his part and on the part of his colleagues.

Keywords: Anti-corruption policy · Corporate fraud · Corruption · Corrupt behavior · Economic crime · Role of a leader

1 Introduction

Corruption is one of the most serious problems in modern Russia. It is a serious obstacle to the development of the state. This problem is detrimental to the economic security of the country, as well as its social stability. It reduces the degree of public confidence in both the authorities and representatives of financial institutions. In this regard, the subject is very relevant under modern conditions, and the authors proved this through the analysis.

Recently, one can see a clear trend that the fight against corruption has intensified. Governments take effective measures to combat corruption in all spheres of public life. First of all, new anti-corruption laws were adopted (for example, on November, 8, 2016, the French Parliament passed Law No 2016-1691 "On Transparency, Fight

S. I. Ashmarina and V. V. Mantulenko (Eds.): IPM 2020, LNNS 161, pp. 702–712, 2021.
https://doi.org/10.1007/978-3-030-60926-9_87

against Corruption and Modernisation of Economic Life" [9]), amendments were made to existing acts regulating the activities of the state anti-corruption system, and new anti-corruption standards were adopted (ISO 37001 "Anti-bribery management systems" [7]) [4].

Awareness of this trend from a global perspective is due to the activation of the legislative activities of many international organisations whose primary purpose is to establish common standards for the fight against corruption through the enforcement of anti-corruption laws, the introduction of criminal liability for corruption-related crimes and to prevent violations of anti-corruption legislation at the national level [1].

Price Waterhouse Coopers' study on anti-corruption compliance in Russia showed the industry that is the most vulnerable to abuse (telecommunications) and the one that is most protected (automotive sector) [13]. However, companies of all sectors are involved in some kind of criminal corruption. The anti-corruption index of Russian companies is growing. Its average value is 68%. At the same time, the index value for various industries ranges from 60% (automotive sector) to 86% (telecommunications sector) [13]. In the framework of generally accessible information, a comparison of anti-corruption compliance in Russian companies and world practice in this area was made: the presence of a compliance control department, a hotline, local policies, the number of litigations, etc.

The loss of financial and tangible assets is a large-scale, but not the only consequence of corporate fraud, including theft of property, distortion of financial statements and corruption. Moreover, for the company, the cost of the investigation may be more than double the direct loss caused by the illegal act. Companies should not underestimate their negative impact on the effectiveness of operating activities, as well as on the moral and psychological climate in the organisation and its reputation. Corruption is an economic crime and acts as one of the risk factors for corporate fraud in the activities of an economic entity. According to PWC, both in Russia and in the world, the negative consequences of economic crimes primarily, affect the morale of the employees, image, business relations, relationships with regulators and regulatory bodies (Table 1).

Table 1. Negative effects of economic crime

Sphere of influence	In Russia	In the world
Company reputation	33%	37%
Relationships with regulators	36%	30%
Business relationship	41%	38%
Employees morale	50%	48%

Source: authors based on [13].

In the study of corporate risk problems, the authors revealed that in Russia, economic crimes in business entities are mainly committed by middle managers (50%) and senior managers (40%) [13] (Table 2).

Table 2. The subjects of internal corporate fraud

Level of Management	In Russia	In the world
Junior level executives	26%	14%
Middle managers	47%	37%
Top executives	39%	24%

Source: authors based on [12].

As noted earlier, in order to prevent the risks of corporate fraud in general, Russian and global companies actively use mechanisms and tools such as internal control and audit, compliance control, hotlines, etc. Table 3 presents signals for determining the facts of corporate fraud.

Table 3. The system of "red flags" of corporate fraud

Risk area	Symptoms		
	Analytical	Documentary	Behavioral
Corruption and bribery	– extremely profitable trade, loss-making or zero trade, – a large number of discounts, – involving the company in major litigation	– loss and substitution of individual documents, – complaints of counterparties about untimely repayment of accounts payable	– unethical behavior; – violation of business standards, – acceptance of expensive gifts, – close relations with external partners and contractors
Conflicts of interest	– imbalance of costs, – unreasonable bonus payments, – frequent change of management teams, – major transactions with nearest counterparties	– coincidence of the names of managers and partners, – delays in the repayment of receivables, – entries made shortly before the end of the reporting period, – renegotiation of the supply contract	– closed work style, – unwillingness to go on vacation and take their own affairs, – close relations with external partners and contractors, – unofficial conflict of interest information
Financial Reporting and Documentation Fraud	– revenue growth with a decrease in inventories, – increase in income with a decrease in cash inflows,	– loss of primary documents, – strange labels, – providing copies, – documentation delays,	– frequent job changes, – spending large amounts of money in excess of income,

(*continued*)

Table 3. (*continued*)

Risk area	Symptoms		
	Analytical	Documentary	Behavioral
	– stock growth with a decrease in accounts payable, – rapid increase in expenses compared to income	– postings without documents, – making records by unauthorized persons	– complaints and suspicions, – the presence of personal debts, – addiction to risky transactions and gambling
Asset mis-appropriation	– write-off of funds ahead of schedule, – payroll increases with staff reductions, – investment growth in the absence of depreciation growth, – "dead souls"	– entries in accounting registers without documentary evidence, – corrected documents; – presentation of copies of documents, – doubtful handwriting on documents	– spending large amounts of money in excess of income, – complaints and suspicions, – personal debt, – negative biographic signs

Source: authors.

Due to the unfavourable situation with corruption, it is necessary to strengthen Russia's fight against corruption. According to the Corruption Perception Index for 2019, Russia ranks 137th out of 180 scoring 28 points out of 100 like Papua New Guinea, Lebanon, Dominican Republic, Kenya, Liberia, Mauritania, Paraguay, and Uganda [16].

Moreover, the Russian lawmaker's attention to the issues of pursuing anti-corruption policies by business entities is increasing every year. The National Anti-Corruption Plan 2018-2020 separately allocates a series of measures to improve anti-corruption measures in the business sector, including protection of business entities from the abuse of power by officials [2].

Therefore, in the context of modern legislation, it is particularly important to effectively prevent and combat corporate entity corruption. It is necessary to change the attitude of society towards corruption. Although most citizens condemn corruption and advocate the fight against it, sometimes they themselves are willingly involved in corruption schemes to solve professional and personal issues, participating in it on their own behalf or as an intermediary of an economic entity or its official [11]. COSO Enterprise Risk Management is a three-layer model of internal control and risk management system, which was improved in 2017. It pointed out that corporate culture should be the main prevention mechanism for any illegal behaviour (including corruption-oriented behaviour) of any company's employee [3].

The areas that need more attention from the point of view of further improving the system of internal control and risk management of an economic entity are the development of an anti-corruption culture and the increasing role of leadership in anti-corruption propaganda.

2 Methodology

The study was carried out in two stages. At the first stage, the study was devoted to understanding whether the anti-corruption programs of economic entities are relevant and in demand in their activities. The analysis was based on reports from leading consulting companies using quality research methods: 2018 PWC study "Fraud Prevention: What measures are companies taking? A study of economic crimes in Russia" [14] and 2017 PWC study "Anti-Bribery and Corruption Index (ABCI) in Russia – 2017" [12]. Since the main respondents in these studies are middle and senior managers responsible for audit, compliance, risk management or a financial or other function, it was possible to identify and analyse the current business needs and key trends in the development of anti-corruption measures at the company level, as well as the role leadership in the corrupt behaviour of employees.

In the second stage, the study was aimed at finding the most effective practice of the authority of the leader in the work of the anti-corruption system of the company. This stage was based on the study of compliance standards, such as ISO 19600 "Compliance management systems" [6], ISO 37001 "Anti-bribery systems" [7] and analysis of "best practices", experience of implementing anti-corruption practices by large Russian companies.

Transparency International defines corruption as an abuse of entrusted powers for personal gain. Corruption affects all aspects of society, leading to significant inefficiencies and creating obstacles to economic development. The anti-corruption issue has attracted more and more attention from national governments, international institutions and the private sector. New tougher anti-corruption standards continue to evolve around the world [16]. Corruption is a problem in all states. It is closely related to other types of offenses, for example, fraud and monopoly. Corruption also causes the growth of the shadow economy and the outflow of capital from countries [10]. Owing to the fact that the assessment takes place only on identified cases, the results of such studies cannot be considered fully representative, however, they demonstrate that employee participation in corruption schemes remains a common and expensive problem for the company [15].

Studies show that every time a person crosses the moral threshold, it is easier for him to go to the next step. Therefore, it usually begins with minor corruption crimes and gradually evolves into larger corruption crimes until they become so large that they can no longer hide the fact. The American professor and anti-corruption expert Klitgaard proposed a formula to explain the occurrence of corruption:

$$K = M + D - A, \quad (1)$$

where K is corruption; M is a monopoly; C – discretion; P – accountability [8].

In accordance with the views of Klitgaard, the one who has a monopoly independently decides who will buy his goods or use his service [8]. This person is not required to report on their decisions. It is this state of affairs that contributes to increased corruption.

3 Results

Obviously, corruption risks, from the point of view of corporate fraud, have turned from an operational problem into a strategic challenge for business, which requires an active approach to managing at the highest level of the company. Fraud committed by senior executives creates serious problems for companies: such economic crimes are difficult to detect, they destroy the corporate culture and set a negative “tone from the top”. It is very important to prohibit the reduction of employee motivation and loyalty, which means to exclude and prevent the following situations:

- uncertainty in social matters,
- low or unreasonably differentiated wages,
- unrealistic financial indicators (for example, excessively high or not controlled by the contractor),
- injustice and inequality in the team,
- management does not communicate with employees, is indifferent to personnel, does not pursue an open policy,
- there is no social program to support members of the labour collective that makes it easier for them to overcome external circumstances,
- demonstrative luxury, to which management is inclined, etc.

Therefore, the formation of a corporate culture as a tool for managing the company’s risks with full awareness and responsibility on the part of senior management will lead the domestic business to a positive transformation of the short-term and long-term operational effectiveness. A favourable working atmosphere as well as a developed system of internal control, combined with a law-abiding compliance with the norms of the anti-corruption program of the company, will prevent the generation of motive, pressure, opportunity, and justification of corrupt behaviour on the part of employees.

The honesty of an employee largely depends on the integrity of company executives. As for the prevention of corruption and the implementation of the anti-corruption program, it is worth noting that the main role in ensuring the anti-corruption system is played by company executives [5].

The personal impact of management on company employees is very important. Company executives serve as team leaders. Their organisational capabilities, their ability to practice innovation, and their ability to prevent and resolve conflicts of interest significantly contribute to the company’s successful realisation of the declared anti-corruption values. Approved anti-corruption measures and uniform ethical standards will affect the behaviour and professional knowledge of all employees in the field of anti-corruption.

The role of leaders at any level of management is not only to manage, but also to provide the most favourable conditions for their subordinates so as to prevent them conducting any violations on their own. Managing activities of colleagues and other individuals as well as cooperating with legal entities and individuals in carrying out business activities, further helps at reducing violations and corrupt behavior or decisions that might be taken. In other words, managers should create a system of psychological, spiritual and material incentives that encourage each employee to think and act in the interests of the company, observing the restrictions and prohibitions, fulfilling the obligations established in the labour contract and other local regulations on anti-corruption issues.

The list of management responsibilities in the local acts of the company, as well as in the employment contract, must include participation in the development and implementation of anti-corruption measures, which are to prevent and overcome corrupt behaviour, corruption-dangerous situations, and their consequences. Any situation in the employees' activities that leads to the possibility of violating the prohibitions and restrictions specified in the local regulatory acts of the company that set up the requirements for employee behavior in accordance with Russian law can be attributed to a corruption-dangerous situation.

The study showed that preventive anti-corruption measures are more difficult to implement in a company's activities than punitive measures. However, the practice of implementing anti-corruption measures has proved that if the management is organised, they will be more effective and efficient, and will achieve high results.

The main areas of company management in the prevention of corrupt behaviour and the prevention of corruption and other offenses include:

- the organization of ensuring compliance by all employees of the company with requirements to prevent and resolve conflicts of interest,
- implementing organisational, administrative, explanatory and other measures aimed at attracting company employees to participate in anti-corruption measures,
- ensuring the access of interested persons to information on the activities of a company that is not a commercial secret, especially in matters of preventing and combating possible corrupt offenses,
- strengthening control over the appeals of employees and partners on corruption issues, as well as decisions on them,
- the organisation of measures to increase the responsibility of employees for not taking measures to eliminate the causes of corruption in accordance with the anti-corruption policy of the company.

The main responsibility of management at any level is to prevent employees from corrupt behavior and the possible offense in the company's divisions, as well as to ensure the activities of the divisions are in compliance with the established restrictions, prohibitions, obligations, and labour rules. It is necessary to exclude a formal approach to the implementation of the anti-corruption policy of the company. The management should be interested in the direct and full implementation of the company's anti-corruption raft, especially in the prevention of corruption.

It is worth paying attention to the role of HR departments. These units develop and implement educational and enlightening work that is preventive in the fight against corruption. They also carry out activities aimed at both legal anti-corruption education and the formation in their minds of intolerance to corruption in the performance of their duties and at home. Preventive discussions should be held on the inadmissibility of corruption, the procedure for a notification of incidents and of incitement to commit them.

When developing and communicating the following information, the specific circumstances of the relationship where corruption risk may occur should be considered:

- specific, inherent in their field of activity, knowledge,
- information distribution channels,
- depth and methods of supplying educational materials.

For this reason, the completeness and frequency of the anti-corruption information provided should differ significantly. It is proposed that workers be divided into three groups. Newly hired employees can be offered adaptation programs for teaching legal knowledge; to experts - a knowledge base equipped with tools for interaction, as well as a newsletter informing all participants about changes in the knowledge base; and, finally, to all employees - online cases based on real events, including practical tasks, incidents. Such a separation is logical, since the legal field and the information environment on anti-corruption education is an extensive material that changes quite dynamically. As a preventive measure, the company's management must inform the subordinates of cases that entail the application of disciplinary measures against their colleagues for non-compliance with anti-corruption laws.

4 Discussion

From a psychological point of view, corruption is interpreted as a deviant behaviour, that is, it violates the role of members of society under the direct influence of private interests. There are three main aspects of the occurrence of corruption: tolerance (the attitude as a normal phenomenon ("everyone takes bribes"), which cannot be eradicated and avoided); expressed condemnation not of the fact of corruption itself, but only of the prohibitive size of bribes; inconsistency.

In this case, inconsistency is a system of double standards. That is, their own corrupt behaviour, as well as the corresponding behaviour of loved ones, is perceived as a forced response to circumstances arising in the external environment. This behaviour has nothing to do with deviant behaviour, so it will not be negatively evaluated by emotions. On the other hand, a similar behaviour of third parties is considered negative and is accompanied by censure by the individual.

All three aspects are reflected in the company's anti-corruption policy, as well as the conflict of interest policy, and have a direct relationship. In addition, in order to meet the legal and ethical requirements of owners, management, employees, partners, customers and society as a whole, it is necessary to comply with national and international legislative standards. Thus, the issue of corruption has become a familiar agenda for the companies' board of directors. An effective anti-corruption program not only helps to prevent the implementation of corruption risks and other abuses in the short and long term, but also serves as the basis for reducing sanctions when violations occur.

Returning to the issue of leveling the risk of corruption by increasing the level of corporate culture, it is worth noting the positive impact on the part of personnel security and its motivational component, which form a working atmosphere of honesty, openness and mutual assistance. Such an atmosphere is aimed at people who are not corrupt (about 80% of the staff), but can commit atrocities under the influence of external circumstances. Creating a microclimate to prevent fraud in the team is the main goal of the person responsible for risk management and the company's violations and the internal control system. The management's promotion of anti-corruption behaviour leads to a number of serious effects in the company's activities (Table 4).

Table 4. The relationship of factors and indicators of economic efficiency

Element of anti-corruption behavior	Directions of influence (effects)	Performance indicator
High responsibility on the part of the heads of divisions and operational personnel for the implementation of the goals and objectives "from the top"	Improving the overall level of management quality, improving KPI	Strengthening personal and collective position to improve the efficiency of the company
Understanding and recognition by management of the key role of corporate culture in the implementation of goals and objectives	Increasing team coherence, the complexity of management and arrhythmia in business processes	Productivity growth
A clear position of the company in relation to any fact of corruption behavior and awareness of the inevitability of penalties by management and staff	Decrease in factors of motive (pressure), possibilities and justification of illegal actions, negative consequences of violations	Reducing the facts of abuse and misconduct, reducing complaints and increasing the efficiency of the company
Improving internal control systems	Eliminating deficiencies in corporate control systems and reducing the level of openness for illegal actions in someone's favor	Productivity growth
Documenting rules and regulations for managing corruption risks	Strengthening control systems and reducing the degree of staff ignorance of what is allowed and what is forbidden to do	Reduce violations, reduce complaints and increase efficiency

Source: authors.

The objectives of the in-depth examination are to verify the validity of the economic calculations and forecasts contained in the investment proposal, the correctness of calculations, as well as to verify the estimates of the preliminary examination. In-depth examination is performed by one or more experts from the expert group appointed by the members of the expert council. If necessary, a special examination is carried out with the involvement of external specialists. In the conditions of digitalisation, it is possible to build a sufficiently reasonable decision-making algorithm for optimising the inclusion of the project in the investment program. If the evaluation of individual criteria was adjusted during the in-depth examination, a new offer rating is calculated. Proposals that have passed an in-depth examination are considered for a meeting with the expert council.

The atmosphere of integrity, openness and honesty is formed through the creation of mechanisms to counter corruption schemes in all aspects of activity, the adoption and observance by all employees of the "code of conduct", and, first of all, the company's management, which counteracts the formation of ideas of self-justification for those who are inclined and show corrupt behaviour in their activities.

When implementing the above measures in the activities of an economic entity, the following positive consequences can be expected:

- increasing the reputation of the company,
- improvement of key performance indicators of the company (increase in net profit, profit from sales, revenue or savings on reducing elements of costs of sales),
- building long-term partnerships with all interested parties.

5 Conclusion

The role of the management of each division of the company is not only to manage, but also to provide the most favourable conditions for their subordinates in order to motivate them in preventing any offenses on their own, and within the framework of the work of colleagues. Top management is recommended to refer to the company's employees to remind them of the need to comply with the company's anti-corruption policy, and, if necessary, hold appropriate meetings. First of all, they should be carried out in connection with the assessment of corruption risks. "Tone from the top", that is, management's commitment to a "culture of compliance", not only on paper but also on a personal example. It is very important to maintain such a culture throughout the company. In other words, managers must create a system of psychological, spiritual, material incentives that encourages each employee to think and act in the interests of the company, observing restrictions and prohibitions, fulfilling the obligations established by the contract and local acts on combating corruption, including preventing conflicts of interest.

References

1. Cherepanova, V.A.: Combining efforts against corruption: directions for the development of regulation in Russia. Implementation of best anti-corruption practices and the exchange of experience in combating corruption in Russia. INFRA-M, Moscow (2018)
2. Decree of the President of the Russian Federation: On the National Anti-Corruption Plan for 2018–2020 (2018). http://www.garant.ru/products/ipo/prime/doc/71877694/#ixzz5ckS88Qj1. Accessed: 22 June 2020

3. Deloitte: the conceptual framework for risk management COSO (2018). https://www2.deloitte.com/content/dam/Deloitte/ru/Documents/risk/russian/rules-of-game-changing.pdf. Accessed: 22 June 2020
4. Dobrynina, A.K.: Anti-corruption compliance as a necessary condition for the development of Russian business. Young Sci. **14**(148), 350–353 (2017)
5. Feshina, S.S., Konovalova, O.V.: Prospects of counteracting petty corruption in the digital economy in Russia. Econ. Taxes Law **11**(5), 92–99 (2018)
6. ISO 19600: Compliance management systems (2014). https://www.iso.org/ru/standard/62342.html. Accessed 22 June 2020
7. ISO 37001: Anti-bribery management systems (2017). https://www.iso.org/iso-37001-anti-bribery-management.html. Accessed 22 June 2020
8. Klitgaard, R.E.: Controlling Corruption. University of California Press, California (1988)
9. Law No. 2016-1691: On Transparency, Fight against Corruption and Modernisation of Economic Life (2016). https://www.legifrance.gouv.fr/eli/loi/2016/12/9/2016-1691/jo/texte. Accessed 22 June 2020
10. Nikolaev, D.A.: The impact of the shadow economy on international capital flows under modern conditions. Econ. Taxes Law **11**(5), 108–117 (2018)
11. Prasolov, V.I., Minkina, D.A.: Analysis of corruption perception index dependence on GDP and level of investment in Russia. Mod. Sci. **5**, 289–292 (2019)
12. PWC: Anti-Bribery and Corruption Index (ABCI) in Russia – 2017 (2017). https://www.pwc.ru/en/publications/anticorruption-2017.html. Accessed 22 June 2020
13. PWC: Russian economic crime survey (2018). https://www.pwc.ru/ru/publications/recs-2018.html. Accessed 22 June 2020
14. PWC: What measures are companies taking? A study of economic crimes in Russia (2018). https://www.pwc.ru/ru/forensic-services/assets/PwC-recs-2018-rus.pdf. Accessed 22 June 2020
15. Sharova, S.A.: Anti-corruption Regulation in Russia and the USA. Law, Moscow (2016)
16. Transparency: Corruption perception index (2020). https://transparency.org.ru/research/indeks-vospriyatiya-korruptsii/. Accessed 22 June 2020

Intellectual Capital and Its Role in the Development of the Company

O. Y. Kuzmina[1(✉)], M. E. Konovalova[1], and A. V. Larionov[2]

[1] Samara State University of Economics, Samara, Russia
pisakina83@yandex.ru, mkonoval@mail.ru
[2] LLC New City, The Official Representative of FINAM in Samara, Samara, Russia
alarionov@corp.finam.ru

Abstract. The study considers the process of forming the intellectual capital of the company. Special attention is paid to the evolution of the theory of the intellectual capital to clarify the content of this socio-economic phenomenon. Various scientific approaches to the definition of the intellectual capital and its structure are studied, its main features are identified, the conditions and factors for the formation of the intellectual capital of the company are determined. The authors analyze various methods for assessing the intellectual capital, identifying the advantages and disadvantages of existing approaches to calculating the intellectual capital of the company. Due to the lack of a single calculation variation, some criteria are proposed that ensure the selection of the most productive method for assessing the intellectual capital of the company. Having studied institutional conditions for the formation of the intellectual capital in the modern economy of Russia, the authors identify priority measures to improve the Russian institutional environment to develop the intellectual capital.

Keywords: Efficiency · Human capital · Intellectual capital · Investments · Q-Tobin coefficient

1 Introduction

At this stage of economic development, the intellectual capital is equated by specialists with the number of key economic categories, factors of production, strategic development assets of subjects of post-industrial, innovative economy [5]. Under the influence of post-industrialization, innovative development, it is obvious that intelligence, information, knowledge, computer and information technologies have become the basis of competitive advantages for subjects of all levels of the economy. Economists argue that in the 21st century the basis of competition will not depend on the optimal use of traditional factors of production, but on the consumption of the human mind, information technology and accumulated knowledge presented in the form of intellectual property [4, 8].

The transformation of innovation into a key source of economic growth and competitiveness of an organization, the increasing role of the intellectual capital in ensuring the efficient operation of companies are important trends in the economy.

S. I. Ashmarina and V. V. Mantulenko (Eds.): IPM 2020, LNNS 161, pp. 713–719, 2021.
https://doi.org/10.1007/978-3-030-60926-9_88

In other words, to maintain their competitiveness and effectiveness, firms must carry out active innovation and build up their intellectual capital.

2 Methodology

The traditional methods used to assess tangible assets when studying the intellectual capital of the company, the most important property of which is “immateriality”, are not applicable. When choosing existing methods for assessing the intellectual capital, it is necessary to consider the goals, the availability of available information, the level of assessment and the fact who will receive the final information. It should be noted that at this stage in the development of economic science there is no single methodology for assessing the intellectual capital. Existing methods for assessing the intellectual capital can be combined into four groups based on whether they are used at the corporate level or in relation to individual elements of the intellectual capital and considering the monetary or non-monetary nature of the assessment [11]. The characteristics of each group are presented in Table 1.

Table 1. A brief description of methods for assessing the intellectual capital

Methods for assessing the intellectual capital	Result of assessment	Applied level of assessment
Direct Intellectual Capital methods	Financial	Assessment of individual items
Market Capitalization methods	Financial	Corporate
Return on Assets methods	Financial	Corporate
Scorecard methods	Non-financial	Assessment of individual items

Source: authors.

The first group includes methods of direct assessment of the intellectual capital - Direct Intellectual Capital methods (DIC). Together, these methods focus on the monetary assessment of each component of the firm’s intellectual capital, followed by a general integrated assessment of this indicator. Finding the latter should not occur by simply adding up cost parameters, it is a more complex assessment of the synergistic interaction of various parts of the intellectual capital with each other. However, in practice it is very difficult to do. There are methods that allow us to assess the contribution of each component to the value of the intellectual capital. Such methods are: broker technology, FiMIAM, the cost of the invention. Using these methods, we ensure transparency and validity of the sources of formation of the intellectual capital.

The second group is Market Capitalization methods (MCM). These methods allow you to calculate the total value of the intellectual capital of the company in monetary terms. These methods use the assumption that the real value of the company corresponds to its market value. Accordingly, if we subtract its tangible assets from the market price of the company, then intangible assets remain – the intellectual capital [1].

The resulting indicator is often interpreted as goodwill, i.e. the amount more than the value of the tangible assets of the company, which is paid upon its acquisition.

Return on Assets methods (ROAM), belonging to the third group of methods, allow you to assess the intellectual capital of the company at the level of the entire organization using standard financial indicators, namely, by comparing the company's ROA and the ROA of the entire industry. Using this method, the value of the intellectual capital can be found as the difference between the value of the entire company, which is the sum of the book value of all its assets and the discounted stream of residual operating profit, and the value of tangible assets. The latter is determined by the carrying value of these tangible assets and the discounted residual income stream corresponding to the average rate of return for the industry. This approach allows companies to be compared both within the same industry and those operating in different industries. The most popular methodology in the framework of Return on Assets methods is the intellectual value-added coefficient. The basis of this coefficient is the idea to assess the effectiveness of the value of the company, not intangible assets.

Return on Assets methods most accurately and reliably allow you to assess the intrinsic value of the intellectual capital. Nevertheless, it should be borne in mind that in the long run the results are not always fair in relation to the real state of affairs: statistically significant dependencies weaken over time, uncertainty increases, and overall market behavior does not fall under the laws of normal distribution [3].

The disadvantages of this method are the following features. ROA is calculated based on accounting data that does not always reflect real economic efficiency. In addition, this method displays the value of intangible assets of the company, which causes difficulties in analyzing the value of individual components of the intellectual capital. There may also be a problem in determining the sample of companies for the study, especially if you are not limited to any industry.

Another group of methods used to assess the intellectual capital are Scorecard methods (SCM). The principle of their use is based on identification of various components of the intangible assets or the intellectual capital, which are then generated into indices and indicators by scoring. In this case, it is not intended to receive a monetary value of the intellectual capital. The analysis can use both financial and non-financial inputs. This allows you to display hidden factors that affect the value of the company. This group of methods includes the largest number of methods. The most famous of them are the following: Balanced scorecard, Intangible asset monitoring, Scandia Navigator, National intellectual capital index and others.

SC methods allow you to determine the potential and ability of the company to develop. They also contribute to increasing the effectiveness of strategic management of the intellectual capital. As part of the study, various approaches to calculating the intellectual capital of the company were implemented and it was concluded that their choice largely depends on the size of the company and its industry sector.

3 Results

A comprehensive analysis conducted in the study allows us to conclude that the scientific concept of the intellectual capital is in the process of formation. There are many definitions of the intellectual capital, which is explained by two points. On the one hand, the category of the intellectual capital has penetrated various fields of scientific knowledge, sociology, psychology, economics, and, therefore, its content is filled with the scientific context from which it is interpreted. On the other hand, the phenomenon under study is constantly evolving, socio-economic conditions of life are changing, and the intellectual capital is "expanding", including all new elements. Within the framework of the generalized interpretation, the category of the "intellectual capital" can be considered as a logical stage in the evolution of society, a form of manifestation of a person's creative abilities in the information economy, the most important aspects of which are new knowledge, competitive behavior, flexibility, information mobility, know-how, advanced decision-making methods which are accumulated, capitalized and applied to ensure sustainable innovative growth of the company [2, 10].

The process of formation of the human capital is fundamentally influenced by the system of national education and the policy of the company in relation to the additional professional education of its employees. The organizational capital is formed under the influence of such an external factor as the level of development of the entrepreneurial culture in the country, the main internal factor is the level of readiness for innovations in organizational business processes of both the company staff, its directors, and top management [6]. The formation of the client (relative) capital is greatly influenced by the company's goodwill, which is supported by effective partnerships, the clients' predisposition to the company's brand, and many other factors that shape the company's image.

When choosing existing methods for assessing the intellectual capital, it is necessary to consider the objectives of the evaluation and the level of information availability. The creation of the universal method for assessing the intellectual capital will eliminate the problem of its accounting both from the investment point of view and from the position of accounting. Unfortunately, at this stage, there is no single methodology for assessing the intellectual capital. It is necessary to apply the above methods that exist within the framework of the above four groups of methods in accordance with the purposes of the study and the conditions for its implementation to obtain correct results. So, for example, for such large enterprises as PJSC MMC Norilsk Nickel, two models for assessing the intellectual capital are effective: Market over Book Value model and q-Tobin coefficient. The advantages of Market Capitalization methods are simplicity of obtaining data from financial statements of companies whose shares are traded on the stock exchange, and relative simplicity of the analysis, given that these methods are not expensive.

Both applied models gave the same result, according to which it can be concluded that PJSC MMC Norilsk Nickel effectively uses its intellectual capital, which demonstrates a steady upward trend. Over the entire period under review, market capitalization of PJSC MMC Norilsk Nickel significantly exceeded its book value, which made the intellectual capital a positive value. From 2014 until 2018, there was a gradual increase in both the company's capitalization and its intellectual capital. Let us assess the intellectual capital of PJSC MMC Norilsk Nickel using q-Tobin coefficient (q). Q-Tobin coefficient is calculated by the formula:

$$q = K/A, \quad (1)$$

where q is Tobin coefficient; K - market capitalization of the company; A - book value of the company.

The data for the calculation and calculation results are presented in Table 2.

Table 2. Calculation of q-Tobin coefficient, in million rubles

Indicator	2014	2015	2016	2017	2018
Market price	1278623	1447946	1601760	1716963	2068743
Book value	469324	528970	632576	557353	728880
Q-Tobin coefficient	2,72	2,74	2,53	3,08	2,84

Source: authors.

Q-Tobin coefficient of more than one indicates the efficiency of the company using its intellectual capital, and the higher it is, the better, at least from the point of view of the company's investment attractiveness. If this ratio is greater for the company than for the closest competitors in the industry, this shows that using its intellectual capital, the company generates more profit and is more innovative. If the coefficient is less than unity, then it cannot be argued that the company does not have intellectual capital, we can only conclude that the company uses its intellectual capital inefficiently, or that the company is underestimated in the market [9]. It is established that, on average, q-Tobin coefficient is quite stable over time, and companies with its high value usually produce unique products or have unique production factors.

Turning to Table 2, we can state that q-Tobin coefficient of PJSC MMC Norilsk Nickel is of constant value. We can conclude that market capitalization reflects assets that are not recorded in the balance sheet of PJSC MMC Norilsk Nickel, and the level of the organizational structure contributes to the growth of capital, rather than its decrease. In 2016, q-Tobin coefficient decreased in comparison with the same indicator in other years. This can be explained by the fact that this year the book value of PJSC MMC Norilsk Nickel increased by 20% compared to the previous period. Such dynamics is due to a rather high share of non-current assets in the total assets of the company.

4 Discussion

At the present stage of development of economic relations, there is a limitation in the ways of using traditional resources and usual sources of economic growth. These restrictions occur in the system in which the driving force of economic development is economic interests of agents seeking to defend their position. Considering these circumstances, it becomes relevant to ensure economic growth using resources such as information, knowledge, because they are limited only by the formation environment. The effectiveness of these resources is possible only when forming special institutions that ensure enforcement mechanisms. However, institutions do not always ensure realization of economic interests of individuals, creating institutional traps. Acting as a centrifugal force, economic interests can disrupt the stability of the economic system, while institutional interests are concentrated on equilibrium and balance and set the system centripetal development vector, form the rules of behavior of business entities, organize them, control and bring the system to rest and equilibrium.

The interconnection of economic relations and the institutional environment excludes their confrontation, forms the basis of any economic process, including the processes of formation and reproduction of the intellectual capital. Speaking about institutions involved in the formation and development of the intellectual capital, their features should be noted. The relationship between the intellectual capital and institutions of its formation is dual. On the one hand, knowledge is a thorough characteristic of all types of institutional relations, on the other hand, all stages of its reproduction are institutionalized. In other words, institutions influence the production of knowledge, and new knowledge affects the development of institutions by modifying them. It follows from this that knowledge as a means of production is necessary for the formation of new knowledge. Being the main component of the intellectual capital, past knowledge affects all aspects of reproduction of the intangible assets of the company [7]. But for the implementation of these processes, functional formal and informal institutions are needed that will create the conditions for generation, transition and use of new knowledge in production and the formation of a new innovative economy.

The main condition for the effective functioning of the intellectual capital is the optimal state of the institutional environment in terms of density and clarity in building its hierarchical structure for the effective environment. The Russian state of institutions responsible for the development of the intellectual capital is in an inefficient state and needs reforms to build the innovative economy in which the intellectual capital is the main element. First, it is necessary to carry out reforms in institutes of science and innovative research, since they act as the main generator of the innovative potential of the country. Attention should also be paid to institutions supporting innovative projects and venture investment institutions. It is necessary to create conditions under which motivation of agents to participate in the creation and completion of the final innovative product will grow. It is proposed to increase motivation by reforming the tax system as a way of state support in the field of innovation, thereby stimulating an increase in the number of innovative projects, which in turn will be regulated by specialized institutions for monitoring innovation.

5 Conclusion

In conclusion, it is important to note that in our country special attention should be paid to the process of creating conditions for the transition to the innovative economy and its further development, the formation of the intellectual capital, which will determine the structure of the economy, and will contribute to an increase in the quality of products and services, as well as will increase the efficiency of the economy at all its organizational levels. The degree of development of intellectual labor and its participation in production processes are the most important factors determining the competitiveness of both the country and individual economic entities.

References

1. Asiaei, K., Jusoh, R.: Using a robust performance measurement system to illuminate intellectual capital. Int. J. Acc. Inf. Syst. **26**, 1–19 (2017)
2. Barrena-Martínez, J., Cricelli, L., Ferrándiz, E., Greco, M., Grimaldi, M.: Joint forces: towards an integration of intellectual capital theory and the open innovation paradigm. J. Bus. Res. **112**, 261–270 (2020)
3. Grant, R.: Toward a knowledge-based theory of the firm. Strateg. Manag. J. **17**, 109–122 (1996)
4. Kianto, A., Sáenz, J., Aramburu, N.: Knowledge-based human resource management practices, intellectual capital and innovation. J. Bus. Res. **81**, 11–20 (2017)
5. Lee, C.-C., Lin, C.-K.: The major determinants of influencing the operating performance from the perspective of intellectual capital: evidence on CPA industry. Asia Pac. Manage. Rev. **24**(2), 124–139 (2019)
6. McDowell, W.C., Peake, W.O., Coder, L.A., Harris, M.L.: Building small firm performance through intellectual capital development: exploring innovation as the "black box". J. Bus. Res. **88**, 321–327 (2018)
7. Pınar, S., Fındıklı, M.A., Köse, A.M.: The mediating roles of solidarity and intellectual capital on the relationship between resource dependency sub-dimensions and innovation performance. Procedia Comput. Sci. **158**, 557–564 (2019)
8. Roslender, R., Nielsen, C.: Lessons for progressing narrative reporting: learning from the experience of disseminating the Danish Intellectual Capital Statement approach. Acc. Forum **41**(3), 161–171 (2017)
9. Secundo, G., Ndou, V., Vecchio, P.D., Pascale, G.D.: Sustainable development, intellectual capital and technology policies: a structured literature review and future research agenda. Technol. Forecast. Soc. Chang. **153**, 119917 (2020)
10. Steenkamp, N., Kashyap, V.: Importance and contribution of intangible assets: SME managers' perceptions. J. Intell. Capital **11**(3), 368–390 (2010)
11. Sveiby, K.E.: The intangible assets monitor. J. Hum. Resource Costing Acc. **2**, 73–97 (1997)

Public Investment in Human Capital as a Factor of Social Progress

A. Kh. Malikova(✉), J. Z. Malikov, and K. A. Temir-Bulatov

Samara State University of Economics, Samara, Russia
malikova_ax@mail.ru, malikovjz@gmail.com,
kamtemirbulatv@rambler.ru

Abstract. The relevance of the problem under study is due to the importance of high quality of human capital for the formation of a strong, economically developed, civilized state. In this regard, this article is aimed at revealing the overdue understanding of the limited natural resources, which led to an increasing need for human capital development. The leading method for the study of this problem is the general scientific method of knowledge, which allowed to comprehensively justify the thesis that the degree of efficiency of public investment in human capital, primarily in the provision of education, primarily affects the development of socio-economic state as a whole. The article reveals the legislation of the Russian Federation in the field of education, presents the author's consolidation with the opinion on the systematization of national legislation in this area, in particular, the possibility of adopting the Federal Law "On Education in the Russian Federation", which would absorb the entire array of normative educational law. Materials of the article are of practical value for law makers, politicians, researchers in the field of development of state and legal phenomena.

Keywords: Education · Human capital · Investment · Socio-economic state

1 Introduction

In recent years, according to the United Nations Development Programmes (UNDP) on Human Development Report 2019, Russia ranked only 49th among 189 countries in the world [15]. The Russian education system can take its rightful place on the world market of educational services by modernizing them while maintaining the best traditions of Soviet education. The Soviet Union, the world's most scientific civilization, was the first and in fact the only society in the 20th century where high culture became the culture of the masses. This phenomenon is directly related to the formation of a universal system of compulsory, free education and enlightenment, and the ideology of consistent universal accessibility of any educational level and direction.

As early as in the 19th century, von Stein, a German philosopher, economist, lawyer, and author of the concept of a social state, repeatedly emphasized in his works that the first major step that the state must take in order to eradicate poverty, and homelessness is the introduction of compulsory, free, "at least elementary mental education" [16, p. 21]. At the present stage, intangible resources such as knowledge,

S. I. Ashmarina and V. V. Mantulenko (Eds.): IPM 2020, LNNS 161, pp. 720–724, 2021.
https://doi.org/10.1007/978-3-030-60926-9_89

skills and work skills are also crucial for economic development. An example is the economies of Europe, Japan and other developed countries. With low levels of natural resource endowment, the high level of national wealth in these countries is due to the development of human capital. In recent years, therefore, the functions of the social state have expanded significantly in advanced states. Above all, this concerns the increase in expenditure on human development, which is based on education.

The great scientist categorically noted that there is no doubt that education has the highest price for any man, and it is a condition and a consequence of any success, the size and depth of its essence measure and value of any individual person [16]. It is now recognized that education is not a guarantee of success in life, but it has a significant impact on its probability and, having both symbolic and practical significance, to a large extent determines the increase of opportunities, raising the status of citizens. The more educated an individual is, the more opportunities he or she has to improve his or her own status, that of his or her family and that of the state as a whole. The realization of the right to education by each individual is an assessment of one of the most important indicators of the state of society as a whole. The problem of access to education is inextricably linked to the problem of its quality.

Modern scientific research indicates that the level of education affects a person's health more than his or her standard of living or position in society. Education brings more benefits to a person in terms of health, life expectancy than a good, well-paid job, as educated people generally tend to make better decisions about their behaviour and lifestyle. Von Stein also pointed out that people need to be highly educated in order to recognize the importance of health and to fulfill their demands [16]. By increasing spending on education, ensuring its accessibility and quality, the state protects and ensures the health of its citizens.

2 Methodology

The following methods were used in the research process:

- universal methods, in particular metaphysics (considering the state and law as eternal and immutable institutions deeply disconnected from each other and other social phenomena) and dialectics,
- general scientific methods, which include analysis, synthesis, induction, deduction, system approach, functional approach, analogy, abstraction, historical method,
- private scientific (or special) methods: statistical, cybernetic, psychological methods,
- private legal methods: formal legal, comparative legal, legal forecasting.

3 Results

The importance of education in public life and the human right to education is reflected in the attention paid to it by the entire world community. Thus, the Unitied Nations has declared the third millennium era of education. And this is understandable because, according to the UNDP's estimates based on data from 192 countries, 64% of the world's wealth is human capital, 21% - physical capital, and only 15% - natural resources [15]. The provision on the increasing role of human capital as the main factor of economic development is also set forth in the Concept of long-term social and economic development of the Russian Federation for the period until 2020 [2].

The level of efficiency of state investments in human capital, primarily in education, has a primary impact on the development of the socio-economic state as a whole. Given that the function of the state to provide education has a higher value, the state enshrines and guarantees it in legal acts. The legal basis for the implementation of the function of the state to provide education in Russia is its Constitution, which proclaims in Article 43: Everyone has the right to education. General availability and free of primary education [3].

The legislation of the Russian Federation in the field of education includes the Federal law "On Education in the Russian Federation" from 29.12.2012 № 273-FL [5], the Federal law "On Science and the State Scientific and Technical Policy" from 23.08.1996 № 127-FL [6], other laws and other standard legal acts of the Russian Federation accepted according to them, laws and other standard legal acts of subjects of the Russian Federation in the field of education, and also the international law ratified by Russia which is a part of its legal system, first and foremost, International covenant on economic, social and cultural rights [7].

Education is a prerequisite for an innovative economy. The world experience shows that the higher the level of education the population has, the higher the quality of life it aspires to. "Educated, not indifferent people always act as generators of ideas that benefit not only themselves but also others, which makes it possible to form new partnerships in the business that ensures the growth of the quality of life" [11, p. 2]. The scientific and technical potential of any country is the most important resource, one of the bases of industrial development, renewal of social and scientific life. Undoubtedly, the more the state spends on research and development, the higher the level of its economic growth and living standards are.

4 Discussion

The growth of society's education contributes to the development of the socio-economic state. The provision of educational opportunities contributes not only to the free development of everyone in the state. Without quality education, highly qualified specialists, universal employment, it is impossible to achieve material prosperity for everyone and high economic development rates of the state as a whole. We concur with the position of scientists who consider human capital to be one of the main resources of any country, the main source of national wealth, which determines its market stability, efficiency of operation, the prospects of strategic development [14], the ways to increase which is formal education and accumulation of practical experience [9]. Given the low level of natural resources availability, the high level of national wealth in European countries is primarily due to adequate development of human capital. Modern competition is won by "skill rather than number" [10]. Expenditure on social development, especially education and basic research, is considered a highly effective investment in human capital. With the formation of a society based on global knowledge, national intellectual capital becomes the basis of economic well-being, a factor of political power for a country. In this regard, the optimal solution of strategic tasks of statehood, including ensuring a decent life and free development, is possible only through the use of the intellectual potential of its society.

5 Conclusion

Educational service is not the sphere of exclusively public-power structures. It may be provided by private individuals and non-governmental corporations [13], however, education should be primarily subject to state paternalism. This sphere, at least at the present stage, should be freed from market mechanisms. A society in which market relations dominate in all spheres and not only in the sphere of commercial activity is a 'bazaar', not a 'socium'. Only the economy should have a market nature [4]. For the sake of the development of the economy, for the benefit of the whole society, one should not give education, science, healthcare, art, and culture to the undivided power of the market.

Education must be made accessible to all, as it is one of the main social values - the intellectual potential of society. A social state must create conditions for education for as many of its citizens as possible. Digital, distance education can help to learn everywhere, everything and from anyone. The activities of public authorities in the field of education, which aim at its commercialization and thus destroy the foundations of physical, mental and moral development of citizens, are ultimately harmful to the whole state. Ensuring individual access to quality education regardless of social and property status, territorial or other differences will not only help to resolve possible future social conflicts, but also address issues of national security, state stability and state authority in the world system [8]. An approach to the education system based on a misunderstanding of its essence, role and place in the process of market-based economic reforms can be a major risk. In no case should the education system be turned into a direct market economic mechanism. It is also unacceptable to save money in this area [11].

Investments in human capital today undoubtedly include expenses for basic research and development. Social development has always been linked to scientific advances [1]. The state, capable to carry out the function of implementing the results of scientific research, gives the country the opportunity to develop and create modern highly productive mechanisms, develop technology, infrastructure, if it concerns the results of scientific research in the field of law, improve the legislation of the country, which is a priority in the formation of a strong, economically developed, civilized state.

Even von Stein noted that public investment in human capital could be richly rewarded, because it reduces sources of poverty, lack of order and education, and that reduces the tax on the poverty [16]. Strengthening the influence of the state in education, giving science and education a high state and public status, a significant increase in budget financing and investment in human capital are now necessary conditions for building a strong state with an effective system of public administration, with a modern army, with a highly developed high-tech economy [12]. Based on the foregoing, there is a need to adopt the Federal Law "On Education in the Russian Federation", which would absorb the entire array of normative educational law [5]. This law should set forth mandatory state standards of education, including the possibility of realizing the right of everyone to modern education, because state, accessible, high-quality education is the main condition for improving the living standards of citizens, leading to the progress of modern society.

References

1. Cobb-Clark, D.A., Salamanca, N., Zhu, A.: Parenting style as an investment in human development. J. Populat. Econ. **32**(4), 1315–1352 (2018). https://doi.org/10.1007/s00148-018-0703-2
2. Concept of long-term social and economic development of the Russian Federation for the period until 2020 (2018). http://www.consultant.ru/document/cons_doc_LAW_82134/28c7f9e359e8af09d7244d8033c66928fa27e527/. Accessed 22 June 2020
3. Constitution of Russian Federation from 12.12.1993 (1993). http://www.constitution.ru/. Accessed 22 June 2020
4. Diniz, F., Leitão, H.V.: Entrepreneurship and social innovation in training and human capital development: The case of the bank palmas. Econ. Region **12**(3), 865–874 (2016)
5. Federal law "On Education in the Russian Federation" from 29.12.2012 N273-FZ (2012). http://www.consultant.ru/document/cons_doc_LAW_140174/. Accessed: 22 June 2020
6. Federal law "On Science and the State Scientific and Technical Policy" from 23.08.1996 № 127-FZ. http://www.consultant.ru/document/cons_doc_LAW_11507/. Accessed 22 June 2020. (1996)
7. International covenant on economic, social and cultural rights (1976). https://www.ohchr.org/EN/professionalinterest/pages/cescr.aspx. Accessed 22 June 2020
8. Kalugina, Z.I.: Deficit in human potential as a threat to modernization. Regional Res. of Russia **2**(3), 225–233 (2012)
9. Kascheev, P.O., Belgorodskiy, V.S., Radko, S.G.: Human capital as an integral factor in the growth of labor productivity. Izvestiya Vysshikh Uchebnykh Zavedenii, Seriya Teknologiya Tekstil'Noi Promyshlennosti **3**, 15–22 (2018)
10. Laplane, A., Mazzucato, M.: Socializing the risks and rewards of public investments: Economic, policy, and legal issues. Res. Policy X **2**, 100008 (2020)
11. Nosyk, O.M.: Capital formation function of subjects investing in human capital: the framework of economic choice. Actual Probl. Econ. **164**(2), 60–66 (2015)
12. Malikova, A.K.: Problems of forming a social state in Tajikistan. Donish, Dushanbe (2013)
13. Ryazanova, O.E., Nazarova, E.V.: Foreign experience of social investment for usage and development of human capital. J. Adv. Res. Econ. **8**(1), 186–191 (2017)
14. Shafuda, C.P.P., De, U.K.: Government expenditure on human capital and growth in Namibia: a time series analysis. J. Econ. Struct. **9**(1), 21 (2020)
15. United Nations Development Programmes: Human development report 2019 (2019). http://hdr.undp.org/sites/default/files/hdr2019.pdf. Accessed 22 June 2020
16. von Stein, L.: Teaching about management and the law of management, comparing literature and laws of France, England and Germany. Printing House of V.V. Pratz, St. Petersburg (1884)

Personnel Analysis of Municipal Authorities

E. S. Nedorezova(✉), K. N. Ermolaev, and F. F. Salamov

Samara State University of Economics, Samara, Russia
nedlen63@yandex.ru, ermolaevkn@yandex.ru,
farrux_sies@mail.ru

Abstract. Today, the state civil and municipal service is the central link in improving the efficiency of the public administration system. One of the most important directions to reform the municipal service is the formation of an effective personnel policy, since the level of implementation of government decisions depends on professional employees who have necessary qualities of the public service, and the effectiveness of their official activities. The solution of the problems associated with the formation and development of the personnel of the municipal service under current economic and institutional conditions variability, justifies the topic of the study and its relevance. The formation of the personnel reserve in the state civil service has some difficulties. They are associated with a lack of regulation at the legislative level. This applies, for example, to the procedure for preparing the personnel reserve, entering the state civil service out of competition, the time spent in the reserve, and holding a competition.

Keywords: Municipal authorities · Personnel reserve · Personnel management

1 Introduction

The formation of the personnel reserve of a municipal organization often causes difficulties, which are most often associated with employee discontent due to a lack of information regarding regulatory regime, the absence of competitive procedures in regulatory legal acts, and insufficiently serious attitude of managers to the selection of candidates [2]. At the federal level, the issue of allocating funds for the training of candidates has not been resolved, which does not provide an opportunity for development. This issue can be solved by attracting candidates to various events, for example, city competitions, festivals, and so on. It is necessary to highlight one more problem - the participation of the personnel service in the formation of the base of candidates for civil service. The personnel department sometimes does not have a clear understanding of professional activities of other structural units, which leads to a poor selection of candidates at the initial stage. Excessive bureaucratization for the formal procedure or, on the contrary, the lack of understanding of tasks at each stage of the formation introduces employees who do not have enough knowledge or professional qualities. HR officers must be both highly competent and ready for innovation.

S. I. Ashmarina and V. V. Mantulenko (Eds.): IPM 2020, LNNS 161, pp. 725–730, 2021.
https://doi.org/10.1007/978-3-030-60926-9_90

2 Methodology

To predict the necessary need for the personnel, the state body or its structural unit applies special methods and techniques. The methods are applied individually or comprehensively, depending on the specifics of state body activities. Let us single out the methods necessary for determining the number of personnel reserves:

1. The method of complexity. Its essence is to fix in time the tasks and actions of a civil servant to determine the significance and expediency of these operations. The result may be the rejection of any tasks in favor of more significant ones, or the reduction in the number of civil servants and the redistribution of tasks. Designed for mid-term recruitment planning.
2. The method of expert assessments. Based on the opinion of the heads of structural divisions, their intuition and professional experience. Suitable for strategic personnel planning, since subjective assessments of several experts gives an objective picture of the results of personnel assessments.
3. The method of extrapolation. It implies the transfer of the existing situation to the planning period, considering the specifics of functions, financial changes, etc. Suitable for a short period of time, implying that most of the conditions will remain the same [1].

Creating a completely new model of the public service process, it is imperative to use the elements of updating the existing personnel. This will be effective when using the potential of the personnel. For the successful implementation of tasks and goals set for the state apparatus, it is necessary to professionally train the personnel to quickly replace the newly formed or vacated key office positions.

The current legislation defines the structure of the personnel management reserve in the form of a group of persons, including the current employees of the municipal service, formed based on principles of professionalism, morality and a few other principles, the application of which by participants of the personnel reserve is necessary for the effective and sustainable development of the state administrative apparatus, and its updating.

3 Results

When developing the personnel management process to provide municipal authorities with competent employees, it is necessary:

1. Develop a selection procedure for candidates based on specific criteria that allow an objective approach to assess a future candidate to include him in the personnel reserve.
2. Strengthen the motivational criteria for the attractiveness of the public service, which can have a positive impact on the influx of competitive young and talented specialists into the sphere of municipal authorities.
3. Carry out a policy aimed at increasing the status of the civil service, its attractiveness for young, but quite experienced specialists.

There are certain factors that negatively affect the attitude of people to the municipal public service. Among these, one can single out an opaque mechanism for selecting employees for state authorities, which are very specific and sometimes formal, at this stage of personnel management development, and preferences from being in the public service. The above factors create risks of reducing the degree of public confidence in the government. In our opinion, the process of creating a talent pool among young people is the most valuable and effective.

For these purposes, the development of internal regulatory documentation and legal acts is of paramount importance. Despite the above principles of forming the personnel reserve among the most experienced employees, the issue of recourse to the youth personnel reserve can be put at the forefront. Since it is much easier for young people to understand and accept the current system of public administration, to become part of the system, to be imbued with its principles [8]. Therefore, considerable attention must be paid to the estimated qualification requirements for young specialists, since most of them do not have professional knowledge and experience in solving complex managerial problems. We single out the most important criteria for the selection of young candidates to include them in the personnel reserve: organizational and leadership abilities, developed creative and communication skills, significant educational outcomes, non-monetary motivation to perform the public service, a high level of patriotism. When organizing the process of filling vacant posts, it is recommended that the heads of municipal authorities concretize the criteria for assessing the level of professional knowledge, professional qualities required in the selection process for filling public service posts.

To study the attitude of the population towards the public service and prospects of its choice as the main type of professional activity in several regions of the Russian Federation in 2018, Rosstat specialists conducted a study of the opinions of civil servants who had ever been in the personnel reserve and who received or had not received the opportunity to fill a vacant position [5]. According to the results of this study, mixed conclusions were obtained: most of the respondents surveyed participated and were in the personnel reserve no more than once, about a little more than 23% indicated that they were in it two or more times. The main motivator for enrolling in the reserve was the opportunity to occupy a high position in the civil service.

More than half of the respondents indicated that the expectation of transferring or enrolling in the staff of the municipal authority was not justified. The reason for this was called the inoperative mechanism of the personnel reserve and the process of transferring officials. Also, about 60% of the respondents indicated that the declared training in the personnel reserve is not carried out, or this process is taking place formally. Figure 1 shows a visualized summary of the study described above.

When trying to find out why the personnel reserve mechanism does not work, many respondents answered that when choosing a candidate for a vacant position, preference is mainly given to employees who are on friendly or family ties with the management. An important factor behind such survey results was the insufficient use of information technology in attracting, selecting, and adapting candidates. A significant part of the respondents said that the employer is likely to choose an employee to fill a vacant position from among current civil servants as persons with experience in civil government work.

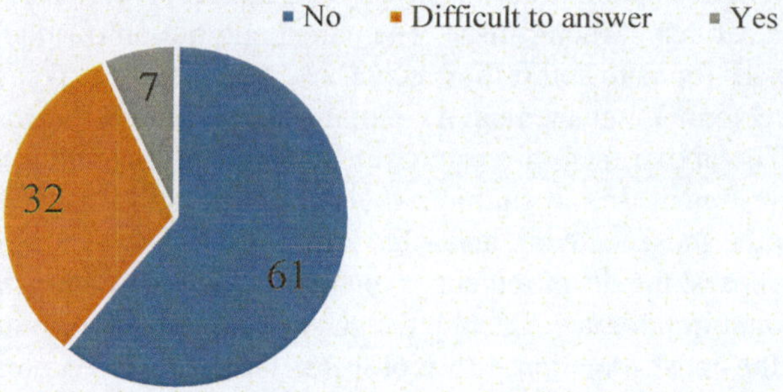

Fig. 1. Distribution of respondents' answers to the question "Did the personnel reserve help you?", % (Source: authors).

4 Discussion

There is a set of problems identified in the above study that impede the development of personnel management in the field of the civil municipal service: a significant level of the impact of family ties on decisions made in the field of staff rotation; frequent changes in the criteria for the correspondence of the position to employees who are on the list of candidates; a small proportion of young specialists both in the reserve and in the process of appointing them to a vacant or new civilian municipal post. We offer solution methods necessary to eliminate the identified problems: strengthen information technology components of competitions for the selection of candidates for the personnel reserve; increase the information awareness of the population; consider the issue of transferring functions of the personnel department to the outside - to outsourcing companies; except lobbying related or friendly candidates when nominating employees from the personnel reserve for vacant positions; realize the technological possibility of tracking the emerging vacancies through the Internet and searching for your candidacy in the list of persons admitted to the competition for filling vacant positions.

In fact, the implementation of these measures is a rather complicated procedure. For example, outsourcing the functions of the personnel department is very difficult, due to the presence of many legal restrictions. When creating a database of persons participating in the competitive selection process, obtaining the consent of each of the participants to process personal data and open part of this information to the public will become rather complicated and time-consuming [4, 6, 7].

Human resources management should be developed in a way of promoting the growth of the number of young people in ongoing competitions. As mentioned above, this aspect has two sides. The obvious one is the lack of professional experience in the implementation of economic and social projects. However, such a contingent (young members of the personnel reserve) is, in our opinion, the most preferable in the context of the planned fulfillment of vacancies. It is much easier to "educate" a young specialist in accordance with the accepted philosophy and principles of work in the organization or in the municipal unit of executive authorities. The consciousness of such a person is not "blinkered"; he can, without question and promptly, accept all instructions directed to him. In addition, the transaction costs of the organization for retraining an experienced professional are reduced, in part of trainings held regarding him [3]. When creating and testing the personnel management mechanism, as a rule, they are guided by the key points of Deming's "management cycle", which includes the following structured algorithm of actions: plan - execute - control - improve, as well as a set of specific tools involved in ensuring its implementation.

5 Conclusion

As a result, we can conclude that the existing range of problems needs to be addressed at the legislative level. For this, it is necessary to clearly prescribe the decision-making process on the selection of candidates for the reserve and their selection from it to fill the vacant position substantiating their professional skills, experience, talent, desire for self-development. It is also important to transfer the control functionality of such a mechanism to the internal control service of the municipal authority to comply with the requirements of transparency, openness, equality in the formation of the personnel reserve. It should be noted that a regulatory framework has been developed in all regions of the country and it is regularly being updated to select candidates to include them in the personnel reserve. Such legislative acts determine approaches to the selection of candidates, their evaluation and regulate the procedure for their assumption of office from the position of the candidate. Therefore, summarizing the process of studying the category of the personnel reserve, we indicate that this category is formed, based on the competitive component, of a group of citizens or civil servants with certain professional qualities required at a certain level to fill public service posts.

References

1. Bach, T., Verhoest, K., Wynen, J.: The interaction of administrative tradition and organisational characteristics: the case of agency personnel management autonomy. International Review of Administrative Sciences (2020, in Press). https://doi.org/10.1177/0020852319889674. Accessed 15 June 2020
2. Gibbs, D.: Civil service reform, self-selection, and bureaucratic performance. Econ. Polit. **32**(2), 279–304 (2020)
3. Gunadi, G.: Sustaining e-government website services: an investigation of dynamic relationships of organizational factors in a government agency. Adv. Intell. Syst. Comput. **1159**, 517–527 (2020)
4. Kaye-Essien, C.W.: Performance reporting delay in local government: a global south view. Int. J. Public Sector Manag. **33**(4), 477–496 (2020)
5. Rosstat: "Professional Development of Federal Civil Servants" (2020). https://gks.ru/folder/471. Accessed 15 June 2020
6. Simonova, M.V., Kolesnikov, S., Spravchikova, N.: Interregional aspects of employment as a factor in the formation of the labor potential. In: Solovev D. (Ed.) Smart Technologies and Innovations in Design for Control of Technological Processes and Objects: Economy and Production. FarEastCon 2018. Smart Innovation, Systems and Technologies, vol. 139, pp. 425–434. Springer, Cham (2019)

7. Troshina, E.P., Mantulenko, V.V.: Influence of digitalization on motivation techniques in organizations. In: Ashmarina, S.I., Vochozka, M., Mantulenko, V.V. (eds.) ISCDTE 2019. LNNS, vol. 84, pp. 317–323. Springer, Cham (2020). https://doi.org/10.1007/978-3-030-27015-5_38
8. Yu, H.H.: Glass ceiling in federal law enforcement: an exploratory analysis of the factors contributing to women's career advancement. Rev. Public Person. Admin. **40**(2), 183–201 (2020)

Creating a Human Capital Model Based on Global Indexes

V. D. Orekhov[1](✉), O. S. Prichina[2], and A. V. Blinnikova[3]

[1] International Institute of Management LINK, Zhukovsky, Russia
vorehov@yandex.ru
[2] Russian State Social University, Moscow, Russia
olgaprichina@mail.ru
[3] State University of Management, Moscow, Russia
allarest@mail.ru

Abstract. The purpose of this work is to form a multi-factor model of indicative diagnostics of human capital in relation to GDP per capita based on 15 aggregated global indexes. The study using regression and correlation analysis allowed us to form the innovative human capital index (IHC), focused on predicting GDP per capita, and to determine the complex optimal Predictor for IHC diagnostics. The main components of this Predictor are the Global Competitiveness Index (0.4), Mean Years of Schooling (0.3), GDP per capita (0.15), and Corruption Perception Index (0.05). The regression error of the IHC index with the optimal Predictor is 2% for a sample of 24 major economies and 5.6% is the average value for five samples of 6–72 economies.

Keywords: Correlation · Forecasting · GDP per capita · Global index · Human capital · Regression analysis

1 Introduction

Human capital (HC) is one of the most important modern socio-economic systems, which contains up to 80% of the world's wealth. The study of HC has become one of the most important areas of economic research. Traditional methods of calculating the value of human capital are focused on obtaining its financial assessment, which allows further including it, along with physical and natural capital, in the equations of economic dynamics [2]. The financial assessment of HC can be obtained by taking into account investments in human capital [8], based on the income received from it [6], or by using the discount method proposed by World Bank. It should be noted that all these methods have a number of significant disadvantages, and estimates of the value of HC using them differ significantly.

The indicator approach is based on the measurement of natural indicators of HC, such as the number of years spent for training of employees [3]. The disadvantage of this method is the use of the average number of training years, although the results of these studies show that with the growth of the training duration, the GDP per capita generated by the corresponding human capital increases exponentially. Consequently, higher levels of education contribute significantly more to the national wealth.

S. I. Ashmarina and V. V. Mantulenko (Eds.): IPM 2020, LNNS 161, pp. 731–740, 2021.
https://doi.org/10.1007/978-3-030-60926-9_91

Moreover, Badinger and Tondl show that economic growth in different regions of the European Union is only sensitive to tertiary education, and differences in the level of secondary education do not have a statistically significant impact on economic development [1].

The authors of this work have developed a method of indicative diagnostics [12] of the educational component of HC, which takes into account the contribution of various educational levels of HC to GDP growth in more detail. For most countries, it provides a low forecast error, but for some countries with a high educational level (Russia, Israel, and South Korea), the forecast error reaches 40%. This makes it necessary to develop assessment methods that use a wider range of information about human capital.

Due to the importance of the HC concept, several human capital indexes have been developed recently, in particular indexes proposed by the World economic forum [15], and the World Bank [17]. These two HC indexes provide a non-financial score that is not directly related to the number of training years. They estimate HC in fractions of a unit or as a percentage. Accordingly, the question arises how to link these estimates to financial units and/or to the GDP per capita generated by them.

It is also important that significantly different indicators (components) are used to calculate these two indexes. Schwab takes into account mainly the competence of the staff, namely, the education level of employees, their accumulated skills, the growth of educational components and qualifications, as well as skills that are used at work (know-how) [15]. The Human Capital Index focuses mainly on the health of HC and takes into account: the survival rate of employees under the age of 60, the probability of living until the age of 5 for children, the proportion of children without developmental disabilities, and the expected number of school years before the age of 18.

It may seem that Global Human Capital (GHC) indicators look more closely related to the essence of human capital than the Human Capital Index (HCI). However, the regression-correlation analysis gives, in general, a different result [13]. It is possible that there is a hidden predictor that correlates better with HCI. For example, good health care and education can be an important factor in the formation of high-performance human capital, although this factor is not measured, but it is closely related to HCI. Can this be identified and how does it relate to the presence of two alternative HC indexes?

Another question is whether it is possible to predict the HC value determined based on HCI and GHC and using a sufficiently large number of aggregated global indexes, and what is the composition of the corresponding Predictors. Thus, the use of new indexes for evaluating human capital brings with it a significant number of issues that will be explored in this paper. The purpose of this work is to form a multi-factor model of indicative diagnostics of human capital in relation to GDP per capita based on a block of global indexes.

2 Methodology

The methodological basis of this work is a systematic approach. It is used for regression and correlation analysis of the system of socio-economic activity based on human capital, the result of which is GDP per capita (in accordance with purchasing power parity – PPP). The authors use a system of global aggregated indexes presented in Table 1, where, in comparison with the work [9], the following indexes are added: CPI, SPI, LEI and KIG, which is relevant, since the Social Progress Index significantly reduced the regression error with GDP/C.

Table 1. Indexes used in the work

i	Complete Name	Abbreviation
1.	Index of Economic Freedom [5]	IEF
2.	Ease of Doing Business Ranking	EDB
3.	Worldwide Governance Indicators [7]	WGI
4.	Global Competitiveness Index [15]	GCI
5.	Mean Years of Schooling [20]	MYS
6.	RandD Expenditure [9]	ERD
7.	World Happiness Index [4]	WHI
8.	GDP per capita (World Bank) [17]	GDP/C
9.	The Legatum Prosperity Index [10]	LPI
10.	Corruption Perception Index [19]	CPI
11.	The Social Progress Index [16]	SPI
12.	Life Expectancy Index [20]	LEI
13.	KOF Index of Globalization [14]	KIG
14.	Global Human Capital [15]	GHC
15.	Human Capital Index [17]	HCI

Source: authors.

Because of the difficulty of comparing countries that differ significantly in GDP size, a series of samples was used that included a different number of economies ranked by the GDP size. Depending on the number of countries, they were designated: G6, G12, G24, G48, G72. The optimization of GDP/C Predictors was based on the average value of the regression error ($\Delta R^2 = 1 - R^2$), which was indicated by the index *m* or *mid.*

Optimal complex Predictors were formed from indexes given in Table 1 by linear composition, with the sum of non-negative weighting coefficients $\Sigma k_i = 1$. Predictors were optimized in order to minimize the average value of the error of the ΔRm^2 regression over five samples using the gradient descent method with cyclic variable replacement.

3 Results

The Global Human Capital approach to measurement allows us to apply it not only to countries, but also to individual corporations, industries, and regions. The Human Capital Index can only be used for larger socio-economic structures such as states or regions. Therefore, it is important to understand their interrelation and areas of application. Considering the statistical dependence of the two human capital indexes on each other, it is shown that the regression error is quite large for them and for the G24 sample is $\Delta R^2 \approx 17\%$. The GHC regression from HCI is approximately linear, GHC values about half as low as HCI. This means that the Human Capital Index differentiates countries significantly more, which is a positive factor. The regression line is weakly dependent on the sample, and the regression error increases with the sample size.

Since the concept of human capital is important in terms of creating national wealth per capita, we will consider the ability of two HC indexes to provide forecasts of GDP per capita (GDP/C), both individually and as part of optimized pair Predictors. Table 2 shows data on the regression error of the GDP/C dependence on these two indexes (power trend), as well as on the optimal pair Predictors and the complex optimal Predictor CP_1, components of which are given in Table 3.

Table 2. The regression error of the GDP/C from different Predictors, %

	G6	G12	G24	G48	G72	mid
GHC	11.3	18.6	18.6	33.5	33.6	23.1
HCI	16.6	11.4	8.9	28.3	22.0	17.4
0.42 GHC + 0.58 SPI	4.5	4.1	7.4	18.9	14.6	**9.9**
0.35 HCI + 0.65 SPI	7.5	6.2	6.4	17.3	13.2	10.1
Predictor CP_1	2.4	2.6	5.8	13.5	9.9	**6.8**

Source: authors.

Table 3. Components of the optimal complex Predictor for GDP/C

IEF	EDB	GCI	MYS	ERD	WHI	LPI	CPI	SPI	KIG	GHC	HCI
0.03	0.06	0.02	0.12	0.01	0.15	0.04	0.01	0.26	0.07	0.08	0.15

Source: authors.

The dependence of GDP/C on the optimal pair Predictor is shown in Fig. 1. In its structure, the share of human capital (GHC) is 42%. The best results are provided by a power trend with a degree of 4.3, although the exponential and power trends are very close. In Fig. 1, the G6 sample points are marked with special icons – from left to right in Fig. 2: India, China, Russia, USA, Japan, Germany. The point corresponding to Russia (square) is close to the trend line, and the USA (circle) is much higher.

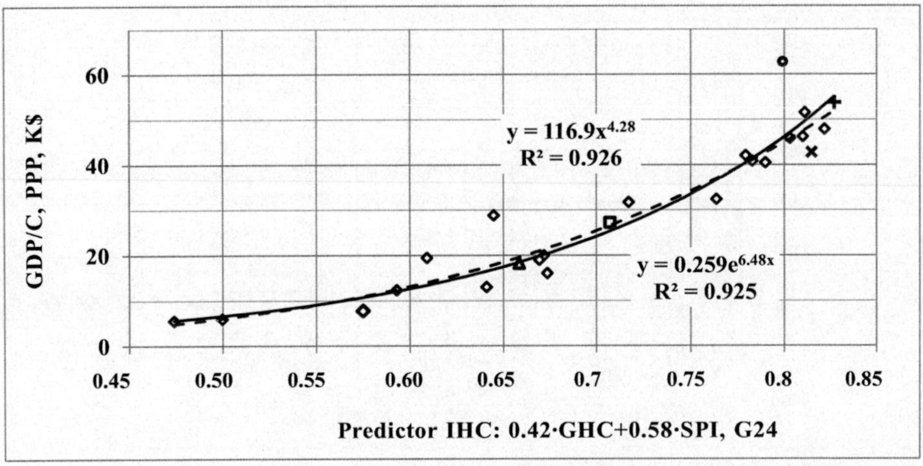

Fig. 1. GDP/C regression from the optimal pair Predictor (Source: authors).

The share of human capital in the optimal complex Predictor is 23%, but the Mean Years of Schooling index [20], whose share is 12%, also reflects characteristic of human capital. Together they contribute 35% to the GDP/C Predictor. However, other indexes that affect GDP/C have components related to the formation of HC too. Thus, the Social Progress Index [16] contains among 12 indicators medical care and nutrition, sanitation and water supply, medicine and health, access to basic and higher education.

Table 2 shows that the regression error of the studied HC indexes varies significantly depending on indexes with which they interact in their impact on GDP/C. To study the influence of other Predictors on the Human Capital Index, it is necessary to create a comprehensive index of HC. Taking into account the obtained data, it can be assumed that the complex HC Index (IHC) can be calculated using the formula (1).

$$\mathrm{IHC} = 0.4\ \mathrm{GHC} + 0.6\ \mathrm{HCI} \tag{1}$$

This index characterizes the relative quality of human capital. In order to calculate the total value of the country's HC, you need to multiply the IHC by the number of employees. On average, the share of the employed population in the 20 largest economies is 44% [18]. Since HC is implemented not only in production, we can assume that the number of people who make up human capital is equal to half of the population of countries. Corresponding to this approximation, the values of the number of HC for the largest economies are shown in Fig. 2.

It is characteristic that this dependence is well approximated by the hyperbolic equation ($R^2 = 0.96$). China's human capital is 23% of the world's, Russia's is 2.5%, and the US – 6.2%. This definition of human capital rather characterizes the intellectual development potential of countries. In market conditions, the accumulated scientific and technological potential and competitive position in the markets are very important. All this affects the amount of GDP (and GDP/C) that countries receive.

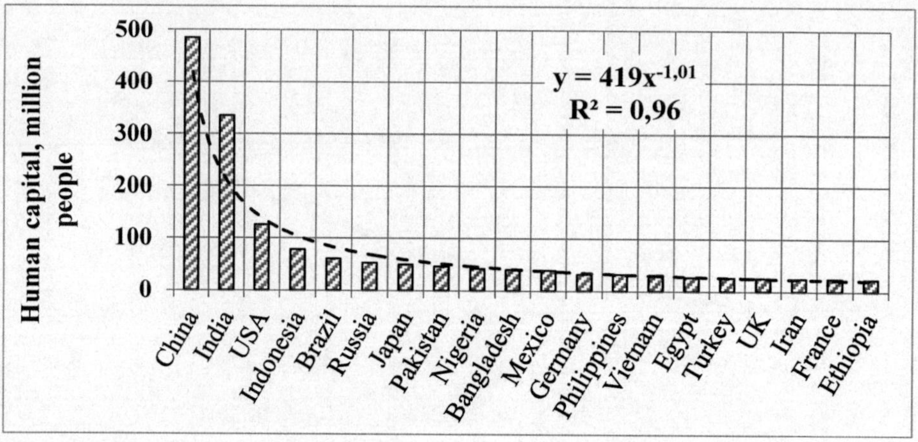

Fig. 2. Number of human capital in the largest economies (Source: authors).

Let's consider factors affecting the human capital index. In order to find out what Predictors and how they affect the complex Human Capital Index (1), we first determine the IHC regression error from each of the indexes listed in Table 1 individually. The corresponding results with a trend in the form of a cubic polynomial are shown in Table 4 (for all samples, except G6, a quadratic polynomial is sufficient). The lowest regression error is provided by the GDP/C indexes (ΔR^2 = 11%) and MYS (14%).

Table 4. The error of IHC regression from global indexes, %

	G6	G12	G24	G48	G72	mid		G6	G12	G24	G48	G72	mid
IEF	0.8	53	40	47	47	38	GDP/C	0.7	6.8	7.2	22	19	**11**
EDB	11	26	28	32	39	27	LPI	15	26	14	18	15	17
WGI	17	18	29	36	33	27	CPI	0.2	28	14	9	36	18
GCI	10	17	13	22	20	16	SPI	0.9	25	12	26	21	17
MYS	3.7	11	11	24	20	**14**	LEI	14	30	18	25	27	23
RDE	12	34	29	15	39	26	KIG	6.7	23	21	29	25	21
WHI	1.7	46	54	49	45	39							

Source: authors.

Table 5 shows the regression error for the best optimal Predictors paired with GDP/C and MYS. It can be seen that the Global Competitiveness Index has the strongest impact on human capital in pairs [15]. Thus, the regression error of the Predictor 0.4 × MYS + 0.6 × GCI was ΔR^2 = 6.2%, which is almost half as much as that of the best individual index of influence on human capital – GDP/C, for which ΔR^2 = 11%. It should be noted that the GCI, in addition to indicators of economic competitiveness, also contains components characteristic for human capital: health, education and the labor market.

Table 5. Error of IHC regression from paired Predictors, %

Pairs		k8/k5	G6	G12	G24	G48	G72	mid
GDP/C	EDB	0.8	0.5	4.7	5.4	19.7	16.8	9.4
	GCI	0.6	0.6	3.9	3.6	18.2	15.3	**8.3**
	LPI	0.65	0.1	7.8	5.2	17.2	14.5	9.0
	CPI	0.9	0.0	4.7	5.8	21.9	17.8	10.0
	LEI	0.8	0.2	7.1	7.1	19.6	16.8	10.2
	KIG	0.8	1.1	7.4	6.9	19.8	16.8	10.4
MYS	EDB	0.75	3.4	9.3	4.8	17.0	13.3	9.6
	GCI	0.4	0.9	3.2	2.9	13.5	10.7	**6.2**
	ERD	0.7	1.3	3.0	7.0	17.5	13.2	8.4
	CPI	0.7	1.1	2.5	5.5	18.2	14.8	8.4

Source: authors.

Figure 3 shows the dependence of IHC on the optimal pair Predictor with a quadratic trend for the G24 sample. It is characteristic that with the growth of the Predictor above the value of 0.8, IHC goes to the "shelf" as a result of the growth of the negative quadratic term. In Fig. 1, we saw that GDP/C is growing like a fourth-degree polynomial. Thus, in order of magnitude, GDP/C will grow proportionally to the square of the paired Predictor linearly associated with Mean Years of Schooling [20].

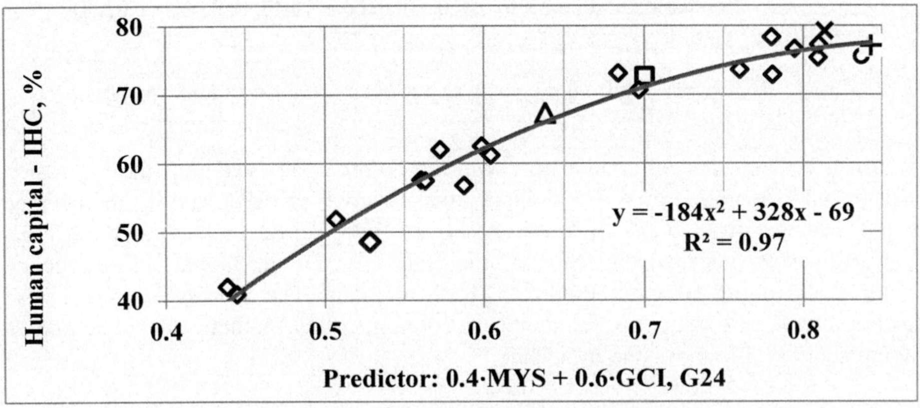

Fig. 3. IHC Regression from the optimal pair Predictor for G24, % (Source: authors).

Obtaining the characteristics of the IHC regression index with paired Predictors allows us to search for the optimal complex Predictor – CP for human capital. Optimization was performed by varying k_i coefficients and search for their values that provide the minimum average of the regression error values for five samples. The obtained optimal values of CP coefficients are shown in Table 6.

Table 6. Components of the optimal complex Predictor for IHC

Weighting coefficients	k_2	k_4	k_5	k_6	k_7	k_8	k_{10}	k_{12}
Indexes	EDB	GCI	MYS	ERD	WHI	GDP/C	CPI	LEI
Optimal values	0.03	0.4	0.3	0.02	0.03	0.15	0.05	0.02

Source: authors.

Figure 4 shows the regression dependence of IHC on the optimal complex Predictor CP for the G24 sample.

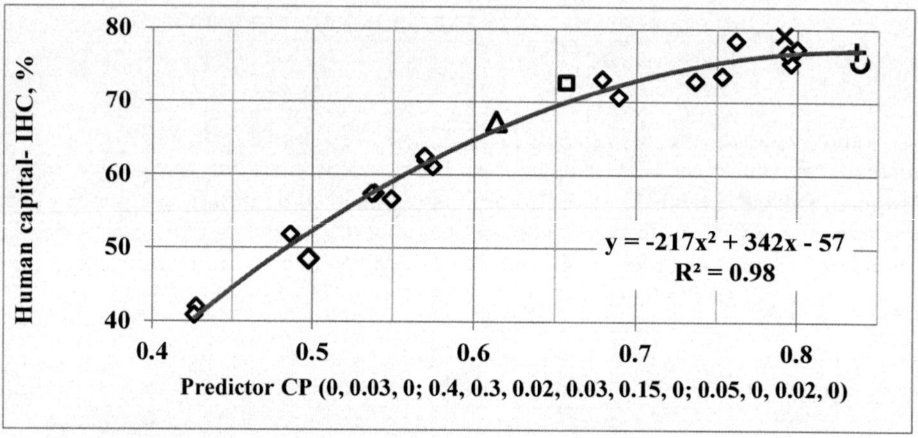

Fig. 4. IHC Regression from the complex Predictor CP, G24 (Source: authors).

It can be seen that, in comparison with the optimal pair Predictor, the regression error for G24 decreased by one and a half times (from 3% to 2%), and for the average of five samples from 6.2% to 5.6%. As in the best paired Predictors, the Global Competitiveness Index, Mean Years of Schooling, and GDP per capital indexes occupy a key place in the complex Predictor (85% in total). The regression line is also quadratic with an output "on the shelf". According to this Predictor, the IHC human capital score is 3% above the trend line.

4 Discussion

In connection with the development of the World Bank [17] and the World Economic Forum [15] two alternative human capital indexes (HCI and GHC, respectively), which are fundamentally different from each other and from the methods of assessing human capital that were used previously. A number of questions and opportunities arose in connection with the use of these unique tools. The difference between these two indexes can be interpreted as a fact that one or both of them are incorrect to varying degrees. At the same time, the argument that HCI is aimed at assessing the future HC is not very convincing, since for it the error of synchronous regression with GDP/C is significantly less than for GHC. However, the presence of two indexes allows them to form Predictors on a statistical basis and in combination with other global indexes for more accurate forecasting of GDP/C, on the one hand, and the formation of a model for indicative diagnostics of human capital, on the other.

Comparison of these two indices and statistical estimation of their regression with GDP/C allowed us to propose a total IHC index aimed at predicting the value of GDP/C. The HCI and GHC indices characterize the relative quality of human capital, but they can be used to estimate the absolute size of human capital of different countries, which is very important from the point of view of investors' understanding of labor markets. The presence of such a total index of human capital also allowed us to form a model of indicative diagnostics of human capital. Somewhat unexpected was the result of the analysis, which showed that indexes related to human capital do not dominate in the impact on the value of GDP/C and a significant contribution is made by World Happiness Index and Social Progress Index, although they also have components related to HC. Another unexpected aspect was that the strong impact on of Global Competitiveness Index on the human capital index (IHC). These results require additional research, in particular, using the method of cognitive modeling, since it is clear that in such a complex, weakly structured system, the relations are quite complex and nonlinear. Additional research is also needed on the fact that IHC's dependence on the Predictor of human capital is "on the shelf", since this may contradict the results of Mincer [11], according to which labor results depend exponentially on the number of training years.

5 Conclusion

Regression and correlation analysis of the impact of 14 global indexes on GDP per capital allowed us to form the IHC human capital index, which is focused on forecasting GDP per capita, and to determine the complex optimal Predictor for IHC diagnostics. It is shown that, according to the IHC, the largest human capital in the world (in millions of people) is possessed by: China (484), India (335), the USA (125), Indonesia (76), Brazil (61), Russia (53) and Japan (50).

The optimal pair Predictor for IHC includes Mean Years of Schooling (0.4) and Global Competitiveness Index (0.6). The regression error ($\Delta R^2 = 1 - R^2$) for it is less than 3.2% for samples that do not exceed 24 countries, and 6.2% is the average value for samples from 6, 12, 24, 48 and 72 samples. The main components of the optimal complex Predictor are the Global Competitiveness Index (0.4), Mean Years of Schooling (0.3), GDP per capita (0.15), and Corruption Perception Index (0.05). The regression error of the IHC index with the optimal Predictor is 2% for a sample of 24 major economies and 5.6% - the average value for five samples of 6–72 economies.

Acknowledgements. The reported study was funded by the RFBR, project No. 19-29-07328.

References

1. Badinger, H., Tondl, G.: Trade, human capital and innovation: the engines of European regional growth in the 1990-s. In: Fingleton, B. (Ed.) European Regional Growth, pp. 215–239. Springer, Berlin
2. Barro, R.J., Lee, J.W.: International data on education and attainment updates and implications. https://scholar.harvard.edu/barro/files/p_jwha.pdf. Accessed 22 June 2020 (2000)
3. Barro, R.J.: Economic Growth. The MIT Press, London (2004)
4. Helliwell, J., Layard, R., Sachs, J.: World happiness report 2019. Sustainable Development Solutions Network, New York (2019)
5. Heritage Foundation: 2020 index of economic freedom (2020). https://www.heritage.org/index/pdf/2020/book/index_2020.pdf. Accessed 27 June 2020
6. Jorgenson, D.W., Fraumeni, B.M.: The accumulation of human and nonhuman capital, 1948–84. In. Lipsey, R.E., Tice, H.S. (Eds.) The Measurement of Saving, Investment, and Wealth, pp. 227–282. University of Chicago Press, Chicago (1989)
7. Kaufmann, D., Kraay, A., Mastruzzi, M.: The worldwide governance indicators: Methodology and analytical issues (2010). https://openknowledge.worldbank.org/handle/10986/3913. Accessed 22 June 2020
8. Kendrick, J.W.: Total capital and economic growth. In: Kendrick, J.W. (Ed.) The Formation and Stocks of Total Capital, pp. 111–125. Cambridge University Press, Cambridge (1976)
9. Knoema: RandD Expenditure as a share of GDP (2017). https://knoema.com/atlas/topics/Research-and-Development/RandD-Expenditure/RandD-expenditure-as-a-share-of-GDP. Accessed 23 June 2020
10. Legatum Institute: The Legatum prosperity index, 2019 (2019). https://www.prosperity.com/rankings. Accessed 22 June 2020
11. Mincer, J.: Schooling, Experience and Earnings. Columbia University Press for the National Bureau of Economic Research, New York. (1974)
12. Orekhov, V.D., Prichina, O.S., Blinnikova, A.V., Panfilova, E.A., Shchennikova, E.S.: Indicative diagnostics of the educational component of human capital based on mathematical modeling. Opción. Año **35**(20), 2337–2365 (2019)
13. Orekhov, V.D., Prichina, O.S., Gizyatova, A.S., Blinnikova, A.V., Kukharenko, O.G.: Development of the indicative system for assessing GDP per capita using cumulative indexes, including human capital. Journal of Advanced Research in Dynamical and Control Systems **12**(05), 1139–1152 (2020)
14. Savina, G., Haelg, F., Potrafke, N., Sturm, J.E.: The KOF globalisation index – revisited. Rev. Int. Organ. **14**(3), 543–557 (2019)
15. Schwab, K.: The global competitiveness report 2019 (2019). http://www3.weforum.org/docs/WEF_TheGlobalCompetitivenessReport2019.pdf. Accessed 22 June 2020
16. Stern, S., Wares, A., Epner, T.: Social progress index methodology report (2018). https://www.socialprogress.org/assets/downloads/resources/2018/2018-Social-Progress-Index-Methodology.pdf. Accessed 22 June 2020
17. The World Bank: World development report 2019 (2019). https://www.worldbank.org/en/publication/wdr2019. Accessed 22 June 2020
18. Trading Economics: Employed population (2020). https://ru.tradingeconomics.com/country-list/employed-persons. Accessed 22 June 2020
19. Transparency International: Corruption perception index (2020). https://www.transparency.org/en/cpi. Accessed 22 June 2020
20. UNDP: Human development indexes and indicators: 2018 statistical update (2018). http://hdr.undp.org/sites/default/files/2018_human_development_statistical_update.pdf. Accessed 23 June 2020

Value Basis of Personnel Development in Digital Economy

G. V. Serebryakova(✉), I. V. Nezamaykin, and T. B. Shramchenko

State University of Management, Moscow, Russian Federation
sega62@mail.ru, {iv_nezamaykin, tb_shramchenko}@guu.ru

Abstract. The material presented in this article is results of a research aimed at justifying the need to use value design in the process of developing the organization's personnel, which required solving the following tasks: to substantiate the prerequisites and conceptual foundations of value design, to offer technologies of forming the value basis for the development of specialists in the digital environment. The development of personnel in modern conditions should be based on value-oriented management, aimed at enhancing the consistency between value orientations of specialists and the value system of the organization. The use of value design in the process of HR development allows implementing a system of value standards for specialists, taking into account requirements to the professional culture in the digital economy.

Keywords: Value design · Value matching · Value standard · Value system

1 Introduction

The modern world is characterized by a high level of variability, uncertainty, complexity and ambiguity caused by the speed of changes in the factors of society development. The increasing complexity of political, economic and social processes in the society creates a demand for a new generation of specialists. Digitalization of all spheres of our society makes adjustments to the interaction of socio-economic systems. Many of tasks currently performed by employees in various economic sectors will be automated or disappear due to changes in the way the society is organized [1]. A number of organizations are transferring employees to remote work already [8], which transforms not only requirements that employees have to meet, but also processes of their professional responsibilities implementation. The new economy will require specialists with new competencies. Despite the penetration of artificial intelligence in various areas, a number of tasks will not be digitized - human resource management, social interactions, and activities associated with a high level of unpredictability. For these reasons, the development of digital economy defines the ability to adapt quickly in difficult conditions, prepare and make decisions correctly in an uncertain environment, and possess technologies for working with groups of people and individuals as key characteristics of future specialists.

The digitalization process changes the social basis of people's lives. On the one hand, additional opportunities for obtaining new knowledge and speeding up the processes of their transfer open up, on the other hand, the number of real opportunities

S. I. Ashmarina and V. V. Mantulenko (Eds.): IPM 2020, LNNS 161, pp. 741–747, 2021.
https://doi.org/10.1007/978-3-030-60926-9_92

for social communication decreases. These features determine new rules and requirements to behavioral models of specialists. According to the researchers' forecast for the year 2030, one of the most important qualities of specialists that will determine the success of socio-economic systems functioning are soft skills (functional literacy, erudition, readiness to learn and update knowledge all the time).

To constantly master new knowledge, it is necessary to develop the need for self-development and skills of independent search and mastering innovative learning technologies. Continuous self-improvement is possible only if the values of responsibility, integrity, and proactivity necessary for the quality performance of professional duties are developed in the process of personnel training.

Environmental conditions constantly make adjustments to organizations' functioning. The spread of coronavirus has changed the way of life for entire countries, not to mention individual organizations. Rapid response, the ability to change management processes, the introduction of new communication systems [7], the transition to new technologies of social interaction – this has become not just a characteristic of the current situation, but has passed into the category of mandatory conditions for functioning of all socio-economic systems.

Even before the pandemic, government agencies began actively working on digitalization of the country's economy. But no initiative will be implemented if we do not prepare specialists with necessary knowledge, skills and desire, regardless of the prevailing conditions, to perform their professional duties efficiently. Moreover, this is not just online technology training, it is a complex process, primarily related to the transformation of the psychological foundations of behavior. The basic competencies that modern specialists should master are critical thinking, creativity, empathic skills in building interactions and relationships, and cognitive flexibility. It should be emphasized that it is not possible to develop a specialist only by transferring modern knowledge and technologies to him. The key aspect in the process of managing the HR development, in our opinion, is the value aspect.

2 Methodology

In conditions of digital economy, the role of value training of a specialist is increasing, which, in our opinion, should be based on value design. Value engineering is a new type of design technology, which is a process of forming and developing the structure and elements of the value system of specialists, ensuring the most effective implementation of their professional activities. The methodological basis of value-based design is value-oriented management. First of all, the basis for the value development of personnel within the framework of the new technology should be based on increased commitment to the interests of the society. In the absence of the formed guidelines, almost all other activities aimed at the development of a specialist are doomed to failure. The German Empire's Chancellor Otto von Bismark said that "Man kann die Russen nicht besiegen, das haben wir über Jahrhunderte gelernt. Aber man kann den Russen falsche Werte einbläuen, dann besiegen sie sich von selbst" [6, p. 4].

In modern conditions, when organizing the process of specialists' development, it is impossible to ignore the phenomenon of "value corruption", which is aimed at destroying generally accepted universal values. It is necessary to do everything so that this value pathology does not penetrate into the sphere of professional activity, otherwise, there may be a value substitution in the system of value orientations of a specialist, which will not allow him to realize his professional duties at the proper level, and lead to a value distortion of his behavior model. For these reasons, it is necessary to actively use the technology of value screening, aimed at conducting advanced diagnostics of the value system of employees in order to identify prerequisites for the development of value distortions.

3 Results

Requirements to modern specialists are based on the moral principles of conscientious performance of official duties, honesty, and social responsibility, which is of particular importance in the context of the penetration of digital technologies in all spheres of our society. For their formation in the process of professionals' development, we should actively use not so much traditional management technologies [2], but innovative approaches to the development of the value basis of a specialists, for example, the technology of informal learning, the main goals of which include the formation of behavioral patterns, the expansion of social ties, increased levels of employees' loyalty and increased level of their civic consciousness. The readiness level of a specialist to work in the digital economy should be assessed by the formation level of an active civil position, which is a basis for maintaining a positive image of the organization to which the employee belongs to. An important place in the process of value development of an employee should be given to strengthening his historical memory [5], which contributes to the formation of respect for moral traditions and social customs, the corporate history and culture of the organization.

The target orientation of employees' development is related to the formation of socio-psychological components of their personality. To ensure that this process is of a continuous nature, it is necessary periodically to scan values. And the results of this procedure should be compared with the specialist's value profile, tailored in compliance with requirements to professional culture and development level of value-based environment of the society. This enables to identify discrepancies in the value characteristics of a particular employee and select the most appropriate tools to bring them in line with the organization's value system.

The professional activity of any specialist should be based on norms and principles of professional ethics. It's worth noting that the digital economy increases the transparency of any socio-economic system, so the reputation level of the entire organization depends on the ability to build processes of social interaction of employees, and the level of their information culture. However, it is impossible to acquire all the qualities and values required for effective and professional behavior very quickly. The process of forming and maintaining the value system necessary for a specialist is continuous. Moreover, we should pay attention to the manageable nature of this activity. An effectively organized process of personnel development should be based on well-structured procedures of value design.

Moral responsibility, active citizenship, honesty, and social communication are mandatory elements of the value standard of a modern specialist. Effective implementation of employees' professional activities becomes possible only if these moral concepts are accepted and deeply shared by them. You cannot impose a behavior model on a specialist, each time making a decision about actions in a certain situation, his choice is based on internal foundations and accepted value orientations. For these reasons, it is necessary to use "action learning" as the main form of professional development, as a result of which it is necessary to evaluate not the acquisition of new knowledge, but the reaction and change in behavior patterns. The development of specialists should take into account the system of relationships in which they participate, the level of their communicative competence, the level of their ability to anticipate and neutralize conflict situations. The choice of forms of capacity-building should be based on the needs of all interaction participants. It involves a variety of training forms, flexible programs, high interest and activity of specialists in the development process.

Value characteristics form the employee's attitude to himself, to his place in society, in the organization. It is necessary to make a transition from point-based learning to a holistic, continuous process of personal development, from standard to adaptive problem-oriented programs. For effective development of a specialist, actively use a trajectory approach is needed that takes into account the employees' personal characteristics. Competence-based approach should be supplemented with the development of employees' behavioral manifestations of competencies. For example, it is not possible to form the "team work" competence only by accumulating knowledge about forms and methods of organizing team activities, without constantly increasing the development level of such qualities as the ability to work in a team, cooperation and partnership [3, 4]. An important tool for achieving this goal is the organization of team work on project tasks, for successful completion of which students should organize network teams and competently implement group interaction in the design process.

Useful is the case study technology applied in groups. Studying situations from the experience of practical activities of organizations is especially helpful by giving this technology a value coloring. The use of this method in modern conditions cannot be limited to problem analysis and solution development, it is necessary to supplement the study of each situation with the diagnosis of participants' value systems, modeling of their behavior, identifying and assessing consequences of the chosen behavior pattern. To change attitudes and behavior of specialists, it is effective to use not just traditional game technologies, but such as a metaphorical game aimed at finding solutions to the problem based on the choice of a metaphor. For example, it is necessary to find new approaches to interacting with problem population groups. At the end of the game, you should identify the most appropriate behavior models based on the value system of different groups, and transfer them to real situations.

For the acquisition of interpersonal interaction skills, an effective means of value design is role-playing, which requires from participants to interact with each other and develop empathic management skills. In terms of the development of the digital economy, there is increasing demand to improve communicative competence: a growing number of information links, this increase occurs not only quantitatively, but also qualitatively (they become more intense and diverse). The interaction engages new participants with different level of interests and social status. For successful implementation of professional activities in these conditions, the ability to quickly and adequately respond to current situations comes first. Game situations allow specialists to “try on” different roles, justify and understand the behavior patterns of participants in different situations, reveal the value discrepancy of their actions, and determine the most appropriate ways to eliminate problems. Game technologies are the most appropriate means of developing moral principles of behavior among specialists. For example, when considering a game situation for building a system of interaction with clients’ problem groups, a specialist should design a structure of a constructive dialogue where he should show maximum correctness, self-control and the presence of skills to neutralize resistance, despite the possibility of using provocative communication technologies by interaction participants.

In order to eliminate such organizational pathologies that can potentially manifest themselves in specialists’ activities, such as civil selfishness, social indifference, and finally, corruption, it is not enough to adopt legal acts punishing their manifestation. We should constantly develop and strengthen such value orientations as integrity, decency, honor, they are antipodes of bureaucracy, bribery, and social indifference. It is the constant increase in the priority of these values that serves as the basis for creating the moral foundations of a harmonious combination of personal and organizational interests. In order to develop this group of values, it is possible to actively use behavioral modeling technology, the purpose of which is the formation of interpersonal communication skills and changes in value settings. Behavioral modeling focuses on the formation of a value system necessary for the quality performance of professional activities. To do this, the following steps are needed:

- designing a “behavioral model” based on the values system of professional behavior, which is proposed to be mastered;
- practical activity when it is necessary to apply the designed “behavioral model” in the situation under consideration;
- feedback sessions where behavioral models are analyzed in order to identify the most successful ones, those fully meet requirements of the professional value standard.

4 Discussion

Increasing the information activity of the environment requires finding new forms of personnel development management, which should be based on the shared value-oriented behavior. Inconsistency of employees’ value systems is a source of conflicts, disorganization of interaction, destruction of relationships, which are often impossible to restore because of the loss of trust both inside and outside the organization. For these reasons, in modern conditions, the process of personnel development should be considered only from the position of interdisciplinarity, based on the integration of behavioral, socio-economic, competence and value approaches developed by representatives of different scientific disciplines.

The development of employees in a digital environment is possible only on an interdisciplinary basis [9], an integrated approach, a link of which is value-based design. The development of the digital space leads to a change in the basis for building relationships between people. The reputation of organizations increasingly depends on the loyalty level of their employees, the development degree of information culture and "computer hygiene".

Multi-aspect manifestations of personnel's value development reveals the relation between the results of the participants' performance and their value orientations, affects the duration and strength of interaction in the socio-economic system. Forecasting staff behavior based on the priority values of joint functioning is a priority element of the value management of its development. The value concept of employees' development management forms a reliable basis for increasing the value of their joint work within the organization.

5 Conclusion

In the process of staff development, it is important to achieve a respectful attitude to the organization by employees. The specialist should understand that his aim is to respond as closely as possible to requests of the external environment. The damage that can be caused by actions of an employee with a distorted moral position, characterized by the satisfaction of personal interests, can lead to the destruction of the organization's reputation where this specialist works. Each specialist has not only to master the knowledge of generational theory and social psychology, but also develop a value basis for their behavior that recognizes the importance of each person and citizen. To reduce the alienation level, the specialist should build a system of social interaction, be able to respond correctly to demands of the external environment, understanding that his behavior is an element of the organization's image as a whole. Actions of a specialist in the conditions of digital economy development should be subject to the "ethics of responsibility" (actions based on the awareness of personal responsibility) [10]. Therefore, in the development process, you should try to achieve a balance between belief and responsibility of a specialist, that his activity is not a formal implementation of official duties, but contributes to the development of the entire organization.

Within the framework of value design, it is necessary not only to form value standards of behavior, to select the most appropriate technologies for training and development of specialists, but also to establish a system for evaluating the compliance with these standards in professional activities, which serve as criteria for the professional suitability of an employee to work in this organization. In the process of value diagnostics, the development of moral consciousness and adherence to the accepted system of values in extreme and conflict situations should be evaluated. When evaluating his performance, the priority should be given not only to the knowledge of professional foundations, but also to moral characteristics taking into account specifics of employees' activity. It is proven to train the employees' knowledge and skills is much easier than to instill in him a pattern of behavior, consistent with principles of professional ethics. So, if the company management does not start an active transition to the specialists' value development, the organization will not be able to achieve strategic goals, and the process of employee development becomes irrational.

Acknowledgements. The materials presented by the authors and conclusions based on the research results were obtained thanks to the support of the State University of Management. This support made it possible to conduct the research work, and the discussion of considered issues in the scientific community. The authors thank the publishing house for the opportunity to present research materials to familiarize the broad scientific community with them and expand the scientific discussion on the studied aspects.

References

1. Aptekman, A., Kalabin, V., Klintsov, V., Kuznetsova, E., Kulagin, V., Yasenets, I.: Digital Russia: The new reality (2017). https://www.mckinsey.com/ru/~/media/mckinsey/locations/europe%20and%20middle%20east/russia/our%20insights/digital%20russia/digital-russia-report.ashx. Accessed 25 June 2020
2. Katkalo, V., Volkov, D.: Corporate Training for the Digital World. Eksmo, Moscow (2018)
3. Koulogeorge, P.: Develop your staff through team-building and team-learning opportunities. Forbes Commun. Council Commun. Voice **24**, 254–270 (2018)
4. Nezamaykin, I.V., Serebryakova, G.V., Shramchenko, T.B.: Features of value management in educational activities. In: Sukiasyan, A.A. (Ed.) Proceedings of the International Scientific and Practical Conference "Pedagogic. Psychology and education", pp. 125–129. AMI, Sterlitamak (2018)
5. Nezamaykin, I.V., Serebryakova, G.V., Sycheva, S.M., Shramchenko, T.B.: (2018) Problems of team formation and functioning in modern conditions. In: Mantulenko, V.V. (Ed.) Proceedings of the International Scientific Conference "Global Challenges and Prospects of the Modern Economic Development". The European Proceedings of Social and Behavioural Sciences, vol. 57, pp. 893–904. Future Academy, London (2019)
6. Otto von Bismark Stiftung: Bismarck und die russischen Werte (2016). https://www.bismarck-stiftung.de/2016/01/22/bismarck-und-die-russischen-werte/. Accessed 22 June 2020
7. Parks, E.: Communicative criticality and savoir se reconnaitre: emerging new competencies of criticality and intercultural communicative competence. Lang. Intercult. Commun. **18**(1), 107–124 (2017)
8. Stroeva, O., Zviagintceva, Yu., Tokmakova, E., Petrukhina, E., Polyakova, O.: Application of remote technologies in education. Int. J. Educ. Manag. **33**(3), 503–510 (2019)
9. Tuguskina, G.N., Rozhkova, L.V., Taktarova, S.V., Salnikova, O.V.: (2018) The role of human capital in the digital economy. In: Mantulenko, V.V. (Ed.) Proceedings of the International Scientific Conference "Global Challenges and Prospects of the Modern Economic Development". The European Proceedings of Social and Behavioural Sciences, vol. 57, pp. 953–959. Future Academy, London (2019)
10. Weber, M.: Selected works. Moscow (1971)

Proactive Leadership in Team Building and Investment Evaluation in Human Capital

N. M. Tverdola[1](✉), O. L. Belova[1], and R. V. Aghgashyan[2]

[1] State University of Management, Moscow, Russia
{nm_tverdola, ol_belova}@guu.ru
[2] National Polytechnic University of Armenia, Yerevan, Armenia
rubag@seua.am

Abstract. The purpose of this article is to identify the role of proactive leadership in the formation of a team based on a metaprograms approach to assessing the human capital of an organization and the formation of a methodology for evaluating the effectiveness of investments made in it. This goal is achieved by solving two tasks: first, the analysis of the concepts of leadership and proactive leadership, as well as the related concepts of human capital and human asset, and second, studying the relationship between the concepts of human capital, proactive leadership and organizational effectiveness. The authors have analyzed the concept of the leadership concept, focused on the phenomenon of proactive leadership. Qualitative characteristics of proactive leadership have been highlighted, and the approach to personnel evaluation based on metaprograms and behavioral patterns has been analyzed. The author's view on the structure of human capital and methods for evaluating the effectiveness of investment in its development are proposed for discussion.

Keywords: Effectiveness · Human asset · Human capital · Leadership · Metaprograms · Proactivity

1 Introduction

In professional and scientific terminology, such concepts as proactivity, leadership, human capital, human asset, and effectiveness have become firmly established. However, the relationship, concept and significance of these phenomena continue to cause controversy in the scientific and business environment. What are the features of a modern effective leader? What is the concept and structure of human capital? Is it possible to evaluate the human capital of organizations without the human potential of the leader? After the death of Steve Jobs, the value of the American Corporation Apple, a manufacturer of personal and tablet computers, has slightly changed. Does this mean that the role of one, even such an inspiring leader, does not solve anything at the company level? Is it possible to measure and manage staff performance and investment in his development? Many organizations want to build a high-quality work with staff and are looking for the best way to evaluate the contribution of their employees in terms of resources, and want to understand how effectively investments in human assets work. On the other hand, in practice, the use of these concepts is superficial and

S. I. Ashmarina and V. V. Mantulenko (Eds.): IPM 2020, LNNS 161, pp. 748–756, 2021.
https://doi.org/10.1007/978-3-030-60926-9_93

does not necessitate the deep understanding of their concepts. The attitude to staff in many organizations is based on the principle of the costs source, not investment. Methods for evaluating investment in human capital are complex, and/or do not give an idea of the true state of the subject.

One of the possible reasons for this problem is the increasing volume and speed of information flows, the high dynamics of changes in the world and the complexity of interpretations, the lack of clear formulations as interconnected elements of a single system that do not require large time and other resources of methodologies measurement. When speaking on the investment evaluating in personnel, there is a desire to find a single magic formula. However, until now there was an opinion that it is possible to evaluate investment, but there is no single methodology or universal set of indicators of human capital efficiency that could be used [9, 14]. They depend on the goals, needs, and specifics of each organization. The article offers for consideration: one of the possible, very accessible and transparent ways to reflect the systemic relationship of ideas: leadership, proactive leadership, human capital; method of universal assessment of investment in human capital.

2 Methodology

The purpose of this study is to identify the role of proactive leadership in the formation of a team, based on a meta-programmatic approach to assessing the human capital of the organization and the formation of a methodology for evaluating the effectiveness of its investment. Based on the purpose of this article several questions have been formulated: How are the concepts and content of such ideas as leadership and proactive leadership perceived these days? How adequate are modern methods of personnel assessment? What are the opportunities and prospects for a metaprogram approach to staff assessment? What is the relationship between the effectiveness of human capital and proactive leadership? What is the peculiarity of human capital in terms of human resource management? How is it possible to reflect the relationship of these ideas in the methodology for evaluating the effectiveness of investment in the development of human capital of the organization?

These issues have been solved on the basis of modern concepts of human capital. However, there is still no single view on the structure of human capital, which from our point of view can be considered from the perspective of the personnel structure. Evaluating the effectiveness of human capital is also under constant discussion. In this research, "The Human Capital Monitor®" investment assessment model by Andrew Mayo has been used as a starting point [9]. The metaprogram approach of Merlevede [11], Collins' ideas [5], as well as the theory of spiral dynamics, and others have been used to discuss approaches to personnel assessment. All these concepts and theories have been used by the authors to create an integrated approach to tasks solving.

3 Results

3.1 Features of Understanding the Concepts and Content of the Leadership and Proactive Leadership Ideas These Days

Firstly, you need to define what includes leadership and "proactive leadership", and what this means in terms of team building and its investment evaluating. The term "leader" comes from the English word "lead - to lead someone or something in/to/in the direction, go ahead." There are two versions of its translation [3]:

- a person who manages a company, group, country, or situation,
- someone or something who wins during a competition.

It is not difficult to see that the idea of "leader" is broader than "director". It is assumed that the director is always the leader. Is this really so? Next, it is important to analyze the differences in the content of the three terms - "leader", "leadership" and "leading". The "leader" position is a role in a system. "Leadership" is usually attributed to the competence of a person to influence. Those who are not formally the leader can also lead. And, at the same time, we know a number of examples when a person who has a formal role of a leader was not able to lead people other than through the use of administrative levers.

"Leading" is the result when a person uses both their role and leadership abilities to influence others [15]. Let's narrow the scope of research to the idea of "leader", as a synonym for the term "manager" or "director". If the "director" has the competence to influence, in other words, to be a leader.

Now let's talk about the leader's proactivity. The key components of proactivity are independence, awareness, activity and responsibility. The leader's proactivity can be considered as a manifestation of professional and managerial subjectivity. Or else its intention to become the subject of action, and not its object. Thus, we suggest to consider proactive leadership as a manifestation of the director's ability to make an informed choice - to make decisions and implement them, taking responsibility for this choice, including, first of all, the formation of a highly productive team for implementation of the company's goals [6].

Dr. Adizes compares the role of leader with the thumb much more often than with the forefinger [7]. Only he can work with all the fingers and form the hand, in our case, the team. Its purpose is for different people to work together, "pollinate" and enrich each other, together create or generate something completely new that goes beyond the abilities of each individual member of the group. Make sure that the performance of the group is higher than of each of its members.

3.2 The Human Capital Structure

Today, the vast majority of company owners do not doubt the existence of a direct link between the quality of human capital management and the financial results of companies. However, as it was mentioned above, understanding does not exclude "blind" following in practice the outdated attitude to personnel and investment in it as a cost center and use of appropriate methodologies for evaluating the quality of personnel management. The last one does not correspond to proactive leadership by virtue of its definition. In addition, a conscious and responsible attitude to the asset "personnel" and its investment requires close attention and concentration of control over the effectiveness of this asset' management in the hands of the company's top officials. Only if there is proactive leadership at this level, we can expect its culture at the lower levels of management in the company [1].

In order to ensure that the attitude towards personnel and the management system is changed in the mind, from the costs source to asset management, it is useful to analyze the structure of human capital [4] in the company in terms of interdependent and complementary components (Fig. 1).

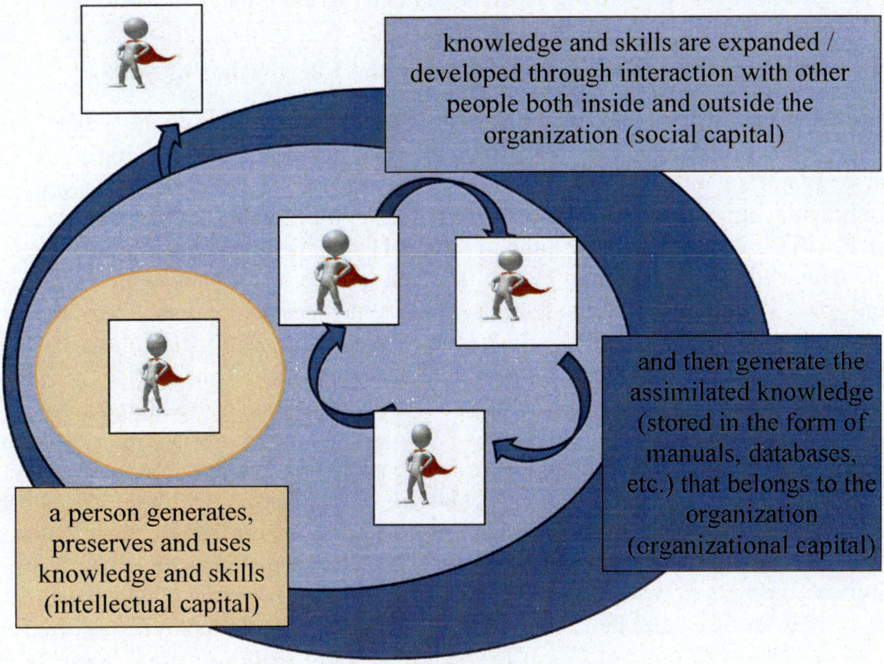

Fig. 1. The structure of human capital in the company (Source: authors).

Thus, the structure of human capital can be represented as 3 components:

- an individual employee who has, multiplies, uses and preserves competencies – intellectual capital,
- employees interact and influence each other's competencies, changing existing ones and creating new ones – social capital,
- the company environment and organizational culture generates further assimilation of competencies, accumulates them and formalizes them in the form of databases and various rules – organizational capital.

Then we see that by analogy with financial asset management, it is extremely important for a director - a proactive leader-to implement three tasks:

- select the “right” employee in the company and in a specific team that requires minimal investment to achieve its “design capacity” in the shortest possible time»,
- to create a culture in the company that will promote a synergistic effect in the work of the team, multiply the return on investment in the development of its members, and, as a result, ensure the implementation of the company’s goals and increase the profitability of the business,
- to measure and evaluate the effectiveness of investment and make decisions based on the results, since without clear measurements, the statement that employees are the greatest asset of the company remains nothing more than just rhetoric.

3.3 Metaprogram Approach to Evaluating the Effectiveness of Staff

There are many different approaches to hiring employees and forming the “right” team, as well as the proposed models for evaluating investment in human capital [2, 8, 13]. The proposed solution is based on the “success model”, “perfection modeling”, the metaprogram approach of Patrick Merlevede [11] and evaluation of the candidate that matches it, the idea of Collins [5], the investment evaluation model of Andrew Mayo – “The Human Capital Monitor®” [9], as well as the theory of spiral dynamics. Employees who are successful in their positions sometimes do not realize how and why they do a good job, which makes them more successful than their colleagues. The reason is that a key part of their actions is automated and located unconsciously. Hence, the usual approaches to designing successful candidate criteria have a high margin [10].

The success modeling approach involves testing and/or interviewing a certain number of the most successful employees who occupy the same position, during which we should find out: their motivation, beliefs they hold, knowledge they consider the most important, studying what skills they use to achieve high results, their levels of values and cultural patterns, manifestations of signs of a particular type of corporate culture according to a typology based on spiral dynamics. From the obtained data, it is necessary to identify common criteria and their values for all employees selected for modeling. They will point out what is crucial for good performance and serve as the basis for a “model of excellence”.

The model can be built on the basis of a competence approach, or go further and find the correspondence of competencies to certain metaprograms, or from the very beginning analyze the key metaprograms of the best employees as criteria, since they form the motivation of a person to master certain competencies and are in fact a prerequisite for specific competencies and competence. In other words, metaprograms, in contrast from the behavioral profile expressed in terms of competencies, can tell us a lot not only about the situation “today”, but also about what we can expect from the employee in the future, about his potential. And from the point of view of proactivity in investing, we will have a clear idea of what competencies training for a particular employee will require more or less investment, and how much they will be justified.

Metaprograms are filters that a person applies to all incoming information. This is what drives a person, their choice of strategies, motivation, their interaction with the world and causes a certain habitual, “studied “ style of thinking [16]. It is important to remember that the models obtained in one company are not universal, that is, they may not be the same for illogical positions in another company, because there are always differences in the corporate culture. The last ones can be supplemented by the specifics of the industry and activity.

Metaprograms of a person are determined during the interview and observations without the use of special automated test methods. Of course, this requires some theoretical and practical training of the employee conducting the assessment. The accuracy of such an assessment, however, may have some margin due to the background of the evaluator's metaprogram profile. It reduces the accuracy of the assessment. The use of special automated test methods allows you to exclude errors. So, not so long ago, the following two JobEQ products entered the market, but they have already proven themselves well:

- iWAM questionnaire that determines motivation and attitude to work,
- VSQ questionnaire that measures values and cultural patterns based on the typology of integral dynamics.

After doing some work, we get a model that will allow us to minimize the risks of choosing the "wrong" candidate, as well as protect us from unnecessary investment in training the wrong people by the wrong things. To do this, we evaluate the compliance of pre-selected candidates who are in the final pool of candidates for this position, as well as all remaining employees who are not included in the list of the "best" employees. The purpose is to analyze how far the degree of deviation from the ideal profile is within acceptable values. As a result, we employ only those candidates who have shown a match or acceptable deviation from the ideal profile, and put under control employees whose profile for one or more position' key criteria significantly exceeds the acceptable values. Based on the obtained systematized data, we make all subsequent decisions on career moves and investment in personnel: hiring, promotion, training, inclusion in the personnel reserve, and employee's termination [12].

3.4 Methodology for Evaluating the Effectiveness of Personnel Taking into Account the Made Investment

Another important step in investment management will be to evaluate the performance of all subsequent and cumulative investment in personnel [8]. For these purposes it is logical to use the formula for evaluating the return on investment:

$$\text{ROI} = \text{profit} * \text{for the reporting period/all personnel expenses of the corresponding category for the reporting period of the} \times \text{discounting}$$

We offer to divide all the company's personnel into categories. The total ROI for all personnel is calculated as the arithmetic average of the received values. We use the usual structure of personnel for financial and accounting reporting:

- main production personnel,
- commercial personnel,
- employees of the administrative and managerial apparatus.

*For each category, the formula includes a specific profit indicator:

- for the main production personnel it will be the difference between revenue and direct expenses, or the margin income,
- for the commercial personnel it is the difference between the profit value for the main production personnel and a part of indirect expenses that, in our opinion, have significantly affected the success of sales, for example, advertising expenses,
- profit value included in the formula for the employees of the administrative and managerial apparatus; – profit for calculating the efficiency of investment for commercial personnel minus all remaining indirect costs.

In addition, we can calculate the individual ROI for a specific employee. In this case the formula will look like this:

$$\text{Employee ROI} = ((\text{profit} * / \text{average number of employees}) / \text{employee expenses}) \\ \text{of the} \times K_{ind}$$

K_{ind} is an average weighted score for the following indicators:

- matching the ideal metaprogram profile;
- potential for inclusion in the personnel reserve;
- contribution, degree of influence on profit;
- matching values with the company's values.

In the K_{ind}, we take into account the employee category by specifying the corresponding indicator "profit*". A measurement scale should be defined for each of the coefficients. It is also important to describe the criteria for awarding a particular point to an employee.

4 Discussion

Nowadays, there are no clear criteria of what concepts used in the discussion of personnel management issues: a competence or a metaprogram. Sometimes the same indicator is called competence and metaprogram at the same time. For example, the target orientation. It is necessary to make a clearer division and get a basic list, which can be further supplemented by second-order metaprograms. In addition, as it was mentioned above, there is only one product on the market that allows you to automate the study of the employee's metaprogram profile. Special training is also required to work with this tool.

Despite the report generated automatically by the system, a large proportion of conclusions require direct participation and analysis of the profile by a specialist working with it. A narrow offer dictates the dependence of companies on the sufficiently established high cost of the product and its training. From the point of view of minimizing investment in personnel evaluation, it would be useful to investigate the strategy of analyzing the metaprogram profile, perform its modeling, and create a basis for the subsequent full automation of the process. There are several other points in which we should continue our research. The specifics of the company's activity may suggest some nuances in calculating the profit value involved in the proposed calculation formula for 3 categories of employees. It is also reasonable to study the criteria and the model building of process in more detail.

5 Conclusion

From our point of view, the proposed methodology for evaluating the effectiveness of investment in personnel, developed on the basis of proactive leadership and a metaprogram approach, is quite convincing and optimal for practical application. Such a methodology could become the basis for changing the approach that is common in the Russian business community to personnel management in terms of costs and switching to the position of human capital theory. The proposed methodology was first tested before the current global economic crisis at one of the small business enterprises operating in the territory of the Russian Federation. And it proved itself well in terms of the ratio of time and financial costs with the received result. The methodology and tools proposed in the article were fully adopted. The proactive approach of the owner and his willingness to look for and use effective tools for managing investment in personnel allowed the company to minimize the impact of the crisis and continue working almost without losses.

References

1. Bernerth, J.B., Hirschfeld, R.R.: The subjective well-being of group leaders as explained by the quality of leader–member exchange. The Leadership Quarterly **27**, 697–710 (2016)
2. Call, M.L., Nyberg, A.J., Thatcher, S.: Stargazing: an integrative conceptual review, theoretical reconciliation, and extension for star employee research. J. Appl. Psychol. **100**, 623–640 (2015)
3. Cambridge Dictionary (2020). https://dictionary.cambridge.org/ru/. Accessed 10 May 2020
4. Coff, R.W., Raffiee, J.: Toward a theory of perceived firm-specific human capital. Acad. Manag. Perspect. **29**, 326–341 (2015)
5. Collins, J.C.: Good to great: Why Some Companies Make the Leap And Others Don't. Mann, Ivanov and Ferber, Moscow (2018)
6. Covey, S.R.: The 7 Habits of Highly Effective People. Alpina Publisher, Moscow (2015)
7. General Manager: the Theory of the thumb from the guru of management Ichak Adizes (2017). https://www.gd.ru/articles/9197-teoriya-bolshogo-paltsa. Accessed 17 Mar 2020
8. Heimann, A.L., Ingold, P.V., Kleinmann, M.: Tell us about your leadership style: a structured interview approach for assessing leadership behavior constructs, The Leadership Quarterly. (2019, In press). https://doi.org/10.1016/j.leaqua.2019.101364. Accessed 15 May 2020
9. Mayo, A.: The Human Value of the Enterprise: Valuing People as Assets – Monitoring, Measuring, Managing. Nicholas Brealey, London (2001)
10. Massaro, M., Handley, K., Bagnoli, C., Dumay, J.: Knowledge management in small and medium enterprises: a structured literature review. J. Knowl. Manag. **20**(2), 258–291 (2016)
11. Merlevede, P.: Talent management: a focus on excellence managing human resources in a knowledge economy (2014). https://hvtc.edu.vn/Portals/0/files/635834387511001885talent-management-a-focus-on-excellence.pdf. Accessed 20 July 2019

12. Nelissen, J., Forrier, A., Verbruggen, M.: Employee development and voluntary turnover: testing the employability paradox. Hum. Res. Manag. J. **27**(1), 152–168 (2017)
13. Podsakoff, P., MacKenzie, S., Lee, J., Podsakoff, N.: Common method biases in behavioral research: a critical review of the literature and recommended remedies. J. Appl. Psychol. **88** (5), 879–903 (2003)
14. SCRIBD: Human capital reporting – CIPD sesson 7 (2005). https://ru.scribd.com/document/241986928/Human-Capital-Reporting-CIPD-Sesson-7. Accessed 15 April 2020
15. Tverdola, N.M.: Leadership and tools for developing leaders in the VUCA environment. Hum. Resources Intell. Resources Manag. Russia **7**(6), 14–18 (2018)
16. Zaharia, C., Reiner, M., Schutz, P.: Evidence-based neuro-linguistic psychotherapy: a meta-analysis. Psychiatria Danubina **27**, 355–363 (2015)

Strategic Features of Forming the Personnel Potential of Land Reclamation in Russia

A. A. Ugryumova and L. E. Pautova(✉)

Federal State Research Institution All-Russia Scientific and Research Institute for Irrigation and Farming Water Supply Systems "Raduga", Kolomna, Russia
feminaa@mail.ru, cosidanie35@yandex.ru

Abstract. The relevance of the research topic is associated with the need to study the characteristics of the formation of the land reclamation's personnel potential as a leading factor in the long-term socio-economic results of the reclamation sector's functioning of the agro-industrial complex of the Russian Federation and the development of individual territories of the Russian Federation. The goal is to study and determine the theoretical, methodological, scientific and practical features of the personnel potential of land reclamation. Tasks are to define the concept of land reclamation's personnel potential, to clarify its structurally meaningful elements, to analyze the elements and factors of the personnel potential of land reclamation, to determine the direction of its development. Methods are system-theoretical analysis, mathematical and statistical processing of primary data, system, logical and correlation analysis, processing and generalization of research results. Results: the concept of "personnel potential for land reclamation", its structurally-substantive elements, methodological and scientific-practical principles for the formation of personnel potential for land reclamation are defined, a scientific and practical study of the personnel potential of 52 Federal State Budget Institutions of all federal districts of the Russian Federation is identified, and strategic features of formation personnel potential of land reclamation.

Keywords: Personnel potential · Personnel structure · Reclamation

1 Introduction

In the context of the implementation of the Resolution of the Government of the Russian Federation of 14 July 2012, No. 717 "On the State Program of Development of Agriculture and Regulation of Agricultural Products, Raw Materials and Food Markets" [20], the issues of studying, identifying and solving the problems of building human resources in the agricultural production sector are relevant. In particular, the problem of formation of the agro-industrial complex reclamation sector's personnel potential is especially relevant.

During the implementation of the personnel policy of Land Reclamation Department of the Ministry of Agriculture of Russia it is possible to identify the main conditions, principles and trends of the land reclamation personnel potential formation and development features. Identification of these characteristics is especially important

S. I. Ashmarina and V. V. Mantulenko (Eds.): IPM 2020, LNNS 161, pp. 757–766, 2021.
https://doi.org/10.1007/978-3-030-60926-9_94

when: 1) taking into account the sectoral and territorial features of the land reclamation systems at the level of federal districts, 2) setting labor standards for specialists and managers of reclamation enterprises, 3) developing and managing the effective use of the labor potential of the industry.

Determination of strategic features of the land reclamation personnel potential formation is aimed at identifying personnel risks and threats in labor relations of the industry, as well as developing of effective measures for successful implementation of the personnel policy and the critical areas of social and economic functioning of the reclamation branch of agricultural industry.

2 Methodology

Theoretical, methodological and strategic features of the formation of the personnel potential of land reclamation in this study are based on:

- concepts of human potential management by Ghasemi, Keykha, Nezhad, Niya [6]; Kalachev [9], Kibanov [11], Kuzmina [14], Maksimova [16], Shamarova [22],
- the concept of regional staffing by Gulyaeva, Buraeva, and Grishaeva [7]; Vartanova [23],
- concepts of innovative development of agriculture by Kozlov [12], Noskova [17]; Primyshev, Cheremisina, and Skaranik [19],
- concepts of industry-specific development of the personnel potential of the agro-industrial complex by Khlusova and Khlusov [10], Senchenko [21], Primyshev, Cheremisina, and Skaranik [19],
- the concept of a system-based point scoring method of the personnel potential's formation by Kalachev [9].

The category of "personnel reclamation potential" is not sufficiently studied and disclosed in industry research despite the noted results of the above authors. From our point of view, it is necessary to determine the scientific, methodological and substantive-structural components of the personnel potential of land reclamation [1]. Research methods are system-theoretical analysis, mathematical and statistical processing of primary data, system, logical and correlation analysis, processing and generalization of research results.

3 Results

It is advisable to highlight the main trends in its formation and development which are based on the results of a system-theoretical and scientific-practical analysis of the concept of personnel potential:

- personal-professional and socio-cultural orientation,
- structurally professional and competency orientation.

These trends in the formation of personnel potential make it possible to determine the totality of its main structurally-substantive elements:

- individual, mental and physiological,
- socio-cultural and demographic,
- professional competence and technology,
- industry, production and innovation,
- self-organization, self-education and self-development.

The results of a theoretical analysis of the personnel potential's concept and the factors of its formation allow us to highlight a certain set of scientific and methodological approaches at the mega-, macro- and micro-levels of socio-economic and industry interaction (Fig. 1).

It is necessary to determine the value of the "human factor" as one of the most capacious categories of content in any production sphere in the context of a strategic approach to the study of personnel potential and management. According to Krechetnikov, personnel potential is a set of qualitative and quantitative characteristics of personnel, including, apart from the number and composition, physical, psychological,

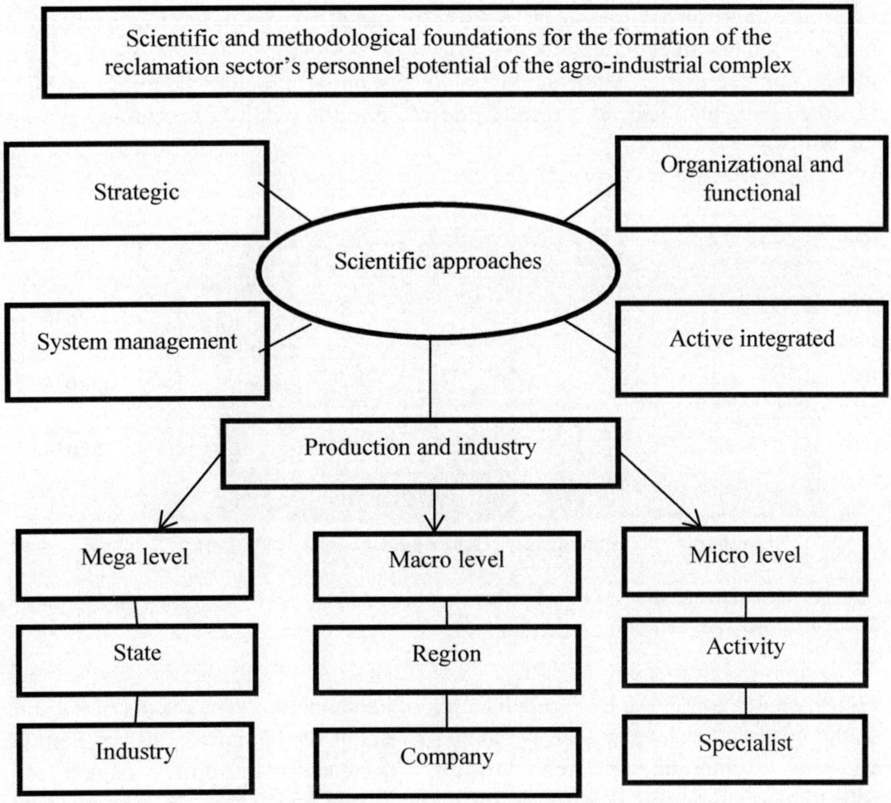

Fig. 1. Scientific and methodological foundations of the personnel potential's formation of the reclamation industry of Russia's agro-industrial complex (Source: authors)

intellectual, creative, professional, qualifying, communicative and other abilities [13]. In addition, in modern conditions, it seems appropriate to distinguish the socio-environmental components of the potential, which determine the features of solving modern socio-economic and sectoral problems [5].

The study of the personnel potential of land reclamation for the period 2015–2017 was held at the Federal State Budgetary Institutions (FSBIs) which are subordinated to the Department of Land Reclamation of the Ministry of Agriculture of the Russian Federation. The study involved FSBIs from all federal districts (hereinafter FD) of the Russian Federation. The total sample of the personnel potential's study of land reclamation by professional categories amounted to 31,533 people.

In our study we studied the following features which are given the selected characteristics of the personnel potential of land reclamation:

1. Structural personnel - structural parts of personnel [8]. In our research we define the following main groups of the reclamation personnel: managers, specialists, other employees, and workers. Dynamics of structural parts of personnel for the period 2015–2017 according to the Federal State Budgetary Institution "Management" Meliodovodkhoz—and the federal districts are presented in Fig. 2. The results of the indicators' analysis of the structural parts of the frames in Fig. 2 allow us to determine that for the period 2015–2017 in the FSBIs the largest decrease in the number of personnel relates to the groups of "specialists" and "managers". This means that there is a deformation of the personnel structure in terms of highly qualified personnel and, as a result, a decrease in the quality of reclamation works (activities).

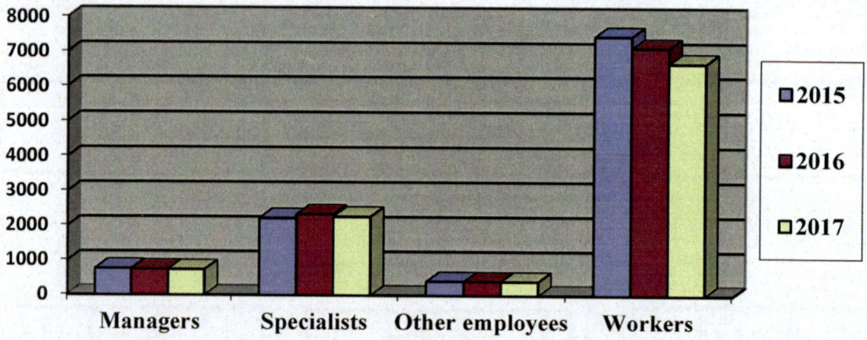

Fig. 2. The dynamics of the personnel structure of the Federal State Budgetary Institutions for 2015–2017,% (Source: authors)

2. Educational feature - the level of education of personnel [4]. The results of the study of the educational level of the personnel potential of the agro-industrial complex allow us to determine the trend towards a decrease in the professional level of education of agricultural workers by main profile specialties. In particular, this applies to the workers with higher and secondary vocational education.

We have studied the indicators of personnel potential - professional growth and self-education in relation to the industry specifics of land reclamation.

3. Professional growth and self-education - advanced training and retraining in all categories of personnel [3]. The features of the dynamics of professional growth and education of personnel potential for land reclamation in our study are determined by the indicators of advanced training and retraining of specialists of the Federal State Budget Institutions for the period 2015–2017. Minimum, maximum and average values of the average annual percentage of employees trained in the system of additional professional education (APE), their age structure, the number of employees studying in the programs and areas of APE for 2015–2017 in all federal districts of Russia are presented in Table 1.

Table 1. The average annual percentage of employees of the FSBIs for 2015–2017 and their average growth,%

Indicators	Percentage of employees trained in the APE system			
	Managers	Specialists	Other employees	Workers
CFD				
Minimum average annual percentage	0	8,33	0	0
Maximum average annual percentage	50,71	28,52	11,11	33,33
Average annual percentage	26,97	18,92	1,85	8,71
The scope of the average annual percent	50,71	20,19	11,11	33,33
Average increase of average annual percent	5,07	4,03	1,85	5,61
NWFD				
Minimum average annual percentage	3,7	6,06	0	0
Maximum average annual percentage	50	40	25	21,98
Average annual percentage	28,86	22,16	9,38	3,66
The scope of the average annual percent	46,3	33,94	25	21,98
Average increase of average annual percent	−10,65	3,87	−8,33	3,57
NCFD				
Minimum average annual percentage	0,00	2,13	0,00	0,00
Maximum average annual percentage	47,33	22,07	26,14	1,42
Average annual percentage	19,94	9,22	8,71	0,71
The scope of the average annual percent	47,33	19,94	26,14	1,42
Average increase of average annual percent	10,17	2,60	0,40	−0,44

(*continued*)

Table 1. (*continued*)

Indicators	Percentage of employees trained in the APE system			
	Managers	Specialists	Other employees	Workers
SFD				
Minimum average annual percentage	29,36	16,14	0,00	0,00
Maximum average annual percentage	45,59	26,92	38,46	10,38
Average annual percentage	37,46	20,12	14,50	5,55
The scope of the average annual percent	16,23	10,78	38,46	10,38
Average increase of average annual percent	−3,78	1,03	2,46	−0,33
VFD				
Minimum average annual percentage	9,26	3,81	0,0	0,0
Maximum average annual percentage	83,33	60,44	26,67	3,69
Average annual percentage	41,16	22,58	7,98	0,99
The scope of the average annual percent	74,07	56,63	26,67	3,69
Average increase of average annual percent	0,44	8,03	2,50	0,81
UFD				
Minimum average annual percentage	4,76	3,45	0,00	0,00
Maximum average annual percentage	12,50	15,28	0,00	2,38
Average annual percentage	8,63	9,37	0,00	1,19
The scope of the average annual percent	7,74	11,83	0	2,38
Average increase of average annual percent	−6,70	3,13	0,00	1,79
SibFD				
Minimum average annual percentage	8,57	3,70	0,00	0,00
Maximum average annual percentage	133,33	116,67	16,67	23,81
Average annual percentage	40,36	43,71	5,21	4,03
The scope of the average annual percent	124,76	112,97	16,67	23,81
Average increase of average annual percent	−4,31	12,32	0,45	1,73
FEFD				
Minimum average annual percentage	20,83	0,00	0,00	0,00
Maximum average annual percentage	55,56	42,86	43,08	2,26
Average annual percentage	40,84	16,81	10,58	0,84
The scope of the average annual percent	34,73	42,86	43,08	2,26
Average increase of average annual percent	−3,16	−0,62	2,17	−0,28

Source: authors' calculations which are based on data received from 52 Federal State Budgetary Institutions.

According to (Table 1) it follows that the smallest average annual percentage of all employees trained in the APE system is zero, and the largest percentage of managers and specialists trained in the APE system was observed in the Siberian Federal District, other employees in the Far Eastern Federal District and workers in the Central Federal District.

It makes possible to identify a tendency toward an increase in the professional level of the personnel potential of the FSBIs according to the data which was obtained from the analysis of training indicators in the APE system: managers and specialists, except of the Siberian Federal District; employees, except of the Far Eastern Federal District, and workers, except of the Central Federal District [18]. In its turn, it will contribute to improve the professional level of the personnel potential of the reclamation industry as a whole.

The authors investigated factors of formation and development of the personnel potential of land reclamation:

1. Social-demographic - the age of employees and their trends [2]. The results of the age categories' analysis of the land reclamation's personnel potential shows that in FSBIs for 2015–2017 the main share of workers is between the ages of 50 and 60 years old - 32%, at a more able-bodied age - workers between the ages of 40 and 50 years, makes up 20%, between the ages of 60 and 65 years - 17%. The problem is that the category of young workers under 30 is not significant enough, accounting for only 7% of the total number of employees. Thus, according to the socio-demographic indicator "age", the personnel potential of land reclamation needs "rejuvenation" [15]. In this regard, it seems advisable to identify and include, during the implementation of the Departmental Program "Development of the Land Reclamation Complex of Russia", sectoral programs for the formation and development of human resources at all levels (from higher education to support in real activities).
2. Industry - the dependence of the personnel's number on the fixed area of the reclaimed land. Linear correlation coefficients for the FSBIs in federal districts were calculated to identify the dependence of the number of FSBIs' workers on the fixed area of the reclaimed land in 2015, 2016 and 2017 (Table 2).

Table 2. Coefficients of linear correlation of reclaimed land's areas which are assigned to the FSBIs and the number of FSBIs' employees in 2015–2017

FD	Reclaimed land area, thousand ha		
	2015	2016	2017
CFD	−0,09	−0,11	−0,04
NWFD	0,97	0,98	0,97
NCFD	0,98	0,98	0,98
SFD	0,86	0,86	0,83
VFD	0,64	0,64	0,64
UFD	1	1	1
SibFD	0,71	0,71	0,74
FEFD	0,63	0,63	0,64

Source: authors

Analysis of (Table 2) shows that the relationship between the area of reclaimed land and the number of employees of the FSBIs: in the Central Federal District is absent in 2015 and 2017; in the North-West Federal District, North Caucasian Federal District and the Ural Federal District, direct (directly proportional) and very strong; in the Southern Federal District, direct and strong; in the Volga Federal District, Siberian Federal District and the Far Eastern Federal District, direct and noticeable.

It's based on the presented correlation analysis, we can conclude that in all federal districts, except for the Central Federal District, a direct correlation relationship is observed between the irrigated area of agricultural land and the number of sectoral personnel, which corresponds to the existing regulatory approaches in staffing the Federal State Budget Institution for Land Reclamation.

3. Normative legal - legislative and normative acts on personnel policy of the industry, regulation of managerial work and workers. The authors identified the regulatory and methodological, legislative and industry specifics of the formation and development of the personnel potential of the Land Reclamation's Department.

4 Discussion

To sum up the results of the study of the personnel potential's features of land reclamation, it is necessary to highlight the following strategic features of its formation and development: 1) lack of interconnection in the regulatory documents of the Department of Land Reclamation on the formation and development of human resources, which determines the need to update the regulatory, methodological, legislative and sectoral support of issues on the personnel potential of land reclamation and agro-industrial complex, in general; 2) the deformation of the personnel's sectoral structure in terms of highly qualified personnel and, as a consequence, a decrease in the quality of reclamation work (activities); 3) a direct correlation between the irrigated area of agricultural land and the number of sectoral personnel was determined; 4) a decreasing trend in the specific gravity of the age group of employees of the FSBIs "up to 30 years" in the number of age groups of agricultural workers and an increasing trend in the specific gravity of groups above 50 years was determined [24, 25].

5 Conclusion

On the basis of the presented research, the following conclusions can be drawn:

- the modern understanding of the category "personnel potential" for land reclamation needs both a methodological and a legislative study ensuring the continuity of this concept in the most important strategic industry documents;
- it seems necessary to identify and neutralize personnel threats that could narrow the industry's personnel potential;
- it is required to ensure a scientific approach to the management of human potential and human resources of the industry, since these resources are the key to the intensive development of reclamation organizations.

The authors identified the regulatory and methodological, legislative and industry specifics of the formation and development of the personnel potential of the Department of Land Reclamation.

References

1. Baharom, R., Hamid, Z.A., Yassin, I.M., Wahab, N.A., Hamzah, N., Yussoff, Y.M.: Implementation of industrial training course: Enhancement of processes and procedures. Paper presented at the IEEE Region 10 Annual International Conference, Proceedings/TENCON, 2017-December, 2712–2715 (2017). https://doi.org/10.1109/tencon.2017.8228322
2. Busari, A.H., Khan, S.N., Abdullah, S.M., Mughal, Y.H.: Transformational leadership style, followership, and factors of employees' reactions towards organizational change. J. Asia Bus. Stud. **14**(2), 181–209 (2019). https://doi.org/10.1108/JABS-03-2018-0083
3. De Guimarães, J.C.F., Severo, E.A., Campos, D.F., El-Aouar, W.A., Azevedo, F.L.B.: Strategic drivers for product and process innovation: a survey in industrial manufacturing, commerce and services. Benchmarking **27**(3), 1159–1187 (2019). https://doi.org/10.1108/BIJ-12-2018-0403
4. Donate, M.J., Ruiz-Monterrubio, E., Sánchez de Pablo, J.D., Peña, I.: Total quality management and high-performance work systems for social capital development: effects on company innovation capabilities. J. Intell. Capital, **21**(1), 87–114 (2019). https://doi.org/10.1108/jic-07-2018-0116
5. Fan, W.: Internal control, board structure and executive compensation - Performance sensitivity. Paper presented at the Proceedings - 2019 International Conference on Economic Management and Model Engineering, ICEMME 2019, pp. 669–673 (2019). https://doi.org/10.1109/icemme49371.2019.00137
6. Ghasemi, M., Keykha, I., Nezhad, A.A., Niya, H.M.: The strategies and challenges of human resources management. Fundament. Appl. Res. Modern World **12–5**, 152–157 (2015)
7. Gulyaeva, T.I., Buraeva, E.V., Grishaeva, OYu.: Staffing the agricultural sector of regional agro-industrial complex: analysis of condition and areas of improvement. Econ. Anal. Theory Pract. **14**(31), 26–38 (2015)
8. Hensel, R., Visser, R.: Does personality influence effectual behaviour? Int. J. Entrepreneurial Behav. Res. **26**(3), 467–484 (2019). https://doi.org/10.1108/IJEBR-06-2019-0343
9. Kalachev, A.M.: An integrated system of HR scoring in the organization. The Manager **910**(37–38), 66–69 (2012)
10. Khlusova, I.A., Khlusov, V.N.: Industry specifics of formation and development prospects of the personnel potential of the agro-industrial complex. Econ. Res. 2 (2017). https://cyberleninka.ru/article/n/otraslevaya-spetsifika-formirovaniya-i-perspektivy-razvitiya-kadrovogo-potentsiala-agropromyshlennogo-kompleksa. Accessed 17 July 2020
11. Kibanov, A.Ya.: Methodology of motivation and stimulation of organization's personnel. Manag. Personn. Intell. Resources Russia, **3**(1(10)), 5–10 (2014). https://doi.org/10.12737/2619
12. Kozlov, A.V.: Staffing of agricultural industry in the context of innovative development. Doctor thesis. Moscow (2015)
13. Krechetnikov, K.G.: Features of the formation and development of the personnel potential of the organization. Syst. Values Modern Soc. **37**, 168–172 (2014)
14. Kuzmina, N.V.: The system of human resources management in integrated corporate structures. Doctor thesis. Moscow (2004)

15. Liu, M.-L., Lin, C.-P., Chen, M.-L., Chen, P.-C., Chen, K.-J.: Strengthening knowledge sharing and job dedication: the roles of corporate social responsibility and ethical leadership. Leadership Organ. Dev. J. **41**(1), 73–87 (2019). https://doi.org/10.1108/LODJ-06-2019-0278
16. Maksimova, L.V.: The components of the personnel potential of an organization. New Word Sci. Pract. Hypothes. Approb. Res. Results **12**, 178–183 (2014)
17. Noskova, M.V.: Key factors in managing the development of human resources in agriculture. Economics in agro-industrial complex. Bull Altai State Agrarian Univ. **66**(4), 109–112 (2010)
18. Olgarenko, G.V., Ugryumova, A.A., Zamahovsky, M.P., Pautova, L.E.: Methodological approaches to the formation of the meliorative complex's personnel support in the Russian Federal districts. Paper presented at the International Multidisciplinary Scientific GeoConference Surveying Geology and Mining Ecology Management, SGEM, **19**(3.1), 369–377 (2019). https://doi.org/10.5593/sgem2019/3.1/s12.048
19. Prismyshev, I.N., Cheremisina, S.G., Skaranik, S.S.: Modern state of productive and skilled potential of the agricultural complex of Crimea. Russian J. Entrepreneurship **19**(5), 1597–1610 (2018)
20. Resolution of the Government of the Russian Federation of 14 July 2012, No. 717 "On the State Program of Development of Agriculture and Regulation of Agricultural Products, Raw Materials and Food Markets" (as amended on 31 March, 2020) (2012). http://docs.cntd.ru/document/902361843. Accessed 27 July 2020
21. Senchenko, E.V.: The personnel potential of the industry: basic general theoretical aspects. Econ. Manag. Anal. Trends Dev. Prospects **22**, 71–78 (2015)
22. Shamarova, G.M.: Issues of human potential management. Personnel Management, **8** (2008). http://www.top-personal.ru/issue.html. Accessed 17 July 2020
23. Vartanova, M.L.: Regional aspects of the development of agro-industrial complex, rural areas and food security. Russian J. Entrepreneurship **18**(5), 869–886 (2017). https://doi.org/10.18334/rp.18.5.37649
24. Winkowski, C.: Classification of forecasting methods in production engineering. Eng. Manag. Product. Serv. **11**(4), 23–33 (2019). https://doi.org/10.2478/emj-2019-0030
25. Zamachowski, M.P., Pautova, L.E., Tyurina, L.M., Sosnovskikh, I.D., Yezhikova, T.S.: Research of the influence of personnel structure in irrigated agriculture on the system of professional development and retraining of personnel. RandD report AAAA-A19-119090290014-6 (2019)

Methodology of Labor Remuneration Differentiation in the Context of Human Capital Assessment

E. Yu. Cherkesova, N. E. Demidova(✉), and D. D. Mironova

Institute of Service and Entrepreneurship (Branch)
"Don State Technical University", Shakhty, Russia
cherkesova.elvira@yandex.ru, ndemidoffa@rambler.ru,
mironova06-86@inbox.ru

Abstract. The article analyzes the author's methods of differentiation of base salaries and base salary rates by professional qualification groups in the budget sphere. These methods allows differentiation between base salaries, both within the professional qualification group, and between the levels of professional qualification groups, depending on the value of the employee's human capital. The proposed integral coefficient reflects the excess of the base valuation of human capital of the employee, the appropriate qualification level compared to the base valuation of human capital of the lower qualifying level of professional qualification group. The coefficient of intellectual and labor personnel efforts provides an opportunity to objectively differentiate base salaries and base salary rates on professional qualification groups. The use of these coefficients provides condition for objective differentiation of base salaries (base official salaries, base salary rates).

Keywords: Base salary · Base salary rate · Professional qualification group · Monetary evaluation of the employee's human capital · Labor efforts of the staff · Expenditures for education

1 Introduction

Within the conditions of rapid renewal and multi-faceted transformation of the technological base of modern society, a person, as a holder of unique knowledge, skills, and accumulated human capital, acquires a completely new meaning in the socio-reproductive process. Because of this, investments in knowledge and intelligence are multiple justified, they bring additional dividends, thereby providing its owner with competitive advantages.

If an employee receives special training in their organization, this means that their salary when moving to another organization will not depend on such training, they will not pay the costs associated with it, so the company must bear the costs. But the firm will also benefit from special training in the form of increased profits due to higher productivity. Then the profitability of special training will be determined by the excess of the discounted profit over the costs incurred, of the profit that is due to the preparation.

S. I. Ashmarina and V. V. Mantulenko (Eds.): IPM 2020, LNNS 161, pp. 767–776, 2021.
https://doi.org/10.1007/978-3-030-60926-9_95

2 Methodology

The authors have developed methods of objective differentiation of base salaries and base wage rates on professional qualification groups of public sector employees. Let's form a condition for objective differentiation in labour remuneration in the j-th professional qualification group, $Sb_{ij.}$ Let's indicate base salary (base official salary, base salary rate) on the *i* qualification level of the *j* professional qualification group, rub. Then:

$$Sb_{(i+1)j} / Eb_{ij} = Eb(_{i+1)j} / Sb_{ij} \text{ and } \quad Eb_{ij/} Eb_{1j} = Sb_{ij} / Sb_{1j} \tag{1}$$

where $i = \overline{1, N}, \quad j = \overline{1, M},$

N is the number of levels in the professional qualification group;

M is a number of professional qualification groups;

Eb_{ij}- basic monetary evaluation of the human capital of an employee whose qualification characteristic corresponds to the *i*-th qualification level of the j-th professional qualification group, rub.

Sb_{1j} - base salary (base official salary, base salary rate) of the first (lowest) qualification level of the *j*-th professional qualification group, rub.;

Eb_{1j} - basic monetary evaluation of the human capital of an employee whose qualification characteristic corresponds to the first (lowest) qualification level of the *j*-th professional qualification group, rub.;

Fulfilling the logical condition (1) means requiring a unidirectional change in the base salary within the professional qualification group in proportion to the base cost of human capital from the minimum to the maximum level. The relationship between the human capital estimates of employees belonging to different qualification levels in a professional qualification group is as follows:

$$Ebi_j = Kint \ * \ Sb_{1j} \tag{2}$$

from here

$$Kint_{ij} = Ebi_j / Eb_{1j} \tag{3}$$

Here $Kint_{ij}$ is an integral coefficient that reflects the excess of the basic value of the human capital of the employee, corresponding to the *i*-th qualification level of the *j*-th professional qualification group in comparison with the basic monetary estimates of the human capital of the lower qualification level of the j-th professional qualification group.

Based on (1) taking into account (3), we get:

$$Sb_{ij} = Sb_{1j} \ * \ Kint_{ij} \tag{4}$$

Thus, the practical meaning of the integral coefficient is to determine the degree of differentiation between base salaries in the professional qualification group, depending on the monetary estimates of the employee's human capital.

3 Results

The higher the qualification level in a professional group, the higher the value of the integral coefficient of differentiation. Let's analyze the evaluation criteria of complexity and performance of personnel (Table 1).

Table 1. Criteria for assessing the complexity and performance of work that corresponds to the*i*-th qualification level of the j-th professional qualification group

Level	Value of integral coefficient	Value of criterial coefficient K_F	Criteria
i	$\text{Kintij} = \prod_{k=1}^{P} K_{Fix}$ $k = 1...P$ number of criteria in the level	1–1,20	1. Education that meets the requirements of professional qualification group
		1–1,25	2. Complexity and variety of performed labour functions
		1–1,3	3. Ability to solve professional tasks independently
		1–1,25	4. Ability to implement innovative approaches in solving professional tasks
		1–1,30	5. The scale of the performed functions and its diversity

Source: authors.

The values of the criterial coefficients shown in Table 1 should be considered as recommended, set to ensure a differentiation between levels in the professional qualification group of at least 20%. The condition for objective differentiation of base salaries (base official salaries, base salary rates) between professional qualification groups follows from the analysis of the intellectual and labor efforts of personnel necessary for attribution to a professional qualification group.

Let's introduce the coefficient of intellectual and labor efforts of the staff of the *i*-th level *j*-th professional qualification group - K_{ILEij}. The practical point K_{ILEij} is that it is possible to objectify the process of differentiation of base salaries (base official salaries, base wage rates) both within the group and amid professional qualification groups.

The coefficient of intellectual and labor effort (K_{ILEij}) can be determined by taking into account the total cost of education required to master the qualification corresponding to the *i*-th qualification level of the j-th professional qualification group. In addition, it is necessary to reflect the possible level of compensation from the society of the staff labor efforts at the current level of economic development.

$$K_{ILEij} = 1 + a\left(K_{EDij} - 1\right), \quad (5)$$

here

$$K_{EDij} = R_{ij} / R_{11} \tag{6}$$

where α is the share of wages in gross domestic product. This value should be taken as one that contributes to the legalization of "gray wages", according to the experience of socially-oriented countries (about 50%);

R_{ij} - the total cost on obtaining the education, necessary to master the qualification corresponding to the *i*-th level of the j-th professional qualification group, rub.;

R_{11}– the total cost on obtaining the education, necessary to perform the work for which the minimum qualification requirements are imposed, rub.

K_{EDij} - a coefficient that reflects the ratio of the total reduced costs for obtaining education, necessary for mastering the qualification corresponding to the *i*-th qualification level of the j-th professional qualification group compared to the minimum necessary qualification for performing simple work.

The total cost on education, taking into account the lost profits, will be justified after a certain time period. The current value of future benefits from investment in education for an employee of the *i*-th qualification level of the j-th professional qualification group is defined as

$$B_{ij} = \sum\nolimits_{t=1}^{n} \frac{B_{t_{ij}}}{(1+r)^{t}}, \tag{7}$$

where B_{tij} is the increment of the individual's income due to the education received in the t-th period;

n - the time period of using the acquired knowledge;

r - discount rate, which takes into account the rate of capital return, the level of risks in the process of obtaining education and the level of inflation.

It is obvious that as the period n increases, the value of future benefits decreases. In modern conditions of dynamic changes in technologies and principles of economic activity, the period of effective use of the received initial knowledges is limited to no more than 10 years.

Thus, it is possible to compare the level of income increment in different professional qualification groups, which will be proportional to K_{EDij}:

$$\frac{\sum\nolimits_{i=1}^{10} B_{tij}(1+r)^{-t}}{\sum\nolimits_{i=1}^{10} B_{t11}(1+r)^{-t}} = \frac{R_{ij}}{R11} = K_{EDij}, \tag{8}$$

The value of future benefits correlates with the amount R_{ij} - the total present costs of obtaining the education necessary to master the qualification corresponding to the *i*-th level of the j-th professional qualification group. Taking into account the levels of education, you can write:

$$R_{ij} = \sum\nolimits_{t=1}^{t_0} C_{t_{0_{ij}}} \beta_t + \sum\nolimits_{t=1}^{t_0} C_{t_{ij}} \beta_t + \sum\nolimits_{t=1}^{t_0} D_{t_{ij}} \beta_t + \sum\nolimits_{t=t_0}^{T} P_{t_{ij}} \beta_t, \tag{9}$$

Where t = 1…T – time periods;

t_o – year of the beginning of employment by the individual, $1 \leq t_o \leq T$;

$\sum_{t=1}^{t_0} C_{t_{0_{ij}}} \beta_t$ – direct costs on general secondary education;

$\sum_{t=1}^{t_0} C_{t_{ij}} \beta_t$ – direct costs on professional education and training corresponding to the *i*-th qualification level of the j-th professional qualification group;

$\sum_{t=1}^{t_0} D_{t_{ij}} \beta_t$ – indirect costs (lost profit), reflecting the non-received income during the period of education to master the qualification characteristics corresponding to the*i*-th qualification level of the j-th professional qualification group;

$\sum_{t=t_0}^{T} P_{t_{ij}} \beta_t$ – direct costs on professional development, professional skills corresponding to the *i*-th qualification level of the j-th professional qualification group.

Here β_t is the discount coefficient in the form

$$\beta_t = (1 + r)^t, \quad (10)$$

if you are bringing past expenses for education to the time you start working or in the form of

$$\beta_t = 1/(1 + r)^t, \quad (11)$$

if the future costs of professional retraining and professional development are brought to the point of starting work.

Thus, the condition for objective differentiation of base salaries (base official salaries, base wage rates) is formulated as follows:

$$S_{Bij} = S_{B11} * K_{IELij} * Kint_{ij}. \quad (12)$$

Here S_{B11} is the base salary (base official salary, base salary rate) of the least qualified employee, rub.

The qualification of the employee of a lower qualification level of higher occupational qualification group (within the macro - occupational qualification group defined by the relevant order of the Ministry of Health) may not be lower than the qualification of high qualification level of the previous occupational qualification group, therefore, the base salary (base salary, base wage rate) the lower category of the higher occupational qualification group should not be less than the base salary (base salary, base wage rate) highest qualification level of the previous occupational qualification group, that is true:

$$S_{B1j} \geq S_{Bim(j-1)}, \quad (13)$$

where i_m is the maximum level (j − 1) of the 2nd professional qualification group within the macro-professional qualification group defined by the relevant order of the Ministry of Health and social development of the Russian Federation.

The base salary (base official salary, base salary rate) corresponding to the maximum qualification level of the (j − 1) professional qualification group within the macro-professional qualification group defined by the relevant order of the Ministry of Health of the Russian Federation can be expressed as:

$$S_{Bim(j-1)} = S_{B11} * K_{ILEim(j-1)} * K_{INTim(j-1)}, \tag{14}$$

where i_m is the maximum qualification level of the (j − 1) professional qualification group.

To ensure objective differentiation of base salaries (base official salaries, base salary rates), a correction mechanism that will allow us to consider a higher level of human capital corresponding to the lower qualification level of the j-th professional qualification group compared to the maximum level of the j (j − 1)-th professional qualification group, should be introduced. In other words, the increment of human capital between the lowest level of the j-th professional qualification group and the maximum level of the (j − 1)-th professional qualification group will be proportional to the ratio:

$$\frac{K_{ILE1j}}{K_{ILEim(j-1)}} \tag{15}$$

Expression (15) reflects the relationship between the base salary (base salary, base salary rate) of the maximum level of the (j − 1)-professional qualification group and the first level of the j-th professional qualification group.

Thus, it is fair to say that the increase in the base salary (base salary, base salary rate) of the first level of the j-th professional qualification group in comparison with the base salary (base salary, base salary rate) of the maximum level of the (j − 1)-professional qualification group is proportional to the increment of human capital. Then, considering (13) and (15), we write:

$$S_{B1j} = S_{Bim(j-1)} * \frac{K_{ILE1j}}{K_{ILEim(j-1)}}, \tag{16}$$

After the conversion, we have:

$$S_{B1j} = S_{B11} * K_{ILE1j} * \prod\nolimits_{m=1}^{j-1} K_{INTim^m}. \tag{17}$$

$m \in z$, $J = \overline{2, M}$, professional qualification groups within the macro-professional qualification group defined by the relevant order of the Ministry of Health of the Russian Federation.

To determine the base salary (base salary, base wage rates) of the *i*-th qualification level *j*-th professional qualification group is necessary in the formula (17) provide the values of the integral gain factor of the intellectual and labour efforts corresponding to the *i*-th qualification level of the j-th professional qualification group:

$$S_{B1j} = S_{B11} * K_{ILE1j} * \prod_{m=1}^{j-1} K_{INTim^{m}} * K_{INTij}. \tag{18}$$

4 Discussion

Let's analyze the aspects of the formation and functioning of human capital from a scientific and methodological point of view. As Becker, the founder of the theory of human capital, wrote [1, 2], initially, the task of his research was to understand the financial return on investment in secondary and higher education (for example, in the United States), how such investments are reflected in labor remuneration. However, the task soon expanded, since there was no methodology for evaluating investment in people, which Becker began to develop. As a result, Becker formulated a number of interesting propositions, which, although they were derived from observations, were theoretically justified by him. Among them:

- the rate of earnings growth decreases with age, but the rate of deceleration depends on the level of education;
- young people are more likely to change jobs than the older generation;
- the degree of natural talent has a positive impact on achieving higher levels of professional training and educational levels;
- impulsiveness is a characteristic of the typical investor in human capital, which leads to more frequent mistakes, compared to the typical investor in "tangible" capital.

A person who invests primarily in himself and understands the importance and profitability of such investments has a more subtle organization of the mental health, as well as due to upbringing, and a tendency to in-depth thinking and understanding not only the processes of the economic field, but also processes of a philosophical and worldview character. Here we come to the first, perhaps the most "subtle", intangible feature of investment in human capital, not even the actual investment, but its precondition.

Becker examines in detail the training on the job, because its example clearly shows the relationship between investment in human capital and the return on them in the form of wages [2]. Increasing the future productivity of the employee is provided by the cost of training in a set of areas: the cost of time and labor effort of the student, the cost of time and, accordingly, the remuneration of those who act as teachers, the cost of raw materials, equipment deterioration. The "costs" and "investments" categories are distinguished in the listed areas: on the one hand, the mentioned costs are the expenditure, as they could do production job instead of using educational process to go on manufacturer products, on the other hand, they also act as investment because they are aimed at future production growth. This, in our opinion, is the second most important feature of investment in human capital [8, 9]. Of course, the amount of expenses, both in their financial and temporary components, directly depends on the complexity of work and the level of required skills.

In our opinion, methodological joint financial return from the human capital investment can be formulated as the sum of the financial return on investment in the human capital for the employee (Fp) and for the employer organization (Fo). The ratio of this model's components depends on a number of circumstances. For example, if the situation on the labor market is tense, there is a decrease in employment, a fall in the number of jobs, firms are implementing a policy of austerity on labor remuneration, the supply of labor is excessive, then the second component of the Fo will prevail over the first. In a situation favorable from the labor demand, employment expansion, creation of new jobs, in the condition of the priority in the corporate policy in the preservation of highly qualified personnel, the first component of the FR will either prevail over the second, or occupy an equal position with it [3].

An important factor is the relationship between the willingness to pay for special training and staff turnover [7]. If trained employees move to other organizations, the firm will lose some of its capital expenditures, they will not bring it financial returns in the future. At the same time, retired employees who have special training will suffer losses, as they will not be able to find a suitable job with a high salary. Another feature of investment in the human capital is revealed here: replacing some individuals who have acquired a high degree of professional qualification with others who do not possess it to the same extent (although potentially capable) leads to a decrease in productivity and income. This is the important difference of the human capital as an asset and material and informational assets (the carrier of which is not a person): personification [6, 10].

As a result of the peculiarities of general and special training, firms are more interested in reducing the turnover of specialists with special training than those who have general training. Personnel with the general qualifications can be easily replaced through the labor market mechanisms, and companies have not spent money on providing such. Another thing is that specialists who have unique skills for a particular company. Firstly, the owners of the companies have invested in their training, and secondly, they can not be easily, faster and cheaply replaced. Companies will be interested in raising wages for such personnel.

We should also consider the situation of training in specialized educational institutions that provide vocational training at the level of secondary or higher education. The student's earnings are usually either absent during the training period, or significantly lower than they could be, since they cannot work full-time. We can assume that the net earnings for the period of training is the difference between the potential earnings and the total, both direct and indirect costs [5].

It is necessary to clarify that the costs associated with education, should be recognize as the investment on the basis that the income (as well as other benefits such as more comfortable working conditions) of an educated worker will be got by him for many years of work and are assumed to be higher compared to an employee who did not receive education of the appropriate level. This is a consequence of the growth of human capital as a result of education.

One of the directions of investment in the human capital is to invest in maintaining and improving the state of physical and psychoemotional health. Investment in health contributes, firstly, to higher productivity due to reduced loss of working time due to illness, secondly, it makes it possible to intensify work without compromising the psychophysical state of the employee, and thirdly, it allows you to work for a longer time, which is especially important amidst the aging population. As a result, the financial return of investment in human capital increases, both for the employee and for the organization [4].

One of the founders of the theory of human capital is Schultz, who, using the example of developing countries, proved that investment in education is the most important factor in economic growth. Schultz refered to the cost of labor the cost of education, the cost of working time that is lost in the process of training and could be paid if the student started working at an earlier age [9]. Accordingly, the forms of investment in human capital were recognized as investments in school education, training in professional skills at the workplace, health care costs, and, importantly, the cost of increasing awareness of the economy state, the availability of jobs, so that you can use this knowledge for employee mobility in the labor market in order to ensure higher labor remuneration.

Schultz considered the cost of human capital formation to be an investment on the basis that the investor receives part of the current income in order to increase future income [8]. The most important, in our opinion, aspect of investment in the human capital, which, although not a distinctive feature of this type of investment, is important for understanding their essence: investment in the human capital serves as a necessary basis for future business, along with investment in real capital, without investment in the human capital, it is impossible to organize a productive business, its stable long-term functioning [3].

5 Conclusion

Thus, an improved methodological approach to the economic assessment of human capital in the field of education and science should be based on an objective differentiation of remuneration rates, which will allow us to give quantitative estimates of the human capital of employees, taking into account the following areas:

- firstly, the socio-economic situation in the region, which is reflected in the establishment of the minimum base salary (base official salary, base wage rate) of the least qualified employee is not lower than the minimum subsistence of the working population in the region,
- secondly, macroeconomic performance indicators, which is reflected in the adjustment of the cost of the labor efforts of the employee compensation by the company, taking into account the share of labor remuneration in the gross domestic product,
- thirdly, evaluation of employee's human capital, relevant to qualifications through the introduction of the coefficient of staff intellectual and labour efforts, expressing the level of training required to perform the work in accordance with the requirements of the professional qualification requirements,

- fourthly, the specifics of working at a specific workplace through a hierarchical system of integral coefficients that reflect the level of human capital through the professional competence of personnel, depending on the qualification level in the professional qualification group,
- fifthly, increasing the economic significance of base salaries (base official salaries, base salary rates) in the system of staff remuneration by integrating into the basic part of the salary increasing coefficients that consider the scale of professional tasks, the variety of performed functions, the ability to innovate and other factors of complexity, intensity and work productivity,
- sixthly, the desire to ensure social stability along with economic justification, which is achieved by focusing of differentiation degree of base salaries (base official salaries, base wage rates) on the experience of countries with a socially oriented market economy.

References

1. Becker, G.S.: Human capital: A theoretical and empirical analysis, with special reference to education. Columbia University Press for NBER, New York, N.Y. (1964)
2. Becker, G.S.: Investment in human capital: a theoretical analysis. J. Political Econ. **5**(2), 9–49 (1962)
3. Cherkesova, E.Y., Breusova, E.A., Savchishkina, E.P., Demidova, N.E.: Competitiveness of the human capital as strategic resource of innovational economy functioning. J. Adv. Res. Law and Econ. Romania **7**(21), 1662–1667 (2016)
4. Collins, N., Chou, Y.-M., Warner, M., Rowley, C.: Human factors in East Asian virtual teamwork: a comparative study of Indonesia, Taiwan and Vietnam. Int. J. Hum. Resour. Manag. **28**(9), 1475–1498 (2017)
5. Cooke, F.L., Liu, M., Liu, L.A., Chen, C.C.: Human resource management and industrial relations in multinational corporations in and from China: challenges and new insights. Hum. Resour. Manag. **58**(5), 455–471 (2019)
6. Demidova, N., Akilina, O., Kirzhetska, M., Lagovskyi, V., Besarab, S.: Accumulation and fulfilment of the human capital potential in order to strengthen the economic security. J. Secur. Sustain. Issues **8**(4), 801–813 (2019)
7. Helfat, C.E., Martin, J.A.: Dynamic managerial capabilities: review and assessment of managerial impact on strategic change. J. Manage. **41**(5), 1281–1312 (2015)
8. Schultz, T.: Investment in human capital. Am. Econ. Rev. **51**(1), 1–14 (1961)
9. Schultz, T.: Investment in human capital: the role of educacation and of research. Am. J. Agric. Econ. **53**(4), 692–693 (1971)
10. Zhu, Y., Yang, H., Bai, G.: Relationship between staff competence and performance of service-oriented manufacturing enterprises in China. Transform. Bus. Econ. **18**(1), 243–249 (2019)

Social Dialogue as a Tool of Entrepreneurship Support

M. A. Andrianova(✉) and N. A. Potapov

Moscow State Institute of International Relations, Moscow, Russia
m.andrianova@inno.mgimo.ru, potapovulgu27@yandex.ru

Abstract. Despite the traditional understanding of social dialogue as a tool of classical labor law only, it seems appropriate to look at it also from the point of view of business law. Since the effectiveness of any business structure is a multi-component phenomenon and depends on many factors, from the availability of sufficient financial assets to the effectiveness of the actions of the governing bodies and the workforce, only social dialogue can ensure the operation of the enterprise as a single organism, where everyone understands a special task and works for it, and where each participant is important. Participation in the social dialogue of the state ensures the normal functioning and relevance of the tool under study, as well as centralized solutions in the field of support of the viability of enterprises.

Keywords: Business law · Labor relations · Social dialogue · Transnational corporations

1 Introduction

Social dialogue is classically considered as instrument of labor law. From the point of view of the doctrine, it appears as part of the subject of labor law and an essential characteristic of its method, sometimes social dialogue is interpreted as a new function of labor law, its principle [11]. The significance of social dialogue as a function of labor law is in the possibility of smoothing social tension between employees and employers [1]. The value of this tool is extremely high. Indeed, one of the four areas of cooperation between Russia and the International labour organization (ILO) is the development of social dialogue [3].

Social dialogue can be developed at the international, national and local levels. Each of levels performs its own function and solves its own tasks: unification-at the international level, creation of regulation at the national level, creation of effective interaction between the employee and the employer at the local level. Depending on the level of social dialogue, there is either trilateral (ILO conventions and recommendations, trilateral social partnership agreements in the Russian Federation) or bilateral interaction (industry agreements, collective bargaining). Within the framework of bilateral interaction, employees and employers (or their representatives) participate in the dialogue; within the framework of trilateral interaction, the state, represented by its bodies, joins the dialogue.

S. I. Ashmarina and V. V. Mantulenko (Eds.): IPM 2020, LNNS 161, pp. 777–781, 2021.
https://doi.org/10.1007/978-3-030-60926-9_96

At the enterprise level, it is possible to track the results of social dialogue in a more applied way. However, we should not conclude from the provision that only the local level is important for business activity, because the need to resort to this tool, the procedure for its application, as well as general decisions on the support of participants in relations can only be formed at the international and national level with the participation of state representatives.

It should be borne in mind that in modern conditions, the local level that affects the interests of large joint-stock companies and their subsidiaries may affect several regions of the state. An example is the general collective agreement of PJSC "Gazprom" [7]. Researchers and practitioners identify the cross-border corporate level of social dialogue. In these cases, the value of social dialogue will go beyond the borders of one state. Such cases will be characterized by the greatest complexity of regulation - on the one hand, and the greatest social significance – on the other.

Focusing on the sources of social dialogue regulation, it should be noted that the issues of social dialogue regulation affect almost all ILO conventions and recommendations. Most conventions contain provisions on the need to address labor issues through the procedures of the studyed Institute. However, special attention has been paid to regulating the activities of transnational corporations in the ILO's normative work, because of its particular economic importance.

In 1977, the Tripartite Declaration of Principles concerning Multinational Enterprises and Social Policy (MNE Declaration) was adopted [8]. Throughout the text of the declaration, there is a need to maintain a social dialogue for the state of origin of MNCs and the host state, employees, employers, their representative organizations, as well as the multinational corporations themselves. It is noteworthy that most of the legal instruments used for the legal regulation of social dialogue, including the already mentioned ILO declarations, as well as numerous internal corporate norms used in the field under study with the highest degree of effectiveness, belong to the category of soft labor law (similar to the category of "soft law" used in international trade): corporate behaviour codes, social responsibility standards, etc.

2 Methodology

To write this article, it was necessary to use both general scientific methods of knowledge: dialectical method, methods of analysis and synthesis, and special methods of legal science: formal legal method, comparative legal method, historical legal method, methods of grammatical and logical interpretation of legal norms, as well as the method of economic analysis. The method of economic analysis is necessary for understanding the position of entrepreneurs, identifying their point of view on labor relations in general, and social dialogue in particular. In the end, it is the method of economic analysis that gives an idea of the effectiveness of a particular institution of labor law. However, taking into account the social component of the research subject, it is extremely impractical to limit the use of the method of economic analysis. This would distort the idea of the labor relations and create an imbalance in the relations between the parties of the labor relationship, and it is for the sake of maintaining this balance that labor law and special regulation in the field under study were created. The use of the historical method is necessary to trace the history of the formation of the institution of social dialogue, to determine its roots. Elements of comparative legal research will help to identify similarities in the understanding of social dialogue in different legal systems, as well as at the level of international legal regulation. The chosen methods allowed us to create a more complete and comprehensive view of the problem under study.

3 Results

This article puts forward the thesis that social dialogue is a multidimensional phenomenon that is important not only for the sphere of work, but also for supporting entrepreneurship. Business activity is classically identified with commercial activity aimed "primarily" at extracting profit. In an uncontrolled market, it would probably be possible to limit this simplified representation. However, over the past decades, there has been an awareness of the importance of entrepreneurship for society and its social function. Even from a purely pragmatic point of view, "profit for profit" works only in the short term, as a result, a serious business, hoping to preserve the enterprise for many years, will build a development strategy based on socially useful and socially significant results.

The norms of labor law appeared when it became obvious that none of the parties of the relationship can exist without regulating social dialogue and the presence of minimum standards for regulating labor relations in the legislation. This is a guarantee of stability in society - for the state, a guarantee of normal life-for workers, a guarantee of production efficiency-for entrepreneurs. The state relies on entrepreneurs as on a component of the state's economic growth. In addition, entrepreneurs assume the function of creating jobs, which contributes to the goal of the state in the fight against unemployment, which cannot be effectively implemented without the help of private capital.

4 Discussion

The most controversial issue in this article is the thesis that social dialogue is not an instrument of state pressure on entrepreneurs in order to protect the interests of employees, but an instrument of soft regulation that creates advantages for all participants of the relationship. From the point of view of economic analysis of law, the goal of a legal entity is to achieve its highest market value [10]. Having studied the work of Coase "The nature of the firm" [2], we can conclude that a legal entity is a way to reduce the costs of participants in business activities using intra—corporate mechanisms. In particular, the use of hired workers who follow the instructions of governing bodies (compared to independent contractors) has the property of reducing costs. Thus, the use of hired labor is a necessity for the enterprise.

Coase's ideas were developed in the works of Demzetz [4]. According to the proposed theory, intra-corporate relations should be considered as agreements between shareholders, management and employees, the terms of which are dictated by market forces. Moreover, all parties interested in long-term development strategies are not inclined to seek immediate satisfaction of their own interests, contrary to the common interests of joint activities.

It is interesting to voice an addition to these views by quoting Dodd's views. According to the researcher, the state authorizes entrepreneurial activity not because it allows investors to profit, but because this activity is associated with the production of goods for society [6]. The author also refers to the experience of the United States in the 1930s, when the American legislator was forced to take into account the interests of corporate employees, their clients, and society as a whole in his regulatory approach due to the crisis. In particular, workers' decent wages and confidence in the stability of their financial situation can increase the consumption of goods produced by enterprises – which is key to the self-preservation of the capitalist economic system.

The issue of the company's interaction with its employees also deserves special attention. Proponents of the liberalist approach are of the opinion that these conditions should be determined by the market, in which – according to a number of researchers – employees and contractors of companies have a negotiating position that allows them to defend their interests, and also have the right to refuse further investments in the company (choose a different commercial partner or leave at their own will) [5]. Meanwhile, the company's profit largely depends on the contribution of interested parties (including the labor of employees and management) [9]. The recognition of social dialogue as an effective mechanism for soft regulation that supports entrepreneurship, which will be particularly evident in the times of crisis that we are all experiencing, can help to reconcile economic and social-oriented labor and legal approaches.

5 Conclusion

Difficult times test the strength of the established algorithms of the state. Performing one of its main functions, providing social support to the population, the state is helpless without an efficient business sector. Entrepreneurial activity becomes most effective when it is supported by a labor collective that shares the ideology of the entrepreneur and the state, which recognizes the importance of the entrepreneur's activity. Thus, at all stages of the state's existence, and especially in times of crisis, the key players are entrepreneurs, workers' representatives, and the state itself as an active regulator.

In our opinion, a tool for supporting entrepreneurship should first of all be understood as a social dialogue, which can create a platform for interaction between all active players involved in the formation of social ties in society, rather than measures related to obtaining a deferred payment for certain types of taxes or subsidizing business. As a classic tool of labor law, social dialogue is currently completely undervalued. It can be used as a modern tool for soft regulation of not only internal corporate relations, but also public relations on a state scale. As long-term experience of the existence of labor law shows, it is a soft regulation based on voluntary agreement of the interests of all participants in the relationship. it is able to ensure the creation of the more effective system.

Negotiations in general and the technique of social dialogue in particular are modeled to inevitably lead to some result. And since even a negative result gives an idea of the presence of a specific problem, after detecting which it is possible to plan ways to solve it, we can conclude that even a negative result of social dialogue can bring positive results. At the same time, the probability of a negative result is extremely low, since initially all sides of the labor relationship are aimed at joint creation and depend on each other.

References

1. Bercusson, B.: The strategy of European social dialogue. Eur. Labour Law **1**, 126–167 (2009)
2. Coase, R.H.: The nature of the firm: origin. J. Law Econ. Organ. **4**(1), 3–17 (1998)
3. Cooperation program between the Russian Federation and the International labour organization for 2017–2020 (2017). http://docs.cntd.ru/document/556323932. Accessed 20 July 2020
4. Demsetz, H.: The theory of the firm revisisted. J. Law Econ. Organ. **4**(1), 141–161 (1988)
5. Dent, G.W.: Stakeholder governance: a bad idea getting worse. Case West. Reserv. Law Rev. **58**(4), 1107–1144 (2008)
6. Dodd, E.M.: For whom are corporate managers trustees? Harv. Law Rev. **45**(7), 1145–1163 (1932)
7. Gazprom Trade Union: Core documents (2020). https://xn–80afnaylbafcido5b6k.xn–p1ai/docs/bazovye_dokumenty. Accessed 20 July 2020
8. ILO: Tripartite Declaration of Principles concerning Multinational Enterprises and Social Policy (MNE Declaration), 5th edn. (2017). https://www.ilo.org/empent/areas/mne-declaration/lang–en/index.htm. Accessed 20 July 2020
9. Keay, A.: Stakeholder theory in corporate law: has it got what it takes. Richmond J. Glob. Law Bus. **9**(3), 249–300 (2010)
10. Keay, A.: Tackling the issue of the corporate objective: an analysis of the United Kingdom's enlightened shareholder value approach. Syd. Law Rev. **29**(4), 577–612 (2007)
11. Kurennoy, A.M.: Issues of effective representation of employees' interests in the sphere of hired labor. Law **11**, 47–56 (2019)

Impact of Innovations on the Dynamic of Human Capital

V. P. Fomin(✉)

Samara State University of Economics, Samara, Russia
fominvp@mail.ru

Abstract. The effective implementation of the potential of human capital requires a careful calculation of its quantitative characteristics. First of all, this concerns the determination of the required value of human capital and its changes under the influence of various factors. It is also important to determine the degree of balance of the values of human capital with other types of physical and financial capital. Indicators of the current and strategic balance of human capital and other types of capital provide great opportunities for managing the sustainable and successful development of a firm.

Keywords: Balance · Factors of changes in the number of employees · Human capital

1 Introduction

Conventionally, methods for calculating the need for production resources can be divided into detailed and enlarged. The former require a well-developed regulatory framework, its constant improvement in accordance with changing conditions and opportunities for scientific and technological progress. The orthodox development of these methods is a virtual model of an enterprise, which allows at an arbitrary moment in time to characterize its current and future state, based on actual or specified parameters of the course of economic processes.

The use of detailed methods presupposes the establishment of the value of endogenous parameters arising from the conditions prevailing at each moment of time in the form of exogenous parameters.

The relationship of cause-and-effect parameters is functional in nature and is set by the system of normative economy. For example, an exogenous value in a cable company is the contractual volume of sales of enameled wire at a specific date. Based on technological, labor and other norms, endogenous indicators of the date (time) of the launch of a given batch of goods, accrued wages, received technological waste and many other quantities are determined.

Aggregated methods for determining the need for production resources and analyzing their balance are more based on stochastic cause-and-effect relationships than functional ones. For example, the market forecast for medium transformers for the next two years (the average payback period for a given production) gives a fifty percent increase in demand for this product. This means that when making a decision on

S. I. Ashmarina and V. V. Mantulenko (Eds.): IPM 2020, LNNS 161, pp. 782–790, 2021.
https://doi.org/10.1007/978-3-030-60926-9_97

responding to such a market situation, it will be necessary to increase the volume of the corresponding production resources by approximately one and a half times.

The factors influencing decision-making on changes in the size and structure of production resources can be divided into current or predictable and spontaneous. The former give the enterprise time to adapt to new challenges or pursue new opportunities. However, modern unstable reality forces us to focus on factors of the second kind.

This is facilitated by fluctuating changes in the production programs of the enterprise associated with sudden (for various reasons) refusals of customers from concluded contracts, the emergence of new orders or financial projects. The probabilistic nature of the formation of a portfolio of orders of enterprises forces them to look for ways of flexible and prompt response to changes in the market situation.

The general criterion for an effective structure of production resources is to obtain the target volume of production at the lowest cost. A rational understanding of the target volume of production implies its compliance with market demand, which is satisfied with the minimum production resources required for this. The creative potential of combining production resources is the higher, the more "technological" the interconnections connecting and realizing the individual potentials of these resources.

It seems that the given value of these potentials is the market demand for certain types of products, works and services. The values of the potentials of certain types of production resources are not specified values, since they depend on many factors. Hence, direct and inverse problems of economic analysis arise to assess, firstly, the sufficiency of individual production resources to meet market demand, and secondly, the production potential of these resources.

The production potential of aggregate resources can be conventionally represented in terms of value of the hypothetical volume of products of a certain type (works or services) that are specific to a particular enterprise, corresponding to market demand at average prices.

2 Methodology

The currently demanded direction in the study of human capital is the determination of a sufficient and balanced size of the company's labor resources in various parameters. Good results in this area of research have been achieved by such scientists as Bruskin, Brezhnev, Dyakonova, Kitova, Savinova, Danko, [2], Dudin, Frolova, Lubents, Sekerin, Bank, Gorokhova [3].

Much credit in revealing the factors that determine the quantitative and qualitative parameters of human capital at the level of individual firms belongs to such scientists as Bazarov [1], Rasskazov, Rasskazov, Deryugin [6].

A special place in scientific research is the assessment of the effectiveness of the use of human capital. Here, directions are highlighted related to financial results that depend on the implementation of the potential of human capital. Also of great importance is the study of factors of labor productivity, measured in physical and value terms. Interesting results in this direction are presented in scientific research of such scientists as Melikhov, Maluev [5].

The dynamics of the value of human capital in the context of individual factors also attracts the attention of scientists. In this direction, the works of such scientists as Glukhova, Syrotyuk, Sherstobitova, Pavlova [4].

3 Results

Algorithms for calculating the number of employees in the context of various types of activity, as a rule, use two main values: the labor intensity of the volume of a specific type of work and the working time fund of one employee. In rare cases, factors related to enthusiasm or emotional motivation for intense work are considered. It seems that when calculating the number of employees required to release the planned volume of production, it is advisable to take into account economic and non-economic factors in a balanced way.

This applies to all interrelated areas of calculations to determine the quantitative parameters of human capital. The basis of such calculations is the definition: the labor intensity of a complex of works for the planned period of time, as well as the establishment of the budget of the working time of individual workers, the intensity of their work per unit of time.

Long-term operating firms with an established staff of workers are more concerned not with the initial calculations of the need for labor resources, but with the adjustment of the existing number of employees. The main reasons for the adjustments are changes in conditions in terms of the volume and structure of products, productivity of means of production, as well as measures to increase labor productivity and the like. Releasing or saving the number of employees due to organizational and technical measures (Eei), which increase production efficiency, is determined by the formula:

$$\text{Eej} = \frac{\sum_{i=1}^{n} (t_i^b - t_i^p) * q_i^p}{F_t^p * Kpn} \quad (1)$$

where

Eej is the relative saving (release) of the number of employees due to the j-th event;

t_i^b, t_i^p - respectively the basic and planned (changed as a result of planned office and technical measures) labor intensity of manufacturing products (or semi-finished products of the i-th redistribution) in normalized hours per unit of the i-th type of product;

q_i^p - the planned volume of the i-th type of product (semi-finished products of the i-th processing);

i - the number of types of products (semi-finished products), which are subject to the action of the j-th office technical measure;

F_t^p - basic fund of working time of one employee in hours;

Kpn - the coefficient of the average skill of workers or the coefficient of the average overfulfillment of production standards.

Firm managers, as a rule, develop various measures to reduce the need for labor resources by eliminating various kinds of losses. These include direct losses resulting from a clear violation of the schedule, as well as hidden losses resulting from the irrational organization of labor and production. The scheme for calculating the release or saving of labor resources for these reasons is as follows.

The total value (in standard hours) of the reduction in the labor intensity of the production program is determined when implementing rationalization proposals that eliminate the identified losses of working time due to the irrational organization of labor and production. Then the amount of these losses of working time is divided by the planned fund of working time of one employee.

For example, a way has been found to combine two sequential technological operations into one operation. This method reduces the production cycle by 0.2 h. Extending this method of rationalization to forty-three thousand five hundred repetitions during the implementation of the annual production program will result in the release of five workers with an average annual fund of 1,740 h for each of them.

The average annual salary of five employees is three hundred thousand monetary units, and the cost of streamlining the production process is one hundred and fifty thousand monetary units. Obviously, the innovation will pay off in six months without taking into account the possibilities of earning additional income from the freed workers.

The elimination of losses of working time is also manifested in an increase in the fund of working time per worker on average. In this case, the saving of labor resources is defined as the ratio of the total increase in the working time fund to the planned working time fund of one employee. For example, vaccinations against infectious diseases increase the average number of working days per worker by two and a half percent. If the number of employees of the enterprise is one hundred and twenty people, then this event will free up three.

The use of simple methods of analysis to calculate the factorial interaction of the volume of output, the number of employees and their labor productivity involves the calculation of the hypothetical value of labor resources in the subjunctive mood. In other words, it is determined what would be the number of workers in the analyzed period, provided that their basic output ($L^{(1)}$) is preserved according to the formulas:

$$L^{(1)} L^{0} * \mathrm{J_Q} = \frac{Q^1}{LP^0} \tag{2}$$

where L^0 - is the base (previous) period, the number of employees;

J_Q - growth rate of production volume, $J_Q = Q1/Q0$;

Q^1 and Q^0 - are the actual and basic, respectively, the volume of production;

LP^0 - basic labor productivity of workers.

The estimated need for labor resources is determined by subtracting from the hypothetical number of employees the savings or release (over expenditure or additional involvement) of labor resources for various factors.

The size of the number of employees is also affected by the change in the structure of products. This is due to changes in the ratio of the specific weights of products with different manufacturing complexity. The calculation results for this change may depend on the product units used. Changes in the labor intensity of products per one monetary unit (in value terms) will differ from changes in the labor intensity of products per one natural unit, for example, a running meter, a piece, and so on. Changes in the structure of manufactured products lead to a corresponding change in the need for the number of employees (ELs):

$$\text{ELs} = \frac{(t_c^p - t_c^b) * q_o^p}{F_t^p * K_{pn}} = \frac{\left(\frac{\sum_{i=1}^{n} T_i^p / q_i^p}{n} - \frac{\sum_{i=1}^{m} T_i^b / q_i^b}{m}\right) * \sum_{i=1}^{n} q_i^b}{F_t^p * K_{pn}} \quad (3)$$

where,

t_c^p, t_c^b - the average labor intensity of manufacturing a unit of measurement of products, respectively, in the planned and baseline periods;

q_o^p, q_i^p, q_i^b - respectively, the volume of production in comparable units of measurement for the total volume in the planning period, for the volume for each i-type of product in the planned and base periods;

T_i^p, T_i^b - is the total labor intensity of manufacturing the i-th type of product, respectively, in the planned and base periods; n, m - the number of types of products, respectively, in the planned and base periods.

The change in the need for the number of employees is influenced by the change in the share of semi-finished products from outside (under outsourcing agreements) in the cost of production. Outsourcing reduces the overall labor intensity of the production program. In this case, headcount savings (Ela) is calculated by dividing the amount of labor intensity reduction by the working time fund of one employee. At the same time, the profitability of replacing our own production with purchased semi-finished products is determined, not only from the point of view of cost reduction, but also from the standpoint of increasing dependence on suppliers and their reliability.

$$\text{Ela} = \frac{\sum_{i=1}^{n} t_{ai} * q_{ai}}{F_t^p * K_{pn}^p} \quad (4)$$

where

t_{ai} - is the specific labor intensity of the manufacture of purchased semi-finished products of the i-th type on their own;

q_{ai} - the number of semi-finished products of the i-th type, replacing their own production.

Improving the organization of labor and production in terms of equipment maintenance, as well as its renewal and modernization lead to a decrease in the need for maintenance workers (Els):

$$\text{Els} = \sum_{i=1}^{n} O_i^p (1/H_{si}^b - 1/H_{si}^p) \quad (5)$$

where

O_i^p - is the number of upgraded equipment of the i-th type, the maintenance rate of which has changed;

H_{si}^b, H_{si}^p - is the maintenance rate (number per worker) of objects of the i-th type, respectively, of the base and planning period.

Reduction of whole-day loss of time (reduction of morbidity, elimination of absenteeism, vacations with the permission of the administration) and intra-shift loss of time (elimination of downtime for various reasons) leads to savings in labor resources (Elt), which is calculated by the formula:

$$\text{Elt} = \frac{\sum_{i=1}^{n} D_{di} * Cm^{p} + \sum_{i=1}^{m} C_{Pi} * D^{p}}{F_t^{p} * K_{pn}^{p}} \tag{6}$$

where

D_{di} - are the day-to-day losses of working time reduced as a result of office technical measures (in days);

Cm^p- the average duration of a work shift in the planned period (hours);

C_{Pi} - the average value of reduced intra-shift losses of working time (in hours);

D^p - the planned number of working days on average per one employee in the planned period.

Similarly to the previous calculations, labor savings are determined by reducing production defects (Elw):

$$\text{Elw} = \frac{\sum_{i=1}^{n} q_{i,fw} t_i + \sum_{i=1}^{m} q_{i,bw}(t_{i,bw} + t_{i,tw})}{F_t^p * K_{pn}^p} \tag{7}$$

Where

$q_{i,fw}$ - is the number of the i-th type of product in final marriage;

t_i - labor intensity of manufacturing the i-th type of product;

$q_{i,bw}$ - the number of the i-th type of product in a correctable marriage;

$t_{i,bw}$, $t_{i,tw}$ - respectively, labor costs lost in a correctable marriage and carried out during its correction.

4 Discussion

The above methods make it possible to find the future needs of an operating enterprise in labor resources, balanced with the expected change in working conditions, by adjusting the existing number of employees for appropriate changes. These methods are based on assessing the influence of factors on the change in the number of employees due to direct changes in working conditions.

However, this approach cannot be considered exhaustive, since it does not fully reflect the behavior of other factors of production. In calculating the values of factors of production, their interaction should be taken into account, which ultimately ensures the corresponding results of economic activity.

Similarly to calculations for labor resources, the change in demand can be determined for other types of resources in the context of planned activities or changes in business conditions. The total savings for individual resources are obtained by summing the savings for each factor. The planned need for production resources obtained by adjusting for the estimated savings will not reflect the balanced structure of these resources, since the labor resources released as a result of organizational technical measures may be insufficient or excessive in comparison with the savings in funds or objects of labor.

The structural features of certain types of production resources also play a role. For example, if opportunities are found to release a certain number of turners, which will not be provided with the corresponding savings in materials and operating time of turning equipment, then an excess of turners will appear.

Methods for determining the needs for the production resources of a firm have been well developed and have reached a high level in the practice of the domestic centralized command economy of the last century. The modern market does not exclude the existing positive developments, but needs their adequate application, development and the emergence of new popular techniques. Firms in market conditions must adhere to a certain way of existence that ensures survival only with a constant increase in the efficiency of the use of production resources.

Determination of the need for production resources in a directive economy was carried out, basically, by the so-called methods "from the achieved", that is, by adjusting the existing values, taking into account the planned activities and anticipated factors. These methods do not fully meet modern requirements for ensuring resource balance with microeconomic instruments. It seems that in a civilized economy, the macro-, mega- and micro-levels of ensuring resource balance should be evenly tense and not have contradictions in solving, in fact, the same task.

To analyze the balance of the production resources of a company, the achieved level of their values, as well as its calculated value, obtained by direct calculation, based on the available funds and market needs for specialized products, can serve as a comparison base. The method of direct calculation of the values of labor resources, objects and means of labor in their structural specifics for a specific production program seems to be more accurate. However, in practice, such a state of calculations is difficult to achieve due to the insufficient level of regulatory support.

The initial stage in the analysis of resource balance is the calculation of the need for specialized production resources, based on market demand for products, taking into account a period comparable to the return on investment in the production of this type of product, as well as the property owned by the company. The value of the firm's funds determines the scale of its activities, expressed in the hypothetical volume of core products (works or services).

One of the main tasks of the balanced scorecard analysis is to assess the possibilities of stable and long-term development of the company by identifying the relationship between the current and future balance of the company. The balanced current state for the long term is ensured by the interconnection of the dynamic characteristics of economic activity.

These characteristics are considered in a sequence that reflects causality. The conditions for the balanced development of a firm are met if the rate of improvement of quality characteristics in terms of the outcome indicators will exceed the rate of change in the indicators of causes. In this case, the presence of potential opportunities for local economic growth is stated, which is the key to increasing the value and property of the company.

5 Conclusion

Effective management of the current and future balance involves the use of appropriate indicators (Table 1). They serve at least two purposes. First, they make it possible, at the required time, to obtain a characterization of balance in its types and levels to test the rationality of new projects for the implementation of current or strategic opportunities for capital gains. Secondly, in the standby mode, they inform about the appearance of unforeseen destabilizing deviations in the balance of the firm's indicators to launch a reflex mechanism to neutralize negative factors or their consequences.

Table 1. Relationship between goals and indicators of business development

Current goals			Strategic goals	
Rapid response indicators			Strategic development indicators	
Weak points of reflex response	Institutional instability of the business environment	Market instability	Promising capital investment projects	Balancing production costs, assortment and quality of products

Source: author.

Achieving these objectives requires five groups of quantitative and qualitative indicators. The first group includes urgent indicators that track vital parameters, deviations in which require immediate correction in order to avoid irreversible negative consequences. This includes identifying the dangers of environmental instability and internal destabilizing factors.

The second group of indicators is responsible for the consequences for the firm's economy resulting from institutional changes (in civil, tax legislation and other rules relating to business organization), which are subject to frequent changes and must be taken into account from a certain date.

The third group of indicators also refers to positions that are generally unstable in the market environment. These indicators monitor spontaneous market forces that change demand, supply, prices, tariffs, etc. A timely response to such changes will avoid unnecessary adaptation time and, accordingly, will reduce losses or lost benefits.

The fourth group of indicators represents a target-oriented system of indicators and is designed to search for directions for the development of a company, the formation of its strategy, based on positioning in the market and the balance of internal controlled factors.

The fifth group of indicators represents manageable indicators that, based on feedback, implement the intended tasks in line with current and future development projects of the company. Here, internal positions are distinguished, associated with production costs, as well as external ones, associated with the range and quality of products, ensuring a sufficient level of competitiveness.

Each group of indicators has specific features, but they are summarized in two directions. The first, second and third are attributed to conditioned reflex functions that ensure viable responses and firm survival in the complex and changing conditions of the business environment. The fourth and fifth groups of indicators are combined into the intellectual function of finding and implementing ways of progressive development of the company.

References

1. Bazarov, T.Yu.: Personnel Management. Academy, Moscow (2017)
2. Bruskin, S.N., Brezhneva, A.N., Dyakonova, L.P., Kitova, O.F., Savinova, V.N., Danko, T.P., Sekerin, V.D.: Business performance management models based on the digital corporation's paradigm. Eur. Res. Stud. J. **20**(4A), 264–274 (2017)
3. Dudin, M.N., Frolova, E.E., Lubents, N.A., Sekerin, V.D., Bank, S.V., Gorohova, A.E.: Methodology of analysis and assessment of risks of the operation and development of industrial enterprises. Qual. Access Success **17**(153), 53–59 (2016)
4. Glukhova, L.V., Syrotyuk, S.D., Sherstobitova, A.A., Pavlova, S.V.: Smart university development evaluation models. In: Uskov, V.L., Howlett, R.G., Jain, L.C. (eds.) Smart Education and e-Learning 2019. Smart Innovation, Systems and Tehnologies, vol. 144, pp. 539–551. Springer, Cham (2019)
5. Melikhov, Yu.E., Maluev, P.A.: Personnel Management. Portfolio of Reliable Technologies. Dashkov and Co, Moscow (2018)
6. Rasskazov, S., Rasskazova, A., Deryugin, P.: Corporate Governance. Infra, Moscow (2020)

Author Index

S. I. Ashmarina and V. V. Mantulenko (Eds.): IPM 2020, LNNS 161, pp. 791–793, 2021.
https://doi.org/10.1007/978-3-030-60926-9